Children with Differences in Sex Development

Simmi K. Ratan
Editor

Children with Differences in Sex Development

Taking a Multidisciplinary Management Approach

Springer

Editor
Simmi K. Ratan
Department of Pediatric Surgery
Maulana Azad Medical College
New Delhi, India

ISBN 978-981-97-1638-8   ISBN 978-981-97-1639-5 (eBook)
https://doi.org/10.1007/978-981-97-1639-5

This Springer imprint is published by the registered company Springer Nature Singapore Pte Ltd.
The registered company address is: 152 Beach Road, #21-01/04 Gateway East, Singapore 189721, Singapore

Foreword

Disorders (or Differences) of Sex Development (DSD) is one of the most challenging topics in modern medicine. It has aspects drawn from nearly all domains of contemporary practice including genetics, embryology and anatomy, endocrinology and physiology, oncology, general surgery, urology, gynecology, and psychology. The treatment options run the gamut from observation to pharmacology and complex reconstructive surgery. The age range spans from the newborn into adulthood. Diagnosis, evaluation, and management require the skills and coordinated care of many specialists. For trainees and new practitioners, the subject of DSD can also be significantly more difficult to grasp due to evolving terminology, classification, and treatment paradigms. It is also not a purely medical topic as increasingly DSD has become an area of discourse in the ongoing discussions concerning gender and sexuality; this has led to a dynamic medico-legal environment where local, regional, national, and international legislation have weighed in and remain in flux.

Prof. Simmi K. Ratan has assembled a team of experts and produced a 68-chapter-long work that should help students and experienced practitioners gain a working knowledge on this complex topic. The work published by Springer Nature is organized into eleven sections. The first addresses what is currently understood to be the molecular, genetic, and embryologic nature of DSD. The second section looks at the concept of gender identity and notes that families, patients, and treating physicians need to be aware of the gender dysphoria which can occur in patients with DSD. This segues naturally into the third section that looks at the current legal environment including Indian and international examples. The fourth section looks at diagnosis and has chapters on terminology and classification, endocrine and cytogenetic evaluation. Section five focuses on DSD conditions due to sex chromosome anomalies. Section six is centered on patients with a 46XY karyotype. Section seven addresses 46XX DSD conditions. The eighth section examines the role of psychological counseling for the family and patients. Surgical options are discussed in section nine including revision surgery, the timing of surgery, and the nature of consent. Long-term outcomes and socialization are covered in section ten including the role of malignancy surveillance and fertility. Finally, section eleven looks at specific concerns such as cloacal exstrophy patients, microphallus, and nephrology issues. The book would be a welcome addition to the collection of any physician involved actively in the care of DSD patients or who may encounter them.

I want to Thank Dr. Dinesh Sarda for availing me this opportunity to look at this book and write a foreword. Dr. Sarda had spent time at our unit in Ann Arbor and we very much enjoyed the time and exchange of experiences and ideas. We are grateful that we have chance to see this great collaborative effort.

Department of Urology, Division of Pediatric Urology
University of Michigan
Ann Arbor, Michigan, USA
Julian Wan

Foreword

"Change Is Around the Corner, We Must Equip Ourselves to Own the Transition with Minimal Disruption"

I am very grateful to Dr Ratan and the team for inviting me to write a foreword for this book titled "Children with Differences in Sex Development: Multidisciplinary Management Approach," which is a collection of very valuable insights that will advance the care of children born with differences in sex development/variations in sex characteristics (DSD/VSC) in India.

I consider myself to be in a unique position in this discussion having trained and practised in India initially and subsequently in Australia. My exposure to two systems with widely different social constructs perhaps places me rightfully on the proverbial fence with the ability to look on either side, both with involvement and independence at the same time.

Thank you to Dr Ratan for letting me browse through the collection. What I see in this book is an exemplary collection of works far more comprehensive than any of its predecessors anywhere in the world as far as my knowledge goes. I place on record my appreciation at the quality of the work produced. The breadth of the topics tackled is vast and the background/expertise of the contributors is awe inspiring. I am particularly pleased with the accommodation of legal and social experts and their pieces. Dr Ratan's leadership in collating this resource hence merits the highest congratulations!

The work is presented and formatted in a very reader-friendly manner with careful thought given to anticipating what the reader would like. There is also much effort evident on avoiding "audience restriction." The collection will be a good reference not only for all medical groups but also for many allied health and relevant advocacy groups.

Wise men say—"*Change is the only permanent thing.*" Time for change in laying out short- and long-term care for children with DSD/ VSC is undeniably around the corner in India. This may likely bring in its wake discussions embedding huge paradigm shifts. The change could easily be rapid, deep, and disruptive. Experts in change management will counsel us to avoid missteps that can unintentionally translate into failure or prolonged delays due to pushbacks. Uncontrolled social disruption may be undesirable. It is hence incumbent upon us that we balance science with social realities and expectations.

This book, I hope, leads the first step of change management as proscribed by Kotter which is to create a climate for THE CHANGE. The editors and contributors have embarked on a worthy albeit challenging task. It may effectively pave the way for future conversations and codesigned processes that are necessary to progress the well-calibrated and tailored process of change. I hope and pray that the various stakeholders engage constructively and take the process of change forward sans friction and unnecessary mistrust.

I will conclude by once again thanking all the contributors of this work and hoping that children are better cared for and their parents/carers/families better supported as a result of this endeavor.

Children's Hospital at Westmead
and University of Sydney
Sydney, NSW, Australia

Aniruddh Deshpande

Preface

A long felt need of bringing the available literature in relation to the management of children with disorder/differences of sex development has led to the compilation of this book.

It is a humble attempt to gather the inputs from stalwarts and experts from across the globe, and working in a variety of fields with respect to this condition under one umbrella; so that the topic could be vividly and lucidly comprehended in a composite manner by the reader.

I pay my extreme gratitude to my mentors and my teachers, who awakened my interest in this relatively less well read and known topic. I appreciate all the co-authors and co-editors of this book, who toiled hard to write and edit chapters on various topics housed in specific sections of the book, addressing key areas (as genetics, embryology, legal, psychological, social rehabilitation and follow-up issues, medical and surgical treatment, and role of support groups)

To the best of my knowledge, no other scientific publication in English gives comprehensive write-ups on a wide variety of relevant fields on this subject on a single platter, and this is the novelty of this book. Updated information for genetic workup and psychosocial rehabilitation are key features of this book and would certainly remain relevant for many years to come

I sincerely feel that students, physicians, surgeons, and researchers who aspire to contribute toward the cause of children with differences of sex development will find this book a very convenient tool to gain knowledge and wider insight about the subject.

Delhi, India

Simmi K. Ratan

Contents

Part IV Diagnosing DSD

Part V Sex Chromosome DSD

Part VI 46XY DSD

Part I

The Science Behind Sexual Differentiation

I whole heartedly thank Dr. Aashima Dabas, Professor Pediatrics Maulana Azad Medical College, New Delhi and Dr. Tanu Anand, Senior Scientist at ICMR, Delhi headquarter for assisting and aiding me in editing this section.

Normal Gonadal Development

1

Sabita Mishra and Surbhi Wadhwa

Abstract

The development of the reproductive system is closely associated with the urinary system in both males and females. Formation and differentiation of gonads in either sex, is dependent on the arrival of the germ cells into the gonadal ridge on the medial side of the intermediate mesoderm. The primordial germ cells for a male fetus carry XY chromosome, while XX chromosome for female fetus. Lateral to the developing gonads and the mesonephric ducts, the coelomic epithelium invaginates in the urogenital ridge to form Müllerian/Paramesonephric ducts at 5 weeks of gestation. The presence of testosterone in the male fetus causes the paramesonephric duct to degenerate except where it forms the appendix of the testis. In female fetuses, the presence of estrogen and absence of testosterone favors the formation of the internal reproductive organs of females. The chapter describes the development of male and female reproductive system in detail.

Keywords

Chromosomes · Primordial germ cells · Estrogen · Testosterone

1.1 Development of Gonads

Although the genetic sex of the embryo is determined at the time of fertilization, the gonads are undifferentiated till the seventh-week post-fertilization. The sex determination is a complex process as the male embryo carries the

S. Mishra (✉) · S. Wadhwa
Department of Anatomy, Maulana Azad Medical College and Lok Nayak Hospital, New Delhi, India

S. K. Ratan (ed.), *Children with Differences in Sex Development*,
https://doi.org/10.1007/978-981-97-1639-5_1

sex-determining region Y protein (SRY) gene which is responsible for the formation of the male gonad. This gene is on the short arm of Y chromosome (Yp1). The protein produced by the gene is a transcription factor that causes a number of cascading events to form the male gonad. In the absence of SRY gene, the female gonad develops.

The development of the reproductive system is closely associated with the urinary system in both males and females. The intermediate mesoderm on either side of the dorsal body wall forms urogenital ridge. The gonads appear on the medial side of the ridge as a pair of longitudinal thickenings, named the genital or gonadal ridge. Formation and differentiation of gonads in either sex are dependent on arrival of the germ cells into the gonadal ridge. The germ cells enter the ridge in the seventh week [1].

1.1.1 Origin of the Primordial Germ Cells

The primordial germ cells originate from the epiblast and migrate through the primitive streak at the third week of gestation to reside among the endodermal cells in the wall of the yolk sac. During the fourth month of gestation, they enter the dorsal mesentery of the hindgut and move by ameboid movement toward the genital ridge. They arrive at the gonadal ridges by fifth week of gestation and start invading the mesoderm of the gonadal ridge by sixth week. When the primordial germ cells fail to reach the gonadal ridge, the gonads do not develop. They remain as two longitudinal ridges of mesenchyme in the dorsal body wall called streak gonads [1].

1.1.2 Formation of the Male Gonad Testis

The primordial germ cells for a male fetus carry XY chromosomes. Under the influence of the SRY gene, the testis determining factor influences the formation of the male sex cords. As the primordial germ cells invade the mesoderm, they form longitudinal cords which dip into the mesoderm. The surface epithelium of the gonadal ridges is the follicular cells that line the cords. The cords are elongated and are called the medullary cords (testicular cords) of the primitive testis. Toward the medial side of the ridge, the cords break up into a network of tiny strands forming the rete testis. This site forms the hilum of the testis. A thick coat of connective tissue the tunica albugenia separates the testis cords from the surface epithelium (Fig. 1.1).

In fourth month, the testis cords form horseshoe loops and are connected to the medial side of the gonad, the rete testis. These cords are formed by the primordial germ cells and the follicular cells, derivatives of the surface epithelium of the gonadal ridge. The Sertoli cells with the spermatogonia form the seminiferous tubules which are solid cords as they have no lumen. The follicular cells form the Sertoli cells which are interposed between the germ cells and function as the supporting cells. They support the germ cells and their side processes lodge the

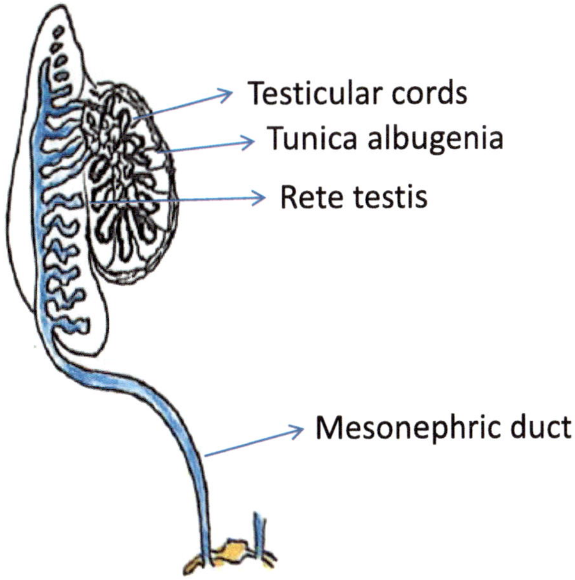

Fig. 1.1 Genital duct in the male at fourth month

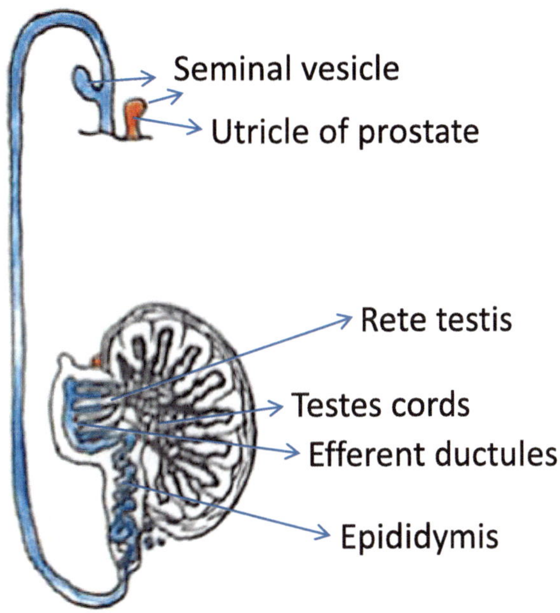

Fig. 1.2 Genital ducts after descent of the testis

spermatogonia during different stages of maturation from puberty onward. At puberty, the seminiferous tubules are canalized to form a lumen. Once they are canalized they join the rete testis network of tubules which enters the efferent ductules to connect the seminiferous tubules to the mesonephric ducts (Fig. 1.2). The mesonephric ducts are responsible for the formation of the male ductal system described in the next section. The original mesenchyme of the gonadal ridge gives rise to the Leydig cells. They produce testosterone which is responsible for the formation of the internal and external genitalia. The testis develops in the posterior abdominal wall. For effective spermatogenesis, the temperature has to be lower than the body temperature [1, 2].

1.1.3 Descent of the Testes

The testes develop in the posterior abdomen behind the peritoneum and move caudally to reach the scrotal sac. It passes from the abdomen to the scrotum via the *inguinal canal*, which in adults is approximately 4 cm long and lies just superiorly to the medial half of the inguinal ligament. Entry to the canal is by the deep (internal) inguinal ring, and exit is by the superficial (external) ring near the pubic tubercle.

By the end of the second month, the *urogenital mesentery* connects the testis and mesonephros to the posterior abdominal wall. Following the degeneration of the mesonephros, the urogenital mesentery serves as a mesentery for the gonad to move caudally and it becomes ligamentous known as the *caudal genital ligament*. From the caudal pole of the testis, there is a mesenchymal condensation, the *gubernaculum*. Prior to descent of the testis, this band of mesenchyme terminates in the inguinal region between the differentiating internal and external abdominal oblique muscles. Later, as the testis begins to descend toward the internal inguinal ring, an extra-abdominal portion of the gubernaculum forms and grows from the inguinal region toward the scrotal swellings. When the testis passes through the inguinal canal, this extra-abdominal portion contacts the scrotal floor. In females gubernaculum remains rudimentary [3].

1.1.4 Development of the Female Gonad Ovary

The germs cells invading the female gonad have XX chromosomes. The absence of Y chromosome causes non-formation of the medullary cords, instead, the germ cells are scattered in isolated clusters in the mesenchyme of the gonadal ridge. The surface epithelium of the female gonad starts proliferating and forms cortical cords which have no germ cells. They penetrate into the parenchyma but remain close to the surface. In the third month, the cords dissociate into isolated clusters. Cells in the clusters continue to proliferate. The follicular cells start surrounding the primitive gem cells called the oogonia. Each oogonium is surrounded by a layer of epithelial cells and is called the follicular cells. The oogonia with its lining follicular cells are primordial follicles [1].

1.2 Ductal System of the Human Fetus

1.2.1 Mesonephric Duct and Its Fate

The development of kidney commences in the late third to early fourth week of intrauterine life with the formation of the pronephros. This transitionary and non-functional excretory system is equivalent to the functional kidneys of primitive fish. They develop bilaterally as tubules or vesicles at the cranial end of the nephrogenic cord of the intermediate mesoderm in the cervical regions and proceed toward the thoracic region. While the pronephric tubules are advancing caudally, simultaneously degeneration sets in the cranial vesicles during fourth week. The question arises that why did these structures develop? The answer lies in its ductal system. The transitory vesicles also induce the formation of the pronephric duct which forms in the urogenital ridge due to epithelial–mesenchymal interactions. The pronephric duct advances toward the developing cloaca. While the tubules degenerate the pronephric ducts survive.

In the thoracic and lumbar region concurrent to the degeneration of the pronephros the interim kidney, mesonephros develops in the nephrogenic cord in the thoracic and lumbar regions. Approximately 40 primitive glomeruli and tubules differentiate in a cranial-caudal direction which drain into the persisting pronephric duct. This duct now is known as the mesonephric or Wolffian duct that drains into the cloaca. The mesonephric kidney excretes hypotonic urine up to 12 weeks of intrauterine life by when the definitive metanephric kidney takes over. A wave of cranial-caudal degeneration takes place in both sexes in the mesonephric tubules while the caudal ones are developing the cranial ones start to degenerate. However, in the differentiating male fetus under the influence of testosterone, a few caudal tubules and the mesonephric duct persist while they dissolute in the female fetus. The persisting tubules in the male reproductive system form the efferent ductules of the testis and the duct forms the genital ductal system of the males called as epididymis and the ductus deferens, and an outpouching the seminal vesicles and beyond as the ejaculatory duct [3]. The cranial portion of the duct also forms the appendix epididymis [2].

In females, in the absence of testosterone and presence of estrogen, the mesonephric tubules and duct form some vestiges as epithelial cysts of the epoöphoron and the paroöphoron within the broad ligament and the Gartner duct cysts in the walls of the vagina.

1.2.2 Paramesonephric Ducts and Its Fate

1.2.2.1 Development of Uterus and Uterine Tubes

Lateral to the developing gonads and the mesonephric ducts at about 5 weeks of intrauterine life in both sexes, the coelomic epithelium invaginates in the urogenital ridge to form Müllerian/Paramesonephric ducts. It is postulated that the mesonephric duct provides the necessary cues for the caudal migration of the developing duct.

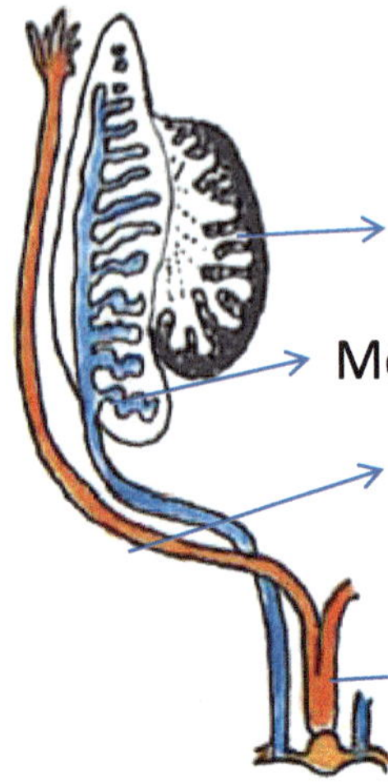

Fig. 1.3 Genital ducts in female second week

The duct crosses the mesonephric duct ventrally to fuse with its opposite counterpart in the midline at about 8 weeks to form the uterovaginal canal which opens in the urogenital sinus [3]. The caudal tip of the duct forms a müllerian tubercle or a sinus tubercle, a swelling, at its entry into the urogenital sinus (anterior part of the cloaca) while the cranial end of the ducts open into the developing peritoneal cavity via a funnel-shaped opening (Fig. 1.3).

The presence of testosterone in the male fetus causes the paramesonephric duct to degenerate except where it forms the appendix of testis. In female fetuses, the presence of estrogen and absence of testosterone favors the formation of the internal reproductive organs of females.

The cranial unfused part of the ducts that open in the peritoneal cavity forms the fallopian tube and its Ostia. Over the next few months, the mucosa of this part becomes highly folded, especially in the regions of the ampulla and the infundibulum and the surrounding mesenchyme differentiates into the smooth muscle. The caudal midline fused uterovaginal canal forms the uterine canal which initially at 8 weeks is separated by a septum that disappears in about a week. Failure to do so results in a variety of defects. While the cranial portion of the uterovaginal canal develops into the corpus and the cervix of the uterus, the caudal part forms the vagina. The fusion of the ducts also causes a reorientation of the developing pelvic cavity which develops the uterovesical and uterorectal pouches and two laterally running peritoneal folds the broad ligament. The splanchnic mesenchyme around the developing uterus and cervix forms the musculature—myometrium and the surrounding parametrium (Fig. 1.4).

1.2.2.2 Development of Vagina

The development of the vagina is, however, debatable with two schools of thought. The vagina is thought to be of dual origin with its upper smaller part being formed by the uterovaginal canal and the remaining larger part from the outgrowth

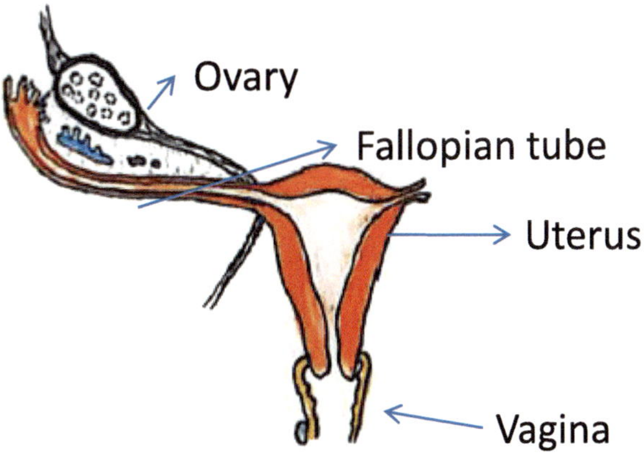

Fig. 1.4 Development of female reproductive system

sinovaginal bulb which is an outgrowth from the endoderm of the urogenital sinus induced by the sinus tubercle. The second school of thought believes that the entire vagina is formed from the outgrowths of the sinovaginal bulbs and is endodermal in origin [3].

Acknowledgment Mr. Ravi Kumar, Artist, Department of Anatomy, Maulana Azad Medical College, New Delhi for drawing all figures in this text.

References

1. Development of reproductive system. In: Schoenwilf GC, Bleyl SB, Brauer PR, Francis-West PH, editors. Larsen's human embryology. China: Churchill Livingstone; 2015.
2. Development of urinary system. In: Schoenwilf GC, Bleyl SB, Brauer PR, Francis-West PH, editors. Larsen's human embryology. China: Churchill Livingstone; 2015.
3. Urogenital system. In: Mishra S, editor. Langman's medical embryology. Harayana: Wolter Kluver; 2019.

Molecular Regulation of Gonadal Development

2

Sunil Kumar Polipalli and Seema Kapoor

Abstract

The process of gonad and genital development is strictly regulated. Placental, fetal, and maternal hormones are among those factors that affect its development. A few number of genes play a crucial role in the gonad development process. The sex-determining region-Y protein, which controls the development of male gonads, is one of the most significant genes. Important genes, including SRY, SF-1, Wilms' tumor related 1 (WTI), GATA4, and SOX9, will be briefly discussed in this chapter along with the intricate connections between these genes during gonadal development.

Keywords

Gonads · Development · Genitalia

2.1 Introduction

In the development of mammals, the formation of a sexually dimorphic phenotype is a crucial moment. The fundamental idea underlying this significant process is that the Y chromosome's presence or absence during fertilization determines genetic sex, which in turn controls the gonads of the developing embryo to develop into either the testes or the ovaries. The testes then release hormones that set off a developmental sequence in males that results in phenotypic sexual differentiation [1, 2]. Differentiation progresses in the female pathway in the absence of the testes and their products of metabolism. Because they trigger the expression of crucial target

S. K. Polipalli · S. Kapoor (✉)
Pediatrics Research and Genetic Lab, Department of Pediatrics, Maulana Azad Medical College, New Delhi, India

S. K. Ratan (ed.), *Children with Differences in Sex Development*,
https://doi.org/10.1007/978-981-97-1639-5_2

genes, a number of transcription factors—including Wilm's tumor-associated 1 (WT1), steroidogenic factor 1 (SF-1), SOX9, and GATA4—are necessary for gonadal development and sexual differentiation. Research has also discovered DAX-1, a different transcriptional regulator that prevents all these aforementioned transcriptional activators from inducing the target gene [3, 4]. After giving a brief introduction regarding gonadal development and sexual differentiation, this chapter examines the research that went into identifying and isolating these genes from the gonad and how they interact to control key processes in sexual differentiation.

2.2 Overview of Sexual Differentiation

The ovaries and testes are referred to as diploid or indifferent gonads because they are physically identical before sexual differentiation occurs. The urogenital rim, which gives rise to the adrenal cortex, gonads, and kidneys, is first formed by these paired structures, which have their origins in the intermediate mesoderm. Separate gonads are formed as a result of mesonephros cell migration and cochlear epithelial growth. The testis is anatomically divided from the ovary and arranged into the testicular cord following sex determination [5]. Primordial germ cells and embryonic Sertoli cells are found in these testicular cords, which are the progenitors of the seminiferous tubules. These germ cells migrate from locations outside the urogenital crest to the gonads. Peritubularmyoid cells, which are located in the basal layer and divide the testicular cord from the steroidogenic Leydig cells in the interstitial region, encircle the cord. An amorphous "ground glass" look is caused by the ovaries' relatively low differentiation prior to birth. They subsequently form primordial follicles, which are functionally equivalent to the testicular cell group and contain distinct cell groups. These include granulosa cells, which aid in the maturation of germ cells, germ cells (oocytes), and stromal cells, which create connective tissue. Theca cells are known to contain the steroidogenic compartment. The Mullerian (paramesonephric) and Wolffian (mesonephric) ducts, which are paired ducts that are initially indistinguishable in male and female embryos, are the source of the internal genitalia. The testes generate three distinct hormones necessary for male development in order to initiate the male pathway. The glycoprotein hormone anti-Müllerian hormone (AMH) is produced first by Sertoli cells and results in the Mullerian ducts that would normally form the uterus, upper vagina, and oviducts regressing. Soon after, the Wolffian ducts differentiate into the seminal vesicles, epididymis, and vas deferens due to the production of testosterone by Leydig cells. Insulin-like peptide 3 (INSL3), a member of the insulin peptide family, is also produced by the Leydig cells later in gestation and is responsible for the testes falling to their characteristic location in the scrotum [6].

The external genitalia, which include the genital tubercle, urethral folds, urethral groove, and genital swellings, originate from structures that are initially shared by both sexes, just like the bipotential gonad and internal genitalia. In contrast to the internal structures that originate from the Wolffian ducts, testosterone, and its 5a-reduced derivative dihydrotestosterone are both necessary for the virilization of the external genitalia and complete male differentiation.

2.3 Essential Genes for Early Gonadal Development

Anomalies in male sexual differentiation result from mutations in genes that control the production of testicular hormones, since these hormones are crucial regulators of the male developmental pathway. Additionally, mutations in genes necessary for the bipotential gonad's differentiation will impair the male pathway. On the other hand, gonadal hormones have no effect on what appears to be normal female development. As a result, XX offspring without gonads will appear normal at birth.

Transcription factors critical for the development of the bipotential gonad have been identified by multiple groups (Fig. 2.1). One of these factors, NR5A1/steroidogenic factor 1 (SF-1), controls the expression of the enzymes that synthesize testosterone in Leydigtumor cell lines, which is why it was initially connected to gonadal function. Analysis of human subjects with abnormalities in the normal pathway of sexual differentiation led to the initial identification of others.

Six of these genes—SRY, NR5A1/SF-1, Wilms' tumor related 1 (WTI), GATA4, SOX9, and DAX1, among others—are reviewed in this chapter, and their intricate interactions with gonadal development are discussed (Table 2.1).

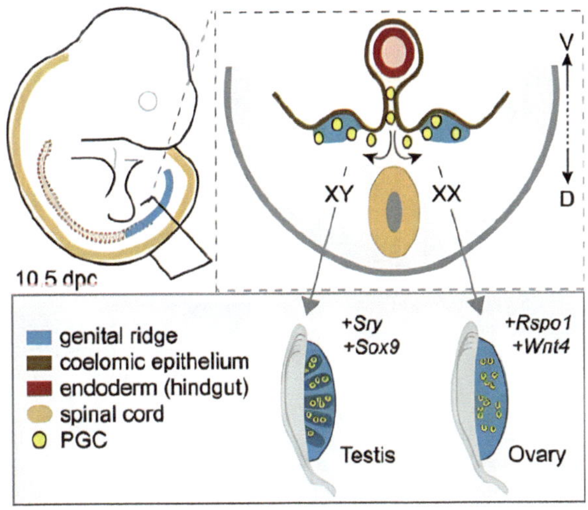

Fig. 2.1 Bipotential gonadal ridges establishment and sex determination. The blue genital ridges are an outgrowth that can be seen in the coelomic epithelium cavity along the mesonephros. In mice, cells are drawn from the surrounding coelomic epithelium (brown) to differentiate it from E10. The dorsal mesentery allows primordial germ cells (yellow) to enter the genital ridges (arrows) after exiting the hindgut (red). Genital ridges are bipotential at this stage, meaning they can differentiate in the testis or the ovaries based on genetic signals. Testis differentiation is triggered by Sox9 expression, which is initiated by Sry expression in the XY genital ridges at E10.5. When Sry is absent, XX gonads start Wnt4 and Rspo1 signaling and grow in the direction of the ovaries, (V) ventral; (D) dorsal

Table 2.1 Sex-differentiating genes

Gene	Chromosomal localization	Expression	Function
GATA4 (GATA-binding protein 4)	8p23.1	Widespread	Regulation of coelomic epithelium thickening
SRY	Yp11.3	Supporting cells of the genital ridges, directing them toward the fate of a Sertoli cell	This protein is involved in male-typical sex development, which usually follows a certain pattern based on an individual's chromosomes
SOX9	17q24-q	Predominantly in mesenchymal condensations throughout the embryo before and during the deposition of cartilage	Plays a critical role during embryonic development, especially important for development of the skeleton and plays a key role in the determination of sex before birth
LHX1 (LIM homeobox gene 1)	17q12	Primitive streak, prechordal and intermediate mesoderm, brain, thymus, tonsil	Differentiation and development of the head, neural and lymphoid tissues and urogenital structures
LHX9 (LIM homeobox gene 9)	1q31.3	Central nervous system, forelimb and hind limb mesenchyme, and urogenital system	Activation of SF1 in gonadal primordia
NR5A1 (Nuclear receptor subfamily 5, group A, member 1, also *SF1:* Steroidogenic factor 1, or *AD4BP*: Adrenal 4 binding protein, or *FTZF1*: Fushi tarazu factor homolog 1)	9q33.3	Gonadal ridges, adrenal gland primordia, hypothalamus, and pituitary	Stabilization of intermediate mesoderm, and transcriptional regulation of several genes (StAR, steroid hydroxylases, aromatase, AMH, DAX1, and many others)
NRG1 (Neuregulin 1)	8p12	Widespread, including progenitors of somatic gonadal cells	Progenitor cell proliferation in the gonads
PAX2 (Paired box gene 2)	10q24.31	Mesonephros, metanephros, adrenals, spinal cord, hindbrain, and optic and otic vesicles	Regulation of WT1 expression and of mesenchyme-to-epithelium transition
SIX1 / SIX 4 (Sine oculis homeobox 1 and 4)	14q23.1	Urogenital ridge derivatives	Regulation of gonadal precursor cell proliferation, and of Fog2 and Sf1
WT1 (Wilms tumor-associated gene 1)	11p13	Urogenital ridge derivatives	DNA- and RNA-binding protein with transcriptional and post-transcriptional regulating capacity

2.4 SRY

A gene called SRY, or Sex-determining Region-Y chromosome, is situated right next to the pseudoautosomal region of the short arm of the Y chromosome and is the critical initiator of testis development [7]. While there is no denying SRY's involvement in the male pathway, the mechanisms by which it works are still poorly understood, and it is not even certain that SRY directly stimulates gene transcription [8]. The majority of the SRY structure is conserved in a region that is similar to high mobility group DNA-binding proteins across various mammalian species, and mutations affecting DNA-binding hinder SRY's capacity to start testes development. These results supported the first theory that SRY stimulates the transcription of downstream genes, which in turn affects male development. However, SRY's specific target genes are still unknown. Several groups have attempted to identify additional genes that are significant in the early stages of gonadal development and sexual differentiation in an attempt to expand our understanding of sexual differentiation.

2.5 Steroidogenic Factor 1 (SF-1)

The adrenal cortex and gonads' cytochrome P450 steroid hydroxylases were found to be significantly regulated by SF-1, also known as adrenal 4-binding protein and formally designated NR.5A1 [9–11]. The transcription factor was found to be a member of the nuclear hormone receptor family proteins, which mediate transcriptional activation by a variety of hydrophobic ligands including bile acids, vitamin D, thyroid hormone, steroid hormones, and retinoids. This was demonstrated by the subsequent isolation of a cDNA encoding SF-1. Since its first description, SP-1 has been identified as a global regulator of gonadal and adrenal steroidogenesis as well as reproduction. It affects the expression of genes including 3B-hydroxysteroid dehydrogenase, steroid hydroxylases, prolactin, corticotropin, luteinizing hormone, and follicle-stimulating hormone receptors, steroidogenic acute regulatory protein, AMH, INSL3, inhibin, and various gonadotropin subunits in the anterior pituitary. Furthermore, SF-1 transcripts were found in the VMH, or ventromedial hypothalamic nucleus. All things considered, these results point to the possibility that SF-1 functions in other capacities at other levels of the endocrine axis and indicate that it controls the expression of the hormones—androgens and AMH—that are essential for male sexual differentiation.

While the exact mechanisms by which SF-1 controls the expression of target genes remain unclear, recent research indicates that SF-1 interacts with several other transcription factors to control the expression of genes essential for gonadal development and sexual differentiation.

2.6 Wilms Tumor Related 1 (WT1)

Analysis of patients with familial Wilms' tumors, an embryonic kidney tumor resulting from aberrant proliferation of the metanephricblastema, led to the initial isolation of the WTI gene. While Wilms tumors are usually sporadic, a small percentage of them are familial, indicating that heritable tumors similar to retinoblastoma may have been caused by a mutation in a tumor suppressor gene. The discovery of a gene on chromosome 11q15 that encoded a potential zinc-finger transcription factor and was named Wilms' tumor related I (WTI) was the result of positional cloning efforts utilizing these kindreds.

WT1 produces up to 24 distinct isoforms of a zinc finger DNA-binding protein through alternative splicing, alternative translation start sites, and RNA editing. This protein is thought to regulate transcription by interfering with particular DNA recognition sequences upstream of target genes. In addition to exhibiting slightly different DNA-binding preferences, WTI isoforms that either contain (+KTS) or lack (−KTS) the amino acids Lys-Thr-Ser also exhibit differential nucleus localization, with +KTS isoforms interacting with spliceosomes and—KTS isoforms displaying more diffuse nuclear staining. Mutations that specifically impact one or the other isoform reveal overlapping and distinct functions of WT1, despite the fact that the molecular basis for this differential localization is still poorly understood.

While WII's expression pattern hinted at potential roles for the gene in gonadal development, the discovery that point mutations in the WTI gene causes urogenital anomalies such as hypospadias and cryptorchidism provided the first concrete proof of these roles. Analyses of WT1 knockout mice established a definitive role for WT1 in gonadal development [12–15]. These WT1 knockout mice not only had renal agenesis but also had impaired adrenal development and no gonads [16]. The internal and external genitalia developed along the female pathway due to gonadal dysgenesis, which occurred before the hormonal mediators of male differentiation were produced. Male and female WTI knockout mice exhibited early adrenal, gonadal, and renal agenesis, indicating that WTI controls the expression of target genes necessary for gonadogenesis and other early phases of urogenital ridge development. While a comprehensive examination of the clinical syndromes resulting from W'T'I mutations is outside the purview of this chapter, a particular area of interest is the Frasier syndrome, which is a condition that affects a subset of patients with WT1 mutations. Patients with Frasier syndrome present with male pseudohermaphroditism, focal glomerular sclerosis, and gonadal dysgenesis instead of Wilms tumors. Their glomerulopathy is comparatively mild; renal insufficiency is not evident until after the age of four, and renal function is partially preserved until adolescence or early adulthood. The WTI mutations that result in Frasier syndrome are located within the WT1 gene's intron 9 and appear to selectively obstruct the synthesis of WT1's + KTS splice variants [17]. This observation suggested that the WT1 + KTS and −KTS isoforms might have non-overlapping roles in the processes involved in gonadal development. Mice have provided elegant confirmation of this model, with selective impairment in the production of either the +KTS or the −KTS isoforms produced by targeting genes in embryonic stem cells [18]. These mice demonstrated that the two isoforms appear to have redundant functions in the heart, adrenal glands, spleen, and early stages of renal development. On the other hand, the successful development of tests specifically requires the +KTS isoform. As will be covered in more detail below, the +KTS isoform's mandatory function in the male pathway may be explained by its function as a crucial upstream regulator of SRY.

Although WT1 has unquestionably demonstrated important roles in the development of the gonadal regions in both humans and mice, little is currently known about the precise target genes through which it carries out these functions [19]. Research conducted on cultured cells has demonstrated that the KTS isoform of WT1 can interact with SF-1 to control the activity of the AMH promoter in relation to gonadal development [20, 21]. The orphan nuclear receptor DAX-1 inhibits this positive interaction, offering a possible method for fine-tuning AMH expression. DAX-1 and SRY itself [22] are additional putative WTI target genes that are known to control gonadal development. Research primarily concentrating on the development of the kidneys has discovered several additional WII target genes, such as growth factors and growth factor receptors and also some molecules that impede apoptosis.

2.7 SOX9

Due to the homology of SOX9's DNA-binding domain with SRY's DNA-binding region, SOX9 was first isolated. Later research revealed that SOX9 haploinsufficiency was linked to 46, XY sex reversal in conjunction with campomelic dysplasia, an autosomal dominant form of skeletal dysplasia [23]. Low levels of SOX9 expression are seen in the indifferent gonads of both sexes during development. Sox9 expression is markedly up-regulated in the testes after SRY initiates male sex determination, but it virtually vanishes in the ovaries. According to this expression pattern, SOX9 may be important in controlling the male developmental pathway [24].

This model is supported by two distinct research avenues. The genetic evidence that SOX9 expression is necessary to initiate the male developmental pathway is the most startling. The first example of this was the discovery that SOX9 is located on chromosome 17q23–24, which was duplicated in humans and resulted in 46, XX sex reversal [25].

The other line of evidence comes from promoter analyses in transgenic mice and cultured cell lines, which offers some insights into the mechanisms by which SOX9 controls gonadal differentiation. According to these studies, Sox9 regulates the AMH promoter crucially [26–28], working in concert with SF-1 to sustain high levels of AMH expression. It appears that DAX-1 can also prevent SOX9 and SF-1 from cooperating, as will be covered in more detail below.

2.8 DAX-1

While our attention has been directed toward the cooperative interactions between transcription factors that promote gene expression, there are other interactions that are known to suppress the expression of genes involved in sexual development. The orphan nuclear receptor DAX-1 is the most well-characterized inhibitor [29]. X-linked adrenal hypoplasia congenital is a condition in which patients present with adrenal insufficiency due to impaired adrenocortical development and subsequently manifest hypogonadotropic hypogonadism due to combined hypothalamic and pituitary defects. Positional cloning of the gene causing this condition allowed for the isolation of DAX-1. The Xp21.3 region, to which DAX-1 maps, is linked to sex reversal from male to female and impaired testicular development when duplicated, indicating that DAX-1 may block the male developmental pathway. Many studies in cultured cells and one study in transgenic mice [30] suggest that DAX-1 can inhibit the expression of several genes that mediate male sexual development, most notably by blocking SP-1's activation of those genes. However, the exact mechanism underlying these effects is still unknown.

2.9 Summary and Perspectives

As mentioned, during a crucial developmental window, a complex series of events in the appropriate tissues are necessary for sexual differentiation and determination. In summary, it has been demonstrated that a number of transcription factors are essential for the development of the testes and male sexual differentiation. Furthermore, a novel idea suggests that both antagonistic interactions with DAX-1 and cooperative interactions among these proteins precisely regulate the expression of the target genes needed for sexual differentiation.

2.10 Key Messages

- SF-1 and WT1 are essential for the development of the indifferent gonad in both sexes.
- SF-1, WT1, SOX9, and GATA4 work together to control the expression of target genes that mediate male sexual differentiation, such as AMH, Ins13, and the steroid hydroxylases.
- DAX-1, a negative regulator of the male developmental pathway, prevents SF-1, WT1, SOX9, and GATA4 from activating important target genes.

References

1. Jost A. Studies on sex differentiation in mammals. Rec ProgHorm Res. 1951;8:379–418.
2. Just A, Vigier B, Prepin J, Perchellet J. Studies on acx differentiation in mammals. Rec ProgHorm Res. 1973;29:1–41.

3. Wilhelm D, Palmer S, Koopman P. Sex determination and gonadal development in mammals. Physiol Rev. 2007;87:1–28.
4. Eggers S, Ohnesorg T, Sinclair A. Genetic regulation of mammalian gonad development. Nat Rev Endocrinol. 2014;10:673–83.
5. Byskov AF. Differentiation of the mammalian embryonic gonad. Physiol Rear. 1986;66:71–117.
6. Zimmermann S, Steding G, Emmen JMA, Brinkmann AO, Nayernia K, Holstein AF, et al. Targeted disruption of the Inal gene causes bilateral cryptorchidism. Mol Endocrind. 1999;13:681–91.
7. Goodfellow PN, Lovell BR. SRY and sex determination in mammals. Annu Rev Genet. 1993;27:71–92.
8. Capel B. Sex in the 90he SRY and the switch to the malepathway. Annu Rev Physiol. 1998;60:497–523.
9. Parker KL, Schimmer BP. Steroidogenic factor 1: a key determinant of endocrine development and function. Endocrine Rev. 1997;18:361–77.
10. Moruhashi KI, Omura T. Ad4BP/S 1, a transcription factor essential for the transcription of steroidogenic cytochrome P450 genes and for the establishment of the reproductive function. FASEB J. 1996;10:1569–77.
11. Hammer GD, Ingraham HA. Steroidogenic factor 1. Its role in endocrine organ development and differentiation. Front Neuroendocrinol. 1999;1999(20):199–221.
12. Sadovsky Y, Dorn D. Function of steroidogenic factor 1 during development and differentiation of the reproductive system. Rev Repro. 2000;5:136–42.
13. Achermann JC, Ito M, Hindmarsh PC, Jameson JL. A mutation in the gene encoding, steroidogenic factor 1 causes XY sex reversal and adrenal failure in humans. Nat Genet. 1999;22:125–6.
14. Biason-Lauber A, Schoenle EJ. Apparently normal ovarian differentiation in a prepubertal girl with transcriptionally inactive steroidogenic factor 1 (NRSA1/SF 1) and adrenocortical insufficiency. Am J Hum Genet. 2000;67:1563–8.
15. Hastie ND. The genetics of Wilms' tumor a case of disrupted development. Annu Rev Genet. 1993;28:523–58.
16. Kreidberg JA, Sariola H, Loring JM, Maeda M, Pelletier J, Houseman D, et al. WT 1 is essential for early kidney development. Cell. 1993;74:679–91.
17. Moore AW, McInnes I, Kreidberg J, Hastie ND, Schedl A. YAC complementation shows a requirement for Wil in the development of epicardium, adrenal gland and throughout nephrogenesis. Development. 1999;126:1845–57.
18. Barbaux S, Niaudet P, Gubler MC, Grunfeld JP, Jaubert F, Kuttenn F, et al. Donor splice site mutations in WTI are responsible for Frasier syndrome. Nat Genet. 1997;17:467–70.
19. Hammes A, Guo J-K, Lutsch G, Leheste JR, Landrock D, Ziegler U, et al. Two splice variants of the Wilm's tumor 1 gene have distinct functions during sex determination and nephron formation. Cell. 2001;106:319–29.
20. Scharnhorst V, van der Erb AJ, Jochemsen AG. WTI proteins: functions in growth and differentiation. Gene. 2001;273:141–61.
21. Nachtigal MW, Hirokawa Y, EnjeartVanHouten DL, Hlanagan JN, Hammer GD, Ingraham HA. Wilms' tumor 1 and Dax 1 modulate the orphan nudear receptor SF 1 in sex specific gene expression. Cell. 1998;93:445–54.
22. Hossain A, Saunders GF. The human sex determining gene actually regulate gene expression within the developing gonads. SRY is a direct target of WTI. J BiolChem. 2001;276:16817–23.
23. Koopman P. Sry and sox? Mammalian testis determining genes. Cell Mol Life Sci. 1999;55:619–56.
24. Morais de Silva S, Hacker A, Harley V, Goodfellow P, Swain A, Lovell Badge R. Sox expression during gonadal development implies a conserved role for the gene in testis differentiation in mammals and birds. Nat Genet. 1995;14:62–8.
25. Huang B, Wang S, Ning Y, Lamb AN, Bartley J. Autosomal XX sex reversal caused by duplication of SOX9. Am J Mad Genet. 1999;32:349–53.

26. Bishop CE, Whitworth DJ, Oin Y, Agoulnik IU, Harrison WR, Behringer RR, et al. A transgenic insertion upstream of sox is associated with dominant XX sex reversal in the mouse. Nat Genet. 2000;26:490–4.
27. Vidal VP, Chaboissier MC, de Rooij DG, Schedl A. Sox9 induces testis development in XX transgenic mice. Nat Genet. 2001;28:216–7.
28. de Santa BP, Bonneaud N, Boizet B, Desclozeaux M, Moniot B, Sudbeck P, et al. Direct interaction of SRY related protein SOX9 and steroidogenic factor 1 regulates transcription of the human anti Müllerian hormone gene. Mol Cell Biol. 1998;18:6653–65.
29. Molkentin JI. The zinc finger containing transcription factors GATA 45, and 6 ubiquitously expressed regulators of tissue specific gene expression. J BiolChem. 2000;275:38949–52.
30. Swain A, Narvaez S, Burgoyne P, Camerino G, Lovell BR. DAXI antagonizes SRY action in mammalian sex determination. Nature. 1998;391:761–7.

Hormonal Regulation During Gonadal Development

3

Sudha Rao Chandrashekar
and Rachna Mohandas Keshwani

Abstract

Although the karyotype of the embryo is decided during fertilization, many other crucial intrauterine events must take place to engender a normal male or female phenotype at birth. These events are classified as phases of sex determination and sex differentiation. Complex interactions of multiple genes, transcription factors, proteins, and hormones are responsible for these intricate changes. The phase of sex differentiation is chiefly brought about by peptide and steroid hormones produced by the developing gonad. Different developing cell types in the fetal testis express different hormones in a pre-programmed manner resulting in the male phenotype. Broadly, Antimullerian Hormone (AMH) causes regression of Mullerian structures, Testosterone stabilizes the Wolffian duct and Dihydrotestosterone (DHT) brings about masculinization of external genitalia. Descent of the testes into the scrotum occurs as a two-stage process influenced by Insulin-like peptide 3 (INSL3), testosterone, and calcitonin gene-related peptide (CGRP). Female sex differentiation occurs in the absence of AMH and testosterone. Ovarian estrogen does not have significant effects on intrauterine sex development. An appreciation of the genetic and hormonal mechanisms of sex development provides insights regarding the causes of differences in sex development (DSDs) and guides investigations and management of this perplexing group of disorders.

S. R. Chandrashekar (✉)
Department of Pediatrics, Division of Pediatric Endocrinology, Bai Jerbai Wadia Hospital for Children, Mumbai, India

Pediatric Endocrinology, Tata Memorial Hospital, Consultant Pediatric Endocrinology, Jupiter Hospital, Thane, India

R. M. Keshwani
Bai Jerbai Wadia Hospital for Children, Parel, Mumbai, India

S. K. Ratan (ed.), *Children with Differences in Sex Development*,
https://doi.org/10.1007/978-981-97-1639-5_3

Keywords

Gonadal differentiation · Gonadal determinationcells

3.1 Introduction

Sex development, a dynamic process necessitating a sequential, well-orchestrated interaction between a multitude of genes, proteins, endocrine, and paracrine factors [1–4], begins at 4–5 weeks post-conception, and is completed with achievement of secondary sexual characteristics and reproductive capability. Knowledge of the biology of sex development provides important insights into the various causes of Disorders of Sex Development, also referred to as Differences of Sex Development (DSD).

In intrauterine life, sex development is essentially a sequential three-step process of sexually undifferentiated organogenesis of gonadal or genital primordia which is bipotential, *sex or gonadal determination* wherein the undifferentiated bipotential gonad gets determined as either testes or ovary due to influence of sex chromosomes and transcription factors, and the process of *sex differentiation* of the gonad into Sertoli and Leydig cells in testes and Granulosa cell and Theca cell in ovary. This process of differentiation of external and internal genitalia is influenced by the peptide hormones and sex steroids produced by the developing gonad.

Gonadal sex denotes the type of gonad present in the individual established after the process of gonadal determination (Testis or Ovary). Genital/Anatomical sex refers to the type of external genitals and internal reproductive structures present in the individual established after the process of gonadal differentiation (Male or Female).

The seminal works of Professor Alfred Jost (1916–1991) provided crucial insights into the hormonal regulation during gonadal development [5]. He demonstrated that castrating a male rabbit fetus before sex differentiation, resulted in a female phenotype with persistent mullerian ducts and no Wolffian structures. He also showed it is testosterone that stimulates Wolffian development and a separate factor (now known as Antimullerian hormone, AMH) secreted by the testis is responsible for mullerian regression.

Broadly, sex determination occurs through transcriptional factors, whereas factors responsible for sex differentiation are the hormones acting through their receptors.

3.2 Hormones in Gonadal Development and Sex Differentiation

Further to the gonadal determination, the sex differentiation is unimorphic and is almost exclusively hormone dependent. Male sex differentiation is a complex, active process regulated by androgens and Antimullerian Hormone (AMH) and their receptors in target tissues. Insulin-like peptide 3 (INSL3) secreted by fetal Leydig cells causes the transabdominal phase of testicular descent. Together these bring about differentiation of internal and external genitalia. In fetal life, the presence or absence of ovaries does not influence phenotypic sex differentiation.

3.3 Male Sex Differentiation

By seventh embryonic week, the coelomic epithelium, mesonephric mesenchyme, and primordial germ cells segregate into testicular cords and interstitial tissue. Coelomic cells form Sertoli cells which gather together and enclose primordial germ cells in testicular cords. The differentiating Sertoli cells express AMH which causes regression of mullerian structures. Sex-determining region Y protein (*SRY*) expression and timing are deemed extremely crucial for normal Sertoli cell differentiation. Sertoli cells and germ cells increase exponentially in the second trimester. After the testicular cords have completely formed, interstitial tissue differentiation into Leydig cells happens by eighth embryonic week which produces androgens and Insulin-like peptide 3 (INSL-3). Sertoli cells are important for maintaining testicular differentiation, including the development of the germ cells and Leydig cells. Functional differentiation of Sertoli cells relies on placental hCG in the first trimester and by fetal pituitary LH in the second and third trimesters with both hormones acting through the same receptor.

3.4 Hormones in Male Sex differentiation

3.4.1 Antimullerian Hormone

Antimullerian Hormone (AMH), also aptly known as Mullerian Inhibiting Substance (MIS), to denoting its principal function, is secreted by the immature Sertoli cells. A glycoprotein homodimer of the Transforming Growth Factor B (TGFB) family, encoded by a 2.75-kb size, 5 exons long, gene located on chromosome 19q13.3 [6–8], AMH is responsible for cranio-caudal regression of the Mullerian ducts by 8–12 weeks of gestational age. All biological effects of AMH are mediated through the membrane-bound AMH type 2 receptor (AMHR2), present in gonads and Mullerian ducts of both sexes [9–11]. AMHR 2 is a serine/threonine kinase encoded by an 8.7-kb long 11 exon gene located on chromosome 12q12-q13. The type 2 receptor binds to the ligand, forms a complex with type 1 receptor, phosphorylating them and initiating a cascade of intracellular signals which by a paracrine mechanism induces epithelial cell apoptosis of mullerian ducts [12].

AMH secretion by fetal Sertoli cells starts at 7–8 weeks post-conception under the control of transcription factors (SOX9, SF1, WT1, GATA4) [13, 14]. It reaches peak concentrations between 9 and 12 weeks. The paracrine action of AMH on AMHR 2 causes regression of Mullerian structures which form the anlagen for the uterus, cervix, and upper 2/3 of the vagina [15]. During the second and third trimesters of fetal life and after birth, AMH production is increased by pituitary FSH which causes Sertoli cell proliferation and upregulates AMH transcription. However, the Mullerian structures appear to be maximally sensitive to AMH between 9 and 12 weeks after conception, when the testis is producing highest concentrations of AMH. Sertoli cells continue to make high amounts of AMH until puberty making it a useful marker of Sertoli cell function in childhood and adolescence. In the developing ovary, AMH is produced slightly later, by the granulosa cells of the preantral and early antral follicles of the late fetal and postnatal ovary.

Males with AMH or AMHR 2 mutations present with Persistent Mullerian Duct Syndrome which can present with undescended testis held tight by the fallopian tubes, but with normal male genitalia. Gonadal dysgenesis with 46, XY can present with the persistence of Mullerian structures due to faulty Sertoli cell development and AMH production. In contrast, 46 XY DSD due to Leydig cell steroidogenesis defects does not show persistence of Mullerian structures, because Sertoli cell production of AMH remains unaffected.

3.4.2 Inhibin B

Inhibin B, a glycoprotein, is a dimer of alpha and beta subunits, produced by the Sertoli cells in prepubertal testes and exerts negative feedback on pituitary FSH secretion. It is unclear whether Inhibin B has any role in gonadal development. Low but detectable Inhibin B levels are seen in cord blood, which increase to the adult range as early as in the first week after birth in a male newborn. These levels then continue to rise up to levels greater than those observed in adults by 3–6 months of age followed by a progressive decline to reach their nadir toward 3–6 years (Fig. 3.1).

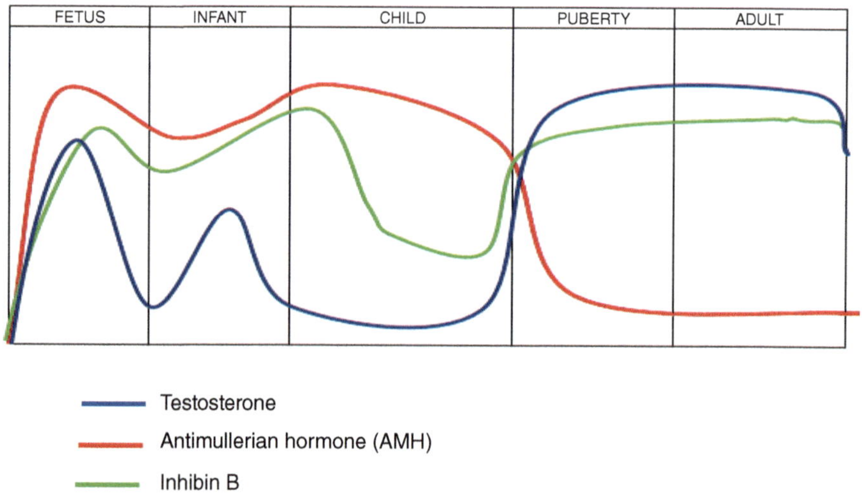

Fig. 3.1 Representative diagram of the levels of serum Testosterone, AMH, and Inhibin B during different phases of life

Inhibin B is postulated to be a marker of the presence of the testis in case of genital ambiguity with cryptorchidism but there are no definitive studies confirming the same. Inhibin B levels are monitored during treatment with recombinant FSH in prepubertal hypogonadal boys.

3.4.3 Androgens

Androgens, are C19 steroids responsible for the male phenotype not only because of induction of secondary sexual characters by their action during puberty, but also due to their role in the developing embryo.

Androgens are secreted by Leydig cells from 9 weeks post-conception age under the feedback regulation of placental hCG initially and by fetal pituitary LH later. Placental Human Chorionic Gonadotropin (HCG) is the principal circulating gonadotropin in the first trimester responsible for fetal Leydig cell steroidogenesis. From about 16–20 weeks of gestation, the developing hypothalamo–pituitary–gonadotropic system producing LH controls testosterone production by the Leydig cell of the testis [16]. Both these hormones (hCG and LH) act on the same LH/CG receptor. In the male fetus, androgen levels peak between 15 and 18 weeks of gestational age and are responsible for the differentiation of internal and external genitalia.

Testosterone, the primary sex hormone is secreted by the fetal Leydig cells and is responsible for Wolffian duct differentiation, the virilization of the urogenital sinus and the external genitalia, and the descent of the testes into the scrotum. The high local concentration of testosterone causes stabilization and proliferation of Wolffian ducts (epididymis, vas deferens, and seminal vesicle).

Testicular testosterone biosynthesis pathway overlaps with adrenal steroidogenesis pathway. In the Leydig cell, cholesterol is taken up from circulation by LDL or HDL receptor or is generated inside the cell by cholesterol synthesis pathways. Steroidogenic acute regulatory protein (StAR) facilitates movement of cholesterol from the outer to the inner mitochondrial membrane [17]. The first step in steroid hormone synthesis involving cleavage of the cholesterol side chain to generate pregnenolone is catalyzed by the P450scc enzyme (CYP11A1) (rate-limiting step). Pregnenolone is converted to progesterone by the microsomal enzyme 3β-hydroxysteroid dehydrogenase type 2 (3βHSD2), or it can undergo 17α-hydroxylation by P450c17 (CYP17) to yield 17-hydroxypregnenolone. 17α-Hydroxylation by the same enzyme also converts progesterone to 17-hydroxyprogesterone. Besides its hydroxylase activity, CYP17 also has 17,20-lyase activity, which breaks the C17,20 carbon bond of 17-hydroxypregnenolone to generate dehydroepiandrosterone (DHEA). The P450 oxidoreductase (POR) and the cytochrome b5 are essential for optimal 17,20 lyase activity.

In the Leydig cell, the major pathway for androgen production is through conversion of 17-hydroxypregnenolone to DHEA rather than through conversion of 17-hydroxyprogesterone (17-OHP) to androstenedione [18]. Subsequent testosterone production can occur through conversion of DHEA to androstenedione by

3βHSD2, followed by the 17β-hydroxysteroid dehydrogenase action of 17βHSD3 to generate testosterone or through the intermediate compound androstenediol.

Dihydrotestosterone (DHT) is responsible for virilization of external genitalia and the urogenital sinus. Dihydrotestosterone (DHT) is the more potent non-aromatizable androgen. In the classic pathway of DHT biosynthesis, testosterone is converted to DHT in target tissues by the 5 alpha-reductase type 2 enzyme. The alternative pathway or the "backdoor" pathway, does not require testosterone as a precursor. The high-affinity action of DHT on the same androgen receptors in the genital skin is required for the formation of prostate and prostatic urethra, a fusion of the urethral folds to form the shaft of penis surrounding the penile urethra, fusion of the labioscrotal folds to form the scrotum and development of the genital tubercle forming glans penis.

Androgen action, at cellular and molecular level, is brought about by varying affinity of androgens to Androgen Receptors (AR). The androgen receptor is 919 amino acid protein encoded by a gene known as AR or NR3C4 located on the X chromosome (Xq11-q12) [19]. It is a member of a superfamily of nuclear receptors that exert their action by acting as transcription factors. In the unbound state, the AR receptor is located in the cytoplasm and on ligand binding is translocated to the nucleus where it binds to target DNA sequences causing expression of effector molecules.

Disorders along the pathway of androgen synthesis and target organ action can result in impaired androgenization and 46, XY, DSD. Since many steps of gonadal and adrenal steroidogenesis overlap, some of the enzyme deficiencies that cause androgen deficiency can also cause glucocorticoid and/or mineralocorticoid deficiency. In all these conditions, because Sertoli cell function and AMH secretion are intact, Mullerian structures are absent.

3.5 Insulin-Like 3 (INSL3) and GREAT

Migration of testes from the lower pole of kidney to the scrotal sac starts at 8 weeks and is completed by the middle of the third trimester. It occurs in 2 phases: transabdominal migration and inguino-scrotal descent [20]. Testicular descent starts at 8 weeks and is completed by the middle of the third trimester. INSL3, a relaxin-like factor mediates the transabdominal stage of testicular descent (8–15 weeks) via action on its G protein-coupled receptor, GREAT [21]. It causes contraction and thickening of the gubernacular ligament and degeneration of the cranio-suspensory ligament.

The trans-inguinal stage of testicular descent (25–35 weeks) is mediated by androgens and the neurotransmitter, calcitonin gene-related peptide secreted by the genitofemoral nerve. The processus vaginalis forms a channel through which the testis descends from the abdomen to the scrotum.

3.6 Female Sex Differentiation

Unlike the male sex differentiation, formation of female sex does not involve significant morphologic changes in the external genitalia. In the absence of AMH, Mullerian structures persist to form the uterus, fallopian tubes, and upper two-third of the vagina. In the absence of local testosterone, Wolffian structures regress. Urogenital sinus develops into urethra and vagina. The genital tubercle forms the clitoris, the urogenital swellings form the labia majora and urogenital folds form the labia minora.

Although fetal ovary can produce steroids and express aromatase activity to produce estrogen, the amount of estrogen produced is insignificant and does not have significant effects on sex development until the time of puberty [22]. In contrast to the testis, the developing ovary does not express LH/HCG receptors until week 16 and the first primary follicles form under the influence of FSH only at around 20 weeks of gestation [23].

3.7 Conclusion

The development, differentiation, and maturation of internal and external genitalia are timed sequential events under the influence of genes, transcription factors, and hormones. The production and action of these hormones is in turn regulated during the prenatal period and later life, with varying levels during different periods of growth. The assessment of these hormones becomes pertinent in investigating a case of DSD.

References

1. Smith P, Wilhelm D, Rodgers RJ. Development of mammalian ovary. J Endocrinol [Internet]. 2014;221(3). Available from: https://pubmed.ncbi.nlm.nih.gov/24741072/
2. Eggers S, Ohnesorg T, Sinclair A. Genetic regulation of mammalian gonad development. Nat Rev Endocrinol [Internet]. 2014;10(11):673–83. Available from: https://pubmed.ncbi.nlm.nih.gov/25246082/
3. Brennan J, Capel B. One tissue, two fates: molecular genetic events that underlie testis versus ovary development. Nat Rev Genet [Internet]. 2004;5(7):509–21. Available from: https://pubmed.ncbi.nlm.nih.gov/15211353/
4. Svingen T, Koopman P. Building the mammalian testis: origins, differentiation, and assembly of the component cell populations. Genes Dev [Internet]. 2013;27(22):2409–26. Available from: https://pubmed.ncbi.nlm.nih.gov/24240231/
5. Josso N. Professor Alfred Jost: The builder of modern sex differentiation. Sexual Development [Internet]. 2008;2(2):55–63. Available from: https://www.karger.com/Article/FullText/129690
6. Massagué J. TGF-beta signaling: receptors, transducers, and Mad proteins. Cell [Internet]. 1996;85(7):947–50. Available from: https://pubmed.ncbi.nlm.nih.gov/8674122/
7. Cate RL, Mattaliano RJ, Hession C, Tizard R, Farber NM, Cheung A, et al. Isolation of the bovine and human genes for Müllerian inhibiting substance and expression of the human gene in animal cells. Cell [Internet]. 1986;45(5):685–98. Available from: https://pubmed.ncbi.nlm.nih.gov/3754790/

8. Cohen-Haguenauer O, Picard JY, Mattéi MG, Serero S, van Cong N, de Tand MF, et al. Mapping of the gene for anti-müllerian hormone to the short arm of human chromosome 19. Cytogenet Cell Genet [Internet]. 1987;44(1):2–6. Available from: https://pubmed.ncbi.nlm.nih.gov/3028714/
9. di Clemente N, Wilson C, Faure E, Boussin L, Carmillo P, Tizard R, et al. Cloning, expression, and alternative splicing of the receptor for anti-Müllerian hormone. Mol Endocrinol [Internet]. 1994;8(8):1006–20. Available from: https://pubmed.ncbi.nlm.nih.gov/7997230/
10. Mishina Y, Rey R, Finegold MJ, Matzuk MM, Josso N, Cate RL, et al. Genetic analysis of the Müllerian-inhibiting substance signal transduction pathway in mammalian sexual differentiation. Genes Dev [Internet]. 1996;10(20):2577–87. Available from: https://pubmed.ncbi.nlm.nih.gov/8895659/
11. Baarends WM, van Helmond MJL, Post M, van der Schoot PJCM, Hoogerbrugge JW, de Winter JP, et al. A novel member of the transmembrane serine/threonine kinase receptor family is specifically expressed in the gonads and in mesenchymal cells adjacent to the müllerian duct. Development [Internet]. 1994;120(1):189–97. Available from: https://pubmed.ncbi.nlm.nih.gov/8119126/
12. Shi Y, Massagué J. Mechanisms of TGF-β signaling from cell membrane to the nucleus. Cell [Internet]. 2003;113(6):685–700. Available from: https://pubmed.ncbi.nlm.nih.gov/12809600/
13. Teixeira J, Maheswaran S, Donahoe PK, Jost A. Müllerian inhibiting substance: an instructive developmental hormone with diagnostic and possible therapeutic applications. Endocr Rev [Internet]. 2001;22(5):657–74. Available from: https://academic.oup.com/edrv/article/22/5/657/2424291
14. Josso N, Rey RA, Picard JY. Anti-müllerian hormone: a valuable addition to the toolbox of the pediatric endocrinologist. Int J Endocrinol [Internet]. 2013;2013. Available from: https://pubmed.ncbi.nlm.nih.gov/24382961/
15. Josso N, Belville C, di Clemente N, Picard JY. AMH and AMH receptor defects in persistent Müllerian duct syndrome. Hum Reprod Update [Internet]. 2005;11(4):351–6. Available from: https://pubmed.ncbi.nlm.nih.gov/15878900/
16. Huhtaniemi IT, Korenbrot CC, Jaffe RB. HCG binding and stimulation of testosterone biosynthesis in the human fetal testis. J Clin Endocrinol Metab [Internet]. 1977;44(5):963–7. Available from: https://pubmed.ncbi.nlm.nih.gov/192755/
17. Papadopoulos V, Miller WL. Role of mitochondria in steroidogenesis. Best Pract Res Clin Endocrinol Metab. 2012;26(6):771–90.
18. Flück CE, Miller WL, Auchus RJ. The 17, 20-lyase activity of cytochrome p450c17 from human fetal testis favors the delta5 steroidogenic pathway. J Clin Endocrinol Metab [Internet]. 2003;88(8):3762–6. Available from: https://pubmed.ncbi.nlm.nih.gov/12915666/
19. Hughes IA, Davies JD, Bunch TI, Pasterski V, Mastroyannopoulou K, Macdougall J. Androgen insensitivity syndrome. Lancet [Internet]. 2012;380(9851):1419–28. Available from: https://pubmed.ncbi.nlm.nih.gov/22698698/
20. Hutson JM, Southwell BR, Li R, Lie G, Ismail K, Harisis G, et al. The regulation of testicular descent and the effects of cryptorchidism. Endocr Rev [Internet]. 2013;34(5):725–52. Available from: https://pubmed.ncbi.nlm.nih.gov/23666148/
21. Zimmermann S, Steding G, Emmen JMA, Brinkmann AO, Nayernia K, Holstein AF, et al. Targeted disruption of the Insl3 gene causes bilateral cryptorchidism. Mol Endocrinol [Internet]. 1999;13(5):681–91. Available from: https://pubmed.ncbi.nlm.nih.gov/10319319/
22. Fowler PA, Anderson RA, Saunders PT, Kinnell H, Mason JI, Evans DB, et al. Development of steroid signaling pathways during primordial follicle formation in the human fetal ovary. J Clin Endocrinol Metab [Internet]. 2011;96(6):1754–62. Available from: https://academic.oup.com/jcem/article/96/6/1754/2834319
23. Kaplan SL, Grumbach MM. Pituitary and placental gonadotrophins and sex steroids in the human and sub-human primate fetus. Clin Endocrinol Metab [Internet]. 1978;7(3):487–511. Available from: https://pubmed.ncbi.nlm.nih.gov/215355/

Brain Imprinting During Gonadal Development

4

Deepa Anirudhan and Parvathy Lalitha

Abstract

The development of sex differences in human beings and other mammals is primarily triggered by the presence of X and Y sex chromosomes. Exposure to gonadal hormones at various stages of development contributes to sex-specific behaviors. Notably, various organs, including the brain, liver, and immune system, exhibit sex differences under the influence of genetic factors and hormonal milieu. Sexual differentiation of the human brain is primarily determined by surge of sex steroids during three critical periods of development namely, embryogenesis, the perinatal period, and puberty, where sexually dimorphic peaks of gonadal hormones are found. Gonadal hormones act on the fetal brain, producing either an activation effect, resulting in a transient modification, or an organization effect, leading to a permanent modification of neuronal function. Testosterone exposure during the critical periods of sexual development induces changes in the sexually dimorphic nuclei of the hypothalamus, contributing to male gender identity. In females, genetic factors play a more significant role, and hormonal influence on the brain is comparatively less pronounced. Recent research has highlighted the role of transcription factors, signaling molecules, and the immune system in the sexual differentiation of the brain. Understanding the concept of sexual differentiation in both brain structure and function is crucial for making informed decisions regarding sex assignment and the clinical management of infants with ambiguous genitalia.

D. Anirudhan (✉)
Govt Medical College, Thrissur, India

P. Lalitha
Aster Medcity, Kochi, India

S. K. Ratan (ed.), *Children with Differences in Sex Development*,
https://doi.org/10.1007/978-981-97-1639-5_4

Keywords

Sexual dimorphism · Brain imprinting · Hypothalamus

4.1 Introduction

The initiation of mammalian sexual differentiation revolves around two primary sexes, dictated by distinct sex chromosomes (X or Y). Apart from the obvious external differences, sexual variations exist in various organs and systems, including the liver, immune system, and brain [1, 2]. While sex chromosomes serve as the primary determinant for sex identification, the majority of sex differences in mammals are believed to stem from differential exposure to sex steroid hormones secreted by gonads during development. In male mammals, the expression of the *SRY* gene on the Y chromosome triggers the formation of testes, triggering a molecular cascade that directs gonadal primordia toward testicular development [3]. Subsequently, testes secrete key hormones, notably the steroid hormone, testosterone, and the peptide Anti-Mullerian Hormone (AMH), playing a crucial role in the development of the male phenotype [4]. It is currently understood that ovaries develop in the absence of *SRY*, with emerging insights suggesting a critical genetic cascade for ovarian development [3].

The human fetal brain exhibits sexual dimorphism similar to primitive gonads. Research across different mammalian species consistently indicates that, the brain, like the rest of the reproductive system, tends to develop as a female structure by default. However, to manifest structural and functional characteristics typical of his species, the male fetal brain must be exposed to testicular hormones during critical periods of development. Much like any other mammals, human beings are most likely to be subjected to this process of hormone-dependent sexual differentiation of the brain.

The sexual differentiation of the brain takes place within a restricted critical window during embryonic life [5]. Essential roles in this process are played by sex steroids, particularly testosterone and its aromatized metabolite estradiol, which exert profound influences on brain development and plasticity throughout the lifespan [6]. In rodents, the metabolite estradiol tends to exert a greater influence on the masculinization of the developing brain, while in primates, including humans, the precursor testosterone may play a significant role in masculinization [7, 8]. Regardless of the species, testosterone is converted to estradiol with the help of the enzyme aromatase, occurring locally in specific regions of the brain [9, 10]. Furthermore, androgen and estrogen receptors are present in discrete brain areas throughout life, contributing to the anatomical specificity of potential responses [11].

More than 60 years ago, Phoenix and colleagues (1959) proposed that gonadal hormones impact the brain in two distinct ways: an activation effect, resulting in a transient modification of neuronal function, or an organization effect, leading to a permanent modification of neuronal function. During embryonic life, the presence of a functional testis triggers the regression of Mullerian structures, promotes the growth of Wolffian structures, and contributes to the masculinization of external genitalia. In contrast, for females, where genetic factors take precedence, the impact of hormones on the brain is relatively less pronounced [12].

The hypothalamus plays a crucial role in the reproductive system regulating the release of gonadotropic hormones (LH and FSH) from the pituitary and the orchestration of the Luteinizing Hormone (LH) surge, a crucial factor for ovulation. Furthermore, gonadal hormones act on the brain to facilitate reproductive behavior.

The concept of sexual differentiation in brain structure and function has the potential to significantly influence clinical decisions regarding sex assignment and the management of infants with ambiguous genitalia. The prenatal hormonal environment may have resulted in lasting alterations to the structure and functional capacity of the brain.

4.1.1 Sexually Dimorphic Regions of the Human Brain

Sex differences in the hypothalamus and closely interconnected limbic structures are considered to be the basis of sex differences in reproduction and sexual behavior, such as the menstrual cycle in women, gender identity, and sexual orientation (including homosexuality and heterosexuality). Certain regions of the hypothalamus display sexual dimorphism and undergo changes in both volume and cell count in response to exposure to sex steroids. The critical stages of development during which sex steroids likely determine the sexual differentiation of the human brain are most likely the three periods marked by sexually dimorphic peaks in gonadal hormone levels: the first half of gestation (when genitalia are formed), the perinatal period, and puberty [13].

The sexually dimorphic nucleus of the preoptic area (SDN-POA) was initially identified in rats by Gorski et al. in 1978 [14]. Immunocytochemical studies provide evidence supporting the homology between the SDN-POA in rats and humans. In male rats, this area is 3–8 times larger than in females, and lesions of the SDN-POA have a significant impact on male sexual behavior [15]. Furthermore, there is a positive correlation between the volume of the SDN-POA and both testosterone levels and male sexual activity in rats [16].

Other sexually dimorphic nuclei include the preoptic anterior hypothalamus (AVPV), ventromedial hypothalamus (VMH), the bed nucleus of the stria terminalis (BNST), mammillary body complex, and the suprachiasmatic nucleus (SCN), where males typically exhibit larger volumes compared to females. Exposure of the developing brain to testicular hormones fosters the development of functional, neurochemical, and neuroanatomical characteristics that are typical of the males of a given species in these areas [17].

During adolescence, the brain undergoes a secondary organization influenced by gonadal steroid hormones secreted during puberty. The presence of testicular hormones during this period not only enhances the activation of male social behaviors but also contributes to both masculinization and defeminization of behavioral responses in adulthood, resembling the effects observed with exposure to testosterone during perinatal brain development [18]. Other research indicates that the overt sexual differentiation of the human hypothalamus occurs much later than traditionally assumed, between approximately 4 years of age and adulthood. This delayed differentiation prompts the consideration that underlying processes may have been programmed during mid-pregnancy or the neonatal period, potentially as a result of exposure to male or female steroid hormones [19].

While sex steroid hormones play a primary role in the majority of brain sexual differentiation processes, recent literature has brought forth crucial inquiries into the direct involvement of genes on sex chromosomes, a role distinct from the actions of sex steroids [20].

4.2 A New View of Sexual Differentiation of the Brain

4.2.1 Role of Transcription Factors

In conjunction with the sex chromosome complement, there exists significant potential for sex differences in transcription regulatory networks (TRNs). Transcription factors (TF) engage directly with DNA or through co-activators and co-repressors within a complex transcriptional system. Steroid receptors, recognized as classical TFs [21], play a role, and steroids can also influence TRNs by regulating the activity of DNA methylating enzymes [22] and histone-modifying enzymes [23]. Further investigations are warranted to determine whether TRNs are established during the developmental critical period and subsequently maintained throughout the lifespan.

4.2.2 Role of Signaling Molecules and Immune System

The preoptic area stands as a pivotal region in shaping sexual behavior and is a key determinant of sex differences in synaptic patterning and cell death rates within the brain. Neurons in this area are rich in estrogen and androgen receptors. The initiation of synaptic pattern differentiation, specifically the development of denser dendritic spine synapses in males, is orchestrated by two immune cell types, namely microglia and mast cells [24]. Mast cells, responsible for producing histamine, serotonin, prostaglandins (PG), and cytokines, are abundantly found in the preoptic area during the critical period of sexual differentiation, with a higher presence in males than females. They continuously degranulate in males, releasing histamine.

During this critical period, nearby microglia respond to the released histamine by increasing prostaglandin production, notably PGE2, with signaling capacity through G-protein coupled receptors linked to adenyl cyclase, cAMP production, and activation of protein kinase A (PKA). Activated PKA, in turn, phosphorylates a subunit of the α-amino-3-hydroxy-5-methyl-4-isoxazolepropionic acid (AMPA) glutamate receptor. This process promotes the trafficking of the receptor to the membrane, facilitating the formation and stabilization of excitatory dendritic spine synapses. This increase in dendritic spines persists into adulthood and correlates positively with the expression of male sexual behavior [25]. Supporting this finding, studies have shown that selectively depleting microglia during the critical period completely eliminates the capacity for mating in mature males rats, but has no impact on females [26].

In the preoptic area, a second nucleus, the anteroventral periventricular nucleus (AVPV), plays a crucial role in regulating luteinizing hormone (LH) release from the pituitary by controlling the activity of gonadotropin-releasing hormone (GnRH) neurons. The AVPV is also a sexually dimorphic nucleus, larger in females compared to males. In this region, inhibitory GABAergic neurons express estrogen receptors, established during the critical period for sexual differentiation, similar to the sexually dimorphic nucleus (SDN). However, in contrast, estradiol here appears to induce cell death, leading to a reduction in the number of surviving cells in males. This process involves signaling pathways typically associated with inflammation, such as TNFα [27].

The amygdala is a complex collection of subnuclei involved in a variety of social, sexual, and emotional behaviors, displays sexual dimorphism in the medial amygdala. This dimorphism arises due to increased microglia in newborn males compared to females [28]. Consequently, the neuronal-immunological system emerges as the key determinant of neural circuit construction, significantly impacting sex differences in physiology and behavior as the animal matures [25].

4.3 Gender Dysphoria in Cases of Atypical Genitalia

The majority of individuals with 46,XX and 21-hydroxylase deficiency (21-OHD) who were assigned female at birth reportedly exhibit a female gender identity. Notably, studies have found no significant correlations between gender identity and virilization scores of the external genitalia [29, 30]. Interestingly, only 5% of 46,XX individuals assigned as adult females experienced gender dysphoria, a prevalence significantly higher than that observed in the general population. Conversely, more than 10% of 46,XX individuals assigned as male at birth reported experiencing gender dysphoria [29].

Identifying the factors that distinguish the majority of females with congenital adrenal hyperplasia (CAH) who maintain a female gender identity from the small minority with a male gender identity could prove valuable in predicting gender identity in challenging cases of intersexuality, such as partial androgen insensitivity. It can also shed light on the causes of extremely atypical gender identity. Factors contributing to this differentiation may encompass hormones (e.g., heightened sensitivity to androgens and low levels of ovarian hormones), genes (especially *SRY*), social environment (e.g., inconsistent rearing environments), or individual characteristics (e.g., perceived pressure to conform to gender roles) [30].

Likewise, insights from other conditions, like 5-alpha-reductase-2 deficiency, indicate a substantial rate (56–63%) of gender role change from female to male during adolescence and adulthood. This suggests the influence of sex steroids in early childhood, shaping future gender identity. In addition to these neural circuits, the mutations in the estrogen receptor have been linked to later gender dysphoria in individuals with apparently normal anatomical development [31, 32]. An analysis of four polymorphisms at the promoter of the ESR1 gene (C1 = rs9478245, C2 = rs3138774, C3 = rs2234693, and C4 = rs9340799) in a population of 273 trans women, 226 trans men, and 537 cis gender controls revealed a higher occurrence of ESR1 mutations in trans men. This suggests an ongoing role of neural and genetic factors in the development of gender identity and roles throughout childhood [32].

4.4 Conclusion

The manifestation of sex-specific behaviors in humans stems from exposure to gonadal hormones at various stages of development. Testosterone, particularly during fetal life, influences the sexually dimorphic nuclei of the hypothalamus, contributing significantly to the development of male gender identity. The sexual differentiation of the brain, mediated by steroids, involves the participation of various transcription factors, signaling molecules, and the immune system. Notably, there is a distinct role of genes on sex chromosomes in the developing brain, distinct from the actions of sex steroids, which is particularly pronounced in females, where hormonal influence on the brain holds less significance.

Furthermore, mutations in the estrogen receptor have been associated with later gender dysphoria, even in individuals with apparently normal anatomical development. Considering these diverse factors is essential when assigning gender to a baby with atypical genitalia to mitigate the risk of gender dysphoria emerging in adulthood.

References

1. Carrer HF, Cambiasso MJ. Sexual differentiation of the brain: genes, estrogen, and neurotrophic factors. Cell Mol Neurobiol. 2002;22:479–500.
2. Cooke B, Hegstrom CD, Villeneuve LS, Breedlove SM. Sexual differentiation of the vertebrate brain: principles and mechanisms. Front Neuroendocrinol. 1998;19:323–62.
3. DeFalco T, Capel B. Gonad morphogenesis in vertebrates: divergent means to a convergent end. Annu Rev Cell Dev Biol. 2009;25:457–82.
4. Gilbert SF. Developmental biology. Subderland, Massachusetts: Sinauer Associates, Inc.; 1994.
5. Blaustein JD, McCarthy MM. Phoenix, Goy, Gerall, and Young, endocrinology, 1959: 50 years young and going strong. Endocrinology. 2009;150:2501.
6. Lenz KM, McCarthy MM. Organized for sex - steroid hormones and the developing hypothalamus. Eur J Neurosci. 2010;32:2096–104.
7. Galani A, Kitsiou-Tzeli S, Sofokleous C, Kanavakis E, Kalpini-Mavrou A. Androgen insensitivity syndrome: clinical features and molecular defects. Hormones. 2008;7:217–29.
8. Smith EP, Boyd J, Frank GR, Takahashi H, Cohen RM, Specker B, et al. Estrogen resistance caused by a mutation in the estrogen-receptor gene in a man. N Engl J Med. 1994;331:1056–61.

9. Balthazart J, Ball GF. New insights into the regulation and function of brain estrogen synthase (aromatase). Trends Neurosci. 1998;21:243–9.
10. Roselli CE, Ellinwood WE, Resko JA. Regulation of brain aromatase activity in rats. Endocrinology. 1984;114:192–200.
11. Lauber AH, Romano GJ, Pfaff DW. Gene expression for estrogen and progesterone receptor mRNAs in rat brain and possible relations to sexually dimorphic functions. J Steroid Biochem Mol Biol. 1991;40:53–62.
12. Phoenix C, et al. Organizing action of prenatally administered testosterone propionate on the tissues mediating mating behavior in the female Guinea pig. Endocrinology. 1959;65:369–82.
13. Swaab DF, Gooren LJ, Hofman MA. The human hypothalamus in relation to gender and sexual orientation. Prog Brain Res. 1992;93:205–19.
14. Gorski RA, Gordon JH, Shryne JE, Southam AM. Evidence for a morphological sex difference within the medial preoptic area of the rat brain. Brain Res. 1978;148:333–46.
15. Turkenburg JL, Swaab OF, Endert E, Louwerse AL, Van de Poll NE. Effects of lesions of the sexually dimorphic nucleus on sexual behaviour of testosterone-treated female Wistar rats. Brain Res Bull. 1988;21:215–24.
16. Anderson RH, Fleming DE, Rhees RW, Kinghorn E. Relationships between sexual activity, plasma testosterone, and the volume of the sexually dimorphic nucleus of the preoptic area in prenatally stressed and non-stressed rats. Brain Res. 1986;370:1–10.
17. Gorski RA. Hypothalamic imprinting by gonadal steroid hormones. In: Zderic, et al., editors. Advances in experimental medicine and biology. Kluwer Academic/Plenum Publishers; 2002. Pediatric Gender Assignment: A Critical Reappraisal. p. 57–73.
18. Schulz KM, Molenda-Figueira HA, Sisk CL. Back to the future: the organizational-activational hypothesis adapted to puberty and adolescence. Horm Behav. 2009;55:597–604.
19. Swabb DF, Chung WCJ, Kruijver FPM, Hofman MA, Ishunina TA. Sexual differentiation of the human hypothalamus pediatric gender assignment – a critical appraisal, vol 511, p. 72–103.
20. Becker JB, Berkley KJ, Geary N, Hampson E, Herman JP, Young E. Sex differences in the brain: from genes to behavior. Oxford: Oxford University Press; 2007.
21. Barrett TJ, Thomas C. Steroid receptors at the nexus of transcriptional regulation. J Cell Biochem Suppl. 1998;30:188–93.
22. Nugent BM, Wright CL, Shetty AC, Hodes GE, Lenz KM, Mahurkar A, et al. Brain feminization requires active repression of masculinization via DNAmethylation. Nat Neurosci. 2015;18:690–7.
23. Matsuda KI, Mori H, Nugent BM, Pfaff DW, McCarthy MM, Kawata M. Histone deacetylationduring brain development is essential for permanent masculinization of sexual behavior. Endocrinology. 2011;152:2760–7.
24. Lenz KM, Nugent BM, Haliyur R, McCarthy MM. Microglia are essential to masculinization of brain and behavior. J Neurosci. 2013;33:2761–72.
25. McCarthy MM. A new view of sexual differentiation of mammalian brain. J Comp Physiol. 2020;206(3):369–78.
26. VanRyzin JW, Yu SJ, Perez-Pouchoulen M, McCarthy MM. Temporary depletion of microglia during the early postnatal period induces lasting sex-dependent and sex-independent effects on behavior in rats. eNeuro. 2016;3:3.
27. Forger NG. Cell death and sexual differentiation of the nervous system. Neuroscience. 2006;138:929–38.
28. VanRyzin JW, Marquardt AE, Argue KJ, Vecchiarelli HA, Ashton SE, Arambula SE, et al. Microglial phagocytosis of newborn cells is induced byendocannabinoids and sculpts sex differences in juvenile rat social play. Neuron. 2019;102:435 49.
29. Dessens AB, Slijper FME, Drop SLS. Gender dysphoria and gender change in chromosomal females with congenitaladrenal hyperplasia. Arch Sex Behav. 2005;34:389–97.
30. Berenbaum SA, Bailey JM. Effects on gender identity of prenatal androgens and genital appearance: evidence from girls with congenital adrenal hyperplasia. J ClinEndocrinolMetab. 2003;88:1102–6.

31. Fernández R, Delgado-Zayas E, Ramírez K, Cortés-Cortés J, Gómez-Gil E, Esteva I, Almaraz MC, Guillamon A, Pásaro E. Analysis of four polymorphisms located at the promoter of the estrogen receptor alpha ESR1 gene in a population with gender incongruence. Sex Med. 2020;8:490–500.
32. Cortés-Cortés J, Fernández R, Teijeiro N, Gómez-Gil E, Esteva I, Almaraz MC, Guillamón A, Pásaro E. Genotypes and haplotypes of the estrogen receptor α gene (ESR1) are associated with female-to-male gender dysphoria. J Sex Med. 2017;14(3):464–72.

Part II

Gender Identity

Special thanks to Dr. Manju Mehta, Senior Consultant, Psychology, Holy Family Hospital, Delhi and Dr. Jagdish Kaur, Public Health Specialist, WHO, for assisting and aiding me in editing this section.

Gender Identity and Role: Revisiting the Psychosocial Mosaic

5

Ishita U. Bharadwaj, Meharmeet Kaur, and Simmi K. Ratan

Abstract

This chapter illustrates the various shades of psychological bodies around gender, and sexuality. The key endeavor of this segment is to deconstruct the notions of moving beyond the gender binary, the most contemporary paradigm adopted. A nuanced understanding of gender, especially in the context of sexual development is crucial when handling clinical concerns around the (ab)normalcy around sexual development of children. It intends to render some psychological insights into notions around gender and sexuality. Appreciating the varied socio-cultural nuances of gendering is pivotal for developing a multi-dimensional approach to building psychotherapeutic interventions when dealing with cases of DSD and their caregivers.

Keywords

Gender · Sexuality · Sexual development · Socio-cultural

I. U. Bharadwaj · M. Kaur
Department of Psychology, University of Delhi, New Delhi, India

S. K. Ratan (✉)
Department of Pediatric Surgery, Maulana Azad Medical College, New Delhi, India

S. K. Ratan (ed.), *Children with Differences in Sex Development*,
https://doi.org/10.1007/978-981-97-1639-5_5

5.1 Introduction

In simplified terms, gender refers to everything a person says or does to identify to oneself and to others to fit in status, which traditionally was restricted being "male" or "female." In modern times with greater delving into this area, newer terminology such as "gender fluidity" or being "asexual" or "ambivalent" have also cropped up.

Since a person's gender is both expressed privately and also has public expression, a more lucid definition is put across for the terms, "gender identity" and "gender role" "Gender identity" as 'sameness, unity and persistence of one's individuality as male, female, ambivalent, especially as it is expressed in self-awareness and behavior.

On the other hand, "Gender role" is everything that a person says and does to indicate to others or to self the degree that one is either male or female or ambivalent and includes but is not restricted to sexual arousal and response. It is interesting to know that in some subjects with differences in sex development (DSD) or even otherwise, gender identity may be ambiguous or fluid whereas gender role may be normal. Hence these terminologies are better used separately [1].

Gender as a theory and practice has been a fairly recent induction in the existing academic discourses across disciplines. It has given an intersectional lens for addressing structural and functional aspects of human behavior. There has been a distinct movement from biological determinism flagging the essentialist belief to deconstructing gender as a category. Contestation on gender has been a long-going debate. On one hand, *Doing Gender* by West & Zimmerman in their work in 1987, which highlighted the process of reproduction of gender through interactions [2] scholars have also argued on the possibility of undoing gender [3] through the voluminous research in this area has contributed to contouring important trends; assertion beyond binaries hence bringing forth the emerging concept of fluid identity(he/she) in place. Butler in 1990 propounded, "gender is what we do and not what we are". She emphasizes on the situatedness of gender through a series of ritualized repetition rather than singular act. There is sufficient scientific literature suggesting a clear difference between gender and sex as categories.

The most critical question is on the possibility of (un)doing of gender. One side of the argument consents, suggesting its diffusion in social interaction [3, 4]. It critiques the traditional doctrines on gender binaries and their role in gender oppression, thereby stressing upon initiating a dialogue on "undoing" of gender for augmenting the feminist endeavor of annihilating gender inequality [3]. In turn, scholars like West and Zimmerman, in 2009, have argued that gender can never be undone, rather it should be redone. The gender structures can be repositioned but not extinguished [5].

Transgender individuals quash the presumption that, sex, sex category and gender are aligned. While cis-gender, or non trans-identified people are ascribed specific sex during birth, are situated in the analogous gender norms, the sex and/or gender categories do not align with their ascribed sex. In a theoretical sense, this disturbance spurs opportunity for (un) and (re) doing gender. A close examination of queer individuals' experiences, incites the pertinent question of processes involved in undoing gender. It makes one contemplate ways through which transgender individuals negotiate gender binaries in their daily routine course [3, 4]. Some suggest their engagement in (re) doing gender through broadening or changing the norms pertaining to gender [5]. In the context of "Hegemonic gender norms" which conform to male dominance, work spaces become one of the crucial domains for examining the limitations to normative gender performance. Transgenders have recently started venturing the formal work spaces, owing to the constant efforts of transgender activists [6]. Looking back in history, feminists have either levied transgender individuals for retaining the oppressive appreciation of gender [7–10] or have viewed them as an agency in resisting the "gender hegemony" [11–15]. Both arguments are substantial, however, neither of them have given sufficient heed to ways through which transgendered people construe their effectiveness of presence in social and institutional spaces. West and Zimmerman, in their work draw significantly from the case assignment of Garfinkel for formulating a legit understanding of gender as a phenomenon in an interactionist context.

5.2 A Case Illustration

The early researchers on transsexual experience conducted lab-like conditions mostly in hospitals and gender identity clinics. The classic Garfinkel's study on Agnes initiated the unethical and dehumanizing practice of using trans people for lab experiments [16, 17].

Feminist Analyst Rogers, in 1992 through the case of Agnes, of 1992 extensively tabbed on the power dynamics between the researcher and the participant. Agne was lured for gender reassignment surgery on conditions of being "appropriately feminine." These situations further significantly impacted on the gender performance of Agnes and Garfinkel's interpretation of Agnes which was a reflection of his own understanding of gender normativity. Assuming that Agnes and other trans individuals might have been motivated by the desire of being represented as normatively gendered (via gender reassignment surgery), the understanding of gender from such works calls for a critical thought.

5.3 Critical Reviews

5.3.1 Etymological Standpoints on Gender and Sexuality

Barbara Risman, in the 2009 symposium on Gender and Society that was themed on, "Doing Gender" deliberated around the *conceptual confusion* induced by uniform practices on "doing gender" in sociological research. She stressed on feminist foundational lens to understand the gendering process. Risman, 2009 further posited that, undoing Gender exists when the essentialism *of binary distinctions between people based on sex category is challenged* [3]. The issue of Language of *undoing gender* was highlighted by West and Zimmerman as a response to Risman. They perceived her thought as a dismissal of accountability to sex category, which according to them, is nearly impossible. The scholars further argued that gender is "*not so much undone as redone*" [5] through repositioning in "accountability structures," that retain gender in interaction(s). In simpler words, accountability structures may alter in order to accommodate limitations to sex category.

Previous researches have demonstrated that transmen, specifically have an increased awareness toward gender inequality in the work domain after this sex change. Schilt, 2006 in his study demonstrated that women who undergo sex change develop good insights into the gendered structure of work owing to the shift among "*gendered worker positions.*"

5.3.2 Gender and Sexual Dysfunctionality

Academic discourses around sexual health of transgender over the last 20 years have been centered around three domains mainly; (1) orgasmic potential for transgender people who have undergone gender-affirmative genital surgery (2) sexual desire in relation to gender-affirmative hormone therapy, and (3) HIV-related risk [18].

Research studies have extensively focused on transgender individuals' sexual orientation as a medium of anatomies of transgender identities (like homosexual, heterosexual, and so on). Contemporary research does not appreciate the diversity of sexual identities of transgender individuals, moreover, it is silent on the subjective and relational shades of sexuality that entail pleasure [19].

Bauer and Harmond, in 2015 critiqued the potent impact of *cis*-normativity and *cis*-sexism in the construing the research trends around transgender sexual health and there is a significant lacunae in lived experiences of transgender that is being represented in academic research [20]. Research on transgender health does not give significance to sexual pleasure [21].

The body is an agentic force of driving (for instance, incidents of violence like rape) or directing a given behavior (for instance, men are not inborn nurturers). Masculinity in its truest essence has always been thought of emerging from the man's body, an essential constituent and/or expression of the male body. Such beliefs are still a conspicuous part of contemporary gender ideology. Hence, the foremost role of the psychosocial analyst is to appreciate men's bodies and their association with masculinity.

5.3.3 Systemic Restructuring Around the Notions of Gender and Sexuality

The hallmark work of Lober, 1994 conceptualized gender as an "institution" that develops and sustains social inequality, demonstrating the inconsequential likelihood of sex and gender, where sex becomes the starting point for gendering people. Lorber, 1999 tried to detangle "masculinity" and "femininity" from the sex category, creating domains like gender status (biological ascription of sex at the time of birth) gender identity (understanding of self as a man or a woman), and gender display (sense of being masculine or feminine). She affirmed the position of trans sexuals and transvestites conform to gender roles since their goal is to be "feminine woman and masculine man" [22].

Sexuality too, like sex and gender is socially constructed, therefore understanding of sexual behaviors varies temporally and culturally [23]. Herdt's classic study in 1981 on same-sex Fellatio of a tribe in Papua, New Guinea, illustrates that such behaviors cannot be termed as homosexual or pedophilic. Although the connotation varies in the regions of the United States, wherein same-sex behavior is presumed to be an attribute of individuals with gay or bisexual inclination. However, with the onset of AIDS compelled the epidemiologist and other scholars to rethink the gender and sexuality given [23]. Growing numbers of thinkers are seen contributing in the area of FTM's and "female masculinities," Devor's (1989, 1997, 1998, 2004) work remains conspicuous, following Cromwell, 1999, Halberstam, 1998, Prosser, 1998 and Rubin, 2003.

Trans sexuals are being legitly visible in the academic spaces, yet there is still a dearth of research on crucial aspects like gender, sex, and sexuality in the academia, existing researches have understood it only through the gender lens [5, 24, 25]. However, critiques on gender through the lens of trans sexuality have been seen more in non-academic spaces [11, 26, 27]. A large chunk of social scientific research on trans sexuals has been mainly centered on male-to-female trans sexuals [28–30]. Intervention by social scientists is very crucial as it can be instrumental in transition of the individual experiences into the larger social realities. Devor, 1997 in their work captured the lived experiences of 46 FTM's. Though this work is intricately detailed, Devor has refrained from making any explicit interpretations. Further, Prosser, 1998 made significant contributions in the area of "loss of materiality" and "body" in the context of trans sexuals, he asserted that the essence of gender is not merely theoretical; rather it is much more than that and is experienced through the agency of body. Critiques have pointed Proser's overemphasis on body in the "transsexual experience." Somatic attributes gain meaning courtesy social ascriptions attached with it.

Cromwell in the year 1999 located gender and sexuality in the essentialist or constructionist fabric(s). He claimed that trans individuals' constructions of their realities around "bodies, identities and sexualities" are distinct as opposed to "deviant" from the larger gender or sex paradigm.

He further critiqued these frameworks and asserted that such explanations are rooted in the notions of social interactionist lens that infers that gender identity is not functional without social context. He believed that studying trans individuals is crucial since rich insights of a constant gender identity can be gained [31].

Rubin in 2003 affiliated with Cromwell's thought that gender identity is paradoxical; it is socially constructed yet is embodied and "absolutely real" [32].

Prosser in 1998, Cromwell and Rubin question the components of gender identity theories that fail to resonate with lived experiences of trans sexual and trans gendered individuals. The notion of "body" is quite central to the beingness of a transgendered person, although gender identities are understood from constructionist lenses, yet it takes a unique strength to believe that transgendered experiences are not solely reliant on social interactions and constructions. Majority of the gendering practices happen to be (un)reflexive, as in they happen swiftly. These practices have a distinct "emotive" attribute. Deducing practices to mere words usurps the experiences of their emotive elements [33].

A renowned queer social activist and an academician, Vikramaditya Sahai in his popular lecture series on transgender issues, pointed out that feminism is relatively silent and (de)sensitized toward transgender issues and is hence selectively assertive in its endeavors. He suggested a parallel of Annihilation Of Caste (by Ambedkar) to be integrated in gender, Annihilation of Gender that would contribute to the de-stigmatization of genders [34].

5.3.4 Biological Aberrations Around Gender and Sexuality

People who are born with "inter sexed" or "trans sexual" conditions regard their "brains" (gendered self) rather than their sex glands to be the reference point for sexual identification [35].

A popular journal, "Nature" in one of its publications highlighted the importance of the "fetal brain being" agentic, even before the gonadal development [36]. This critical assumption can aid in formulating a "biological tool" that can predict the psychological bodies of these infants who are born with "ambiguous genitalia" [37]. This should guide the medical personnel in the assignment of sex during birth so as to avoid "gender dissatisfaction" in the later years of these individuals [38].

Denis illustrates two case examples of Susan and Tony; Susan being assigned as a female through gender assignment surgery and hormonal therapy and is leading a normal life. However, this was not the case with Tony who was not satisfied with female gender assignment. She later opted for "testosterone treatment" and lived as a man.

Debates pertaining to cerebral grounding of hermaphrodite identities have been prevalent since the 1960s and have been perpetually reaffirmed by the medical personnel ever since. American Academy of Pediatrics (AAP), in one of its conventions stated that, the organic factors are given utmost priority in determining the psychosexual development.

> Psychosexual development is influenced by multiple factors such as exposure to androgens, sex chromosome genes, and brain structure, as well as social circumstances and family dynamics [39].

According to this statement, the sexing process of the fetal brain begins in the garb of "sex chromosomes" and "certain hormones" [36].

This ideology of "intersex/transsexual brain" has been under-appreciated by feminist scholars and transgender thinkers [40]. The attribute of DSD being inborn, calls for planning for long-term management of this illness that involves a team of experts from diverse fields. Primary interventions such as timely diagnosis, surgical treatments, and psychosocial assessments focus on psychosexual development. Hence, the understanding of sex dysmorphic features such as gender identities and sexual orientation are crucial when thinking of gender assignment.

Discontentment toward one's ascribed sex in gender assignment situations in DSD is still not adequately appreciated and thereby the data is scanty. This resentment and gender reassignment have been associated with factors such as parental psychopathology and prenatal hormonal climate [41–43]. Explicit outcomes of sex dysphoric features (i.e., gender identity, gender roles, and sexual orientation) can be measured. Hughes, in 2006 claimed that these features should be treated separately, and gender roles and/or homosexual behaviors are not the essential indicators of faulty gender assignment.

Zucker in the year 2005 has extensively critiqued the available clinical scales that assess "psychosexual differentiation" in children with Gender identity disorder and DSD. When examining trends in psycho-sexual development in the clinically diagnosed children with DSD, intricacies such as the typology and temporality of hormonal exposure, the changes in the social environment of the child are to be integrated in the assessments [44–49]. Although fair progress has been attained in developing a compact understanding of the interplay of biological and environmental attributes that lead to "healthy psychosexual development". However, variations in behavioral outcomes make it dicey of the sex assignment and thereby thoughtful planning is always needed [50].

Furthermore, the complex differentiation of gender identity and role depends not only on the greatly investigated and heeded aspects of prenatal exposure of brain to target areas during critical window period but also to continues postnatal hormonal influence and societal interactions. It is stated that development of a child's juvenile gender identity/role begins approximately around 18 months of age and continues to develop until 5 years of age. One important and critical factor is the genital appearance. The latter influences not only the behavior of others but also of the child to perceive his own image. In children with DSD, parental uncertainty regarding the child's sex may be the most important factor for a child to develop an ambiguous gender identity, especially during this period. Another important period is puberty when hormones play an important role in shaping gender identity and role, pubertal hormones bring about morphological changes in the body increasing one's sexual attractiveness and initiation of sexual behavior. Such changes in hormonal milieu lead to changes in gender identity perceived by children in a few conditions as 5 alpha reductase deficiency from female to male with pubertal spurt. Various studies have pointed to the fact that hormones, especially androgens (pre/postnatal and pubertal) have a greater role in determining gender identity than sex of rearing, though few earlier reports hinted that few children with 46 XY reared as females continued to bear female identity in later years too. Mostly these were children with complete or partial androgen insensitivity syndrome. Hence not only androgen levels but also their subsequent actions and societal interactions intermingle in a complex manner to modulate the perception of gender identity of an individual [1]. The various ways hormones influence an individual's gender identity and role are:

1. Physical energy expenditure.
2. Play rehearsal of parenting behavior.
3. Peer preferences.
4. Grooming behavior.

All of the above has led to an ever greater need for the role of psychologists to evaluate such children and they hold a greater role to guide the attending clinicians and surgeons for definitive management of such patients. Using the array of psychological tests they can in most situations reach these results. The various ways to assess may be:

1. Detailed interviews with the client and their family members
2. Psychometric and projective tests.
3. Sex role inventories.
4. Direct behavioral observation [51].

One easily applicable test is "human figure drawing," an easy tool whereby the psychological bent of a DSD child toward male or female can be easily gauged. As more and more aspects are evaluated, gender identity and role will find an apex guide for medical and surgical management of such children and the role of psychologists will be rated as pivotal in the multidisciplinary team for their management.

References

1. Mazur T. Gender identity/role. In: Josso N, editor. The intersex child, vol. 1. 1st ed. Malades, Paris: S. Karger; 1981. p. 14–20.
2. Connell R. Accountable conduct: "doing gender" in transsexual and political retrospect. In: Gender & society, vol. 1. 23rd ed. 2009. p. 104–11.
3. Deutsch F. Undoing gender. In: Gender & society, vol. 2. 21st ed. 2007. p. 106–27.
4. Risman B. From doing to undoing: gender as we know it. Gend Soc. 2009;23(1):81–4.
5. West C, Zimmerman DH. Accounting for doing gender. Gend Soc. 2009;23(1):11–122.
6. National Gay and Lesbian Task Force. One community. One ENDA 2007. http://www.thetaskforce.org/enda07/enda07.html. Accessed 17 July 2009.
7. Eichler M. Sex change operations: the last bulwark of the double standard. In: Salamon ED, Robinson N, editors. Gender roles: doing what comes naturally? Toronto: Methuen; 1987.
8. Irvine J. Disorder and desire: sex and gender in modern American sexology. Philadelphia: Temple University Press; 1990.
9. Raymond J. Transsexualism: the ultimate homage to sex-role power. Chrysalis. 1977;3:11–23.
10. Raymond J. The transsexual empire: the making of the she-male. London: Women's Press; 1979.
11. Bornstein K. Gender outlaw: on men, women, and the rest of us. New York: Vintage; 1995.
12. Butler J. Gender trouble: feminism and the subversion of identity. New York: Routledge; 1990.
13. Butler J. Undoing gender. New York: Routledge; 2000.
14. Stone AR. The empire strikes back: a posttransexual manifesto. Camera Obscura. 1993;10(2):150–76.
15. Stryker S. Identity theory: developments and extensions. In: Yardley K, Holmes T, editors. Self and identity: psychological perspectives. New York: Wiley; 1987.
16. Meyerowitz J. How sex changed: a history of transsexuality in the United States. Cambridge, MA: Harvard University Press; 2002.

17. Rudacille D. The riddle of gender: science, activism, and transgender rights. New York: Pantheon; 2005.
18. Devor H. Sexual orientation identities, attractions, and practices of female-to-male transsexuals. J Sex Res. 1993;30:303–15.
19. Kuper LE, Nussbaum R, Mustanski B. Exploring the diversity of gender and sexual orientation identities in an online sample of transgender individuals. J Sex Res. 2012;49:244–54.
20. Bauer GR, Hammond R, Travers R, Kaay M, Hohenadel KM, Boyce M, et al. "I don't think this is theoretical; this is our lives": how erasure impacts health care for transgender people. J Assoc Nurses AIDS Care. 2009;20(5):348–61.
21. Higgins JA, Hirsch JS. The pleasure deficit: revisiting the "sexuality connection" in reproductive health. Perspect Sex Reprod Health. 2007;39:240–7.
22. Lorber J. Gender paradoxes. New Haven, CT: Yale University Press; 1994.
23. Parker R, Peter A. Culture, society and sexuality: a reader. Los Angeles: UCLA Press; 1999.
24. Kessler SJ, Wendy M. Gender: an ethnomethodological approach. New York: Wiley; 1978.
25. Nicholson L. Interpreting gender. J Women Cult Soc. 1994;20(1):79–105.
26. Sisson K, Moser C. Book Review: Sex changes: the politics of transgenderism. By Pat Califia. Cleis Press, San Francisco; 1997, 307 pp., $16.95. Archives of Sexual Behavior - ARCH SEX BEHAV. 2004.
27. Feinberg L. Transgender liberation: beyond pink or blue. Boston: Beacon Press; 1998.
28. Bolin A. In search of eve: transsexual rites of passage. South Hadley, MA: Begin and Garvey Publishers; 1988.
29. Ekins R. Male femaling: a grounded theory approach to cross-dressing and sex-changing. London: Routledge; 1997.
30. Wirth R. Lewin/Schein's change theory. 2004.
31. Dozier R. Beards, breasts, and bodies doing sex in a gendered. WORLD. 2005;19(3):297–316.
32. Cromwell J. Transmen and FTMs: identities, bodies, genders, and sexualities. Urbana: University of Illinois Press; 1999.
33. Rubin H. Self-made men: identity and embodiment among transsexual men. Nashville: Vanderbilt University Press; 2003.
34. Kaur M, Menon S. Status of Hijras in India in marginality, minoritisation and identity de/re construction: gender perspective. Cape Comrin Publishers; 2020.
35. Diamond M, Hazel G. The right to be wrong. Sex and gender decisions [web version], University of Hawaii. 2005.
36. American Academy of Pediatrics Section on Urology. Timing of elective surgery on the genitalia of male children with particular reference to the risks, benefits, and psychological effects of surgery and anesthesia. Pediatrics. 1996;97:590–4.
37. American Academy of Pediatrics. Committee on genetics, section on endocrinology, and section on urology. Evaluation of the newborn with developmental anomalies of the external genitalia. Pediatrics. 2000;106:138–42.
38. American Psychiatric Association. DSM-III: diagnostic and statistical manual of mental disorder. 3rd ed. Washington, DC; 1984.
39. Lee PA, Christopher PHS, Faisal A, Hughes IA. In collaboration with the participants in the International Consensus Conference on Intersex organized by the Lawson Wilkins Pediatric Endocrine Society and the European Society for Paediatric Endocrinology, "Consensus Statement on Management of Intersex Disorders". Pediatrics. 2006;118:488–500.
40. Kraus C. Critical studies of the sexed brain: a critique of what and for whom? Neuroethics. 2011.
41. Meyer-Bahlburg HFL, Ehrhardt AA. Prenatal sex hormones and human aggression: a review, and new data on progestogen effects. Aggress Behav. 1982;1982(8):39–62.
42. Zucker KJ, Bradley SJ. Gender identity disorder and psychosexual problems in children and adolescents. New York: Guilford Press; 1995.
43. Bradley SJ, Oliver GD, Chernick AB, Zucker KJ. Experiment of nurture: Ablatio penis at 2 months, sex reassignment at 7 months and a psychosexual follow-up in young adulthood. Pediatrics. 1998;102:102–1.

44. Goy RW, Bercovitch FB, McBrair MC. Behavioral masculinization is independent of genital masculinization in prenatally androgenized female rhesus macaques. Horm Behav. 1988;22:552–71.
45. Grumbach MM, Hughes IA, Conte FA. Disorders of sex differentiation. In: Larsen PR, Kronenberg HM, Melmed S, Polonsky KS, editors. Williams textbook of endocrinology, vol. 10. Philadelphia: Saunders; 2003. p. 842–1002.
46. Hines M. Prenatal gonadal hormones and sex differences in human behavior. Psychol Bull. 1982;92:56–80.
47. Hughes IA, Houk C, Ahmed SF, Lee PA, LWPES/ESPE Consensus Group. Consensus statement on management of intersex disorders. Archives of Diseases in Childhood (online). 2005 hardcopy version in, Archives of Diseases in Childhood. 2006; 91: 554–63.
48. Nordenstrom A, Servin A, Bohlin G, Larsson A, Wedell A. Sex-typed toy play behavior correlates with the degree of prenatal androgen exposure assessed by CYP21 genotype in girls with congenital adrenal hyperplasia. J Clin Endocrinol Metab. 2002;87:5119–24.
49. Wallen K. Hormonal influences on sexually differentiated behavior in nonhuman primates. Front Neuroendocrinol. 2005;26:7–26.
50. Pasterski V. Disorders of sex development and atypical sex differentiation in handbook of sexual and gender identity disorders. Wiley; 2008.
51. Slijper SL, Molenaar JC, de Muinck Keizer Schramer SM. Long term psychological evaluation of intersex children. Arch Sex Behav. 1998;27:125–44.

Gender Dysphoria in DSD Children

6

V. R. Ravikumar

Abstract

Gender dysphoria is a broad term used to describe an individual's discontent with the gender they were assigned at birth. It is thought to have its origins in the brain's sexual differentiation, where the crucial roles of prenatal and postnatal androgens as well as societal interactions are taken into consideration. It has been observed that children with various disorders of sex development might experience GD to varying degrees, as is highlighted in the following chapter.

Keywords

Gender dysphoria · Disorders of sex development · Androgens

The biological sex and gender are based on chromosomal configuration and the external genitalia and thus assign male and female sex. A person's gender identity is determined by how well they conform to the societal standards, behaviours, and characteristics that are typically associated with men and women. The individual's gender identity and biological sex do not coincide in cases of gender dysphoria.

Gender dysphoria results in the desire to be perceived as the other gender. They believe that the sex they were assigned at birth is not the same as their gender. Their ideas and emotions about themselves are at odds with the sexual attributes of their physical bodies.

Children who suffer from gender dysphoria may show signs of longing to be the other gender and demand on toys, clothes, and hairstyles that belong to that gender. As you may remember, Georgina, thc tomboyish girl in Enid Blyton's Famous Five

V. R. Ravikumar (✉)
Coimbatore Medical College, Coimbatore, India

S. K. Ratan (ed.), *Children with Differences in Sex Development*,
https://doi.org/10.1007/978-981-97-1639-5_6

series, wished to be called George. These kids would rather play with kids of the other gender.

However, the GD in Disorders of Sexual Differences (DSD) are different from those individuals born with normal genitalia. These children have congenital defects in development due to some hormonal influences which undergo changes as they attain puberty. In many, the gonadal functions are not normal. These hormonal changes during foetal life produce cerebral imprinting of the sex based on the predominant hormonal profile of the individual.

6.1 Congenital Adrenal Hyperplasia

In congenital adrenal hyperplasia (CAH) due to intrauterine exposure of androgens to the brain, it will be expected to adopt male type of behaviour in children with CAH. With proper treatment and social adjustments, most get adapted to be heterosexual. Ricardo Gonalez et al. reported 9% of their cases showing GD in late adolescence in their study of CAH children [1]. Gastaud evaluated long-term outcome in 35 women with CAH and almost all women were satisfied with their gender assignment. However, many experienced pains during penetrative intercourse, only 17% had children and in 20% homosexual inclinations were noted [2].

Only a small number of people with CAH, who were assigned the female gender at birth and experienced severe gender dysphoria that resulted in gender transition in adulthood. Many of these individuals did not get glucocorticoid replacement [1, 3].

Children with 46XX who were raised as males in CAH experience GD issues. A detailed psychosexual examination conducted on 46 XX adult patients with congenital adrenal hyperplasia who were raised as males indicated well-established male gender identities consistent with their male gender assignments at birth. This study examined the long-term results of these individuals. Every patient who had been raised as a man exhibited behaviours and gender roles that were compatible with male gender identity, including having sex with female partners [4]. Except for sterility, they acquire a masculine gender identity and exhibit normal sexual function [5].

In the Indian scenario, 7 out of 173 children with CAH who were diagnosed at the paediatric Intersex Clinic at AIIMS Delhi presented late and had severe virilization. These children were raised as males and their female adnexa was removed. Six children had male genitoplasty.

After puberty, appropriate hormone supplements were recommended. Socialisation went well overall, except for one person with a more complex mental makeup [6].

6.2 5 alpha Reductase Deficiency

The most frequent cause of 46XY DSD is 5 alpha reductase deficiency (5ARD). In a series, most patients with a diagnosis of 5ARD raised as female and untrivialised were fully compatible with being female and were heterosexual. Patients who did not accept their assigned female gender were more virilized [7].

Nonetheless, the occurrence of gender role changes in 5ARD and 17β-HSD-3 based on recent research indicates that they happen frequently but not always. 56–63% of cases with 5α-RD-2 and 39–64% of cases with 17β-HSD-3 who were raised as girls reported changes in their gender roles. Usually, the adjustments were made in early adulthood and adolescence. The degree of external virilisation at birth in these two syndromes does not appear to be systematically correlated with changes in gender roles [8]. According to the author's own study of 5ARD featuring 46XY siblings, one child was raised as a girl and the other as a boy. Both had grown up, got married and were heterosexual. The boy is azospermic. No gender dysphoria was reported in either of them [9].

The present consensus is that the 5ARD with 46 XY children should be reared as males even if they are grossly under-virilised since they virilise at puberty and have normal sexual activity of a male [7].

6.3 17β-Hydroxysteroid Dehydrogenase-3 Deficiency

17β-Hydroxysteroid dehydrogenase-3 deficiency (HSD17B3) is a rare form of 46 XY DSD, simulates 5ARD and many children are reared as females. Diagnosis is made on Androstenedione–Testosterone ratio following HCG stimulation. Genetic testing shows mutations of HSD17B3 gene [10]. These children virilise at the time of puberty. In a study by Peggy T. et al. on gender Change in 46,XY persons with 17β-Hydroxysteroid Dehydrogenase-3 deficiency gender role changes were observed in 39–64% of cases [8].

6.4 Complete Androgen Insensitivity Syndrome

Complete androgen insensitivity syndrome (CAIS) where 46 XY children are reared as girls since they are fully feminised at birth, gender dysphoria is uncommon except for one report of gender change in literature by Piet Hobeke [11]. A large series revealed that there were no documented cases of gender change in individuals with CAIS ($N = 156$ females) [12].

6.5 Partial Androgen Insensitivity Syndrome

Partial androgen insensitivity syndrome (PAIS) is an X-linked recessive condition, which may be mild, and moderately virilised and are usually reared as males. Nine out of 99 (9.1%) individuals with PAIS changed their gender [12]. Despite being raised as either a male or a female, about 25% of people with PAIS are unhappy with the gender they were assigned [13].

6.6 Mixed Gonadal Dysgenesis

In mixed gonadal dysgenesis, MGD the child has XY/XO cell line and has a streak gonad with Mullerian remnant on one side with functional testis on the opposite side. If the child is reared as female based on external appearance it is imperative to remove the streak gonad and testis to prevent virilisation of external genitalia at puberty producing gender dysphoria. The gender dysphoria in Mixed gonadal dysgenesis is more often due to hurried assigning the female sex in infancy without removing the testis. Late desire to change of male sex has been reported in literature where the female sex has been assigned in infancy without adequate investigations [14].

6.7 Ovo Testicular DSD (OT-DSD)

There are currently few studies on gender dysphoria in this patient population, and there are no clear, well-coded recommendations. Matsui et al. examined cases of OT-DSD from Japanese institutions and described their experience with eight children who had OT-DSD. In their 20 years of experience in Japan, they have frequently been asked to perform early gonadal surgery and genitoplasty, in children even though they have informed families of all available treatment options. None of the patients had gender identity disorder or gender dysphoria, according to an 8.2-year mean follow-up [15, 16].

According to a recent research article about the treatment of OT-DSD, the appearance of the bisexual phenotype and the parents' ongoing anxiety demanded that these cases be managed early, even though ethicists and patient support groups argued that genital surgery should not be needed until the patient was able to give informed consent. The authors concluded that commencing the evaluation process early on was both logical and required [16–18].

A retrospective analysis of 64 OT-DSD cases in South Africa revealed that the most common karyotype was 46 XX 88%, followed by 46XY 8%, 46XY/45X, and 46XX/46XY in 3%. Two-thirds of the subjects were reared as male, and eight patients (11%) at a median age of 6.4 years had gender dysphoria [17]. In their long-term follow up of OT-DSD, two, cases had gender dysphoria, and four cases had neuropsychiatric disorders [18].

6.8 Conclusion

Disorders of sexual differences are a unique challenging congenital anomaly that needs a deep understanding of the psychosocial, and psychosexual aspects of rearing a child with ambiguous genitalia to a particular sex. Understanding gender dysphoria in people without DSD may help in clinical care of children with DSD. The development of protocols and educational materials tailored to individuals with DSD is necessary to help decision-making for both professionals and DSD patients. This will improve psychosocial and psychosexual care in this domain [19].

References

1. Gonalez R, Ludeikowski BM. Gender dysphoria in 46,XX persons with adrenogenital syndrome raised as females: an addendum. Front Pediatrics. 2015;2:140.
2. Gastaud F, Bouvattier C, Duranteau L, Brauner R, Thibaud E, Kutten F. Bougnères PImpaired sexual and reproductive outcomes in women with classical forms of congenital adrenal hyperplasia. J Clin Endocrinol Metab. 2007;92(4):1391–6.
3. Choi JH, Yoo HW. Management issues of congenital adrenal hyperplasia during the transition from pediatric to adult care. Korean J Pediatr. 2017;60(2):31.
4. Khattab A, et al. Long term outcomes in 46, XX adult patients with congenital adrenal hyperplasia reared as males. J Steroid Biochem Mol Biol. 2017;165:12.
5. Apóstolos RAC, et al. Gender identity and sexual function in 46,XX patients with congenital adrenal hyperplasia raised as males. Arch Sex Behav. 2018;47(8):2491–6.
6. Sharma S, Gupta DK. Male genitoplasty for 46 XX congenital adrenal hyperplasia patients presenting late and reared as males. Indian J Endocrinol Metab. 2012;16(6):935.
7. Nascimento RLP, de Andrade Mesquita IM. Gender identity in patients with 5-alpha reductase deficiency raised as females. J Urol. 2018.
8. Peggy T, Cohen-Kettenis PD. Gender change in 46,XY persons with 5α-reductase-2 deficiency and 17β-hydroxysteroid dehydrogenase-3 deficiency. Arch Sex Behav. 2005;34:399–410.
9. Ravikumar VR. 5 alpha reductase deficiency. In: Disorders of sexual differences. 1st ed. Chapter 4; 2020. p. 84–6.
10. Boehmer AL, Brinkmann AO, Sandkuijl LA, Halley DJ, Niermeijer MF, Andersson S, de Jong FH, Kayserili H, de Vroede MA, Otten BJ, Rouwé CW. 17β-hydroxysteroid dehydrogenase-3 deficiency: diagnosis, phenotypic variability, population genetics, and worldwide distribution of ancient and de novo mutations. J Clin Endocrinol Metabol. 1999;84(12):4713–21.
11. Hobeke P, Monstrey S. Male gender identity in complete androgen insensitivity syndrome. J Pediatric Urol. 2010;6(Suppl 1):S86.
12. Mazur T. Gender dysphoria and gender change in androgen insensitivity or micropenis. Arch Sex Behav. 2005;34(4):411–21.
13. Migeon CJ, Wisniewski AB. Ambiguous genitalia with perineoscrotal hypospadias in 46,XY individuals: long-term medical, surgical, and psychosexual outcome. Pediatrics. 2002;110:110.
14. Ocal G, Berberoğlu M, Siklar Z, Bilir P. Gender dysphoria and gender change in an adolescent with 45,X/46,XY mixed gonadal dysgenesis. Clin Endocrinol Diabetes. 2009;117(6):301–4.
15. Matsui F, Shimada K, Matsumoto F, Itesako T, Nara K, Ida S, et al. Long-term outcome of ovotesticular disorder of sex development: a single center experience. Int J Urol. 2011;18:231–6. https://doi.org/10.1111/j.1442-2042.2010.02700.x.
16. Mao Y, Shaoji C, Wang R, Wang X, Qin D, Tang Y. Evaluation and treatment for ovotesticular disorder of sex development (OT-DSD)—experience based on a Chinese series. BMC Urol. 2017;17:21–7. https://doi.org/10.1186/s12894-017-0212-8.

17. Ganie Y, Aldous C, Balakrishna Y, Wiersma R. The spectrum of ovotesticular disorders of sex development in South Africa: a single-centre experience. Horm Res Paediatr. 2017;87(5):307–14. https://doi.org/10.1159/000466693.
18. Scarpa MG. Ovotesticular differences of sex development: surgery or not surgery? That is the question. Front Pediatr. 2017;5:231.
19. Cohen-Kettenis PT. Psychosocial and psychosexual aspects of disorders of sex development. Best Pract Res Clin Endocrinol Metab. 2010;24(2):325–34.

Meta-analysis and Current Literature on Gender Dysphoria

7

Ramesh Babu

Abstract

In children with differences in sexual development (DSD) gender assignment is a crucial decision. Gender identity disorder (GID) previously known as gender dysphoria has been reported in some adults with DSD. Appropriate sex assignment of DSD children may help in prevention of development of GID in later life. While the population risk of gender dysphoria is less than 0.5%, among DSD individuals the risk is variable 3–15% as per reported literature. Gender dysphoria is low conditions with little or no androgen effects reared as female despite the presence of Y chromosome. In conditions like mixed gonadal dysgenesis or partial androgen insensitivity there is a variable risk of gender dysphoria. In 5-alpha reductase deficiency it is high if they are reared as female. Thus, it is essential to arrive at the diagnosis and also be aware of future risk of gender conflict before assigning sex of rearing in DSD children.

Keywords

DSD · Gender incongruence · Meta-analysis · Gender dysphoria

7.1 Introduction

Conditions are known earlier as intersex, pseudo-hermophroditism, etc., were named as disorders of sex differentiation in the year 2006 [1]. They were again renamed into a more inclusive terminology, differences of sex development (DSD) in the consensus statement of the year 2018 [2]. The cosmetic or functional issues

R. Babu (✉)
Sri Ramachandra Institute of Higher Education and Research, Chennai, India

S. K. Ratan (ed.), *Children with Differences in Sex Development*,
https://doi.org/10.1007/978-981-97-1639-5_7

of surgical management have been discussed widely in the literature [3, 4]. However, it is equally important to focus on the psychological and mental aspects of those with DSD as they become adults. The main reason for lack of consensus in DSD surgery in [5] is the potential risk for gender dysphoria or gender identify disorder (GID) as these individuals go through adolescence [6, 7]. Gender decisions in DSD children are one of the most complex issues faced by the multi-disciplinary treatment (MDC) team involved in DSD care [8].

7.2 Sex Versus Gender

Sex refers to anatomical characteristics while gender is the inner feeling of an individual as a male, female, or an alternative gender. So one has to understand that sex assignment is what MDC decides for the child while gender identity is what the grown-up DSD individual decides for them. Gender role refers to the attitudes or behaviors (as female or male) one portrays in society. Gender roles and gender identity are different as one may assume a male gender role despite having an inner female identity for a safe living in society [9]. Sexual orientation, on the other hand, is totally different from gender. It is based on the sex of the person to whom the individual is sexually attracted to (heterosexual, bisexual, etc.). Since homosexual orientation or strong cross-sexual interest are common even in the general population, these tendencies in DSD individuals may not actually mean wrong sex assignment [1].

7.3 Gender Dysphoria

The "optimal gender policy" in infants with DSD has been disputed by several studies [10, 11]. Many studies have shown that DSD patients do not always choose the same gender as the sex of rearing as they grow into adults [8]. Gender dysphoria or GID refers to a conflict between a person's physical or assigned sex and the gender with they choose later. In GID, there is a strong desire to be the other sex, or to live or be treated as the other sex, or the conviction that he or she has the feelings of the other sex [12]. With a sense of being entrapped in a wrong body, individuals with GID often request hormones, surgery, or other procedures to physically alter sexual features to make them in alignment with their gender identity. The notion that "nurture overrules nature" in gender identity is a concept of the past as the environmental gender influence based on the sex of rearing has often been overruled by the gender feelings of the grown-up individual. The prevalence of GID in DSD has been very poorly understood in the literature [10, 11, 13].

7.4 Meta-analysis

In 2020, the author conducted a systematic review and meta-analysis [14] aiming at evaluating the prevalence of GID in adolescents and adults with DSD. The articles describing GID prevalence in Congenital adrenal hyperplasia (CAH); complete androgen insensitivity syndrome (CAIS); partial androgen insensitivity syndrome (PAIS); 5 alpha reductase deficiency (5ARD); 17-hydroxysteroid dehydrogenase deficiency (17HSD); mixed gonadal dysgenesis (MGD); and complete gonadal dysgenesis (CGD) were analyzed. Articles that failed to provide exact subgroups of DSD diagnosis were not included [15, 16]. GID was defined as a strong feeling of dissatisfaction about oneself as male or female [10]. Articles that used only questionnaires or pediatric gender identity tools [17] and those which focused on sexual orientation but not on gender identity were excluded [5, 11]. Duplicates publications and those prior to 2005 were excluded [18–21]. The meta-analysis included articles from the Netherlands, the UK, the USA, Canada, Egypt, France, Turkey, Finland, China, Hong Kong, Germany, Iran, Brazil, Italy, and India [20–39].

7.5 Gender Dysphoria Prevalence

Gender dysphoria prevalence in those with DSD was (178/1167) 15%. Gender dysphoria was high among 5ARD and 17HSD patients raised as females. In CAH patients reared female and 5ARD and MGD patients reared male, the gender dysphoria rate was less than the overall prevalence. Table 7.1 summarizes the GID prevalence of the individual conditions and thus the preferred sex assignment guidelines to prevent later development of GID.

Table 7.1 Recommendations for sex assignment in DSD (based on GID %); adapting recommendations from Houk [1], Yang [8], Fisher [13], Babu [14], Kreukels [16], Bayraktar [40], Lee [42], and Weidler [43]

DSD	Subgroups	Preferred sex of rearing	Reasoning (based on GID %)
46 XX DSD	CAH	Female	GID only 4% if raised as female
			GID risk 15% (Some CAH male with high androgen effect may have less GID)
46 XY DSD	CAIS	Female	GID only 2% if raised as female
	PAIS	Male or Female	GID risk 12% if raised female 25% if raised male
	5ARD	Male	If reared as female, 54% virilize at puberty/develop GID and demand gender change; no GID if reared male
	17HSD	Male	
	Complete gonadal dysgenesis	Female	GID% is almost nil if reared female (poor androgen effect)
Sex chromosome DSD	Mixed gonadal dysgenesis (MGD)	Male > Female	GID risk 22% if reared female as functioning testicular tissue may have androgen effects
	Ovotesticular DSD (TH)	Female > Male	GID risk unclear; menstruation and pregnancy have been reported

CAH—congenital adrenal hyperplasia; CAIS—complete androgen insensitivity syndrome; PAIS—partial androgen insensitivity syndrome; 5ARD—5 alpha reductase deficiency; 17HSD –17-hydroxysteroid dehydrogenase deficiency; MGD—mixed gonadal dysgenesis; TH—true hermophroditism; GID—gender identity disorder

7.6 Sex of Rearing in DSD

Important areas of psychosexual development [1, 2] include gender identity (self-presentation as male or female), gender role (sexually dimorphic behavior in a population such as toy preferences, aggression, etc.), and sexual orientation. Assigned and experienced gender differences can be of different intensity, manifest in different forms, and be associated with distress. Diagnostic and Statistical Manual of Mental Disorders, 5th Edition (DSM-5) of the American Psychiatric Association [16] considers significant distress as a criterion for gender dysphoria or GID. The notion that the "sex-assignment" would later guide the future gender identity of the child has been disproven several times [8, 9, 13, 16]. Thus, it is essential that eventual GID outcomes determine the sex assignment rather than the other way around in DSD individuals. Variable types of questionnaires or gender instruments have been used in GID assessment or sexual orientation [3, 11, 17, 24, 27, 29, 30, 36]. These gender assessment tools can be particularly useful in late presenting DSD adolescents where MDC has to decide on sex re-assignment.

7.7 Congenital Adrenal Hyperplasia

GID prevalence in CAH-reared females has been uniformly low at 1% [16], 3% [18], 5% [21], and 6% [35] in the reported literature. In a meta-analysis on CAH, Almasari et al. [4] reported that female gender identity was seen in 88.7% while 76.2% were heterosexual. In a systematic review, Daae et al. [11] reported that a majority of assigned females with CAH identified themselves as heterosexual. In a recent meta-analysis by the author [14] the prevalence of GID in CAH was 5%. In CAH patients reared female GID was significantly low at 4% compared to 15% seen in male-assigned CAH. The authors favored female sex of rearing in all CAH children in view of the very low incidence of GID [8, 13]. Some CAH adults, however, want a male gender change [29]. Successful male rearing of a fully masculinized Prader 5 CAH has been reported [38]. Lee et al. reported nine cases of male-reared CAH where there was no GID [25]. They even reported that masculinized CAH who were assigned male at birth had male gender identity in adulthood in view of the androgen effects. 46XX fetuses exposed to high levels of androgens during fetal development, may develop complete masculinization of genitalia, androgen-induced brain imprinting and strong masculine behavior [13]. In a select group of 46XX CAH patients (Prader 5/late diagnosed CAH) alone, a male rearing may be okay without GID.

7.8 Complete Androgen Insensitivity Syndrome

CAIS children present with female external genitalia at birth and they are raised as girls usually. The functional intra-abdominal testes in these patients may descend later into labia. At puberty, aromatized estrogen from testosterone leads to female breast development. CAIS patients may be referred for inguinal hernia (containing testis) or with primary amenorrhea at puberty. In a large series [19] no GID was reported in CAIS female. In a recent meta-analysis by the author [14] very low GID was noted in patients with CAIS reared female. Therefore it is safe to rear 46 XY-CAIS children as females.

7.9 Partial Androgen Insensitivity Syndrome

PAIS patients have XY DSD but the phenotype may be male looking with a good size penis (penoscrotal hypospadias, and cryptorchidism), or female looking (labioscrotal swellings with buried micropenis). Gender fluid status also has been reported among persons born with PAIS [13]. Mazur et al. [19] reported a GID prevalence of 20–40% in those with PAIS. In the meta-analysis by the author [14] GID prevalence in those raised as female was 12% while it was 25% in those raised as male. A clear recommendation on sex of rearing is difficult in PAIS. The degree of masculinization is often used as a guide in assigning PAIS children to either male or female sex.

7.10 5ARD and 17HSD

These patients also have XY DSD but many have female looking phenotype at birth. A very high GID prevalence of 63 and 64% have been reported in 5ARD and 17HSD if they were reared as females [20]. In the meta-analysis by the author [14] also 54% GID was reported in these conditions when reared as female. A female-looking external genitalia is often noted at birth in these conditions and they are often raised as girls despite having XY DSD. But patients with these two disorders often virilize at puberty (voice becomes hoarse, scrotum becomes dark with rugosities and intra-abdominal testes descend). This condition is often called in Dominican Republic as Machihembras (men born as women) or Guevedoces (penis at 12). As a high percentage of this group request female-to-male gender reassignment at puberty, a male sex assignment is recommended in these conditions despite feminized external appearance in them during infancy [1, 8, 13]. Thus, it is crucial to identify these conditions at birth with the help of HCG stimulation test and differentiate it from other causes of XYDSD (androgen insensitivity). In children with 5ARD, there will be a significant rise of testosterone (TST) but not dihydrotestosterone (DHT) following HCG stimulation; the ratio of TST:DHT will be above 35 thus aiding the MDC to choose a male sex of rearing.

7.11 Complete Gonadal Dysgenesis (Swyer Syndrome)

Patients with CGD often have female-type phenotypes despite having 46XY chromosome due to lack of androgen effect. In CGD patients raised as females GID prevalence has been reported to be very low [15, 16]. In the meta-analysis by the author [14] a female sex of rearing is favored as they have typical female physical and mental development. They can become pregnant with egg donation and reconstructive surgery is not required as the external genitalia are already consistent with the female gender [13].

7.12 Mixed Gonadal Dysgenesis

Appearances like penoscrotal hypospadias and unilateral cryptorchidism (intra-abdominal streak gonad on one side) can be noted in MGD. A mosaic karyotype of 45X/46XY is often seen. Due to testicular androgen effects a higher GID has been reported [24] in MGD patients raised females. In the meta-analysis by the author [14] a higher GID is reported in MGD females. Multiple factors like internal anatomy, gonadal function, phallic size, etc. have to be considered for sex of rearing in children with MGD [13]. In those with good phallic size, male sex rearing can be attempted. The risk of germ cell tumor risk is lowest (<5%) in patients with complete androgen insensitivity syndrome or 5α-reductase deficiency, while it is highest (15%–60%) in 46,XY gonadal dysgenesis [39]. The streak gonads noted in these children may need laparoscopic excision biopsy sooner than later after MDC discussion due to the malignant risk.

7.13 Ovotesticular DSD

Ovotesticular DSD, also known as "true hermaphroditism," is a condition where there may be the presence of both ovarian and testicular tissues. If ovarian tissue is present it can be functional. Spontaneous puberty and even pregnancy have been reported in some [40, 41]. A female sex of rearing may be considered in ovotesticular DSD based on the potential for fertility as a female [13]. No clear reports on GID prevalence have been reported in the literature.

7.14 Androgen Effects the Current Concept in Gender Identity

Originally, social milieu and sex of rearing were supposed to mold a child's gender identity. Later the role of Y chromosome and brain imprinting was felt to be essential. Of late the androgen effect—a combination of adequate androgen levels in the absence of androgen receptor resistance—is considered essential for male gender choice (Fig. 7.1). In a multi-institutional DSD life study, Kreukels et al. [16] reported a low prevalence of GID in female-reared XY DSD patients—CGD or CAIS where there was no androgen effects (they classified them together as one group—DSD with no androgen effect; XY DSD-F-na). Interestingly, a proportion of XX DSD

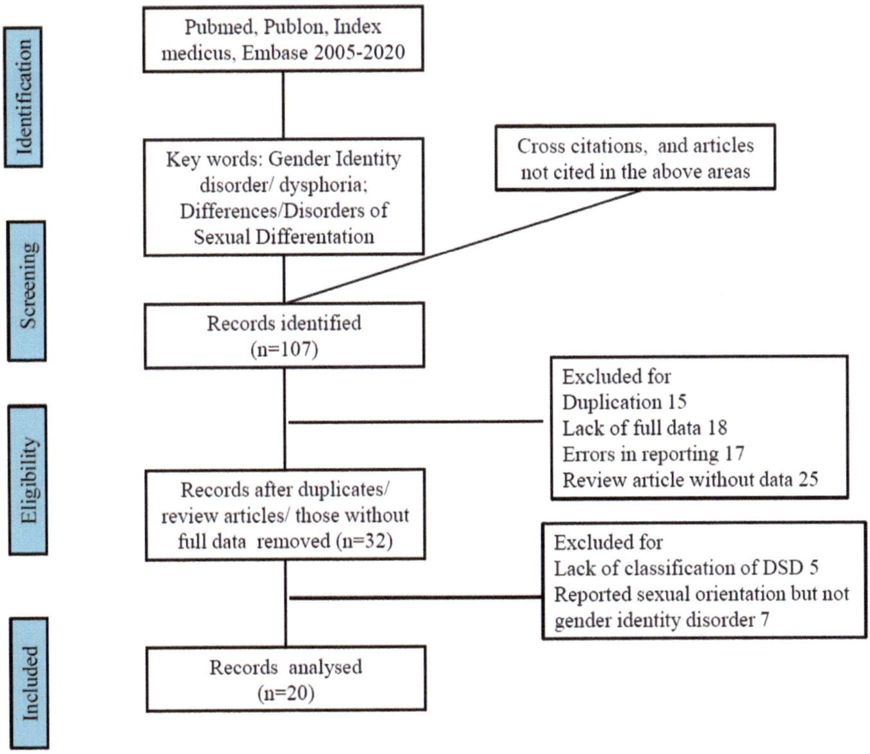

Fig. 7.1 PRISMA flow chart on meta-analysis synthesis

patients (Prader 5/late diagnosed CAH) with strong androgen effects have been successfully reared as male with little GID [25]. It thus appears that androgen milieu in early life is probably more essential in gender orientation than presence/absence of "Y chromosome."

7.15 Preferred Sex of Rearing in DSD Children

Mental well-being is as important as physical functionality. Similarly considering future risk of gender dysphoria is as essential as considering sexual function when making sex-of-rearing decisions in DSD patients. Table 7.1 describes recommendations for sex assignment in DSD based on GID prevalence reported in the recent meta-analysis by the author [14]. Recommendations from various consensus statements [1, 42] have also been incorporated in making these suggestions. Each DSD patient is unique with their own special needs and each one of them warrants multidisciplinary care and long-term psychosexual support.

References

1. Houk CP, Hughes IA, Ahmed SF, Lee PA, Writing Committee for the International Intersex Consensus Conference Participants. Summary of consensus statement on intersex disorders and their management. International intersex consensus conference. Pediatrics [Internet]. 2006;118(2):753–7. Available from: http://www.ncbi.nlm.nih.gov/pubmed/16882833
2. Cools M, Nordenström A, Robeva R, Hall J, Westerveld P, Flück C, et al. Caring for individuals with a difference of sex development (DSD): a consensus statement. Nat Rev Endocrinol [Internet]. 2018;14(7):415–29. Available from: http://www.nature.com/articles/s41574-018-0010-8
3. Crouch NS, Liao LM, Woodhouse CRJ, Conway GS, Creighton SM. Sexual function and genital sensitivity following feminizing genitoplasty for congenital adrenal hyperplasia. J Urol. 2008;179(2):634–8.
4. Almasri J, Zaiem F, Rodriguez-Gutierrez R, Tamhane SU, Iqbal AM, Prokop LJ, et al. Genital reconstructive surgery in females with congenital adrenal hyperplasia: a systematic review and meta-analysis. J Clin Endocrinol Metab. 2018;103(11):4089–96.
5. Mouriquand PDE, Gorduza DB, Gay C-L, Meyer-Bahlburg HFL, Baker L, Baskin LS, et al. Surgery in disorders of sex development (DSD) with a gender issue: if (why), when, and how? J Pediatr Urol [Internet]. 2016;12(3):139–49. Available from: https://linkinghub.elsevier.com/retrieve/pii/S1477513116300122
6. Indyk JA. Disorders/differences of sex development (DSDs) for primary care: the approach to the infant with ambiguous genitalia. Transl Pediatr [Internet]. 2017;6(4):323–34. Available from: http://tp.amegroups.com/article/view/16243/17487
7. Ahmed SF, Achermann JC, Arlt W, Balen AH, Conway G, Edwards ZL, et al. UK guidance on the initial evaluation of an infant or an adolescent with a suspected disorder of sex development. Clin Endocrinol (Oxf) [Internet]. 2011;75(1):12–26. https://doi.org/10.1111/j.1365-2265.2011.04076.x.
8. Yang JH, Baskin LS, DiSandro M. Gender identity in disorders of sex development: review article. Urology [Internet]. 2010;75(1):153–9. Available from: https://linkinghub.elsevier.com/retrieve/pii/S0090429509023784.
9. Hines M. Gender development and the human brain. Annu Rev Neurosci [Internet]. 2011;34(1):69–88. https://doi.org/10.1146/annurev-neuro-061010-113654.
10. de Vries ALC, Doreleijers TAH, Cohen-Kettenis PT. Disorders of sex development and gender identity outcome in adolescence and adulthood: understanding gender identity development and its clinical implications. Pediatr Endocrinol Rev [Internet]. 2007;4(4):343–51. http://www.ncbi.nlm.nih.gov/pubmed/17643082
11. Daae E, Feragen KB, Waehre A, Nermoen I, Falhammar H. Sexual orientation in individuals with congenital adrenal hyperplasia: a systematic review. Front Behav Neurosci [Internet]. 2020;14 https://doi.org/10.3389/fnbeh.2020.00038/full.
12. Cohen-Kettenis PT, Pfäfflin F. The DSM diagnostic criteria for gender identity disorder in adolescents and adults. Arch Sex Behav [Internet]. 2010;39(2):499–513. https://doi.org/10.1007/s10508-009-9562-y.
13. Fisher AD, Ristori J, Fanni E, Castellini G, Forti G, Maggi M. Gender identity, gender assignment and reassignment in individuals with disorders of sex development: a major of dilemma. J Endocrinol Invest [Internet]. 2016;39(11):1207–24. https://doi.org/10.1007/s40618-016-0482-0.
14. Babu R, Shah U. Gender identity disorder (GID) in adolescents and adults with differences of sex development (DSD): a systematic review and meta-analysis. J Pediatr Urol [Internet]. 2021;17(1):39–47. Available from: https://linkinghub.elsevier.com/retrieve/pii/S1477513120306410
15. Jürgensen M, Kleinemeier E, Lux A, Steensma TD, Cohen-Kettenis PT, Hiort O, et al. Psychosexual development in adolescents and adults with disorders of sex development—

results from the German clinical evaluation study. J Sex Med [Internet]. 2013;10(11):2703–14. Available from: https://linkinghub.elsevier.com/retrieve/pii/S1743609515301764.
16. Kreukels BPC, Köhler B, Nordenström A, Roehle R, Thyen U, Bouvattier C, et al. Gender dysphoria and gender change in disorders of sex development/intersex conditions: results from the dsd-LIFE study. J Sex Med [Internet]. 2018;15(5):777–85. Available from: https://linkinghub.elsevier.com/retrieve/pii/S1743609518301437
17. Wallien MSC, Cohen-Kettenis PT. Psychosexual outcome of gender-dysphoric children. J Am Acad Child Adolesc Psychiatry [Internet]. 2008;47(12):1413–23. https://linkinghub.elsevier.com/retrieve/pii/S0890856708601422
18. Dessens AB, Slijper FME, Drop SLS. Gender dysphoria and gender change in chromosomal females with congenital adrenal hyperplasia. Arch Sex Behav [Internet]. 2005;34(4):389–97. https://doi.org/10.1007/s10508-005-4338-5.
19. Mazur T. Gender dysphoria and gender change in androgen insensitivity or micropenis. Arch Sex Behav. 2005;34(4):411–21.
20. Cohen-Kettenis PT. Gender change in 46,XY persons with 5alpha-reductase-2 deficiency and 17beta-hydroxysteroid dehydrogenase-3 deficiency. Arch Sex Behav. 2005;34(4):399–410.
21. Meyer-Bahlburg HFL. Gender identity outcome in female-raised 46,XY persons with penile agenesis, cloacal exstrophy of the bladder, or penile ablation. Arch Sex Behav. 2005;34(4):423–38.
22. Reiner WG. Gender identity and sex-of-rearing in children with disorders of sexual differentiation. J Pediatr Endocrinol Metab. 2005;18(6):549–53.
23. Warne G, Grover S, Hutson J, Sinclair A, Metcalfe S, Northam E, et al. A long-term outcome study of intersex conditions. J Pediatr Endocrinol Metab [Internet]. 2005;18(6). Available from: https://www.degruyter.com/view/j/jpem.2005.18.6/jpem.2005.18.6.555/jpem.2005.18.6.555.xml
24. Szarras-Czapnik M, Lew-Starowicz Z, Zucker KJ. A psychosexual follow-up study of patients with mixed or partial gonadal dysgenesis. J Pediatr Adolesc Gynecol [Internet]. 2007;20(6):333–8. Available from: https://linkinghub.elsevier.com/retrieve/pii/S1083318807001660
25. Lee PA, Houk CP, Husmann DA. Should male gender assignment be considered in the markedly virilized patient with 46,XX and congenital adrenal hyperplasia? J Urol. 2010;184(4 Suppl):1786–92.
26. Ismail SI, Mazen IA. A study of gender outcome of Egyptian patients with 46,XY disorder of sex development. Sex Dev Genet Mol Biol Evol Endocrinol Embryol Pathol Sex Determ Differ. 2010;4(4–5):285–91.
27. Fagerholm R, Santtila P, Miettinen PJ, Mattila A, Rintala R, Taskinen S. Sexual function and attitudes toward surgery after feminizing genitoplasty. J Urol [Internet]. 2011;185(5):1900–4. https://doi.org/10.1016/j.juro.2010.12.099.
28. Maimoun L, Philibert P, Cammas B, Audran F, Bouchard P, Fenichel P, et al. Phenotypical, biological, and molecular heterogeneity of 5α-reductase deficiency: an extensive international experience of 55 patients. J Clin Endocrinol Metab [Internet]. 2011;96(2):296–307. https://doi.org/10.1210/jc.2010-1024.
29. Ercan O, Kutlug S, Uysal O, Alikasifoglu M, Inceoglu D. Gender identity and gender role in DSD patients raised as females: a preliminary outcome study. Front Endocrinol (Lausanne) [Internet]. 2013;4. https://doi.org/10.3389/fendo.2013.00086/abstract
30. Schweizer K, Brunner F, Handford C, Richter-Appelt H. Gender experience and satisfaction with gender allocation in adults with diverse intersex conditions (divergences of sex development, DSD). Psychol Sex [Internet]. 2014;5(1):56–82. https://doi.org/10.1080/19419899.2013.831216.
31. Zhu H, Liu W, Han B, Fan M, Zhao S, Wang H, et al. Phenotypic and molecular characteristics in eleven Chinese patients with 5α-reductase type 2 deficiency. Clin Endocrinol (Oxf) [Internet]. 2014;81(5):711–20. https://doi.org/10.1111/cen.12456.

32. Chan AO, But W, Lee C, Lam Y, Ng K, Loung P, et al. Aetiological bases of 46,XY disorders of sex development in the Hong Kong Chinese population. Hong Kong Med J [Internet]. 2015. Available from: http://www.hkmj.org/abstracts/v21n6/499.htm
33. D'Alberton F, Assante MT, Foresti M, Balsamo A, Bertelloni S, Dati E, et al. Quality of life and psychological adjustment of women living with 46,XY differences of sex development. J Sex Med [Internet]. 2015;12(6):1440–9. Available from: https://linkinghub.elsevier.com/retrieve/pii/S1743609515310365
34. Khorashad BS, Roshan GM, Reid AG, Aghili Z, Moghadam MD, Khazai B, et al. Childhood sex-typed behavior and gender change in individuals with 46,XY and 46,XX disorders of sex development: an Iranian multicenter study. Arch Sex Behav [Internet]. 2018;47(8):2287–98. https://doi.org/10.1007/s10508-018-1281-9.
35. Razzaghy-Azar M, Karimi S, Shirazi E. Gender identity in patients with congenital adrenal hyperplasia. Int J Endocrinol Metab [Internet]. 2017;In Press(In Press). Available from: https://sites.kowsarpub.com/ijem/articles/12537.html
36. Nascimento RLP, de Andrade Mesquita IM, Gondim R, dos Apóstolos RAAC, Toralles MB, de Oliveira LB, et al. Gender identity in patients with 5-alpha reductase deficiency raised as females. J Pediatr Urol [Internet]. 2018;14(5):419.e1–419.e6. Available from: https://linkinghub.elsevier.com/retrieve/pii/S1477513118304996
37. Gangaher A, Chauhan V, Jyotsna V, Mehta M. Gender identity and gender of rearing in 46 XY disorders of sexual development. Indian J Endocrinol Metab [Internet]. 2016;20(4):536. Available from: http://www.ijem.in/text.asp?2016/20/4/536/183471
38. Houk CP, Lee PA. Approach to assigning gender in 46,XX congenital adrenal hyperplasia with male external genitalia: replacing dogmatism with pragmatism. J Clin Endocrinol Metab [Internet]. 2010;95(10):4501–8. https://doi.org/10.1210/jc.2010-0714.
39. Abacı A, Çatlı G, Berberoğlu M. Gonadal malignancy risk and prophylactic gonadectomy in disorders of sexual development. J Pediatr Endocrinol Metab [Internet]. 2015;28(9–10) https://doi.org/10.1515/jpem-2014-0522/html.
40. Bayraktar Z. Potential autofertility in true hermaphrodites. J Matern Neonatal Med [Internet]. 2018;31(4):542–7. https://doi.org/10.1080/14767058.2017.1291619.
41. Nihoul-Fékété C, Thibaud E, Lortat-Jacob S, Josso N. Long-term surgical results and patient satisfaction with male pseudohermaphroditism or true hermaphroditism: a cohort of 63 patients. J Urol [Internet]. 2006;175(5):1878–84. https://doi.org/10.1016/S0022-534728052900934-1.
42. Lee PA, Houk CP, Ahmed SF, Hughes IA, International Consensus Conference on Intersex organized by the Lawson Wilkins Pediatric Endocrine Society and the European Society for Paediatric Endocrinology. Consensus statement on management of intersex disorders. International Consensus Conference on Intersex. Pediatrics [Internet]. 2006;118(2):e488–500. Available from: http://www.ncbi.nlm.nih.gov/pubmed/16882788
43. Weidler EM, Pearson M, van Leeuwen K, Garvey E. Clinical management in mixed gonadal dysgenesis with chromosomal mosaicism: considerations in newborns and adolescents. Semin Pediatr Surg [Internet]. 2019;28(5):150841. Available from: https://linkinghub.elsevier.com/retrieve/pii/S105585861930099X

Models of Psychopathology and Psychiatric Comorbidities in Children and Adolescents with Gender Dysphoria

8

Anurag Jhanjee and Rituja Roli

Abstract

Gender Dysphoria (GD), according to DSM-V, is a marked incongruence between the sex that one feels or expresses and the sex assigned at birth; causing clinically significant discomfort or impairment in social, occupational, or other important functioning areas. The previous term "Gender Identity Disorder" was replaced to avoid pathologizing identity. Not all gender-variant individuals experience gender dysphoria. Transgender person is an umbrella term used for identifying all those individuals who experience discordance or discomfort between natal gender and expressed gender. GD can adversely affect the daily life of a person in various ways. Experiencing disgust towards their own body or feeling another's attention to their "wrong" gendered body may lead them to avoid social activities, sports, sexual encounters, etc.

Understanding the various models of psychopathology helps in the clinical management of GD. If Stroller's model is correct, the underlying systemic psychopathology and dynamics need to be addressed. In case of Coates and Person's, resolution of separation anxiety is the target. In case of social ostracism model, therapeutic aim could be a reduction in the gender-variant behavior expressed by the child and taking measures to provide a safer social environment. In case of GD as an inherent cause of distress, treatment is focused on eliminating GD. Studies show increased rates of psychiatric morbidity in GD children and adolescents, including that of suicidality and self-harm, depression, and eating disorders when compared with their peers. Prevalence of psychiatric comorbid-

A. Jhanjee (✉)
Department of Psychiatry, Maulana Azad Medical College (MAMC) and Govind Ballabh Pant Institute of Postgraduate Medical Education and Research (GIPMER), New Delhi, India

R. Roli
Mental Health Liaison Team, Essex Partnership University NHS Foundation Trust (EPUT), Colchester, United Kingdom

S. K. Ratan (ed.), *Children with Differences in Sex Development*,
https://doi.org/10.1007/978-981-97-1639-5_8

ity among GD children increases with age; the most vulnerable period being adolescence. Various studies show that 58–71% of GD cases exhibit at least one psychiatric disorder, most frequently being depression (50%) followed by anxiety (37.2%) and self-harm (8%). Having awareness of risk of psychiatric comorbidities is essential for clinicians working with GD children.

Keywords

DSD · Gender dysphoria · Gender non-conforming · Transgender

8.1 History

The concept of "gender dysphoria" has been around for many years, although under other names. It was in 1977, that the term "transsexualism" was included in ICD9 within the "Sexual Deviations and Disorders" category, alongside paraphilias, sexual dysfunctions, and others [1, 2].

It first became a part of DSM in DSM-III within the 'Sexual Identity Disorders' in 1980. ICD-10 (1992) and DSM-IV (1994) termed it as "Gender Identity Disorder" and "Sexual Identity Disorder," respectively.

Currently, DSM-V (2013) considers Gender Dysphoria (GD) as a marked incongruence between the sex that one feels or expresses and the sex assigned at birth; causing clinically significant discomfort or impairment in social, occupational, or other important functioning areas. The previous term was rejected to avoid pathologizing identity [3–7]. Not all gender-variant individuals experience gender dysphoria.

ICD-11 (2022) includes the term "Gender Incongruence" within the chapter "Conditions related to sexual health." It defines it as the presence of a marked and persistent disagreement between the gender experienced by the person and the assigned sex [8].

Transgender person is an umbrella term used for identifying all those individuals who experience discordance or discomfort between natal gender and expressed gender [9]. To be identified as a transgender person or gender diverse is considered being a matter of human diversity and not pathology [10].

In the contemporary discussions of gender, there are many questions raised, e.g., what is considered normal expressions of gender and non-pathological variant or gender variance; and who is ultimately responsible for deciding what is normal and what is not. Furthermore, there are some cultures who inherently do not just have binary notions of gender with only "men" and "women" and may have a culturally acceptable third gender, e.g., the Hijras of India or the Native American "two spirit people" [11].

The recent Indian legislation on transgender individuals, The Transgender Persons (Protection of Rights) Act, 2019, defines a transgender person as any person whose expressed gender does not match with the gender assigned to that person at birth and includes trans-man or trans-woman (whether or not such person has undergone Sex Reassignment Surgery or hormone therapy or laser therapy or such other therapy), a person with intersex variations, genderqueer, and person having such socio-cultural identities as Kinner, Hijra, Aravani, and Jogta [12]. This definition is slightly different from that of ICD-10 and DSM-5 as it also includes persons with disorders of sexual development; and shall be used for all issues and matters related to the law [13].

8.2 Background

Children begin to develop a sense of their identity in terms of gender, typically around age three. At this age, they may develop gendered behaviors and interests. Some may also start expressing a desire to be the other gender. Such children are often brought for medical consultations at school age, as it is also the time when they first start interacting with other children closely and come under the scrutiny of adults other than their caregivers. As they approach puberty, many children diagnosed with GD start to experience increased anxiety levels in association with anticipating changes in their bodies.

Studies show that more than half of those diagnosed with gender identity disorder as per DSM-IV, later identify with their birth-assigned gender. Children who identify as transgender in adulthood are found to have had more extreme GD as children. Nevertheless, according to many studies, there are increased rates of gay and bisexual identity among those adults who were gender non-conforming during childhood [14].

GD can adversely affect the daily life of a person in various ways. Experiencing disgust toward their own body or feeling another's attention to their "wrong" gendered body may lead them to avoid social activities, sports, sexual encounters etc. They may also face many social restrictions based on real or feared resistance from other people in an equal way [15].

8.3 Introduction

8.3.1 Diagnosis

8.3.1.1 Gender Dysphoria in Children

DSM 5 diagnosis of gender dysphoria in children requires:

- At least six out of the eight features listed in criterion A; and one of these six must be "a strong desire to be of the other gender or an insistence that he or she is the other gender [or some alternative gender (that is neither conventionally male or female) different from one's assigned gender]." The other symptoms include preferring cross-dressing and rejecting conventional attire of one's assigned gender; preferring cross-gender roles in make-believe play; preferring toys, games, or activities stereotypically associated with the other gender; and

preferring play with children of the other gender; rejecting toys, games, and activities typical of one's assigned gender; a strong dislike of their sexual anatomy and a strong desire for the sex characteristics, primary and/or secondary, of the experienced gender.
- Associated clinically significant distress or impairment in social, school, or other important areas of functioning (Criterion B).

DSM-5 diagnosis also includes a new specifier: "With a DSD." When this specifier is used, both the specific DSD as well as the GD diagnosis should be coded.

The differential diagnosis of GD in childhood:

1. Simple atypical gender behavior in the absence of clinically significant distress or impairment.

   It includes children showing non-conformity to stereotypical gender role and gender-variant behavior like "tom-boyism" in girls or "girly" or "sissy" behavior in boys. The majority of these children feel comfortable with their bodies and assigned genders and continue to remain so in adolescence or adulthood.
2. Epiphenomena of coexisting disorders, including psychotic disorders.
3. Gender concern as a stated desire to be the other gender due to some perceived social advantage of being so or to escape a perceived disadvantage of the assigned gender (e.g., following sexual trauma).

In cases where gender dysphoria is experienced during periods when psychotic symptoms are in remission, both schizophrenia and gender dysphoria may be diagnosed.

The criteria set for GD in Childhood is different than GD in Adolescents and most children with GD in Childhood do not manifest GD in Adolescence and later.

8.3.1.2 Gender Dysphoria in Adolescents

DSM 5 diagnosis of gender dysphoria in adolescents requires:

At least two of the six Criterion A symptoms (listed below), to be present for at least 6 months, and associated clinically significant distress or impairment in function, in social, occupational, or other areas (Criterion B):

1. A pronounced incongruence between one's experienced or expressed gender and the primary/secondary sex characteristics of the assigned or natal gender. In prepubescent adolescents or those in puberty's early stages, it may be expressed as aversive anxiety to anticipated secondary sex characteristics.
2. A strong desire to be rid of the natal gender's primary and/or secondary sex characteristics due to incongruence with one's experienced/expressed gender as opposed to other motives. In adolescents who are prepubescent or in puberty's early stages, it is expressed as a strong desire to prevent further development of anticipated secondary sex characteristics.
3. A strong desire for primary and/or secondary sex characteristics of the other gender.

4. A strong desire to be either of the other gender or some alternative gender, neither conventionally male nor female, rather than one's assigned gender.
5. A strong desire to be treated by others as a member of the other gender or some alternative gender.
6. Finally, a strong conviction that one's feelings and reactions are more typical of the other gender or some alternative gender to the assigned gender.

DSM 5 Specifiers:

- "With a DSD": It requires coding a specific DSD condition, if present, in addition to the GD diagnosis.
- "Post-transition": For those who have transitioned to living full-time in the desired gender, regardless of legal name change, who have undergone or plan to have at least one cross-sex medical or surgical procedure.

The differential diagnosis of GD in Adolescents includes:

1. Nondiagnosable atypical gender behavior.
2. Transvestic disorder (a paraphilia in which a person is aroused by cross-dressing.
3. Body dysmorphic disorder (may have a wish to alter or remove a specific body part perceived to be unattractive or abnormally formed).
4. Schizophrenia and other psychotic disorders (may have delusions of being another gender).

If a person with a psychotic disorder meets the full criteria for GD when the psychosis is in remission, the diagnosis of GD can still be made in case the GD preceded the onset of the psychotic illness. In rare cases, a man may adopt a male eunuch identity, in which they seek castration to ameliorate psychological or physical effects of androgens while still identifying as male [11].

8.4 Models of Psychopathology in Children and Adolescents with Gender Dysphoria

The following models help explain the presence of other types of psychopathology that are seen in some children and adolescents with GD.

1. *Psychodynamic Formulations*
   A. *Stroller's model (1968): No characteristic psychopathology* was found in GD boys. They were normal, except for their extreme GD. Their GD was ego-syntonic and probably the result of certain predispositions (like feminine physical features) interacting with complex family dynamics and psychopathology [16, 17].

B. *Coates and Person's model (1985):* GD was considered to be secondary to *separation anxiety* as the underlying psychopathology; due to its high rate of association to the same [18].

While the two psychodynamic models share the same hypothesis that complex family dynamics lead to GD development, they arrive at totally different conclusions in terms of the associated psychopathology. It was pervasive in the Coates and Person's model. There was none in the Stroller's model.

2. *Psychosocial Factors*

Social Ostracism: GD is conceptualized as secondary to the *experience of negative social reaction*, especially from their peer group, toward their marked gender-variant behavior. This model implies that a supportive environment would significantly reduce the associated psychopathology in GD [19].

Green (1976) interviewed mothers of feminine boys and masculine control boys asking them to classify the peer relations of these boys with other boys into four categories: leader, good mixer, voluntary loner, or rejected. The study showed that in case of a male peer group, the masculine boys were much more likely to be classified as good mixers than were the feminine boys while the feminine boys as either liners or rejected than were the masculine boys [20]. A similar study of ad-recruited "tomboys" and "non-tomboys" was conducted by Green et al. (1982) in peer relations with other girls. It showed that non-tomboys were more likely to be classified as good mixers than were tomboys. On the other hand, tomboys were more likely to be classified as loners than were the non-tomboys. However, tomboys did not get rejected by their female peers, unlike the data from the study on feminine boys [21].

In addition, a study by Wallien et al. (2010) showed a strong relationship between peer relations of GD children and the sex of the peer group. In that, the social position of 28 GD children (14 girls and 14 boys) was examined in their naturalistic environment. Results showed that GD boys, in case of their same-sex peers, were somewhat more likely to be rejected than were the other boys while acceptance was less likely. On the contrary, they were more likely to be accepted by the girls than were the other boys in the class, and less likely to be rejected. In case of GD girls, acceptance by same-sex peers was less likely as compared to the other girls in the class and they were no more or less likely to be rejected. Thus, there is more cross-sex acceptance of both GD girls and boys. Also, GD boys experienced more rejection from same-sex peers than GD girls [22].

Psychologist Diane Ehrensaft also suggested that distress in GD children arises due to stigma and prohibitions within the child's social environment. This was based on her observation that distress in GD often resolves or reduces when these children are in an environment where they are free to express their experienced gender [23].

3. *General Risk Factors*

Psychopathology in GD is associated with the general risk factors that relate to psychopathology in any child or adolescent, like significant family history, socioeconomic status, family discord, etc. [24]. Nevertheless, it is not thought to cause GD per se.

Zucker et al. (2008) showed that maternal psychopathology was an important predictor of behavioral problems in GD children [25].

4. *GD as an Inherent Cause Itself*

   In contrast to the Coates and Pearson model (1985), this model views distress as a direct function of GD, which subsequently causes other types of psychopathology like depression and substance use disorder [19].

   For instance, a study by Vasey and Bartlett (2007) showed that among the *fa'afafine*, a third gender category in Samoa, only a minority experienced GD. It was attributed to the Samoan culture's high degree of social tolerance toward them [26]. There is very little research with regard to this factor.

8.4.1 Summary and Conclusion

An understanding of the various models plays a very important role with regard to clinical management policies and decisions. Whichever model is supported by the empirical evidence, the treatment approach is based accordingly. If Stroller's model is correct, the underlying systemic psychopathology and dynamics need to be addressed. In case of Coates and Person's resolution of separation anxiety is the target. In case of social ostracism model, the therapeutic aim could be a reduction in the gender-variant behavior expressed by the child making it more gender typical. Alternatively, the child could be provided with a safer social environment. Parents may take measures to reduce social stigma and prohibitions in the home of preschool children. However, once the children start school, their increasing interactions with the outside world may prove traumatic and exacerbate dysphoria. Hence, preventive measures need to be taken on a larger social scale. Finally, in case of GD as an inherent cause of distress, treatment is focused on eliminating GD. Some studies of GD adolescents show that biomedical treatment leads to reduction of the associated psychopathology [11, 27–29].

8.5 Psychiatric Comorbidities in Children and Adolescents with Gender Dysphoria

Transgender non-conforming persons face various social adversities including violence, stigma, and discrimination in the healthcare systems making them more vulnerable to the risk of mental illness [30]. Studies on mental health of transgender youth show increased rates of psychiatric morbidity including that of suicidality and self-harm, depression, and eating disorders when compared with their peers [31]. The risk of anxiety, depression, self-harm, and suicide was twofold to threefold higher among transgender youth when compared to cisgender-matched controls in both outpatient as well as inpatient mental health care settings [32]. Studies showed that among all the psychiatric comorbidities, risk of suicide was 20 times higher among transgender individuals when compared to the general population [33].

The prevalence of psychiatric comorbidity among GD children increases with age; the most vulnerable period being adolescence. They have 2–3 times higher rates of psychiatric diagnoses than the non-transgender population. Various studies show that 58–71% of GD cases exhibit at least one psychiatric disorder, most frequently being depression (50%) followed by anxiety (37.2%) and self-harm (8%). The high rate is likely due to the stigma they face related to their gender-variant behavior and identities [32, 34, 35].

8.5.1 Substance Use

Studies show a high prevalence of substance abuse in GD: 40% [cannabis (24.4%), alcohol (21.5%)], with equal prevalence in trans female and male. Prevalence of other illegal drugs reaches upto 11.6%, being more frequent in trans male. Prevalence of alcohol and cannabis in relation to the non-trans population is high (upto 1.25–1.53%). This difference becomes greater when considering the use of ecstasy or cocaine, being upto 6–8 times higher [36–38].

8.5.2 Suicide-Related Events

Suicide Attempts: Prevalence ranges between 16.1% and 50.3%. Regarding the predominance of suicide attempts in either gender, there is no agreement between various studies. Prevalence is 1.6–5.4 times higher than the non-trans population.

Suicidal Ideation: Prevalence was found to range between 30 and –95.5% in a young population, with a preponderance in trans females.

Self-Harm In Non-Suicidal Attempts: Prevalence ranges between 36.8 and –86.9% with predominance in trans males, and exhibits positive association with age [35, 39–43].

8.5.2.1 Depression/Anxiety

Studies show a high prevalence in young and adolescent trans populations; 4–10 times higher than in the non-trans population. Prevalence of depression ranges from 33% to 50.6% and anxiety has a wider range (26.7–63.3%) [32, 34, 35, 39].

8.5.2.2 Eating Disorders

The risk is four times higher than in the non-trans population. Diagnosis of eating disorder is found in 15.8% of trans population. Prevalence of weight manipulations with gender-affirming purposes is 65%. Prevalence of behaviors related to eating disorders, such as binge eating, fasting or vomiting, and laxatives varies between 33 and 50% [44–46].

8.5.2.3 Dissociative Disorders

They have a prevalence of 29.6%. The risk is more than that in non-trans population, with no differences between genders [47].

8.5.2.4 Autism Spectrum Disorders (ASD)

Most studies have found a higher prevalence of ASD in GD. The prevalence of traits related to the ASD ranges between 14.5 and 36.3%. Prevalence of ASD in GD: 14%. This may be related to intrauterine hormone exposure [48–50].

8.5.3 Rapid Onset GD

It was first described by Littman. The two major psychiatric comorbidities seen in these adolescents were found to be: anxiety (46.6%) and depression (39.4%). Average age of identification as trans was 15.2 years. Majority did not present indicators of gender dysphoria in childhood or adolescence. Some agents favoring this could be family conflicts and circles of friendship in which most members are trans or gender variability increasingly represented in social networks and media [51].

8.5.4 Clinical Implications

It is important to remember that GD children and adolescents may also have psychiatric disorders that are independent of their gender concerns. This awareness of the risk of psychiatric comorbidities is essential for clinicians working with GD children, especially regarding internalizing problems; while also realizing that externalizing comorbidity, if present, can make a GD child more vulnerable to social ostracism [52]. In terms of implication on the clinical treatment, even serious psychiatric diagnoses like major depressive disorder, bipolar disorder, and schizophrenia are not an absolute contraindication for gender transition. Nonetheless, these must be adequately treated so as to avoid any interference with adherence to transition regime of treatment. In addition, it is important that manifestation of a psychiatric comorbidity is not mistaken for GD. While GD resolves with transition, a comorbidity will most likely persist. Hence, it is imperative to recognize it and continue the treatment even after transition.

References

1. Fernández Rodríguez M, García-Vega E. Surgimiento, evolución y dificultades del diagnóstico de transexualismo. Rev Asoc Esp Neuropsiquiatría. 2012;32(113):103–19.
2. World Health Organization. International classification of diseases: [9th] ninth revision, basic tabulation list with alphabetic index [Internet]. World Health Organization; 1978 [cited 2023 Jan 25]. https://apps.who.int/iris/handle/10665/39473
3. Mas Grau J. Del transexualismo a la disforia de género en el DSM. Cambios terminológicos, misma esencia patologizante. Rev Int Sociol. 2017;75(2):059.
4. Kendell RE. Diagnostic and statistical manual of mental disorders, 3rd ed. Am J Psychiatry. 1980;137(12):1630–1.
5. World Health Organization. The ICD-10 classification of mental and behavioural disorders: clinical descriptions and diagnostic guidelines [Internet]. World Health Organization; 1992 [cited 2023 Jan 25]. Available from: https://apps.who.int/iris/handle/10665/37958

6. Guze SB. Diagnostic and statistical manual of mental disorders, 4th ed. (DSM-IV). Am J Psychiatry. 1995;152(8):1228.
7. Diagnostic and statistical manual of mental disorders: DSM-5™, 5th ed. Arlington, VA, US: American Psychiatric Publishing, Inc.; 2013. xliv, 947 p. (Diagnostic and statistical manual of mental disorders: DSM-5™, 5th ed).
8. ICD-11 [Internet]. [cited 2023 Jan 25]. Available from: https://icd.who.int/en
9. Papoulias C. Transgender. Theory Cult Soc. 2006;23(2–3):231–3.
10. Evidence supporting the biologic nature of gender identity - PubMed [Internet]. [cited 2023 Feb 1]. Available from: https://pubmed.ncbi.nlm.nih.gov/25667367/
11. Sadock BJ, Sadock VA, Ruiz P. Kaplan and Sadock's comprehensive textbook of psychiatry. 11th Edition, Wolters Kluver; 2020, Alphen aan den Rijn. - References - Scientific Research Publishing [Internet]. [cited 2023 Jan 10]. Available from: https://www.scirp.org/%28S%28lz5mqp453edsnp55rrgjct55%29%29/reference/referencespapers.aspx?referenceid=2723304
12. Coleman E, Radix AE, Bouman WP, Brown GR, de Vries ALC, Deutsch MB, et al. Standards of care for the health of transgender and gender diverse people, Version 8. Int J Transgender Health. 2022;23(Supp 1):S1–259.
13. (PDF) Manual on Mental Healthcare of Transgendered Persons in India [Internet]. [cited 2023 Feb 1]. Available from: https://www.researchgate.net/publication/353906777_Manual_on_Mental_Healthcare_of_Transgendered_Persons_in_India
14. Kaplan & Sadock's Synopsis of Psychiatry [Internet]. [cited 2023 Jan 25]. Available from: https://www.wolterskluwer.com/en/know/synopsis-of-psychiatry
15. Dowers E, White C, Kingsley J, Swenson R. Transgender experiences of occupation and the environment: a scoping review. J Occup Sci. 2019;26(4):496–510.
16. Stoller RJ. Male childhood transsexualism. J Am Acad Child Psychiatry. 1968;7(2):193–209.
17. Stoller RJ. Sex and gender: on the development of masculinity and femininity. Science House; 1968. 306 p.
18. Extreme boyhood femininity: isolated behavior or pervasive disorder? | Semantic Scholar [Internet]. [cited 2023 Jan 9]. Available from: https://www.semanticscholar.org/paper/Extreme-boyhood-femininity%3A-isolated-behavior-or-Coates-Person/c3febc85aca350662adaa3d9d3d614c413530733
19. Zucker KJ, Wood H, VanderLaan DP. Models of psychopathology in children and adolescents with gender dysphoria. In: Kreukels BPC, Steensma TD, de Vries ALC, editors. Gender dysphoria and disorders of sex development: progress in care and knowledge [Internet]. Boston, MA: Springer US; 2014 [cited 2023 Jan 9]. p. 171–92. (Focus on sexuality research). https://doi.org/10.1007/978-1-4614-7441-8_9
20. Green R. One-hundred ten feminine and masculine boys: behavioral contrasts and demographic similarities. Arch Sex Behav. 1976;5(5):425–46.
21. Green R, Williams K, Goodman M. Ninety-nine "tomboys" and "non-tomboys": behavioral contrasts and demographic similarities. Arch Sex Behav. 1982;11(3):247–66.
22. Wallien M, Veenstra R, Kreukels B, Cohen-Kettenis P. Peer group status of gender dysphoric children: a sociometric study. Arch Sex Behav. 2009;39:553–60.
23. Beemyn G. The children should Lead us: Diane Ehrensaft's *gender born, gender made: raising healthy gender-nonconforming children*. J LGBT Youth. 2013;10(1–2):159–62.
24. Connell AM, Goodman SH. The association between psychopathology in fathers versus mothers and children's internalizing and externalizing behavior problems: a meta-analysis. Psychol Bull. 2002;128:746–73.
25. Zucker K. Associated psychopathology in children and adolescents with gender identity disorder. In 2008.
26. Vasey PL, Bartlett NH. What can the Samoan "Fa'afafine" teach us about the Western concept of gender identity disorder in childhood? Perspect Biol Med. 2007;50(4):481–90.
27. Cohen-Kettenis PT, van Goozen SH. Sex reassignment of adolescent transsexuals: a follow-up study. J Am Acad Child Adolesc Psychiatry. 1997;36(2):263–71.

28. de Vries ALC, Steensma TD, Doreleijers TAH, Cohen-Kettenis PT. Puberty suppression in adolescents with gender identity disorder: a prospective follow-up study. J Sex Med. 2011;8(8):2276–83.
29. Smith YL, van Goozen SH, Cohen-Kettenis PT. Adolescents with gender identity disorder who were accepted or rejected for sex reassignment surgery: a prospective follow-up study. J Am Acad Child Adolesc Psychiatry. 2001;40(4):472–81.
30. Radix AE, Lelutiu-Weinberger C, Gamarel KE. Satisfaction and healthcare utilization of transgender and gender non-conforming individuals in NYC: a community-based participatory study. LGBT Health. 2014;1(4):302–8.
31. Connolly MD, Zervos MJ, Barone CJ, Johnson CC, Joseph CLM. The mental health of transgender youth: advances in understanding. J Adolesc Health Off Publ Soc Adolesc Med. 2016;59(5):489–95.
32. Reisner SL, Vetters R, Leclerc M, Zaslow S, Wolfrum S, Shumer D, et al. Mental health of transgender youth in care at an adolescent urban community health center: a matched retrospective cohort study. J Adolesc Health Off Publ Soc Adolesc Med. 2015;56(3):274–9.
33. Maguen S, Shipherd JC. Suicide risk among transgender individuals. Psychol Sex. 2010;1(1):34–43.
34. Reisner SL, Biello KB, White Hughto JM, Kuhns L, Mayer KH, Garofalo R, et al. Psychiatric diagnoses and comorbidities in a diverse, multicity cohort of young transgender women: baseline findings from project life skills. JAMA Pediatr. 2016;170(5):481–6.
35. Peterson CM, Matthews A, Copps-Smith E, Conard LA. Suicidality, self-harm, and body dissatisfaction in transgender adolescents and emerging adults with gender dysphoria. Suicide Life Threat Behav. 2017;47(4):475–82.
36. Rowe C, Santos GM, McFarland W, Wilson EC. Prevalence and correlates of substance use among trans female youth ages 16-24 years in the San Francisco Bay Area. Drug Alcohol Depend. 2015;147:160–6.
37. Gonzalez CA, Gallego JD, Bockting WO. Demographic characteristics, components of sexuality and gender, and minority stress and their associations to excessive alcohol, cannabis, and illicit (noncannabis) drug use among a large sample of transgender people in the United States. J Prim Prev. 2017;38(4):419–45.
38. Johns MM, Lowry R, Andrzejewski J, Barrios LC, Demissie Z, McManus T, et al. Transgender identity and experiences of violence victimization, substance use, suicide risk, and sexual risk behaviors among high school students - 19 states and large Urban School districts, 2017. MMWR Morb Mortal Wkly Rep. 2019;68(3):67–71.
39. Marshall E, Claes L, Bouman WP, Witcomb GL, Arcelus J. Non-suicidal self-injury and suicidality in trans people: a systematic review of the literature. Int Rev Psychiatry Abingdon Engl. 2016;28(1):58–69.
40. Suicidal ideation and attempted suicide amongst Chinese transgender persons: National population study - PubMed [Internet]. [cited 2023 Jan 10]. Available from: https://pubmed.ncbi.nlm.nih.gov/30699856/
41. Thoma BC, Salk RH, Choukas-Bradley S, Goldstein TR, Levine MD, Marshal MP. Suicidality disparities between transgender and cisgender adolescents. Pediatrics. 2019;144(5):e20191183.
42. Kuper LE, Adams N, Mustanski BS. Exploring cross-sectional predictors of suicide ideation, attempt, and risk in a large online sample of transgender and gender nonconforming youth and young adults. LGBT Health. 2018;5(7):391–400.
43. Perez-Brumer A, Day JK, Russell ST, Hatzenbuehler ML. Prevalence and correlates of suicidal ideation among transgender youth in California: findings from a representative, population-based sample of high school students. J Am Acad Child Adolesc Psychiatry. 2017;56(9):739–46.
44. Diemer EW, Grant JD, Munn-Chernoff MA, Patterson DA, Duncan AE. Gender identity, sexual orientation, and eating-related pathology in a National Sample of College Students. J Adolesc Health Off Publ Soc Adolesc Med. 2015;57(2):144–9.
45. Watson RJ, Veale JF, Saewyc EM. Disordered eating behaviors among transgender youth: probability profiles from risk and protective factors. Int J Eat Disord. 2017;50(5):515–22.

46. Avila JT, Park KT, Golden NH. Eating disorders in adolescents with chronic gastrointestinal and endocrine diseases. Lancet Child Adolesc Health. 2019;3(3):181–9.
47. Colizzi M, Costa R, Todarello O. Dissociative symptoms in individuals with gender dysphoria: is the elevated prevalence real? Psychiatry Res. 2015;226(1):173–80.
48. van der Miesen AIR, de Vries ALC, Steensma TD, Hartman CA. Autistic symptoms in children and adolescents with gender dysphoria. J Autism Dev Disord. 2018;48(5):1537–48.
49. Stagg SD, Vincent J. Autistic traits in individuals self-defining as transgender or nonbinary. Eur Psychiatry J Assoc Eur Psychiatr. 2019;61:17–22.
50. Nobili A, Glazebrook C, Bouman WP, Glidden D, Baron-Cohen S, Allison C, et al. Autistic traits in treatment-seeking transgender adults. J Autism Dev Disord. 2018;48(12):3984–94.
51. Littman L. Parent reports of adolescents and young adults perceived to show signs of a rapid onset of gender dysphoria. PLoS One. 2018;13(8):e0202330.
52. Wallien MSC, Swaab H, Cohen-Kettenis PT. Psychiatric comorbidity among children with gender identity disorder. J Am Acad Child Adolesc Psychiatry. 2007;46(10):1307–14.

Part III

Legal Aspects

I duly acknowledge Dr. Sreenivas, Professor Forensic Medicine, Maulana Azad Medical College, New Delhi and Dr. Devendra Yadav, Additional Professor of Paediatric Surgery, AIIMS, New Delhi for assisting and aiding me section.

Legal Aspects of DSD: Indian and International Scenario

9

Ramesh Babu

Abstract

The terms intersex or hermaphroditism have currently been replaced with more inclusive terminology, difference of sex development (DSD). In the past, the medical and surgical outcomes in relation to functionality were the main focus when describing outcomes of those with DSD. Of late gender-related issues and the psychological and social problems associated with DSD have taken the center stage. Both the medical and the social problems start soon after the birth and get extended during the adulthood. In addition to identification of life-threatening conditions, medical and surgical management may be promptly required in some. The timing of medical or surgical intervention has always been a matter of debate. As soon as the baby with DSD is born assignment of sex of rearing becomes a major social emergency. Gender confirming treatments have been a matter of great debate recently. Several states and countries have even gone ahead with their own regulations in DSD management. While the right of the DSD individual to decide treatment as an adult is being upheld legal and ethical guidelines have often cut a balance in not denying the essential lifesaving treatment. While the decision to defer treatment is being called for, one should also consider the need for safe upbringing of the school going child. Medical and legal issues should be addressed side by side without compromising ethics.

Keywords

DSD · Gender incongruence · DSD legal · Gender dysphoria

R. Babu (✉)
Sri Ramachandra Institute of Higher Education and Research, Chennai, India

S. K. Ratan (ed.), *Children with Differences in Sex Development*,
https://doi.org/10.1007/978-981-97-1639-5_9

9.1 Background

The terminology DSD (differences of sex development) has been proposed for all conditions erstwhile known as "disorders of sexual differentiation" or "intersex" as the previously used terms such as "hermaphrodite" or "pseudo-hermaphrodite" were felt to be derogatory and causing a social stigma. The proposed new classification takes into account, the sensitivities of the individuals and families and reflects the molecular understanding of DSD [1, 2]. There is still significant controversy existing in the field, regarding appropriate and respectful nomenclature. Although the term "intersex" has been replaced by the term "DSD" in the medical literature, many activists and patient support groups still prefer to call themselves intersex individuals [3]. Another major issue is including people with "transgender" or LGBT also into the umbrella together as intersex. This has a major implication for legal issues as clubbing multiple conditions under the same umbrella can create incorrect decisions when applying a blanket rule for the treatment. For example, intersex or DSD refers to disorders of genotypic/ phenotypic sex characteristics, referring to a medical condition needing essential medical/ surgical treatment. On the other hand, transgender refers to anatomically unambiguous individuals who later as an adult seek sex reassignment due to their changing gender feelings. The latter is an issue with mind/desire and they may seek/deny a treatment with their own consent as a grown-up adult. But social issues faced by intersex/DSD patients start right from birth, when a sex of rearing is needed by the parents for schooling and socially secure upbringing. Denying them prompt medical treatment, or making them wait for procedures until adulthood, to obtain an informed self-consent, may expose them to many irreversible psychological and social issues during childhood.

9.2 Sex and Gender

The terms "sex," "gender," and "sexuality" are interpreted differently. "Gender" is a social term that refers to the way society reflects the individual identity. It does not take into account the actual "gender identity" ("internal identity") and future "gender role" ("behavioral expressive role") [4]. Gender identity is the basic sense of belonging to a specific gender and the self-identification as male, female, or any another gender [5]. The terminology "gender role" means the attitudes and personality traits that society defines as male or female. Gender roles should be considered separately from gender identity [6]. For example, a person with female gender identity may take a fatherly gender role for the sake of the family or society.

The term "sex" refers to anatomical characteristics (such as phenotypic sex, genotypic sex—karyotype, internal anatomy, and gonadal sex). Sex of rearing means the sex assigned to the child by the caregivers/parents. The word, Sexual orientation, is completely different from sex or gender. Sexual orientation is defined by a person's sexual reaction/response to another person, the sex of the person to whom one is sexually attracted (heterosexual, homosexual, bisexual) [2]. Thus when we discuss further, we would be discussing only sex assignment and not gender assignment, as gender itself is a self-identity while sex of rearing is that assigned by a multi-disciplinary committee (MDC).

9.3 Gender Dysphoria

Several studies [7, 8] have questioned the "optimal gender policy" for infants with DSD. While in past, the idea that "nurture overrules nature" prevailed, recent studies have shown that, children with DSD do not always conform to the sex of rearing when they become adults [9]. Gender dysphoria means a mental conflict between a person's assigned or physical sex and the gender they identify with. The terminology is now recently being changed to gender identity disorder (GID). People with GID have a strong and persistent cross-identification of gender usually manifested in adolescents and adults. It is detected by the following symptoms: desire to be the opposite sex or the desire to live/be treated as the opposite sex, or the conviction that they have the classical feelings and reactions that are usually noted in the opposite sex [10].

9.4 Legal Implications

The term transgender refers to a person, whose sense of personal identity and gender does not match with their sex at birth. It is distinct from DSD in the sense that they have unambiguous sexual anatomy at birth. When they become adults, they seek sex reassignment surgery (SRS) due to gender dysphoria and it is done by adult specialist surgeons with the patient's consent. On the other hand, surgery for intersex is often performed for ambiguous genitalia in children, mainly for medical indications by pediatric surgeons after obtaining parental consent. There is considerable controversy on the timing of irreversible surgery in children with DSD. Many children with DSD may require medical and surgical treatment because of the "medically necessary" or "life-threatening" and "urgent" indication to intervene.

However, the LGBTQIA+ activist groups consider the rights of intersex children along with the rights of transgender people. The 2019 case in the Madurai bench of Madras High Court (India) started with transgender marriage rights [11]. While the case itself started with a denial of transgender marriage the intersex activists pleaded to be heard on surgeries on intersex people. The judge asked the director of medical education to form the guidelines on intersex surgeries while delivering judgment on transgender rights! The judge quoted from Khalil Gibran: "*Your children are not your children. They are the sons and daughters of Life's longing for itself. They come through you but not from you, And though they are with you yet they belong not to you. You may give them your love but not your thoughts, For they have their own thoughts. You may house their bodies but not their souls, For their souls dwell in the house of tomorrow, which you cannot visit, not even in your dreams. You may strive to be like them, but seek not to make them like you.*"

Fig. 9.1 Legal guidelines should balance between rights of child, parent, doctor, and activists

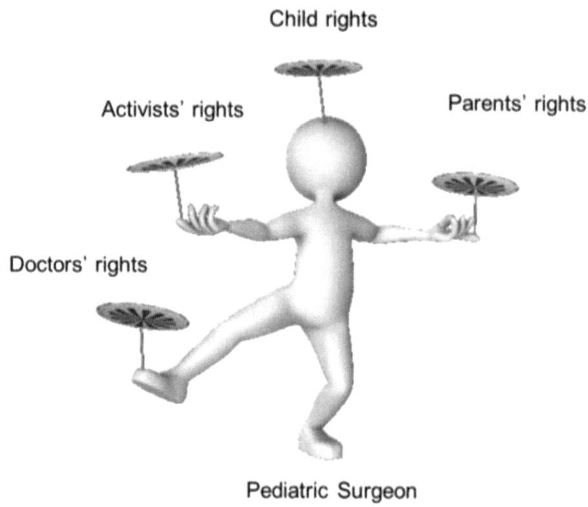

Another hearing in the Madras High Court in 2022 on a transgender marriage complaint (they had sought police protection from their parents!) took up and discussed difficulties faced by LGBTQIA+ community and issued directions to the National Medical Council (NMC) and the State Medical Councils to add the medical and legal updates in their curricula [12]: to avoid pathologizing and criminalizing the non-heterosexual relationships, gender non-conformity, gender incongruence, etc. The NMC and State Medical Councils were asked to form a committee and submit a report on the compliance with the recommendations.

The activists are against any irreversible surgical procedures and feel that patients should grow up and decide themselves the need for surgery after gender identity is established. The legal diaspora too seeps to fall in line, failing to understand that the decision to defer surgery itself is a decision taken without the patient's consent. For example, an adolescent with androgen insensitivity may develop deepening of the voice around puberty and deferring orchidectomy/medical suppression until adulthood may lead to permanent alteration of voice! While formulating legal guidelines one has to balance the child's rights, parents' rights, activists' rights, and doctor's rights (Fig. 9.1).

9.5 Ethical Implications

Human Rights Watch, USA, CAH research/education/support foundation (CARES) and the Advocates for Intersex Youth (InterACT) together with Dr. Dix Poppas, Chief of Pediatric Urology at Cornell-Weil Medical Center, in New York City created an extensive document after interviewing several intersex adults [13]. They argued that assigning a sex of rearing to a child never required surgery and suggested the Society of Pediatric Urology to formulate guidance on surgeries in DSD individuals. They also claimed that intersex surgeries are unethical and must be avoided till the patient can participate actively in decision-making process [13].

Diamond et al. [14] in their article on evidence regarding unnecessary surgery in infants say that "if no effort is made, we will be left, in the next generation, to continue making the same judgment, based on 'experience' and 'expert opinion." However, this is promptly refuted by Lee et al. [15] in their commentary with the following points: (1) current outcome data, shows better outcomes than historical data; (2) "clitoral surgery is done less frequently understanding NV bundle and outcomes of current advances could be better"; (3) "sexual satisfaction may not relate to prior genital surgery as females are often satisfied with their surgical outcome."

Recent legislations in Australia have been attempting to include the hypospadias repair also under the purview of DSD legislation. This may leave a vast population of hypospadiac children uncorrected with their psychological disadvantages during their childhood. In hypospadiac children, the decision to defer surgery in itself is a decision taken without the consent of the child and it has detrimental psychological consequences warned Dr. Snodgrass in a personal communication to the Australian government. He also mentioned that uncorrected hypospadias negatively impacts a man's daily life and his self-image. Surgery is not the same in teens as it is in boys and they will suffer more if surgery is not allowed before they reach an age to provide legal consent.

9.6 Medical Implications

Assigning a sex of rearing is a great responsibility and it should not be rushed up. As many factors must be taken into account, it is imperative to collect all necessary and relevant information while still being as expeditious as possible. When addressing the baby, it is essential to be gender-neutral and to avoid "he" or "she" pronouns that could bias the family inadvertently toward a particular gender that may possibly get changed later. Instead of referring to the baby as "it", warmer terms like "your baby" may encourage greater bonding with parents. Until the decision is finalized, remaining gender-neutral may help prevent misunderstandings and confusion later. However, such decisions should be made with caution as it causes enormous stigma and social pressure on the family.

While there is considerable controversy on the timing of surgery itself, even the activists and legal fraternity agree that sex of rearing could be assigned without active irreversible treatment. The sex of rearing decisions often rests with parents and MDC dealing with these children. It may become a social emergency in certain societies where there is considerable pressure on the parents when the entire family demands to know the sex of the child. Sex of rearing decisions, conventionally, have been dependent on the functional potential of external or internal genitalia. Of late there is a considerable interest in incorporating future gender identity results and gender dysphoria evidence in sex of rearing decisions. Variable types of questionnaires or gender instruments have been used in gender assessment or sexual orientation [8, 16–22]. These gender assessment tools can be particularly useful in late presenting DSD adolescents where MDC has to decide on sex reassignment.

A recent meta-analysis [23] reported that gender dysphoria is low in women with congenital adrenal hyperplasia (CAH), complete androgen insensitivity syndrome (CAIS), and pure gonadal dysgenesis favoring female sex of rearing in them. Gender dysphoria is high in women with 5 alpha reductase deficiency (5ARD)/17 hydroxysteroid dehydrogenase deficiency (17HSD) favoring male sex of rearing in these DSD individuals. Gender dysphoria is variable in partial androgen insensitivity syndrome (PAIS) or mixed gonadal dysgenesis (MGD) and hence, no clear recommendations on sex of rearing could be made in these conditions. In PAIS/MGD children, multiple factors such as local anatomy, hormonal profile and genetics need to be considered. These parents can be given the option to leave the sex undetermined (if they are willing) until their children themselves are able to develop gender identity as they grow through the adolescence period.

9.7 Way Forward

Due to growing concerns, it is essential to understand the intricacies, complexities, and develop some guidelines in each state. Ideally, DSD should be managed only in specific centers with multi-disciplinary committee. The Head/medical director/dean of each tertiary hospital or medical college should act as the lead in the formation of such local multi-disciplinary committees (LMDC). The committee must include at least three specialists: a Pediatrician/Endocrinologist, a Pediatric Surgeon, and a clinical psychologist/psychiatrist. Wherever available it may include a geneticist, social worker, or a support group individual.

Sex assignment must be carried out only at designated centers by LMDC after a thorough evaluation, diagnosis, and discussion with all the involved stakeholders. While a thorough evaluation is indeed necessary, MDC should also remember the social urgency of declaring sex of the newborn in some cultures. In complex situations where the LMDC could not take a decision, there should be three-tier regulatory setup to guide them (Fig. 9.2). A state-level Apex multi-disciplinary committee (SMDC) and an apex multi-disciplinary committee (AMDC) (a premier center for pediatric surgical care in the country) should also be formed to guide the LMDC.

Decisions about the nature and timing of any surgical intervention are made along with the family (or involving the adolescent child), by the LMDC acknowledging and taking into account the significant psychological impact. When a potentially "life threatening" situation arises (like urosepsis/cancer/risk to life) on the intersex child, the well-being of the child is of crucial importance in the decision-making (Fig. 9.2). Those decisions can be made after discussion with a LMDC if urgent: DSD with retention of menstrual or mucinous fluid or recurrent urinary tract infection due to urinary stasis in the common channel—becomes potentially "life threatening" due to urosepsis (in these children vaginoplasty/such procedures to evacuate excretory fluids is justified). In those children with XY/mixed gonadal dysgenesis with streak gonads, the risk of cancer is very high (40–60%) making it a potentially "life threatening condition" [24]. Excision biopsy/removal of the streak gonad in MGD with a high risk of cancer is justified after getting an informed

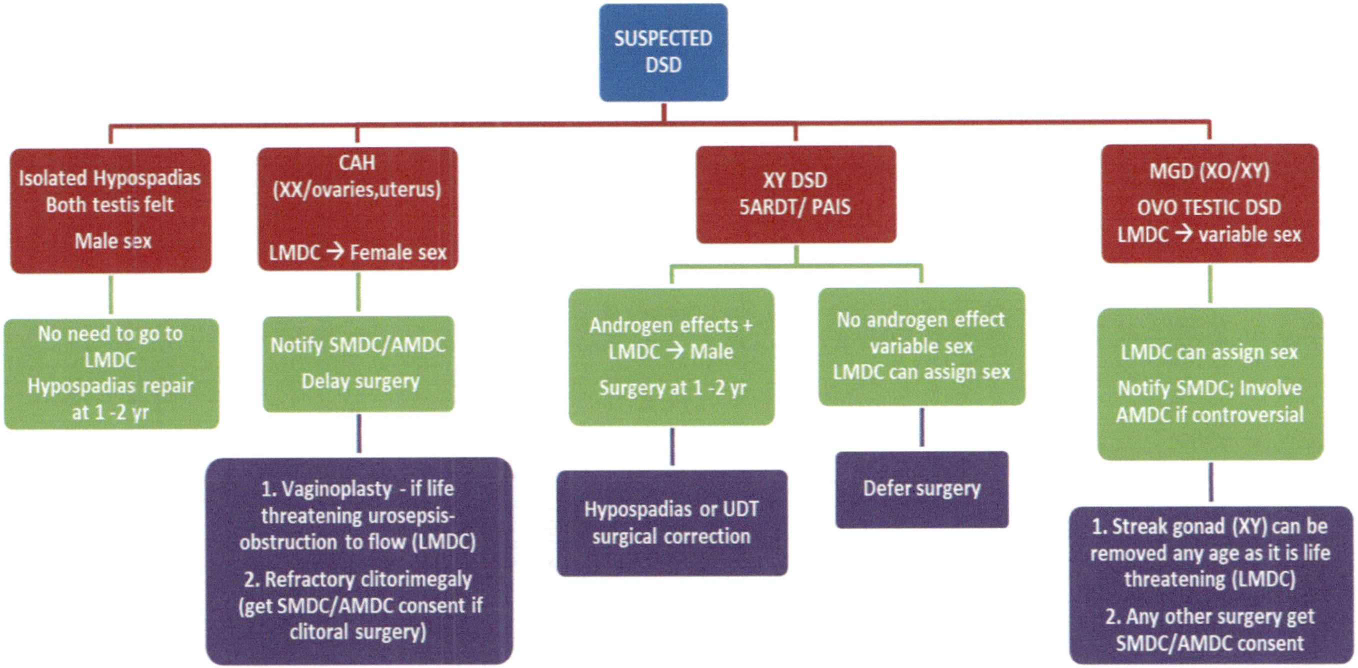

Fig. 9.2 Proposed suggestions for local, state, and apex multi-disciplinary committees (LMDC, SMDC, AMDC). One has to understand that there may be many variations or combinations that may warrant an individualized care plan

consent. LMDC must maintain a record of a signed document from all members and parents about the need for surgical intervention. The same record should also be forwarded to SMDC which should act as a nodal center for record maintenance. The minutes of these meetings could act as a legal document later in case of any dispute.

There are many other conditions like CAH, where surgery is controversial (Fig. 9.2). Clitoral surgery has been known to alter/impair the sensation [17, 25] and any form of clitoral surgery is considered cosmetic. Canning believes that maintaining older techniques is unlikely to reflect the results of current nerve sparing procedures [26]. In most cases of CAH, an adequate hormonal suppression is known to reduce clitorimegaly [27]. Therefore, clitoral reduction procedures are becoming less indicated these days. However, in situations where refractory clitorimegaly is there, one has to get SMDC/AMDC concurrence before clitoral surgery (Fig. 9.2). Similarly, vaginoplasty in CAH has been fraught with lots of failures and painful vaginal dilatations. Many authors recommend delaying vaginoplasty until puberty or to an age when it can be safely performed with the informed consent of the grown-up [14, 28]. However, in conditions where there is retention of fluid/urine in common channel leading to urosepsis, surgical correction is justified (Fig. 9.2).

Two-thirds of caregivers/parents of female infants with CAH reported not regretting their decision on surgical intervention [29]. Most of the females with CAH and their parents believe CAH should be excluded from the intersex category and should be considered in isolation, in legislation regarding childhood genital surgery [30]. While the outcry to delay surgeries for cosmetic purposes to a later date gets louder, intangible effects of having atypical genitalia on children and parents are unclear [31]. Only recently, a small feasibility study on postponing surgery in CAH has reported that girls and their parents did not have significant concerns regarding the ambiguity of genitalia [32]. So when formulating state guidelines, we have to weigh the pros and cons of genital surgery in CAH. Another area of debate is, Prader 5 CAH child reared as male, in whom a late feminizing genitoplasty may be counterproductive as due to androgen effects, they often do not have gender dysphoria as men [33]. These children may also need hysterectomy/gonadectomy surgery, but this should be postponed until puberty, when an informed decision by the patient can be made [34].

In children with CAIS, the germ cell tumor risk is <1% and the risk is more only after second decade of life [24]. Therefore, gonadectomy surgery can be postponed for allowing spontaneous feminization at puberty and hence avoid hormonal replacement [27]. However, in those reared as females, voice change at puberty may force the MDC to resort to surgical/medical methods to prevent the irreversible voice change. This happens around 12–15 years of age, when the individual is still not of legal age of consent. In such situations, there should be legal provision to accommodate the assent of the Gillick competent adolescent to proceed with the necessary medical/surgical treatment.

The least controversial decision of all decisions would be surgery for hypospadias in those with bilateral palpable testes and XY karyotype. The appropriate timing of the isolated hypospadias repair is around 9–18 months of age [27] and for this decision-making, these children need not go through LMDC/DSD pathway

(Fig. 9.2). Even in those DSD kids where male sex of rearing has been decided upon by MDC (mixed gonadal dysgenesis or 5-alpha reductase deficiency) hypospadias repair should be offered at the right age to enable them to use boys' toilet in school without risk of being bullied. Psychological risk of deferring surgery in hypospadias children could be considerable. The decision to wait for hypospadias repair after they become adults and offer informed consent may itself be an incorrect view due to the irreversible psychological damage it could pose during schooling.

9.8 Legal Impediments for Stakeholders

Many activists feel that DSD children who have undergone surgery have been subjected to genital mutilation without their informed consent during their childhood. Some of the grown-up DSD patients have revolted against surgeons for putting them through this trauma. Although they also blame their parents they never file a case against their parents. Instead, they file lawsuits against the treating doctors who performed the procedures in the good interest—for their safe upbringing, after getting informed consent from the parents/guardians.

As there are laws enacted against discrimination or bullying of DSD children/adults in many countries, a few countries/states have gone ahead and enacted laws restricting DSD surgery. It is possible that in future, some activists may even express regret of not having the surgery and even go to court seeking legal sanction to have their surgery done (typical example—a CAIS patient who was reared as girl, wanting gonadectomy prior to voice breaking).

It is essential to meticulously record the minutes of DSD multi-disciplinary meetings and keep them for a prolonged time (25–30 years) at the apex nodal center—like apex MDC office to safeguard the interests of parents and the treating physicians, who may have to demonstrate that procedure performed was medically indicated and was decided after careful and thorough discussion with all stakeholders (parents/ grownup children and specialists).

9.9 Summary

There is a rising concern among the DSD support groups that sex assignment is often performed with binary notions (fitting them into male/female framework), not considering the possible assignment of third/neutral sex. Even though a gender-neutral child upbringing and possible third sex of rearing is accepted by the Law, the parents prefer to have a sex assigned for the child, in order to provide socially secure and safe upbringing. Even in the so-called progressive-minded Western countries where there is social acceptance of a third gender, the psychosocial issues of not choosing a gender are unclear [35] and there is no law yet to regulate their medical management.

The children with DSD completely rely upon the support of their parents, who in turn are usually dependent on their extended family and society (not social welfare/

government support). Denial of the necessary medical attention may lead to child neglect, or even casting their child away. While the court has instructed the school authorities to build toilets for third gender, we are not sure whether this is always available. Even if available using a gender-neutral toilet itself may be a reason for these children to be bullied and teased by their peers. It is quite possible that gender-neutral upbringing in school going children may lead to depression or even self-harm. A rather safer way would be to keep the sex assigned until the individual grows and feels safe to express their own gender identity and then embark upon taking up the chosen gender role.

Imposing laws or rules to ban all DSD surgeries will make the parents turn to quackery to achieve their aims, unless until the views of society are changed via self-realization or via education. The better way, rather is to ensure that DSD children continue to receive the required medical care and support from qualified pediatric surgeons. Having said that, it is indeed prudent to regulate these irreversible surgeries via guidelines, regulations, consensus, and informed decisions involving parents and LMDC. Each patient with DSD is unique and indeed they all warrant a prompt multi-disciplinary care and due psycho sexual support in the long term.

References

1. Lee PA, Houk CP, Ahmed SF, Hughes IA, International Consensus Conference on Intersex organized by the Lawson Wilkins Pediatric Endocrine Society and the European Society for Paediatric Endocrinology. Consensus statement on management of intersex disorders. International consensus conference on intersex. Pediatrics [Internet]. 2006;118(2):e488–500. Available from: http://www.ncbi.nlm.nih.gov/pubmed/16882788
2. Houk CP, Hughes IA, Ahmed SF, Lee PA, Writing Committee for the International Intersex Consensus Conference Participants. Summary of consensus statement on intersex disorders and their management. International intersex consensus conference. Pediatrics [Internet]. 2006;118(2):753–7. Available from: http://www.ncbi.nlm.nih.gov/pubmed/16882833
3. Cools M, Nordenström A, Robeva R, Hall J, Westerveld P, Flück C, et al. Caring for individuals with a difference of sex development (DSD): a consensus statement. Nat Rev Endocrinol [Internet]. 2018;14(7):415–29. Available from: http://www.nature.com/articles/s41574-018-0010-8
4. Mouriquand PDE, Gorduza DB, Gay CL, Meyer-Bahlburg HFL, Baker L, Baskin LS, et al. Surgery in disorders of sex development (DSD) with a gender issue: if (why), when, and how? J Pediatr Urol [Internet]. 2016;12(3):139–149. Available from: https://linkinghub.elsevier.com/retrieve/pii/S1477513116300122
5. Fisher AD, Ristori J, Fanni E, Castellini G, Forti G, Maggi M. Gender identity, gender assignment and reassignment in individuals with disorders of sex development: a major of dilemma. J Endocrinol Invest [Internet]. 2016;39(11):1207–24. https://doi.org/10.1007/s40618-016-0482-0.
6. Hines M. Gender development and the human brain. Annu Rev Neurosci [Internet]. 2011;34(1):69–88. https://doi.org/10.1146/annurev-neuro-061010-113654.
7. de Vries ALC, Doreleijers TAH, Cohen-Kettenis PT. Disorders of sex development and gender identity outcome in adolescence and adulthood: understanding gender identity development and its clinical implications. Pediatr Endocrinol Rev [Internet]. 2007;4(4):343–351. Available from: http://www.ncbi.nlm.nih.gov/pubmed/17643082.

8. Daae E, Feragen KB, Waehre A, Nermoen I, Falhammar H. Sexual orientation in individuals with congenital adrenal hyperplasia: a systematic review. Front Behav Neurosci [Internet]. 2020;14 https://doi.org/10.3389/fnbeh.2020.00038/full.
9. Yang JH, Baskin LS, DiSandro M. Gender identity in disorders of sex development: review article. Urology [Internet]. 2010;75(1):153–159. Available from: https://linkinghub.elsevier.com/retrieve/pii/S0090429509023784
10. Cohen-Kettenis PT, Pfäfflin F. The DSM diagnostic criteria for gender identity disorder in adolescents and adults. Arch Sex Behav [Internet]. 2010;39(2):499–513. https://doi.org/10.1007/s10508-009-9562-y.
11. Justice Swaminathan G. Arun Kumar and Sreeja versus Inspecter General at Madurai Bench of Madras High court. 2019. p. WP(MD)No.4125 of 2019.
12. Justice Anand Venkatesh N. Sushma and Seema versus commissioner of police at Madras High court. 2021. p. W.P.No.7284 of 2021.
13. Human Rights Watch. I want to be like nature made me: medically unnecessary surgeries on intersex children in the US. p. hrw.org/sites/default/files/report-pef/lgbtntersex
14. Diamond M, Garland J. Evidence regarding cosmetic and medically unnecessary surgery on infants. J Pediatr Urol [Internet]. 2014;10(1):2–6. Available from: https://linkinghub.elsevier.com/retrieve/pii/S1477513113003112.
15. Lee PA, Houk CP. Commentary to "evidence regarding cosmetic and medically unnecessary surgery on infants". J Pediatr Urol. 2014;10(1):7.
16. Wallien MSC, Cohen-Kettenis PT. Psychosexual outcome of gender-dysphoric children. J Am Acad Child Adolesc Psychiatry [Internet]. 2008;47(12):1413–23. Available from: https://linkinghub.elsevier.com/retrieve/pii/S0890856708601422
17. Crouch NS, Liao LM, Woodhouse CRJ, Conway GS, Creighton SM. Sexual function and genital sensitivity following feminizing genitoplasty for congenital adrenal hyperplasia. J Urol. 2008;179(2):634–8.
18. Nascimento RLP, de Andrade Mesquita IM, Gondim R, dos Apóstolos RAAC, Toralles MB, de Oliveira LB, et al. Gender identity in patients with 5-alpha reductase deficiency raised as females. J Pediatr Urol [Internet]. 2018;14(5):419.e1–419.e6. Available from: https://linkinghub.elsevier.com/retrieve/pii/S1477513118304996
19. Ercan O, Kutlug S, Uysal O, Alikasifoglu M, Inceoglu D. Gender identity and gender role in DSD patients raised as females: a preliminary outcome study. Front Endocrinol (Lausanne) [Internet]. 2013;4. https://doi.org/10.3389/fendo.2013.00086/abstract
20. Fagerholm R, Santtila P, Miettinen PJ, Mattila A, Rintala R, Taskinen S. Sexual function and attitudes toward surgery after feminizing genitoplasty. J Urol [Internet]. 2011;185(5):1900–4. https://doi.org/10.1016/j.juro.2010.12.099.
21. Schweizer K, Brunner F, Handford C, Richter-Appelt H. Gender experience and satisfaction with gender allocation in adults with diverse intersex conditions (divergences of sex development, DSD). Psychol Sex [Internet]. 2014;5(1):56–82. https://doi.org/10.1080/19419899.2013.831216.
22. Szarras-Czapnik M, Lew-Starowicz Z, Zucker KJ. A psychosexual follow-up study of patients with mixed or partial gonadal dysgenesis. J Pediatr Adolesc Gynecol [Internet]. 2007;20(6):333–338. Available from: https://linkinghub.elsevier.com/retrieve/pii/S1083318807001660
23. Babu R, Shah U. Gender identity disorder (GID) in adolescents and adults with differences of sex development (DSD): a systematic review and meta-analysis. J Pediatr Urol [Internet]. 2021;17(1):39–47. Available from: https://linkinghub.elsevier.com/retrieve/pii/S1477513120306410
24. Abacı A, Çatlı G, Berberoğlu M. Gonadal malignancy risk and prophylactic gonadectomy in disorders of sexual development. J Pediatr Endocrinol Metab [Internet]. 2015;28(9–10) https://doi.org/10.1515/jpem-2014-0522/html.

25. Creighton S, Chernausek SD, Romao R, Ransley P, Salle JP. Timing and nature of reconstructive surgery for disorders of sex development—introduction. J Pediatr Urol [Internet]. 2012;8(6):602–10. Available from: https://linkinghub.elsevier.com/retrieve/pii/S1477513112002367
26. Canning DA. Commentary on genital sensation after feminizing genitoplasty. J Urol [Internet]. 2005;173(3):982. https://doi.org/10.1097/01.ju.0000152140.37086.4b.
27. Bangalore Krishna K, Kogan BA, Mazur T, Hoebeke P, Bogaert G, Lee PA. Individualized care for patients with intersex (differences of sex development): part 4/5.Considering the Ifs, Whens, and Whats regarding sexual-reproductive system surgery. J Pediatr Urol [Internet]. 2021;17(3):338–345. https://linkinghub.elsevier.com/retrieve/pii/S1477513121000917
28. Ahmed SF, Achermann JC, Arlt W, Balen AH, Conway G, Edwards ZL, et al. UK guidance on the initial evaluation of an infant or an adolescent with a suspected disorder of sex development. Clin Endocrinol (Oxf) [Internet]. 2011;75(1):12–26. https://doi.org/10.1111/j.1365-2265.2011.04076.x.
29. Fisher RS, Espeleta HC, Baskin LS, Buchanan CL, Chan YM, Cheng EY, et al. Decisional regret about surgical and non-surgical issues after genitoplasty among caregivers of female infants with CAH. J Pediatr Urol [Internet]. 2021. Available from: https://linkinghub.elsevier.com/retrieve/pii/S1477513121004654
30. Szymanski KM, Rink RC, Whittam B, Hensel DJ, Braga LH, Donahue KL, et al. Majority of females with a life-long experience of CAH and parents do not consider females with CAH to be intersex. J Pediatr Urol [Internet]. 2021;17(2):210.e1–210.e9. https://linkinghub.elsevier.com/retrieve/pii/S1477513120305362
31. Lee PA, Fuqua JS, Houk CP, Kogan BA, Mazur T, Caldamone A. Individualized care for patients with intersex (disorders/differences of sex development): part I. J Pediatr Urol [Internet]. 2020;16(2):230–7. https://linkinghub.elsevier.com/retrieve/pii/S1477513120300474.
32. Bougnères P, Bouvattier C, Cartigny M, Michala L. Deferring surgical treatment of ambiguous genitalia into adolescence in girls with 21-hydroxylase deficiency: a feasibility study. Int J Pediatr Endocrinol [Internet]. 2017;20(1):3. https://doi.org/10.1186/s13633-016-0040-8.
33. Chowdhury TK, Laila K, Hutson JM, Banu T. Male gender identity in children with 46,XX DSD with congenital adrenal hyperplasia after delayed presentation in mid-childhood. J Pediatr Surg [Internet]. 2015;50(12):2060–2. https://linkinghub.elsevier.com/retrieve/pii/S0022346815005370.
34. Bangalore Krishna K, Kogan BA, Ernst MM, Romao RL, Mohsin F, Serrano-Gonzalez M, et al. Individualized care for patients with intersex (disorders/differences of sex development): part 3. J Pediatr Urol [Internet]. 2020;16(5):598–605. https://linkinghub.elsevier.com/retrieve/pii/S147751312030379X.
35. Ernst MM, Kogan BA, Lee PA. Gender identity: a psychosocial primer for providing care to patients with a disorder/difference of sex development and their families [individualized care for patients with intersex (disorders/differences of sex development): part 2]. J Pediatr Urol [Internet]. 2020;16(5):606–11. https://linkinghub.elsevier.com/retrieve/pii/S1477513120304034

Evolution and History of Law with Regard to DSD

10

M. Sreenivas and Mohit Chauhan

Abstract

Disorder of sexual differentiation or gender dysphoria has been put in a separate category in the Diagnostic and Statistical Manual of Mental Disorder V. This is important for the management of the patients who were previously labelled with a diagnosis of gender identity disorders. It helps to address the mental agony that one may undergo due to an asynchronization between the perceived and assigned gender. These disorders are less common in females than males, with a less common incidence in natal females. Adults and adolescent females present gender dysphoria at an early age. This is nearly always gynephilic when the assigned sex is female in such cases. In India, transgender is duly identified as the third gender. In spite of this, it seems to remain much under-reported in India and other nations for the stigma related to it and along with homosexuality, this issue more sensitively severe in the female population. Diagnosis of gender dysphoria continues to be the most intricate, myriad extremely sensitive topic that is most pervasive to the human life. This chapter is intended to be of benefit to almost all the strata of stakeholder professional population, wherein the authors traverse with conceptual and evolutionary genesis of the increasingly important and much sensitive issues of international concern related to gender dysphorias. With a prologue of landmark Judgment from the Indian context, the sub-topics that transcend through historical perspective in relation to gender dysphorias include but are not limited to the *principles and theories laid down for gender identity (Yogyakarta's), constitutional rights of an individual as per the country's statute*

M. Sreenivas (✉)
Department of Forensic Medicine, University College of Medical Sciences and GTB Hospital, New Delhi, India

M. Chauhan
Department of Forensic Medicine and toxicology, Lady Hardinge Medical College and Associated Hospitals, New Delhi, India

S. K. Ratan (ed.), *Children with Differences in Sex Development*,
https://doi.org/10.1007/978-981-97-1639-5_10

(including right to marry, respect to private and family life), *offences of civil and criminal nature related to the person of transgender, role of specialist and non-specialist health care providers, finer nuances along with the sub-issues of sex-reassignment surgeries, approach to the management, and other variegated psycho-socio-medico-legal challenges faced by persons suffering from gender dysphorias.* The authors also lay down notable cases from the world, and various statutory solutions that have been prescribed available to the persons suffering from gender dysphorias at the level of an individual and as part of various sub-systems involved within the Indian context.

Keywords

Sex · Gender dysphoria · DSM-V · Transgender · Sex-reassignment · Vulnerable · DSD legal

10.1 Historical Background

National Legal Services Authority (NALSA) versus Union of India.

Issues before the apex court—(1) Whether persons who fall outside the male/female gender binary classification can be legally recognized as "third gender" persons. (2) Disregarding non-binary gender identities is a breach of the fundamental rights that have been guaranteed by the Constitution of India [1]. These are the issues decided by the apex court (cannot be amended).

Legal Recourse—The matter was referred to an "Expert Committee on Issues Relating to Transgender" that was constituted under the chairpersonship of the Ministry of Social Justice and Empowerment.

Outcome—The verdict placed one's gender identity as diversity in one's self-expression that is guaranteed as a right to dignity as defined in the Article 21 of the Constitution of India. It also extended the Right to express one's gender "through dress, words, action, or behavior" under the ambit of freedom of expression, again guaranteed as the Right to equality and freedom of expression. The judgment also reiterated that discrimination in any form whatsoever on the ground of "sex" is explicitly prohibited in Articles 15 and 16 of the Indian constitution. It was made further clear by the apex court that gender identity refers to "an innate perception of one's gender by a person, rather than that of mere biological characteristics." The apex court also defined the "third gender" and upheld the right of all persons to self-identify their gender. *Hijras* and eunuchs were legally identified to be called as "third gender." The apex court forbids that no "third gender" person/s should be subjected to any medical examination or biological test and reiterated that this may amount to invasion of the privacy of an individual. The apex court in this case also delineated the thin difference that exists between sex and the gender identity.

This landmark judgment legally recognized non-binary gender identities and uphold the fundamental rights of transgender persons in India, for the first time. It also directed the Central and State governments to take proactive actions in securing the rights of transgender persons.

10.2 Introduction

10.2.1 Definitions

Sex—It refers to the "gender" (based on one's self-perception), in addition to the biological attributes (chromosomes, genitalia + secondary sexual characteristics) carried by a person. "Gender" is also called the psychological sex. The other components of sex are (1) chromosomal sex, (2) gonadal sex, (3) internal anatomic sex, (4) external anatomic sex, (5) hormonal sex, (6) sex of assignment and rearing, and (7) gender role or psychosexual identification.

Gender Identity—It refers to a person's deeply felt internal perception and experience of the gender, which may or may not be concordant with the sex assigned at the time of birth. The term *gender identity* may also involve the freely chosen modification of the bodily appearance or its function, acquired in either case by medical, surgical, or other means. Other expressions of gender, including dress, speech, and mannerisms are also considered within the context of gender identity.

Transgender People—As per "queer theory," transsexual, gender-queer, third sex, androgynous, drag king/queen, transvestite, and/or people who had undergone hormone treatment and/or surgery, to accustom to the body form with their perceived gendered identity are included in the category of *transgender people.* The further sub-types of transgender people are (a) *Trans-man*—A person who is assigned gender female at birth (PAGFB), but identifies with the gender male, irrespective of the undergoing of the sex reassignment surgery/procedures and (b) *Trans-woman*—A person who is assigned gender male at birth (PAGMB), but identifies with the gender female, irrespective of the undergoing of the sex reassignment surgery/procedures.

Stein for the first time stated the definition of gender identity as that, "Someone who has the feeling possessed by a person that they possess a different sex than the one that they were born–to some extent" [2]. In other words, their perception of the sex does not align with the sex assigned at birth. The DSM-IV manual defines gender identity disorders (coined as *transsexualism by Benjamin*) as a heterogeneous group of disorders which share a strong and persistent desire to achieve the status and role of the opposite sex [3]. It is a condition characterized by the perception of being born into a body of wrong sex, as an error of nature. It is accompanied by a strongly grown-up obsessive and intense desire from an early age, to change one's bodily morphology and identity to become a person of the opposite sex (the pursuit of role reversal is a defining characteristic in this context). The patient remains restless he or she gets radical transformation and may even go up to performing self-mutilation. These disorders may manifest verbally, in the assertion that one properly belongs to the opposite sex, or non-verbally, in cross-sex behavior. The condition is a passionate, lifelong conviction that one's psychological gender—that indefinable feeling of maleness or femaleness—is opposite to one's anatomic sex. In adults, the reason for the preponderance of male-to-female transsexualism is (1) the greater publicity given to male-female transsexuals and (2) the greater success of vaginoplasty, in comparison to phalloplasty. There may be a delay by the adult males (up to the age of about 30 years) in seeking treatment in spite of the pre-existence of prodromal features [4]. Transsexualism, along with transvestism is a form of gender role inversion, whereas fetishism, exhibitionism, and impotence are associated with a normal gender role acquisition [4].

With the passage of time, it was felt to have a non-derogatory term for such cases. In 2012, the APA declared that the term "gender identity disorder" needs replacement by a neutral term "gender dysphoria" in the latest version of the Diagnostic and Statistical Manual of Mental Disorders (DSM-V) [5]. The affective component of gender identity disorder is known as gender dysphoria. Principle 3 of the Yogyakarta's Principle, i.e.—The right of equal recognition before the Law is closely applicable in this context [6]. This may be extremely relevant to intersex people who constitute up to 1.7% of the population, and nearly 0.5% of people with identifiable sexual or reproductive variations [7].

10.3 Evolution of the Socio-Medico-Legal Issues in Gender Dysphorias: World Scenario

- *1959–1960*—Phalloplasty as a part of sex change operation in a transsexual patient took place for the first time in the Netherlands. The operative technique had already been described in 1938 and perfected after World War II [8].
- *1955*—Switzerland—A case of sex reassignment surgery happened where a change in sex was laid down ought to be accepted.
- *1971*—The "Ormrod test" found its application where the respondent, a post-operative female transsexual, was transformed into a male. The case found that marriage between members of same sex was legally impossible, and thus ought to be declared null and void. Lord Justice Ormrod—explained that the court was unable to approach a case of transsexualism with the knowledge of medical profession [9]. *Kusum* analyzed these biological phenomena, and opined that medicine as a subject often cannot give polar "yes" or "no" answers. This fundamental conflict is deep-rooted at all the relations between the subjects of medicine and law. *Kusum* extended the *Ormrod test* to determine the sexual identity of a person. It is based on three biological factors: the chromosomal, the gonadal (presence of ovaries/testes), and the genital (internal sex organs at the time of birth) of the individual concerned [10].
- *1972*—Legislations for transgender were considered in Sweden (first country to legislate sex reassignment surgery). The same was considered for Italy, Spain, and Great Britain—1982, and the Netherlands—1984 [11].
- *1990*—Increased transgender societal movement—truth claims from "queer theory" [12].
- *1995*—International Bill of Gender Rights (IBGR) was proposed [13].

10.4 Questions of International Human Concern that Are Required to Be Addressed with Regard to the Cases of Gender Dysphorias

1. Rectification in the birth certificates and other documents of personal identity?
2. Statutory provisions related to transgender were required to be laid down in countries like Sweden, West Germany, Finland, Czechoslovakia, Greece, Italy, Holland, Switzerland, Spain, the UK, and the USA.
3. Courts in many countries relied on the biological sex of a transsexual, who has neither acquired the features of his/her original sex nor of the opposite sex.
4. The practice of the courts to decide on ad hoc basis in each case and simultaneously the difficulty faced by courts to admit the psychological basis (fourth factor) which otherwise is duly accepted by the medical witnesses [9, 11, 14].
5. Legalization of the new sex (decision about fixing the sexual identity) of the transsexual post-surgery—A landmark case in this regard that went up to the European Commission of Human Rights and brought changes in the European Civil Law system. In the case titled—Van Oosterwijk versus Belgium (1989) [15]. The Commission found that by failing to amend the original birth records, the respondent state was violating Article 8 of the constitution, the right of respect of private and family life, and Article 12, the right to marry. However, in another case, titled Rees v. United Kingdom (1985)—the European Court ignored the determination of the sexual identity of the post-operative transsexual person, right at the outset and forced the social system to consider Mr. Rees as a biological female and not as post-operative transsexual male [15].

10.5 Association Bodies for Transgender People

The World Professional Association for Transgender Health (WPATH) was originally named as the Harry Benjamin International Gender Dysphoria Association (HBIGDA). It was formed in the year 1978) [16]. The European Professional Association for Transgender Health (EPATH) dictates unequivocal trans-affirmative clinical guidance and continuing professional development courses for health professionals working for welfare of the transgender people [17]. The Henry Benjamin HBIGDA has also established the essential prerequisite protocol to manage a case of a transsexual.

10.6 Diagnosis of Gender Dysphorias (Laub's Criteria)

1. Two social scientists, one of whom must be a behavioral psychiatrist with aMD degree or an equivalent and the other may be a psychologist with a PhD or an equivalent, must participate in the diagnosis. Their recommendations must be in writing.

2. One of them must have known the patient in a diagnostic or therapeutic relationship for at least 6 months.
3. It must be established that the patient has had uninterrupted and unchangeable feelings of being in the wrong body for more than 2 years.
4. The patient should have had a successful cross-living test over a 1-year period, which includes legal, social, and sexual success.
5. Before any hormonal/surgical gender reassignment is contemplated, an appropriate monitoring of the medical conditions, as well as the effects of the hormones on the various systems, should be accomplished with appropriate documentation. An endocrinologist, preferably a member of the same gender dysphoria team, should be entrusted with this task.
6. When these diagnostic prerequisites have been wholly and completely met, the rearrangement of the body toward its new sexual appearance can be performed on those subjects whose true, or most appropriate, gender has been determined to be different from their present sexual appearance [18].

10.7 Importance of the Assignment of the Sex at Birth and Problems in Ambiguous Cases

Assigned sex—It is an evaluative judgment of the sex with which babies are born with.

Genitalia at birth—If it is not clearly recognizable at the time of birth in absolute terms, it can lead to misrecognition about the sex of the child. The ambiguity may also have discovered at a later point of time (in approximately 1.7% of the cases—intersex mixed chromosomal composition, gonadal dysgenesis, impaired hormone levels, and/or the genital duct system malformations).

10.8 Stewart's Philosophical Theory About Gender Identity

It is the most idealistic theory that refers to a conjunction of the epistemic and political foundations referred to as the "Queer Theory" and "Existentialism" [19]. The *Queer Theory* describes gender to be a non-binary entity that exists on a spectrum range of genders. The concept of *Existentialism* is directed at social justice for young people so that they can move more freely and do what they feel is right for themselves. The underlying principle of compatibility has an important role to play in biological essentialism.

10.9 Role of Health Care Providers in Dealing with Cases of Gender Dysphorias

They should adopt a fair and conscious approach in the following manner while attending the cases of gender Dysphorias.

1. *Non-discriminative approach in rendering health care services—Transgender Person's (Protection of Rights Act, 2019*) [20]—Section 3 (d) of the Chapter 2 of this act states that there should be "No denial or discontinuation of, or unfair treatment in, health care services; it prescribes for prompt treatment—There should be a provision for the prompt treatment in cases of sexual violence to transgender because non-treatment of the victim (transgender) may attract—166 B IPC [21]. *Clinical Establishment Act 2010.* As per this act, a female patient is to be examined in the presence of a female attendant or by a female registered medical practitioner only—the provisions of this act in furtherance of ensuring the consent, dignity, safety, equality, autonomy of the transgender people may be applicable as per the sections of this act that may be found relevant in this reference [22].
2. *Recognition and Identity by the way of Medical certification* (Chapter—3 of the *Transgender Rights Act*)—Section 4 of the act—This states about the right to be recognized as a transgender and to enjoy the self-perceived gender identity by a person. Section 5 of the act—This states that an application for a certificate of identity as a transgender person when submitted to the office of the District Magistrate if accompanied by the report of a psychologist of a govt. hospital, should be entertained without any medical examination (if a minor child, such application is to be from a parent or guardian of such child). Section 6 of the act—This states that the District Magistrate may (1) issue a certificate of identity as transgender person, (2) Gender can be recorded in all the official documents in accordance with the details incorporated in the certificate, and (3) a person has to be conferred with ~~the~~ rights and ~~the~~ proof of the recognition of identity. A male, female, transgender, male-to-female (MtF) transgender, female-to-male (FtM) transgender, and third sex/gender may see issues in the exercise of civil rights. Since India as a nation has strong affirmative action policies for women, it has codified conditions where a male-to-female transgender person is entitled to be recognized as female as per the land laws of this country. There are multiple options available for the legal recognition of transgender people in India. (1)—Legal recognition of the gender identity of the trans-people as women or men. (2)—Legal recognition of the gender identity of the trans-people as ~~a~~ separate gender ("third gender" or "transgender") *Option 3*—Legal recognition of the gender identity of the trans-people based upon their choice—women/men or ~~a~~ separate gender ("third gender"/"transgender").
3. *Procedure for a certificate of identity—Medical officers to act*—Section 7 of the *Transgender Person's (Protection of Rights Act, 2019)* states that after the issue of a certificate under sub-section (1) of Section 6, if a transgender person wishes to undergo surgery to change gender either as a male or female, it may be applied for the same along with the certificate from the Medical Superintendent or Chief Medical Officer of the medical institution where the surgery had been performed. The office of the District Magistrate may then issue a revised certificate indicating a change in gender upon being satisfied with the correctness of such certificate. The certificate of identity under Section 6 or a revised certificate under sub-section (2) entitles a person to change his first name on the birth certificate

and all the other official documents related to the identity of such person, provided that such change in the gender and the issue of the revised certificate under sub-section (*2*) shall not affect the rights and entitlements of such person under this Act. The Sex Revision Surgery (SRS) providers are legally bound to issue a post-SRS medical certificate. This may be useful in changing sex/gender and the name in official documents such as ration card, voter ID, driving license, passport, and bank passbook.

4. *Knowledge of medical criteria for identification*—The "Kusum—Ormrod test" may not be applicable in the Indian scenario in cases without SRS, but may be considered after SRS for the revised certificate [9, 10].
5. *Examination for Certification (MtF–Male to Female and FtM–Female to Male)*—Tanner Staging may reveal inconsistencies in the distribution of secondary sexual characteristics. The examination of transgender women may show underdeveloped breasts, feminine fat redistribution, diminished muscle mass, sparse or absent body hair, softened thinner skin, and testicles that have diminished in size or retracted completely. The anatomy of neo-vagina or natal vagina is examined with an anoscope. The examination of Transgender men may reveal facial and body hair growth, clitoromegaly, increased muscle mass, masculine fat distribution, and androgenic alopecia. The pelvic exam may be traumatic, and may be avoided. The cervical cancer screenings and Pap smear are advocated in such cases. There may be features suggestive of the use of testosterone and the presence of amenorrhea. There may be a history suggestive of medical intervention in a transgender patient under 18 years of age. An evaluation of the psychosexual orientation of the candidate may be warranted.
6. *Duties of a health care provider as an employee under the State as an authority*—These have been described under Section 15 of the Chapter 6 of the *Transgender Person's (Protection of Rights Act, 2019)*. There should be separate HIV sero-surveillance centers under NACO, a programme led by Government of India. There should be availability of free medical care facilities, along with counseling services for sex reassignment surgery and hormonal therapy to the transgender. There should be wide access to health care facilities in health care institutions, along with a provision to cover medical expenses through an inclusive insurance scheme. There has to be an ensured availability of health manual describing about the sex reassignment surgery in accordance with the guidelines issued by the WPATH. There should be a periodic review of the medical curriculum and research avenues for doctors, so that they can address the specific health issues and needs of the transgender in an efficient manner.
7. *Duties pertaining to SRS and specialist health care providers*—The management of transsexualism is one of the most demanding challenge in clinical sexology. It requires the collaboration of behavioral, endocrinological, and surgical specialists working as a team. It is much unlikely that a transsexual person will seek help to reduce or eliminate the transsexual feelings. The role of the behavioral scientist, (psychologist or a psychiatrist) is most crucial, and also time consuming in this regard. He is the person to identify patients with gender identity disorders. He works together with the physician who recommends hormone

therapy and the surgeon who accepts the patient for gender reassignment surgery, thereby sharing equal responsibility for patient selection and the final outcome of the assigned treatment [23].

10.10 Documents Issued for Identification in Cases of Gender Dysphorias

1. *Federal government*
   (1) Voter ID (Others, Male, Female)
   (2) Unique identification card—Aadhar (Male, Female, Transgender)

The passport and Permanent Account Number PAN card provide only two options for the column of the gender.

2. *State government*
   The ration card issued by the state of Tamil Nadu does mention about the third category as Aravani, for the gender category of an individual. The birth certificate, driving license, and bank passbook do not mention about any other column for a gender other than male or female [24].

10.11 Medico-legal Issues Related to Cases of Gender Dysphorias

i. *Acts of violence*
   1. *Violence against and by transgender/individuals with gender dysphorias*—It is "the intentional use of physical force or power, threatened or actual, against oneself, another person, or against a group or community, that either results in or has a high likelihood of resulting in injury, death, psychological harm, maldevelopment, or deprivation" [25].

Rules prescribed for the medical examinations are to be followed in cases falling under the following categories for stages of trials under the relevant procedural and penal codes.

(a) Transgender as a victim or survivor or accused of sexual violence.
(b) Transgender victim under Protection of Children from Sexual Offences Act, 2012 (POCSO Act, 2012).
(c) A juvenile transgender offender.
(d) False claim as a Transgender [26].

2. *Offences against the body and health of Transgender*—The Section 18 (d) of the Chapter 8 of the *Transgender Person's (Protection of Rights Act, 2019)* describes that any act that harms or injures or endangers the life, safety, health, or well-

being, whether mental or physical, of a transgender person or tends to acts including causing the physical abuse, sexual abuse, verbal and emotional abuse, and economic abuse, shall be punishable with imprisonment for a term which shall not be less than 6 months but which may extend to 2 years and with fine.

ii. *Considerations while performing medical or surgical interventions*

- Principles of Autonomy, Benevolence, Non-Maleficence, and Justice to be strictly adhered to in cases of sex reassignment surgeries (SRSs).
- The treatment should be aimed to alleviate the feeling of depression that might have pre-existed due to a feeling of inadequacy in the subject in relation to his or her physical and psychiatric state. This is important to deal it as an issue, which otherwise can lead to suicide or self-mutilation.
- Right to refuse medical treatment should be honored in dealing with transgender patients.
- Capacity and competency of children and young people to give assent and consent must be determined.
- Disputes over young people's access to medical treatment must be addressed.
- Application of psycho-legal integration in gender reaffirming and certification in various types of dysphorias is to be facilitated.

iii. *Sex Reassignment Surgery (SRS)*

1. SRS should be avoided in intersex infants.
2. There should be provisions for adequate and proper medical care facilities, counseling, availability of manuals, and published material for transgender health and that related to SRSs.
3. There should be provisions for the coverage of medical expenses through a comprehensive insurance scheme for people with gender dysphorias including transgender.
4. A lack of thorough examination by Psychiatrist/Clinical Psychologist, Endocrinologist, Urologist, Plastic Surgeon/Surgeon, and Gynecologist before a surgical procedure is likely to attract issues of Human Rights Violation.
5. The type of Sex Reassignment Surgery that has to be performed should be on the case-specific basis, and also should be the one that has been recommended by the expert committee to the Ministry of Social Justice and Empowerment, Govt. of India).
6. SRS is to be performed only at the time of the attainment of majority under documentary proof.
7. The patient's immunity status should be tested preoperatively. However, a seropositive patient should not be discriminated against solely on this basis because his or her expectation of life, and particularly the quality of it, cannot be truly assessed and anticipated for future. However, in the case of a patient with AIDS, surgical intervention may not be done as a part of management.

8. The type of sex to be revised is the one with which the person has lived continuously <= 12 months in a gender role that is congruent with their gender identity.
9. The preoperative factors that indicate a favorable outcome are:
   (A) Reasonable degree of psychological stability, without any history of psychosis.
   (B) Successful adaptation in the desired role for at least 1 year, with convincing physical appearance and behavior.
   (C) Sufficient understanding of the limitations and consequences of the surgery.
   (D) Preoperative psychotherapy in the context of a gender identity programme.

iv. *Parental disputes before or after SRS*
1. The surgeons may face complex dilemmas in cases where there is an active opposition from the parents of transgender persons, who are otherwise above 18 years of age and are fully competent with a right to choose the type of SRS they want to undergo.
2. Sometimes, the doctors under anticipation and to avoid future litigation/s do not agree to perform SRS in Trans-people without an explicit consent from the parents of otherwise competent adults.
3. Court may indirectly guide the transgender persons to exercise a choice to undergo SRS provided when they are >18 years of age and fulfil the case requirements for an appropriate surgery by the doctors.
4. Minors—There are reported cases where police arrested trans-community persons (and/or quack doctor who performed emasculation) based upon the complaints filed by parents that trans-people forced their son to undergo SRS or emasculation [27].
5. *Gillick's Competence or Mature Minor Test* (Maturity, Intelligence, Autonomy, Critical analysis)—The clinical team may assess the mental capacity of a child under the age of 16 years regarding the ability to furnish consent to receive hormone treatment or refuse otherwise, after completely understanding the likely consequences, both the positive and the negative outcomes of the indicated form of treatment, along with available alternatives (including no treatment at all), and their consequences. It calls to avoid any undue influence from outside in deciding about the form of treatment. The child's autonomy, existentialism and self-determination should be the priority of the adult oversight and state obligation. However, the legal sanctity of Gillick's competence is still equivocal in the Indian context.
6. *Measures against parental disputes*

(A) *Constitution of support groups*
   (i) The members of these groups may help the parents to tide-over the misconceptions, understand the difference between sex, sexuality, gender, and intersex variations, as manifested in their child.
   (ii) Realize that they as parents will be provided support.

(iii) That the perceived gender nonconformity of their children is not their fault
(iv) Help them to cope up through the negative feelings of anger, disbelief, shock, or disappointment in a safe space.
(v) Understand human rights and impact of its violation.

10.12 Socio-Medico-Legal Challenges Faced by Persons Suffering from Gender Dysphorias [27]

1. Increased HIV risk.
2. Structural vulnerability (Sex work, Sexual Violence, Poverty).
3. Discrimination.
4. Barriers to accessing health and social protection services.
5. Lack of legal recognition of gender status of hizras and other transgender people.
6. Denial of the legal rights of hizras/transgender people.
7. Ambiguity in the legal status of the sex reassignment surgery and post-surgical gender recognition.

10.13 Safeguards for Health Care Providers

It is a better option to get these consent forms converted into affidavits by paying the appropriate court or notary fee and getting these attested by a magistrate or notary. This may serve as a future safeguard. Here, it can be impleaded in future that "the state" was a witness to this agreement [28].

10.14 Socio-legal Issue in Gender Dysphorias

1. *Sexual inclination*—Studies reported that nearly all transsexual women were homosexually inclined. However, those having homosexual relations do not regard these relations as deviant. This is because they identify themselves so strongly associated with the opposite sex, thereby giving rise to another socio-legal issue [29, 30].
2. *Right to contest election*—The Election Commission has introduced the option of "other" in the voter's identity card and indicated that "hijras" can vote or contest as "other" [31]. However, the legal validity of this executive order on the "*right to contest*" is not clear. Hijras had contested elections in the past. It has been documented that the victory of a transgender woman who contested election on a seat reserved for women was overturned for this reason, with reference a judgment of the Madhya Pradesh High Court in this context that the referred person was not a woman but a "hijra" [32].
3. *Inheritance of the property*—A hijra woman was allowed to receive property from her Guru, as the court accepted that the property cannot be transferred to

anyone outside the community. The court via this ruling explicitly acknowledged the existence of a distinct "eunuch" class with its own customs and rituals, that must be respected [33].

4. *Legal classification as "backward" community*—The C.S. Dwarakanath Backward Classes Commission Report of 2010 recommended that the people from transgender community (including Hijras, Kothis, Jogappa, Female to Male [transgender people], and Mangalamukhis) should be included in the category of more backward communities (2-B) that would then entitle them to government benefits [34].
5. *Economic justice*—The Karnataka High Court appointed C. Anu, a transgender person, to the post of an administrative job in the Group D category [35].
6. *Identification*—The Karnataka Government instituted an initiative entitled *Lingathwara Alpasankhyathara Yojana* (Gender Minorities Programme). It proposed to implement trainings, loans, and subsidies through NGOs working in the area (because transgender people may not have permanent addresses and may have issues with identification) [36].
7. *Civil Society*—The Bangalore University has announced reservation for transgender persons in postgraduate courses offered under its statute [37].

10.15 Highlighted Cases from India Related to Gender Dysphorias/Transgender

1. *Naz Foundation vs. NCT Delhi in 2009*—Previously, all types of non-penovaginal sexual relationships among consensual adults were dealt as criminal offences under Section 377 of the Indian Penal Code. One decision of the high court of Delhi, in this case, gave a new interpretation to the existing framework of few Constitutional Rights (Article 14—*equality before the Law*, Article 15—*discrimination on the basis of sexual orientation is not permitted,* and Article—21 *the right to live with dignity and the right of privacy*) to a minority group by decriminalizing consensual sexual activity between adults. This type of an issue had never been deemed worthy of protection of the rights of minority group by the way of judicial consideration [38].
2. *Jayalakshmi versus Government of Tamil Nadu—Year 2007*—The case of a petitioner (Jayalakshmi) was that she along with her husband and her brother Pandian were living in Vyasarpadi. Pandian, who was aged about 18 years, was working as a helper in a vegetable shop belonging to one Ms. Renuka at New Merginpuram at Vyasarpadi. According to the petitioner, her brother Pandian had the traits of a woman and later, became a transgender person. By knowing his affliction, the Police (respondent-3rd in this case) registered a First Information Report against him in a criminal case. This transgender person (Pandian/Pandiammal) was then used to be repeatedly and brutally abused sexually by the police in Chennai upon being called on the pretext of an enquiry, and was threatened on all the occasions with dire consequences so as not to reveal the matter to anyone. The Sub-Inspector of Police, one of the respondents had tor-

tured him by thrusting lathi inside his anus and a few other police personnel (other respondents) forced her to have oral sex. This was going on for the past 2 weeks. When the mother of the petitioner came to know about the same and enquired at the Police Station, she was beaten up. This led the transgender Pandian to immolate herself in front of the police station. A Writ Petition was filed under the Article 226 of the Constitution of India that had a prayer for the issuance of a Writ of Mandamus directing the respondents to pay a sum of Rs.10,00,000/– as compensation to the petitioner for the death of her brother. It also sought directions to the State (the 2nd respondent) to initiate appropriate disciplinary action against respondents numbers 3–8. The ruling of the Madras High Court—under the bench of Chief Justice A. P. Shah and Justice P. Jyothimani—directed the Tamil Nadu Government to provide compensation to the next of kin of the deceased [39].
3. http://articles.timesofindia.indiatimes.com/2012-04 01/Chennai/31269633_1_transgender-parents-change-operation [40].
4. Transgender kills self after parents oppose sex change operation [41].
5. Twenty-year-old Rohtak youth shot deadall family including father, mother, maternal grandmother, and sister as they refused money for gender reassignment surgery [42].

10.16 Scope for Further Evolution

1. Possibilities of additions of sub-sections to the Section 375 IPC (Indian Penal Code).
2. Consent in 12–18 years—Any Indian system similar to Gillick's competency test.
3. Potency disputes in FtM transgender candidates.
4. Implications pertinent to the laws on artificial insemination and surrogacy act.
5. Regulations for the adoption of a transgender child.
6. Deciding upon the applicability of the legal right of a MtF mother over a child.
7. Further issues related to marriage, divorce, and annulment in case of transgender discovered later in life.
8. Inheritance issues of the property by the legal heir or the offspring of a transgender.

10.17 Conclusion

Legal recognition of the surgically changed gender will discharge the obligation on the part of the state to provide a healthy life to its citizens. The wider acceptance of the transsexual individuals by the society has changed the outlook of the "transsexual's new gender" for the better. Many states are amending their laws in accordance with the advances in medical sciences but a lot more remains to be done by many others.

References

1. National Legal Services Authority versus Union of India & Ors on 15 April, 2014 (SC Civil Original Jurisdiction (WRIT PETITION (CIVIL) NO.400 OF 2012) Reportable.
2. Stein E. The mismeasure of desire: the science, theory, and ethics of sexual orientation. New York, NY: Oxford University Press, Inc.; 1999.
3. Benjamin H. Transsexualism and transvestism as psychosomatic and somatopsychic syndromes. Am J Psychother. 1954;8(2):219–30.
4. Sharma: Gender identity disorder and its medico-legal considerations. Med Sci Law. 2007;47(1):31–40.
5. American Psychiatric Association. Diagnostic and statistical manual of mental disorders. 5th ed. Arlington, VA: American Psychiatric Association; 2013.
6. https://yogyakartaprinciples.org/introduction/. Accessed 31 Dec 2018.
7. https://www.americanprogress.org/article/key-issues-facing-people-intersex
8. Haeseker B, Nicolai JP. [The first gender-changing operation from female to male in The Netherlands, 1959/'60]. Ned Tijdschr Geneeskd. 2007;151(9):548–52. Dutch.
9. Arthur Cameron Corbett v. April Corbett (Otherwise Ashley) [1971] P. 83, [1970] 2 All E.R. 33. https://en.wikipedia.org/wiki/Corbett_v_Corbett.
10. Kusum V. The legal implications of a sex change operation. The Times of India. March 23, 1993. p. 5.
11. Gromb S, Chanseau B, Lazarini HJ. Judicial problems related to transsexualism in France. Med Sci Law. 1997;37(1):27–31.
12. Halperin DM. The normalization of queer theory. J Homosex. 2003;45(2–4):339–43.
13. https://www.digitaltransgenderarchive.net/downloads/2z10wq28m.
14. Taitz J. Judicial determination of the sexual identity of post-operative transsexuals: a new form of sex discrimination. Am J Law Med. 1987;13(1):53–69.
15. https://hudoc.echr.coe.int/app/conversion/docx/?library=ECHR&id=001-57549-&filename=CASE%20OF%20VAN%20OOSTERWIJCK%20v.%20BELGIUM.docx&logEvent=False
16. https://www.wpath.org/
17. https://epath.eu/
18. Laub DR. The role of surgery in the treatment of transsexualism (invited comment). Ann Plast Surg. 1984;13:476–81.
19. Shrage LJ, Stewart RS. Philosophizing about sex. Broadview Press; 2015.
20. https://www.indiacode.nic.in/handle/123456789/13091?sam_handle=123456789/1362
21. https://legislative.gov.in/sites/default/files/A1860-45.pdf
22. http://www.clinicalestablishments.gov.in/
23. Harish D, Sharma BR. Medical advances in transsexualism and the legal implications. Am J Forensic Med Pathol. 2003;24(1):100–5.
24. https://transgender.dosje.gov.in/
25. Krug EG, Dahlberg LL, Mercy JA, Zwi AB, Lozano R, editors. World report on violence and health. Geneva: World Health Organization; 2002.
26. https://wcd.nic.in/sites/default/files/POCSO%20Act%2C%202012.pdf
27. Chakrapani V, Narain A. Legal recognition of gender-identity of transgender people in India: current situation and potential options. Policy brief. Page no 13. UNDP.
28. Gupta R, Murarka A. Treating transsexuals in India: history, prerequisites for surgery and legal issues. Indian J Plast Surg. 2009;42(2):226–33.
29. Pauly IB. Male psychosexual inversion: transsexualism. Arch Gen Psychiat. 1965;13:172.
30. Roth M, Ball JRB. Psychiatric aspects of intersexuality. In: Armstrong CN, Marshall AJ, editors. Intersexuality in vertebrates including man. London: Academic Press; 1964. p. 395.
31. http://www.queerty.com/india-asks-voters-male-female-or-transgender-20091113/
32. Kamla Jaan v. Sadik Ali, Civil Revision Petition No 1294 of 2002 on file with the Alternative Law Forum.

33. Illyas and Others v. Badshah, alias Kamla. AIR 1990MP 334.
34. CS Dwarakanath backward classes commission report, 2010.
35. Krishnaprasad, "High Court Strikes a Blow for Transgender", The Hindu, 5 February 2012. Available at http://www.thehindu.com/news/states/karnataka/article2860948.ece accessed on 9 May 2012.
36. Government of Karnataka Budget 2012-2013, 21 March 2012, p. 32. Available at www.finance.kar.nic.in/bud2012/budsp-gen2012-e.pdf. Accessed on 1 October 2012.
37. http://www.dnaindia.com/bangalore/report_transgender-quota-in-post-graduate-courses-has-not-found-takers_1431913 accessed on 1 October 2012.
38. Naz Foundation vs. NCT Delhi (2009) 160 DLT 277.
39. https://indiankanoon.org/doc/1373799/
40. http://articles.timesofindia.indiatimes.com/2012-04-01/chennai/31269633_1_transgender-parents-change-operation
41. https://timesofindia.indiatimes.com/city/chennai/transgender-kills-self-after-parents-oppose-sex-change-operation/articleshow/12488105.cms
42. https://www.hindustantimes.com/cities/others/rohtak-youth-killed-his-family-as-they-refused-money-for-gender-reassignment-surgery-cops-101630871498153.html

Part IV

Diagnosing DSD

Special thanks to Dr. Kanishka Das, Professor Pediatric Surgery, AIIMS, Bhubaneshwar for assisting and aiding me in editing this section.

Terminology and Clinical Evaluation in the Diagnosis of DSD

11

Kanishka Das and Mainak Deb

Abstract

The diagnosis of a suspect DSD is best done in the setting of a multidisciplinary specialty clinic involving experts from various fields under a lead consultant. The clinical evaluation begins with a critical analysis of the prenatal history (including maternal obstetric history) and investigations (imaging, genetic studies), birth and neonatal events, and relevant family history. In later presentations, the prepubertal and pubertal history is recorded. The indications for evaluation for a DSD are reviewed.

Clinical evaluation includes general (e.g., BP, height, and habitus) and systematic review for syndromic associations (e.g., Turner and Kline Felter) and multisystemic involvement (e.g., Cloacal anomaly, hydrometrocolpos, and hydroureteronephrosis), particularly of the closely associated urogenital tract. A detailed assessment of external genital morphology (phallic structure, genital orifice, genital folds, and gonads) with attention to specific details (e.g., Stretched penile length, gonadal symmetry, and anogenital ratio) is done. Findings are serially documented using standard scoring systems (e.g., Tanner, Prader, and External genitalia score) for uniformity and comparison across time periods. Pubertal changes are carefully chronicled and photographic documentation with appropriate consent is advisable in all cases.

The clinical evaluation concludes with making a broad provisional working diagnosis of a sex chromosomal DSD/46 XY DSD/46 XX DSD. A preliminary panel of investigations (serum biochemistry, ultrasonography, karyotype) is

K. Das (✉)
All India Institute of Medical Sciences, Bhubaneswar, India

M. Deb
Rainbow Children's Hospitals, Hyderabad, India

S. K. Ratan (ed.), *Children with Differences in Sex Development*,
https://doi.org/10.1007/978-981-97-1639-5_11

awaited before further specific, directed investigations to clinch a final diagnosis as per the Chicago consensus terminology.

Keywords

History · Clinical evaluation · DSD

The diagnosis of DSD is a carefully orchestrated deductive process and recent insights into the genetics and natural history of these entities have sharpened our earlier perspectives and strategies for an early optimal diagnosis of this medico-social emergency. This chapter elaborates on a methodical and directed approach to the diagnosis and differential diagnosis of DSD. The terminology (Table 11.1) adheres to the international Chicago consensus statement, 2006 [1] to allow for easy comprehension and international comparison. Earlier terminologies that are currently deemed unacceptable are best avoided. While the Chicago consensus is useful, traditional clinical observations and skills that are indispensable for a rational evaluation are also detailed.

Table 11.1 Terminology in DSD

SEX DSD	46, XY DSD	46, XX DS D
1. Turner Syndrome 45, X, and variants *2.Klinefelter Syndrome* 47, XXY and variants *3. Mosaicism* 45, X /46, XY, and variants *4. Chimerism* 46, XX/46, XY	*1.* Disorders of gonadal development: – Gonadal dysgenesis – Testicular regression syndrome – Ovotesticular DSD *2. Disorders of testosterone synthesis* – Leydig Cell hypoplasia or aplasia (variants in LH receptor) – Steroidogenic defects: Smith—Lemli—Opitz syndrome; StAR protein, P450scc, 3βhydroxysteroid dehydrogenase type 2, 17 α - hydroxylase/17–20 lyase, P450 oxidoreductase, cytochrome B5, 17β-hydroxysteroid dehydrogenase type 3 deficiencies *3. Disorders of testosterone action* – Androgen insensitivity syndrome (with or without androgen receptor mutations) – 5α Reductase deficiency *4. Other*: Cloacal exstrophy, persistence of Mullerian duct syndrome, etc.	*1. Disorders of gonadal development:* – Ovarian dysgenesis – Testicular DSD – Ovotesticular DSD *2. Androgen excess* – Steroidogenic defects: 21 α-hydroxylase, 11 β-hydroxylase, 3 β-hydroxysteroid dehydrogenase type 2, P450 oxidoreductase, and aromatase deficiencies – Glucocorticoid receptor mutations – Maternal tumors or exposure to androgenic drugs *3. Other:* Malformations involving Müllerian ducts, cloacal exstrophy

Bailez MM, Costanzo M, Guercio G. Role of minimally invasive surgery (MIS) in different sexual development (DSD). Semin Pediatr Surg. 2021 Aug;30(4):151078

11.1 Clinical Evaluation

The diagnosis and management follow an interdisciplinary approach involving the fetal medicine expert, obstetrician, clinical geneticist, neonatologist – pediatrician, endocrinologist, pediatric surgeon – urologist, pathologist, clinical psychologist–child psychiatrist and medico-social worker. The services are best organized in the ambit of a multidisciplinary specialty clinic where the experts meet and discuss under a lead consultant and arrive at a consensus before any management decision is reached or communication is made with the family or child.

The incidence of DSD is 1 in 5000 newborns [1]. Some children with DSD are born with genital ambiguity; this is a medico-social emergency and can be evaluated on broad lines quickly within 2–7 days, depending on the availability of a diagnostic armamentarium, to offer a tentative diagnosis to the parents. Till such time, the baby is referred to with gender-neutral terms like "it" or "the/your baby" and positive body language. All communication to the parents is provided through a designated spokesperson or member of the multispecialty team (e.g., the treating pediatric surgeon) who crystallizes the consensus opinion of the experts and fields the queries of the family too. The latter involves a continuing consultative process with the team. Since a host of web-based information may be accessed by the parents, it is useful to offer them select reliable sources for basic information.

DSD can manifest at any time from birth and across childhood to puberty. It may be suspected and detected depending on various factors including the clinical perspective of the physician, optimal investigations, and referral patterns in a healthcare system. Late detection and referrals in DSD are related to mild genital ambiguity and/or systemic manifestations, e.g., XY partial gonadal dysgenesis, complete androgen insensitivity, hypogonadotropic hypogonadism, 5 AR2 deficiency, Persistent Mullerian duct syndrome, and nonclassical/atypical CAH [2].

11.2 Historical Review

Prenatal and family histories of several factors may help contribute toward a diagnosis. Maternal administration of hormonal medications (contraceptives, abortifacients, spurious medications for a male issue), especially peri conceptually or in the first trimester is noteworthy. First-trimester miscarriages, multiple miscarriages, and stillbirths suggest genetic defects while parental consanguinity hints at recessive disorders like CAH, 5AR2 deficiency, or PMDS. Points to note include sex of rearing (parental/physician assigned), social identity and interactions of the child, incongruent secondary sexual characters at puberty, failure to attain menarche or delayed menstruation in a presumed female, and infertility. Cyclic abdominal pain and abdominal mass are related to functioning endometrial tissue and obstructed menstruation; cyclic haematuria points to an ovotesticular DSD. Also useful are family history of infertility in presumed females of earlier generation, e.g., in 46 XY DSD—complete androgen insensitivity or any form of familial hypospadias. A history of earlier unexplained sibling deaths and neonatal-infantile occurrence of unexplained sepsis, vomiting—gastroenteritis—dehydration is common with classic CAH due to 21 hydroxylase deficiency. An early pathway steroidal deficiency like 3 β HSD manifests with features of deficiency of cortisol (Addisonian crisis), aldosterone (salt-losing) as well as sex steroids (sexual infantilism or virilization depending on the sex). Features of maternal virilization (hirsutism, acne, clitoromegaly) are forthcoming with rare etiologies like androgen-secreting tumors (ovarian, adrenal) in pregnancy and aromatase deficiency.

Prenatal diagnosis is gaining ground in certain disorders, particularly 46 XX DSD–CAH, especially in the presence of a previous pregnancy being affected or an issue with a specific diagnosed DSD and known inheritance pattern. Where prenatal sex determination and other targeted investigations (imaging of genitalia and associated multisystemic anomalies, fetal karyotype, amniotic fluid hormone studies, etc.) are legally allowed and have been conducted, the results must be reviewed in readiness for an early neonatal diagnosis. Any form of prenatal counseling that has been provided and therapeutic measures (e.g., maternal dexamethasone administration during early gestation in a presumed fetal CAH) are also noted.

Adolescent/peripubertal presentations that require investigation for DSD are—a girl with primary amenorrhoea and variable breast development, a girl with virilization, and a boy with pubertal delay. Several of these are between 13 and 15 years of age and obese [3].

11.3 Clinical Examination

Meticulous clinical examination and documentation are necessary at the first and every examination to register an evolving phenotype. Though repeated examinations of the genitals are inadvisable [4], they may be necessary to discern a typical evolution of changes in certain DSD states, particularly in the peripubertal period. Common physical findings that children are referred with from a primary physician and merit investigations for an underlying DSD inneonatal-infantile referrals include the following [1, 5]:

- Phenotypic female genitalia with atypical features like isolated clitoromegaly, single genitourinary orifice, posterior labial fusion, inguinal, or labial mass.
- Phenotypic male genitalia with atypical features like bilateral undescended testes, micropenis, isolated proximal/perineal hypospadias, or distal hypospadias with undescended testis/testicular ectopia.

11.3.1 General Examination

A detailed general/systemic examination and genital examination are performed sequentially. Other systemic, dysmorphic, and non-genitourinary features that are typical of well-defined clinical entities (e.g., Short stature, neck webbing, and cubitus valgus in classic Turner syndrome; in untreated simple virilizing CAH, affected ones are tall children but short adults) are noted including soft findings linked to obesity, height, typical facial features, and cognitive status. Some DSD may have few, if any genital features, e.g., Male CAH with neonatal cortisol deficiency and shock, but the latter typical non-genitourinary features will clinch the diagnosis.

11.3.2 Genitalia

The gross appearance of all genital orifices and sinuses is detailed. The findings may be pictorially documented with diagrams and photographs with necessary permission. Specific details are relevant to the suspected clinical diagnosis (e.g., BP and cutaneous pigmentation in congenital adrenal hyperplasia) and scoring systems (e.g., Prader grades for a virilised female) are noted. In an older child, a gentle per rectal examination to feel a uterus/Mullerian structure is in order; this may be deferred in young children when an examination under anesthesia is anyway planned.

Familiarity with the normal neonatal male and female external genitalia (Fig. 11.1a, b) and ethnic variations thereof is important. In term males, the urethra is at the tip of the glans, the penis has a stretched flaccid length of 2.5–4 cm and girth/diameters of 0.9–1.3 cm. Both testes are descended into the scrotal sacs and measure 8–14 mm in diameter. The normal female has a 2–6 mm clitoris that may be just visible between the labia in the anatomical position, sometimes maternal hormones may render it prominent but the effect wanes over a few days. The vaginal orifice is a 3–4 mm slit or stellate one with heaped-up mucosa. There is neither posterior labial fusion nor any gonads palpable in the inguinolabial region.

Specific genital anatomic details (Figs. 11.2 and 11.3) that are recorded include:

(a) Phallic structure—SFPL (Stretched flaccid phallic length), CBG (Corporal body girth or diameter), presence and severity of chordee (mild <15 degrees, moderate 15–30 degrees, severe >30 degrees). Significant SFPL is common in virilizing CAH and ovotesticular DSD with a predominant testicular component. Short phalluses are noted in 46XY DSD with proximal testosterone biosynthetic blocks and testicular dysgenesis. An erection or tumescence during the examination or after administration of human chorionic gonadotrophin (hCG)/androgen for various investigations grossly indicates responsive erectile tissue; it is often noted during the mini puberty (birth—4 months) when hormonal levels are peaking and is predictive of expected pubertal changes. Urethral plate anatomy (plate width, waisting) and glans details (navicular fossa width and depth, well-formed/hypoplastic glans, flat/conical) and preputial anatomy (complete, deficient, or hooded) are important for the characterization of hypospadias—a common component in DSD. The anatomy of hypospadias with DSD is more severe than that seen in non-DSD states. A cali-

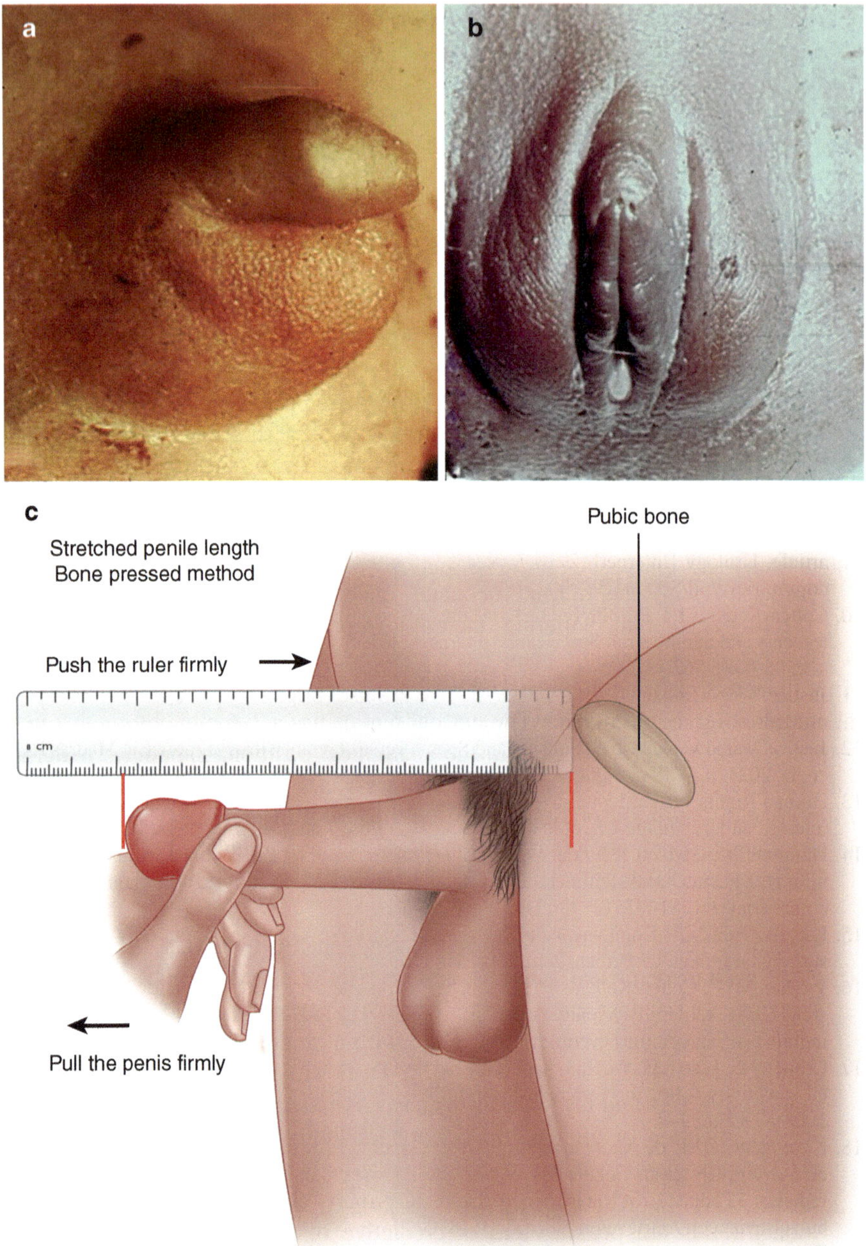

Fig. 11.1 Normal male (**a**) and female (**b**) external genitalia at birth. The stretched penile length is generally measured using the bone-pressed method (**c**)

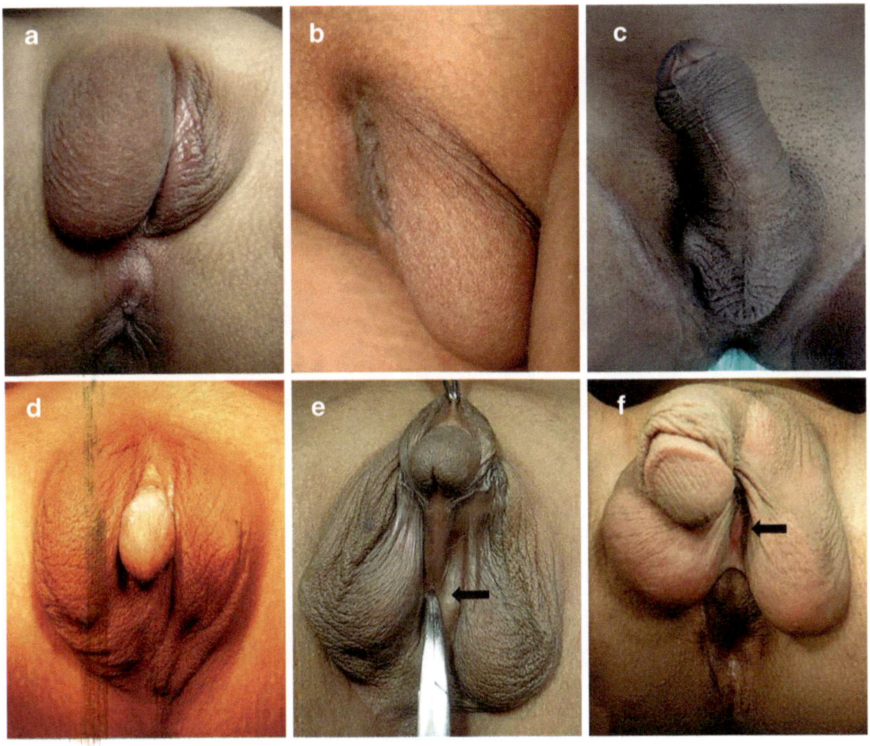

Fig. 11.2 Spectrum of genital phenotype in 46 XY DSD. The various features illustrated are barely visible micropenis (**a**, **b**), asymmetric labioscrotal folds (**a**), bilateral hypoplastic and empty scrotum (**c**), severe chordee (**f**), mid-scrotal hypospadias and unfused/bifid scrotum (**e**, **f**) and varying degrees of penoscrotal transposition (**d**, **e**, **f**)

per is used for specific measurements. The SFPL is measured with a firm ruler/scale pressed against the pubic bone and the phallus aligned in a stretched manner against it (Fig. 11.1c). A micropenis is defined as a short penis, <24–25 mm (i.e., <2.5 SD below the mean) with a normal configuration [6].

(b) Genital orifice—Number, size, diameter, mucosal appearance—flat or mildly protuberant. A flat mucosa is often seen in urogenital sinus openings in 46 XY DSD and gonadal dysgenesis while exuberant heaped-up mucosa implies estrogenic stimulation. In phenotypic females, separate urethral and genital orifices may not be apparent in a cursory OPD examination and need EUA. Stenotic orifices and common urogenital channels must be scrutinized and potential obstructive sequel—hydrocolpos, hydrometrocolpos, hydronephrosis, and their infective states.

(c) Gonads—Their presence or absence in the inguino scrotal/labial region on either side, approximate size or diameter in comparison to an orchidometer, feel-firm and nodular (ovarian) or soft and smooth (testicular) or heteroge-

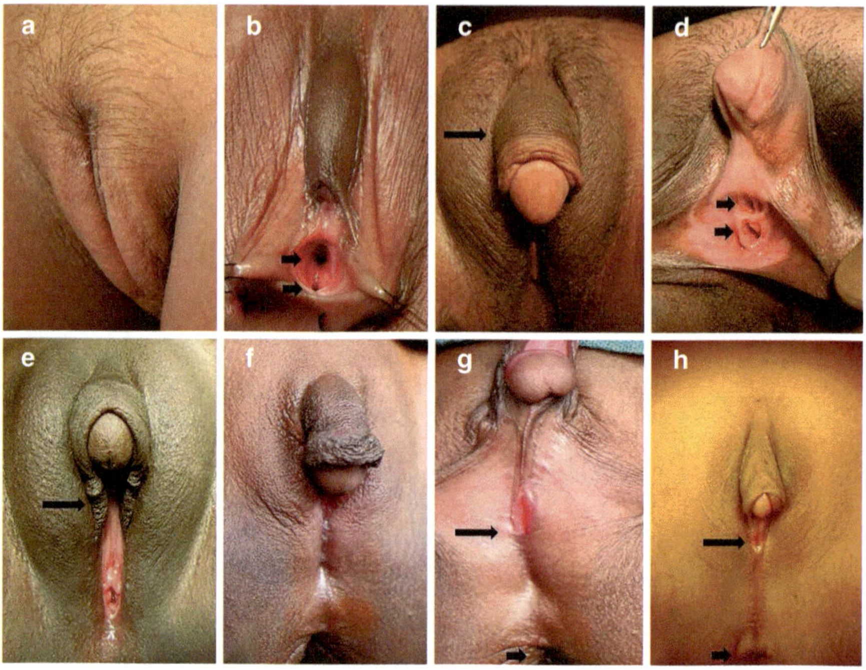

Fig. 11.3 Spectrum of phenotype and virilisation in 46 XX DSD -CAH. In this 3 year old (**a**, **b**) with pubic hair (**a**), the clitoromegaly and posterior labial fusion are not apparent till the rugose labia are separated (**b**). A 6-year old(**c**, **d**) with obvious clitoromegaly (**c**) and more rugose, pigmented labia. Both children have separate recognizable urethral and vaginal orifices (arrows, **b**, **d**) within a funnel-shaped urogenital sinus—Prader 2. Note the hypoplastic labia minora (arrow, **e**), clitoromegaly (**c**, **e**, **f**), and increasing degree of posterior labial fusion(short arrow—anus, long arrow—urogenital orifice) in other children (**e**—Prader 1, **g** and **h**–Prader 3)

neous. The symmetry of the gonadal position refers to the relative gonadal location proximal (impalpable) or distal (palpable) to the deep inguinal ring on either side. Asymmetric gonadal position along the line of testicular descent is common with asymmetric (mixed) gonadal dysgenesis and ovotesticular DSD and results from lateralizing, paracrine effects of the developing gonad during embryogenesis. On the contrary, symmetric gonadal position (intraabdominal ovaries in CAH or inguinal testes in AIS) signifies global systemic hormonal effects. Again gonadal descent (testicular dependent) is less affected in classical 5AR2 deficiency than the development of DHT-responsive areas (urogenital sinus, prostate, urethral folds)—hence a common phenotype of the entity being perineoscrotal hypospadias, urogenital sinus remnant, and inguinal or scrotal gonads. Dysgenetic/streak gonads do not descend and are diminutive, hence impalpable.

(d) Genital folds—Labial, scrotal, or indeterminate (labioscrotal) genital folds are noted along with symmetry, pigmentation, rugosity, and palpable structures (gonads/other genital ductal adnexa) within them. An indeterminate (labioscrotal) nature implies a mixed hormonal (androgenic and estrogenic) influence as in CAH and ovotesticular DSD rather than a predominant one (labial in CAIS or scrotal in PAIS/5AR2 deficiency). The differentiation of labia majora from minora is absent in congenital adrenal hyperplasia and attendant pigmentation points to indirect MSH secretion of elevated ACTH. Labioscrotal folds may harbor patent processus vaginalis/hernial sac with herniated adnexa of Mullerian structures in gonadal dysgenesis or ovotesticular DSD. Likewise, an innocuous hernia in a phenotypic male may carry with it a prolapsing uterus/fallopes (Mullerian remnants) along with the testis (one/both) and vas (mesonephric derivatives) coursing sometimes into the opposite sac as in the spectrum of anomalies in 46 XY DSD - Persistent Mullerian Duct syndrome.

Features of a bifid scrotum and penoscrotal transposition signify increasing degrees of feminization in a male, the latter is graded as partial or complete. Similarly, posterior labial fusion and formation of a common urogenital (urethrovaginal) channel implies the masculinization of a female and is specified by the Prader scale.

(e) Anogenital distances (AGD)—Anogenital distance is the distance between the anterior limit of the anus and the fourchette (females) or posterior limit of the fused scrotum (males); it is twice as large in males than in females [5]. The anogenital distance and ratio are sensitive indicators of masculinization or feminization. In male-term newborns, the mean AGD is 24.7 (4.5) mm. In female-term newborns, the mean AGD is 16.0 (3.2) mm [7]. Anogenital ratio = Distance between the anus and posterior fourchette/distance between anus and base of the clitoris, values >0.5 are noted androgen exposure in the 8–14 weeks masculinizing window of the first trimester. Lately, the measurements are being taken from the center of the anus instead of the anterior border [8].

(f) Pubertal changes (Figs. 11.4, 11.5, and 11.6).

The period from infancy to puberty is usually quiescent. However, the onset of puberty with changes in the hormonal milieu and tissue responsiveness manifests as phenotypic changes. Previously diagnosed and characterized cases of DSD are reviewed annually to document evolving changes in general anthropometry/habitus and genital anatomy; they are reviewed closely at the onset of puberty as a host of secondary sexual characters set in—thelarche, pubarche, and menarche. These offer vital clues to diagnosis in DSD, some cryptic and milder forms of DSD may be obvious to the unsuspecting clinician only at puberty, e.g., complete AIS, reared as female, who does not attain menarche. The classic example of pubertal virilization seen with 5AR2 deficiency and 17ßHSD deficiency and incongruent phenotype—sex of rearing instances with ovotesticular DSD makes these moving targets for clinical review. Close clinical differentials like non-classic 5AR2 deficiency and PAIS that often exhibit overlapping phenotypes may become clearer with features more specific to one than the other, e.g., breast development but a relative lack of axillopubic hair is commoner with PAIS than 5AR2 deficiency at puberty. Psychological changes are equally varied and need evaluation alongside somatic changes.

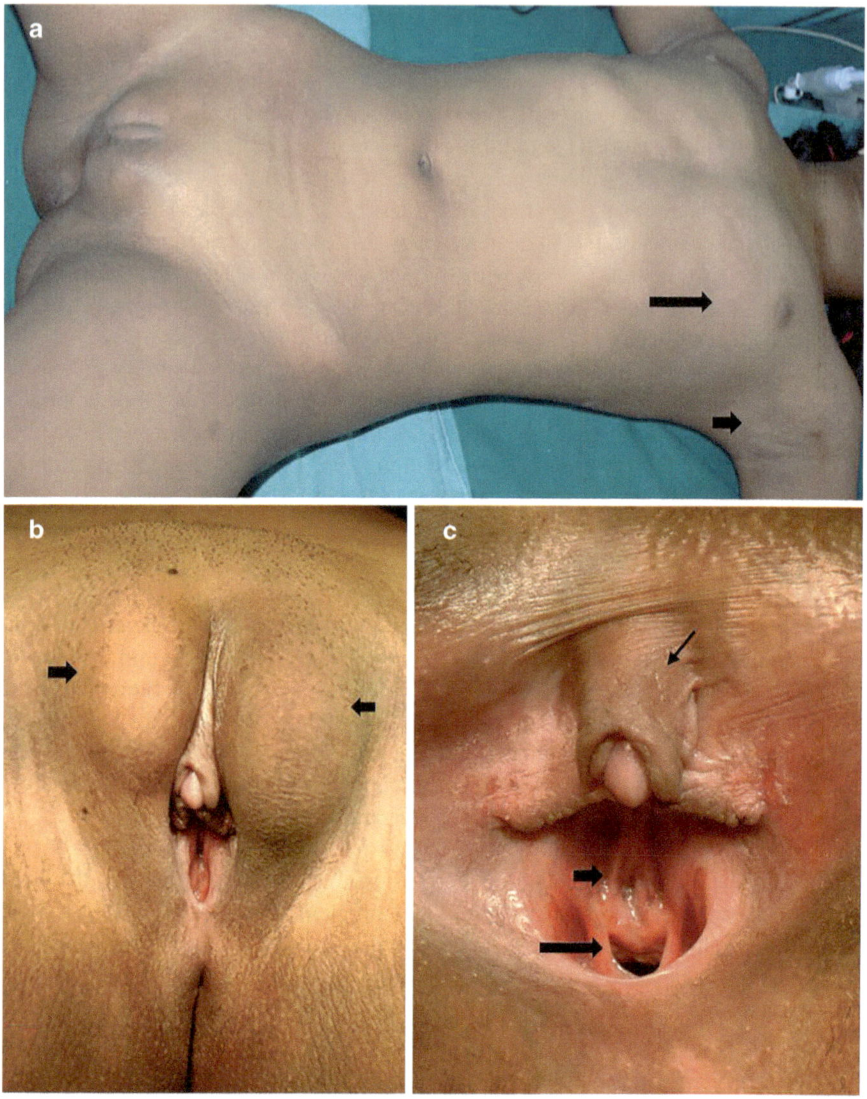

Fig. 11.4 : Pubertal presentation at 13 years with atypical genitalia and recent change in voice in 46XY DSD-Partial Androgen Insensitivity Syndrome Note the feminine habitus, poor breast development (long arrow, and sparse axillopubic hair (short arrow, **a**; **b**). Both gonads were palpable in labioscrotal folds (arrows, **b**), and the phallic structure was 2.3 cm (oblique arrow, **c**). There were two orifices—the external meatus (short arrow, **c**) and a blind vaginal pouch (long arrow, **c**). The EGS was 5.5

The nature of genital hair distribution or escutcheon at puberty differentiates the male (triangle with the base at the pubes, apex at the umbilicus) from a female (triangle with the base at the pubes, apex at the fourchette) pattern. The

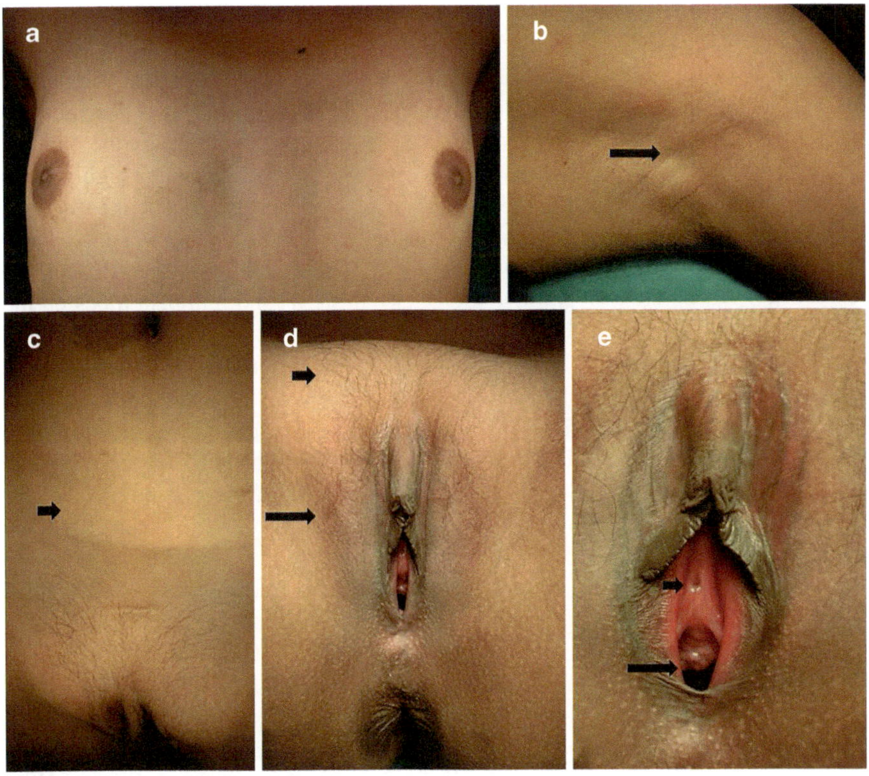

Fig. 11.5 Pubertal presentation at 15 years with primary amenorrhoea and poorly developed secondary sexual characters in 46XX DSD Gonadal Dysgenesis. Note the modest breast development (**a**) and lack of axillopubic hair (short arrow, **b**, **c**)—Tanner stage 2. The mons pubis (short arrow, **d**) and labia (long arrow, **d**) were hypoplastic. There were two orifices—the external meatus (short arrow, **e**) and a vaginal orifice (long arrow, **e**)

nature of the hair is further classified as per Tanner stage. The tough, curly pubic hair in an untreated toddler with virilizing CAH with elevated androgens is typical. Body hair distribution in non-genital areas is also noted. Breast and areolar development is described as per Tanner staging too (Annexure 1). Breasts are relatively well developed in ovotesticular DSD, milder in PAIS, and almost absent in pure ovarian gonadal dysgenesis.

Yet several minute details may not be apparent or cannot be evaluated satisfactorily in an outpatient examination of a small child with anxious parents and a thorough examination under anesthesia (EUA) and requisite endoscopy may be scheduled before a final diagnosis is made.

Several scoring systems are used to allow for accurate documentation and comparison in specific clinical situations. The Prader's scale (1–5) is used to categorize the degree of genital masculinization in 46 XX DSD/CAH from the mildly virilized

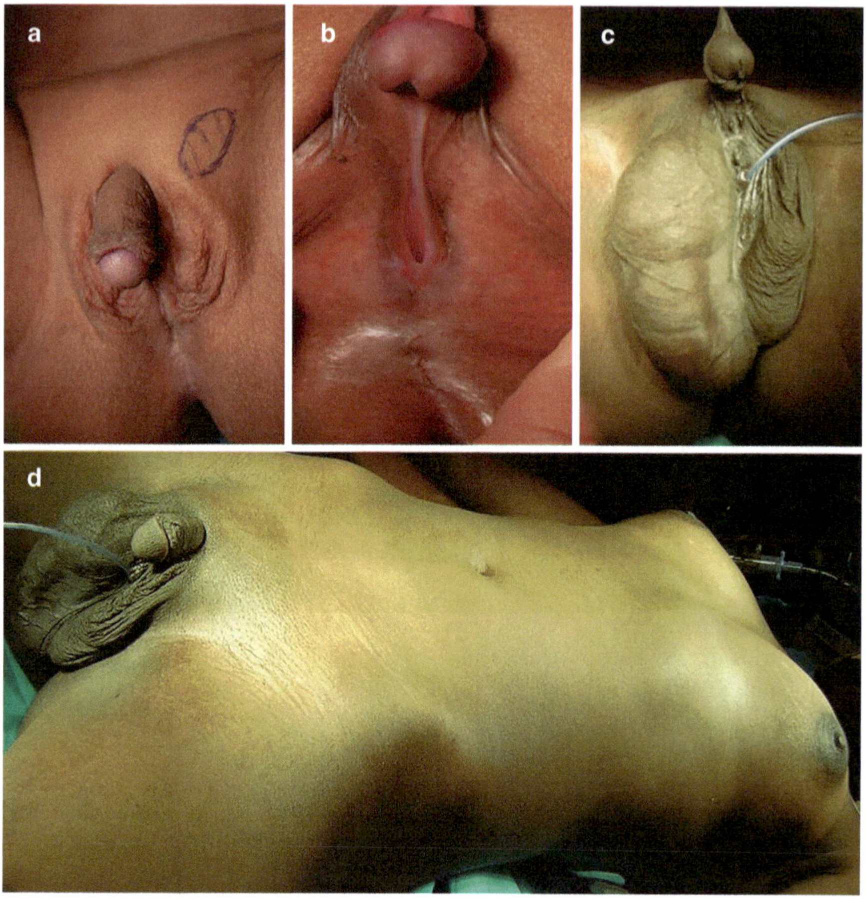

Fig. 11.6 Spectrum of phenotype in 46 XX ovotesticular DSD. A 4-year child (**a**, **b**) reared as male had a left, palpable inguinal gonad (marked, **a**) and inguinal hernia; the right gonad was impalpable. Note the severe chordee (>30 degrees) and mid-scrotal meatus (**b**). This 16-year-old male (**c**, **d**) presented with a mix of masculine (rugose scrotum, penis) and feminine (habitus, breast development–Tanner stage 3, proximal hypospadias)features. The right labioscrotal fold was large and contained a palpable gonad and several cystic structures while the left was empty

female to the male phenotype (Annexure 2). The External Masculinization score (0–12) includes scoring on the testicular position in maldescent, meatal position in hypospadias, presence/absence of micropenis, and scrotal fusion in a feminized male at birth [9]. The EGS (External Genitalia Score) is a recently validated scoring system that improves on the EMS and captures the appearance of the genitalia across the phenotypic spectrum in both sexes [8]. It uses a non-binary, gradual scale from female to male (range, 0–12) of the same anatomical landmarks as the EMS and can be used in premature and term babies till 2 years of age (Annexures 3, 4).

An EMS of less than 11 or EGS of less than 10.5 has been used to select infants for detailed evaluation [9]. These scoring systems have their limitations too; the Prader and EMS do not reflect any correlation between the degree of external virilization and the level of the confluence between the vagina and urethra in common urogenital sinus anatomy. Though these scores can be used for precise documentation instead of photography which requires additional consent and ethical clearance, descriptive notes must be added where required.

After a detailed historical review and physical examination, it must be clear whether there is enough ground for a directed investigation of DSD. The other team members are consulted for clinical inputs and a scheme of investigations planned The sequence generally begins with a preliminary karyotype and non-invasive imaging that help to select the more specific investigations (hormonal profile, genitogram, genitoscopy, laparoscopy, etc.) that are elaborate, expensive, and invasive in an algorithmic manner. Examination and other investigations needing anesthesia are scheduled last with a clear plan of surgical interventions if any to minimize the number and duration of anesthesia exposure in the course of the management calendar.

Annexures

Annexure 1: Tanner Scale for the Onset and Progression of Puberty

Because the onset and progression of puberty are so variable, Tanner has proposed a scale, now uniformly accepted, to describe the onset and progression of pubertal changes (Fig. 9-24). Boys and girls are rated on a 5-point scale. Boys are rated for genital development and pubic hair growth, and girls are rated for breast development and pubic hair growth.

Pubic hair growth in females is staged as follows (Fig. 9-24, B):

- Stage I (Preadolescent)—Vellos hair develops over the pubes in a manner not greater than that over the anterior wall. There is no sexual hair.
- Stage II—Sparse, long, pigmented, downy hair, which is straight or only slightly curled, appears. These hairs are seen mainly along the labia. This stage is difficult to quantitate on black-and-white photographs, particularly when pictures are of fair-haired subjects.
- Stage III—Considerably darker, coarser, and curlier sexual hair appears. The hair has now spread sparsely over the junction of the pubes.
- Stage IV—The hair distribution is adult in type but decreased in total quantity. There is no spread to the medial surface of the thighs.
- Stage V—Hair is adult in quantity and type and appears to have an inverse triangle of the classically feminine type. There is a spread to the medial surface of the thighs but not above the base of the inverse triangle.

The stages in male pubic hair development are as follows (Fig. 9-24, B):

- Stage I (Preadolescent)—Vellos hair appears over the pubes with a degree of development similar to that over the abdominal wall. There is no androgen-sensitive pubic hair.
- Stage II—There is sparse development of long pigmented downy hair, which is only slightly curled or straight. The hair is seen chiefly at the base of the penis. This stage may be difficult to evaluate in a photograph, especially if the subject has fair hair.
- Stage III—The pubic hair is considerably darker, coarser, and curlier. The distribution is now spread over the junction of the pubes, and at this point that hair may be recognized easily on black-and-white photographs.
- Stage IV—The hair distribution is now adult in type but still is considerably less than seen in adults. There is no spread to the medial surface of the thighs.
- Stage V—Hair distribution is adult in quantity and type and is described in the inverse triangle. They can be spread to the medial surface of the thigh.

In young women, the Tanner stages for breast development are as follows (Fig. 9-24, C):

- Stage I (Preadolescent)—Only the papilla is elevated above the level of the chest wall.
- Stage II (Breast Budding)—Elevation of the breasts and papillae may occur as small mounds along with some increased diameter of the areolae.
- Stage III—The breasts and areolae continue to enlarge, although they show no separation of contour.
- Stage IV—The areolae and papillae elevate above the level of the breasts and form secondary mounds with further development of the overall breast tissue.
- Stage V—Mature female breasts have developed. The papillae may extend slightly above the contour of the breasts as a result of the recession of the aerolae.

The stages for male genitalia development are as follows (Fig. 9-24, A):

- Stage I (Preadolescent)—The testes, scrotal sac, and penis have a size and proportion similar to those seen in early childhood.
- Stage II—There is an enlargement of the scrotum and testes and a change in the texture of the scrotal skin. The scrotal skin may also be reddened, a finding not obvious when viewed in a black-and-white photograph.
- Stage III—Further growth of the penis has occurred, initially in length, although with some increase in circumference. There also is increased growth of the testes and scrotum.
- Stage IV—The penis is significantly enlarged in length and circumference, with further development of the glans penis. The testes and scrotum continue to enlarge, and there is a distinct darkening of the scrotal skin. This is difficult to evaluate on a black-and-white photograph.
- Stage V—The genitalia are adult with regard to size and shape.

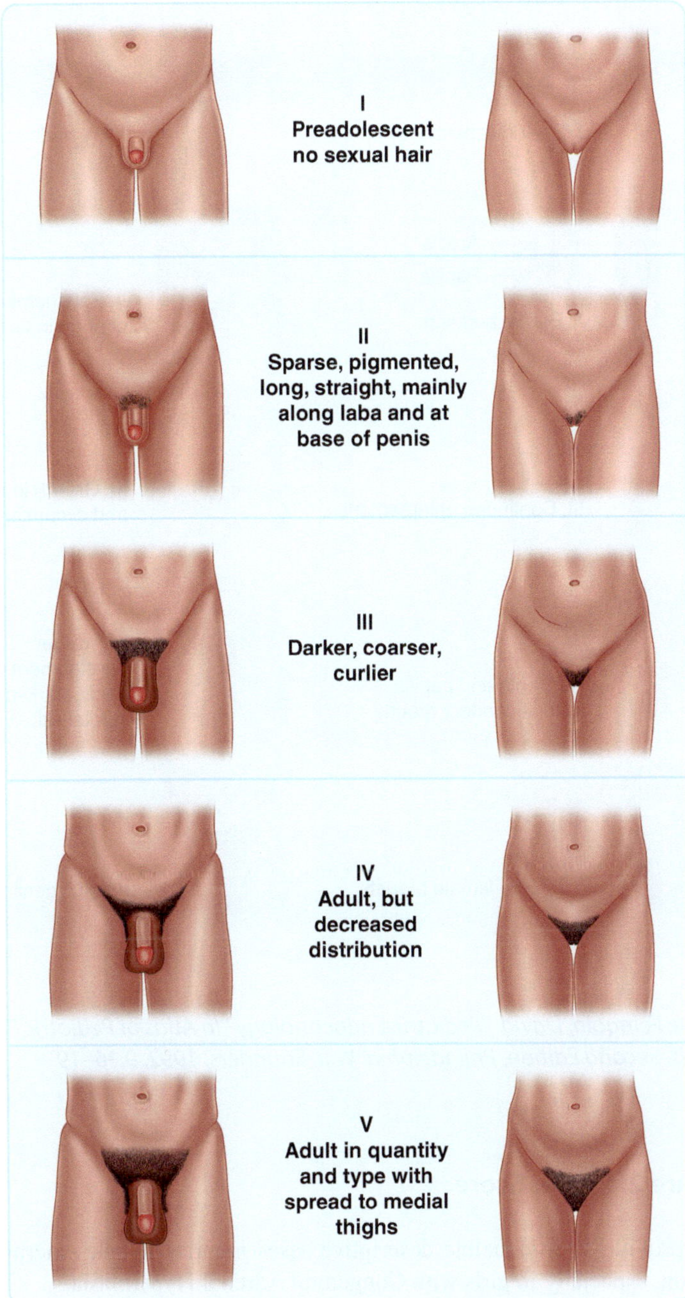
I
Preadolescent
no sexual hair
II
Sparse, pigmented,
long, straight, mainly
along laba and at
base of penis
III
Darker, coarser,
curlier
IV
Adult, but
decreased
distribution
V
Adult in quantity
and type with
spread to medial
thighs

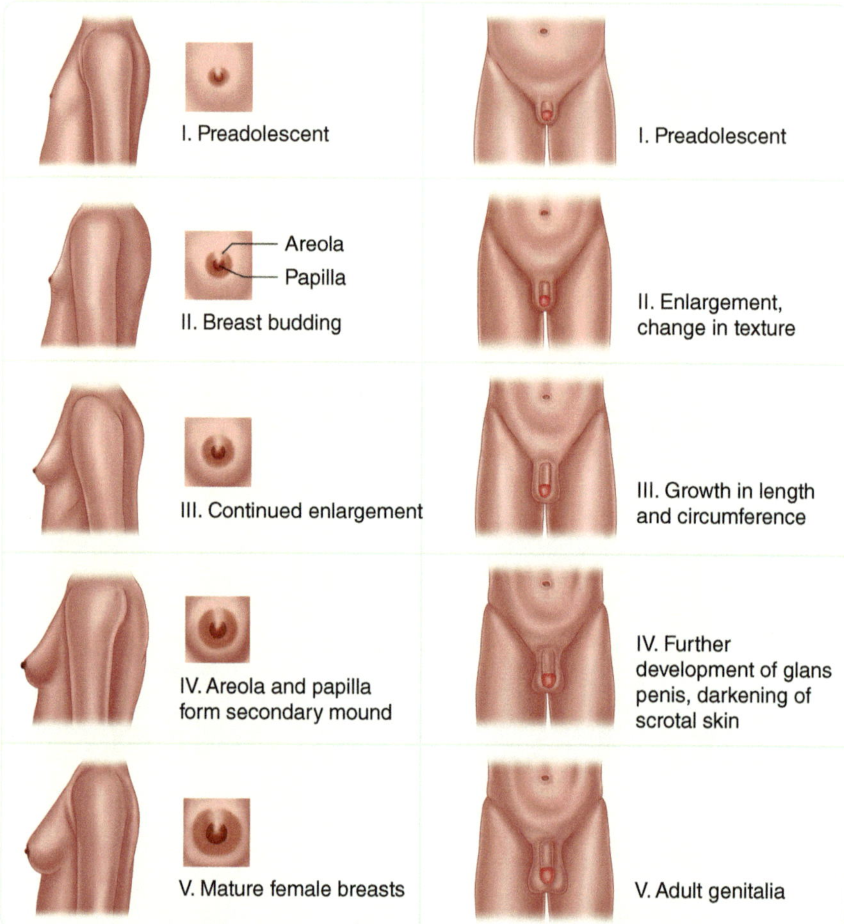

Reference: Feingold, David. "Pediatric Endocrinology" In Atlas of Pediatric Physical Diagnosis, Second Edition, Philadelphia. W.B. Saunders, 1992, 9.16–19

Annexure 2: Prader Score

Used to provide a reproducible descriptive assessment of female external genital virilization, especially in girls with Congenital Adrenal Hyperplasia.

Described by Prader and Andreas in 1954.

Prader score	Genital description
Stage 0	Normal female genital phenotype
Stage 1	Isolated clitoral hypertrophy may not be prominent. Separate urethral and vaginal orifices
Stage 2	Noticeable phallic hypertrophy. Separate urethral and vaginal orifices—close together. Smaller vaginal orifice. Posterior labial fusion may be seen
Stage 3	Further phallic hypertrophy. Single urogenital sinus opening. Posterior labial fusion
Stage 4	Penile proportions of enlarged phallus, usually with chordee. Urogenital sinus in a perineoscrotal hypospadiac location. Complete fusion of labioscrotal folds. Impalpable gonads
Stage 5	Complete masculinization with urogenital sinus opening at the tip or near the tip of phallus. Scrotal appearance of fused labioscrotal folds. Gonads are impalpable. Normal Mullerian structures are present

Reference: Ogilvy-Stuart, A. L. Early assessment of ambiguous genitalia. Archives of Disease in Childhood, 2004; 89(5), 401–407

Annexure 3: External Masculinization Score/Internal Masculinization Score (EMS/IMS)

The EMS/IMS Scores provide a format to objectively summarize findings of genital examination in newborns with suspected genital ambiguity.

Feature	Score for Yes/Noor condition
EMS	
Scrotal fusion	3/0
Micropenis	0/3
Urethral meatus	
Normal	3
Glandular	2
Penile	1
Perineal	0
Right and left gonad (score for each)	
Scrotal	1.5
Inguinal	1
Abdominal	0.5
Absent	0
IMS	
Uterus	0/3
Fallopian tube (right and left score each)	0/2
Epididymis (right and left score each)	2/0
Vas deferens (right and left score each)	2/0

N.B. Micropenis is defined as Penile length < 2.5 cm

The Masculinization scores can help determine the need for a further extensive evaluation of an under-masculinized male.

Infants with an EMS Score <11 warrant further investigation for DSD.

Reference: Ahmed S.F, Khwaja O, Hughes I.A. The role of a clinical score in the assessment of ambiguous genitalia. BJU International. 2000; 85, 120–124

Annexure 4: External Genitalia Score (EGS)

The External Genitalia Score (EGS) aims to overcome the male bias in the EMS and provide a more gender-neutral score to objectively document genital examination findings in newborns.

The score uses phenotypic features at five anatomic areas and assigns a specific score to each finding as in the table below.

EGS	Labioscrotal fusion	Genital tubercle Length (mm)	Urethral meatus	Right gonad	Left gonad
3	Fused	>31	Top of the GT		
2.5		26–30	Coronal glandular		
2			Along the GT		
1.5	Posterior fusion	21–25	At the GT base	Labioscrotal	Labioscrotal
1		10–20	Labioscrotal	Inguino-Labioscrotal	Inguino-Labioscrotal
0.5				Inguinal	Inguinal
0	Unfused	< 10	Perineal	Impalpable	Impalpable

Abbreviation: GT—Genital Tubercle

The EGS correlates well with the measurements of Anogenital distance in both boys and girls as a mark of prenatal androgen exposure. Full-term infants with an EGS >0 and ≤10.5 warrant evaluation by a DSD team. Preterm infants with EGS scores <10th percentile for gestation need evaluation.

Reference: Van der Straaten, S., Springer, A., Zecic, A., Hebenstreit, D., Tonnhofer, U., et al. The External Genitalia Score (EGS): A European multicenter validation study. J Clin Endocrinol Metab, March 2020, 105(3):1–9

References

1. HughesIA HC, Ahmed SF, Lee PA. Consensus statement on management of intersex disorders. Arch Dis Child. 2006;91:554–63.
2. El Beck MS, Germano CW, Barros BA, Andrade JG, Guaragna-Filho G, Paula GB, et al. Why pediatriciansneed to know the disorders of sex development: experience of 709 cases in a specialized service. J Pediatr. 2020;96:607–13.

3. Ahmed SF, Achermann J, Alderson J, Crouch NS, Elford S, Hughes IA, et al. Society for Endocrinology UK Guidance on the initial evaluation of a suspected difference or disorder of sex development (Revised 2021). Clin Endocrinol (Oxf). 2021;95(6):818–40.
4. Malone PS, Hall-Craggs MA, Mouriquand PDE, Caldamone AA. The anatomical assessment of disorders of sex development (DSD). J Pediatr Urol. 2012;8(6):585–91.
5. Alkhzouz C, Bucerzan S, Miclaus M, Mirea AM. Miclea D.46, XX DSD: developmental, clinical and genetic aspects. Diagnostics. 2021;11:1379. https://doi.org/10.3390/diagnostics11081379.
6. Lee PA, Mazur T, Danish R, Amrhein J, Blizzard RM, Money J, Migeon CJ, Micropenis I. Criteria, etiologies and classification. Johns Hopkins Med J. 1980;146(4):156–63.
7. Sathyanarayana S, Grady R, Redmon JB, Ivicek K, Barrett E, Janssen S, Nguyen R, Swan SH, TIDES Study Team. Anogenital distance and penile width measurements in the infant development and the environment study (TIDES): methods and predictors. J Pediatr Urol. 2015;11(2):76.e1–6.
8. van der Straaten S, Springer A, Zecic A, Hebenstreit D, Tonnhofer U, Gawlik A, et al. The external genitalia score (EGS): a European multicenter validation study. J Clin Endocrinol Metab. 2020;105(3):dgz142. https://doi.org/10.1210/clinem/dgz142.
9. Ahmed SF, Khwaja O, Hughes IA. The role of a clinical score in the assessment of ambiguous genitalia. BJU Int. 2000;85(1):120–4.

12 Cytogenetic and Molecular Genetic Investigations in DSD

Dinesh Kumar and Sunil Kumar Polipalli

Abstract

An early confirmation of the genetic sex in DSDs is the key to its diagnosis and management. Genetic diagnosis of DSDs requires consideration of causes and undestanding sequential processes of normal sex development and differentiation. Karyotyping results are the first and most defining, therefore, mandatory genetic test and help classify the three major DSD subclasses. The involved principal genes are classified as Testicular promoting genes (SOX3, SOX9, FGF9, PGD2) and Ovarian promoting genes (DAX1, WNT4, FOXL2, RSPO1, and β-catenin). Considering the wide and varied spectrum of DSDs, success in ascertaining a genetic cause in XY DSD remains challenging. The standard approach in genetic evaluation of children with DSDs should be to first determine the genetic constitution followed by determining the affected genes.

Keywords

Gonadal differentiation · Genetic defects · Chromosomal defects

Children with DSDs and their families are extremely distressed and seek medical advice which depends heavily on genetic investigations for diagnosis and prognosis. To administer such investigations requires an in-depth understanding of the normal gonadal development and the roles of candidate genes and transcription factors.

D. Kumar (✉)
Department of Anatomy, Maulana Azad Medical College, New Delhi, India

S. K. Polipalli
Division of Genetics, Department of Pediatrics, Maulana Azad Medical College, New Delhi, India

S. K. Ratan (ed.), *Children with Differences in Sex Development*,
https://doi.org/10.1007/978-981-97-1639-5_12

12.1 Early Stages of Development

At conception, the genetic sex is established, once a spermatozoon (with X or Y sex chromosome) fertilizes an oocyte, which results in a diploid 46, XX or 46, XY cell. During the first weeks of embryonic life, sexually indifferent gonads and genitalia start to develop. By the fourth week, development of the urogenital sinus and the adreno-gonadal ridge occurs. The male sex and female characteristics are determined during fetal life by complex biological processes involving cascades of gene expression which involve the proteins with highly specific functions in location and time. The development of the bipotential gonad involves well-characterized human genes: *EMX2, CBX2, NR5A1, GATA4*, and *WT1* [1–3]. From the sixth week, the presence of a Y chromosome and its *SRY* gene (located at Yp11) directed activity triggers a series of genes that mediate development of undifferentiated gonad into a testis [4] and inhibits those that induce development into an ovary [1, 3, 5, 6].

Complex interactions between antagonistic genes regulate the processes of differentiation into a testis or an ovary and therefore are critical to gonadal development [7].

To summarize, we can classify the key genes in the gonadal development as: Testicular promoting genes (SOX3, SOX9, FGF9, PGD2) and ovarian promoting genes (DAX1, WNT4, FOXL2, RSPO1, and β-catenin).

12.2 Male Sex Differentiation

Adequate secretion of testicular hormones, in a specific time sequence regulates differentiation of both internal and external genitalia. Internal male genitalia development requires secretion and action of testosterone (T), which causes Wolffian ducts to develop into epididymis and vas deferens. The anti-Mullerian hormone (AMH) induces Mullerian duct regression and its activity is mediated by its receptor (AMHR2) [9]. Conversion of testosterone into dihydrotestosterone (DHT) by 5alpha-reductase type 2 directs development of the prostate and external genitalia [11] (Fig. 12.1).

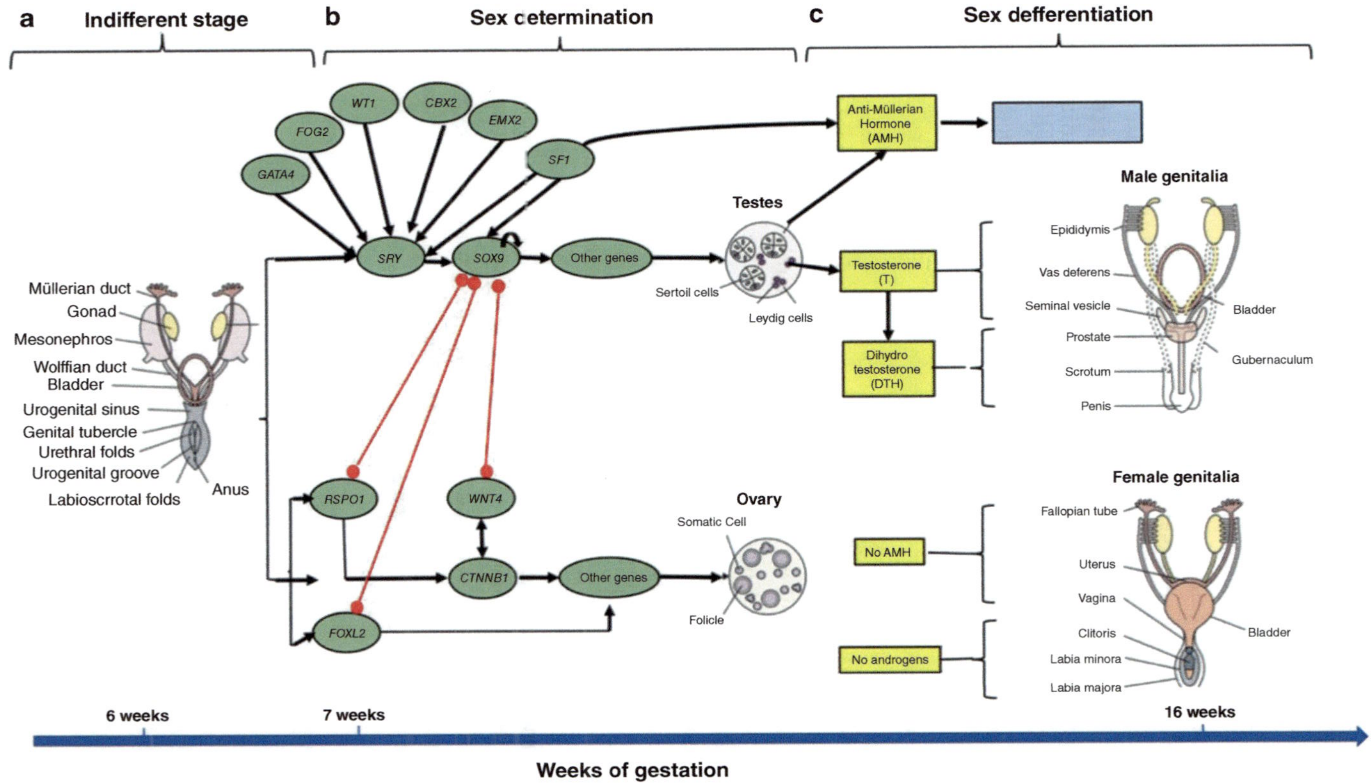

Fig. 12.1 Human sex development during fetal life. Diagrammatic representation of interplay of Genes in gonadal differentiation (Black arrows indicate gene activation and red lines indicate gene repression) [Courtesy 8]

12.3 Female Sex Differentiation

Differentiated female genitalia is result of absence of AMH and elevated androgen concentrations (T and DHT) (Fig. 12.1). Murine model of estrogen receptor (ER) knockout suggests that estrogens could feminize the genital tubercle by mediation of ER [12], however, morphology of external genitalia of a female newborn with complete resistance to estradiol (E2) has not yet been described.

Any alteration in developmental stages may disrupt normal sex development, resulting in differences in sex development (DSD) [13]. DSDs arise when this highly synchronized multipronged process is disrupted as a result of genetic mutations, interfering with either the development of the testes or ovaries or the actions of endocrine and local factors in extra-gonadal tissues [14, 15].

According to Chicago Consensus [13], DSDs have been classified into three major groups:

1. Sex chromosome DSD
2. 46, XX DSD with female karyotype
3. 46, XY DSD with male karyotype

Each of these groups is divided into subgroups. Groups 2 and 3, involve a broad range of genes, which has increased over the years. With the exception of Group 1 DSD, especially Klinefelter syndrome with 47, XXY karyotype and Group 3 DSD, 46,XY DSD male infants born with congenital hypospadias, the population frequency of Groups 2 and 3 DSD is so low that they are considered rare diseases, with a population frequency < 1/2000.

12.4 Environmental Factors and DSDs

DSDs in the absence of DNA sequence defects are likely to be induced by environmental factors which cause DNA methylation, Histone protein modifications, or presence of non-coding RNA. Therefore, the role of epigenetics cannot be ignored.

12.5 Population Affected

The incidence of 46, XY DSD has been estimated to be 1 in 20,000 births. Ovotesticular DSDs have been estimated to occur in 1 of 100,000 live births [6]. The frequency of testicular or mixed gonadal dysgenesis is estimated at 1:10,000 [6]. The worldwide incidence of 46,XX DSD, consisting primarily of CAH (mostly 21-hydroxylase deficiency) has been estimated to be 1 in 14,000–15,000 live births [8] but has ethnic differences induced gene mutation frequency. CAH and mixed gonadal dysgenesis constitute about half of all DSD patients presenting with genital ambiguity [9].

The rate may be as high as 1:200 to 1:300, when all congenital genital anomalies (including cryptorchidism and hypospadias) are considered [9]. Among patients with hypospadias and cryptorchidism, currently the diagnosis of specific DSD conditions is generally limited to those with proximal hypospadias with cryptorchidism. The overall incidence estimations also include those with Klinefelter syndrome (estimated in 1:500 to 1:1000 live births) and Turner syndrome (about 1:2500 live births). These estimates provide an important perspective to the burden on the society and health resources.

Genetic diagnosis of DSDs requires an understanding of the causes and processes of normal sex development and differentiation. Considering the varied features of DSDs, the success in ascertaining a genetic cause in XY DSD remains relatively low.

After physical examination, the standard protocol for confirming the diagnosis and characterizing the DSDs includes (a) Hormonal testing—testosterone, estrogen, and progesterone; (b) Genetic testing—Chromosomal analysis; (c) Imaging studies—Ultrasound or MRI; (d) Biopsy; and (e) Psychological evaluation.

12.5.1 Genetic Testing

Genetic testing is the cornerstone of diagnosis in all patients with DSDs, especially in children where gender assignment weighs heavily on the parents.

Genetic causes of DSDs include (1) Numerical or structural variations in sex chromosomes and (2) Variations in genes involved in gonadal and/or genital development (leading to inactivation or activation). Sex assignment in these children continues to be a challenging diagnostic and therapeutic problem.

Genetic testing should be performed along with biochemical investigations, or before even before performing detailed endocrine investigations [16] but not later to at best decision making position.

Candidates for genetic testing DSD can include fetuses, neonates, infants or prepubertal children or adolescents/adult persons either in cases of family history of DSD or because the phenotype suggests a DSD.

Apart from chromosomal variations—deletions, duplications, translocations, both in autosomes and sex chromosomes, DSD may also involve copy number variations (CNV) which are especially suspected when the phenotype includes additional anomalies [17–20].

1. *Cytogenetics*:

Karyotyping is essential to define major diagnostic DSD subclasses and therefore is the mandatory first investigation.

Chromosomal analysis is performed for:

(a) Detection of abnormalities in the individual's sex chromosomes. This can help to identify conditions such as Turner syndrome, Klinefelter syndrome, and other genetic disorders.
(b) DSD categorization into one of the three diagnostic groups based on the sex chromosomes found.

Overall, karyotyping is an important tool in the evaluation of DSDs, as it can help to provide a definitive diagnosis and guide medical management.

Cytogenetics is still the gold standard in this modern era with techniques like array-complementary genomic hybridization (array-CGH) techniques are being increasingly used [21] especially, for deletions/translocations smaller than 20 kb but with greater experience and technological support, one can identify chromosomal abnormalities up to 5–10 Mb in size in metaphase spreads of traditional karyotyping.

Limitations: The turnaround time for results is 7–10 days. Limited by resolution of 5 Mb.

2. *FISH* (Fluorescence in situ hybridization)

FISH is another genetic diagnostic test that can be used in the evaluation of DSD, which uses fluorescent probes to bind to specific regions of the chromosomes and produce a visual signal.

A comprehensive panel of genes (Table 12.1), known to be affected sites in human 46, XY and 46, XX DSD (including gonadal dysgenesis, primary and secondary gonadal insufficiency, disordered steroidogenesis, androgen resistance, isolated urogenital anomalies and syndromic conditions associated with ambiguous genitalia) is reported along with their chromosomal locations: it includes 62 genes in 46, XY and 61 in 46, XX [22]. Considering the number of genes is large, therefore specific suspected FISH probes post-karyotyping can be used as confirmatory evidence (Table 12.2).

Using FISH, one can detect presence or absence of SRY region of Y chromosome or loss or gain of specific segments in suspected chromosomal segments of genes. It can also be used to detect the presence of specific genetic mutations or deletions that may be associated with dysmorphic sex disorders. e.g., deletion in the X chromosome with Turner syndrome.

Overall, FISH is a valuable diagnostic tool that can provide additional information to help diagnose and manage dysmorphic sex disorders. However, it is often used in conjunction with other diagnostic tests, such as karyotyping, to provide a comprehensive evaluation of these conditions (Table 12.3).

Limitations: Only known chromosomal imbalances can be determined by specific probes. Isolated probes are expensive.

Molecular genetics has an important role in the diagnosis of disorders of sex development (DSD). For example, mutations in the genes that regulate gonadal development and hormone production can cause DSD. By sequencing these genes, geneticists can identify specific mutations or variants that may be responsible for an individual's DSD.

Table 12.1 Role of genes in gonadal differentiation

Developmental stage	XX Gonads	XY Gonads	Age (week)
(A) *Indifferent stage:* By 5 weeks of life—Bipotential gonads has-, two pairs of genital ducts (Mullerian and Wolffian ducts). The common external genitalia includes: Genital tubercle, the urethral folds, urogenital groove, and labioscrotal folds [8], at different stage, male and female embryos share two genital ducts (Mullerian and Wolffian ducts) and external genitalia (genital tubercle and labioscrotal folds) [9]	SF-1 responsible for adrenal and gonadal steroids. SF1 continues in testes but is repressed in the ovary by SRY. WT1 promotes the secondary development of bipotential gonads		5–6
(B) Sex determination: Depending on the activation or repression of signalling pathways, somatic cells and gonocytes in the bipotential gonad begin to differentiate into testicular or ovarian cells. They acquire the destined morphological character	Absence of SRY, which results in decreased SOX9 expression	Insulin receptors are responsible for gonadal proliferation before SRY expression Later differentiation depends on whether SRY binds to the enhancer region of SOX 9	6–7
(C) Sex differentiation: Presence or absence of testicular hormones (anti-Mullerian hormone [AMH] and testosterone [T] and dihydrotestosterone [DHT]) are principal players in internal and external genitalia differentiation [10]	RSPC1, WNT4, and FOXL2 in the absence of SRY (during meiosis) cause ovarian differentiation, to form granulosa cells. RSPC1, WNT4 stimulate β-catenin which in turn suppresses SOX 9 Male reversion with SRY negative can occur due to either a pathogenic variant of loss of function in the ovarian pathway or function gain of a pathogenic variant in genes mimicking SRY. In 10%, SRY expression is seen in XX due to Y-X translocation	Failure of SRY expression results in male-to-female reversion. This can result in pure or mixed gonadal dysgenesis Gene targets of SRY gene influence SOX9, CBLN4, and ER71/ETV2	8

Table 12.2 Defect—features DSD matrix

1	Palpable testes	Absent	Present	Absent	Maybe present
2	Mullerian derivatives	Maybe present	Absent	Present	Present
3	Chromosomal complement	45X, 46XY	46XY	46 XX	46 XX/46 XY
4	Diagnosis	Testes agenesis/ dysgenesis at very early stage	Gonadal dysgenesis	– 21 α-hydroxylase deficiency, CAH – Maternal ingestion of androgens	Ovotesticular/ anti-Mullerian inhibitor

Table 12.3 Frequently associated chromosomal defects with DSD

S. no.	Chromosomal defect	Diagnosis
1	46 XX	Congenital adrenal hyperplasia 21 α-hydroxylase deficiency 11β-hydroxylase deficiency Ovotesticular DSD
2	45 X	Gonadal dysgenesis
3	46 XX/46XY, 45 X/46 XY, 46 XX/47XXY	Ovotesticular DSD Mixed Gonadal dysgenesis
4	46 XY	Agenesis/Dysgenesis Persistent Mullerian duct syndrome
5	46 XX	Y chromosomal translocation
6	47 XXY	Klinefelter syndrome

In some cases, genetic testing can also help to identify the specific type of DSD a person has. For example, a genetic test can identify whether an individual has 46, XX DSD or 46, XY DSD, which are two of the most common types of DSD. This information can be useful for determining the appropriate course of treatment and management for the individual's condition.

Molecular genetics can also be used to identify whether a person has an underlying genetic condition that may be causing their DSD, such as Turner syndrome or Klinefelter syndrome. This information can be important for identifying potential health risks and providing appropriate medical care.

3. *Reverse Transcription Polymerase Chain Reaction*

With the cloning of the genes that code for proteins, known to be altered in the clinical and biochemical phenotype, the most frequent monogenic causes of DSD have been identified. This is especially useful for the determination of enzyme deficiencies causing disorders of adrenal and gonadal steroidogenesis (Fig. 12.2), both in 46, XX and 46, XY DSD and in cases suspected of complete androgen insensitivity. The genes involved in the differentiation and development of male and female

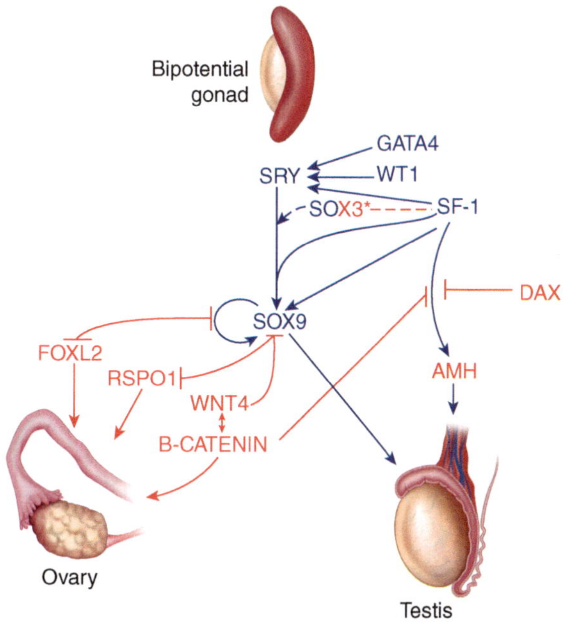

Fig. 12.2 Proposed scheme of gene regulation in gonadal development. Courtesy: [11]

gonads are being progressively detected on the basis of family studies, animal models, and functional studies in vitro [7, 20].

Limitations: Reaching a diagnosis in 46, XY DSD is relatively low [23]. This is most likely due to the heterogeneous clinical and hormonal presentation of many DSDs, the number of known DSD-related genes being large (Table 12.1) and often with low genotype–phenotype correlation and the increasing evidence that a significant proportion of the underlying pathogenesis may be multifactorial [24, 25].

4. *Microarray [array Comparative genomic hybridization (aCGH)/Single nucleotide polymorphisms (SNP)]*: SNP array has wider diagnostic potential compared to aCGH—polyploidy, chimerism, mosaicism, and regions of homozygosity can be detected.

Limitations: aCGH cannot detect balanced chromosomal rearrangements and may not detect mosaicism unless combined with conventional cytogenetic techniques; some CNVs identified by microarray are challenging to interpret in relation to the degree of pathogenicity.

5. *Sanger sequencing* involves PCR amplification of coding and flanking regions of the promoter region. A large number of genes involved in the development of DSD encode the regulatory transcription factors of other genes (*AR, DAX1, DMRT1, FOXL2, NR5A1, SOX3, SOX9*, and *SRY*). The literature shows evidence of mutations in noncoding regulatory regions, therefore testing noncoding DNA regions will be useful to identify the cause of some DSDs in which molecular

diagnosis could not be previously performed [26]. The high-throughput DNA sequencing allows whole exome testing (coding regions) or screening a panel of candidate genes.

Targeted panel approach of multiple DSD-related genes has the advantage of being time efficient and cost effective in comparison to a single gene approach, providing a greater diagnostic yield [27, 28]. However, panel development is time consuming, labor intensive, and flexibility is limited for the addition of new DSD-related genes in comparison to alternative high-throughput screening techniques such as whole genome/exome sequencing.

In addition, broad expressivity in some DSD phenotypes could be explained by an oligogenic origin, in which the interaction of multiple genes could give rise to a phenotype unique to each individual [17]. Thus, testing a specific gene or one of its regions will progressively be limited to the diagnosis of a new patient who is a relative of a well-characterized case [13, 19–22, 26, 29].

Limitation: A large amount of template DNA is needed for each read and is time consuming. It has low throughput and NGS has overtaken Sanger sequencing due to this also has a much lower cost per base.

6. *Whole exome and genome sequencing (WES and WGS)*

Genome sequencing is helpful in prognosticating and counselling. For example, genome sequencing can identify variants that may increase the risk of families with certain health conditions. Whole exome sequencing helps examine the protein coding regions (~1–2% of the genome) and whole genome sequencing does not require targeted enrichment.

Limitation: High cost.

The following general algorithm can be followed for genetic testing in the diagnosis of DSD:

1. *Karyotype analysis*: For patients with suspected DSD. Karyotyping involves examining the number and structure of a person's chromosomes. It can detect abnormalities such as extra or missing chromosomes, which can indicate conditions such as Turner syndrome or Klinefelter syndrome.
2. *Molecular genetic testing*: If the karyotype analysis is normal or inconclusive, additional molecular genetic testing may be performed. This may include targeted gene sequencing, chromosomal microarray analysis (CMA), or whole exome sequencing (WES). These tests can identify mutations or changes in specific genes that are associated with DSD.

Genetic investigations are essential for diagnosing and understanding disorders of sex development, and can provide important information for treatment and management of these conditions.

DSDs are complex not only for treating doctors but for the patient and the family, especially due to potentiality of the primitive gonad under the influence of complex interplay of various gene products. It is important to note that genetic testing alone may not always provide a definitive diagnosis or answer all questions related to DSD. An optimal approach in children with DSD should be cytogenetics to determine the genetic constitution followed by molecular biology tests to determine the affected player genes.

A multidisciplinary approach, which shall include a range of specialists, including endocrinologists, geneticists, and psychologists, is often necessary for the most accurate diagnosis and effective management of DSD.

References

1. Eggers S, Sinclair A. Mammalian sex determination—insights from humans and mice. Chromosom Res. 2012;20(1):215–38.
2. Munger SC, Natarajan A, Looger LL, Ohler U, Capel B. Fine time course expression analysis identifies cascades of activation and repression and maps a putative regulator of mammalian sex determination. PLoS Genet. 2013;9:e1003630.
3. Eggers S, Ohnesorg T, Sinclair A. Genetic regulation of mammalian gonad development. Nat Rev Endocrinol. 2014;10:673–83.
4. Berta P, Hawkins J, Sinclair A, Taylor A, Griffiths BL, Goodfellow PN, et al. Genetic evidence equating SRY and the testis determining factor. Nature. 1990;348:448–50.
5. Harley VR, Clarkson MJ, Argentaro A. The molecular action and regulation of the testis-determining factors, SRY (sex determining region on the Y chromosome) and SOX9 [SRY-related high-mobility group (HMG) box 9]. Endocr Rev. 2003;24:466–87.
6. Sekido R, Lovell-Badge R. Sex determination and SRY: down to a wink and a nudge? Trends Genet. 2009;25:19–29.
7. Bashamboo A, Eozenou C, Rojo S, McElreavey K. Anomalies in human sex determination provide unique insights into the complex genetic interactions of early gonad development. Clin Genet. 2017;91:143–56.
8. Granada ML, Audí L. The laboratory in the multidisciplinary diagnosis of differences or disorders of sex development (DSD). Adv Lab Med. 2001;2(1):468–80.
9. Zhao F, Yao HH. A tale of two tracts: history, current advances, and future directions of research on sexual differentiation of reproductive tracts. Biol Reprod. 2019;101:602–16.
10. León NY, Reyes AP, Harley VR. A clinical algorithm to diagnose differences of sex development. Lancet Diabetes Endocrinol. 2019;7:560–74.
11. García-Acero M, et al. Disorders of sexual development: current status and progress in diagnostic approach current urology. 2020;13(4):169–78.
12. Yang JH, Menshenina J, Cunha GR, Place N, Baskin LS. Morphology of mouse external genitalia: implications for a role of estrogen in sexual dimorphism of the mouse genital tubercle. J Urol. 2010;184(4 Suppl):1604–9.
13. Lee PA, Nordenström A, Houk CP, Ahmed SF, Auchus R, Baratz A, et al. Global DSD update consortium. Global disorders of sex development update since 2006: perceptions, approach and care. Horm Res Paediatr. 2016(85):158–80.
14. Nordenvall AS, Frisen L, Nordenstrom A, Lichtenstein P, Nordenskjold A. Population based nationwide study of hypospadias in Sweden, 1973 to 2009: incidence and risk factors. J Urol. 2014;191:783–9.
15. Pang SY, Wallace MA, Hofman L, Thuline HC, Dorche C, Lyon IC, Dobbins RH, Kling S, Fujieda K, Suwa S. Worldwide experience in newborn screening for classical congenital adrenal hyperplasia due to 21-hydroxylase deficiency. Pediatrics. 1988;81:866–74.
16. Ahmed SF, Alimusina M, Batista RL, Domenice S, Lisboa Gomes N, McGowan R, Patjamontri S, Mendonca BB. The use of genetics for reaching a diagnosis in XY DSDSexDev. 2022;16:207–24.

17. Ledig S, Hiort O, Scherer G, Hoffmann M, Wolff G, Morlot S, et al. Array-CGH analysis in patients with syndromic and nonsyndromic XY gonadal dysgenesis: evaluation of array CGH as diagnostic tool and search for new candidate loci. Hum Reprod. 2010;25:2637–46.
18. White S, Ohnesorg T, Notini A, Roeszler K, Hewitt J, Daggag H, et al. Copy number variation in patients with disorders of sex development due to 46,XY gonadal dysgenesis. PLoS One. 2011;6(3):e17793.
19. Kon M, Fukami M. Submicroscopic copy-number variations associated with 46,XY disorders of sex development. Mol Cell Pediatr. 2015;2(1):7.
20. Croft B, Ohnesorg T, Sinclair AH. The role of copy number variants in disorders of sex development. Sex Dev. 2018;12(1–3):19–29.
21. Audi L, Ahmed SF, Krone N, Cools M, McElreavey K, Holterhus PM, et al. Genetics in endocrinology: approaches to molecular genetic diagnosis in the management of differences/disorders of sex development (DSD).
22. Yatsenko SA, Witchel SF. Genetic approach to ambiguous genitalia and disorders of sex development: what clinicians need to know. Semin Perinatol. 2017;41:232.
23. Alhomaidah D, McGowan R, Ahmed SF. The current state of diagnostic genetics for conditions affecting sex development. Clin Genet. 2017;91(2):157–62.
24. Werner R, Monig I, Lunstedt R, Wunsch L, Thorns C, Reiz B, Krause A, Schwab KO, Binder G, Holterhus PM, et al. New NR5A1 mutations and phenotypic variations of gonadal dysgenesis. PLoS One. 2017;12:e0176720.
25. Camats N, Fernandez-Cancio M, Audi L, Schaller A, Fluck CE. Broad phenotypes in heterozygous NR5A1 46,XY patients with a disorder of sex development: an oligogenic origin? Eur J Hum Genet. 2018;26(9):1329–38.
26. Baetens D, Mendonça BB, Verdin H, Cools M, De Baere E. Noncoding variation in disorders of sex development. Clin Genet. 2017;91(2):163–72.
27. Dong Y, Yi Y, Yao H, Yang Z, Hu H, Liu J, et al. Targeted next generation sequencing identification of mutations in patients with disorders of sex development. BMC Med Genet. 2016;17:23.
28. Eggers S, et al. Disorders of sex development: insights from targeted gene sequencing of a large international patient cohort Genome Biol. 2016;17(1):243.
29. Bashamboo A, McElreavey K. Human sex-determination and disorders of sex-development (DSD). Semin Cell DevBiol. 2015;45:77–83.

Imaging and Endoscopy in the Evaluation of DSD

13

Mainak Deb and Kanishka Das

Abstract

The diagnosis of Differences in Sex Development (DSD) in a child is associated with considerable anxiety in the parents and patients alike and arriving at an early diagnosis is important. A thorough clinical evaluation with a preliminary panel of karyotyping, biochemical, and hormonal assay is accompanied by non-invasive imaging. Cystogenitoscopy and laparoscopy are further specific adjuncts that help in treatment planning and prognostication.

Ultrasonography, a widely available and non-invasive modality, is extensively used in the prenatal and postnatal periods. Phenotypic abnormalities in the external genitalia—genitourinary tract in prenatal scans and discordance of findings with the fetal karyotype lead to a suspicion of DSD. A fetal MRI may aid further in imaging the fetal pelvic anatomy and adrenomegaly.

The ultrasonogram is the preferred initial investigation of choice postnatally to assess Mullerian structures, gonads, rectal and bladder abnormalities, and allows for serial observation and follow-up in these patients. A genitogram is a contrast study that delineates the lower genital tracts in the evaluation of a persistent urogenital sinus or a persistent cloaca. It is difficult to perform and is being replaced by genitoscopy where feasible.

Magnetic Resonance Imaging is superior to ultrasonography in imaging Mullerian structures and localizing intra-abdominal gonads but is limited by the need for general anesthesia in young children.

M. Deb
Rainbow Children's Hospitals, Hyderabad, India

K. Das (✉)
All India Institute of Medical Sciences, Bhubaneswar, India

S. K. Ratan (ed.), *Children with Differences in Sex Development*, https://doi.org/10.1007/978-981-97-1639-5_13

Direct visualization of the lower genital tracts by genitoscopy offers objective information that aids surgical reconstruction of the urogenital anomaly. Laparoscopy is superior to both USG and MRI in assessing Mullerian and Wolffian structures and gonads and obtaining crucial biopsies. In addition, it allows multiple therapeutic interventions including gonadectomy, orchidopexy, and genital reconstruction.

Imaging and endoscopic investigations are tailored to the individual patient to allow quick and cost-effective evaluation with those requiring anesthesia combined with planned therapeutic interventions.

Keywords

Imaging · Endoscopy · DSD

13.1 Introduction

The evaluation of a neonate with atypical genitalia is a medical and social emergency. The diagnosis is a source of considerable parental anxiety, hence an immediate need to assign a gender of rearing to conform to societal norms. Where the diagnosis presents in the adolescent years as abnormal or discordant sexual maturation, the patient's mental and psychological distress further adds to the complexity of the clinical situation. Besides, some children with DSD may present with apparently normal genitalia (e.g., male CAH, persistent Mullerian duct syndrome with a normal male phenotype) or emergent systemic presentations (e.g., dyselectolytemia, dehydration, and shock in CAH) requiring urgent diagnosis and management.

Previous chapters have stressed the need for early evaluation by a multidisciplinary team in order to establish a diagnosis, counsel the caregivers, allay anxiety, and offer a reasonable management calendar with outcomes. This process requires a quick but comprehensive assessment. A detailed clinical history and examination, karyotype, and a biochemical-hormonal evaluation can largely provide a working diagnosis. Preliminary non-invasive imaging coupled with a select invasive endoscopic evaluation thereafter provides adjunctive data to aid immediate counseling and provide data for long-term prognostication. It is imperative to avoid radiation in this young group of patients who may need repeat evaluations over years, hence non-ionizing modalities such as US and MRI are preferably employed over others (CT, isotope scans, etc.).

13.2 Role of Imaging in Prenatal Diagnosis

Prenatal diagnosis of DSD during routine screening is possible. In India, the disclosure of the sex of the fetus is prohibited under the provisions of the Pre-Conception and Pre-Natal Diagnostic Techniques (PCPNDT) Act which complicates prenatal diagnosis, disclosure to the parents, and related counseling.

The fetal karyotype may be determined as early as 8 weeks of gestation by assessing cell-free fetal DNA in maternal blood. Discordance of fetal chromosomal sex and sonologic appearance of the fetal genitalia can raise suspicion of DSD [1]. More frequently, abnormalities detected in other organ systems rather than the external and internal genitalia may point toward a DSD. These are common, the presence of structural bladder anomalies, cystic lesions in the lower abdomen, fetal rectal abnormalities, etc. Their presence prompts a detailed survey of the pelvic structures, kidneys, adrenals, and gonads [1].

On sonography, fetal ovaries are not normally visualized, and hence a visualization/enlargement implies pathology. The uterus is identifiable from the 19th week growing linearly through pregnancy [2]; its presence is also deduced from indirect signs like a space between the bladder and rectum or posterior bladder indentation. Yet agenesis of Mullerian structures in MRKH is usually missed! Intra-abdominal streak gonads and testes are difficult to identify. Various scenarios may induce suspicion of DSD prenatally. In the male fetus with a short rectovesical space (no uterus), these include abnormalities of the phallic structure (abnormal shape/size), genital folds (absent/bifid scrotum), or gonad (absent/undescended testes). Similarly, in the female fetus with an identifiable uterus / enlarged rectovesical space, an enlarged phallic structure, abnormal/fused labia, or alternatively any discordance between external and internal fetal gender. Specific findings and the accuracy of detection vary with the various.

A fetal MRI is less operator dependent and can aid in a more detailed assessment of fetal pelvic structures and ureterovaginal abnormalities, especially with oligohydramnios, where sonology may be less sensitive. It is reserved for situations that remain inconclusive after an ultrasonography, particularly with coexisting urinary tract anomalies.

In at-risk families, CAH may occasionally be diagnosed prenatally by the presence of enlarged hypoechoic adrenals in the fetal USG. The adrenal enlargement may be seen only on the left. This may be confirmed by fetal MRI as an iso or hypointense adrenalomegaly in T2 images [1].

Severe hypospadias can be diagnosed on the basis of an abnormally shaped phallus, the presence of a chordee leading to a downturned phallus, and visualization of the urinary stream arising from a hypospadiac location during fetal micturition. The Tulip sign is the demonstration of a bifid scrotum with a downturned phallus between the scrotal folds and is associated with severe hypospadias. Ascertaining the position of the gonads during the third trimester may help in narrowing down the differential diagnoses. Hypospadias coexisting with pelvic cysts is suggestive of Mullerian remnants and necessitates a more intensive fetal and postnatal evaluation [1, 3].

The prenatal diagnosis of cloacal anomalies is suspected if an over-distended bladder, midline retrovesical cystic lesions suggestive of a hydrometrocolpos or rectal abnormalities are seen in isolation or coexisting in various combinations. Urorectal fistulae may be present when enteroliths are demonstrated in the fetal rectum. Enteroliths are echogenic luminal foci on USG, hyperintense on T1, and hypointense on T2 images on fetal MRI. The presence of enteroliths is a reliable predictor of urine mixing with meconium within the bowel lumen. Hydroureteronephrosis is commonly seen in conjunction with these conditions [1].

13.3 Imaging Modalities in the Postnatal Period

A detailed history and a careful clinical examination are the cornerstone in the postnatal evaluation of a child with DSD. Imaging remains the primary modality to assess the internal genital anatomy, mainly the uterus and the vagina. Ultrasound can be used for gonadal localization and assessing the nature of the gonads. Evaluation of other associated anomalies like renal and adrenal anatomy and malformations of the terminal spine is also facilitated by detailed imaging.

13.3.1 Ultrasonography

Ultrasonography is the primary modality for assessing the internal pelvic organs as also the inguinal, perineal, subperitoneal, and anal regions. It is widely available, can be performed without the need for sedation, and does not expose the child to ionizing radiation. The quality of information gained on a sonogram is highly operator dependent, a practical limitation. Both transabdominal and transperineal sonography are contributory; the latter providing a closer view of the urethral anatomy.

The identification of normal contoured or distended (hydrometrocolpos) Mullerian structures - upper third vagina, cervix, uterus and fallopian tubes on the USG aids in narrowing the diagnostic possibilities. Simultaneous characterization of gonads when combined with the available clinical details and karyotype, can clear the path to a specific diagnosis.

Neonatal uterus and ovaries are easily identified as a result of enlargement due to maternal hormonal stimulation. Neonatal ovaries may have multiple anechoic ovarian follicles and a volume of barely 1 ml [3]. The follicles are few and inconspicuous in the interim and appear again at puberty with an increase in gonadal volume to about 8 ml. USG can identify the uterus in 90% and ovaries in approximately 40% of cases. The echogenic endometrial stripe is distinctive of the uterus at all ages, though neonates have a larger cervix and the pubertal uterus a larger fundus. The prostatic utricle represents the Mullerian remnant in males and may be large in certain DSD (Fig. 13.1).

Normal testes appear as homogenous isoechoic or hyperechoic masses in newborns. They have a distinctive mediastinum testis and increase in volume from 1 ml at birth to 20 ml by puberty. The sensitivity (~75%) and specificity (~40%) of USG in the identification of testes remain low and several structures including lymph nodes are mistaken. Ovotestis is visualized as a heterogeneous structure with solid testicular echotexture juxtaposed to a relatively cystic ovarian part with follicles, the latter more evident at puberty. The arrangement may be at the ends of the gonadal structure (polar) or otherwise. Identification of streak gonads is usually not feasible

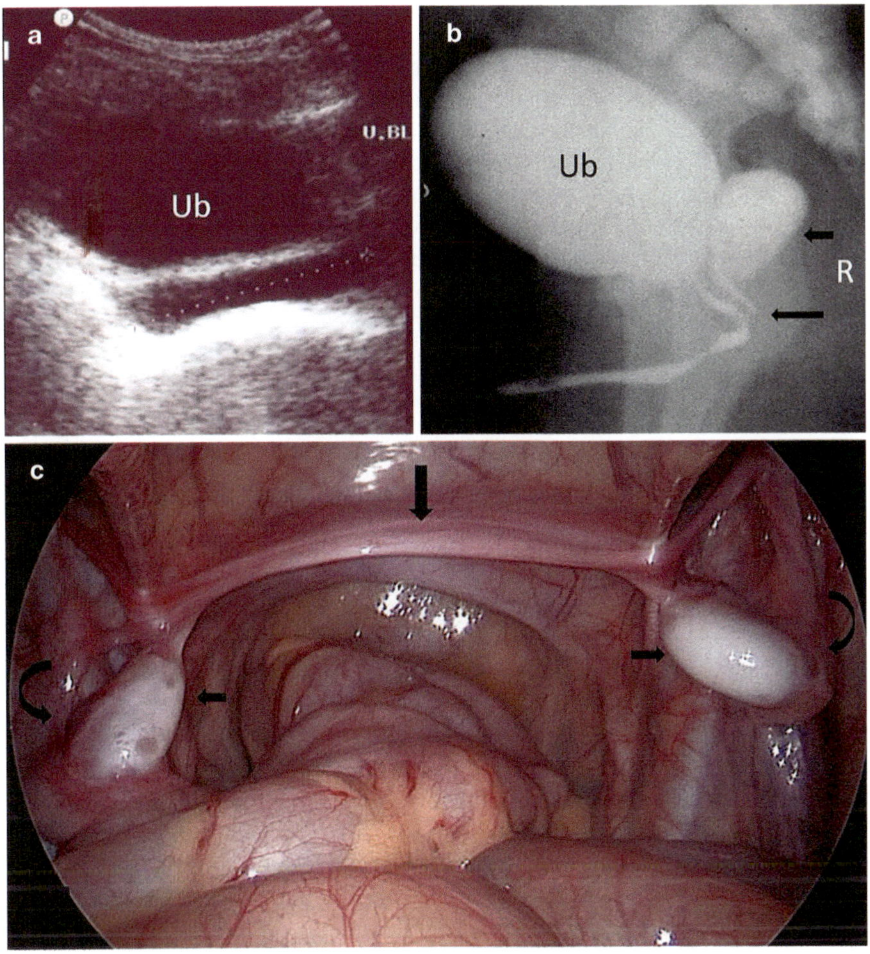

Fig. 13.1 Persistent Mullerian Duct Syndrome. A Mullerian derivative may be imaged on an ultrasonogram (**a**) as a thick-walled, linear cavity (dotted line) posterior to the bladder (Ub) In the voiding cystourethrogram (**b**), note the ovoid Mullerian remnant taking off from the prostatic urethra (long arrow) and wedged between the bladder (Ub) and the rectum (R). Laparoscopic view of the pelvis in an infant with bilateral undescended testis and normal penis showing intrabdominal testicular gonads (short arrows), Mullerian derivatives—uterus (long arrow), bilateral fallopian tube, and fimbriae(curved arrows)

on USG; they are small, elliptical, ectopic, and bland and composed largely of fibrous tissue [3, 4].

In addition, USG is an important tool in assessing the status of the kidneys and the adrenals. The presence of obstructive hydroureteronephrosis is common in cloacal anomalies. Many may have associated structural renal anomalies. The adrenal gland may or may not be enlarged in CAH. Altered adrenal texture, lobulated surface (cerebriform), and increase in adrenal limb length >20 mm and width >4 mm suggest a diagnosis of CAH [5].

Serial USG screening is also useful in patients of DSD with a propensity to develop gonadal, renal, or adrenal neoplasia. Children with dysgenetic gonads and a Y chromosome have an increased risk of developing gonadoblastomas. These tumors appear as echogenic calcified pelvic masses. Echogenic foci in inguinal and labioscrotal gonads suggest microlithiasis, are of controversial significance, and are followed up rigorously. Patients of DSD with concurrent glomerulopathy/aniridia will require long-term surveillance for developing Wilms' tumors.

13.3.2 Genitography

The assessment of the urethra and vaginal tracts and their relation to each other is essential in the management of children with DSD. Genitography is used to delineate the urethra, vagina, cervix, and uterus and the configuration of the urethrovaginal confluence. Performing a genogram requires determination and perseverance to obtain satisfactory results. It is best interpreted when performed by the treating surgeon in collaboration with a skillful radiologist.

The genitogram is a retrograde contrast study performed by injecting water-soluble contrast into the visible perineal orifices. If a single urogenital sinus orifice is present, lateral and oblique views with the hips flexed at 90° accurately delineate the urethrovaginal confluence level and its distance from the bladder neck and the external meatus. A sagittal image gives the clearest view. The level of the rectal confluence is noted in cloacal anomalies. If the urethral and vaginal orifices are separately seen, images are obtained after contrast is instilled into both orifices. Vaginal films may also help assess the anatomy of the cervix and uterus. The degree of virilization of the urethra is assessed by the ratio of the lengths of the horizontal anterior urethra to the vertical posterior urethra; the normal male ratio being 3:2. Presence of a large prostatic utricle and rectourethral communication may also be demonstrated [3].

Contrast spills during the study can be messy and lead to suboptimal images. Hence, it is important to obtain a good seal during instillation of the contrast. This can be achieved with a Foley catheter of appropriate size inserted into the meatus with the bulb inflated outside and pushed against the external meatus to prevent backflow of the contrast. Occasionally, direct injection with a blunt-nosed syringe and the hub pressed against the meatus can achieve a similar seal.

Shopfner's criteria described in 1964 are sometimes used to categorize urogenital sinus anatomy on a genogram. This classification system describes six types of perineal anatomy ranging from normal female type with separate urethral and vaginal orifices to normal /hypospadiac single perineal orifice with a normally virilized male urethra [6].

The genitourinary passages and confluences can also be visualized during a VCUG alone or when combined with a genitogram and provide adjunctive information about bladder size and morphology, presence of VUR, etc. which may contribute to the treatment plan. Generally, the catheter tip slips preferentially into the bladder; alternatively, a curved tip/coude catheter directs it toward the anteriorly placed urethral channel and bladder.

The genitogram is commonly a fluoroscopic one as described above; genitosonography with saline/contrast agents requires more expertise and less practice [7]. Finally, genitoscopy can add further refined and objective information to imaging the urogenital sinus anatomy and aid in management.

13.3.3 Magnetic Resonance Imaging

Magnetic Resonance Imaging (MRI) is not the first-choice modality of imaging in children with DSD. The superior soft tissue resolution, multiplanar image reconstruction, absence of ionizing radiation, and low operator dependence make it an attractive option. It is useful in older, cooperative children where it can be performed without anesthesia. The logistic difficulties of administering safe anesthesia during the MRI study make it less preferred in small children. Alternatively, a feed and wrap technique is used to obtain the study without sedation.

MRI is more sensitive than USG in detecting Mullerian derivatives. Its accuracy in detecting intrabdominal gonads is similar to that of USG but the localization of intrabdominal gonads is better done by MRI. A normal test is has a homogenous high signal intensity on T2-weighted and intermediate signal intensity on T1-weighted images Ectopic gonads, testes, and immature ovaries appear hyperintense on T2-weighted images. Streak or dysgenetic gonads, though as difficult to detect by MRI as in the USG, appear as hypointense stripes on T2-weighted images. MRI is accurate in diagnosing malignant transformation of streak gonads by demonstrating hyperintense signals on T2-weighted images within streak gonads [8].

MRI may be useful in imaging masses and tumors that appear concurrently in other organs in these patients, e.g., Wilms' tumors in patients of DSD with aniridia, glomerulopathy, and WT1 gene mutations [1, 3, 4].

13.4 Role of Endoscopy and Laparoscopy in Diagnosis

As discussed, the diagnosis of DSD is made on the basis of a good clinical assessment, genetic and biochemical investigations. Non-invasive imaging facilitates an accurate diagnosis and effective treatment planning in most cases. Endoscopy and laparoscopy are more invasive tools that require anesthesia and are employed to overcome the lacunae in information gleaned about the internal genitalia and gonads thus far and refine surgical treatment and long-term counseling. The process begins with a thorough examination under anesthesia—abdominal, genital, and per rectal and then proceeds to genitoscopy and laparoscopy as appropriate. Endoscopy or genitoscopy, in the context of the assessment of DSD, denotes direct visualization of the urogenital sinus, urethra, vagina, bladder, and rarely the rectum with the use of a cystourethroscope. Laparoscopy is the direct visualization of intrabdominal/

pelvic structures, especially the gonads and Mullerian derivatives. The expected yield of information from these tests, possible changes in the treatment plan and therapeutic options like gonadal biopsy or gonadectomy should be discussed with the patient and caregivers. This will reduce diagnostic surprises and the need for repeat anesthesia/procedures.

Given the current limitations of MRI and other imaging modalities in the assessment of intrapelvic structures in small children with DSD, laparoscopy remains the investigation of choice in this regard. Both MRI and laparoscopy require the administration of anesthesia in young children, but the more invasive laparoscopy offers a few advantages over MRI. Superior, direct, and magnified visualization allows accurate characterization of Mullerian structures, identification, and localization of gonads and streak gonads. Therapeutic procedures like gonadal biopsies, streak gonadectomies, orchidopexies, and excision of Mullerian structures can be performed simultaneously if indicated.

13.4.1 Genitoscopy

Genitoscopy is primarily used in the assessment of the lower urogenital tract and its malformations in children with DSD. It may be used to complement a genitogram or as an alternative with the advantage of providing more accurate and objective information [9]. An appropriate-sized video cystourethroscope with a working channel is used. The magnified view of the distended tubular structures allows a detailed examination of the urogenital sinus, urethra, vagina, and bladder and their spatial relationship [10].

Common indications for the genitoscopic assessment in DSD include severe/proximal hypospadias, a urogenital sinus in CAH, or cloacal anomalies. It allows for the assessment of the length and course of the native urethra, the verumontanum, and the prostatic utricle in a male urethra. Mullerian derivatives are located at the prostatic utricle on the verumontanum and may/may not have a cervical impression, the latter indicating a uterus cranially. They are encountered in ovotesticular DSD and gonadal dysgenesis. A pseudovaginal pouch is a urogenital sinus derivative located distal to the veru/site of the prostatic utricle as in some forms of 5AR deficiency. The presence of either a true Mullerian remnant or a urogenital sinus derivative in DSD influences the working diagnosis and directs investigation (Fig. 13.2). In severe hypospadias, a pseudovaginal pouch or a prostatic utricle is evaluated and suitably managed prior to urethroplasty; the latter invariably restricts access to the region thereafter.

Assessment of the urogenital sinus is mandatory before surgical reconstruction in CAH. Accurate measurements of the length of the urogenital sinus, the position of the vaginal confluence, and the length of the proximal urethra from the bladder neck to the vaginal confluence can be obtained. These are more accurately measured on the genitoscopy than the genitogram [11] and have been combined with the Prader scoring and phallic length in Rink's PVE classification (Annexure).

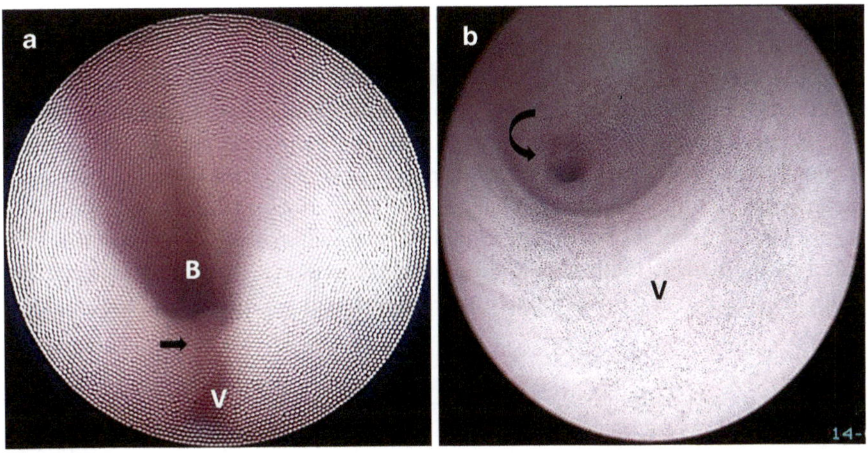

Fig. 13.2 Genitoscopy of a common urogenital sinus (**a**) showing the confluence/septum (arrow) between the anterior urethra (U) and posterior vagina (V). Note the transverse folds (**b**) in the capacious vagina, the cervical dimple (curved arrow) at its cranial end differentiates a true Mullerian remnant from a pseudovaginal pouch

The level of rectal confluence too can be assessed in patients with persistent cloacal anomalies. A cystogenitoscopy helps identify coexistent urological abnormalities like ectopic ureters, bladder trabeculations, urethral narrowing, diverticula, and rectourethral communication. It may be combined with an intraoperative retrograde contrast study to opacify smaller orifices not accessible to the scope. Placement of a catheter in the vagina under genitoscopic guidance aids intraoperative identification.

Genitoscopy is an invaluable tool in the assessment of postoperative results of genital reconstruction. It is a part of the diagnostic and therapeutic armamentarium in common sequels like vaginal stenosis and urethral strictures.

13.4.2 Laparoscopy

The structural evaluation of the internal genital organs is important. The presence or absence of gonads, their concordance or discordance with chromosomal gender, and the associated presence of Mullerian/Wolffian remnants are major determinants in the management of these patients.

Non-invasive modalities like USG and MRI provide useful information regarding internal genital ductal and gonadal anatomy but have their limitations. USG has a low sensitivity in identifying both intra-abdominal gonads and Mullerian structures while MRI is superior in both aspects. However, the identification and localization of dysgenetic streak gonads are equally poor with both modalities. Like MRI in small children, laparoscopy requires the administration of anesthesia but offers direct, magnified visualization of the internal pelvic organs. Apart from the accurate delineation of the genital ducts and gonads, the option of a simultaneous biopsy and therapeutic intervention is available [12].

Laparoscopy offers a superior identification of Mullerian structures and their relation to the ipsilateral gonad (Fig. 13.3). Localization of intra-abdominal gonads is accurately done by following the gonadal vessels from the lumbar area into the inguinal canal and various regions of the pelvis. A streak gonad is best visualized at laparoscopy and any neoplastic change is reliably assessed based on appearance (Fig. 13.4). The presence of an open inguinal ring with a potential inguinal hernia is clearly identified. A unilateral development of the uterus (hemiuterine horn) and fallopian tube or a vasdeferens in chromosomal DSD-like mixed gonadal

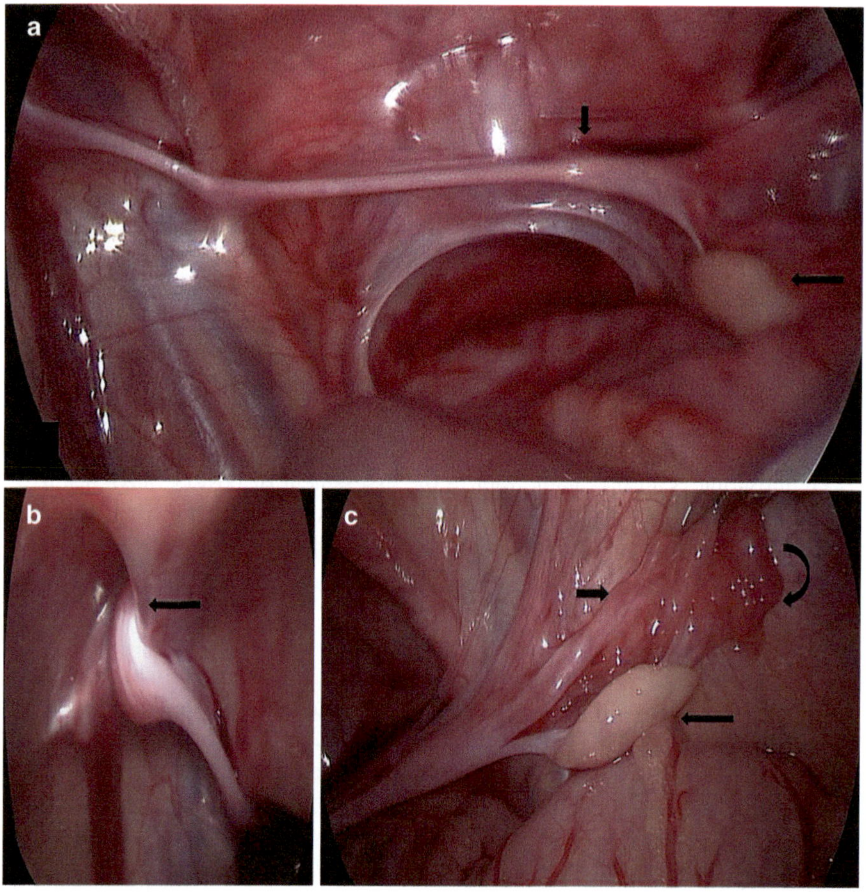

Fig. 13.3 Ovotesticular DSD: Laparoscopic view of the pelvis in an infant with atypical genitalia and a left inguinal hernia with a palpable gonad. The right half of the uterus was larger (**a**, long arrow). The fallopian tube entered the lt inguinal canal (arrow, **b**) with the vas and testicular gonad. In c, note the right ovary (long arrow) with adjacent fallopian tube (short arrow) and fimbria (curved arrow)

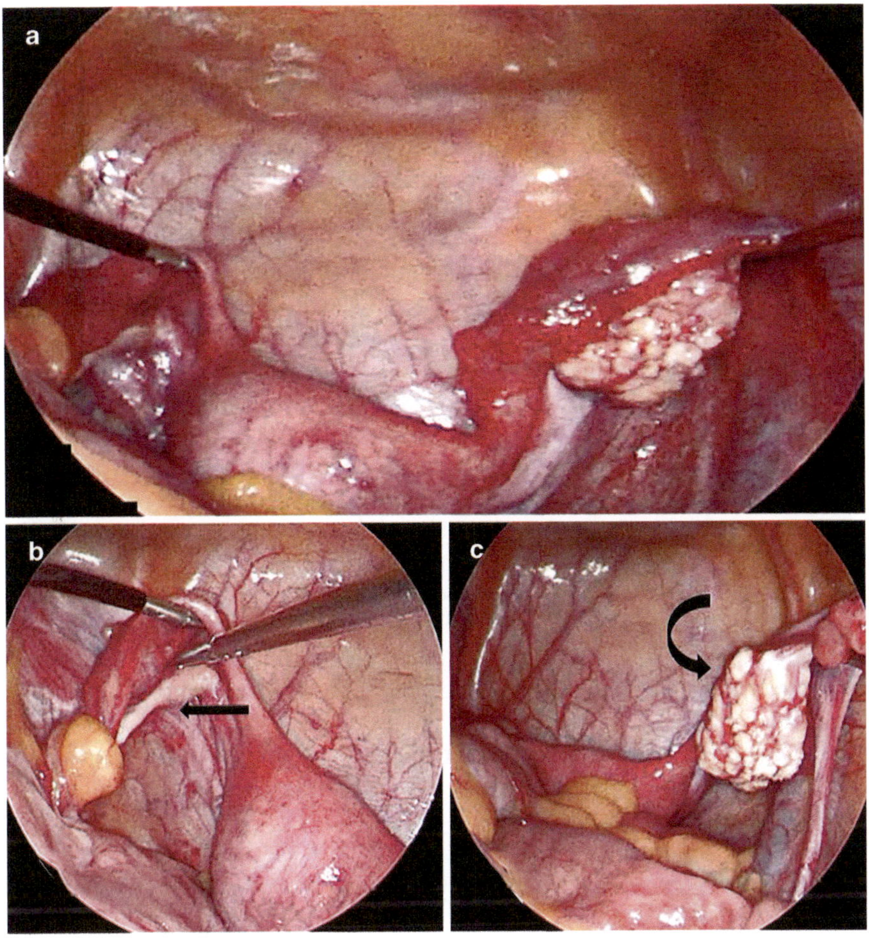

Fig. 13.4 Laparoscopic view of the pelvis (**a**) in a child with gonadal dysgenesis and bilateral streak gonads. Note the slender streak on the left side (**b**, arrow) and a calcific mass (**c**, arrow) replacing the rt. gonad (gonadoblastoma)

dysgenesis/ovotesticular DSD is noted. The visualization of intra-abdominal Wolffian structures is optimal with laparoscopy. Perhaps the most important for diagnosis is a gonadal biopsy in confusing cytogenetic reports (rare karyotypes/specific deletions/translocations, etc.), inconclusive preoperative imaging, and macroscopically indeterminate gonads—especially where ovotesticular DSD or gonadal dysgenesis is suspect. In ovotesticular DSD, depending on polar/hilar/other arrangements of heterogeneous gonadal tissue, longitudinal biopsies, or bipolar biopsies from representative areas of one or both gonads are performed. These may be done laparoscopically or laparoscopic-assisted, delivery per port open biopsy.

Ultrasonographic assistance and frozen section evaluation may be judiciously used. At all costs, the biopsy must be meticulous, representative, and preserve gonadal tissue till a decision is made for the removal of part/whole.

Apart from diagnostic capabilities, laparoscopy is widely used as an adjunctive therapeutic tool in children with DSD. Therapeutic interventions include gonadectomies, excision of Mullerian/Wolffian remnants, and unilateral or bilateral orchidopexies as indicated in the therapeutic plan. These are discussed in detail elsewhere [13, 14].

The obvious advantages offered by laparoscopy over imaging in the evaluation of intrabdominal internal genitalia in children with DSD have earned wide acceptance as a procedure of choice.

In conclusion, the evaluation of the internal genital anatomy remains an important part of the workup in a patient with DSD. It should ideally begin with a non-invasive and widely available screening tool like the USG, which in experienced hands can provide reliable information aiding diagnosis, counseling, and prognostication. More detailed information can be obtained by specific features on an MRI or genitography. Invasive investigations like genitoscopy and laparoscopy offer the advantage of superior visualization, high sensitivity and specificity, and simultaneous therapeutic intervention. These modalities require anesthesia and are undertaken after meticulous planning and discussion with the multidisciplinary team and the patient's family to combine diagnosis and definitive therapeutic procedures or genital reconstruction with minimal exposure to anesthesia and return to the operating room.

The combination of investigation modalities is carefully tailored so that most information is obtained as quickly and cost-effectively as possible with minimal possible morbidity.

Annexure 1

The PVE Classification of Urogenital Sinus (UGS) Anatomy

The PVE classification of UGS anatomy is aimed at an objective element to the classical Prader classification and aids in a more informed surgical treatment plan.

The elements measured in the PVE classification are:

P—The *P*hallic length.
V—The measured position of the *V*aginal confluence from both the external UGS meatus and bladder neck.
E—The *E*xternal genital appearance.

P—Stretched Phallic length (flaccid phallus) measured at the time of genitoscopy. Expressed in centimeters to the nearest 0.1 cm.

V—The distance of the vaginal confluence is measured from the external meatus and from the bladder neck with the aid of a calibrated ureteral catheter passed alongside the cystourethroscope into the bladder.

Expressed as $V_{x,y}$ where x stands for the distance of vaginal confluence from bladder neck and y, the distance from the UGS external meatus.

E—As in the Prader classification from Types E 1–5 with the addition of a Type E 0 for a normal phenotype and a Type E 6 for a rare or uncharacterizable phenotype.

Reference: Rink R.C, Adams M.C, Misseri R. A new classification for genital ambiguity and urogenital sinus anomalies. BJU Int. 2005 Mar;95(4):638–42

References

1. Goncalves LF, Hill H, Bailey S. Prenatal and postnatal imaging techniques in the evaluation of disorders of sex development. Semin Pediatr Surg. 2019;28(5):150839. https://doi.org/10.1016/j.sempedsurg.2019.150839.
2. Soriano D. Development of the fetal uterus between 19 and 38 weeks of gestation: in-utero ultrasonographic measurements. Hum Reprod. 1999;14(1):215–8.
3. Moshiri M, Chapman T, Fechner PY, Dubinsky TF, Shnorhavorian M, Osman S, et al. Evaluation and management of disorders of sex development: multidisciplinary approach to a complex diagnosis. Radiographics. 2012;32:1599–618.
4. Guerra-Junior G, Andrade KC, Barcelos IHK, Maciel-Guerra AT. Imaging techniques in the diagnostic journey of disorders of sex development. Sex Dev. 2018;12:95–9.
5. Eapen A, ChandramohanA SB, PuttaT JR, Kekre A. Imaging evaluation of disorders of sex development. Journal of gastrointestinal and abdominal. Radiology. 2020;3:181–92.
6. Shopfner CE. Genitography in intersexual states. Radiology. 1964;82(4):664–74.
7. Hryhorczuk AL, Phelps AS, Yu RN, Jeanne S, Chow JS. The radiologist's role in assessing differences of sex development. Pediatr Radiol. 2022;52(4):752–64.
8. Jennifer K, Son JK, Ali S, Al Khori N, Lee EL. MR imaging evaluation of pediatric genital disorders: MR technologic overview and interpretation. Magn Reson Imaging Clin N Am. 2019;27(2):301–21.
9. Ahmed SF, Achermann J, Alderson J, Crouch NS, Elford S, Hughes IA et al. Society for Endocrinology UK Guidance on the initial evaluation of a suspected difference or disorder of sex development (Revised 2021). Clin Endocrinol (Oxf). 2021;95(6):818–40.
10. Tafazzoli K, Wünsch L, Bouteleux M, Lindert J, Schulz T, Birnbaum W, Marshall L, Hiort O, Tüshaus L. Endoscopy and laparoscopy in disorders of sex development. Sex Dev. 2018;12:100–5.
11. Rink RC, Adams MC, MisseriR. A new classification for genital ambiguity and urogenital sinus anomalies. BJU Int. 2005;95(4):638–42.
12. Steven M, O'Toole S, Lam JPH, MacKinlay GA, Cascio S. Laparoscopy versus ultrasonography for the evaluation of Mullerian structures in children with complex disorders of sex development. Pediatr Surg Int. 2012;28(12):1161–4.
13. Burgmeier C, de Leriche C. Laparoscopy in the surgical treatment of disorders of sexual development. J Laparoendosc Adv Surg Tech A. 2016;26(9):730–3.
14. Bailez MM, Costanzo M, GuercioG. Role of minimally invasive surgery (MIS) in different sexual development (DSD). Semin Pediatr Surg. 2021;30(4):151078.

Endocrine Profiling in DSD

14

Belinda George and Ganapathi Bantwal

Abstract

Differences in sex development (DSD) are conditions in which there is an inconsistency between chromosomal, gonadal, and genital/anatomical development. Among the tools available for their evaluation, endocrine profiling is the least reliable owing to the dynamic nature of hormone secretion and limitations in the currently used methods for assay. Nevertheless, when employed appropriately, it provides immense insight into the possible etiology. With the introduction of liquid chromatography-tandem mass spectroscopy, the assay sensitivity and specificity have improved significantly; however, a standardization of laboratory values and normal reference ranges is needed. The source of androgens in humans is either the adrenal gland or the gonads and the pathology of most DSDs originates from either of these. In suspected isolated hypofunctioning of the testes, the work-up focuses on the assessment of the presence and functionality of testicular tissue. As the steroidogenic pathway is similar in the adrenal and the gonads, deficiency of any of the enzymes involved can present with both gonadal and adrenal failure. If adrenal involvement is evident, a complete adrenal steroid profile is the preferred approach. In a 46 XX infant with atypical genitalia, the commonest etiology being 21 hydroxylase deficiency, the initial work-up is geared toward ruling out adrenal insufficiency and confirming hyperandrogenism. In this chapter, we discuss common clinical scenarios at birth, beyond mini-puberty and at puberty; and suggest appropriate timing and choice of hormonal evaluation.

Keywords

Hormonal profile · Adrenal · Gonad

B. George (✉) · G. Bantwal
Department of Endocrinology, St. John's Medical College Hospital, Bengaluru, India

S. K. Ratan (ed.), *Children with Differences in Sex Development*,
https://doi.org/10.1007/978-981-97-1639-5_14

14.1 Introduction

Differences in sex development (DSD) are defined as conditions in which there is an inconsistency between chromosomal, gonadal, and genital/anatomical development. DSDs are complex genetic disorders with a varied clinical presentation that poses significant diagnostic and management challenges for the clinician [1]. Overlapping phenotypes, inaccessible and unreliable hormonal profiling, genetic heterogeneity, and complex psychosocial issues involving gender assignment and the need for surgical intervention make the evaluation and management of this disorder extremely difficult and challenging [2].

In females (46 XX), the presentation of DSD is usually due to the effect of excess androgens leading to the virilization of external genitalia. In males (46 XY), the presentation is usually due to total or partial absence of androgen effect leading to the under-virilized external genitalia. The source of androgens in humans could be either the adrenal gland or the gonad; hence, the majority of the DSDs represent a pathology occurring either in the adrenal gland (especially in females) or the gonads. DSDs may present at birth with obvious ambiguity of genitalia or later in childhood with premature adrenarche or at puberty with the absence of or poor development of secondary sexual characters or with contra-sexual pubertal development. The phenotype of atypical genitalia can range from mild hypospadias or isolated clitoromegaly to complete male or female genitalia with a discordant karyotype. A thorough history, physical examination, imaging studies, and cytogenetic tests including karyotype and FISH for the presence of the Y chromosome are mandatory initial tools that have been discussed elsewhere in the book. In this chapter, we discuss the utility of hormonal profiling along with its limitations and appropriate interpretation that enables and guides us to an accurate diagnosis of the etiology of DSD.

14.2 Physiology

14.2.1 Adrenal Steroidogenesis

An understanding of the androgen biosynthetic pathway that occurs in the adrenal gland and gonads is essential to accurately assess and interpret the hormonal profile seen in these disorders. Figure 14.1 illustrates the simple steroidogenesis pathway with multiple enzymes that are involved in the pathogenesis of DSD. Complete or partial deficiency of any of these enzymes can lead to DSD [3]. The commonest cause for DSD, especially in 46 XX females, is 21 hydroxylase enzyme (21OH) deficiency which leads to the accumulation of precursors like 17 hydroxy progesterone, which then becomes the substrate for increased androgen synthesis and production leading to virilization of female external genitalia [4]. As the machinery for androgen biosynthesis is similar in both adrenals and gonads, deficiency of any of these enzymes in 46 XY males including Star, side chain cleavage, 17 alpha-hydroxylase, P 450 oxidoreductase, and 3 beta-HSD can all lead to reduced androgen synthesis and under virilization of male genitalia.

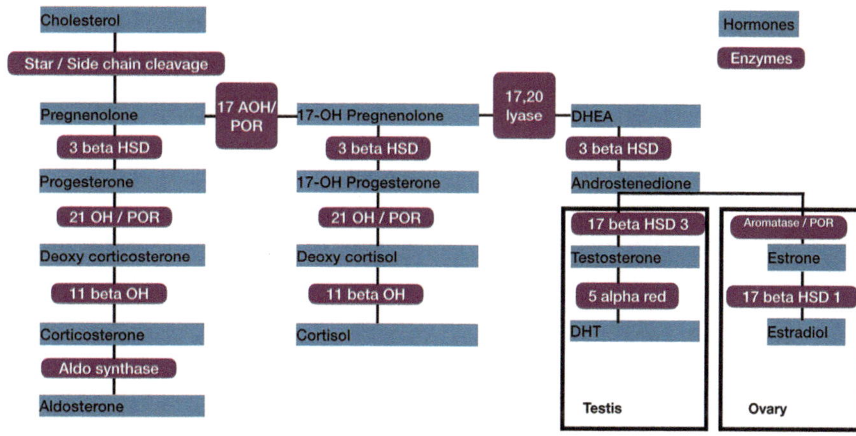

Fig. 14.1 Steroidogenesis pathway in adrenas and gonads

14.2.2 Gonadal Function and Mini-puberty

The hypothalamic–pituitary–gonadal (HPG) axis is quiescent at birth and the levels of follicle-stimulating hormone (FSH), luteinizing hormone (LH), and testosterone are low at birth. These levels increase by the first week of life and peak at around 3 months of age before becoming undetectable again by 6 months of life. This period between 1 week and 6 months of life is referred to as "mini-puberty"; this time-frame provides an opportunity to assess the functioning of the HPG axis without stimulation [5]. Figure 14.1 illustrates the steroidogenesis pathway including the final steps that occur only in the testes, leading to the formation of testosterone and dihydrotestosterone (DHT), the predominant androgens in circulation. DHT is the active androgen that acts upon the androgen receptor bringing about the effects of androgenization and is necessary for the complete development of male external genitalia. Among the 46 XY DSDs, the most common cause is androgen insensitivity syndrome, where the androgen receptor is resistant to the action of both testosterone and DHT [6]. 5 Alpha-reductase enzyme deficiency which catalyzes the conversion of testosterone to DHT is a close differential diagnosis.

14.3 Hormonal Assessment

I. *At birth*

A. *46 XX DSD*

A newborn presenting with genital ambiguity, no palpable gonads, and the presence of Mullerian structures on imaging will likely have a 46 XX DSD, with 21 hydroxylase deficiency CAH being the most probable cause. Hormonal assessment in these cases must be done at 48–72 h of life (to avoid the adrenal hormone surge seen at birth). As the main substrate for the 21 hydroxylase enzyme is 17 hydroxy progesterone (17-OHP), this is assessed to rule out 21 hydroxylase deficiency. Plasma renin levels may also be tested. An additional serum sample should ideally be collected and saved before therapy is initiated. If the results are not suggestive of 21 hydroxylase deficiency or another enzyme deficiency is suspected based on the clinical manifestations, additional steroid hormone precursors may be tested as a panel. Many programs for newborn screening also include serum 17-OHP to rule out CAH, as this would enable early detection and prevent a possibly fatal adrenal crisis [7]. This strategy is particularly useful in male infants as genital ambiguity will not be in evidence to alert one toward the diagnosis. 21-Deoxycortisol (produced by 11b-hydroxylation of 17OHP) is normally not secreted in large amounts (even in preterm infants), and thus elevated levels are highly specific for 21OHD. This test is being recommended instead of 17-OHP to diagnose 21 hydroxylase deficiency. if available, 21 deoxy cortisol levels can be used to confirm CAH in those with elevated 17-OHP levels on screening. The presence of maternal hormones is generally protective in the first 1–2 weeks of life and most infants with CAH present in the second week of life with failure to thrive and dyselectrolytemia (hyponatremia and hyperkalemia) suggestive of cortisol and aldosterone deficiency. An ACTH stimulation test is necessary in the absence of dyselectrolytemia to confirm cortisol deficiency.

B. *46 XY DSD*

A newborn presenting with genital ambiguity, palpable gonads, and no Mullerian structures on imaging will likely have a 46 XY DSD. FISH for Y chromosome/karyotyping may also be helpful in identifying these infants, especially in those with undescended or absent testes. FSH, LH, and testosterone should be measured in these cases to assess testicular function. The ideal period for assessing testicular function in infants should match the timing of mini-puberty, which ranges from the second week of life to 6 months of age.

If FSH, LH, and testosterone are detectable, additional androgens including androstenedione and DHT should be measured to rule out 17 beta-HSD 3 and 5 alpha-reductase deficiency, respectively. An elevated testosterone (T) level with a low T:DHT ratio is suggestive of androgen insensitivity syndrome [8].

Markedly elevated FSH and LH levels with undetectable testosterone are suggestive of primary testicular dysfunction, gonadal dysgenesis, or anorchia. Measurement of anti-Mullerian hormone (AMH) and inhibin B could be additional diagnostic tests that are useful in this scenario [9].

If adrenal involvement is suspected based on a clinical picture (hyperpigmentation, hyponatremia, hyperkalemia, hypotension), adrenal profiling is added. Plasma glucose, serum electrolytes, cortisol, and 17-OHP levels are the first-line investigations in this scenario. Paired adrenocorticotropin hormone (ACTH) and cortisol levels evaluate for cortisol deficiency. In addition, an entire steroid panel including pregnenolone, 17-OH pregnenolone, progesterone, 17-OHP, 11-Deoxy cortisol, and dehydroepiandrosterone sulfate (DHEAs) if feasible, will help ascertain the level of the blockade in the steroidogenesis pathway [10]. The defect can be anywhere starting from the LH/HCG receptor culminating at the androgen receptor in the end with all the enzyme deficiencies placed in between (Star, Side chain cleavage, 17 alpha-hydroxylase, 17–20 lyase, P450 oxidoreductase, 3 beta-HSD type 2, 17 beta-HSD type 3, 5 alpha-reductase and aromatase).

Undescended testes and micropenis with a urethral opening at the tip of the phallus and normal scrotal development may be suggestive of hypogonadotropic hypogonadism, which is not categorized as DSD. Evaluation should focus on pituitary hormone assessment including FSH, LH, and testosterone beyond 2 weeks of life.

C. *45X/46XY DSD (mixed gonadal dysgenesis)*

If the karyotype reveals 45X/46XY mosaicism, and/or if there is marked asymmetry in the appearance of the external genitalia, the diagnosis could be mixed gonadal dysgenesis [11]. The functioning status of the gonad needs to be assessed using FSH, LH, testosterone, inhibin B, and AMH levels.

II. *Beyond Mini-Puberty*

In several countries, it is not uncommon for children with atypical genitalia present since birth to seek medical attention at a later date. The approach to endocrine profiling is similar to those cases that have been detected at birth. 46 XX children without gonads need adrenal profile testing including serum electrolytes, 17-OHP, ACTH, and Cortisol. As mini-puberty would have subsided beyond 6 months, assessment of testicular function in 46 XY children can only be done after stimulating the HPG axis using human chorionic gonadotropin (HCG) hormone [12].

A. *Premature Adrenarche*

Girls with simple virilizing forms of CAH that were missed at birth could present with hyperandrogenism. Boys with a simple virilizing form of CAH could also present in childhood with the development of pubic hair, growth spurt, and advanced bone age [13]. Adrenal profile including serum electrolytes, 17-OHP, ACTH, and cortisol could be measured in these cases, with an additional steroid panel if available in selected situations.

B. *Bilateral Inguinal Hernias (female babies)*

Girls presenting with bilateral inguinal hernias should always be examined for the presence of gonads in the inguinal region. Complete androgen insensitivity syndrome (CAIS) can rarely present in this age group with bilateral inguinal masses [14]. Basal or stimulated LH, FSH, and testosterone along with karyotyping are usually adequate for confirming the diagnosis. Additional androgens including androstenedione and DHT could be useful, especially if there is mild ambiguity on thorough clinical examination to rule out partial androgen insensitivity syndrome (PAIS).

C. *Transverse testicular ectopia*

These children may seek medical attention for undescended testes, unilateral inguinal hernia with gonads within, or rarely, testicular torsion [15]. Transverse testicular ectopia (TTE) usually occurs in cases of Persistent Müllerian Duct Syndrome (PMDS); evaluation should include AMH levels to assess for deficiency of AMH or resistance to AMH.

III. *Puberty*

A. *Virilization at Puberty*

Children with 46 XY DSD reared as females may present with virilization at puberty, a typical presentation of 5 alpha-reductase deficiency. Other differentials include partial androgen insensitivity syndrome (PAIS), 17 beta-HSD 3 deficiency, and partial gonadal dysgenesis. Androgen-secreting adrenal and gonadal tumors may also present with virilization; evaluation to rule out these sinister conditions may be necessary if clinical suspicion is high. Biochemical testing should include FSH, LH, testosterone, DHT, and androstenedione. Additional tests like DHEAs and tumor markers may be done if an adrenal tumor or malignant transformation of a dysgenetic gonad is suspected, respectively [16].

B. *Feminization At Puberty*

In a child reared as a male, feminization at puberty in the form of breast development could suggest androgen insensitivity syndrome or ovotesticular DSD [17]. The presence of cyclical abdominal pain or hematuria will provide additional clues to the presence of functioning ovarian and endometrial tissue. Biochemical testing should include FSH, LH, testosterone, and estradiol levels; HCG stimulation to assess testicular component and FSH stimulation to assess ovarian component could be done if the child is still in the pre-pubertal phase. In childhood, the presence of atypical genitalia with/without asymmetrical external genital appearance and the presence of Mullerian structures (including hemi-uterus) should raise suspicion of ovotesticular DSD. Ovotesticular DSDs can present with either 46XX, 46XX/46XY, or 46XY karyotype.

C. *Primary Amenorrhea*

The most common cause for primary amenorrhea with sex chromosome DSD is Turner's syndrome, where they present with short stature, typical skeletal features, and the absence of secondary sexual characteristics. Rarely, primary amenorrhea could be the first manifestation of a 46 XY DSD [18], the differentials being CAIS, pure gonadal dysgenesis, and 17 alpha-hydroxylase deficiency. Biochemical testing should include FSH, LH, testosterone, estradiol, and serum electrolytes (particularly with hypertension). Adrenal profile including ACTH, cortisol, renin, and the entire steroid panel may be done if 17 alpha-hydroxylase is suspected clinically.

IV. *Infertility*

Adults with DSD and normal-appearing genitalia may present later in life with infertility. The most common etiology for males presenting with infertility and azoospermia is Klinefelter syndrome, a sex chromosome DSD. Occasionally, the karyotyping may reveal 46 XX testicular DSD in otherwise normal-appearing males presenting with azoospermia and infertility [19]. Biochemical testing should include FSH, LH, and testosterone for these scenarios. Rarely, a child with 46 XX simple virilizing CAH reared as male may present in adulthood with infertility: 17-OHP or complete adrenal profile confirms the diagnosis.

14.4 Interpreting Hormone Assays

Steroid hormones can be measured either using liquid chromatography/tandem mass spectrometry (LC-MS) or the more widely used immunoassay. The European Cooperation in Science and Technology recommends the use of either of these methodologies in the diagnosis and follow-up of DSDs [20]. LC-MS has better specificity when compared to immunoassay, and it can also analyze multiple metabolites in the same sample. Immunoassays also suffer from a high degree of cross-reactivity between various steroid precursors, particularly when they are elevated as is seen in DSDs. Hence, the preferred methodology when evaluating DSDs is LC-MS, if available. However, this is not widely available, is expensive, cumbersome for laboratories, needs expert personnel, and suffers from a lack of standardization. Most of the cut-offs mentioned in guidelines and reviews are applicable to immunoassays. An understanding of the methodology that is being used and available is essential for accurate interpretation of the hormonal assays. Apart from the methodology, the timing of the assay is also important in interpretation, e.g., a test done during the quiescent phase of puberty should not be misinterpreted as poor testicular function and gonadal failure. The clinical phenotype, patient's chronological age, and other ancillary investigations should be considered in toto while interpreting these dynamic hormonal assays.

14.5 Tests for Assessing Testicular Function

FSH, LH, AMH, inhibin B, testosterone, and DHT are measured as the first-line investigations for the evaluation of 46 XY DSD to assess the presence and functionality of testicular tissue. In primary testicular failure, FSH and LH levels are usually markedly elevated. A low FSH and LH levels do not rule out testicular dysfunction particularly if the child is outside the mini-puberty period. AMH and inhibin B are two hormones produced by the Sertoli cells that are excellent markers of Sertoli cell function during childhood. AMH is secreted exclusively by the Sertoli cells, whereas inhibin B is secreted predominantly by Sertoli cells and other testicular cell types. The concentration of both AMH and inhibin B are low at birth and thereafter increase gradually till the second year of life [21]. AMH levels remain stable for the rest of childhood and start reducing once the child enters puberty. Inhibin B concentrations, on the other hand, decline after the second year of life and tend to increase when the child enters puberty. Blood levels of another hormone called insulin-like 3 protein (INSL-3) reflect Leydig cell function.

If baseline testosterone levels are not detectable, and gonadotropin levels are not markedly elevated, HCG stimulation is advisable to assess and confirm testicular dysfunction. HCG is administered at doses of 1000–1500 units per day subcutaneously or intramuscularly (daily or every other day) for one to 5 days. Blood sampling for testosterone is done 24 h after the last injection. Hormone assessments post-HCG should include testosterone, dihydrotestosterone, and androstenedione. The HCG test not only helps us determine if testicular tissue is present and

Table 14.1 Hormonal assessment in common 46 XY DSDs

	FSH	LH	Testosterone (basal or stimulated)	DHT	AMH	Steroid panel
Gonadal dysgenesis	Markedly elevated	Elevated	Low	Low	Low	Low
LHCGR defect	Elevated/ Normal	Markedly elevated	Low	Low	Normal	Low
Adrenal steroidogenic defects	Normal	Elevated	Low	Low	Normal	Variable based on level of defect
17 beta HSD 3 deficiency	Normal	Elevated	Low	Low	Normal	T:A < 0.8
5 alpha reductase deficiency	Normal	Normal	Normal/ Elevated	Usually low	Normal	T:DHT > 10
PAIS	Normal/ Elevated	Markedly elevated	Normal/ Elevated	Normal	Normal	Normal

functional, but it also helps us to decide the etiology in case of 46 XY DSD which includes 17 beta-HSD 3 deficiency, 5 alpha-reductase deficiency, and partial androgen insensitivity syndrome (Table 14.1). A testosterone value of 1.58 nmol/L (0.45 ng/ml) following HCG administration has excellent predictive value to suggest the presence of functional testicular tissue [9]. A T:DHT ratio more than 10:1 is suggestive of 5 alpha-reductase deficiency; and a testosterone to androstenedione (T:A) ratio less than 0.8 is suggestive of 17 beta-HSD 3 deficiency [6].

14.6 Adrenal Profile

The classic picture seen in 21 hydroxylase deficiency, which is the most common form of 46 XX DSD, is elevated levels of all precursors in the serum including 17-OHP, androstenedione, and testosterone. In all patients with suspected 46 XX DSD presenting with ambiguous genitalia, it is essential to rule out life-threatening congenital adrenal hyperplasia. Plasma glucose, serum electrolytes, and 17-OHP are the first-line investigations to be sent in this scenario. As there is an increase in adrenal hormone synthesis at birth, measurement of 17-OHP should be delayed for at least 48–72 h after birth. If congenital adrenal hyperplasia is suspected, additional serum samples should be collected before initiating the child on steroid replacement therapy. Additional hormonal assays that can be ordered include other precursors of the steroidogenesis pathway like deoxy cortisol, pregnenolone, progesterone, and their 17-hydroxylated products. Renin is a useful test in patients with suspected congenital adrenal hyperplasia with cortisol deficiency. Paired ACTH and cortisol levels or an ACTH-stimulated cortisol level are useful in identifying the presence of adrenal insufficiency. There is no clear normative data for serum cortisol reference range at birth and early infancy, hence a paired sample with unequivocally elevated

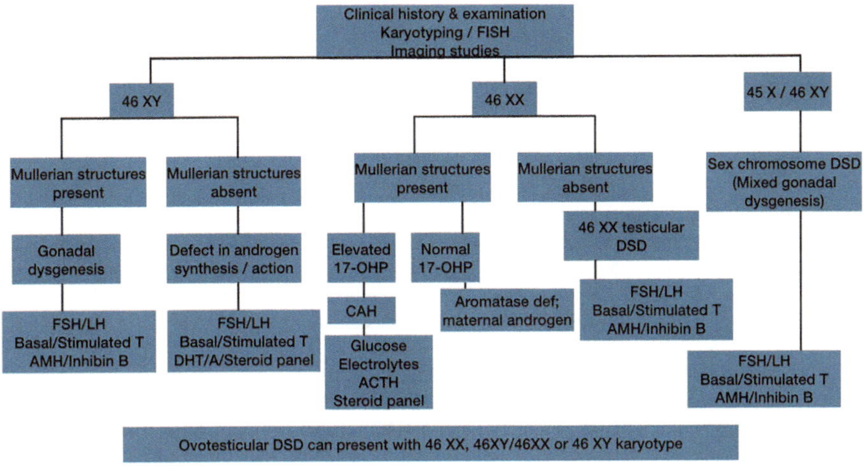

Fig. 14.2 Approach to infant/child with atypical genitalia

ACTH is highly suggestive of adrenal insufficiency. The cut-off values for all the above tests will vary depending on the child's chronological age, presence of premature birth, and timing of testing. 17-OHP values above 10 ng/ml are highly suggestive and diagnostic of 21-hydroxylase enzyme deficiency [4]. Other differentials including aromatase deficiency, maternal androgen exposure, and 46 XX testicular disorders are rarely encountered in clinical practice (Fig. 14.2).

14.7 Conclusion

Among the investigative modalities available for the diagnostic evaluation of DSDs, hormonal profiling is the least accurate and least reliable tool in its current form. Endocrine profiling comes with a lot of caveats. The hormonal values can show wide physiological fluctuations depending on fetal maturity, circadian rhythms (which are poorly established in the neonatal period), and stress associated with birth, making their interpretation extremely challenging. Accurate interpretation of hormonal assays also necessitates a clear understanding of the limitations of assays, and the patient's clinical scenario with regard to the presence of mini-puberty/puberty at the time of sampling. Finally, putting the overall clinical picture together including history, physical examination, imaging, and cytogenetic studies along with the hormonal assays will help us to arrive at an accurate diagnosis. Therefore, though endocrine profiling should be carried out in all cases of suspected DSD, their interpretation should be cautiously undertaken. Standardization of LC-MS testing and reporting will certainly improve the diagnostic accuracy of currently available assays; clear and accurate capture of the clinical phenotype is highly essential for ensuring appropriate interpretation.

References

1. Lee PA, Nordenström A, Houk CP, Ahmed SF, Auchus R, Baratz A, Baratz Dalke K, Liao LM, Lin-Su K, Looijenga LH 3rd, Mazur T, Meyer-Bahlburg HF, Mouriquand P, Quigley CA, Sandberg DE, Vilain E, Witchel S; Global DSD Update Consortium. Global Disorders of Sex Development Update since 2006: perceptions, approach and care. Horm Res Paediatr. 2016;85(3):158–80. https://doi.org/10.1159/000442975
2. León NY, Reyes AP, Harley VR. A clinical algorithm to diagnose differences of sex development. Lancet Diabetes Endocrinol. 2019;7(7):560–74. https://doi.org/10.1016/S2213-8587(18)30339-5
3. Miller WL, Auchus RJ. The molecular biology, biochemistry, and physiology of human steroidogenesis and its disorders. Endocr Rev. 2011;32(1):81–151. https://doi.org/10.1210/er.2010-0013
4. Claahsen-van der Grinten HL, Speiser PW, Ahmed SF, Arlt W, Auchus RJ, Falhammar H, Flück CE, Guasti L, Huebner A, Kortmann BBM, Krone N, Merke DP, Miller WL, Nordenström A, Reisch N, Sandberg DE, Stikkelbroeck NMML, Touraine P, Utari A, Wudy SA, White PC. Congenital adrenal hyperplasia-current insights in pathophysiology, diagnostics, and management. Endocr Rev. 2022;43(1):91–159. https://doi.org/10.1210/endrev/bnab016.
5. Kurtoğlu S, Baştuğ O. Mini puberty and its interpretation. Turk Pediatri Ars. 2014;49(3):186–91. https://doi.org/10.5152/tpa.2014.2038.
6. Wisniewski AB, Batista RL, Costa EMF, Finlayson C, Sircili MHP, Dénes FT, Domenice S, Mendonca BB. Management of 46,XY differences/disorders of sex development (DSD) throughout life. Endocr Rev. 2019;40(6):1547–72. https://doi.org/10.1210/er.2019-00049.
7. Grosse SD, Van Vliet G. Challenges in assessing the cost-effectiveness of newborn screening: the example of congenital adrenal hyperplasia. Int J Neonatal Screen. 2020;6(4):82. https://doi.org/10.3390/ijns6040082.
8. Gulía C, Baldassarra S, Zangari A, Briganti V, Gigli S, Gaffi M, Signore F, Vallone C, Nucciotti R, Costantini FM, Pizzuti A, Bernardo S, Porrello A, Piergentili R. Androgen insensitivity syndrome. Eur Rev Med Pharmacol Sci. 2018;22(12):3873–87. https://doi.org/10.26355/eurrev_201806_15272.
9. Jespersen K, Ljubicic ML, Johannsen TH, Christiansen P, Skakkebaek NE, Juul A. Distinguishing between hidden testes and anorchia: the role of endocrine evaluation in infancy and childhood. Eur J Endocrinol. 2020;183(1):107–17. https://doi.org/10.1530/EJE-20-0041.
10. Sánchez-Guijo A, Oji V, Hartmann MF, Traupe H, Wudy SA. Simultaneous quantification of cholesterol sulfate, androgen sulfates, and progestagen sulfates in human serum by LC-MS/MS. J Lipid Res. 2015;56(9):1843–51. https://doi.org/10.1194/jlr.D061499
11. Das DV, Jabbar PK. Clinical and reproductive characteristics of patients with mixed gonadal dysgenesis (45,X/46, XY). J Obstet Gynaecol India. 2021;71(4):399–405. https://doi.org/10.1007/s13224-021-01448-3.
12. Honour JW. Diagnosis of diseases of steroid hormone production, metabolism and action. J Clin Res Pediatr Endocrinol. 2009;1(5):209–26. https://doi.org/10.4274/jcrpe.v1i5.209
13. Bonfig W, Schwarz HP. Growth pattern of untreated boys with simple virilizing congenital adrenal hyperplasia indicates relative androgen insensitivity during the first six months of life. Horm Res Paediatr. 2011;75(4):264–8. https://doi.org/10.1159/000322580.
14. Tyutyusheva N, Mancini I, Baroncelli GI, D'Elios S, Peroni D, Meriggiola MC, Bertelloni S. Complete androgen insensitivity syndrome: from bench to bed. Int J Mol Sci. 2021;22(3):1264. https://doi.org/10.3390/ijms22031264.
15. Mukhtar MU, Niazi SAK, Sarwar MZ, Naqi SA. Transverse testicular ectopia with persistent Mullerian duct syndrome: an operative eureka. Int J Surg Case Rep. 2020;71:338–340. https://doi.org/10.1016/j.ijscr.2020.04.035

16. Santi M, Graf S, Zeino M, Cools M, Van De Vijver K, Trippel M, Aliu N, Flück CE. Approach to the virilizing girl at puberty. J Clin Endocrinol Metab. 2021;106(5):1530–9. https://doi.org/10.1210/clinem/dgaa948.
17. Syryn H, Van De Vijver K, Cools M. Ovotesticular difference of sex development: genetic background, histological features, and clinical management. Horm Res Paediatr. 2021:275–284. https://doi.org/10.1159/000519323
18. Eren E, Saglam H, Cakir ED, Tarim O. Etiological evaluation of adolescents with primary amenorrhea. Indian J Pediatr. 2014;81(9):861–5. https://doi.org/10.1007/s12098-013-1266-6.
19. Sharma A, Minhas S, Dhillo WS, Jayasena CN. Male infertility due to testicular disorders. J Clin Endocrinol Metab. 2021;106(2):e442–59. https://doi.org/10.1210/clinem/dgaa781.
20. Kulle A, Krone N, Holterhus PM, Schuler G, Greaves RF, Juul A, de Rijke YB, Hartmann MF, Saba A, Hiort O, Wudy SA, Action EUCOST. Steroid hormone analysis in diagnosis and treatment of DSD: position paper of EU COST Action BM 1303 'DSDnet'. Eur J Endocrinol. 2017;176(5):P1–9. https://doi.org/10.1530/EJE-16-0953.
21. Josso N, Rey RA. What does AMH tell us in pediatric disorders of sex development? Front Endocrinol (Lausanne). 2020;11:619. https://doi.org/10.3389/fendo.2020.00619.

Pathology of Differences in Sex Development

15

Suravi Mohanty and Uttara Chatterjee

Abstract

Various DSD diagnoses are usually evident from clinical evaluation and preliminary investigations including karyotype, imaging and hormonal studies. However, some may require histopathological examination of the gonads and genital ducts to clinch a final diagnosis. The role of the pathologist extends beyond establishing a diagnosis to identifying premalignant and malignant lesions in these gonads.

The commonest DSDs are congenital adrenal hyperplasia and mixed gonadal dysgenesis. Gross and microscopic features of the gonads and genital ducts in various DSD, especially different forms of gonadal dysgenesis, Persistent Mullerian Duct syndrome, Androgen Insensitivity Syndrome and Ovotesticular DSD are discussed. The aberrant genesis of a streak gonad during ovarian or testicular differentiation and the pathology of an ovotestis are detailed. A partial, asymmetric, and ipsilateral development of Mullerian /Wolffian ducts is typical of both gonadal dysgenesis and ovotesticular DSD.

Premalignant and frankly malignant gonadal lesions are often detected at diagnosis itself. Germ Cell Neoplasia In Situ (GCNIS) and gonadoblastomas can develop in a dysgenetic gonad, gonadoblastomas being the precursor lesions of germ cell tumors. The risk of malignancy is higher in mixed gonadal genesis, PAIS (partial androgen insensitivity syndrome), Fraiser and Denis Drash syndrome; and lower in ovotesticular DSD and in CAIS (complete androgen insensitivity syndrome).

S. Mohanty
Department of Pathology, Kalinga Institute of Medical Sciences, Bhubaneswar, India

U. Chatterjee (✉)
Department of Pathology, Institute of Postgraduate Medical Education and Research, Kolkata, India

S. K. Ratan (ed.), *Children with Differences in Sex Development*, https://doi.org/10.1007/978-981-97-1639-5_15

The histology of the gonad in DSD is often challenging and a final diagnosis is made with careful deliberation on paraffin sections. However, frozen section diagnosis, immunohistochemistry, and molecular studies like FISH are useful in select situations.

Keywords

Pathology · DSD · Gonadal dysgenesis · Neoplasia

Differences in sex development (DSD) refer to conditions of incomplete or disordered genital or gonadal development leading to discordance between genetic sex, gonadal sex, and phenotypic and behavioral sex [1]. Although the majority of DSDs are congenital adrenal hyperplasia and mixed gonadal dysgenesis, the commonest biopsy specimens received are from testicular regression syndrome and androgen insensitivity syndrome. Most specimens related to DSD are that of excised gonads or gonadal ducts and occasionally diagnostic gonadal biopsy (Figs. 15.1 and 15.2). The pathologist's work involves the interpretation of morphological findings in light of clinical, karyotypic, and biochemical parameters in order to confirm the final diagnosis and evaluate for premalignant and malignant changes. There is sparse

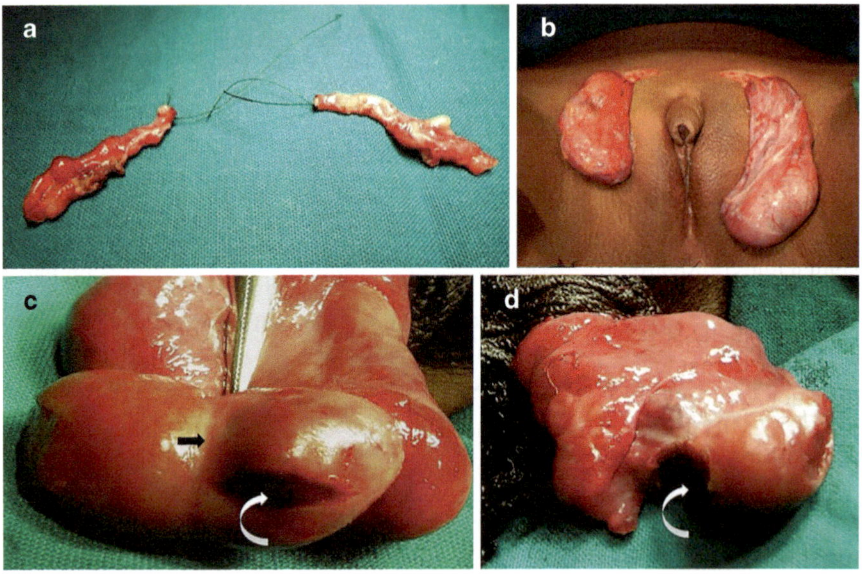

Fig. 15.1 Gross appearance of various gonads in DSD. The slender, pale, streak of gonadal dysgenesis (**a**), the inguinal testis in 46XY DSD–Androgen insensitivity syndrome (**b**) and the bipotential polar gonad in ovotesticular DSD (**c**, **d**). A groove (black arrow) demarcates the polar ovarian and testicular components. A linear intraoperative biopsy identified the components for selective excision of the discordant gonad (white arrows)

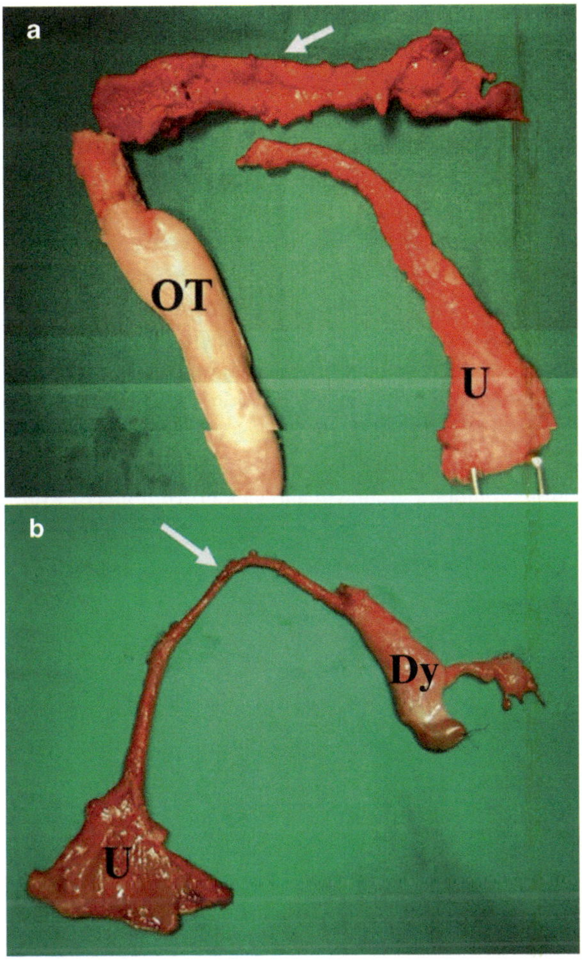

Fig. 15.2 Asymmetric genital ducts. (**a**) Ovotesticular DSD. The uterine horn (U) and Fallopian tube (arrow) were on the side of the partially descended rt. ovotestis (OT). (**b**) Mixed gonadal dysgenesis. The uterus (U) and left Fallopian tube (arrow) lead to a dysgenetic gonad (Dy) at its fimbrial end

literature on the pathology of DSD. A brief review of the pathology of the common DSD is presented with an emphasis on diagnostic aspects, associated risk of gonadal neoplasia and therapeutic implications.

15.1 Congenital Adrenal Hyperplasia

Congenital Adrenal Hyperplasia (CAH) is the commonest cause of DSD overall and of virilization with a 46 XX karyotype. Male CAH with 46XY is rare and often missed. It is an autosomal recessive disorder of the steroid biosynthesis pathway; 21-Hydroxylase deficiency accounts for more than 90% of cases. The degree of virilization varies depending on the severity of the enzyme defect, the location of the blockade and the timing of the onset of the endocrine effects.

Clinical features and phenotype: In 21-Hydroxylase defect, females present with ambiguous genitalia with clitoromegaly, posterior labial fusion, penile urethra or hypospadias or PCOS. The males have normal genitalia, precocious puberty and advanced bone age. In 11 β-hydroxylase deficiency, virilized females have hypertension. Males have normal genitalia and precocious puberty.

Gross: Stimulation of the adrenal cortex by the increased ACTH leads to marked enlargement of the adrenal cortex with a cerebriform appearance and a tan-brown color. *Microscopy:* There is diffuse cortical hyperplasia, particularly in the zona reticularis (Fig. 15.3). The cortical cells are lipid depleted. However, in lipoid congenital adrenal hyperplasia, the adrenal glands are pale yellow and are characterized by vacuolated cells, with cholesterol clefts and giant cell reaction [2, 3]. *Risk of Malignancy:* Adrenal cortical adenomas and carcinomas are rare. Testicular tumors resemble adrenal tissue morphologically and functionally hence called "Testicular Adrenal Rest Tumors" [4]. These lesions are bilateral, located in the hilar region of the testes are dependent on elevated levels of ACTH. They contain abundant eosinophilic granular cytoplasm and lack crystalloids of Reinke. These tumors may lead to gonadal dysfunction and infertility. Ovarian steroid cell tumors and giant and bilateral myelolipomas also occur.

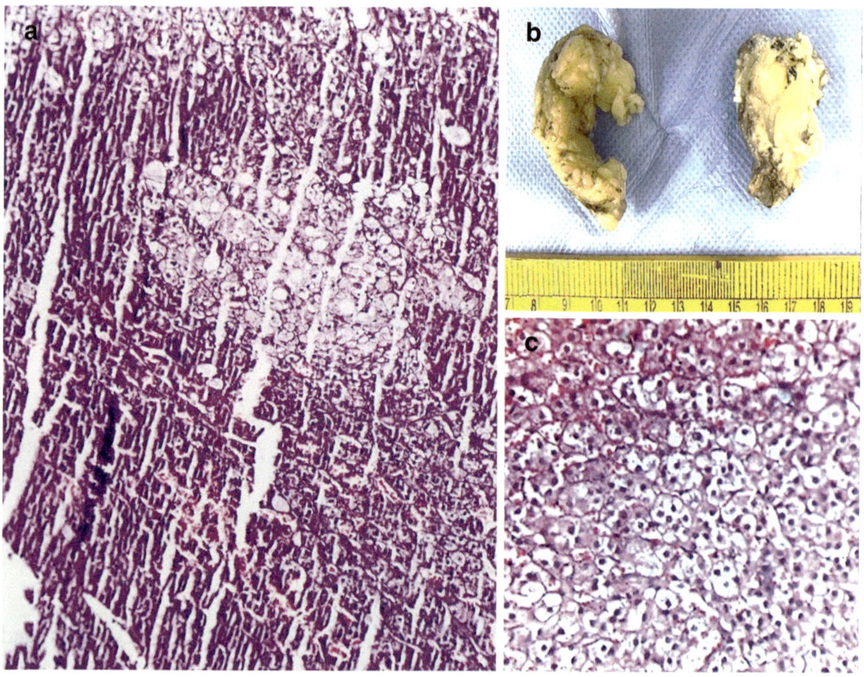

Fig. 15.3 Congenital adrenal hyperplasia. (**a**): Photomicrograph of adrenal cortex showing hyperplasia involving mainly the zona reticularis (H&E ×100). (**b**) Excised bilateral adrenal glands. (**c**) Photomicrograph of testicular adrenal rest tumor (TART) composed of large nests of cells with abundant clear cytoplasm at hilum of testis (H&E ×400)

15.2 Testicular Regression Syndrome

It is seen in phenotypic males with XY karyotype but with one or both testes rudimentary or absent. In most, the left testis is preferentially absent.

Clinical features and phenotype: The clinical presentation depends upon whether the testicular injury was unilateral/bilateral and the timing thereof. With early gonadal regression before the testis releases androgenic or anti-Mullerian hormones, the testis is absent, and the genitals atypical with retained Mullerian structures. In late afflictions with the damage after mid/late fetal stage, testosterone, and MIS were elaborated and they have a male phenotype with infantile to nearly normal male genitalia, regressed Mullerian structures, and differentiated Wolffian-derived structures.

Gross: In most, the vas deferens end blindly in a circumscribed, 0.5–1.5 cm fibrous nodule/button—the "nubbin" representing the regressed testis. It may lie in the retroperitoneum, iliac fossae, or scrotum and has hypoplastic vascular attachments. *Microscopy:* The testis may be absent or represented by rudimentary epididymal tubules in half. Other findings include fibroconnective tissue with scattered thick-walled blood vessels/ nerve bundles, hemosiderin deposits, calcification, or giant cells near the residual mesonephric ductular derivatives—vas deferens or epididymis (Fig. 15.4). Rarely, isolated clusters of Leydig cells may be found. Outlines

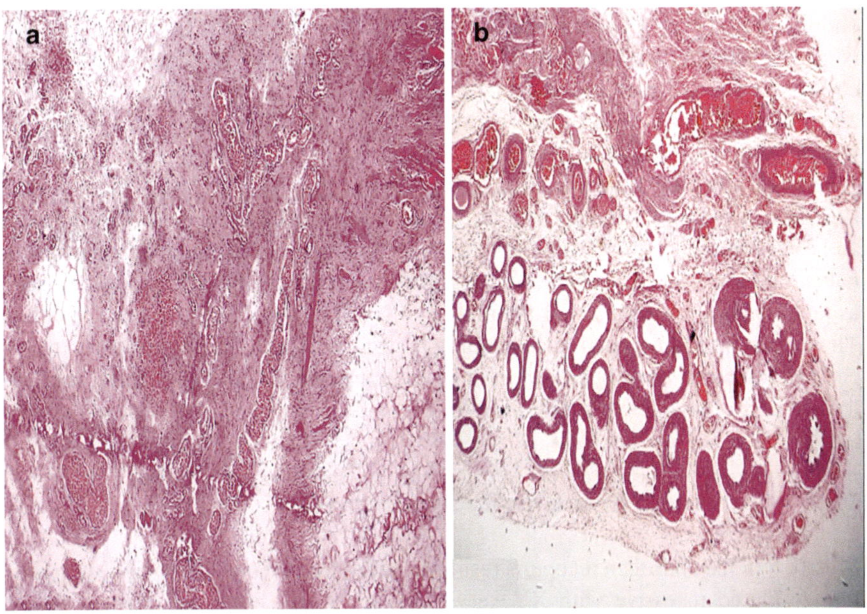

Fig. 15.4 Testicular regression syndrome. (**a**) Low power view of testicular nubbin with vascularized fibroconnective tissue and hemorrhage (H&E ×40). (**b**) Low power view of gonadal remnant showing epididymal tubules and the proximal end of the spermatic cord at far right (H&E ×40)

of seminiferous tubules are characteristically absent [5]. In unilateral afflictions, the contralateral testis is usually normal. *Risk of Malignancy:* Unlike cryptorchidism, there is no increased risk of malignancy as there is little gonadal tissue.

15.3 Klinefelter Syndrome

Klinefelter syndrome (KS) is one of the most common causes of prepubertal delay and primary hypogonadism in males, with a prevalence of 1 in 500–1000 live newborn males. In the majority, the karyotype is 47XXY due to the nondisjunction of chromosomes during the meiosis of paternal or maternal gametes. Other karyotypes include 46XY/47XXY, 48XXXY, 49XXXXY, and rarely 47XXY/46XX [2].

Clinical features and phenotype: Typical features include a tall man with narrow shoulders, broad hips, sparse body hair, gynecomastia, small testicles, and azoospermia. They present with sequelae of androgen deficiency like infertility, erectile dysfunction and low bone mineral density. Infants with KS may have normal male external genitalia and hence a delayed diagnosis; associated findings are hypospadias, micropenis, and cryptorchidism. They have hypogonadotropic hypogonadism due to intrinsic defects in Leydig cell function with low serum testosterone and elevated serum LH and FSH.

Gross: The testes are smaller and softer. *Microscopy*: The seminiferous tubules start showing degenerative changes in fetal life which progresses with age; by late childhood, the primary spermatogonia are reduced in numbers. In adults, the testis is largely atrophic with hyalinized seminiferous tubules and prominence of Leydig cells in the form of pseudo-adenomatous clusters. Germ cells are absent and tubules are lined by Sertoli cells only. There is a similarity to post-pubertal cryptorchid testis; however, excessive peritubular elastic tissue is lacking. Also, unlike the pseudo-Leydig cell hyperplasia in post-pubertal cryptorchid testis, there is true Leydig cell hyperplasia. *Risk of Malignancy:* KS can be associated with gonadal and extragonadal germ cell tumours. In the testis, seminoma, teratoma, and embryonal carcinoma have been observed. Extragonadal tumors occur in mediastinum such as teratoma and embryonal carcinoma. Leydig cell tumors have also been reported. They are more predisposed to developing breast carcinoma. A range of hematological malignancies have been reported including acute leukemia, chronic myeloid leukemia, and malignant lymphomas [6–8].

15.4 Androgen Insensitivity Syndromes

Mutation in the androgen receptors results in androgen insensitivity syndrome (AIS). It is X-linked recessive with XY karyotype, a variably feminized phenotype and bilateral testicular gonads. Androgen insensitivity syndrome is of two types—complete (CAIS) or partial (PAIS), they are differentiated by phenotype and gene studies.

Pathology: In AIS, the testes may be intra-abdominal (usually in complete forms) or inguinolabial (usually in incomplete forms). They are small for age; the sectioned

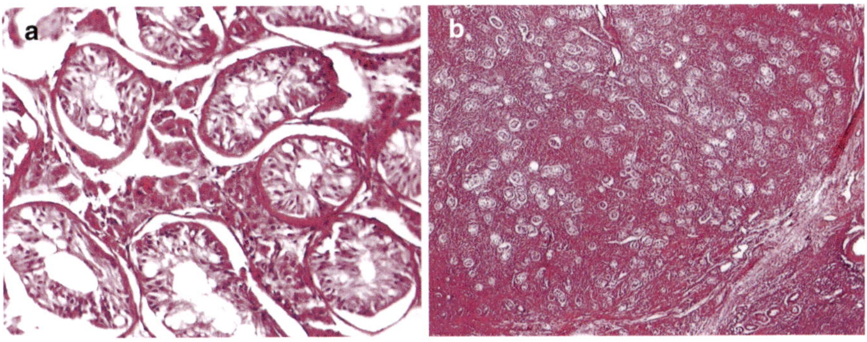

Fig. 15.5 46XY testicular DSD. (**a**) High power view of seminiferous tubules showing Sertoli cells only (H&E ×400) and (**b**) Scanner view of Sertoli cell adenoma (H&E ×40)

surfaces are tan-brown with yellow specks. On microscopy, seminiferous tubules are small and lined by immature Sertoli cells; germ cells are rarely encountered. Leydig cells appear hyperplastic; crystals of Reinke are rare. The abundant stroma resembles ovarian stroma. Hamartomatous nodules are found in the majority and are bilateral. *Risk of neoplasia:* The prevalence of germ cell tumors is 5.5%. In PAIS, the risk is higher as there is a larger germ cell population in PAIS compared to CAIS where they are lost after infancy [2, 9] (Fig. 15.5).

15.5 Persistent Mullerian Duct Syndrome

Persistent Mullerian Duct Syndrome (PMDS) is characterized by the persistence of Mullerian ductal derivatives in 46 XY males with near normal phenotype or mild atypical genitalia.

Clinical features and phenotype: The age at diagnosis ranges from a neonate to the fourth decade. It is a heterogeneous group of disorders caused mainly by a defect in the Mullerian inhibiting substance (MIS) gene or an abnormal Anti Mullerian hormone type II receptor resulting in end-organ insensitivity to biologically active MIS. Clinically, the spectrum varies from unilateral or bilateral cryptorchidism, to normal or near normal male external genitalia with inguinal hernia harboring a prolapsed infantile uterus and fallopian tubes.

Gross: Remnant of the upper vagina, infantile uterus, and fallopian tubes are found. The vas is embedded in the wall of the lower uterine and upper vaginal structure in most. *Microscopy:* The uterus has a well-developed myometrium lined by shallow, non-functioning type of endometrium due to low levels or absence of estrogen and progesterone (Fig. 15.6). The testis and other Wolffian structures may be essentially normal in childhood but by adulthood, show features of dysgenesis with progressively hyalinized seminiferous tubules, mixed atrophy and Leydig cell hyperplasia. Sometimes, the testes can have a low tubular fertility index and decreased tubular diameter [2, 10]. *Risk of neoplasia:* The risk of malignant neoplasia (5–15%) is associated with cryptorchidism such as germ cell neoplasia in situ

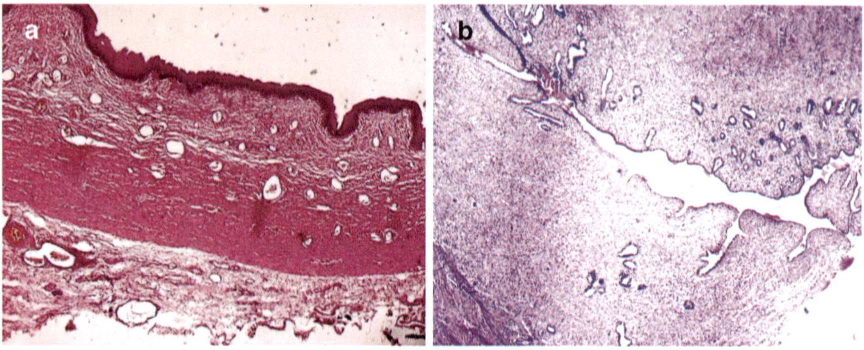

Fig. 15.6 Persistent Mullerian Duct Syndrome. (**a**) Low power view of the cervix lined by stratified squamous epithelium (H&E ×100). (**b**) Scanner view of endomyometrium in another case of PMDS. The endometrium is shallow and non-functional (H&E ×40)

(GCNIS), intra-tubular germ cell neoplasia (ITGCN), and seminoma [11]. These tumors usually develop in the cryptorchid testis and in the pexed testis several years after surgical descent and may be bilateral.

15.6 Gonadal Dysgenesis

These are disorders in which there is asymmetrical and misdirected differentiation of an undifferentiated gonad during gonadal development with one or both gonads being poorly developed as "streak" gonads. When the differentiation is along ovarian lines, it may arrest as a streak gonad with or without ovarian follicles or a hypoplastic ovary. If differentiation is along testicular lines, an accident leads to a streak gonad with epithelial cords or a dysgenetic testis. The gonad may be completely absent too. Any "undifferentiated gonadal tissue" (UGT) arrested at the fetal stage in the dysgenetic gonad or ovotestis may transform into gonadoblastoma.

Gonadal dysgenesis can be either pure (46XX or 46XY) with bilateral streak gonads or mixed with mosaic karyotypes (commonly 46XY, 45XO) and combinations of one streak gonad and one relatively normal gonad. A variant with two dysgenetic testes is termed partial gonadal dysgenesis.

15.6.1 Turner Syndrome

The prototype of gonadal dysgenesis is Turner syndrome. They bear the typical Turner phenotype and are short-statured females with infantile external genitalia and normal internal genitalia. By adolescence, they have primary amenorrhea.

Pathology: The fetal ovaries are histologically normal till about 16–18 weeks of gestation. The germ cells enter meiosis without the second X chromosome and progressively vanish by adulthood A classical streak gonad is seen with germ cells in primordial follicles in a background of fibrous connective tissue or ovarian-like stroma. In adults, the follicles are atretic, the hilum shows cortical cysts, hilar cells, or clear cells in groups. IHC shows positivity of the germ cells for DDX4 and C-Kit and negative for OCT3/4, PLAP, and TSPY (testis-specific protein Y-encoded). *Risk of neoplasia*: As there are neither germ cells nor Y chromosomes in patients with pure Turner syndrome, germ cell malignancies are rare. These ovaries can develop epithelial ovarian tumors, the incidence being similar to healthy females [2, 10].

15.6.2 Mixed Gonadal Dysgenesis

Mixed gonadal dysgenesis (MGD) is one of the most frequent causes of male sexual ambiguity in which there is asymmetry of the external gonads and an asymmetrical, variably feminized internal genital tract.

Clinical features and phenotype: There is asymmetry with the development of a testis and a streak gonad. The loss of testicular functions leads to the persistence of Mullerian duct derivatives, incomplete differentiation of Wolffian duct structures, and atypical male external genitalia. Testicular maldescent is common and few have Turner-like features.

Pathology: Usually there is a differentiated gonad on one side and a streak gonad or streak testis on the other side. The commonest type is testis and a contralateral streak gonad; variations include testis and contralateral gonadal agenesis, bilateral hypoplastic gonads with rudimentary tubules in one, or streak gonad with contralateral tumor. A unilateral macroscopic testis is found in over half the cases, it may be normal in size or smaller; small pale foci on the surface imply poorly formed tunica. On the side of the streak gonad, genital ducts of both Wolffian (vas deferens) and Mullerian (rudimentary uterus, hemi uterus, fallopian tube) types can be seen side by side, often ending blindly. An intra-abdominal or inguinal cryptorchid testis or streak/dysgenetic testis without an accompanying fallopian tube is present on the contralateral side. A germ cell tumor on one side with contralateral agenesis is also considered to be MGD. *Risk of Malignancy:* Tumors develop in 10% of patients with MGD, 25%–30 of these are malignant. The most common tumor in MGD (75%) is a gonadoblastoma, other GCT reported are germinoma, yolk sac tumor, teratoma, embryonal carcinoma, and choriocarcinoma. These may present unilaterally or bilaterally. Occasionally Sertoli cell tumors and Sertoli cell tumor-like proliferations of sex cord elements have been reported. Early gonadal resection is done to prevent the development of invasive germ cell tumors [10, 12–14].

15.6.3 Pure Gonadal Dysgenesis

The individuals could be 46XX, 46XY or 45XO and are phenotypically females, sometimes with clitoromegaly. Primary amenorrhea is the usual clinical presentation. Both gonads are streak gonads with ovarian-like stroma. Mullerian structures like uterus and fallopian tubes are also present in most cases. *Risk of neoplasia:* Pure Gonadal Dysgenesis (PGD) patients with 46XX karyotype rarely develop gonadal tumors, which include germ cell tumors and mucinous epithelial tumors. Those with 46XY karyotype are at a higher risk of developing gonadoblastoma and other germ cell tumors (10%–25%) [10].

Patients with partial gonadal dysgenesis have bilateral dysgenetic testes, persistent Müllerian duct structures, cryptorchidism, and inadequate virilization.

15.6.4 The Streak Gonad and Genital Ducts

Gross: Streak gonads are small elongated, flattened, pale gonads. It measures 2–3 cm in length and 0.5 cm in width in adults. *Microscopy:* Streak gonads show variable amounts of dense fibrous tissue and ovarian-type stroma with wavy elongated spindle cells in a background of connective tissue (Fig. 15.7). The hilus may be well developed and feature hilus cells, rete ovarii or mesonephric tubules. Streak gonads usually lack germ cells and are devoid of tunica albuginea. Scattered germ cells can, however, be present in streak gonads and show immunopositivity for DDX4 and C-kit but negative for OCT3/4, PLAP, and TSPY (testis-specific protein

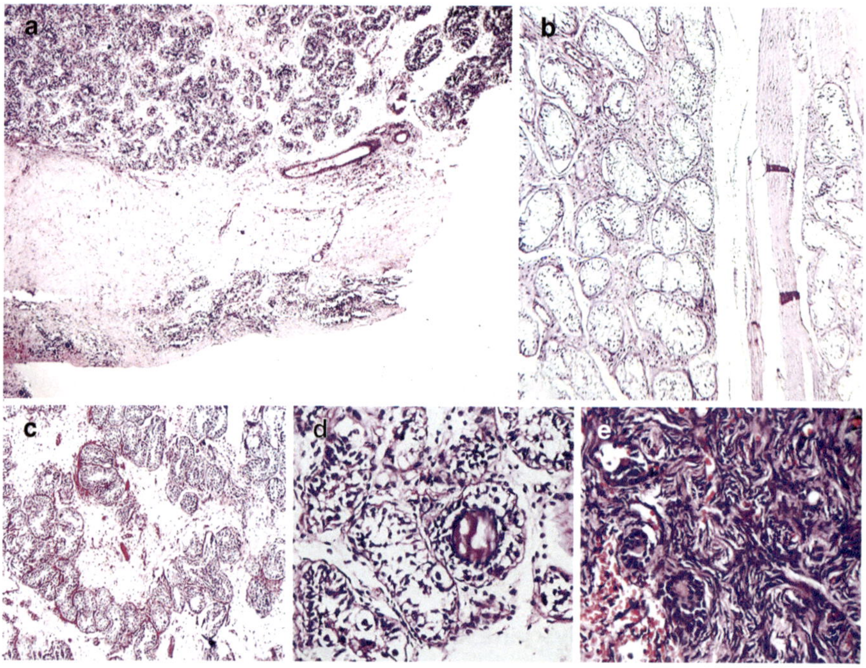

Fig. 15.7 Mixed Gonadal Dysgenesis. (**a**) Scanner view of testis showing poorly collagenized tunica with seminiferous tubules within (H&E ×40). (**b**) Another case showing branching of seminiferous tubules in tunica (H&E ×100). (**c**) Scanner view of dysgenetic gonad with irregularly distributed seminiferous tubules in edematous stroma (H&E ×40). (**d**) High power view of the same showing seminiferous tubules lined mainly by Sertoli cells and a reduced number of germ cells. A focus of microcalcification is noted in the center (H&E ×400). (**e**) High power view showing ovarian stroma (H&E ×400)

Y-encoded). Streak gonads sometimes show cord-like structures composed of two types of cells. Most of the cells are small round to oval cells with hyperchromatic nuclei and are immature Sertoli/granulosa cells forming cord-like structures. They are strongly positive for cytokeratin AE1/AE3 and show moderate staining with inhibin and D2-40, and weak staining with anti-Mullerian hormone (AMH). The basal membrane of these epithelial cords has variable thickness, with nodular projections into the cords in the thickest areas. Thickened hilar vessels and Leydig/hilar cell hyperplasia can be present. Unlike ovotesticular DSD, primordial follicles with oocytes are lacking. By puberty, the germ cells present in a streak gonad degenerate and disappear resulting in a gonad composed of fibrous tissue [3, 10].

The *dysgenetic testis* shows varying degrees of disorganization together with progressive abnormalities of cryptorchidism. The tunica albuginea or cortex is thin, irregular and poorly formed. Often seminiferous tubules are identified within the tunica—an important clue for underlying gonadal dysgenesis in orchidectomy for cryptorchidism. Widely spaced seminiferous tubules, with ovarian-like stroma or immature primary sex cords indeterminate between male and female structures may be present. The seminiferous tubules frequently are branched, with variable diameters. Most of the tubules lack germ cells. Leydig cells may be present as small clusters.

A streak is often excised for biopsy; an indeterminate gonad may be biopsied to establish the type of gonad. Biopsy specimens are obtained along the long axis of the gonad to be most representative.

15.6.4.1 Germ Cell Neoplasia In Situ (GCNIS) and Gonadoblastomas (Gb) in Gonadal Dysgenesis

Depending on the degree of development in the dysgenetic gonad, the lesion may develop into a GCNIS or gonadoblastoma, the latter being a precursor of an invasive germ cell tumor. Early gonadoblastomatoid changes can be often seen in the dysgenetic gonads.

GCNIS is a common precursor of all seminomas and most of the post-pubertal germ cell tumors (GCTs). Although typically seen in post-pubertal testis, in DSD it may be found in pre-pubertal testis. On microscopic examination, GCNIS consists of atypical gonocytes with abundant clear cytoplasm and large nuclei with prominent nucleolus and coarse chromatin. The cells are typically located in the spermatogonial niche in a single layer resembling a string of beads (Fig. 15.8). Tubular lumina can still be identified. Gain of chromosome 12p is not present in GCNIS unlike most of the invasive GCTs. Isolated GCNIS may be found in the workup of male infertility, cryptorchidism and DSD. The term ITGCN (intra-tubular germ cell neoplasia) is used for extensive GCNIS which fill up the entire seminiferous tubules without any discernible lumina and represents an intermediate stage between GCNIS and invasive GCT. Tumors encountered in these patients are seminoma in the testis, dysgerminoma in the ovary, gonadoblastomas and nonseminomas. The immunoprofiles of GCNIS and seminoma are similar and show positivity for PLAP, OCT4, SOX17, AP-2y, NANOG, LIN28A, and podoplanin. c-Kit expressed in normal spermatogonia is useful because it is overexpressed in GCNIS and seminomas

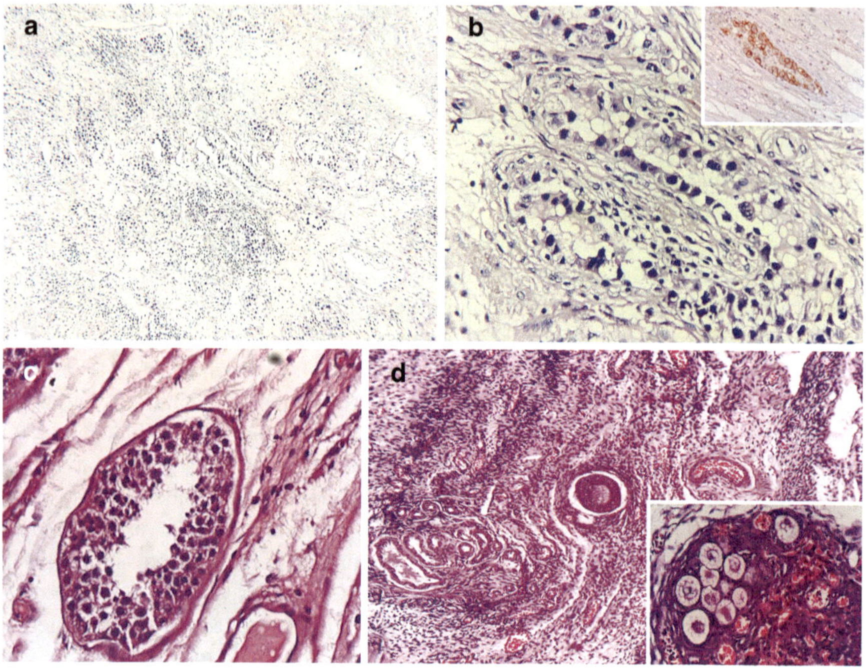

Fig. 15.8 Germ Cell Neoplasia In Situ. (**a**) Scanner view of germ cell tumor in testis showing GCNIS and ITGCN (H&E ×40). (**b**) High power view of GCNIS comprising a single row of atypical germ cells in the spermatogonial niche (H&E ×400). Inset shows the same on c-Kit IHC (×400). (**c**) High power view of GCNIS (H&E ×400). (**d**) Hypoplastic and dysgenetic ovary with an abnormal follicle in the center surrounded by a uniform coat of granulosa cells. The stroma is edematous and disorganized (H&E ×100). The inset shows abnormal closely spaced follicles

[11, 12]. It is important to diagnose GCNIS accurately because of the associated increased risk of developing invasive GCTs.

Gonadoblastomas were considered to be mixed-sex cord/germ cell tumors. In the recent 2022 WHO classification, they have been reclassified under in situ form of malignant germ cell tumors and the sex cord component is considered to be reactive proliferation. The key feature is that sex cord cells surround germ cell nests in peripheral palisades or radially surround small round deposits of eosinophilic, PAS-positive material, resembling Call-Exner bodies. The primitive germ cells resemble those of dysgerminoma and the sex cord stromal cells morphologically and immunohistochemically resemble immature Sertoli and granulosa cells. Sex cord cells with globoid deposits of PAS-positive, hyaline basement membrane material in variable proportions are usually present (Fig. 15.9). The basement membrane material frequently develops laminated calcifications resembling psammoma bodies. The germ cell component of gonadoblastoma may overgrow the stromal elements and form a germinoma, or rarely other types of GCTs. Overgrowth of the sex cord stromal component with the formation of a Sertoli cell tumor is even rarer. Coarse calcifications in an invasive GCT hint at an origin from a gonadoblastoma and underlying DSD. The immunophenotype includes PLAP, CD117 and podoplanin

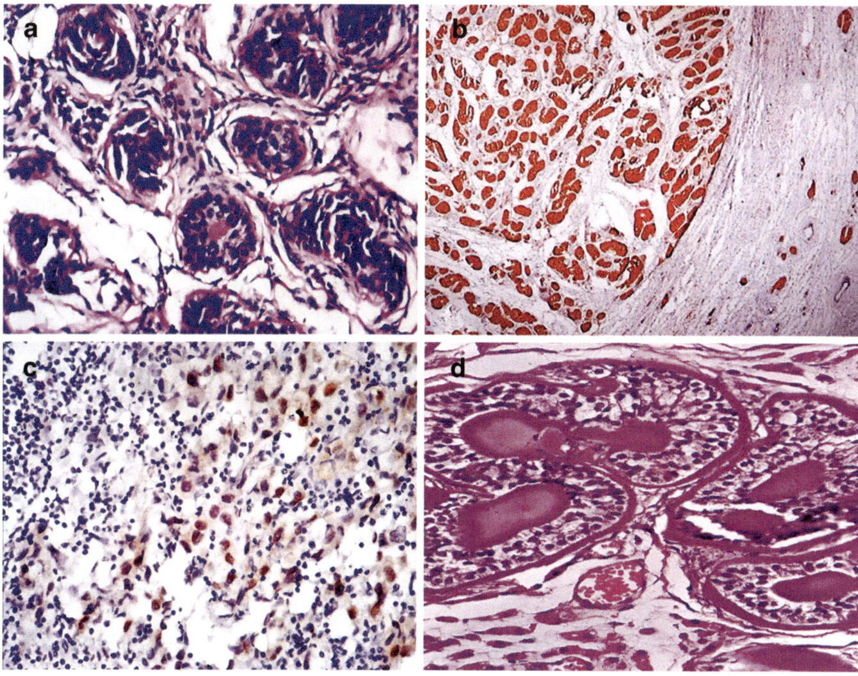

Fig. 15.9 Gonadoblastoma. (**a**) Low power view showing globoid eosinophilic deposits in the center of sex cord-stromal nests (H&E ×100). (**b**) The tumor cell nests show strong positivity for inhibin (H&E ×40). (**c**) Germinomatous component of the same tumor showing large pale clusters of germ cells with nuclear positivity for OCT 4. Lymphocytes are noted in the background (×400). (**d**) High power view of early gonadoblastoma composed of sex cord-stromal cells radially surrounding small deposits of amorphous eosinophilic PAS positive material resembling Call-Exner bodies (PAS ×400)

staining in the germ cells; inhibin, nuclear SOX9 (weak) and FOXL2(strong) in the sex cord cells; and inhibin plus calretinin in the Leydig cells [11, 15].

15.6.4.2 Tumors Associated with Gonadal Dysgenesis

Gonadoblastoma is seen in patients with gonadal dysgenesis with Y chromosome material including 46,XY pure gonadal dysgenesis and mixed gonadal dysgenesis. Those with 46,XX pure gonadal dysgenesis and Turner syndrome are at risk of developing hilus cell hyperplasia and tumors. Genetic mutations such as WT1, SF1, DAX1 and SOX9 in gonadal dysgenesis patients help in assessing the tumor risk. Germ cell tumors are usually discovered in infancy itself.

15.7 Ovotesticular DSD

Though ovotesticular DSD has been considered to be one end of the spectrum of gonadal dysgenesis by some, these gonads are composed of mature ovarian/testicular/both tissues. The genotype is commonly 46 XX, but may be mosaics 46 XX/XY or 46 XY. Similarly, clinical features are spread across the spectrum of atypical genitalia with asymmetric features in the external genitalia as well as the internal gonads and the genital ducts. This is due to the effect of the ipsilateral gonad and a particular combination of gonad on each side influencing the ductal derivatives on that side. Common combinations are ovary—ovotestis, ovary and testis and bilateral ovotestis. The testis is usually descended into the labia/scrotum while the ovary is intra-abdominal. The location of ovotestis depends on the proportion of each, most are intrabdominal. Importantly, patients with uterus menstruate.

Gross: A pure testis or ovary looks unremarkable. The mixed gonad/ovotestis looks heterogeneous with a softer, smoother, testicular part and a firmer, lobated, ovarian part. There may be arranged side by side (polar) or one gonadal component within the hilus of another (hilar). *Microscopy:* The testicular tubules are sclerotic with fewer germ cells at puberty. The ovarian tissue is usually norma, and ovulation may occur (Fig. 15.10). The genital duct accompanying the gonad is specific for the gonad, i.e., the Fallopian tube accompanies the ovary and the vas deferens is next to the testis. In the case of ovotestis, the genital duct may be either Fallopian tube or

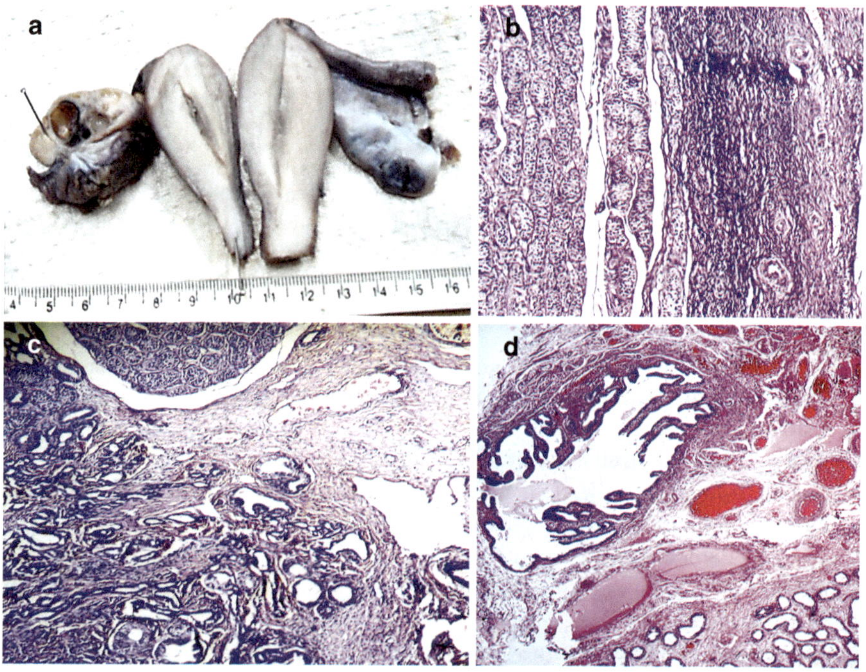

Fig. 15.10 Ovotesticular DSD. (**a**) Uterus and bilateral gonads and genital ducts in an ovotesticular DSD. The pin marks the testicular portion. (**b**) Closely packed seminiferous tubules are noted (left) flanked by ovarian tissue (right) (H&E ×100). (**c**) A case of ovotestis showing testicular portion along with rete testis (H&E ×40). (**d**) Juxtaposition of the Fallopian tube (upper left) and epididymal tubules (lower right) in the same case (H&E ×40)

vas deferens or parts of both. *Risk of neoplasia:* As the tissues are mostly mature, malignant tumors occur in less than 3% of cases—commonly germinoma, but gonadoblastomas and others have been reported [2, 9, 10].

15.8 Premalignant Conditions

It is well known that some patients with a DSD are at an increased risk of malignancy. There are two known premalignant conditions seen in patients with DSD-Gb and GCNIS.

Gb is a distinct type of GCNIS occurring in ovarian-type gonads and originates from germ cells initially developing along the female pathway (ovaries) but arrested in primordial follicle stage. It is generally found at the time of gonadectomy.

GCNIS is often seen in patients with DSD with testicular-type tissue and has the potential to develop into several types of testicular-type GCTs. Like Gb, GCNIS cells are fetal gonocytes in the seminiferous tubules that are arrested during development and fail to develop into mature spermatogonia. These often progress to an invasive GCT over years, 50% by 5 years. Both GCNIS and Gb can be present in the same gonad if there are separate testicular and ovarian elements.

The risk of gonadal malignancy is high (15–60%) in gonadal dysgenesis and PAIS, especially with Y chromosome/expression of TPSY gene and intra-abdominal location of gonads, Frasier syndrome or Denys-Drash syndrome (+Y). It is low (1–3%) in CAIS and ovotesticular DSD [10, 11, 16].

15.9 Role of IHC, Frozen Section, and Pitfalls in Histopathology of DSD

Surgical specimens of DSD need to be interpreted along with clinical, karyotypic, imaging, and biochemical data. Hence, a summary of these findings must be accurately filled in the pathology request form. The age and location of the gonad are important as atrophic changes set in early. Histochemical stains like PAS and elastic stains are useful to highlight these changes in the basement membrane. This is useful in distinguishing KS from post-pubertal cryptorchid testis. CD 117 (c-Kit), PLAP, podoplanin (D2-40), and OCT4 help to pick up GCNIS and ITGCN in cryptorchid testis and in gonadal dysgenesis. These stains are particularly helpful in delineating foci of germinoma in the background of gonadoblastoma. Co-expression of OCT 4 and TSPY in the seminiferous tubules indicates maturation arrest and predates development of GCNIS. Along with increased Kit-ligand expression in the germ cells, this is of pathogenetic relevance for the development of gonadoblastoma in DSD [11]. The distribution of OCT 4 positive germ cells in the tubule, their relationship to the basal lamina, and consideration of the patient's age have been described as useful aids in distinguishing immature germ cells from maturation delay. In cases where dysgenetic testes are retained, biopsy at orchidopexy and post-puberty with appropriate immunohistochemical staining is warranted to identify patients at increased risk of malignancy.

Inhibin, SF1, calretinin, SOX9, and FOXL2 are markers of sex cord and stromal cells. The Sertoli cell-only tubules, immature Sertoli cells, Sertoli cell component of gonadoblastoma, Leydig cells and testicular adrenal rest tumors can all be stained using a combination of sex cord markers. FISH analysis is helpful in demonstrating Y chromosome in tissues of phenotypically female patients. FISH analysis for isochromosome 12p is also important in separating the more aggressive type II from type I GCT which can develop in a background of DSD.

The histology of the gonad in DSD is often challenging and a clear final diagnosis is made with careful deliberation and special staining on paraffin sections. However, frozen section diagnosis may be used in certain situations. Rarely a dysgenetic gonad may not be recognizable macroscopically as a "streak" and an intraoperative frozen may be necessary prior to removal. Mapping the testicular and ovarian areas in an ovotesticular DSD helps in retaining the gonad concordant with the sex of rearing (e.g., the testis in those raised as males) and removing the other. This is feasible in gonads with macroscopically distinct testicular and ovarian areas (e.g., polar or less commonly hilar arrangement) but difficult in others. Enzyme reactivity for alkaline phosphatase has been used to diagnose GCNIS on frozen sections to assist in testis sparing and fertility-preserving surgery.

References

1. Hughes IA, Houk C, Ahmed SF, Lee PA. Consensus statement on management of intersex disorders. J Pediatr Urol. 2006;2:148–62.
2. Husain AN, Dehner LP. Stocker and Dehner's pediatric pathology. Lippincott Williams & Wilkins; 2021 Apr 6.
3. Gwin K, Singh N, Cavello A, Reyes-Mugia M. Reproductive system. In: Cohen M, Scheimberg I, editors. Essentials of surgical pediatric pathology. Cambridge University Press; 2014. p. 323–6.
4. Knape P, Reish N, Dorr HG, et al. Testicular rest tumours (TART) in adult men with classic congenital adrenal hyperplasia(CAH). Urolog A. 2008;1596-7:1599–602.
5. Law H, Mushtaq I, Wingrove K, Malone M, Sebire NJ. Histopathological features of testicular regression syndrome: relation to patient age and implications for management. Fetal Pediatr Pathol. 2006;25:119–29.
6. Scully RE. Gonadoblastoma; a gonadal tumor related to the dysgerminoma (seminoma) and capable of sex-hormone production. Cancer. 1953;6:455–63.
7. Hersmus R, de Leeuw BH, Wolffenbuttel KP, et al. New insights into type II germ cell tumor pathogenesis based on studies of patients with various forms of disorders of sex development (DSD). Mol Cell Endocrinol. 2008;291:1–10.
8. Obsterhuis JW, Looijenga LH. Testicular germ-cell tumors in a broader perspective. Nat Rev Cancer. 2005;5:210–22.
9. Rutgers JL. Advances in the pathology of Intersex conditions. Hum Pathol. 1991;22(9):884–91.
10. Nistal M, Paniagua R, Gonzalez-Peramato P, Reyes-Mugica M. Perspectives in paediatric pathology, Chapter 5. Gonadal Dysgenesis. 2015;18:259–78.
11. Berney DM, Tickoo SK. Tumours of the testis. In: WHO Classification of Tumours Editorial Board, editor. Urinary and male genital tumours. Lyon: IARC Press; 2022. p. 243–57.

12. Looijenga LH, Stoop H, de Leeuw HP, et al. POU5P1 (OCT3/4) identifies cells with pluripotent potential in human germ cell tumors. Cancer Res. 2003;63:2244–50.
13. Cools M, Drop SL, Wolffenbuttel KP, Oosterhuis JW, Looijenga LH. Germ cell tumors in the intersex gonad: old paths, new directions, moving frontiers. Endocr Rev. 2006;27:468–84.
14. Cools M, van Aerde K, Kersemaekers AM, et al. Morphological and immunohistochemical differences between gonadal maturation delay and early germ cell neoplasia in patients with undervirilization syndromes. J Clin Endocrinol Metab. 2005;90:5295–303.
15. Buell-Gutbrod R, Ivanovic M, Montag A, et al. FOXL2 and SOX9 distinguish the lineage of the sex cord-stromal cells in gonadoblastomas. Pediatr Dev Pathol. 2011;14:391–5.
16. Morin J, Peard L, Saltzman AF. Gonadal malignancy in patients with differences of sex development. Transl Androl Urol. 2020;9(5):2408–15. https://doi.org/10.21037/tau-19-726.

Approach to Diagnosis in Children with Differences of Sex Development

16

Kanishka Das

Abstract

The diagnosis of DSD is a deductive process. It begins with a detailed historical review and careful physical examination followed by a preliminary panel of screening investigations (karyotype, ultrasonography, serum biochemistry, serum 17 hydroxy progesterone (OHP)) followed by more specific tests (hormonal studies, imaging—MRI, genitogram) in a reasonable, cost-effective, and stepwise manner. Endoscopic evaluation (genitoscopy/laparoscopy, biopsy) requiring anesthesia is scheduled with planned therapeutic interventions in a calendar that combines diagnosis and management. Various diagnostic algorithms follow these principles to move from a provisional diagnosis along a process of differential diagnosis to a final one as per the Chicago Consensus terminology.

Keywords

DSD · Diagnosis · Algorithm

The diagnosis of DSD is akin to a mathematical deductive process in the majority of common scenarios. After a detailed historical review and physical examination of a case that merits investigation for an underlying DSD, the first tier of investigations is a basic genetic study (karyotype) and noninvasive imaging (ultrasonography/ occasionally an MRI). In the neonate, this preliminary set also includes plasma glucose, serial serum electrolytes, and serum 17 OHP in an attempt to detect the

K. Das (✉)
Department of Paediatric Surgery, All India Institute of Medical Sciences, Bhubaneswar, Odisha, India

S. K. Ratan (ed.), *Children with Differences in Sex Development*, https://doi.org/10.1007/978-981-97-1639-5_16

common 21 hydroxylase deficient Congenital Adrenal Hyperplasia (CAH) that must be immediately recognized and emergently managed.

Based on the phenotype and the results of the first tier (karyotype, imaging, biochemistry) investigations which could return crucial information in 48–72 h with recent advanced methods, further specific investigations are considered in a reasonable, cost-effective, and stepwise manner.(Fig. 16.1).A single or a panel of hormonal studies are carefully selected to diagnose biosynthetic defects in the adrenogonadal steroidal pathway and are based on a specific deficiency/defect leading to a lack of the steroidal end product, excess of precursors, and the effects of feedback on a functioning hypothalamic–pituitary–gonadal axis. Various algorithms have been devised to arrive at a conclusive diagnosis employing the results of the investigations detailed above. Once a broad diagnosis is made, the specific diagnosis is made with further inquiry [1].

The various forms of 46 XX DSD-CAH are differentiated based on their specific clinical features (virilization, hypertension, salt wasting, pigmentation, growth retardation, adrenal crisis) and biochemical profiles (serum 17OHP, DOCA, PRA, etc.) depending on the site of enzymatic block in each form. Rare forms of masculinization in 46 XX DSD (non-CAH) like fetomaternal ovarian/adrenal tumors, placental aromatase deficiency, and exogenous androgenic medications/progestins to mother during pregnancy are elucidated separately with specific features. Other forms of 46 XX DSD bearing Mullerian structures internally—ovotesticular DSD and the rare 46XX testicular DSD are diagnosed with additional hormone

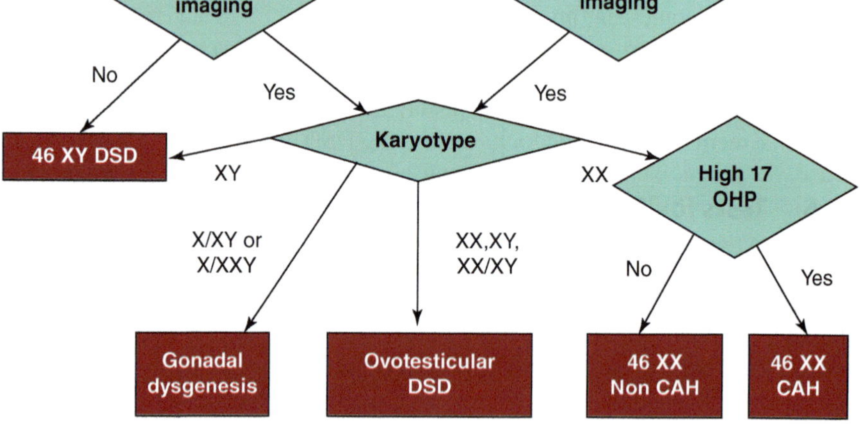

Fig. 16.1 Basic diagnostic algorithm in DSD

evaluation (serum Anti Mullerian hormone (AMH), estradiol, testosterone to identify the type of gonadal tissue) and gonadal biopsy at laparoscopy. Very rarely, 46 XX DSD is associated with complete Mullerian agenesis.

After a diagnosis of 46 XY DSD in a child with bilateral undescended testis and microphallus/unilateral undescended testis with proximal hypospadias and combinations thereof, an hCG-stimulated hormonal profile comprising of serum LH, FSH, androstenedione, testosterone, and dihydrotestosterone will identify the level of defect at various levels: supra gonadal/gonadal/androgen receptor level. A lack of testosterone effects can thus be narrowed down to testicular differentiation defect (hypothalamic–pituitary trophic hormone deficiency, gonadal dysgenesis—regression), testosterone biosynthetic defects (20, 22-demolase, 17, 20-lyase 3b-hydroxysteroid dehydrogenase deficiency), conversion defect (5AR2 deficiency), or an end organ responsiveness (partial or complete Androgen Insensitivity Syndrome (AIS)). Similar phenotypes with 46 XY DSD can be differentiated according to the nature of the gonad, the presence or absence of a Mullerian remnant, and hormonal profiles. The presence of a cervix and a uterus in a 46, XY DSD implies dysgenetic gonads, ovotesticular DSD, or persistent Müllerian duct syndrome. The differentiation would require confirmation with gross visualization and biopsy after imaging, specific hormonal profile (serum AMH, testosterone, estradiol, LH, FSH), and endoscopy /laparoscopy. Pure gonadal dysgenesis (PGD) is often confused with complete AIS with both presenting at puberty with a female phenotype with failure of menarche. However, in PGD, there are usually normal or hypoplastic Müllerian derivatives as the dysgenetic gonads often do not produce MIS, but in complete AIS, the MIS from testis ensures there are no Müllerian derivatives. A trial of testosterone injections and observation of variable androgenic effects in the partial form will distinguish the partial from complete AIS. There may still be entities under 46 XY DSD that may overlap in clinical presentation and hormonal profile like certain forms of 5AR2 deficiency and partial AIS (PAIS). Where available, specific genetic studies and receptor studies will lead to the final diagnosis.

Similarly, in a child with asymmetric gonads and a Mullerian remnant on imaging, the karyotype will help differentiate between a mixed gonadal dysgenesis (commonly 46XY,45XO) and an ovotesticular DSD (commonly 46XX/46 XX-46XY). With inconclusive karyotypes, serum testosterone/AMH and estradiol indicate mature functioning testicular and ovarian tissue, respectively, rather than dysgenetic gonads. Serum AMH is now commonly available and is a marker of testicular tissue. Though the rare pure dysgenesis is clearly picked up by these, results may vary depending on the extent and laterality of dysgenesis in the common mixed gonadal dysgenesis form. In indeterminate cases, the final diagnosis may have to wait till gonadal and internal genital ductal visualization and biopsy is available.

About 20% of DSD may evade a specific diagnosis till some investigations needing anesthesia are completed; similarly specific molecular diagnosis is reached in barely 20% of all DSD. A small Mullerian derivative may go unrecognized till a genitoscopy or a streak gonad is picked up only at laparoscopy/laparotomy. Even on direct visualization at laparoscopy /laparotomy, the gross appearance of some gonads and the ipsilateral genital ducts may be confusing till a histological examination is available on gonadal/genital ductal biopsy. Laparoscopy affords an excellent, minimally invasive means of establishing the diagnosis conclusively. A provisional diagnosis will accordingly need a revision after the results of all investigations are available [2] before a treatment plan is charted out. The diagnosis of DSD can be difficult terrain for the uninitiated and no committed resectional procedure should be attempted before a clear diagnosis is reached.

References

1. Rawal AY, Austin PF. Concepts and updates in the evaluation and diagnosis of common disorders of sexual development. Curr Urol Rep. 2015 Dec;16(12):83. https://doi.org/10.1007/s11934-015-0556-0.
2. Ahmed SF, Achermann J, Alderson J, Crouch NS, Elford S, Hughes IA, et al. Society for Endocrinology UK guidance on the initial evaluation of a suspected difference or disorder of sex development (revised 2021). Clin Endocrinol. 2021 Dec;95(6):818–40.

"Error Traps" in Diagnosis of Disorders of Sexual Development

17

Simmi K. Ratan and Nitin Jain

Abstract

Arriving at a diagnosis in a child with ambiguous genitalia is a challenging task which is associated with many pitfalls and likelihood of error commitment. There could be overlooking of certain aspects which can lead to an erroneous diagnosis. The text here details diagnostic dilemmas or atypical results during work up of patients with DSD. Such situations which are likely to cause suboptimal or erroneous diagnosis may be labelled as the "error traps" and attending clinician or surgeon must keep these in mind so as to prevent making a wrong diagnosis as it has got the far-reaching implications. The wrong diagnosis not only leads to a wrong sex assignment, but it can have long-lasting psychosexual development disorders and in later time can raise a legal issue also. So, the attending clinician has to be very careful about proceeding with the diagnosis of the case and must look into smallest details of the diagnostic investigations undertaken along with the correct interpretation of these investigations with the help of experts.

Keywords

Ambiguous genitalia · Disorders of sexual development · Error traps · Gender assignment · Gonads

Disorders of Sex development (*DSD*) are defined as congenital conditions in which development of chromosomal, anatomical, or gonadal sex is atypical. The estimated frequency is reported to be in the range of 1:2500–1:4500 live births and it is same

S. K. Ratan (✉) · N. Jain
Department of Pediatric Surgery, Maulana Azad Medical College, New Delhi, India

S. K. Ratan (ed.), *Children with Differences in Sex Development*,
https://doi.org/10.1007/978-981-97-1639-5_17

as mentioned in earlier literature [1, 2]. Out of these disorders, congenital adrenal hyperplasia (CAH, 46XX DSD) is the commonest one that constitutes 50% of the reported cases [3].

17.1 Diagnosis and Management: An Overview of the Line of Approach

Diagnosis and management of DSD patients in a timely manner is always a challenge, especially to avoid issue of sex reassignment and its consequences [4]. Early diagnosis and management greatly can help in avoiding problems of gender dysphoria altogether. The different areas of legal angles also go hand in hand. A lot of uproar resultant to a wrongly or sub-optimally arrived diagnosis, thus, has an outreaching bearing not only in providing a rather clearer perception to the family regarding the gender status of their child and its future prospects; but also for avoiding consequent social adjustment problems and psychological issues for the child [5]. Special attention in this matter cannot be underestimated. Needless to say that the construction of family tree and genetic counselling play an important role.

The main goal is to reach a correct diagnosis and this always starts with detailed and appropriate history and clinical examination. It has already been explained in the previous chapters regarding the importance of a complete detailed history which may include unexplained deaths, similar genital ambiguity in the family, infertility, hirsutism, amenorrhea, maternal exposure to exogenous androgens, and consanguinity [6–8]. Similarly examination preferably in frog like position also plays an important role in reaching towards a correct diagnosis parallelly to be followed by the anatomical status of the internal organs, by utilizing the various radiological investigations like sonography of pelvis (USG), genitography, and magnetic resonance imaging (MRI) [9, 10]. Workup of genetic sex (karyotype) is important to arrive at a diagnosis especially in the conditions that may have overlapping diagnosis or have similar presentations.

In the following text, the errors or challenges in the diagnosis of DSD are highlighted so that the reader can realize the diagnostic dilemmas or any bizarre results during work up of patients with DSD. Such situations which are likely to cause suboptimal or erroneous diagnosis may be labelled as the "error traps." These can have far-reaching adverse effects for future prospects of the patients and hence, should be avoided at all costs.

17.2 Error Trap 1: Not Able to Recognize a Condition as of DSD

First and foremost, important thing is to recognize the presence of marker signs while working up patients with DSD. Presentation of DSD with typical or atypical phenotype may be variable in different individuals and may be related to the severity of deviation from normal as well as to age of presentation. So, it is very important to know all different aspects and presentations at different ages [11]. Therefore, a strong index of suspicion is topmost requisite. By and large, a surgeon will come across the setting as stated below and should be able to recognize these as pointers to the possible existence of DSD; failing which it can be labelled as "error trap."

Marker signs for suspecting DSD in younger children include: (1) Overt genital ambiguity (e.g., cloacal exstrophy); (2) Apparent female genitalia with an enlarged clitoris and posterior labial fusion (as in congenital adrenal hyperplasia, i.e. CAH); (3) Apparent male genitalia with bilateral undescended testes, hypospadias, or micropenis; and (4) Discordance between genital appearance and a prenatal karyotype [12].

Also going beyond what has been stated above, reference should also be made to inappropriate behavior history or even psychological stress, suboptimum academic performance, toilet use phobia, withdrawal behavior, as these may be the FIRST POINTERS to the existence of DSD condition in an individual.

In older children, the marker signs for evaluation of DSD also include: (1) Previously unrecognized genital ambiguity; (2) Inguinal hernia in a girl (e.g., Complete androgen insensitivity syndrome (CAIS) and few cholesterol cycle related enzymatic deficiencies); (3) Delayed or incomplete puberty; (4) Primary amenorrhea or virilization in a girl; (5) Breast development in a boy; and (6) Gross or cyclic hematuria in a boy and a suprapubic mass (hematocolpos) with breast enlargement after the age of puberty in a boy with the history of hypospadias and undescended gonads [5, 12].

Hence, the attending clinician and specially the pediatric surgeons must keep a wide scope of knowledge about these variations and presentations at different ages.

17.3 Error Trap 2: Not Able to Differentiate Hypoplastic Phallus from Hypertrophied Clitoris

In most cases, ambiguous genitalia in genetic females has the following features:

- An enlarged clitoris that looks like a small penis.
- The urethral opening can be anywhere along under surface of clitoris, or below it.
- The labia may be fused and look like a scrotum.
- The infant may be thought to be a male with undescended testicles.
- Sometimes a lump of tissue is felt within the fused labia, further making it look like a scrotum with testicles.
- *Hypertrophied clitoris*—Generally clitoris has otherwise good amount of skin hood which almost drapes it and is immediately downwards flanked by labia minora, upper border of which that may be fused or unfused. There under is urethral and vaginal meatus as urogenital sinus opening which is wide and broad. The clitoris has inner epithelialized mucosa underneath, and generally bears pink

color or is less pigmented in comparison to hypoplastic phallus [13]. The enlarged labioscrotal folds on either side of clitoris in virilized female child may be difficult to differentiate from cleft scrotum in males. In our experience, the hypertrophied clitoris lacks adequate skin hood coverage and despite the presence of urethral meatus at the base, the preputial hood may be quite unremarkable in comparison to that seen in males with hypospadias (Fig. 17.1a).

- *Hypoplastic phallus or microphallus*—The stretched phallic length is 2 SD below mean for that population [14] (Fig. 17.1b).

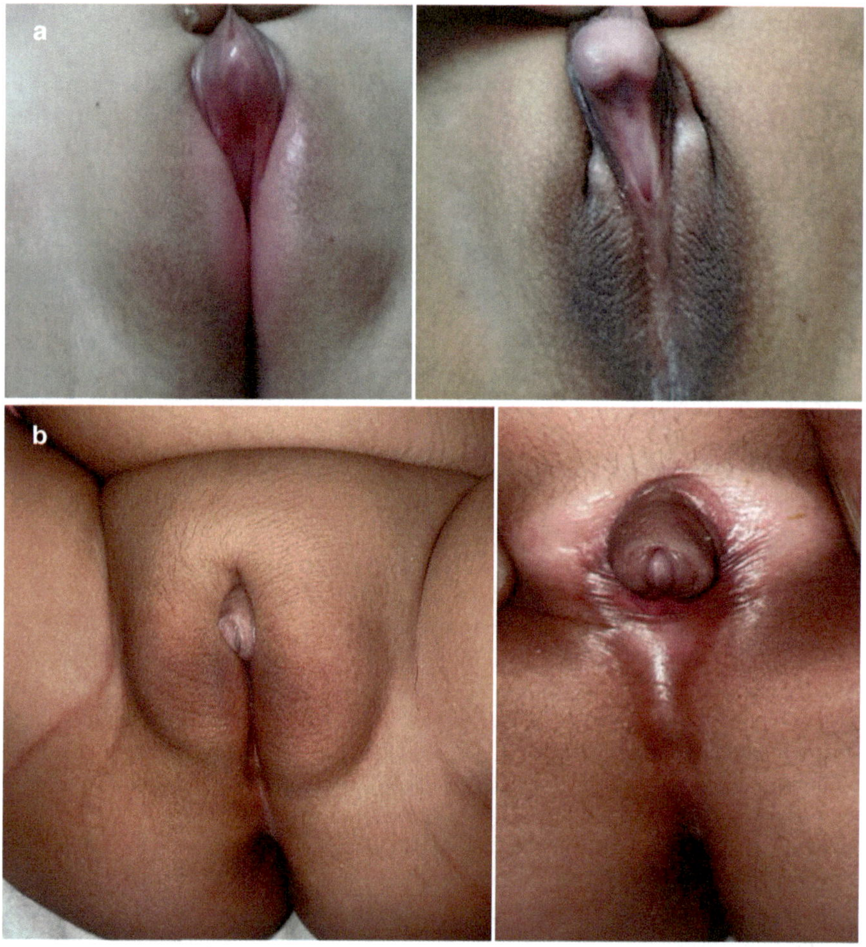

Fig. 17.1 (**a**) Hypertrophied Clitoris in a case of CAH. (**b**) Hypoplastic Phallus in a case of 5α-reductase deficiency. (**c**) Bifid scrotum with penoscrotal transposition in a case of testosterone biosynthetic defects

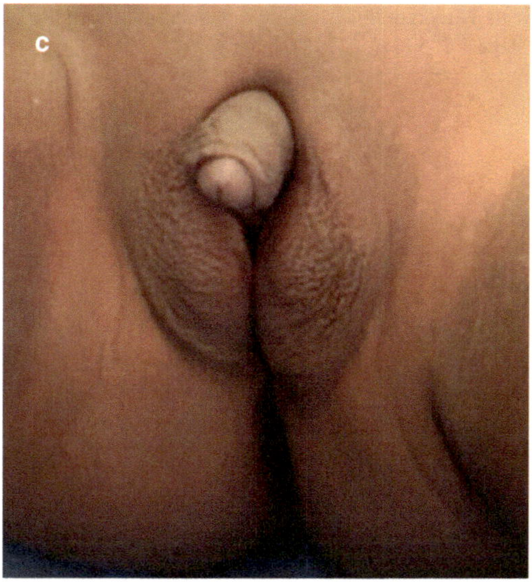

Fig. 17.1 (continued)

- *Bifid scrotum and Penoscrotal transposition* (partial or complete) should be properly noted as these may warrant DSD evaluation [15]. (Fig. 17.1c).

Going by our experience it is always good to label the phallus as hypertrophic clitoris or hypoplastic penis based on overall greater similarity of genitalia towards male or female type.

As we are already aware of steroid hormones synthesis cycle, when there is deficiency of 17,20 lyase only sex hormones are affected. Similarly, when there is deficiency of 17-beta hydroxy steroid dehydrogenase then formation of testosterone is affected exclusively [16].

The evaluation of chromosomal makeup of an individual here plays an important role in reaching conclusive diagnosis in the cases where it is difficult to adjudge whether the phallus is hypertrophic clitoris or hypoplastic penis.

One should be aware of the Prader staging that provides a useful guide to interpret the severity of virilization in girls with CAH. Though it is rare, but it is not impossible to have a child with Prader stage 5 and female genetic makeup. We are showing here Prader stage 4 & 5 in a genetic female which are very confusing in terms of overlapping phenotypic characters (Table 17.1).

Tiny phallus and hypoplastic scrotum can pose a diagnostic challenge. For e.g., tiny phallus is seen in Mixed Gonadal Dysgenesis due to the presence of dysgenetic testis and streak gonad. Hypoplastic scrotum is seen in DMP (dysgenetic male pseudo hermaphroditism). Micropenis and poorly defined secondary sexual characteristics are the features of Kallman syndrome [17].

Table 17.1 Prader scale is a grading system used to measure degree of genital virilization and showing here Prader stage 4 & 5 in a genetic female

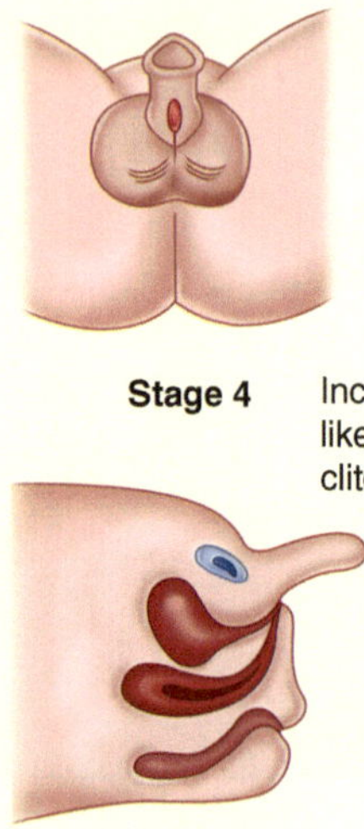	**Stage 4**	Increasingly phallic clitoris, urethra like urogenital sinus at base of clitoris and complete labial fusion
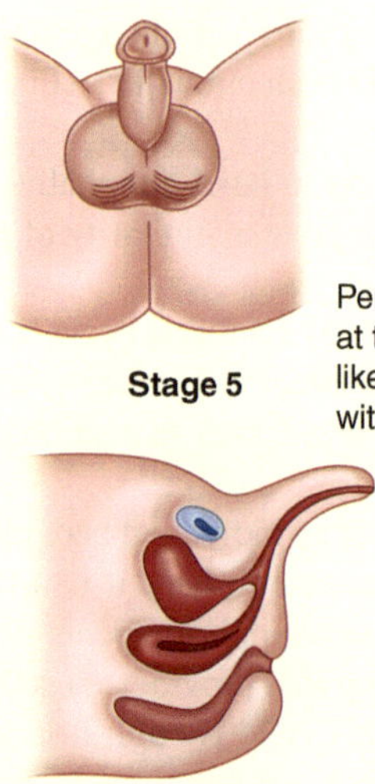	**Stage 5**	Penile ciltoris, urethral meatus at tip of phallus and scortum like labia (appear like male without palpable gonads)

17.4 Error Trap 3: Failure to Appreciate Development and Pigmentation Status of Genitalia

Hyperpigmentation noted in females and some males with CAH due to high concentration of adrenocorticotropic hormone (ACTH) and melanocytestimulating hormone (MSH) mediated by reciprocally low hormonal levels. It is better appreciated under daylight/white light and pigmentation is brownish yellow in character. However, in some conditions, male DSD accompanies with hypopigmentation due to 17 ketosteroid dehydrogenase deficiency. The condition may be unrecognized early in males as only signs are precocious (or delayed puberty) puberty and few systemic complaints.

Ambiguous hypoplastic and sometimes hypopigmented genitalia in male CAH observed with 17,20 lyase, 17-hydroxy steroid dehydrogenase. So, good size phallus means there has been a good testosterone exposure in utero. The enzymatic defects proximal to these in cholesterol cycle may not only result in metabolic syndrome but even female type of genitalia.

Hence, pigmentation status of the genitalia along with the chromosome result is very important in interpreting the type of defect which is seen in a child with ambiguous genitalia.

17.5 Error Trap 4: Failure to Keep in Mind the Incidence of Conditions Leading to DSD and Gender along with Age-linked Impacts, Degree and Extent of Blocks of Hormonal Synthesis

This means when evaluating a case of DSD, pediatric surgeon must keep a wider horizon and open mind as frequency of condition may be different in different geographical areas or among different ethnic groups.

Intersex clinic data from AIIMS, New Delhi, India, Male pseudohermaphrodite (46XY DSD) is the commonest type while, data from rest of the world congenital adrenal hyperplasia is the commonest [18]. This may be because nonclassic form manifests later in subtle form.

Among the MPH (46XY DSD) androgen insensitivity including complete androgen insensitivity syndrome (CAIS) and partial androgen insensitivity syndrome (PAIS) are more common as compared to other testosterone biosynthetic defects like 5-alpha reductase which is far less common [19]. In contrast, data from SKIMS, Kashmir, India, and other countries from Middle East where consanguineous marriages are common, 5-alpha reductase deficiency is frequent [20]. Virilization may occur at puberty due to testosterone surge.

Classical CAH causes hyperpigmentation and virilization in young girls but precocious puberty among boys.

17,20 lyase deficiency and 17 ketosteroid dehydrogenase deficiency can affect male hormones; cause undervirilization but at puberty the substrate from alternate sources can overcome this deficiency.

Worldwide CAH is the commonest followed by mixed gonadal dysgenesis (MGD). Salt-wasting type is the most common type of CAH followed by simple virilizing that is a more moderate form and non-salt wasting is the least common form of CAH (SW > SV > NSW types of CAH) [5, 21]. Among CAH, 21-hydroxylase deficiency is the commonest while 11-hydroxylase is the second commonest. Degree of enzymatic blockade can affect phenotypic outcomes.

Also, certain defects are commoner in some geographic areas and ethnic groups; for e.g., 5-alpha reductase deficiency is seen in Turkey and true hermaphrodite is seen in Bantu tribe of Africa [22]. True hermaphrodite (ovotesticular DSD) is the rarest.

17.6 Error Trap 5: Not Being Able to Recognize and Manage Life-threatening Metabolic Conditions Associated with DSD which Point to the Diagnosis

The metabolic conditions may be the only presenting feature and draw your attention towards genitalia later or while you are working up for ambiguous genitalia and subsequently encountered severe metabolic derangements.

CAH patients are generally associated with severe biochemical abnormalities which can be life-threatening referred to as classical form; however, CAH can present in simple virilizing form which is seen in 25% of cases and has different presentation. If a patient is metabolic stable that does not mean that the patient does not have CAH. The main feature of simple virilizing CAH in newborn females is the abnormal development of the external genitalia, which has varying degrees of virilization [23, 24]. In addition to the salt-wasting and simple virilizing forms of CAH diagnosed in infancy, a mild or "non-classic" form exists, which is characterized by varying degrees of postnatal androgen excess but is sometimes asymptomatic. The nonclassic form may be noticed in late childhood and may lead to accelerated growth, premature sexual maturation, acne, and secondary polycystic ovary syndrome.

It is also worth noting that 17,20 lyase and 17-beta ketosteroid deficiencies are not associated with metabolic problems but are generally associated with undervirilization among males as these largely affect testosterone biosynthesis.

High blood pressure or hypertension is associated with 11-hydroxylase deficiency.

So, in the background of inaccessible gonads, especially with metabolic problems and growth-related problems, it is an important "error trap" if CAH is not recognized on time.

17.7 Error Trap 6: Not Been Able to Recognize Nature of Gonads: Ovary, Testes, or Ovotestes

Despite laparotomy or laparoscopy, faliure to maintain a high degree of promptness in recognizing the nature of gonads can act as yet another error trap, some dictums must be kept in mind while doing work-up in a patient of DSD.

- Any palpable gonad is testis—normal or dysplastic
- Ovaries will never be palpable in the groin at birth
- If both gonads are palpable, it cannot be a case of 46 XX DSD

- Generally, the gonads are accompanied with its duct so, in the presence of fallopian tube-like structure, the attached gonad has to be ovary or streak gonad and if there is vas like structure the gonad attached has to be testis or ovotestis.

Palpable testes indicate that Mullerian ducts would have undergone regression since transabdominal testicular descent and former are controlled by MIS (Mullerian inhibiting substance) [21].

Ovotestis will have dual consistency; softer part being testis and firmer being ovary. Ovarian tissues are polar in location while the testicular tissues are present towards the center although it is a histological diagnosis in gonadal biopsy. The distribution of testicular and ovarian components can be polar or hilar in location [25, 26].

Descent of the gonads depends on testicular amount in such patients and fully descended gonads in ovotesticular karyotype do not rule out presence of ovarian tissue.

Streak gonads (small dysgenetic ovaries without follicle) are generally intra-abdominal while in dysgenetic testis may descend and in latter, testicular tissue is predominant. Such gonads hardly contribute to fertility though may still have some hormonal output.

17.8 Error Trap 7: Not Able to Identify the Presence of Mullerian Structures and Limitation of Investigations

Close inspection and evaluation of genitalia should be done along with identification and characterization of urogenital sinus or meatal opening.

Per rectal examination used to be done earlier by some clinician revealed cord-like structure behind urethra and bladder as uterus and it is well appreciated in neonates. This is generally not resorted to nowadays due to availability of better investigation modalities and chances of errors on plain examination and the chances of pain and injury caused [27].

USG may show well-formed uterus behind bladder, gonads, and ducts but this radiological investigation is operator dependent, and MRI may be utilized in such cases [9, 10].

Ultrasonography (US), a tool most commonly used for the assessment of ambiguous genitalia (AG), and demonstrating internal reproductive organs, identification of the uterus or testes, as well as establishing the presence or absence of gonads and Müllerian derivatives [10]. The sensitivity and accuracy of US depend on probe resolution, as well as an experienced ultrasonographer. Magnetic resonance imaging (MRI) is used as an adjunct modality to evaluate ambiguous genitalia, as it is helpful in clarifying the internal anatomy and searching for internal gonad with excellent soft-tissue contrast and tissue characterization [9, 28]. MRI is now

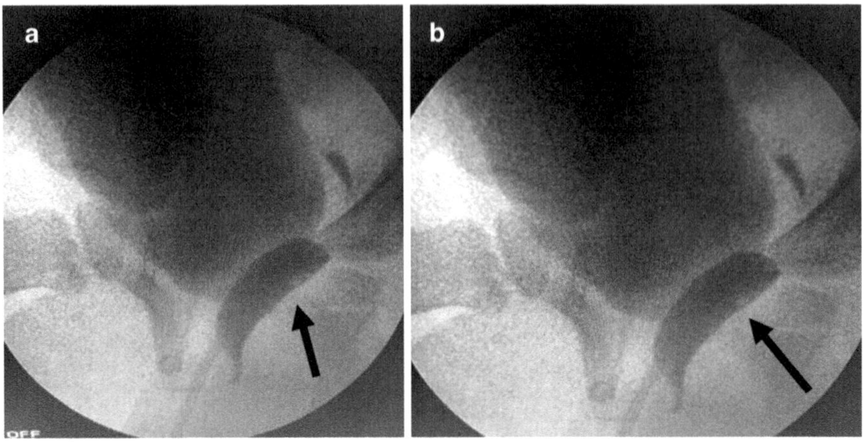

Fig. 17.2 (**a**, **b**) Dye study showing vagina (arrow) as a rounded vault structure

considered as investigation with 100% sensitivity and specificity, whereas 3D USG can match its efficiency in few cases while for detection of normal uterus its sensitivity ranges around 83.3% [29]. MRI can additionally detect renal problems [30]. The above limitations should be kept in mind before interpreting and arriving at a conclusive diagnosis. Appropriately done genitogram is an important tool in evaluation of a case of ambiguous genitalia. Usually, vagina is seen as a capacious streak even in newborn extending above pubic symphysis with convex filling defect at apex and absence of this defect may indicate vaginal pouch or blind ended vagina [31, 32]. Inappropriately done or misinterpreted dye study may miss the common channel or give wrong impression and confuse the interpretation of results.

Genitoscopy/endoscopy and laparoscopy are next line of evaluation. Care must be taken to identify the structures as the anatomy of urogenital system may be variable.

In ovotesticular DSD, uterine cavity and cervical impressions are rudimentary on dye study, whereas in CAH, uterine cavity, cervix are normal, and capacity of vagina would depend upon virilization [19] (Fig. 17.2a, b).

17.9 Error Trap 8: Not Being Able to Evaluate Family History and Look for Additional Syndrome

It is important to keep in mind that many of these disorders are inheritable as these could be X-linked or autosomal recessive. History of unexplained deaths, similar genital ambiguity in the family, infertility, hirsutism, amenorrhea, maternal exposure to exogenous medications, containing androgens, such as synthetic progesterone preparation, consanguinity and abnormal pubertal spurts, or growth in family is important in diagnostic workup.

Disorders like androgen insensitivity syndromes (AIS) are X linked recessive with multiple syndromic associations. Many syndromic names have been given to many of the presentations of AIS, such as Reifenstein syndrome, Goldberg–Maxwell syndrome, Morris' syndrome, Gilbert-Dreyfus syndrome, Lub's syndrome, incomplete testicular feminization, Rosewater syndrome, and Aiman's syndrome. These different presentations were all caused by the same set of mutations in the androgen receptor gene [33].

Family history of other affected individuals related to each other is in a pattern consistent with X-linked inheritance. "Other affected family members" refers to:

- Affected 46,XY individuals
- Manifesting heterozygous females (46,XX). About 10% of heterozygous females have asymmetric distribution and sparse or delayed growth of pubic and/or axillary hair

CAH disorders, the commonest cause of DSD are mostly autosomal recessive [23, 24]. However, the next common cause of DSD, i.e. 46XY DSD profile due to androgen insensitivity syndrome has X-linked inheritance and the case gets clearer delving into the family tree. Family history of other affected individuals related to each other is in a pattern consistent with X-linked inheritance. While affected 46,XY individual is the presenting child with DSD, "Other affected family members" refer to manifesting heterozygous females (46,XX). About 10% of heterozygous females have asymmetric distribution and sparse or delayed growth of pubic and/or axillary hair.

Syndromic features can lead to the diagnosis and such features are seen in MGD (in as high as 30% of patients). MGD is associated with 45XO (Turner Syndrome) or 46XY (XY Mosaicism). Turner syndrome (TS) is a condition characterized by short stature and ovarian dysgenesis [34].

Disorders in genetic male (46XY) with Mullerian structures as persistent Mullerian duct syndrome (PMDS) are inherited as autosomal recessive and so is 5 alpha reductase deficiency. The history of consanguinity may point to these conditions and hence should be enquired into to prevent any error trap.

17.10 Error Trap 9: Failure to Institute Appropriate Investigation and Interpret these Correctly in Terms of Rare Diagnosis

There are many investigations used for basic workup in these disorders to reach at specific problem or diagnosis. We should keep in mind that there could be some limitations of the radiological workup and other investigations as already stated in earlier sections.

Chromosomal analysis: Short arm of "Y" translocation to X causes XX males; while dysgenetic gonads with XY females. Hence, chromosome makeup alone cannot lead to a proper diagnosis.

Radiological investigations: USG is operator dependent. If an ultrasonologist is not able to look Mullerian structures on ultrasound that does not mean that these are absent, they can still be present. MRI is much superior and used in difficult cases. MRI has advantages of being multiplanar, superior tissue characterization and coupled with dye study.

Biochemical investigations: To assess gonadal tissue functionality [11, 12]:

- Entire reproductive hormonal profile including gonadotropins may be required in complex cases
- 17-OH progesterone is raised in CAH but not in few other CAH where androstenedione/testosterone needs to be evaluated; ACTH stimulation tests may be required specially in proximal defects
- Precursors or T/DHT levels; specially in pre-pubertal males, beta hCG stimulation may be required
- Even estriol estimation will be required; in those with TH
- While computing the ratios of the hormones, their units should be properly adjusted, or the results may be misleading
- The samples for hormone estimation have to follow natural circadian rhythm and following waning away of influence of maternal hormones. Suboptimal collection and storage of the samples may lead to genesis of the fallacious results.
- Additional special studies like receptor study that can be demonstrated by study on genital skin biopsy and histopathological study on genital skin biopsy specimens can be taken at the time of endoscopies [35]. The donor site has to be carefully chosen. Similarly, inappropriately done gonadal biopsy can miss hilar location especially in TH, bisecting gonads where thin biopsy is taken [36].

17.11 Error Trap 10: Not Able to See Acquired or Central Causes

Thorough history is very important in diagnostic workup of DSD patients involving any extraneous causes or other central causes responsible for ambiguous genitalia.

- Intake of androgens, for e.g. In 46,XX newborns, exposure to an excess of intrauterine androgen causes different degrees of virilization and genital ambiguity as defined by the Prader Scale
- Mothers with virilizing tumors
- Kallman syndrome—Kallman syndrome and its normosmic variant collectively known as congenital hypogonadotropic hypogonadism are principally characterized by the association of an isolated defect in the secretion (or, less commonly, action) of gonadotropin-releasing hormone (GnRH) and consequent infertility, with anosmia and potentially other associated non-reproductive features [17, 35]
- Idiopathic virilization
- Subtle hormonal biosynthetic defects

This is as far as various error traps showing how vigilant a pediatric surgeon/clinician must be before approaching the investigation map in a case of DSD.

References

1. Blackless M, Charuvastra A, Derryck A, Fausto-Sterling A, Lauzanne K, Lee E. How sexually dimorphic are we? Review and synthesis. Am J Hum Biol. 2000;12:151–66.
2. Sax L. How common is intersex? A response to Anne Fausto-Sterling. J Sex Res. 2002;39:174–8.
3. Thyen U, Lanz K, Holterhus PM, Hiort O. Epidemiology and initial management of ambiguous genitalia at birth in Germany. Horm Res. 2006;66:195–203.
4. ALOmran HI, ALJurayyan NA. Disorders of Sex Development (DSD): diagnostic approach and Management in Infants and Children. Biomed J Sci Techn Res. 2021;36(5):28940–50.
5. Kim KS, Kim J. Disorders of sex development. Korean J Urol. 2012;53:1–8.
6. Cohen-Bendahan CC, van de Beek C, Berenbaum SA. Prenatal sex hormone effects on child and adult sex-typed behavior: methods and findings. Neurosci Biobehav Rev. 2005;29:353–84.
7. Meyer-Bahlburg HF. Gender and sexuality in classic congenital adrenal hyperplasia. Endocrinol Metab Clin N Am. 2001;30:155–71.
8. Wallen K. Nature needs nurture: the interaction of hormonal and social influences on the development of behavioral sex differences in rhesus monkeys. Horm Behav. 1996;30:364–78.
9. Mehdi SA, Shaukat A, Dogar IH, Hussnain N. Imaging of ambiguous genitalia. Professional Med J. 2008;15(4):492–5.
10. Govind CB, Parra DA, Oudjhane K, Miller SF, Babyn PS, Pippi Salle JL. Imaging of ambiguous genitalia: classification and diagnostic approach. Radio Graphics. 2008;28(7):1891–904.
11. Brain CE, Creighton SM, Mushtaq I, Carmichael PA, Barnicoat A, Honour JW, Larcher V, Achermann JC. Holistic management of DSD. Best Pract Res Clin Endocrinol Metab. 2010 Apr;24(2):335–54.
12. Lee PA, Houk CP, Ahmed SF, Hughes IA. International consensus conference on intersex organized by the Lawson Wilkins Pediatric Endocrine Society and the European Society for Paediatric Endocrinology. Consensus statement on management of intersex disorders. International Consensus Conference on Intersex Pediatrics. 2006;118:e488–500.
13. Yerkes EB, Rink RC. Surgical management of female genital anomalies, disorders of sexual development, urogenital sinus, and cloacal anomalies. In: Gearhart JG, Rink RC, Mouriquand PDE, editors. Pediatric urology. Elsevier Inc; 2010. p. 476–99.
14. Cimador M, Catalano P, Ortolano R, Giuffrè M. The inconspicuous penis in children. Nat Rev Urol. 2015 Apr;12(4):205–15.
15. Wong YS, Tam YH, Pang KK, Yau HC. Incidence and diagnoses of disorders of sex development in proximal hypospadias. J Pediatr Surg. 2018 Dec 1;53(12):2498–501.
16. Andersson S, Geissler WM, Wu L, Davis DL, Grumbach MM, New MI, Schwarz HP, Blethen SL, Mendonca BB, Bloise W, Witchel SF. Molecular genetics and pathophysiology of 17 beta-hydroxysteroid dehydrogenase 3 deficiency. J Clin Endocrinol Metabol. 1996 Jan 1;81(1):130–6.
17. Stamou MI, Georgopoulos NA. Kallmann syndrome: phenotype and genotype of hypogonadotropic hypogonadism. Metabolism. 2018;86:124–34.
18. Gupta DK. Ambiguous Genitilia and hypospadias: Pediatric intersex clinic – AIIMS experience. 1st ed. New Delhi: sponsored by WHO/MCI/NAMS; 2000.
19. Gupta DK. Ambiguous Genitalia and hypospadias: disorders of sexual differentiation. 1st ed. New Delhi: sponsored by WHO/MCI/NAMS; 2000.
20. Bashamboo A, McElreavey K. Consanguinity and disorders of sex development. Hum Hered. 2014;77(1–4):108–17.
21. Pagani RL, Abou GR. Disorders of sex determination and development. Medscape; 2018. p. 258–62.

22. Laubscher WM. A male pseudohermaphrodite. S Afr Med J. 1967 Sep 1;41(37):929–31.
23. Pang S, Wallace MA, Hofman L, Thuline HC, Dorche C, Lyon IC, Dobbins RH, Kling S, Fujieda K, Suwa S. Worldwide experience in newborn screening for classical congenital adrenal hyperplasia due to 21-hydroxylase deficiency. Pediatrics. 1988 Jun;81(6):866–74.
24. Gidlöf S, Falhammar H, Thilén A, von Döbeln U, Ritzén M, Wedell A, Nordenström A. One hundred years of congenital adrenal hyperplasia in Sweden: a retrospective, population-based cohort study. Lancet Diabetes Endocrinol. 2013 Sep 1;1(1):35–42.
25. Van Niekerk WA. True hermaphroditism: an analytic review with a report of 3 new cases. Am J Obstet Gynecol. 1976 Dec 1;126(7):890–907.
26. Heier J. A review of anatomical presentation and treatment in true hermaphroditism. Best Integrated Writing. 2014;1(1):12.
27. Mehmood KT, Rentea RM. Ambiguous genitalia and disorders of sexual differentiation. StatPearls [Internet] 2022 May 8. StatPearls Publishing.
28. Ravi N, Kadakola B. Radiological evaluation of ambiguous genitalia with various imaging modalities. Int J Adv Res Technol. 2012;1(2):1–14.
29. Graupera B, Pascual MA, Hereter L, Browne JL, Úbeda B, Rodríguez I, Pedrero C. Accuracy of three-dimensional ultrasound compared with magnetic resonance imaging in diagnosis of Müllerian duct anomalies using ESHRE–ESGE consensus on the classification of congenital anomalies of the female genital tract. Ultrasound Obstet Gynecol. 2015 Nov;46(5):616–22.
30. Kanemoto K, Hayashi Y, Kojima Y, Maruyama T, Ito M, Kohri K. Accuracy of ultrasonography and magnetic resonance imaging in the diagnosis of non-palpable testis. Int J Urol. 2005;12(7):668–72.
31. Hryhorczuk AL, Phelps AS, Yu RN, Chow JS. The radiologist's role in assessing differences of sex development. Pediatr Radiol. 2022 Apr;52(4):752–64.
32. Marei M, Abdelsattar AH, Fares AE, Elkotby M, Elbarbary M, Seif H. Accuracy and predictivity of cystoscopy and genitography to operative findings in persistent urogenital sinus due to virilized female congenital adrenal hyperplasia. Kasr El Aini J Surg. 2014 Jan;15(1):21e5.
33. Pérez-Palacios G, Chávez B, Méndez JP, McGinley JI, Ulloa-Aguirre A. The syndromes of androgen resistance revisited. J Steroid Biochem. 1987;27(4–6):1101–8.
34. Jung JY, Yang S, Jeong EH, Lee HC, Lee YM, Han HS, Yi KH. Mixed gonadal dysgenesis in 45, X Turner syndrome with SRY gene. Ann Pediatr Endocrinol Metab. 2015 Dec;20(4):226.
35. Ogilvy-Stuart AL, Brain CE. Early assessment of ambiguous genitalia. Arch Dis Child. 2004 May 1;89(5):401–7.
36. Nistal M, Garcia-Fernández E, Mariño-Enríquez A, Serrano A, Regadera J, González-Peramato P. Diagnostic value of the gonadal biopsy in the disorders of sex development. Actas Urologicas Espanolas. 2007 Oct 1;31(9):1056–75.

Part V

Sex Chromosome DSD

I whole heartedly thank Dr. Sangita Yadav WHO Fellow, Ex Director Professor Pediatrics, Maulana Azad Medical College, New Delhi, Dr. Amita Sen Consultant Professor Pediatric Surgery, Atal Bihari Vajpayee Institute of Medical Sciences, New Delhi and Dr. Anjan Dhua Additional Professor. Pediatric Surgery, AIIMS Delhi for assisting and aiding me in this section.

Turner Syndrome

18

Preeti Singh and Sangeeta Yadav

Abstract

Turner syndrome (TS) is a chromosomal disorder seen in phenotypic females due to partial or complete loss of the X chromosome. The clinical phenotype in TS varies widely despite similar chromosome complement. The clinical spectrum in girls with TS includes typical phenotypes of varying severity (shield chest with widely spaced nipples, short webbed neck, cubitus valgus, Madelung deformity), short stature, delayed puberty, hypergonadotropic hypogonadism, neurocognitive and behavioral problems, cardiovascular malformations of the heart, osteoporosis, and association with various autoimmune and metabolic disorders. A standard 30-cell karyotype is recommended to diagnose TS and to detect 10% mosaicism with 95% confidence. A multidisciplinary team approach is needed to manage children and adolescents with TS across their lifespans. The karyotype–phenotype and genotype–phenotype correlations in TS may help to prognosticate and develop an individualized care plan. Screening and surveillance of comorbidities are recommended at diagnosis and thereafter for early detection and management of complications. A good transition care plan to adulthood can achieve the best quality of life outcomes.

Keywords

Short stature · Hypergonadotropic hypogonadism · Delayed puberty · Karyotype

---

P. Singh
Department of Pediatrics, Lady Hardinge Medical College and Kalawati Saran Children's Hospital, University of Delhi, New Delhi, India

S. Yadav (✉)
Department of Pediatrics, Maulana Azad Medical College and Assoc. Hospitals, University of Delhi, New Delhi, India

S. K. Ratan (ed.), *Children with Differences in Sex Development*,
https://doi.org/10.1007/978-981-97-1639-5_18

18.1 Turner Syndrome

Turner syndrome (TS) is the most common sex chromosome abnormality in females. It affects phenotypic females with one intact X chromosome and complete or partial absence of the second sex chromosome in association with one or more clinical manifestations. It was first described by Otto Ullrich in 1930 as Ullrich scleroatonic muscular dystrophy, and then in 1938 by Henry Turner [1, 2]. A clinical diagnosis of Turners was made in seven females based on the presence of short stature, sexual immaturity, neck webbing, and cubitus valgus. Therefore, TS is also known as Ullrich–Turner syndrome.

18.2 Incidence and Prevalence

TS affects 1 in 2,500 liveborn females, estimated through epidemiological surveys and reports of newborn screening from Europe, Japan, and the USA [3–5]. The true prevalence of TS is however difficult to ascertain because most patients remain undiagnosed. Some individuals are not diagnosed until late adulthood if their phenotype is mild [6]. The prevalence of TS is roughly similar in various countries and ethnic groups. There has been a decline in the prevalence of TS at birth. This is due to the increased use of prenatal ultrasound for screening TS during pregnancy and hence the termination of affected fetuses [7, 8]. Most fetuses carrying X chromosome monosomy (45, X) do not survive until birth. In 10% of all spontaneous abortions, the genotype reported is 45, X [9, 10].

18.3 Pathophysiology

TS is a chromosomal disorder characterized by complete or partial absence of the second X chromosome. Most (40–50%) girls with TS have the 45, X karyotype, while a varied degree of mosaicism (45,X/46,XX; 45,X/47,XXX; 45,X/46,XX/47,XXX) is observed in 15–25% of them [11, 12]. The aneuploidy in classic Turner's karyotype results from the nondisjunction of the paternal genome in a majority of (60–80%) cases (left with only maternal X chromosome). In mosaics, on the other hand, the nondisjunction of the sex chromosome occurs during postzygotic cell division, which results in a combination of both normal and abnormal cell lines. Structural aberrations in the X chromosome, like isochromosome 46, X,i(Xq), ring 46,X, r(X)chromosomes, and deletion 46,X, del(Xp) have been reported in 20–25% of individuals with TS [11, 12]. Varied amounts of Y chromosome material were found in 10–12% of individuals with TS; 3% may have 45X,46XY karyotype [11]. The clinical phenotype in mosaics depends on the timing of the mosaicism during the developmental stage and its distribution in the affected tissues. A low degree of mosaicism is found in phenotypically normal females. Since mosaicism can occur in all cell lines, it can present a diagnostic challenge if only the peripheral blood is assessed for the karyotype. Women with 45,X/46, and XX mosaicism usually have a milder phenotype with less severe congenital heart disease and are more likely to undergo spontaneous menarche and experience pregnancies [11]. In a recent systematic review, the highest rate of spontaneous thelarche and menarche has been reported in women with 45, X/47, and XXX mosaicism [13]. In TS, the presence of Y chromosomal material is associated with virilization during puberty and the rare development of gonadoblastoma. A standard 30-cell karyotype is recommended to detect 10% mosaicism with 95% confidence [4].

Karyotype–Phenotype Correlation There is a wide variability in the phenotypic expression and severity of comorbidities, despite the presence of similar chromosome complement in individuals with TS [11, 12]. This can be attributed to various mechanisms beyond the genomic imbalance (haploinsufficiency of X-chromosomal genes) that determine the phenotype in TS. Multiple factors including epigenetic mechanisms, imprinting of genes on X chromosome, and the degree of X chromosome inactivation regulate the gene expression across the genome and influence other autosomal genes [11, 14]. The precise impact of these changes and other genomic mechanisms need to be studied and elucidated. The parental origin of the X chromosome can also affect phenotypic diversity in TS [11, 12]. The varied phenotypes associated with karyotypes in TS are given in Table 18.1.

Genotype–Phenotype Correlation Despite the clinical variability, a few genes have been reported to have a definite correlation with specific phenotypes in TS. Short stature homeobox-containing gene (SHOX gene) escapes X-inactivation and its decreased expression can explain the growth deficit observed in TS. Other clinical features associated with SHOX haploinsufficiency include Madelung deformity, high-arched palate, micrognathia, scoliosis, and reduced leg length (but nor-

Table 18.1 Karyotype–phenotype correlation in TS [14]

Karyotype	%	Phenotype characteristics
Monosomy X 45,X	40–50	
Mosaicism 45,X/46,XX	15–25	Milder phenotype Less severe congenital heart disease/lymphatic abnormalities Spontaneous menarche and pregnancy May experience frequent early miscarriages
Mosaicism with "Triple X" 45,X/47,XXX; 45,X/46,XX/47,XXX	3	Spontaneous menarche and thelarche
45,X/46,XY	10–12	Less likely to develop autoimmune disorders such as hypothyroidism; Virilization during puberty, risk for gonadoblastoma
Deletion Xp22.3—46,XX, del(Xp22.3);		Hearing problems
Isochromosome Xq; X 46,X, I(Xq); isodicentric 46,X,idic(Xp)	10	Similar to 45 X, but higher incidence of hearing loss, autoimmunity, and congenital heart disease
Ring X chromosome 46,X,r(X)/46,XX		Learning difficulties, behavioral maladjustment, metabolic syndrome

mal sitting height) [15]. The UTX gene is implicated in repeated episodes of otitis media in TS. The haploinsufficiency of the gene TIMP1 (tissue inhibitor of matrix metalloproteinase 1) that escapes X-inactivation in combination with structural aberrations of the TIMP3 gene (on chromosome 22) increases (>10 fold) the risk of bicuspid aortic valve, aortic dilation and aneurysms in women with TS [16]. The variable phenotype in women with TS is due to the differential methylation of a few escape genes (KDM6A, UBA1, STS, and USP9X) resulting in changes in the RNA expression as compared to those with 46, XX. Differential methylation of the KDM6A (lysine-specific demethylase 6A) gene located on the short arm of chromosome X at position 11.3 is associated with gonadal dysgenesis [17]. Similarly, differential expression of the gene KDM5C (encoding lysine-specific demethylase 5C) can explain the distinct neurocognitive profile of TS [17].

18.4 Clinical Features

The clinical phenotype in TS varies widely among affected females. The spectrum and the severity of clinical presentation vary, while some may develop features over a period of time. Short stature is the most characteristic feature, invariably reported in all individuals with TS. The distinct facial features are epicanthic folds, low set/rotated and deformed ears, fish-like mouth, micrognathia, and narrow high-arched palate. Other characteristic phenotypic features in TS include a "shield" chest with the appearance of widely spaced nipples, a short and webbed neck (hypoplasia of vertebrae), cubitus valgus (increased carrying angle of the elbow), and Madelung deformity of the forearm and wrist (Figs. 18.1 and 18.2). Distinctive clinical clues towards the diagnosis of TS in the newborn include webbing of the neck, narrow and high-arched palate, lymphedema of the hands and feet, nail dysplasia, and short fourth metacarpals and/or metatarsals. The malformation of the ears, neck, chest, nipples, nails, and kidneys is attributed to the lymphatic duct hypoplasia that results in severe edema. Intelligence is usually preserved in TS, but girls may experience deficits in specific neurocognitive domains like visuospatial organization or a non-verbal learning disorder.

The two most common manifestations in individuals with TS are linear growth failure and pubertal delay or premature ovarian insufficiency.

Short stature is seen in all girls with TS. In the absence of treatment, the adult height is approximately 20 cm shorter than the age and sex-matched reference population and correlates well with the parental stature (mid-parental height) [18]. The adult height of Indian Turner's girls was observed to be 17.7 cm shorter than that of the national average in a study published in the western part of India [19]. The birth length of a newborn with TS is 1 SD shorter than a neonate of comparable gestational age, highlighting the onset of the growth failure in fetal life. The post-natal growth velocity is slow during childhood and is further blunted by delayed puberty and a lack of a pubertal growth spurt.

Short stature and other skeletal abnormalities in TS are associated with haploinsufficiency of the SHOX (short stature homeobox-containing gene on the X

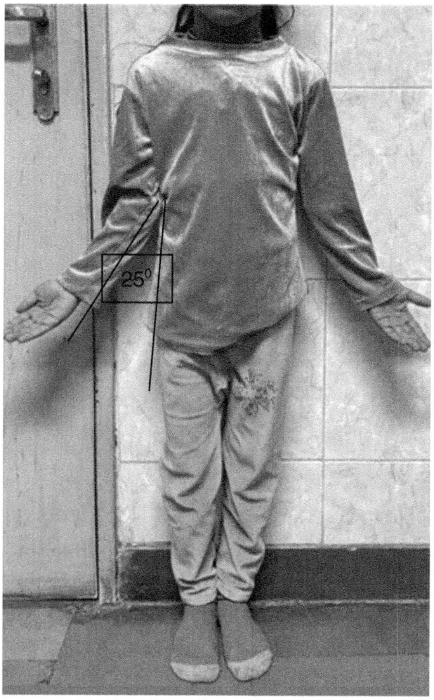

Fig. 18.1 Increased carrying angle and short stature in a 10-year girl with TS. In the normal female population, the carrying angle at the elbow averages 12^0, while it increases to ≥ 20–30^0 in girls with TS

chromosome) gene. The proliferation and maturation of chondrocytes in the growth plate are regulated by the SHOX gene. It controls the expression of growth factors including FGFR3 (encoding fibroblast growth factor receptor 3) and NPPB (natriuretic peptides B). In TS, abnormal growth hormone (GH) secretory dynamics (nocturnal GH secretion) result in functional GH insufficiency; the levels of GH, insulin-like growth factor-1 (IGF-1), and insulin-like growth factor-2 (IGF-2) are within normal range. Therefore, the growth failure in TS is due to the relative resistance of GH action at the growth plate. Further, there is minimal or no permissive release of the GH pulses in the absence of sex steroids, i.e. estrogen during the pubertal phase. There is an increased risk of autoimmune disorders like autoimmune thyroid disease, celiac disease, and inflammatory bowel disease (IBD) that compounds the problem of growth failure and short stature in TS. The characteristic skeletal malformations like short neck and scoliosis also contribute to poor stature in TS.

Hypergonadotropic Hypogonadism and Pubertal Delay The premature ovarian failure in TS can be attributed to haploinsufficiency of genes on the pseudoautosomal region of the X chromosome (BMP-15, FMR 1,2) [11, 12]. Though the exact mechanism is still unclear, accelerated loss of oocyte and progressive increase in connective tissue stroma are observed in the ovaries after 6–8 weeks of gestation in fetuses with TS. This increased rate of apoptosis results in gonadal failure in them. The degree of ovarian dysfunction is variable, ranging from no ovarian function

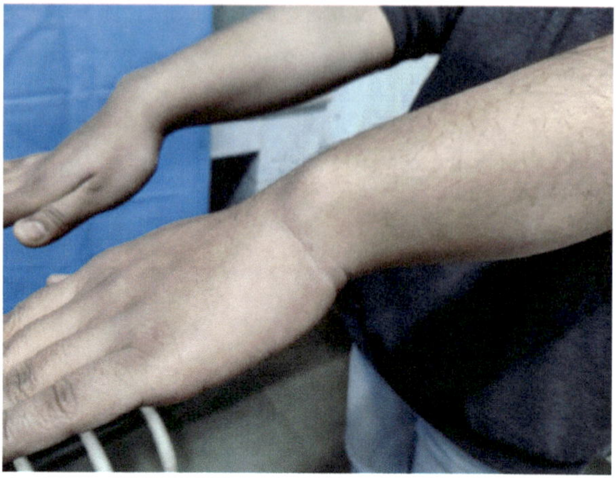

Fig. 18.2 Madelung deformity (prominent ulnar head and apparent volar subluxation of the wrist on the forearm) in TS

("streak gonads" with small amounts of connective tissue without follicles or only a few atretic follicles) to spontaneous thelarche (30%) and menarche (15%) with normal reproductive function for at least a few years after menarche. Girls with TS can present with primary or secondary amenorrhea (after normal pubertal onset and progression), delayed or arrested puberty, or infertility in adult life. It is characterized by delayed bone age and features of hypergonadotropic hypogonadism (raised FSH/LH and low estradiol levels). A small proportion of girls with mosaicism can have normal onset and progression of puberty with regular menstrual cycles [13]. TS should be considered in any girl with unexplained short stature and or delayed pubertal development.

Due to wide variability in gonadal function, predicting the pubertal outcome of girls with TS is difficult. Anti-Mullerian hormone (AMH) is a sensitive marker of prepubertal follicular development and its measurement can help to assess the ovarian function in TS throughout childhood [20]. All women with TS will require hormone replacement at some point, to maintain age-appropriate secondary sexual characteristics, and normal uterine growth, optimize bone mineral health, and improve metabolic health and psychosexual development.

Cardiovascular Abnormalities Cardiovascular diseases are one of the most important life-threatening comorbidities that contribute substantially to increased mortality. There is wide variability in the prevalence based on the referral pattern, imaging studies, and the focus of examination [21]. The spectrum ranges from congenital heart disease [coarctation of the aorta (7–14%), bicuspid aortic valve (14–34%)], aortic aneurysm/dilatation (3–42%), to hypertension (50%) and conduction/repolarization defects attributed to neuroanatomic dysfunction. Other left-sided structural defects observed in TS include subaortic obstruction, coronary anomalies, abnormal pulmonary vein drainage, and mitral valve dysplasia [22]. Cardiac MRI is a more sensitive tool (than echocardiography) to detect various cardiovascular defects, including valvular and systemic or pulmonary venous abnormalities in TS [23]. Aortic dissection (1–2%) is the most dreaded life-threatening complication in TS, usually preceded by dilatation of the aortic root or ascending aorta. Hypertension is a significant risk factor for aortic dissection, stroke, and ischemic heart disease. Various factors implicated in the pathophysiology of hypertension in TS include a dysregulated sympathetic nervous system, increased resting noradrenaline levels, coarctation of the aorta, and estrogen deficiency [24, 25]. A thorough cardiovascular assessment is recommended for all women with TS at diagnosis and during follow-up [14, 26].

Body Composition and Metabolic Disease Women with TS have an increased risk of metabolic disease due to the altered body composition. They have a high total fat mass and visceral fat and decreased total lean body mass compared with normal age-appropriate women [27, 28]. Published studies have observed a high incidence of impaired glucose tolerance with insulin resistance, type 2 diabetes (5–25%), dyslipidemia, and metabolic syndrome in individuals with TS [29]. Type 1 diabetes has been reported in TS from the Western series.

The haploinsufficiency for SHOX genes is associated with intrinsic bony abnormalities in TS. This leads to the development of abnormal trabecular microarchitecture and lower cortical bone porosity in the radius, tibia, metacarpal bones, femoral neck, and lower spine [30]. The women with TS have low bone mineral density (BMD) and are vulnerable to fracture risk. Estrogen deficiency, osteoporosis, and abnormal bone development and modeling can result in increased bone fragility in TS.

Autoimmune Diseases Women with TS are at increased risk of developing a wide range of autoimmune diseases. In a meta-analysis, the pooled prevalence of autoimmune thyroid diseases in TS was 38.6% (95% CI 29.7–47.6%) [31]. Other autoimmune disorders commonly seen in TS include celiac disease, type 1 diabetes, IBD, rheumatoid arthritis, psoriasis, vitiligo, and alopecia areata [32].

Renal Abnormalities Congenital malformations of the renal/urinary tract system have been reported in 30–40% of girls with TS. Collecting duct system malformations (20%), malrotation, and horseshoe kidneys (10%) are the common renal malformations in TS [33]. The obstructive pathologies at the ureteropelvic junction can lead to clinically significant hydronephrosis and risk for pyelonephritis.

Gastrointestinal and Hepatic Disorders The derangement of liver function in TS mainly comprises elevated transaminases and alkaline phosphatase, while the bilirubin levels may remain within the normal range [34]. Dyslipidemia, estrogen deficiency, and developmental abnormalities of the liver are responsible for hepatic dysfunction. Other common comorbidities associated with TS are celiac disease (8%) and IBD (3%).

Psychological, Neurocognition, and Quality of Life The intellectual ability is usually unaffected in TS except in the subgroup with the ring X chromosome.

However, neurocognitive problems are seen in a substantial proportion of girls with TS [35]. More than two-thirds of girls have nonverbal learning difficulties translating to poor mathematical and problem-solving tasks, impaired social cognition, poor psychomotor skills, and poor visual-spatial organization [36]. Girls with TS have an increased risk of attention deficit disorder (ADD) and problems with executive functioning. Hence, early referral to an appropriate psychologist with specific neurocognitive expertise should be done. Untreated adolescents experience social interaction challenges, poor self-esteem, anxiety, and even depression with poor quality of life [37]. Periodic psychological assessment and timely intervention can improve their outcome. Age-appropriate puberty induction with hormone replacement therapy is associated with improved self-esteem and social adaptation in girls with TS.

Ear and Hearing Loss Structural abnormalities of the Eustachian tubes and the cranial base have been observed in TS. This predisposes to recurrent middle ear infections and effusion during childhood, often resulting in conductive hearing loss [38]. Adult women with TS can develop progressive sensorineural hearing loss due to a defect in the outer hair cells of the lower middle coil of the cochlea. It worsens with age and is more commonly associated with 45, X or 45,X/46,i(Xq) karyotype [39].

18.5 Diagnosis

The prenatal diagnosis of TS is usually an incidental finding detected on amniocentesis or chorionic villus sampling, performed for unrelated reasons [40]. Physicians involved in the care/management should be well versed/aware of the complications, prognosis, and recent advances in managing children with TS. Nuchal thickening, lymphangiectasia (fetal hydrops), septated cystic hygroma, cardiac defects, kidney anomalies, or a short femur are important clues towards TS on fetal ultrasonography. In suspected cases, prenatal testing is done using FISH or GTG karyotype analysis performed on either chorionic biopsy (11–13 weeks), amniocentesis (15–19 weeks), or cordocentesis (>19 weeks). A repeat karyotype analysis of the newborn is recommended at delivery (either cord blood or neonatal blood) to confirm the diagnosis [14].

A standard 30 metaphase karyotype is recommended during childhood and adolescence in suspected phenotypes at varied time points, to confirm the diagnosis of TS [3, 14]. In the neonatal period and early infancy, the clinical clues for TS include edema of hand and feet, nuchal folds, low hairline, short fourth metacarpal, distinct facies, and left-sided cardiac anomalies (coarctation of aorta/hypoplastic left heart syndrome). Karyotype is indicated for all girls with unexplained short stature, poor growth velocity, delayed puberty or arrest, primary or secondary amenorrhea, multiple pigmentary nevi, and cubitus valgus. Cases with a normal karyotype but with a high clinical suspicion for TS should be evaluated with adjuvant FISH analysis and/or reverse transcriptase polymerase-chain-reaction (PCR) assay to identify low-level mosaics and the presence of Y chromosome material, respectively [41, 42]. In case of inconclusive results, a second karyotype should be performed using a different tissue such as skin (fibroblasts), bladder epithelial cells from a urine sample, or buccal mucosa cells [43].

18.6 Management

A multidisciplinary team approach is required to manage children and adolescents with TS across their lifespan. Once the diagnosis of TS is established, a baseline evaluation and monitoring for associated abnormalities, including cardiovascular disease, renal pathology, thyroid disease (and other autoimmune conditions like celiac disease), hearing and eye abnormalities, and learning disabilities is part of the management plan. This involves a close partnership of the primary care physician with a team comprising of an endocrinologist, cardiologist, gynecologist, psychologist, clinical geneticist, and timely referral to other subspecialty experts like a gastroenterologist, otorhinolaryngologist, nephrologist, etc. Education and counseling of the child and family with TS are essential and form the cornerstone of management. The need for a healthy lifestyle and regular physical activity is stressed at each visit to prevent cardiovascular diseases and metabolic complications. Regular and continued surveillance throughout the life course helps in early recognition of the comorbidities and complications during clinical follow-up. Psychological support and behavioral therapy are integral to managing these children and adolescents. Turner's syndrome support groups play a vital role in improving the well-being and quality of life of these children and their families. Pharmacological therapy in TS is directed towards augmentation of stature, age-appropriate puberty induction, and management of other comorbidities.

18.6.1 Augmentation of Stature

Children and adolescents with TS do not have GH deficiency, so its assessment is not indicated. GH therapy is recommended in children with TS who have short stature or are at high risk of short stature and low growth velocity. Besides maximizing the height potential, GH therapy helps improve body proportions, body composition, and attainment of peak bone mass, metabolic health, cardiovascular risk, psychosocial adaptation, and improved quality of life. However, the response to GH therapy is varied (7–15 cm over the projected adult height) as it depends upon the age of initiation, dose, and duration of GH therapy, and timing of induction of puberty [44]. Baseline height before initiation of GH therapy and mid-parental height are the other non-modifiable predictors of final adult height. Growth monitoring in TS should be done by plotting height on growth curves specific to this condition.

The optimal age for initiating GH therapy is 4–6 years, as it results in more significant height gain during childhood and allows for age-appropriate induction of puberty (minimum interval of 4 years before puberty induction). Therapy may be continued until the girl is satisfied with her height or until little growth potential remains (bone age ≥14 years and height velocity <2 cm/year). The dose recommended for GH therapy (40–50 μg/kg/day or 1.3–1.5 $mg/m^2/d$) in TS is higher than that used in GH deficiency alone, administered once daily by subcutaneous injection before bedtime [14]. Children are monitored (at least every 6 months) for treatment efficacy and undesirable side effects. An annual assessment of serum IGF-1 is recommended to titrate the dose of GH therapy. Maintaining IGF-1 concentrations below 2SD above the mean for age and/or Tanner stage of pubertal development is desirable but ideally above the mean. Screening for adverse effects like benign intracranial hypertension, scoliosis, slipped capital femoral epiphysis, and abnormalities in lipid and glucose metabolism is routinely done in children receiving GH therapy.

Low-dose oxandrolone @0.03mg/kg/d (<0.05mg/kg/d) alone or combined with GH is recommended to augment height in girls with TS. It is indicated where there is a delay in diagnosis (≥10 years) and/or if adult height outcome is likely to be unsatisfactory with standard GH alone [14]. Oxandrolone is a synthetic anabolic steroid derived from dihydrotestosterone, that increases final adult height by its direct action on the growth plate and increase in IGF-1 concentration. Its androgenic effect increases protein synthesis, lean body mass, and bone mineral density. Children treated with oxandrolone and GH for 2–3 years achieve an adult height of 2–5 cm taller than GH therapy alone (due to synergistic effect) [45, 46]. Girls are monitored for dose-dependent virilizing side effects (acne, hirsutism, clitoromegaly) and delayed puberty [47].

18.6.2 Treatment of Gonadal Failure

Hormone replacement therapy is needed for all individuals with TS at some point in life for induction of puberty, augmentation of growth, maintenance of secondary sexual characteristics, attaining peak bone mass, adequate uterine growth, improved metabolic profile, cardiovascular function, and psychosocial and neurocognitive outcome. During adolescence, girls with TS may experience signs of premature ovarian failure reflected by high gonadotropins (high FSH and LH), low estrogen, and low AMH. Children may present late in adolescence with clinical signs & symptoms of delayed puberty or pubertal arrest after having normal or late pubertal onset. Current practice guidelines recommend initiating a low-dose estrogen replacement (one-eighth or one-tenth of the adult dose) by 11–12 years to mimic the normal progression of puberty while maximizing the growth potential and minimizing risks [14]. A delay in pubertal induction beyond 12 years can adversely impact bone mineral health, cognition, and psychosocial effects. Small (25–50%) and gradual increments in estrogen dose are made (every 6 months to 1 year) to reach the adult dose in 2–2.5 years. This low and gradual building up of estrogen levels will preserve the normal shape of breast development. The timing of pubertal induction and titration of the dose are further individualized based on the clinical response.

Transdermal natural estrogens or oral 17 β-estradiol preparations (estradiol valerate) are preferred over older ethinyl estradiol or conjugated equine estrogen, as the former is more physiological. The studies comparing similar doses of oral versus transdermal estrogen preparations have not found a difference in the outcome related to body composition, lipids, HbA1c, and bone mineralization [48]. Following the commencement of breakthrough/menstrual bleeding or after 2 years of estrogen therapy, a gestagen is added to establish monthly menstrual cycles. Progesterone therapy should be introduced after ensuring optimal uterine growth with mature endometrium and adequate breast development. Cyclical therapy with medroxyprogesterone acetate 5–10 mg/day is added for 10–12 days per month while continuing estrogen to reduce the risk of endometrial hyperplasia and cancer.

18.6.3 Management of Other Comorbidities

A comprehensive cardiovascular evaluation by a pediatric cardiologist is recommended for all children with TS at diagnosis. Blood pressure (both the upper and lower extremities) measurement is recommended at every clinical visit for early detection and management of hypertension. ECG (prolonged QTc interval) and echocardiography (screen structural abnormalities such as coarctation of the aorta and bicuspid aortic valve) are done for all children at diagnosis and repeated during follow-up as needed. Cardiac magnetic resonance (CMR) is utilized in older children (usually after 12 years) for both screening and surveillance of cardiovascular abnormalities like dilated ascending aorta, an elongated transverse aortic arch (other aortic phenotypes), and partial anomalous pulmonary venous return. Renal ultrasonography is recommended for all children at diagnosis to identify congenital malformations of the renal/urinary system. Annual developmental and behavioral screening is recommended until adulthood, with specialized referrals as needed. Regular neuropsychologic assessments are performed at critical transitional stages in school, at entry into elementary, middle, and high school. Addressing specific learning disabilities will help improve these children's quality of life. Regular audiology and ophthalmologic evaluation (at diagnosis and every three years) are recommended for all girls in TS. Screening for other comorbidities like celiac disease, autoimmune thyroiditis, type 1 diabetes, liver disease, and risk factors for metabolic syndrome (dyslipidemia, type 2 diabetes) is performed regularly after 10 years of age in TS (Fig. 18.3).

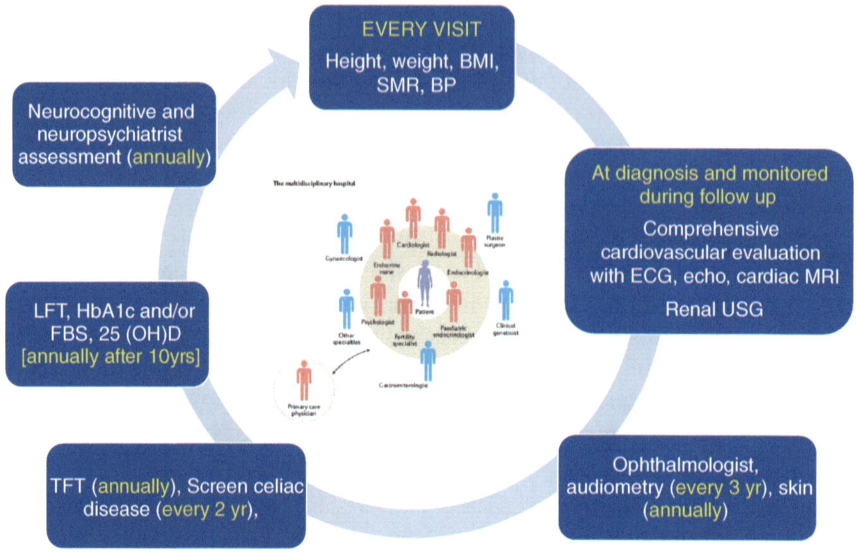

Fig. 18.3 Screening recommendations for comorbidities in TS [12]

18.7 Fertility and Pregnancy

Only 5–7% of women with TS (with mosaic karyotype) with normal FSH and AMH concentrations experience spontaneous pregnancies [49]. However, the rate of spontaneous miscarriage is high among these women [49]. Adoption or oocyte donation from 46, XX can be considered a viable alternative to achieve pregnancy in women with TS. Cryopreservation of ovarian tissue and/or oocytes from selected preadolescent or adolescent girls with TS is a potential upcoming technique (under investigation) that might permit reproduction later in life [50]. Pregnancy should, however, be planned after careful assessment of cardiovascular risk (especially aortic dissection) and other comorbidities.

18.8 Transition of Care

A planned and staged healthcare transition, beginning in early adolescence is recommended for girls with TS [14, 51]. A comprehensive adult care roadmap should be developed by the treating physician/pediatrician/pediatric endocrinologist before transitioning to adult care. During the transition phase, the management plan gradually shifts from optimizing stature to timely initiation and compliance with sex hormone replacement therapy in girls with TS. Regular screening for comorbidities and complications must be ensured before transfer. A cardiovascular assessment including an electrocardiogram, transthoracic echocardiogram, or cardiac MRI is recommended to screen for cardiac anomalies and aortic disease. Implementing lifestyle changes with regular weight-bearing exercises to prevent obesity and metabolic abnormalities and maintaining adequate bone health assume importance. All of these need to be initiated while the child is still under the care of a pediatric endocrinologist. It is important to discuss and counsel the adolescent and her family about the prospects of fertility (spontaneous versus assisted pregnancy). They should be sensitized about various options available for reproductive techniques like oocyte preservation [52]. Assessment of psychosocial problems like poor self-esteem, body image issues, impaired visual-spatial processing, perceptual abilities, and motor coordination should be actively pursued during adolescence. Psychological comorbidities pose a significant challenge in girls with TS. Adequate motivational support from family and peer groups during the transition phase can improve behavioral and emotional well-being in girls with TS [53]. It is important to establish collaboration between primary and other specialty providers before/during the transition, to address the multifaceted care issues specific to TS.

Fostering social interactions through TS advocacy groups encourages peer-to-peer (and parent-to-parent) contact and provides resources for education or employment. They are vital in enhancing knowledge and confidence during the transition, reducing stress and distress, and promoting the reciprocal sharing of experiences. Valuable resources to facilitate transition include "Turner Syndrome Pediatric to Adult Care Transition Toolkit" available at https://www.endocrine.org/improving-practice/transitions/turner-syndrome, and "Transition to adulthood" guidelines by the Turner Syndrome Foundation available at https://turnersyndromefoundation.org/what_is_turner_syndrome/transition.

References

1. Turner HH. A syndrome of infantilism, congenital webbed neck, and cubitus valgus. Endocrinology. 1938;28:566.
2. Kinderheilk Z. Über typische Kombinationsbilder multipler Abartungen. Eur J Pediatr. 1930;49:271.
3. Bondy CA. Turner Syndrome Study Group. Care of girls and women with Turner syndrome: a guideline of the Turner syndrome study group. J Clin Endocrinol Metab. 2007;92:10.
4. Cockwell A, MacKenzie M, Youings S, Jacobs P. A cytogenetic and molecular study of a series of 45,X fetuses and their parents. J Med Genet 1991;28:151.
5. Nielsen J, Wohlert M. Chromosome abnormalities found among 34,910 newborn children: results from a 13-year incidence study in Arhus. Denmark Hum Genet. 1991;87:81.
6. Gunther DF, Eugster E, Zagar AJ, et al. Ascertainment bias in Turner syndrome: new insights from girls who were diagnosed incidentally in prenatal life. Pediatrics. 2004;114:640.
7. Baena N, De Vigan C, Cariati E, et al. Turner syndrome: evaluation of prenatal diagnosis in 19 European registries. Am J Med Genet A. 2004;129A:16.
8. Iyer NP, Tucker DF, Roberts SH, et al. Outcome of fetuses with Turner syndrome: a 10-year congenital anomaly register based study. J Matern Fetal Neonatal Med. 2012;25:68.
9. Hassold T, Pettay D, Robinson A, Uchida I. Molecular studies of parental origin and mosaicism in 45. X conceptuses Hum Genet. 1992;89:647.
10. Romero ST, Geiersbach KB, Paxton CN, et al. Differentiation of genetic abnormalities in early pregnancy loss. Ultrasound Obstet Gynecol. 2015;45:89.
11. Viuff M, Skakkebaek A, Nielsen MM, Chang S, Gravholt CH. Epigenetics and genomics in Turner syndrome. Am J Med Genet C Semin Med Genet. 2019;181(1):68–75.

12. Gravholt CH, Viuff MH, Brun S, Stochholm K, Andersen NH. Turner syndrome: mechanisms and management. Nat Rev Endocrinol. 2019;15(10):601–14.
13. Dabrowski E, Jensen R, Johnson EK, Habiby RL, Brickman WJ, Finlayson C. Turner syndrome systematic review: spontaneous Thelarche and menarche stratified by karyotype. Horm res Paediatr 2019;92(3):143–149.2.
14. Gravholt CH, Andersen NH, Conway GS, Dekkers OM, Geffner ME, Klein KO, Lin AE, Mauras N, Quigley CA, Rubin K, Sandberg DE, Sas TCJ, Silberbach M, Söderström-Anttila V, Stochholm K, van Alfen-van derVelden JA, Woelfle J, Backeljauw PF; International Turner syndrome consensus group. Clinical practice guidelines for the care of girls and women with Turner syndrome: proceedings from the 2016 Cincinnati international Turner syndrome meeting. Eur J Endocrinol 2017;177(3):G1-G70.
15. Marchini A, Ogata T, Rappold GA. A track record on SHOX: from basic research to complex models and therapy. Endocr Rev. 2016;37(4):417–48.
16. Corbitt H, Morris SA, Gravholt CH, Mortensen KH, Tippner-Hedges R, Silberbach M, Maslen CL, Investigators GTACR. TIMP3 and TIMP1 are risk genes for bicuspid aortic valve and aortopathy in Turner syndrome. PLoS Genet. 2018;14(10):e1007692.
17. Trolle C, Nielsen MM, Skakkebæk A, Lamy P, Vang S, Hedegaard J, Nordentoft I, Ørntoft TF, Pedersen JS, Gravholt CH. Widespread DNA hypomethylation and differential gene expression in Turner syndrome. Sci Rep. 2016;6:34220.
18. Brook CG, Mürset G, Zachmann M, Prader A. Growth in children with 45. XO Turner's syndrome Arch Dis Child. 1974;49(10):789–95.
19. Khadilkar VV, Karguppikar MB, Ekbote VH, Khadilkar AV. Turner syndrome growth charts: a Western India experience. Indian J Endocrinol Metab. 2020;24(4):333–7.
20. Lunding SA, Aksglaede L, Anderson RA, Main KM, Juul A, Hagen CP, Pedersen AT. AMH as predictor of premature ovarian insufficiency: a longitudinal study of 120 Turner syndrome patients. J Clin Endocrinol Metab. 2015;100(7):E1030–8.
21. Mortensen KH, Young L, De Backer J, Silberbach M, Collins RT, Duijnhouwer AL, Pandya B, Gravholt CH, Lopez L, Roos-Hesselink JW. Cardiovascular imaging in Turner syndrome: state-of-the-art practice across the lifespan. Heart. 2018;104(22):1823–31.
22. Mortensen KH, Gopalan D, Nørgaard BL, Andersen NH, Gravholt CH. Multimodality cardiac imaging in Turner syndrome. Cardiol Young. 2016;26(5):831–41.
23. Silberbach M, Roos-Hesselink JW, Andersen NH, Braverman AC, Brown N, Collins RT, Backer D, Kim HK, Gottliebson W, Hor K, Backeljauw P, Gutmark-Little I, Salisbury SR, Racadio JM, Helton-Skally K, Fleck R. Cardiovascular anomalies in Turner syndrome: spectrum, prevalence, and cardiac MRI findings in a pediatric and young adult population. AJR Am J Roentgenol. 2011;196(2):454–60.
24. Gravholt CH, Hansen KW, Erlandsen M, Ebbehøj E, Christiansen JS. Nocturnal hypertension and impaired sympathovagal tone in Turner syndrome. J Hypertens. 2006;24(2):353–60.
25. Zuckerman-Levin N, Zinder O, Greenberg A, Levin M, Jacob G, Hochberg Z. Physiological and catecholamine response to sympathetic stimulation in turner syndrome. Clin Endocrinol. 2006;64(4):410–5.
26. Silberbach M, Roos-Hesselink JW, Andersen NH, Braverman AC, Brown N, Collins RT, De Backer J, Eagle KA, Hiratzka LF, Johnson WH Jr, Kadian-Dodov D, Lopez L, Mortensen KH, Prakash SK, Ratchford EV, Saidi A, van Hagen I, Young LT; American Heart Association council on cardiovascular disease in the Young; council on genomic and precision medicine; and council on peripheral vascular disease. Cardiovascular health in Turner syndrome: a scientific statement from the American Heart Association. Circ Genom precis med. 2018;11(10):e000048.
27. Gravholt CH, Weis NR. Reference values for body proportions and body composition in adult women with Ullrich-Turner syndrome. Am J Med Genet. 1997;72(4):403–8.
28. Gravholt CH, Hjerrild BE, Mosekilde L, Hansen TK, Rasmussen LM, Frystyk J, Flyvbjerg A, Christiansen JS. Body composition is distinctly altered in Turner syndrome: relations to glucose metabolism, circulating adipokines, and endothelial adhesion molecules. Eur J Endocrinol. 2006;155(4):583–92.

29. Freriks K, Timmermans J, Beerendonk CC, Verhaak CM, Netea-Maier RT, Otten BJ, Braat DD, Smeets DF, Kunst DH, Hermus AR, Timmers HJ. Standardized multidisciplinary evaluation yields significant previously undiagnosed morbidity in adult women with Turner syndrome. J Clin Endocrinol Metab. 2011;96(9):E1517–26.
30. Hansen S, Brixen K, Gravholt CH. Compromised trabecular microarchitecture and lower finite element estimates of radius and tibia bone strength in adults with turner syndrome: a cross-sectional study using high-resolution-pQCT. J Bone Miner Res. 2012;27(8):1794–803.
31. Mohamed SOO, Elkhidir IHE, Abuzied AIH, Noureddin AAMH, Ibrahim GAA, Mahmoud AAA. Prevalence of autoimmune thyroid diseases among the Turner syndrome patients: meta-analysis of cross sectional studies. BMC Res Notes. 2018;11(1):842.
32. Bonamico M, Pasquino AM, Mariani P, Danesi HM, Culasso F, Mazzanti L, Petri A, Bona G. Italian society of pediatric gastroenterology Hepatology (SIGEP); Italian study Group for Turner Syndrom (ISGTS). Prevalence and clinical picture of celiac disease in Turner syndrome. J Clin Endocrinol Metab. 2002;87(12):5495–8.
33. Tukel T, Bas F, Kilic G, Basaran S, Gunoz H, Apak M. Frequency of renal malformations in Turner syndrome: analysis of 82 Turkish children. Pediatr Nephrol. 2000 Oct;14(12):1111–4.
34. El-Mansoury M, Berntorp K, Bryman I, Hanson C, Innala E, Karlsson A, Landin-Wilhelmsen K. Elevated liver enzymes in Turner syndrome during a 5-year follow-up study. Clin Endocrinol. 2008;68(3):485–90.
35. McCauley E, Feuillan P, Kushner H, Ross JL. Psychosocial development in adolescents with Turner syndrome. J Dev Behav Pediatr. 2001;22(6):360–5.
36. Sybert VP, McCauley E. Turner's syndrome. N Engl J Med. 2004;351(12):1227–38.
37. Reis CT, de Assumpção MS, Guerra-Junior G, de Lemos-Marini SHV. Systematic review of quality of life in Turner syndrome. Qual Life Res. 2018;27(8):1985–2006.
38. Hamberis AO, Mehta CH, Dornhoffer JR, Meyer TA. Characteristics and progression of hearing loss in children with turner's syndrome. Laryngoscope. 2020;130(6):1540–6.
39. Hultcrantz M, Sylvén L, Borg E. Ear and hearing problems in 44 middle-aged women with Turner's syndrome. Hear Res. 1994;76(1–2):127–32.
40. Swauger S, Backeljauw P, Hornung L, Shafer J, Casnellie L, Gutmark-Little I. Age at and indication for diagnosis of Turner syndrome in the pediatric population. Am J Med Genet A. 2021;185(11):3411–7.
41. Binder G, Koch A, Wajs E, Ranke MB. Nested polymerase chain reaction study of 53 cases with Turner's syndrome: is cytogenetically undetected Y mosaicism common? J Clin Endocrinol Metab. 1995;80(12):3532–6.
42. Akcan AB, Boduroğlu OK. Y chromosome material in Turner syndrome. Cureus. 2021;29;13(11):e19977
43. Azcona C, Bareille P, Stanhope R. Lesson of the week: Turner's syndrome mosaicism in patients with a normal blood lymphocyte karyotype. BMJ. 1999;318(7187):856–7.
44. Li P, Cheng F, Xiu L. Height outcome of the recombinant human growth hormone treatment in Turner syndrome: a meta-analysis. Endocr Connect. 2018;7(4):573–83.
45. Sheanon NM, Backeljauw PF. Effect of oxandrolone therapy on adult height in Turner syndrome patients treated with growth hormone: a meta-analysis. Int J Pediatr Endocrinol. 2015;2015(1):18.
46. Mohamed S, Alkofide H, Adi YA, Amer YS, AlFaleh K. Oxandrolone for growth hormone-treated girls aged up to 18 years with Turner syndrome. Cochrane Database Syst Rev. 2019;2019(10):CD010736
47. Sas TC, Gault EJ, Bardsley MZ, Menke LA, Freriks K, Perry RJ, Otten BJ, de Muinck Keizer-Schrama SM, Timmers H, Wit JM, Ross JL, Donaldson MD. Safety and efficacy of oxandrolone in growth hormone-treated girls with Turner syndrome: evidence from recent studies and recommendations for use. Horm Res Paediatr. 2014;81(5):289–97.
48. Klein KO, Rosenfield RL, Santen RJ, Gawlik AM, Backeljauw PF, Gravholt CH, Sas TCJ, Mauras N. Estrogen replacement in Turner syndrome: literature review and practical considerations. J Clin Endocrinol Metab. 2018;103(5):1790–803.

49. Mortensen KH, Rohde MD, Uldbjerg N, Gravholt CH. Repeated spontaneous pregnancies in 45. X Turner syndrome Obstet Gynecol. 2010;115(2 Pt 2):446–9.
50. Borgström B, Hreinsson J, Rasmussen C, Sheikhi M, Fried G, Keros V, Fridström M, Hovatta O. Fertility preservation in girls with turner syndrome: prognostic signs of the presence of ovarian follicles. J Clin Endocrinol Metab. 2009;94(1):74–80.
51. Bernard V, Donadille B, Le Poulennec T, Nedelcu M, Martinerie L, Christin-Maitre S. Management of endocrine disease: transition of care for young adult patients with Turner syndrome. Eur J Endocrinol. 2019;180(1):R1–7.
52. Grynberg M, Bidet M, Benard J, Poulain M, Sonigo C, Cédrin-Durnerin I, Polak M. Fertility preservation in Turner syndrome. Fertil Steril. 2016;105(1):13–9.
53. Oris L, Seiffge-Krenke I, Moons P, et al. Parental and peer support in adolescents with a chronic condition: a typological approach and developmental implications. J Behav Med. 2016;39(1):107–19. https://doi.org/10.1007/s10865-015-9680-z.

Klinefelter Syndrome

19

Kanika Shanker and Sangita Yadav

Abstract

The diagnosis of Klinefelter syndrome often goes unidentified due to a lack of recognition of clinical features. The recognition of this is important for the treatment and prevention of consequences and complications, respectively. In this chapter, we will review the age-based presentation and discuss treatment options.

Keywords

Disorders of Sexual Differentiation · Klinfelters Syndrome · Delayed Puberty · Infertility · Gynecomastia

- History and Background: Epidemiology

Klinefelter syndrome is the most common cause of primary hypogonadism. It is seen in approximately 1 to 2.5 per 1000 men [1]. Importantly, 25 to 50% of them are diagnosed in their second decade of life.

- Pathogenesis and Genetic Abnormality

K. Shanker
Division of Pediatric Endocrinology, Goryeb Children's Hospital, Atlantic Health System, Morristown, NJ, USA

S. Yadav (✉)
Department of Pediatrics, Maulana Azad Medical College and assoc. Hospitals, University of Delhi, New Delhi, India

S. K. Ratan (ed.), *Children with Differences in Sex Development*,
https://doi.org/10.1007/978-981-97-1639-5_19

The genetic abnormality of Klinefelter syndrome is the presence of an extra sex chromosome. It occurs due to meiotic nondisjunction during the formation of either eggs or sperms. This defective process has equal chances to occur in either mother or father.

Due to the genetic abnormality, there is progressive degeneration, hyalinization, and fibrosis mediated by myoid cells of both the germ cells and the Leydig cells. The damage to the seminiferous tubules leads to decreased sperm production and the damage to the Leydig cells leads to hormonal imbalance leading to decreased testosterone production. Abnormalities of spermatogenesis generally precede and are more severe in comparison to abnormalities in testosterone production [2, 3].

- Age-based Clinical Presentation
  1. Neonate and infancy: It is very important to have clinical suspicion of Klinefelter syndrome in a newborn presenting with a shorter size of penis, hypospadias, and or undescended testis.
  2. Children: Many boys with Klinefelter syndrome are often diagnosed during the normal duration of puberty. Either they do not start puberty or have stalled puberty, meaning failure to complete pubertal development. It usually presents with age inappropriate size (smaller) of the testes, less pubic hair and body hair. In later years, they can present with persistent gynecomastia [2].
  3. Boys of pubertal age with Klinefelter syndrome are taller in comparison to their mid-parental height. In terms of body habitus, the legs are much taller and therefore the lower segment is more than the upper segment. This leads to lower upper segment to lower segment ratio. Their leg length is on average, 4 to 8 cm higher than men without Klinefelter syndrome. This difference varies by the severity of the Klinefelter syndrome phenotype and presence of other hormone issues.
  4. Adults usually present with primary infertility and gynecomastia (Fig. 19.1).

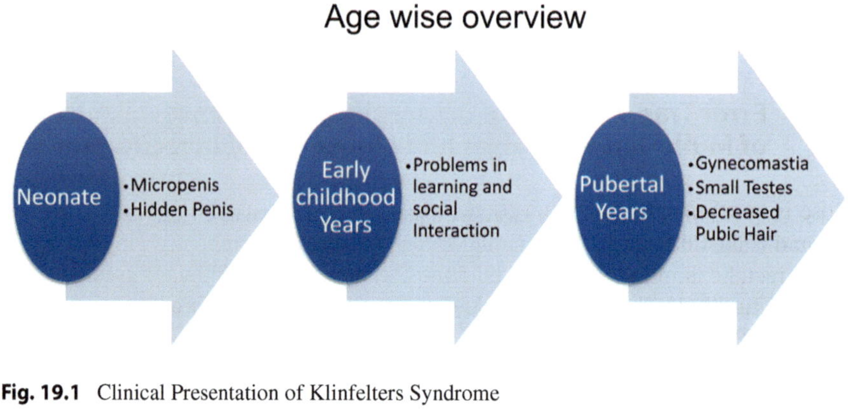

Fig. 19.1 Clinical Presentation of Klinfelters Syndrome

19.1 Comorbidities

Behavioral abnormalities, cognitive and learning disabilities may present with behavioral concerns and learning delay. Most individuals with Klinefelter syndrome have near normal or slightly low cognitive function, but good overall wellbeing. It is very important that the diagnosis of Klinefelter syndrome is made early and important resources such as speech and reading therapy are started without delay [2].

Dentofacial abnormalities—Taurodontism, defined as an enlargement of the pulp of molars on dental radiography may be associated with an increased risk of dental infections, mandibular prognathism, delayed or absence of eruption of permanent teeth [3, 4]. The relationship of the dental and gum abnormalities absolute and relative risk are currently being studied [3].

Cardiovascular—Increased association of ischemic heart disease due to dysfunction and incompetence, mitral valvular prolapse, lower varicose veins, deep venous thrombosis, and pulmonary blood clots [2]. The increased risk of deep venous thrombosis and pulmonary embolism is likely due to Klinefelter syndrome and not a side effect of testosterone therapy [5].

Pulmonary—Chronic obstructive pulmonary disease and increased lung infections are more commonly diagnosed in men with Klinefelter syndrome [6].

Musculoskeletal—Men may have decreased muscle mass and weakness due to lower testosterone values. In addition, there is an increased lifetime risk of fractures due to lower bone and osteoarthritis later in life [2].

Neurologic—There is an increased risk (25%) of essential tremors, Parkinson-like syndrome presenting as shakiness of hands [2]. Occasionally tremors can be a presenting complaint of the syndrome and therefore present in the Neurologist's office.

Cancer—The risk of breast cancer is up 50-fold higher in men with Klinefelter syndrome [2]. Men with Klinefelter syndrome account for 4% of all cases of male breast cancer.

Extragonadal germ cell tumors (especially in the mediastinal area) and non-Hodgkin lymphoma also appear to occur at a higher frequency, but the absolute risk is small [2]. The lifetime risk of prostate cancer seems to be lower than the general adult male population [2].

Autoimmune disorders—Autoimmune diseases such as rheumatoid arthritis, systemic lupus erythematosus, and hormone related autoimmune diseases such as Hashimoto's thyroiditis and type 1 diabetes mellitus are relatively more common in men with Klinefelter syndrome. Similar predisposition to autoimmune diseases is also seen with girls in Turner syndrome [2].

- Differential diagnosis
  1. Pituitary disorder presents as secondary hypogonadism, with biochemical hormonal evaluation showing low LH, FSH, and low testosterone values
  2. Kallmann syndrome, which presents with a decreased sense of smell and secondary hypogonadism
  3. Constitutional delay of puberty

4. Systemic causes such as celiac and inflammatory bowel disease
5. Testicular issues such as anorchia, luteinizing hormone receptor gene mutation, radiation, chemotherapy, trauma, surgery, and inflammation of the testis

- Diagnosis

Definitive diagnosis is based on chromosomal analysis showing one extra X chromosome with a Y chromosome (46 XXY). The biochemical features are indicative of primary hypogonadism presenting as high FSH and LH. The rise in LH and FSH is due to positive feedback from the pituitary gland to the testicular tissue. The elevation in FSH is more than LH. In the initial stages, testosterone values are near normal or low normal and then gradually fail to rise as puberty progresses; in other words, the values of testosterone are inappropriate for the stage of puberty.

- Management
  - Counseling

The first step in treatment is to communicate with the patient or the patient's family the diagnosis in a sensitive manner. The discussion regarding the diagnosis must be provided in a gentle, sensitive manner that openly discloses the genetic basis and the potential health consequences. The clinician must explain that Klinefelter syndrome is not inherited per se and that there is no fault of either patients or the patient. When initially discussing the potential consequences, it is important to stress the broad range of variability of symptoms and focus on the features that are universal: a very high likelihood of eventual permanent androgen deficiency and infertility. Because of the decreased absolute risk of other consequences of Klinefelter syndrome, it can be discussed later [4] (Table 19.1).

- Testosterone therapy

For pubertal boys (13 to 18 years) the goal of testosterone therapy is to cause gradual virilization to include penile growth, facial and pubic hair, and increased muscle mass without causing early closure of bone epiphyses and decreased adult height. Relatively lower doses of testosterone in the range of 50 to 100 mg either

Table 19.1 Age appropriate of Management of Klinfelters Syndrome

Age	Management
Prenatal diagnosis	Parental education about the evaluation and management
Neonate	Testosterone therapy for micropenis and surgical management of hypospadias [7, 8]
1–12 years	Speech and reading therapy
13–18 years	Testosterone therapy
18–50 years	Testosterone therapy and artificial reproductive techniques
>50 years	Evaluation and management of comorbidities including consideration of bone health by optimizing vitamin D and calcium intake

intramuscular testosterone enanthate or cypionate until growth ceases or until age 18 to 19 years give a good response.

For adults, initiation of testosterone therapy should be started at lower doses such as 25% to 50% of adult replacement, followed by gradual escalation to a full adult replacement dose over a few months so that the patients do not experience abrupt behavioral and mood changes (e.g., due to increased libido).

– Male infertility

Assisted reproductive technologies—The two most effective treatments for infertility in men with Klinefelter syndrome are adoption and utilization of donor sperm. However, patients with Klinefelter syndrome who are interested in having biological children should be informed about the microscopic testicular extraction of sperm (microTESE). In about 50% of men with Klinefelter syndrome liver sperms can be extracted by microTESE. These sperms can be used in assisted reproductive techniques such as intracytoplasmic injection of sperm into an ovum [9].

– Gynecomastia

For boys who have completed puberty and men with cosmetically concerning gynecomastia, reduction mammoplasty can be done. This surgery should be delayed after testosterone therapy has been initiated because untreated hypogonadism will cause breast tissue growth to reappear. The extent of surgery depends upon the severity of the breast enlargement and whether there is also excess adipose tissue present [10].

19.2 Prognosis

Overall health expectancy is good for them. The risks of adverse effects of Klinefelter syndrome appear to be low, and these risks do not justify deviation from published guidelines for preventive health care in males. Although many adverse health consequences are associated with Klinefelter syndrome, there is no evidence that specific preventive or screening measures should differ from the general population.

References

1. Herlihy AS, Halliday JL, Cock ML, McLachlan RI. The prevalence and diagnosis rates of Klinefelter syndrome: an Australian comparison. Med J Aust. 2011;194:24.
2. Gravholt CH, Chang S, Wallentin M, et al. Klinefelter Syndrome: integrating genetics, neuropsychology, and endocrinology. Endocr Rev. 2018;39:389.
3. Belling K, Russo F, Jensen AB, et al. Klinefelter syndrome comorbidities linked to increased X chromosome gene dosage and altered protein interactome activity. Hum Mol Genet. 2017;26:1219.

4. Tremblay I, Van Vliet G, Gonthier M, Janvier A. Partnering with parents to disclose Klinefelter syndrome to their child. Acta Paediatr. 2016;105:456.
5. Chang S, Biltoft D, Skakkebæk A, et al. Testosterone treatment and association with thrombin generation and coagulation inhibition in Klinefelter syndrome: a cross-sectional study. Thromb Res. 2019;182:175.
6. Bojesen A, Juul S, Birkebaek NH, Gravholt CH. Morbidity in Klinefelter syndrome: a Danish register study based on hospital discharge diagnoses. J Clin Endocrinol Metab. 2006;91:1254.
7. Flannigan R, Patel P, Paduch DA. Klinefelter Syndrome. The effects of early androgen therapy on competence and behavioral phenotype. Sex Med Rev. 2018;6:595.
8. Ross JL, Kushner H, Kowal K, et al. Androgen treatment effects on motor function, cognition, and behavior in boys with Klinefelter Syndrome. J Pediatr. 2017;185:193.
9. Gies I, Oates R, De Schepper J, Tournaye H. Testicular biopsy and cryopreservation for fertility preservation of prepubertal boys with Klinefelter syndrome: a pro/con debate. Fertil Steril. 2016;105:249.
10. Gikas P, Mokbel K. Management of gynaecomastia: an update. Int J Clin Pract. 2007;61:1209.

Mixed Gonadal Dysgenesis

20

Amita Sen and Vishrut Narang

Abstract

Mixed Gonadal Dysgenesis (MGD), a disorder of sexual differentiation and is the most frequent chromosomal abnormality presenting with ambiguous genitalia, second only to congenital adrenal hyperplasia (CAH) in newborns. The majority have a 45X/46XY karyotype. Patients have variable internal and external anatomy, with some cases presenting as virilized females. Somatic signs of Turner's syndrome may be present. Patients may have various degrees of undescended testis, labial fusion, clitoromegaly, epispadias, or hypospadias. Along with this varied clinical spectrum, the chromosomal disposition also shows a wide variation (e.g., 45,X/46,XY or 46,XX/46,XY, and others). Patients are infertile and there is a notable vulnerability towards malignancy which increases with age. Treatment depends on whether the child has been reared as a male or a female and the genetic imprint on the child. Patients also need screening for other associated malignancies like Wilms tumor. Treating Physicians should consider subjective threat, formulate an personalized plan best suited to the patient, and avoid doing irreversible surgeries. A comprehensive plan for long-term management allowing children to do well and thrive in childhood should be executed by a multidisciplinary team.

A. Sen (✉)
Department of Pediatric Surgery, Atal Bihari Vajpayee Institute of Medical Sciences and Dr. Ram Manohar Lohia Hospital, Delhi, India

V. Narang
Child Care Centre, Sonipat, Haryana, India

S. K. Ratan (ed.), *Children with Differences in Sex Development*,
https://doi.org/10.1007/978-981-97-1639-5_20

Keywords

Mixed gonadal dysgenesis · Ambiguous genitalia · Streak gonad

20.1 Introduction

Phenotypic characters in both sexes develop due to a precise sequence of genetic, molecular, and morphological events. Mixed gonadal dysgenesis (MGD) also known as asymmetric gonadal dysgenesis is the most common chromosomal abnormality, classified as per the Chicago consensus on the management of Intersex disorders 2005. The phrase MGD was refined with the 2006 consensus statement in which chromosomal conditions were separately classified [1]. This consensus received a further update in 2016 including "chimerism" under the category listing MGD [2]. MGD is second to CAH in the genesis of ambiguous genitalia in newborns. It was first described by Pich in 1937 and the term MGD was coined by Sohval in 1963 [3]. Some synonyms of MGD are 45, X/46, XY MGD, 45, X0/46, XY MGD, asymmetrical gonadal differentiation, etc.

20.2 Epidemiology

The prevalence of this disorder is largely unknown. Approximately 40% of patients have 46 XY karyotypes, while others have mosaicism (mosaic could be 45 XO/46XY). The overall gonadal situation is between pure gonadal dysgenesis and a true hermaphrodite (Fig. 20.1).

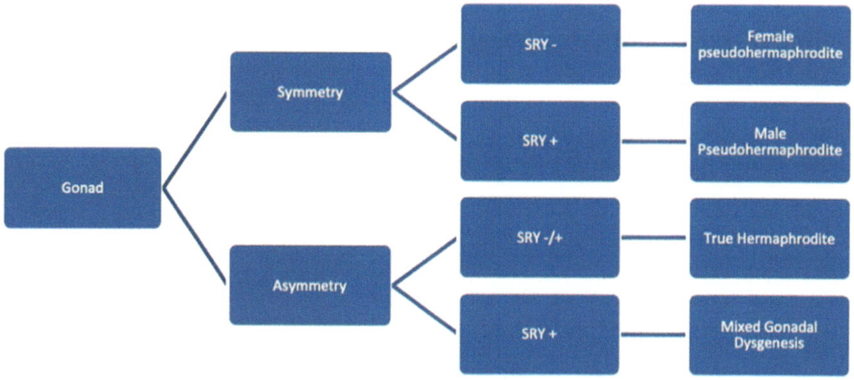

Fig. 20.1 The overview of disorders of sexual differentiation

20.3 Pathophysiology

MGD is a disorder of sexual differentiation. There is low testosterone, Mullerian inhibiting substance levels are reduced, the passage of vagina to the perineum stalls and joins the urethra to form UGS (urogenital sinus), usually adjacent to the bladder neck (incomplete differentiation of testis resulting in incomplete inhibition of Mullerian development because of lag during anaphase) [4]. Pathogenesis is due to the inadequacy of a Y chromosome-related inductive factor in the testis.

20.4 Deviation

A subgroup of MGD (45X0/XY) having near normal external genitalia, who are raised as boys, are a gamut of this disease known as DMP (dysgenetic male pseudohermaphroditism). MGD and DMP share similar features. In DMP, all cells have an XY cell line and both gonads are testis but are streak gonads. In DMP, there is the presence of bilateral dysgenetic testis and defective masculinization internally and externally [5].

20.5 Genetic Aspects and Histology

The majority cases with MGD have a 45X/46XY karyotype, with the phenotype being clinched by the fraction of monosomic cells. It has been seen that Y chromosome rearrangements are associated with a 45, X cell line like dicentric and ring Y chromosome. All cases are sporadic. The genotype–phenotype correlations seen are as follows:

1. Partial expression of SRY gene
2. Existence of gonadoblastoma locus (TSPY1) on the Y chromosome in females
3. Dosage loss of the SHOX gene leads to dwarfism in patients

The uterus develops due to a deficiency of output of the anti-Mullerian hormone (AMH).

X-linked DAX-1, which is thought to subdue differentiation of male gonads might have a part to play in the genesis of MGD. This variant of DSD has the minimal understanding at the molecular level. Some patients with Denys–Drash syndrome have MGD.

Three types of gonads seen: dysgenetic testis, streak testis, and streak gonads. The laterality of gonads in this disorder has remained an enigma.

Dysgenetic testis have no spermatids, and Leydig cells are immature. Interstitial fibrosis and areas of dysgenetic tissue are common. The seminiferous tubules reach a fairly complex state, but in the hilar zone, they become disorganised, and the tubules undergo atrophy and hyalinization. The tunica albuginea is poorly collagenized which evokes a picture of the stroma of the ovarian cortex than that of the testicular tunica which contains epithelial cords or seminiferous tubules with isolated germ cells. Sertoli cells in the dysgenetic testis fail to produce anti-Müllerian hormone (AMH), which is essential for the collagenization of the tunica albuginea and for maintaining a clear boundary between the albuginea and the testicular parenchyma. This deficiency also hinders the regression of Müllerian derivatives. Paradoxically the dysgenetic testis can respond to gonadotropins and secrete testosterone in average quantity at puberty.

Streak testis—It may occur due to a series of developmental mishaps originating from flaws of the H-Y antigen revelation or function that leads to abnormal differentiation of the indifferent urogenital ridge and, furthermore, to anomalous fabrication of Mullerian inhibiting substance and testosterone. The gonads bearing significant resemblance to a testis are commonly linked with the presence of a fallopian tube because these streak testes are anomalous and secrete minimal Mullerian inhibiting substance or secrete it behind schedule to affect the developing tissue [6]. Histologically there is a loosely associated collection of tubules with a fibrovascular stroma consistent with a rudimentary epididymis and seminiferous tubules. The streak testis can have adequate functional tissue to produce satisfactory testosterone levels.

Streak gonads—These gonads were noticed to be arranged in a stroma resembling that of a normal ovarian cortex. In older patients, the entire population of germ cells in streak gonads disappear; Robboy et al., therefore, suggested that patients with MGD should be raised as girls. Timely gonadectomy will prevent the development of malignancy. In case the uterus is preserved and the patient receives exogenous estrogen, care should be taken to detect early signals of development of endometrial carcinoma or its precursor [7].

20.6 Clinical Presentation

The phenotype in MGD is heterogeneous. External genitalia is masculinized to some extent and a blend of features of both sexes can be found in whom both gonads are abnormal. Patients may present with various degrees of undescended testis, labial fusion, clitoromegaly, epispadias, or hypospadias (Fig. 20.2). Despite these variations, MGD cases can usually be grouped into three broad clinical groups: -

(a) Female phenotype with ambiguous genitalia
(b) Male phenotype
(c) Female phenotype with Turnerian stigmata (wide neck, low set ears, broad chest with broadly spaced nipples, arms turned outwards at the elbows)

Patients with female phenotype with ambiguous genitalia have a normal ipsilateral hemiscrotum with testis, indicating sufficient production of testosterone. The contralateral gonad is a streak gonad that is not able to produce either Anti-Mullerian Hormone (AMH) or testosterone, and the external genitals are predestined to a

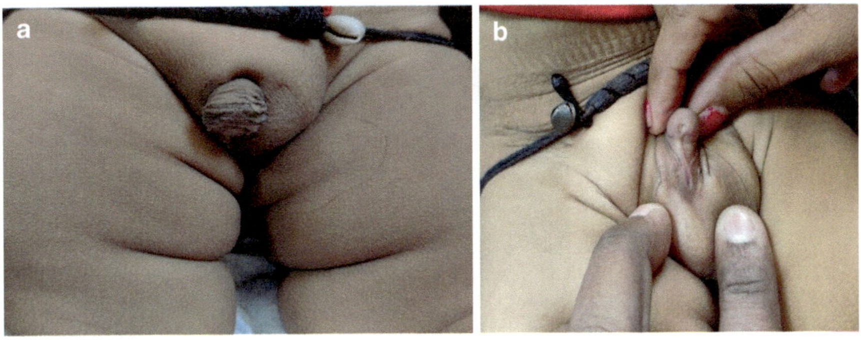

Fig. 20.2 A patient with (**a**) right undescended testis, and left intrascrotal gonad and (**b**) proximal hypospadias. On evaluation, it was found to be a case of MGD

female type and have rudimentary Mullerian derivatives (uterus and fallopian tube without a lumen). Some cases may have hernia uterine inguinale. These findings fit into the typical type of mixed gonadal dysgenesis.

Patients with phenotypic characteristics of males, likely due to the predominance of the 46, XY line of cells, can have bilateral testicles. MGD is generally diagnosed during the evaluation for hypospadias, cryptorchidism, or infertility. These physical findings mimic the features of male dysgenetic pseudohermaphroditism.

Female phenotype with Turnerian stigmata has an influence of X0 line and would be difficult to differentiate from patients of 45,X0 gonadal dysgenesis. Gonads in these cases are streak gonads bilaterally.

With this in the background, the age at diagnosis of MGD can range theoretically from the antenatal period till adulthood. Such cases can be antenatally diagnosed if an increased nuchal fold is noted, neonatal, or in infancy if there are ambiguous external genitalia. Later in childhood, MGD can be diagnosed during evaluation for short stature, in adolescence if there is delayed onset of puberty, or in adults, undergoing workup for infertility [8].

Infertility is the rule, and there is a heightened risk for malignant transformation in the gonads (due to exposed germ cells and exceptionally high and prolonged gonadotropin stimulation). Some researchers have reported an elevated risk of germ cell tumors in patients with a DSD having Y chromosome, with the risk proportionate to the amount of Y material present [9]. The risk of neoplasia increases with age, reported to be around 15–30 percent [10, 11].

Wolffenbuttel, in his study, found that gonadal histology gives an improved comprehension of the likelihood of germ cell malignancy in patients with disorders of sex development (DSD) and proof-based tactics are evolving. Especially in 46,XY DSD and 45,X/46,XY DSD, which are typified by gonadal dysgenesis, the risk of germ cell malignancy increased multifold. The authors sum up the advances made over the past decade in malignancy risk assessment in patients with DSD and its ramification for optimal timing of surgery [12].

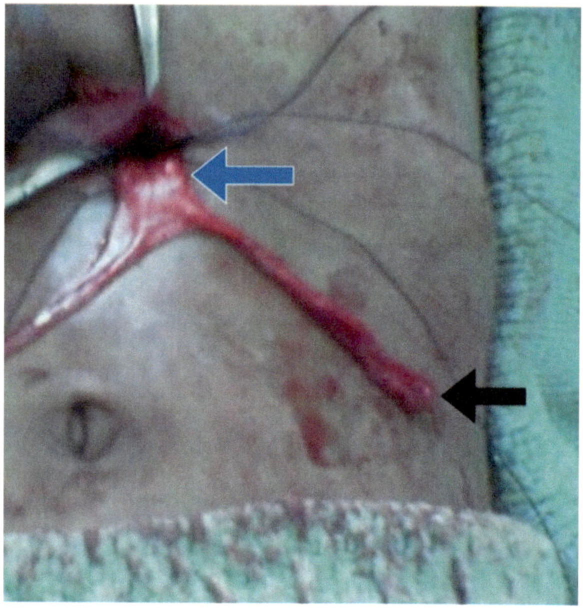

Fig. 20.3 An intraoperative picture of a streak gonad with a rudimentary uterus that was excised. (Black arrow-gonad, blue arrow-rudimentary uterus)

Our experience: A patient with penoscrotal hypospadias and right undescended testis presented to us in the outpatient department. Upon evaluation, he was found to have a 45X0/46XY mosaicism and a rudimentary uterus on the right side. He underwent right gonadectomy in 2005 followed by histological confirmation of a streak gonad (Fig. 20.3). He was subsequently planned for hypospadias repair since the parents wanted male sex rearing. He underwent chordee correction followed by Thiersch–Duplay urethroplasty in subsequent stages. After the urethroplasty, the rudimentary uterus was also removed. Until the last follow-up, the patient was doing well, the left testis was present in the left hemiscrotum, with no evidence of a mass lesion.

20.7 Investigations

Karyotyping-Standard G-banding (Giemsa, Leishman's, or variant) karyotyping evaluates the chromosomal pattern in around 20–30 cells during the metaphase (in this phase, the chromosomes assume their characteristic condensed and discrete shape). The 2016 consensus statement on the Management of Intersex Disorders defines MGD as 45,X/46,XY or 46,XX/46,XY, diagnosed by karyotyping [2]. A mosaic chromosome analysis evaluating a minimum of 50 cells to be done. In fluorescence in situ hybridization (FISH) analysis, the cells are evaluated in 100 cells at interphase and specific short sequences of single-stranded DNA probes are used to detect the X and Y probes, thereby yielding quicker results (within 48 hours) than standard karyotyping (2 weeks) as the FISH technique does not require cell culturing [13].

Ultrasound of the abdomen and pelvis is the first imaging usually done. It provides information about the urinary tract, the status of the Mullerian derivatives, and the gonads. If the uterus is identified, it rules out the possibility of a 46XY disorder because of metabolism anomalies of the male hormone and defects in androgen action. Instead, the following differentials are kept:

1. MGD
2. Ovo testicular DSD "(Internally, gonads of both sexes are found. They can be an amalgamation of the ovaries, testes, or combined ovotestes). The external genitalia are commonly ambiguous but a range from normal male to normal female can be seen.
3. Virilized female
4. Persistent Mullerian duct syndrome (PMDS): External genitalia appear typical, usual presentation is with bilateral cryptorchidism or inguinal hernias, Mullerian structures are discovered during surgery for the same

A retrograde genitogram is done to delineate the anatomy of the urogenital sinus.

MRI is recommended as a baseline investigation in patients undergoing orchiopexy especially when gonads were not visible on ultrasound (US). Nakhal et al. in their research compared MRI imaging with histologic findings in gonads post-gonadectomy. They concluded that MRI and histology of macroscopic findings tallied with each other and could portray the gonads sufficiently [14]. In another paper by Alaniz et al., the usefulness of MRI to ultrasound for a broad group of DSD patients was performed. They underwent either of the two before a prophylactic gonadectomy and found no evidence in the imaging characters to indicate premalignant or malignant lesions. The instigators could not find a difference between USG and MRI in identifying the gonads [15].

Diagnostic Laparoscopy is useful as a first surgical intervention in identifying the anatomy and any conditions that require immediate attention and action in a minimally invasive way. Cystoscopy and vaginoscopy performed along with laparoscopy can help us understand pelvic anatomy hence a multidisciplinary team comprising pediatric surgeons, urologists, and gynecologists can take necessary action. The 2012 series by Farrugia et al. pursues this surgery at an age of 2–3 months. In contrast, Weidler et al. preferred to reach an age of sixth months to hearten the family bond in the neonatal period and to take the focus away from the genitalia as the most important part of the infant's care [16, 17].

20.8 Treatment/Management

Management of this disorder has been described as a complex situation as many factors need to be evaluated before deciding on a treatment pathway including family issues, the presence of reproductive organs of both sexes, risks of developing malignancy, infertility, gender identity, and dysphoria, family dynamics, social adaptation, and psychological endurance skills. On that account, a standard globally agreed-upon treatment algorithm is not present. Broadly, it depends on the sex of rearing and genetic imprinting [16, 18] and entails the following aspects:

1. Gender assignment
2. Appropriately timed early gonadectomy
3. Proper screening for Wilms tumor

Previously, these patients were frequently subjected to procedures that assigned them female gender in the neonatal period. It was suggested that patients with a Y chromosome should be assigned male gender or that irreversible procedures should be delayed [6, 17]. Fertility is not considered during gender assignment [19]. Borer et al. preferred female gender assignment. The male gender allocation was a legitimate option in extreme virilization and a descended testis [20].

If the patient is brought up as a girl, she would need estrogen and progesterone therapy and surgery for clitoral recession, vaginal exteriorization, labioscrotal reduction, and gonadectomy (risk of tumor). The most common tumor seen is gonadoblastoma which is a steroid-secreting tumor. In many parts of the world, there is an inclination towards delayed reconstructions and avoiding irreversible gonadectomy [11].

If the patient is engendered as a boy, he would need the removal of the streak gonad and regular screening of the testis. Orchiopexy and hypospadias repair are usually required with presurgical testosterone stimulation [21].

Growth hormone treatment is needed if the patient has a short stature. On the basis of the external masculinization score (EMS), Cools et al. classified the patients into 3 categories:

1. Minimal virilization needing orchiopexy, biopsy, and surveillance with self-examination
2. Patients with ambiguous genitalia in whom there is a decreased threshold to perform a gonadectomy
3. Phenotypically female in which an early planned gonadectomy is needed

The EMS gives consideration to gonadal differentiation and probability of tumor in patients with 45,X/46,XY. Testosterone production is often at a desired level in boys, but timely surveillance is needed because of malignancy risk, which is conversely associated to EMS. In girls, tumor risk is low, but gonads are not working; therefore, a gonadectomy is the most sensible treatment modality.

International experience is limited to studies with small sample sizes, considering the rarity of this disorder. In a study from Boston Children's Hospital, the authors concluded that in cases where the neonatal brain is exposed to testosterone, a balanced approach involving both vaginoplasty and phallic preservation could help meet parental desires while maintaining anatomical options. This approach allows for the potential reconstruction of either a male or female phenotype at a later date, respecting the evolving gender identity of the child and the rights of both parents and the patient [22].

20.9 Follow-up/Prognosis

Clinical and psychological results hinge on the quality of care and level of support provided. For patients being followed up without gonadectomy or with partial gonadal tissue removal, annual visits are necessary and should be done by a multidisciplinary team. In cases where orchiopexy has been done, self-exam is a must every 3 months and should have annual imaging with any abnormality found warranting a biopsy of retained gonads. Patients and families should be encouraged to enroll in social and peer support groups and should also be screened for depression or, anxiety [22].

Psychological Therapy/Genetic Counseling Genetic counseling should be offered to the mother and father with a prenatal diagnosis of 45 X/46 XY but it is often limited by the broad spectrum that are linked with this situation [23].

References

1. Lee PA, Houk CP, Ahmed SF, Hughes IA. International consensus conference on intersex organized by the Lawson Wilkins Pediatric Endocrine Society and the European Society for Paediatric Endocrinology. Consensus statement on management of intersex disorders. International consensus conference on intersex. Pediatrics. 2006 Aug;118(2):e488–500.
2. Lee PA, Nordenström A, Houk CP, Ahmed SF, Auchus R, Baratz A, et al. Global disorders of sex development update since 2006: perceptions, approach and care. Horm Res Paediatr. 2016;85(3):158–80.
3. V R Ravikumar. A handbook on disorders of sexual differentiation.
4. Nathalie J. The intersex child. Basel: Karger; 1981.
5. Makhija D, Shah H, Tiwari C, Jayaswal S, Desale J. Mixed gonadal dysgenesis with an unusual "inverted" Y chromosome. Dev Period Med. 2016 Jan 1;20:178–80.
6. Donahoe PK, Crawford JD, Hendren WH. Mixed gonadal dysgenesis, pathogenesis, and management. J Pediatr Surg. 1979 Jun;14(3):287–300.
7. Robboy SJ, Miller T, Donahoe PK, Jahre C, Welch WR, Haseltine FP, et al. Dysgenesis of testicular and streak gonads in the syndrome of mixed gonadal dysgenesis: perspective derived from a clinicopathologic analysis of twenty-one cases. Hum Pathol. 1982 Aug;13(8):700–16.
8. Rosa RFM, D'Ecclesiis WFB, Dibbi RP, Rosa RCM, Trevisan P, Graziadio C, et al. 45,X/46,XY mosaicism: report on 14 patients from a Brazilian hospital. A retrospective study. Sao Paulo Med J Rev Paul Med. 2014 Dec;132(6):332–8.
9. Gonadoblastoma, testicular and prostate cancers, and the TSPY gene - PubMed [Internet]. [cited 2023 Jan 22]. Available from https://pubmed.ncbi.nlm.nih.gov/10090875/
10. Kurose Y, Nagai T, Shigematsu K, Uotani T, Akahori T, Takai Y, et al. Chance diagnosis of mixed gonadal dysgenesis in an adult case of malignant gonadal germ cell tumor: a case report. J Med Case Rep. 2021 Mar 28;15(1):165.
11. Cools M, Pleskacova J, Stoop H, Hoebeke P, Van Laecke E, Drop SLS, et al. Gonadal pathology and tumor risk in relation to clinical characteristics in patients with 45,X/46,XY mosaicism. J Clin Endocrinol Metab. 2011 Jul;96(7):E1171–80.
12. Wolffenbuttel KP, Hersmus R, Stoop H, Biermann K, Hoebeke P, Cools M, et al. Gonadal dysgenesis in disorders of sex development: diagnosis and surgical management. J Pediatr Urol. 2016 Dec;12(6):411–6.

13. Sinclair A. Genetics 101: cytogenetics and FISH. CMAJ Can Med Assoc J. 2002 Aug 20;167(4):373–4.
14. Nakhal RS, Hall-Craggs M, Freeman A, Kirkham A, Conway GS, Arora R, et al. Evaluation of retained testes in adolescent girls and women with complete androgen insensitivity syndrome. Radiology. 2013 Jul;268(1):153–60.
15. Alaniz VI, Kobernik EK, Dillman J, Quint EH. Utility of ultrasound and magnetic resonance imaging in patients with disorders of sex development who undergo prophylactic gonadectomy. J Pediatr Adolesc Gynecol. 2016 Dec;29(6):577–81.
16. Weidler EM, Pearson M, van Leeuwen K, Garvey E. Clinical management in mixed gonadal dysgenesis with chromosomal mosaicism: considerations in newborns and adolescents. Semin Pediatr Surg. 2019 Oct;28(5):150841.
17. Farrugia MK, Sebire NJ, Achermann JC, Eisawi A, Duffy PG, Mushtaq I. Clinical and gonadal features and early surgical management of 45,X/46,XY and 45,X/47,XYY chromosomal mosaicism presenting with genital anomalies. J Pediatr Urol. 2013 Apr;9(2):139–44.
18. Mouriquand PDE, Gorduza DB, Gay CL, Meyer-Bahlburg HFL, Baker L, Baskin LS, et al. Surgery in disorders of sex development (DSD) with a gender issue: if (why), when, and how? J Pediatr Urol. 2016 Jun;12(3):139–49.
19. Anand A, Gupta NP, Singh MK, Mathur SR, Nayyar R. Mixed gonadal dysgenesis with normal karyotype: a rare case report. Indian J Pathol Microbiol. 2010;53(2):313–5.
20. Borer JG, Nitti VW, Glassberg KI. Mixed gonadal dysgenesis and dysgenetic male pseudohermaphroditism. J Urol. 1995 Apr;153(4):1267–73.
21. Proto G, Bartolomei P, Mazzolini A, Grimaldi F, Torossi I, Bertolissi F. Mixed gonadal dysgenesis (MGD). Description of a case. Minerva Endocrinol. 1991;16(4):203–6.
22. Chen D, Hidalgo MA, Leibowitz S, Leininger J, Simons L, Finlayson C, et al. Multidisciplinary care for gender-diverse youth: a narrative review and unique model of gender-affirming care. Transgender Health. 2016;1(1):117–23.
23. Intersex-affirming hospital policy guide: providing ethical and compassionate health care to intersex patients | Lambda Legal [Internet]. [cited 2023 Jan 22]. Available from https://www.lambdalegal.org/publications/intersex-affirming

46,XX/46,XY Chimeric Ovotesticular DSD 21

Sangita Yadav and Sohini Pradhan

Abstract

Disorder of sex development is a condition of congenitally derived atypical development of chromosomal, gonadal, or anatomic sex. Ovotesticular DSD has testicular tissue with seminiferous tubules as well as ovarian tissue with follicles in the same or opposite gonad and is characterized by karyotype—46,XX/46,XY chimerism. The most common form of ovotesticular DSD is associated with 46,XX karyotype. Most commonly present as ambiguity of genitalia or significant hypospadias. The internal genitalia is extremely variable in these patients.

Diagnosis of chimeric ovotesticular DSD depends on karyotyping—46,XX/46,XY. Other measures—hormonal assay, radio-imaging to visualize the internal genitalia, laparoscopy for detailed information of internal structures, and biopsy for testicular/ovarian tissue.

Gender designation—most challenging decision. Female designation for 46,XX karyotype with uterus and minimal virilization. Male designation given if clitoromegaly with functional testicular tissue and poor Mullerian structure. Gonadotrophin Relaesing Hormone (GnRH) agonists and/or aromatase inhibitors used till puberty if more time is needed for decision-making. Sex hormone supplementation is required. There is an increased risk of malignancy in these patients. Parental counseling is one of the key measures.

S. Yadav (✉)
Department of Pediatrics, Maulana Azad Medical College and assoc. Hospitals, University of Delhi, New Delhi, India

S. Pradhan
Maulana Azad Medical College, University of Delhi, New Delhi, India

S. K. Ratan (ed.), *Children with Differences in Sex Development*,
https://doi.org/10.1007/978-981-97-1639-5_21

Keywords

Ovotesticular DSD · True hermaphroditism · 46,XX/46,XY chimerism

21.1 Introduction

Disorder of sex development is defined as a condition of congenitally derived atypical development of chromosomal, gonadal, or anatomic sex. DSD has been classified by karyotype: Sex chromosome DSD, 46,XX DSD, and 46,XY DSD. Ovotesticular DSD, previously known as true hermaphroditism, has testicular tissue with seminiferous tubules and ovarian tissue with follicles in the same or opposite gonad and is characterized by karyotype–46,XX/46,XY chimerism. The most common form of ovotesticular DSD is associated with 46,XX karyotype [1]. The gonads in ovotesticular DSD can be ovary, testis, and/or ovotestis, with the most common histology being ovotestis. One of the most common differentials of ovotesticular DSD is mixed gonadal dysgenesis, which is discussed further in this chapter, and how to differentiate both these conditions.

21.2 Epidemiology

DSDs are rare, with a reported incidence varying from 1 in 4500 to 1 in 5500. Ovotesticular DSD is the rarest form of DSD. It has an approximate incidence of less than 1/20,000 globally, comprising only 3–10% of all cases of DSD, with only 500 cases reported till now worldwide [1, 2].

21.3 Etiopathogenesis

Most cases are sporadic. However, familial as well as genetic basis also exist in some cases. Among the most common form of DSD, i.e. 46XX ovotesticular DSD, there is a mutation involving a specific residue of NR5A1 (steroidogenic factor 1) that has recently been identified responsible for initiating testis development pathway in 46, XX karyotype [3–5]. Other mechanisms identified are the disruption of nuclear receptor NRN2F2 and homozygous mutation of RSPO1 (ovarian-testis repressor gene) encoding R-spondin 1 [6–8]. 46,XX/46,XY chimeric ovotesticular disorder (true hermaphroditism) that has presence of both testicular and ovarian tissue in the patient is a form of tetragametic chimerism in which two zygotes fuse during early embryonic development, resulting in a single individual with cell lines from both zygotes [9].

Patients with ovotesticular DSD have asymmetrical development of gonads and often been classified as per various types and location of gonads [10].

1. Unilateral (50%): Ovotestis on one side and ovary/testis on the other side
2. Bilateral (30%): Testis and ovary present on both sides
3. Lateral (20%): Testis on one side and ovary on the other side

21.4 Clinical Features

These patients present most commonly with early complaints of ambiguous genitalia or significant hypospadias. The internal genitalia is extremely variable in patients of chimeric ovotesticular DSD [11, 12].

Cryptorchidism is common, but at least one gonad is often palpable, either in the inguinal canal or in the labio-scrotal fold. They are often associated with inguinal hernia. These individuals can experience estrogenization during puberty and develop cyclic hematuria if uterus is present. It is also important to note that Y chromosome is necessary for sperm production; therefore, 46,XX patients with ovotestis will be infertile. Spermatogenesis is rare and testicular interstitial fibrosis is common.

Individuals with chimeric ovotesticular DSD also develop breast during puberty and menses have been noted in significant patients when uterus is present. They have also been reported to undergo ovulation and pregnancy when ovary is present. However, these patients at the same time have been reported to undergo androgenization leading to voice changes and clitoral enlargement of left untreated during adolescence.

21.5 Diagnosis

After clinical evaluation, ovotesticular DSD should be considered as a diagnosis in all patients with ambiguous genitalia. The diagnosis of chimeric ovotesticular DSD depends on karyotyping—46,XX/46,XY which is strongly suggestive of the same. Hormonal assay in the form of basal testosterone, Anti Mullerian Hormone (AMH), and inhibin B in the first month of life and basal AMH levels thereafter, as well as ovarian hormonal assay with serum estradiol, inhibin A levels, and follicular response to recombinant human Follicle Stimulating Hormone (FSH) have been found to be useful in diagnosis [13]. Patients should also undergo pelvic ultrasound as the first radiological investigation, followed by magnetic resonance imaging (MRI) to visualize the internal genitalia. Surgical techniques—laparoscopy for detailed information of internal structures and biopsy for testicular and ovarian tissue sometimes help confirm the diagnosis [14, 15].

21.6 Differentiating Chimeric Ovotesticular DSD from Mixed Gonadal Dysgenesis (MGD)

Both chimeric ovotesticular DSD and mixed gonadal dysgenesis can have similar karyotype and ambiguous genitalia. It is difficult to differentiate both and requires detailed investigations to reach a diagnosis.

Mixed gonadal dysgenesis (MGD) is a condition in which individuals have chromosomal mosaicism as well as dysgenetic gonads and variable internal and external reproductive anatomy [16]. This has unremarkable similarity to chimeric ovotesticular DSD especially, when patients have 46,XX/46,XY karyotyping. However, it is rare to find dysgenetic gonads in chimeric ovotesticular DSD patients.

Clinical presentation in mixed gonadal dysgenesis can be on a spectrum with varying chromosomal arrays (e.g., 45,X/46,XY or 46,XX/46,XY and others) as well as varying gonadal characterizations (e.g., streak gonads or normal ovaries/testes) [16, 17].

The treatment for MGD and chimeric ovotesticular DSD is identical and no difference has been seen.

21.7 Treatment

The management of chimeric ovotesticular DSD is very complex and depends on the following:

1. Age of diagnosis
2. Stage of genital development
3. Internal genitalia
4. Reproductive ability

The decision for designating the gender of the child is an extremely difficult one. Individuals with 46,XX karyotype with uterus, minimal virilization, and functional ovarian tissue shall have female designation as appropriate. These individuals may undergo functional testicular tissue removal if androgenization is not desired. These patients are then monitored post-operatively by measuring serum AMH and testosterone levels. GnRH agonists can be used in these patients till puberty if more time is needed for decision-making [18].

Similarly, a male gender designation is more appropriate if there is significant clitorophallic development and functional testicular tissue with absent or poorly formed Mullerian structures [18]. GnRH agonists and/or aromatase inhibitors might be considered if more time is required for decision-making. Remnant mullerian structures shall be removed if there are associated problems in the form of hematuria, urinary retention, or any disruption to the blood supply of the testis. Often, sex hormone supplementation may be required to complete pubertal development [19].

Parental counseling is one of the key measures in managing patients with ovotesticular DSD. Parents should also be explained the possible risk of malignancy in the form of gonadal tumors, which are quite common in these patients. Gender identity is an important consideration in patients with ovotesticular DSD, often witnessed during androgenization in girls and estrogenization in boys.

References

1. Krob G, Braun A, Kuhnle U. True hermaphroditism: geographical distribution, clinical findings, chromosomes and gonadal histology. Eur J Pediatr. 1994;153:2–10.
2. Nistal M, Paniagua R, Gonzalez-Peramato P, Reyes-Mugica M. Perspectives in Pediatric pathology, Chapter 7. Ovotesticular DSD (true hermaphroditism). Pediatr Dev Pathol. 2015;18:345–52.

3. Baetens D, Stoop H, Peelman F, Todeschini AL, Rosseel T, et al. NR5A1 is a novel disease gene for 46, XX testicular and ovotesticular disorders of sex development. Genet Med. 2017;19:367–76.
4. Bashamboo A, Donohoue PA, Vilain E, Rojo S, Calvel P, et al. A recurrent p.Arg92Trp variant in steroidogenic factor-1 (NR5A1) can act as a molecular switch in human sex development. Hum Mol Genet. 2016;25:3446–53.
5. Swartz JM, Ciarlo R, Guo MH, Abrha A, Weaver B, et al. A 46,XX ovotesticular disorder of sex development likely caused by a steroidogenic Factor-1 (NR5A1) variant. Horm Res Paediatr. 2017;87:191–5.
6. Bashamboo A, Eozenou C, Jorgensen A, et al. Loss of function of the nuclear receptor NR2F2, encoding COUP-TF2, causes testis development and cardiac defects in 46, XX children. Am J Hum Genet. 2018;102(3):487–93.
7. Tomaselli S, Megiorni F, De Bernardo C, et al. Syndromic true hermaphroditism due to an R-spondin (RSPO1) homozygous mutation. Hum Mutat. 2008;29:220–6.
8. Naasse Y, Bakhchane A, Charoute H, et al. A novel homozygous missense mutation in the FU-CRD2 domain of the R-spondin 1 gene associated with familial 46,XX DSD. Sex Dev. 2017;11(5–6):269–74.
9. Tate JW, O'Hara BF, Wei S. 46,XX/46,XY Chimerism & Human Sexual Development. OBM Genetics. 2022;6(2):12.
10. Grumbach MM, Hughes IA, Conte FA. Disorders of sex differentiation. In: Williams textbook of endocrinology, vol. 2. 10th ed. Philadelphia: Saunders; 2003. p. 842–1002.
11. Damiani D, Fellous M, McElreavy K, et al. True hermaphroditism: clinical aspects and molecular studies in 16 cases. Eur J Endocrinol. 1997;136:201–4.
12. Sircili MH, Denes FT, Costa EM, et al. Long term followup of a large cohort of patients with ovotesticular disorder of sex development. J Urol. 2014;191:1532–6.
13. Steinmetz L, Rocha MN, Longui CA, et al. Inhibin a production after gonadotropin stimulus: a new method to detect ovarian tissue in ovotesticular disorder of sex development. Horm Res. 2009;71:94–9.
14. Steven M, O'Toole S, Lam JP, et al. Laparoscopy versus ultrasonography for the evaluation of Mullerian structures in children with complex disorders of sex development. Pediatric Surg Int. 2013;28:1161–4.
15. Moriya K, Morita K, Mitsui T, et al. Impact of laparoscopy for diagnosis and treatment in patients with disorders of sex development. J Pediatr Urol. 2014;10:955–61.
16. Weidler EM, Pearson M, van Leeuwen K, Garvey E. Clinical management in mixed gonadal dysgenesis with chromosomal mosaicism: considerations in newborns and adolescents. Semin Pediatr Surg. 2019;28(5):150841.
17. Martinerie L, Morel Y, Gay CL, et al. Impaired puberty, fertility, and final stature in 45,X/46,XY mixed gonadal dysgenetic patients raised as boys. Eur J Endocrinol. 2012;166:687.
18. Lee PA, Houk CP, Ahmed SF, Hughes IA. Consensus statement on management of intersex disorders. International Consensus Conference on Intersex. Pediatrics. 2006;118:e488–500.
19. Van Bever Y, Wolffenbuttel KP, Brüggenwirth HT, Blom E, de Klein A, Eussen BHJ, et al. Multiparameter investigation of a 46,XX/46,XY tetragametic chimeric phenotypical male patient with bilateral scrotal ovotestes and ovulatory activity. Sex Dev. 2018;12:145–54.

Part VI

46XY DSD

I duly acknowledge Dr. Samir Kant Acharya, Professor, Pediatric surgery Vardhaman Mahavir Medical College, New Delhi for assisting and aiding me in editing this section.

Disorder of Testicular Development

22

J. D. Rawat and Sudhir Singh

Abstract

The development of gonads and their differentiation is a complex procedure that occurs in a fetus during 4–5th week of intrauterine life. Multiple genes and steps are involved in the differentiation and development of gonads and the determination of the sex of the embryo. The 46 XY DSD (a disorder of sex development) can occur due to defects in testis development, inappropriate production, and the action of testosterone. The clinical feature of the condition varies from female external genitalia to micropenis, although Mullerian structures may or may not be present. Testicular tissue is found in some cases, occasionally unilateral or bilateral streak gonads can be found. Patients can present either in the neonatal period or later during puberty with amenorrhea. Detailed clinical examination with relative prenatal, antenatal, and family history should be taken. A complete physical examination of cases, hormonal evaluation including LH, FSH, and testosterone, radiological evaluation for better delineation of anatomy, and karyotyping are required as a part of the workup. Some cases may even require a gonadal biopsy to reach a diagnosis. After confirmation of the diagnosis, parental counseling should be done. Sex rearing of the child is decided by treating team after detailed parent counseling. The chapter includes Swyer syndrome, partial gonadal dysgenesis, and testicular regression. Patients treated well usually have a good quality of life in adulthood.

Keywords

Pure gonadal dysgenesis (Swyer syndrome) · 46XY ovotesticular DSD · Androgen receptor defect · 5-alpha-reductase deficiency

J. D. Rawat (✉) · S. Singh
Department of Paediatric Surgery, King George's Medical University, Lucknow, Uttar Pradesh, India
e-mail: jdrawat@kgmcindia.edu

S. K. Ratan (ed.), *Children with Differences in Sex Development*,
https://doi.org/10.1007/978-981-97-1639-5_22

Embryology of Testicular Development During embryonic development, around the fourth week of gestation, intermediate mesoderm develops on both sides of the dorsal mesentery in the coelomic cavity [1]. From the intermediate mesoderm, a structure called a genital ridge develops, which later forms the testis. The coelomic epithelial cells, present on the ventromedial part of mesonephros, contain the Wolffian duct. The primordial germ cells originate from the area around the yolk sac [2]. Around the 5 weeks of gestation, primordial germ cells begin to form gonads. The bipotential primitive gonads will become ovary by default. Only by the presence of the Y chromosome, the stimulation by the SRY gene causes the development of primitive gonads into a testis [3]. MIS (Mullerian inhibiting substance) produced by the Sertoli cells prevents the further development of Mullerian structures [4]. SF1 (steroidogenic factor 1) is a transcription factor encoded by the NR5A1 gene, located on the long arm of chromosome 9, which regulates the SRY gene [5]. SOX 9 is a transcription factor responsible for producing MIS in Sertoli cells [6]. In the Leydig cells, SOX 9 induces the production of testosterone. It also activates FGF9 (Glia-activating factor) which causes the development of testicular structures [7]. Suppression of the WNT4 gene found on chromosome 1, is required for male sex development [8]. Leydig cells produce testosterone which helps in the development of epididymis, seminal vesicles, and vas deferens, which are the Wolffian ducts derivatives. Enzyme 5-alpha-reductase converts testosterone to dihydrotestosterone (DHT), which is required for the development of male external genitalia. Around the 5–6th week of gestation, the gender of the embryo is thus determined.

Testicular Descent Testicular descent occurs in two stages: transabdominal and inguinoscrotal stage. The descent of the testis, initially located adjacent to the kidney reaching up to the internal inguinal ring is the abdominal phase of testicular descent and starts around eight weeks of gestation [1]. The abdominal phase of testicular descent relies on the gubernaculum, cranial suspensory ligament, abdominal pressure, and hydraulic pressure by the processus vaginalis. The inguinoscrotal descent is an androgen-dependent stage starting around 26 weeks of gestation [9]. At term, the testis lies in the scrotum.

Histology of Testis The testis contains Sertoli cells, Leydig cells, and germ cells [10]. The Sertoli cells, found inside the seminiferous tubules, support the germ cells and provide nutritional support for the progress of spermatogenesis [11]. Hormones, such as FSH and testosterone, act on the Sertoli cells. The Sertoli cells also secrete the hormone Inhibin B, which decreases FSH levels and promote spermatogenesis [12]. The Leydig cells present in the seminiferous tubule produce steroidal hormones, including androgen [13].

Important Functions of the Testis The main functions of the testis are spermatogenesis and the production of androgen [14]. The endocrine function of the testis is under neuroendocrine control [15]. The Leydig cells produce androgen under the influence of the luteinizing hormone (LH) secreted from the anterior pituitary. Spermatogenesis is regulated by Sertoli cells under the influence of FSH. The LH and FSH secretion occur under the control of gonadotrophin-releasing hormone

(GnRH) by the anterior hypothalamus [12]. For proper functioning, the testis requires a temperature, 2 degrees lower than our core body temperature. Anatomical adaptations for the regulation of testicular temperature include presence of thin scrotal skin which lacks adipose tissue, presence of pampiniform plexus, response of cremasteric muscle to temperature, and presence of sweat glands in the scrotal skin [1].

22.1 Abnormalities of Testicular Development Leading to a Disorder of Sexual Differentiation (46 XY DSD)

Complete (Pure) Gonadal Dysgenesis (Swyer Syndrome) This is a rare condition, with a documented incidence of 1: 1,00,000 in literature [16]. The characteristic clinical feature in this condition is presence of a female phenotype with sexual infantilism. The child presents with clinical features of delayed puberty, amenorrhea, and absent breast development. This occurs due to insufficient virilization of the fetus, with undifferentiated gonads, occurring due to abnormal sexual differentiation during embryonic development. The condition occurs due to a decreased production of testosterone and Anti-Mullerian hormone (AMH). The serum concentration of gonadotropins is noted to be abnormally high. A high level of serum LH leads to increase androgen leading to virilization. The karyotype in these cases is 46XY [17]. The external genitalia in these cases are female, the uterus is small, and gonads are dysgenetic and non-functional. These gonads are fibrous structures, 2–3 cm long and 0.5 cm wide located in the broad ligament. Histopathologically, there is presence of fibrous stroma but no ovarian tissue.

There are differences noted between Turner's syndrome and 46XY gonadal dysgenesis. There is a notably high chance of gonadal malignancy in complete gonadal dysgenesis [18]. In these cases, there is a 35% risk of malignancy, occurring bilaterally, up to the age of 30 years, although earlier presentation has also been noted to occur in infancy [19]. The tumor development in complete gonadal dysgenesis is linked with the H-A antigen [20]. Gonadoblastoma is the most common tumor and is non-metastatic [21]. Foci of other tumors like metastatic dysgerminoma may also be present [22]. Gonadoblastoma tissue produces testosterone and estrogen, causing virilization through ectopic hormone production. The cause of 46 XY complete gonadal dysgenesis may be attributed to the SRY gene. The absence of testicular development would permit ovarian differentiation. The management option in these cases is the removal of both streak gonads and cyclical estrogen and progesterone replacement.

22.1.1 Partial Gonadal Dysgenesis (Male Pseudohermaphroditism)

The spectrum of 46, XY partial gonadal dysgenesis (46, XY PGD) encompasses individuals ranging from phenotypically normal females to those exhibiting clitoromegaly, as well as from phenotypically normal males to those presenting isolated hypospadias. Patient with bilateral testicular regression may also present with ambiguous genitalia. A gonadal tumor can also develop in these cases. In certain instances, cases might manifest adrenal insufficiency due to a defect in the SF1 gene, which arises from NR5A1 variants [23]. In this condition, there may be a partial defect in the supporting Leydig and Sertoli cells. This condition may also be seen in children affected with Denys–Drash syndrome. Management of these patients requires a multi-disciplinary team in a designated DSD center [24].

Testicular Regression In this condition, there is the absence of one or both testis, with partial or complete absence of testicular tissue. The phenotype in these cases is a normal male with unilateral or bilateral non-palpable testis, though phenotypic males with micropenis, to phenotypic females, are other presentations [25]. The clinical presentation depends on the extent and timing of the intrauterine accident during the process of sexual development and differentiation. The condition clinically presents as nonpalpable testis, with surgical exploration revealing blind-ending spermatic vessels or a nubbin lacking normal morphology. This accounts for one-third of cases of non-palpable undescended testis. Compensatory hypertrophy of the contralateral testis is seen in these cases. There is a debate regarding the removal of nubbin of testicular tissue due to the potential of malignancy, although according to some recent reports few authors do not recommend this [24].

Ovotesticular Disorder (46XY Ovotesticular DSD) Ovotesticular DSD is the rarest disorder of sex development and only 500 cases have been reported so far [26]. The gonads can be a combination of the ovary, testes, or combined ovary and testis (ovotestis) on either side in the same individual. The external genitalia is usually ambiguous, but can range from normal male to normal female. The condition is also known as true gonadal intersex or true hermaphroditism. Karyotype in these cases is usually 46XX, which normally results in female sexual development. There is the translocation of the *SRY* gene from the Y chromosome to the X chromosome or another chromosome leads to the development of this condition [27], however, XY karyotype has also been noted to occur in few cases. Other causes such as deletions of *DMRT1*, mutations of *SRY*, and mutations of *MAP 3 K1* have been reported [28].

Testicular Number Abnormalities: (Monorchidism and Polyorchidism) Trauma, torsion, or malignancy may cause the loss of one testis, described as monorchidism. The condition can also be congenital due to vascular insult in intrauterine life [29] or can also occur with DSD. Generally, one testis is sufficient for hormonal and reproductive function. The condition of more than one testis in the scrotum or presence of bilaterally normally descended testis with cryptorchidism occurs due to abnormal embryological development. This condition may vary from a separate testis along with epididymis and vas to an extra testis common epididymis and vas [30].

Androgen Receptor Defect This disorder of sexual differentiation (DSD) occurs in a child with a normal male karyotype. The androgen production is normal but is unable to act on the target tissue receptor. The condition can vary from complete insensitivity to androgen, which leads to a female phenotype, to mild androgen insensitivity in which normal male external genitalia is present. The condition occurs due to a genetic defect on the X chromosome that is responsible for the androgen receptor [31].

5-Alpha-Reductase Deficiency Testosterone acts after being converted to dihydroxy testosterone by the enzyme 5-alpha-reductase [32]. Since the active form of testosterone is required for the development of the male phenotype, deficiency of this enzyme leads to development of DSD with a normal XY chromosome. The patient presents with a female phenotype or a less virilized male [33]. Clinically there is ambiguous genitalia with fused labio-scrotal fold and a small phallus that looks like a clitoris. The internal genitalia in these cases is male, with vas and epididymis present.

22.1.2 Management Approach

Counseling the parents of affected children is important. The determination of sex in these children depends on the diagnosis, clinical findings, parents' acceptance, and social traditional culture. In those with female phenotypes, hormonal replacement is needed at the time of puberty in the form of cyclical estrogen and progesterone. In those with male gender assignment, androgen replacement is often required. Surgical correction may often be required in these cases, which can have medicolegal bearing. Surgery should ideally be delayed till the baby can take part in the decision-making process. To avoid the risk of malignancy in these gonads, patients should be kept on close follow-up for early detection, alternatively gonadectomy can be considered. Sex assignment in cases of pure gonadal dysgenesis is generally female. Other conditions like testicular dysgenesis, partial gonadal dysgenesis, and 46XY ovotesticular DSD may have a variable sex assignment, depending on case-to-case basis.

References

1. Patni P, Mohanty SK, Singh R. Embryonic development of the testis. In: Singh R, Singh K, editors. Male infertility: understanding causes and treatment. Singapore: Springer; 2017.
2. Witchel SF, Lee PA. Ambiguous genitalia in sperling pediatric endocrinology. 5th ed. Philadelphia, PA: Elsevier; 2021. p. 123–74.
3. Fechner PY. The role of SRY in mammalian sex determination. Acta Paediatr Jpn. 1996 Aug;38(4):380–9. https://doi.org/10.1111/j.1442-200x.1996.tb03512.x.
4. Brunello FG, Rey RA. AMH and AMHR2 involvement in congenital disorders of sex development. Sex Dev. 2022;16:138–46. https://doi.org/10.1159/000518273.

5. Ferraz-de-Souza B, Lin L, Achermann JC. Steroidogenic factor-1 (SF-1, NR5A1) and human disease. Mol Cell Endocrinol. 2011 Apr 10;336(1–2):198–205. https://doi.org/10.1016/j.mce.2010.11.006.
6. Li X, Yu H, Wang Y, Liu X, Liu Y, Qu J, Wang X. Roles of two Sox9 genes during gonadal development in Japanese flounder: sex differentiation, spermatogenesis and gonadal function maintenance. Int J Mol Sci. 2018 Feb 8;19(2):512. https://doi.org/10.3390/ijms19020512.
7. Sánchez L, Chaouiya C. Primary sex determination of placental mammals: a modelling study uncovers dynamical developmental constraints in the formation of Sertoli and granulosa cells. BMC Syst Biol. May 2016;10(1):37. https://doi.org/10.1186/s12918-016-0282-3.
8. Kim Y, Kobayashi A, Sekido R, DiNapoli L, Brennan J, Chaboissier MC, Poulat F, Behringer RR, Lovell-Badge R, Capel B. Fgf9 and Wnt4 act as antagonistic signals to regulate mammalian sex determination. PLoS Biol. June 2006;4(6):e187.
9. Achermann JC, Hughes IA. Textbook of endocrinology (thirteenth edition). In: Melmed S, Kenneth S, editors. Pediatric disorders of sex development in Williams. Philadelphia, PA: Elsevier; 2016. p. 893–963.
10. McLachlan RI, Rajpert-De Meyts E, Hoei-Hansen CE, de Kretser DM, Skakkebaek NE. Histological evaluation of the human testis—approaches to optimizing the clinical value of the assessment: mini review. Hum Reprod. 1 January 2007;22(1):2–16.
11. Houda A, Nyaz S, Sobhy BM, Bosilah AH, Romeo M, Michael JP, et al. Seminiferous tubules and spermatogenesis. In: Wu W, editor. Male reproductive anatomy. London: IntechOpen; 2021. https://doi.org/10.5772/intechopen.98917.
12. Smith LB, Walker WH. The regulation of spermatogenesis by androgens. Semin Cell Dev Biol. 2014 Jun;30:2–13. https://doi.org/10.1016/j.semcdb.2014.02.012.
13. Zirkin BR, Papadopoulos V. Leydig cells: formation, function, and regulation. Biol Reprod. 2018 Jul 1;99(1):101–11. https://doi.org/10.1093/biolre/ioy059.
14. Wang RS, Yeh S, Tzeng CR, Chang C. Androgen receptor roles in spermatogenesis and fertility: lessons from testicular cell-specific androgen receptor knockout mice. Endocr Rev. 2009 Apr;30(2):119–32. https://doi.org/10.1210/er.2008-0025.
15. Victor E, Osarugue I, Obukohwo OM, Kingsley NE, Obidike NA. Endocrine functions of the testes. In: Wu W, editor. Male reproductive anatomy. London: IntechOpen; 2021. https://doi.org/10.5772/intechopen.101170.
16. Jakovleva A, Kovalova Z. Complete gonadal dysgenesis analysis in the population of Latvia: malignant outcomes and a review of literature. Med Pharm Rep. 2022 Jan;95(1):47–53. https://doi.org/10.15386/mpr-2064.
17. Guidozzi F, Ball J, Spurdle A. 46,XY pure gonadal dysgenesis (Swyer-James syndrome)—Y or Y not?: a review. Obstet Gynecol Surv. 1994 Feb;49(2):138–46. https://doi.org/10.1097/00006254-199402000-00026.
18. Da Silva RS, Monteiro IC, Braz Dos Santos LG, Caldas NG, Chen AC, Chen JR, Silva HS. A case of Swyer syndrome associated with advanced gonadal dysgerminoma involving long survival. Case Rep Oncol. 2015 Mar 31;8(1):179–84. https://doi.org/10.1159/000381451.
19. Olsen MM, Caldamone AA, Jackson CL, Zinn A. Gonadoblastoma in infancy: indications for early gonadectomy in 46XY gonadal dysgenesis. J Pediatr Surg. 1988 Mar;23(3):270–1. https://doi.org/10.1016/s0022-3468(88)80738-3.
20. Amice V, Amice J, Bercovici JP, Riviere D, Corolleur MJ. Gonadal tumor and H-Y antigen in 46,XY pure gonadal dysgenesis. Cancer. 1986 Apr;57(7):1313–7. https://doi.org/10.1002/1097-0142(19860401)57:7<1313::aid-cncr2820570710>3.0.co;2-k.
21. Scully RE. Gonadoblastoma. A review of 74 cases. Cancer. 1970 Jun;25(6):1340–56. https://doi.org/10.1002/1097-0142(197006)25:6<1340::aid-cncr2820250612>3.0.co;2-n.
22. Troche V, Hernandez E. Neoplasia arising in dysgenetic gonads. Obstet Gynecol Surv. 1986 Feb;41(2):74–9. https://doi.org/10.1097/00006254-198602000-00002.
23. Ribeiro G, de Andrade J, Marques-de-Faria AP, Fabbri HC, et al. Long-term follow-up of patients with 46,XY partial gonadal dysgenesis reared as males. Int J Endocrinol. 2014;2014:480724.

24. Nataraja RM, Asher CM, Nash R, Murphy FL. Is routine excision of testicular remnants in testicular regression syndrome indicated? J Pediatr Urol. 2015 Jun;11(3):151.e1–5. https://doi.org/10.1016/j.jpurol.2015.01.018.
25. Pirgon Ö, Dündar BN. Vanishing testes: a literature review. J Clin Res Pediatr Endocrinol. 2012 Sep;4(3):116–20. https://doi.org/10.4274/Jcrpe.728.
26. Özdemir M, Kavak RP, Yalcinkaya I, Guresci K. Ovotesticular disorder of sex development: an unusual presentation. J Clin Imaging Sci. 2019;9:34. Published 2019 Jul 12. https://doi.org/10.25259/JCIS_45_2019.
27. Scarpa MG, Grazia MD, Tornese G. 46,XYovotesticular disorders of sex development: a therapeutic challenge. Pediatr Rep. 2017;9(4):7085. Published 2017 Dec 11. https://doi.org/10.4081/pr.2017.7085.
28. Ledig S, Hiort O, Wünsch L, Wieacker P. Partial deletion of DMRT1 causes 46,XYovotesticular disorder of sexual development. Eur J Endocrinol. 2012 Jul;167(1):119–24. https://doi.org/10.1530/EJE-12-0136.
29. Belman AB, Rushton HG. Is the vanished testis always a scrotal event? BJU Int. 2001;87:480–3. https://doi.org/10.1046/j.1464-410X.2001.00101.x.
30. Savas M, Yeni E, Ciftci H, Cece H, Topal U, Utangac M. Polyorchidism: a three-case report and review of the literature. Andrologia. 2010;42:57–61. https://doi.org/10.1111/j.1439-0272.2009.00957.x.
31. Ovidiu B, Marcu DR, Mischianu DLD, Poiana C, Diaconu CC, Bungau SG, Tit DM, Cumpanas A, Bohiltea R. The challenges of androgen insensitivity syndrome. Arch Med Sci. 2021 Mar 15;18(4):881–9. https://doi.org/10.5114/aoms/125584.
32. Azzouni F, Godoy A, Li Y, Mohler J. The 5 alpha-reductase isozyme family: a review of basic biology and their role in human diseases. Adv Urol. 2012;2012:530121. https://doi.org/10.1155/2012/530121.
33. Okeigwe I, Kuohung W. 5-alpha reductase deficiency: a 40-year retrospective review. Curr Opin Endocrinol Diabetes Obes. 2014 Dec;21(6):483–7. https://doi.org/10.1097/MED.0000000000000116.

46XY Complete Gonadal Dysgenesis

23

Anjana Sreedharan and Samir Kant Acharya

Abstract

Complete gonadal dysgenesis or Swyer syndrome is a type of 46XY DSD where various genetic abnormalities lead to failure of development of the testis resulting in bilateral streak gonads. As the streak gonads have no functional tissue, there is a lack of testosterone and anti-Mullerian hormone (AMH) in the fetal period, resulting in the female phenotype. The patient typically presents in the adolescent period to the physician with complaints of primary amenorrhea and failure of development of secondary sexual characteristics. A well-formed vaginal orifice and cervix is present with a small sized uterus. Investigations include genetic testing, assessment of hormonal levels, and imaging studies. Management is along the lines of complete ovarian failure. As there is a high risk of development of gonadoblastoma, bilateral gonadectomy is recommended.

Keywords

Swyer syndrome · Female phenotype · Gonadoblastoma

23.1 History

Complete gonadal dysgenesis is also called as Swyer syndrome after it was described by G. Swyer in 1955. He described it as a new form of "male pseudohermophroditism." The two females presented to him with primary amenorrhea with tall stature, female external genitalia (one had clitoromegaly) with hypoestrogenic cervix,

A. Sreedharan · S. K. Acharya (✉)
Department of Paediatric Surgery, Vardhman Mahavir Medical College and Safdarjung Hospital, New Delhi, India

S. K. Ratan (ed.), *Children with Differences in Sex Development*,
https://doi.org/10.1007/978-981-97-1639-5_23

vagina, and a rudimentary uterus [1]. On evaluation, karyotyping revealed XY chromosome. It was later identified that the cause of this syndrome was because of dysgenetic gonads.

23.2 Background

The exact incidence of complete gonadal dysgenesis (CGD) is unknown. However, one estimate placed it at 1 in 80,000 [2]. It is also known as pure gonadal dysgenesis or gonadal dysgenesis, XY female type. The chromosomal makeup is always XY. But they appear phenotypically as a female with female external genitalia. Due to various genetic mutations, they lack ovaries and develop what has been described as streak gonads. The streak gonads do not develop properly (aplasia) and are replaced by fibrous tissue with no function. In the absence of a testis, there is no testosterone and no AMH; therefore, the XY genotype will not produce any features of masculinity and will allow for the development of internal female genitalia. However, the non-functioning streak gonads fail to produce adequate estrogen which leads to poor development of secondary sexual characteristics.

In 15–20% of the patients with CGD, it occurs due to the mutation of the SRY gene (sex determining region Y), present on the Y chromosome. Other mutations noted are MAP3K1, NROB1 gene mutation on X chromosome and the desert hedgehog gene (DHH) located on chromosome 12. In a few rare cases, mutations of steroidogenic factor 1 (SF1) gene, protein WNT4 gene, CBX2, GATA4, and WWOX genes have been implicated [3–5].

23.3 Introduction

23.3.1 Normal Embryology (Fig. 23.1)

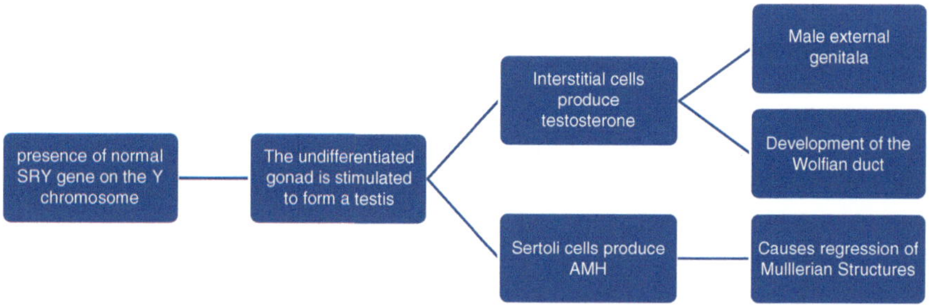

Fig. 23.1 Embryology in a normal male child

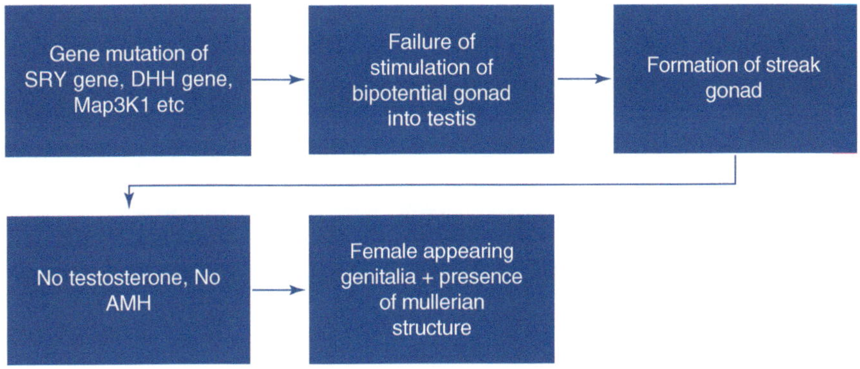

Fig. 23.2 Etiopathogenesis of complete gonadal dysgenesis

23.3.2 Etiopathogenesis of CGD (Fig. 23.2)

CGD comes under 46XY disorders which comes under a larger umbrella of disorders of sexual differentiation. The genotype of these patients is always 46XY and they are always phenotypically female. The SRY gene is the sex determining region of the Y chromosome also called the "switch gene." In the absence of any mutation, this gene directs the bipotential gonad to develop into a testis. The testis in turn will produce testosterone (resulting in the development of male external genitalia) and anti-Mullerian hormone (resulting in the absence of the Mullerian structures, i.e. fallopian tubes, uterus, and upper 2/3rd of the vagina) [6].

The exact cause behind the development of CGD is not known. It has been linked to mutations of SRY gene, DHH gene, MAP3K1, and others. Failure to signal the bipotential gonad into a testis leads to the development of a streak gonad. On microscopy, streak gonads show features similar to ovarian stroma but does not contain follicles [7]. The streak gonads are incapable of producing testosterone, estrogen, or anti-Mullerian hormone. This results in a child with female phenotype with development of uterus, fallopian tubes, and vagina. However, during puberty, there is no production of estrogen. This results in poor breast development, rudimentary uterus, and primary amenorrhea.

23.4 Clinical Features (Fig. 23.3)

The classic presentation is at puberty, with the median age of diagnosis being 16 years [2]. A phenotypically female presents with a history of primary amenorrhea to the gynecologist. They are also taller than women of their age. The presence of the Y chromosome has been postulated as the cause for this height.

On general examination, there is absence of secondary sexual characteristics and poor breast development.

On local examination or examination under anesthesia, typically there is a well-formed vaginal orifice and cervix. Histopathology will show it is hypoestrogenic.

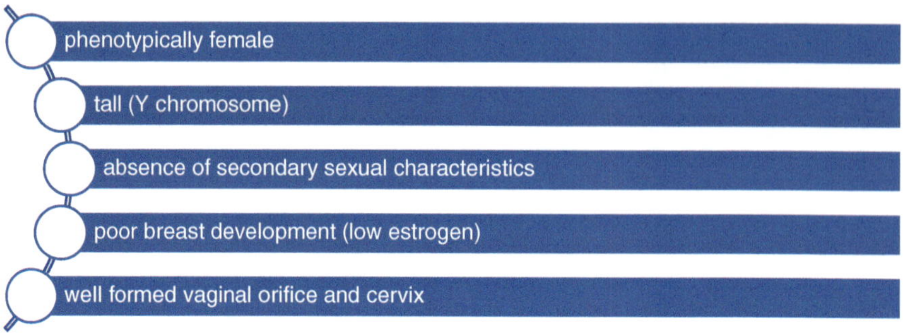

Fig. 23.3 Clinical features of CGD

Fig. 23.4 CGD phenotype

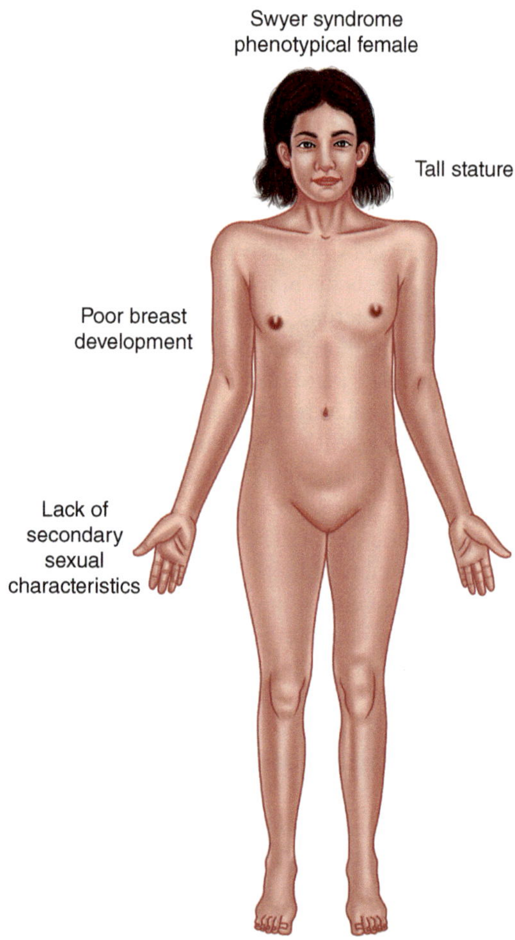

This type of DSD does not present at the neonatal period as a social emergency as the newborn child looks typically female and is usually brought up as a female. The complaints begin at puberty and are similar to complete ovarian failure (Fig. 23.4).

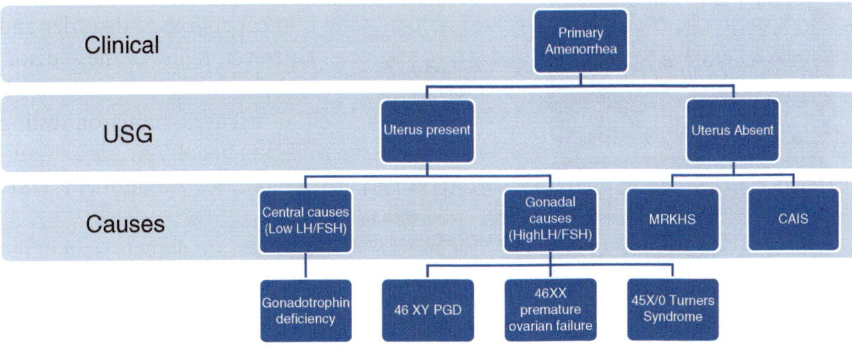

Fig. 23.5 Diagnostic algorithm for a patient presenting with primary amenorrhea

23.5 Differential Diagnosis

The patient typically presents at adolescence with primary amenorrhea. The differentials include hypogonadotrophic hypogonadism (Kallmann syndrome), complete androgen insensitivity, 46XX premature ovarian failure, and Mayer–Rokitansky–Kuster–Hauser syndrome. The algorithm is explained in the figure.

Primary evaluation with an ultrasound will identify the presence or absence of a uterus.

If a uterus is present, an elevated LH and FSH level suggests a gonadal problem such as 46XX premature ovarian failure, 45X/O Turner's syndrome, or CGD. They can be differentiated with the help of karyotyping. If a uterus is present, low levels of LH/FSH suggest that the problem exists either in the pituitary or the hypothalamus. Differentials, in case of an absent uterus, include Complete Androgen Insensitivity Syndrome (CAIS) and Mayer Rokitansky Kuster Hauser Syndrome (MRKHS) [8] (Fig. 23.5).

23.6 Management

23.6.1 Investigations (Fig. 23.6)

A. **Genetic testing**
   1. Chromosomal Analysis—Karyotyping and array CGH. A 46XY karyotype helps in narrowing the diagnosis to a 46XY DSD. Further analysis can then be done to differentiate if the fault lies in the gonadal development, the synthesis or the action of testosterone.
   2. FISH for SRY—Fluorescence in situ hybridization is used for identification of mutations in the SRY gene which is the cause in approximately 15% of the cases of CGD.
   3. Gene sequencing—Several other genetic mutations have been linked to SOX9, GATA4, FOG2, NR5A1, and WT1. 13–18% of CGD has also been found to have a mutation in MAP3K1 gene.

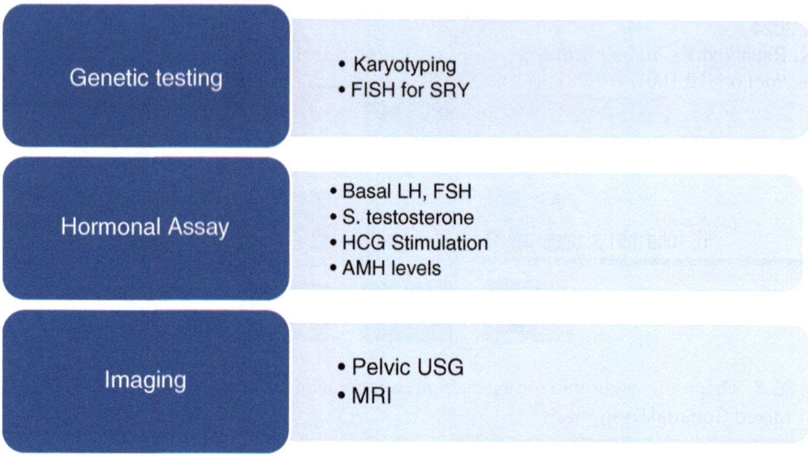

Fig. 23.6 Investigations

Table 23.1 Normal serum testosterone levels according to age [9]

Age group	Testosterone levels (ng/dL)
Newborn	75–400
Prepubertal	3–10
Puberty 10–11 years	1–48
12–13 years	5–619
14–15 years	100–320
16–17 years	200–970
>18 years	350–1080

B. **Hormone levels**

1. Basal LH and FSH levels—It remains low in the prepubertal aged children and is elevated in the post-pubertal age group.
2. Serum testosterone—Lower than normal (Table 23.1).

HCG stimulation test—There will be no increase in the testosterone levels following HCG injections.

3. HCG stimulation test—There will be no increase seen in serum testosterone levels following HCG stimulation. Many HCG stimulation test protocols have been described. One method is to administer HCG intramuscularly on three consecutive days as per age (<1 year –500 units; 1–10 years—500 units; >10 years—1500 units). A serum testosterone level > 200 ng/dL is indicative of the presence of functional Leydig cells.
4. AMH is almost undetectable [10] (Table 23.2).
5. Tumor markers such as Alpha Feto Protein (AFP), Lactate Dehydrogenase (LDH), and beta-Human Chorionic Gonadotropin (HCG) if there is a gonadal mass on imaging or discordant pubertal characteristics. This is not routinely recommended (Figs. 23.7 and 23.8).

Table 23.2 Normal serum AMH range as per age

Age	Serum AMH (ng/ml)
<14 days	35–140
15 days–6 months	55–120
6 months–2 years	85–320
2–9 years	55–250
9–18 years Tanner 1	35–200
Tanner 2	10–140
Tanner 3	4–55
Tanner 4	4–22
Tanner 5	4–21
Adults	3–18

References levels taken from Refs. [11, 12]

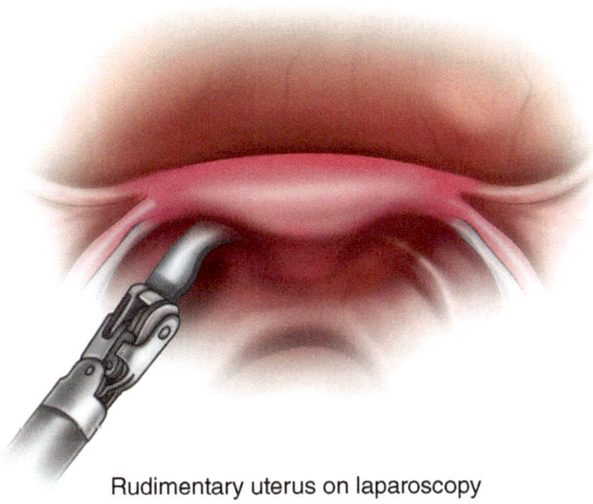

Fig. 23.7 Laparoscopic view of rudimentary uterus

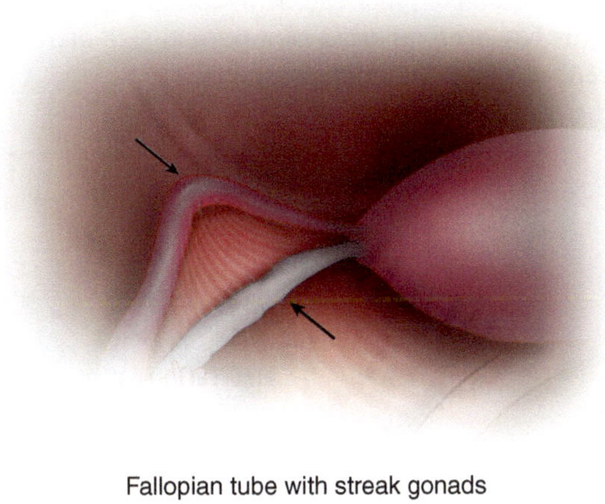

Fig. 23.8 Laparoscopic view of streak ovary

C. **Imaging**

Pelvic ultrasound or MRI—To look for Mullerian structures and visualize gonadal structures [13]

23.6.2 Treatment (Fig. 23.9)

Sex of rearing is always female due to female appearing external genitalia and absence of testosterone.

The management includes

1. Bilateral gonadectomy
2. Induction of puberty by estrogen therapy
3. Long-term combined estrogen and progesterone replacement therapy

In complete gonadal dysgenesis, there is usually no doubt regarding gender due to unambiguously female external genitalia. The patient commonly identifies as female and is raised as a female and treated along the lines of complete ovarian failure.

23.6.3 Risk of Malignancy

The presence of Y chromosome increases the risk of tumor development, the commonest being gonadoblastoma. Gonadoblastoma is a slow growing tumor but malignant dysgerminomas have also been known to occur in streak gonads. The risk of development of gonadoblastoma and dysgerminoma has been estimated to be around 15–30%. For the above reason, bilateral gonadectomy is essential.

Induction of puberty with estrogen is necessary for the development of secondary sexual characteristics as well as for treatment of of primary amenorrhea. However, unlike other causes of ovarian failure like Turner's syndrome where the uterus has been reported as a normal size uterus, Swyer syndrome patients have been found to have a smaller uterus.

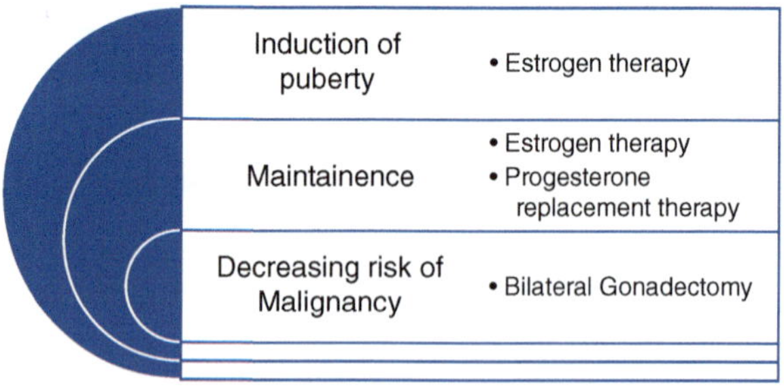

Fig. 23.9 Treatment of CGD

References

1. Swyer GIM. Male pseudohermaphroditism: A hitherto undescribed form. Br Med J. 1955;2(4941):709–11.
2. Michala L, Goswami D, Creighton SM, Conway GS. Swyer syndrome: Presentation and outcomes. BJOG An Int J Obstet Gynaecol. 2008;115(6):737–41.
3. Pearlman A, Loke J, Le Caignec C, White S, Chin L, Friedman A, et al. Mutations in MAP3K1 cause 46,XY disorders of sex development and implicate a common signal transduction pathway in human testis determination. Am J Hum Genet [Internet]. 2010;87(6):898–904. https://doi.org/10.1016/j.ajhg.2010.11.003.
4. Baetens D, Stoop H, Peelman F, Todeschini AL, Rosseel T, Coppieters F, et al. NR5A1 is a novel disease gene for 46,XX testicular and ovotesticular disorders of sex development. Genet Med. 2017;19(4):367–76.
5. Canto P, Söderlund D, Reyes E, Méndez JP. Mutations in the Desert hedgehog (DHH) gene in patients with 46,XY complete pure gonadal dysgenesis. J Clin Endocrinol Metab. 2004;89(9):4480–3.
6. Du H, Taylor HS. Development of the genital system [Internet]. Principles of developmental genetics. 2nd ed. Elsevier; 2015. p. 487–504. https://doi.org/10.1016/B978-0-12-405945-0.00027-2.
7. German J, Simpson JL, Chaganti RSK, Summitt RL, Reid LB, Merkatz IR. Genetically determined sex-reversal in 46,XY humans. Science (80-) [Internet]. 1978;202(4363):53–6. Available from: http://www.jstor.org/stable/1746109
8. Porsius E, Spath M, Kluivers K, Klein W, Claahsen-van der Grinten H. Primary amenorrhea with apparently absent uterus: A report of three cases. J Clin Med. 2022;11(15)
9. Chernecky CCBB. Testosterone levels in males. In: Laboratory tests and diagnostic procedures – E-book. Elsevier Health Sciences; 2012. p. 1059–62.
10. Rey RA, Belville C, Nihoul-Fékété C, Michel-Calemard L, Forest MG, Lahlou N, et al. Evaluation of gonadal function in 107 intersex patients by means of serum antimullerian hormone measurement. J Clin Endocrinol Metab. 1999;84(2):627–31.
11. Edelsztein NY, Grinspon RP, Schteingart HF, Rey RA. Anti-Müllerian hormone as a marker of steroid and gonadotropin action in the testis of children and adolescents with disorders of the gonadal axis. Int J Pediatr Endocrinol [Internet]. 2016;2016(1):1–10. https://doi.org/10.1186/s13633-016-0038-2.
12. Josso N, Rey RA. What does AMH tell us in pediatric disorders of sex development? Front Endocrinol (Lausanne). 2020;11(September):1–12.
13. McCann-Crosby B, Mansouri R, Dietrich JE, McCullough LB, Sutton VR, Austin EG, et al. State of the art review in gonadal dysgenesis: challenges in diagnosis and management. Int J Pediatr Endocrinol [Internet]. 2014;2014(1):1–17. Available from: International Journal of Pediatric Endocrinology

Partial Gonadal Dysgenesis

24

Anjana Sreedharan and Samir Kant Acharya

Abstract

The term partial gonadal dysgenesis is used to denote a condition in children with 46XY chromosome with incomplete gonadal differentiation. Prior to 2006 consensus meeting, the differentiation between PGD and mixed gonadal dysgenesis (MGD) was done on the basis of histopathology. Earlier, a patient with a dysgenetic testis on one side and streak gonad on the other was referred to as mixed gonadal dysgenesis, whereas a child with bilateral streak gonads was diagnosed as partial gonadal dysgenesis or dysgenetic male pseudohermophroditism. On further studies, it was found that the prognosis between these two similar diseases varied despite a similar phenotype. Following the 2006 consensus meeting, a child with 45 XO/46XY mosaic cell line presenting with incomplete gonadal differentiation was diagnosed to have MGD.

The clinical presentation of children with PGD is variable ranging from predominantly female to ambiguous genitalia to predominantly male. The management includes genetic testing, hormonal evaluation, and imaging studies.

The management is variable and has to be tailored to the needs of the individual patient by a multidisciplinary team after a complete discussion with the patient and the patient's family.

Keywords

Partial gonadal dysgenesis · Ambiguous genitalia · 46XY

A. Sreedharan · S. K. Acharya (✉)
Department of Paediatric Surgery, Vardhman Mahavir Medical College and Safdarjung Hospital, New Delhi, India

S. K. Ratan (ed.), *Children with Differences in Sex Development*,
https://doi.org/10.1007/978-981-97-1639-5_24

24.1 History

In the 1960s, Sohval in New York, described a type of intersex which he felt required to be classified separately from the existing nomenclature as it did not fit into the gonadal histology of male hermaphrodites (bilateral testis), female hermaphrodites (bilateral ovaries), or true hermaphrodites (presence of testicular and ovarian tissue). It is to be noted that the terms "pseudohermaphrodites" and "true hermaphrodites" are no longer used due to the stigmatizing nature of this term. Sohval described the presence of a rudimentary "streak gonad" on one side and a testis on the other side and used the term mixed gonadal dysgenesis to denote this histological description. In 2 years, there were five reports suggesting that this was not a rare condition as they previously thought. Bergada et al. from Johns Hopkins University also described patients with gonadal asymmetry. It reflected a lack of awareness of the pathological entity called a streak gonad that was described in 1944 by Wilkins and Fleischmann [1–3].

After its initial description, the term MGD was used to refer to patients with dysgenetic testis on one side and a streak gonad on the other side with no differentiation using karyotype. While, the term partial gonadal dysgenesis (PGD) was used to describe individuals with bilateral dysgenetic gonads [4]. This, however, causes confusion, as the prognosis is different in both groups.

Since the 2006 Consensus Statement on DSD, the term PGD has been applied to 46XY DSD patients who have incomplete gonadal differentiation and the term MGD is reserved for patient with mosaicism with 45X cell line in addition to the 46XY cell line. This differentiation uses karyotyping and does not take histology into account [5].

24.2 Background

In 2005, the Chicago Consensus on Management of Intersex Disorders proposed to substitute the terms intersex, hermaphroditism, and pseudohermaphroditism with the terms Disorders of Sexual Differentiation and its various classifications, to reduce the stigma associated with this condition.

In 2006, The International Consensus Conference on Intersex organized by Lawson Wilkins International Society and the European Society for Paediatric Endocrinology proposed a broad classification that separated DSD with sex chromosome abnormalities (Sex Chromosome DSD) from those with a normal chromosome complement (46,XX DSD and 46,XY DSD). Disorders of gonadal development which include ovotesticular DSD (previously termed as true hermaphrodite) can be found in all the groups. The classification into these three categories is decided by the karyotyping. In the event that gonadal dysgenesis is caused by an abnormality of the sex chromosome (for e.g., Turner's syndrome, Klinefelter's syndrome, or chimeric ovotesticular DSD) it is classified into Sex Chromosome DSD [5, 6].

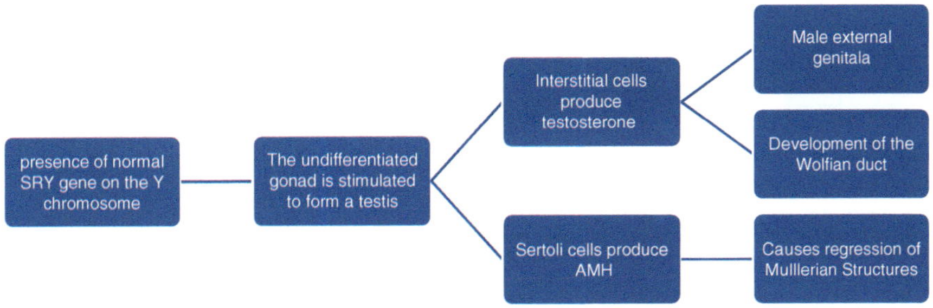

Fig. 24.1 Embryogenesis in a normal male child

24.3 Introduction

Embryogenesis in the normal male child (Fig. 24.1)

24.4 Etiopathogenesis (Fig. 24.2)

In a normal male, the SRY (Sex determining region of Y chromosome) is responsible for stimulation of the undifferentiated gonad to form a testis. The interstitial cells of the testis produce testosterone which promotes the development of the Wolffian duct and upon conversion to dihydrotestosterone, it allows the development of the male external genitalia.

The Sertoli cells, on the other hand, produce anti-Mullerian hormone (AMH) which stimulates the regression of the Mullerian structures.

In a child with partial gonadal dysgenesis, there can be variable failure in gonadal development.

Historically, a child with bilateral streak gonads presented as a phenotypical female and was classified into 46,XX or 46, XY complete gonadal dysgenesis based on the karyotype.

In addition, the children with the histological finding of a dysgenetic testis associated with a contralateral streak gonad was termed as mixed gonadal dysgenesis. Patients with bilateral dysgenetic testis were diagnosed as partial gonadal dysgenesis. This classification was used prior to the Chicago Consensus [5, 7, 8].

Further research has shown that dysgenetic testis maybe bilateral or associated with a contralateral streak gonad in subjects with a 46,XY or a 45,X/46,XY karyotype. The histological picture of the dysgenetic testis can range from having few tubular structures with fibrous tissue to only a reduction in the tubular size and number of germ cells.

Depending on the histology, the external genitalia of these patients can range from predominantly male to predominantly female to classical ambiguous genitalia. This is associated with the persistence of the Mullerian structures which is in turn dependent on the amount of AMH that is produced by the dysgenetic gonad [7, 9].

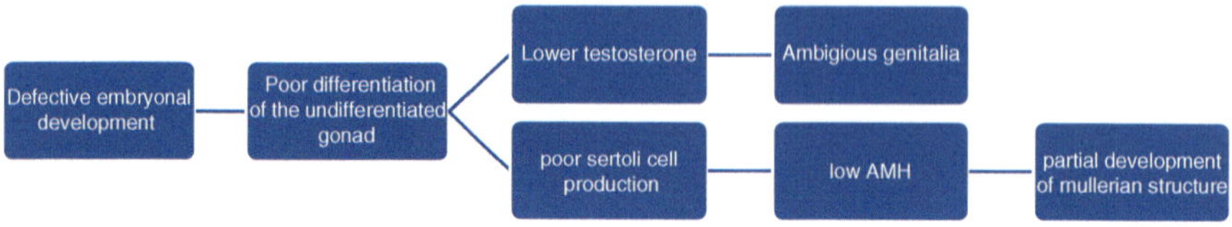

Fig. 24.2 Etiopathogenesis of partial gonadal dysgenesis

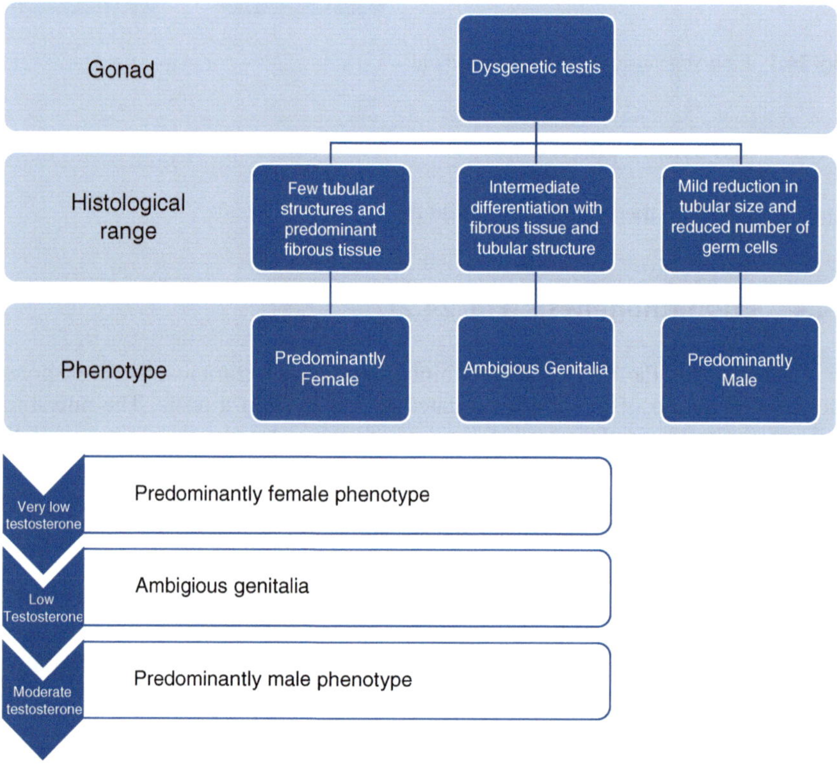

Fig. 24.3 Variations in histopathology and phenotype in PGD

24.4.1 Karyotyping

In partial gonadal dysgenesis, there is no mosaicism. All cell lines contain 46XY (Fig. 24.3).

To further emphasize, the term MGD is reserved for cases with testicular dysgenesis with a 45,X/46,XY mosaic while PGD is used for testicular dysgenesis with 46,XY chromosome constitution, irrespective of the histological picture. In patients with MGD, the histological picture of dysgenetic testis with contralateral streak gonad is found more often than bilateral dysgenetic testes. In PGD, the frequency of dysgenetic testis occurring with contralateral streak is similar to that of bilateral dysgenetic testis [7]. The details of MGD are given under a separate chapter under sex chromosomal DSD.

24.4.2 Histopathology

In PGD, the dysgenetic testis is characterized by variable degrees of immaturity or regressive changes. This, in turn, results in a wide range of clinical ambiguity [10]. It consists of a thin albuginea with underlying seminiferous tubules separated by wide intertubular spaces. The pattern seen mimics the initial stages of testicular embryogenesis and this suggests a failure in completion of full embryogenesis.

A streak gonad appears as a white and narrow strip on gross morphology. Histologically, it is characterized by dense ovarian stroma covered by coelomic epithelium.

24.5 Clinical Features

External Genitalia

The external genitalia can vary from predominantly male to ambiguous genitalia to predominantly female (Fig. 24.4). This spectrum is seen due to the varying degree of dysgenesis of the testis. The degree of virilization is dependent on the testosterone output by the fetal testis. Therefore, the patient can present as a child with hypospadias with undescended testis or ambiguous genitalia or as a female child with partial virilization. The presence of at least one descended testis in the labioscrotal region is usually associated with spontaneous onset of puberty [8].

Evaluation of a child with DSD, especially ambiguous genitalia, is extremely crucial. It is a social emergency more than a medical emergency.

Malignancy

Tumors are likely to arise from a dysgenetic testis. The risk ranges from 15–35% and it increases with increased dysgenesis and intra-abdominal testis. Tumors can also arise from streak gonads if they are associated with the presence of a Y chromosome. [6, 11]

Gender Dysphoria

Studies show that patients raised as female did not have a strong female gender identity on long-term follow-up as compared to PGD patients who were raised as male (Fig. 24.5). They also appeared to have more social problems and difficulty adapting to the female gender-associated hobbies and occupations. These patients also had displayed psychosexual disturbances and a low sex drive [12] (Table 24.1).

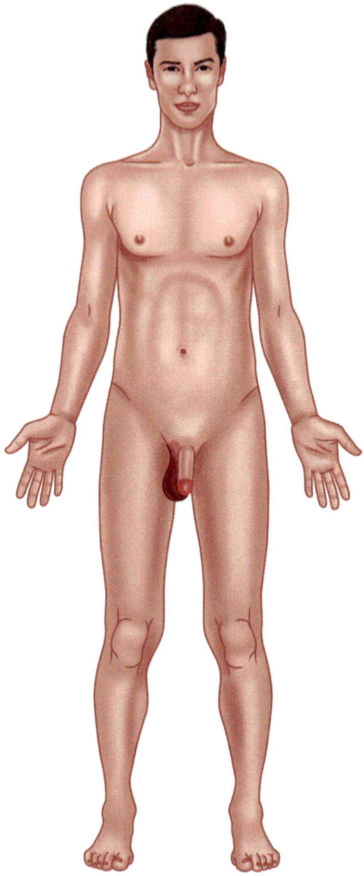

Fig. 24.4 Predominantly male phenotype in PGD

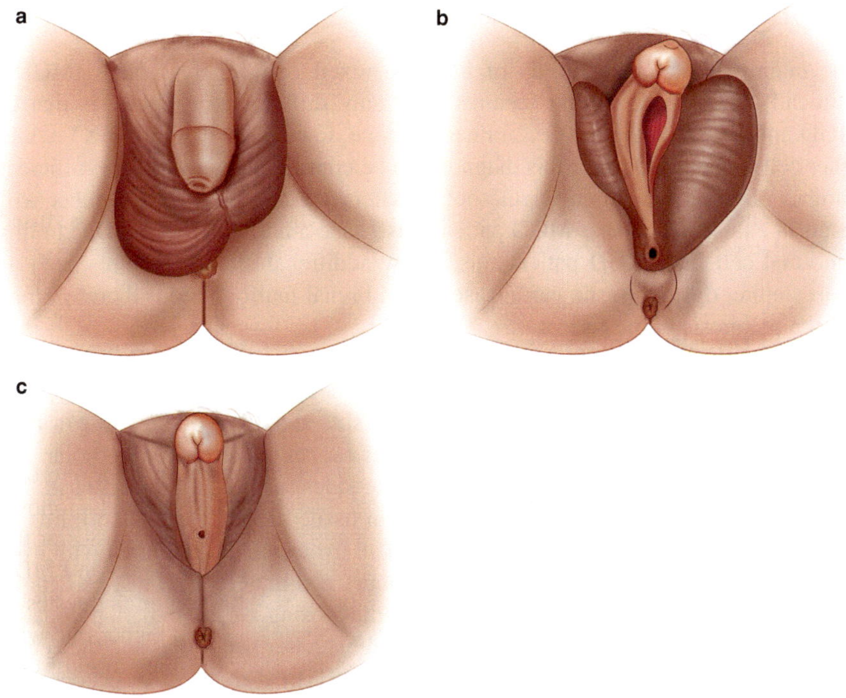

Fig. 24.5 (**a**) PGD appearance—Small phallus with unilateral undescended testis and fused labioscrotal folds. (**b**) PGD appearance—Unilateral undescended testis with unfused labioscrotal folds and proximal penile hypospadias. (**c**) PGD appearance—Bilateral undescended testis with unfused labioscrotal folds, small phallus, and proximal penile hypospadias

Table 24.1 Comparison of CGD, PGD, and MGD

Diagnosis	Karyotype	External genitalia	Risk of malignancy	Location of gonads
CGD	46 XY	Female	Present	Bilateral intra-abdominal streak gonads
PGD	46XY	Male/female or ambiguous genitalia	Present	Can present with dysgenetic or streak gonads
MGD	46XY/ 45X/O	Male/female or ambiguous genitalia	Present	Can present with dysgenetic or streak gonads

24.6 Differential Diagnosis

The main differential diagnosis of PGD is mixed gonadal dysgenesis (MGD). Clinically, they share similar features. The only differentiating factor is that in MGD, mosaicism with 45,XO cell lines exist, i.e. few cells will show 46XY while other cells show 45XO. For this diagnosis to be confirmed, analysis of a sufficient number of cells is required to rule out mosaicism. Analysis of less number will result in a false negative as MGD patients also have a normal cell line. Due to this abnormal cell line, MGD patients also show additional clinical features such as short stature, dysmorphism, cardiovascular and renal malformations thereby altering the prognosis of the patient.

The other differential diagnosis could be ovotesticular DSD (OT DSD). It must be kept in mind that 7% of the OT-DSD comprises patients with a karyotype of 46XY DSD. The gonads could be testis on one side and ovary on the other or they can coexist in the same gonad called an ovotestes. An ovotestes can be polar, i.e. testicular tissue and ovarian tissue on opposite sides, or it can be concentric, i.e. with testicular tissue inside covered by ovarian tissue or vice versa. Two-thirds of children with ovotesticular DSD present with an ovotestes. The children with ovotestestis usually present with ambigious genitalia and therefore it must be kept in mind as a differential diagnosis when one is diagnosing a case as partial gonadal dysgenesis. A wedge biopsy of the gonad which includes both poles and the center is necessary to diagnose ovotesticular DSD [13, 14]. In such children, gender dysphoria is a possibility. Therefore, the possibility of gender dysphoria and the status of the external genitalia must be taken into account while coming to a decision regarding gender assignment.

24.7 Management

24.7.1 Investigations

Patients with suspected 46XY PGD

A) Genetic testing:
   a. Chromosomal analysis—Karyotyping and FISH for SRY. There is a decreased expression of Y chromosome.
   b. Sequence analysis for SRY, NR5A1, and DHH.
B) Hormonal evaluation
   a. LH and FSH: It will be low to normal in the prepubertal age group but elevated in the postpubertal age group. Elevation is secondary to the low testosterone causing a feedback loop and stimulating the pituitary gland [15]
   b. Rule out CAH—Adrenal cause for ambiguous genitalia
   c. Serum testosterone—Variable
   d. AMH—Low

Table 24.2 Normal range of testosterone according to age

Age group	Testosterone levels (ng/dL)
Newborn	75–400
Prepubertal	3–10
Puberty 10–11 years	1–48
12–13 years	5–619
14–15 years	100–320
16–17 years	200–970
>18 years	350–1080

Table 24.3 Normal ranges of AMH according to age

Age	Serum AMH (ng/ml)
<14 days	35–140
15 days–6 months	55–120
6 months–2 years	85–320
2–9 years	55–250
9–18 years Tanner 1	35–200
Tanner 2	10–140
Tanner 3	4–55
Tanner 4	4–22
Tanner 5	4–21
Adults	3–18

Table 24.4 Hormonal levels in CGD, PGD, and MGD

Hormonal levels	CGD	PGD	MGD
LH	Basal level—Low Postpubertal—High	Basal level—Low Postpubertal—Normal to high	Basal level—Low Postpubertal—Normal to high
FSH	Basal level—Low Postpubertal—High	Basal level—Low Postpubertal—Normal to high	Basal level—Low Postpubertal—Normal to high
Testosterone	Low	Low to normal	Low to normal
AMH	Low	Low to normal	Low to normal

e. HCG stimulation test—Inadequate rise of testosterone will be seen (suggesting insufficient functional Leydig cells producing testosterone) (Tables 24.2 and 24.3)

C) Imaging: Pelvic USG or MRI

a. For internal organs: the findings on imaging can range from a fully developed Mullerian structures (i.e. a well developed uterus with fallopian tubes) to absent uterus and fallopian tubes. A moderate production of AMH can result in a rudimentary uterus. The development of the uterus and fallopian tubes depends on the level of AMH produced.

b. Gonads maybe located in the abdomen, in the inguinal region, or in the scrotum [16] (Table 24.4)

It must be kept in mind that the variation in hormonal levels in both PGD and MGD is dependent on the degree of gonadal dysgenesis. A streak gonad may produce very low testosterone and low AMH and in turn result in high levels of LH and FSH in the postpubertal period secondary to the negative feedback loop. On the other hand, a mildly dysgenetic testis will produce testosterone and AMH within the lower limits of normal, resulting in a normal or mildly elevated gonadotrophin levels.

24.8 Treatment

The management of such children is dependent on the clinical features which are in turn dependent on the degree of testicular dysgenesis.

A) **Female phenotype**: Gonadectomy is recommended in view of gross dysgenesis of the testis and high risk of malignancy. It must be noted that there is a risk of gender dysphoria in such children. Genitoplasty might be required to give the appearance of a normal female external genitalia.
B) **Ambiguous genitalia**
   a. If the child is to be raised as a female, then gonadectomy has to be done. A feminizing genitoplasty maybe required.
   b. In case of male sex assignment, the decision should be based on the position and appearance of the testis. If it appears normal and palpable, orchidopexy and biopsy with close monitoring can be done. However, a low threshold must be kept for gonadectomy.
   c. Monitoring includes yearly follow-up including USG with the onset of puberty.
C) Mildly undervirilized male—If testis appears to be normal on palpation and on imaging, routine yearly examination will suffice. In case of undescended testis—Orchidopexy and biopsy is recommended.
D) The above treatment protocol can be followed even for a fully virilized male (46XY PGD) with undescended testis. If the testis is descended, then biopsy is not recommended and routine testicular self-examination will be sufficient [16].

Risk of germ cell malignancy—is high among patients with gonadal dysgenesis, particularly those with Y chromosome and in the children with intra-abdominal testis.

The risk of malignancy is 15–35%. Due to this high risk, gonadectomy has been recommended at the time of diagnosis. The associated risk increases with age.

The testicular biopsy can be sent for IHC staining. Studies have shown that OCT ¾ positive cells located in the basal lamina, TSPY positivity and SCF positive have a high risk of development of malignancy. It must be kept in mind that biopsies can miss out on a malignancy due to limited sampling and hence patients have to be monitored closely.

For patients assigned the male sex, retaining the gonadal tissue preserves the source of testosterone even if it is limited [16, 17].

While the above are the recommendations, the decision for gonadectomy should be taken when the benefits outweigh the risks and has been discussed by a multidisciplinary team with the patient and the family.

24.8.1 National Experience

The Delhi Commission for Protection of Child Rights (DCPCR) has given it recommendation to ban medically unnecessary sex-selective surgeries on intersex infants except in cases of life-threatening situations. The decision by the government is pending.

Tamil Nadu government has been the first state in India to place a prohibitive ban on sex-selective surgeries in infants.

24.8.2 International Experience

The United Nations has condemned intersex surgeries on infants multiple times. However, only three nations—Malta, Chile, and Germany have actually banned this practice in Infants.

References

1. SOHVAL AR. "Mixed" gonadal dysgenesis: a variety of hermaphroditism. Am J Hum Genet. 1961;1963(15):155–8.
2. Wilkins L, Walter F. Ovarian agenesis. J Clin Endocrinol. 1944;4:8.
3. Bergada C, Cleveland WW Jr, HWJ. Gonadal dysgenesis : relation to theories by. Acta Endocrinol. 1962;40:493–520.
4. Lipay MVN, Bianco B, Verreschi ITN. Disgenesias gonadais e tumores: aspectos genéticos e clínicos. Arq Bras Endocrinol Metabol. 2005;49(1):60–70.
5. Lee P, Houk C, Ahmed SF, Hughes IA. Consensus statement on management of intersex. Pediatrics. 2006;118(2):488–500.
6. Hughes IA, Houk C, Ahmed SF, Lee PA. Topic collections Consensus statement on management of intersex disorders Management of intersex disorders. Apr Arch Dis Child [Internet]. 2006;91:554–63.
7. de Andrade JGR, Guerra-Júnior G, Maciel-Guerra AT. 46,XY and 45,X/46,XY testicular dysgenesis: similar gonadal and genital phenotype, different prognosis. Arq Bras Endocrinol Metabol. 2010;54(3):331–4.
8. Gabriel Ribeiro De Andrade J, Marques-De-Faria AP, Fabbri HC, De Mello MP, Guerra-Júnior G, Maciel-Guerra AT. Long-term follow-up of patients with 46,XY partial gonadal dysgenesis reared as males. Int J Endocrinol. 2014;2014:1–9.

9. Andrade JGR, Fabbri-Scallet H, Dos Santos AP, Cools M, Werner R, Hiort O, et al. Clinical findings and follow-up of 46,XY and 45,X/46,XY testicular dysgenesis. Sex Dev. 2020;13(4):171–7.
10. Chemes HE, Muzulin PM, Venara MC, Muhlmann MDC, Martínez M, Gamboni M, et al. Early manifestations of testicular dysgenesis in children: pathological phenotypes, karyotype correlations and precursor stages of tumour development. APMIS. 2003;111(1):12–24.
11. Matsumoto F, Shimada K, Ida S. Tumors of bilateral streak gonads in patients with disorders of sex development containing y chromosome material. Clin Pediatr Endocrinol. 2014;23(3):93–7.
12. Szarras-Czapnik M, Lew-Starowicz Z, Zucker KJ. A psychosexual follow-up study of patients with mixed or partial gonadal dysgenesis. J Pediatr Adolesc Gynecol. 2007;20(6):333–8.
13. Aleck KA, Argueso L, Stone J, Hackel JG, Erickson RP. True hermaphroditism with partial duplication of chromosome 22 and without SRY. Am J Med Genet. 1999;85(1):2–4.
14. Ledig S, Hiort O, Wünsch L, Wieacker P. Partial deletion of DMRT1 causes 46,XY ovotesticular disorder of sexual development. Eur J Endocrinol. 2012;167(1):119–24.
15. Donahoe PK, Crawford JD, Hendren WH. Mixed gonadal dysgenesis, pathogensis, and management. J Pediatr Surg. 1979;14(3):287–300.
16. McCann-Crosby B, Mansouri R, Dietrich JE, McCullough LB, Sutton VR, Austin EG, et al. State of the art review in gonadal dysgenesis: challenges in diagnosis and management. Int J Pediatr Endocrinol [Internet] 2014;2014(1):1–17. Available from: International Journal of Pediatric Endocrinology.
17. Barros BA, Moraes SG, Coeli FB, Assumpção JG, De Mello MP, MacIel-Guerra AT, et al. OCT4 immunohistochemistry may be necessary to identify the real risk of gonadal tumors in patients with Turner syndrome and y chromosome sequences. Hum Reprod. 2011;26(12):3450–5.

Disorders of Testosterone Synthesis 25

Nidhi Sugandhi

Abstract

Disorders of testosterone synthesis are rare causes of 46XY DSD. Testosterone is the main driver of male sexual differentiation. Absence of testosterone due to defects in synthesis affects the formation of both external and internal genitalia in the fetus and development of secondary sexual characteristics later. Since there is a common metabolic pathway for synthesis of glucocorticoids, mineralocorticoids, and testosterone, most of these disorders are associated with deficiency of glucocorticoids and mineralocorticoids. This could lead to life-threatening complications like hypotension and salt wasting at birth. Majority of the children with these disorders have rudimentary external and internal genitalia and behave as females. These children may only be diagnosed at puberty due to amenorrhea. Due to absolute deficiency of testosterone, male pattern brain imprinting and virilization potential are minimal. Feminizing genitoplasty and estrogen supplementation can help them function as a sexually active female despite infertility. These children may also need to be supplemented with glucocorticoids and mineralocorticoids to prevent life-threatening symptoms due to hypertension and salt wasting. Diagnosis is reached by low level of testosterone, not increasing in response to gonadotropin levels along with higher level of the precursor substrates which differ according to the level of the block. Genetic testing for these disorders is confirmatory but difficult due to incomplete characterization of all genes involved, high costs, and limited availability.

Keywords

Lipoid congenital adrenal hyperplasia · StAR deficiency · Smith–Lemli–Opitz syndrome · Cholesterol metabolism · Leydig cell hypoplasia

N. Sugandhi (✉)
Department of Pediatric Surgery, Vardhaman Mahavir Medical college, New Delhi, India

S. K. Ratan (ed.), *Children with Differences in Sex Development*,
https://doi.org/10.1007/978-981-97-1639-5_25

25.1 Introduction

Disorders of testosterone synthesis are relatively infrequent cause of 46XY DSD. Their incidence has not been successfully estimated as they are difficult to diagnose. They need a high index of suspicion in addition to a battery of hormonal and molecular tests which may not be universally available. As in the other 46XY DSD disorders, the severity of the hormone synthesis defects falls on a spectrum and hence also the very variable phenotype. In general, the disorders of testosterone synthesis are characterized by severe deficiency of testosterone synthesis starting in the fetus itself thus interfering even in embryogenesis of the male internal and external genitalia. Most of these children have rudimentary external and internal genitalia and appear and behave as females. Male secondary sexual characteristics do not develop at puberty. Gender dysphoria is less common in these children as compared to Partial Androgen Insensitivity Syndrome (PAIS) and 5-alpha-reductase deficiency as developing brain is usually not exposed to testosterone.

25.2 Pathophysiology

Testosterone and its derivatives are the main drivers of male sexual differentiation. Testosterone starts being secreted by Leydig cells by 7th week of embryonal life and induces differentiation of the Wolffian duct to form the male internal genitalia including epididymis, vas deferens, and seminal vesicles. Also, it acts on the genital tubercle and folds after conversion to 5 dihydrotestosterone to form the phallus and scrotum. At puberty, a surge in testosterone induces the male secondary sexual characteristics.

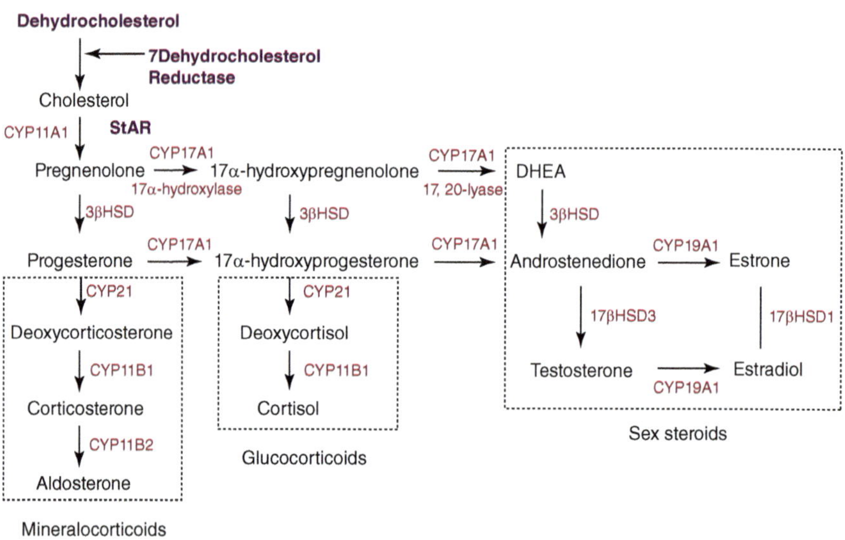

Fig. 25.1 Flowchart showing cholesterol metabolism enzymes and pathway leading to biosynthesis of testosterone, glucocorticoids, and mineralocorticoids

Figure 25.1 outlines the pathway through which cholesterol is converted to the sex steroids, glucocorticoids, and mineralocorticoids. This pathway takes place in testis, adrenals, and some peripheral tissues such as adipose tissue, brain, skin, thymus, and liver [1–4]. Any deficiency in the enzymes of this pathway affects the testosterone biosynthesis and also secretion of mineralocorticoids and glucocorticoids. Thus, these patients in addition to DSD may also have different metabolic abnormalities.

25.3 Smith–Lemli–Opitz Syndrome (SLOS)

This is a rare autosomal recessive syndrome where the enzyme 7-dehydrocholesterol reductase coded by 7DHCR gene is deficient leading to non-conversion of dehydrocholesterol to cholesterol. Thus, the substrate for further enzyme synthesis, that is cholesterol, is deficient. It leads to multiple anomalies including microcephaly, distinct facial features (anteriorly facing nostrils), behavioral and intellectual problems, growth restrictions, 2–3 toe syndactyly, and also may be associated with kidney, heart, and adrenal problems [5].

The ambiguous genitalia occur in 20–25% of affected individuals because of deficiency of testosterone to varying severity. It is characterized by severe undervirilization with undescended gonads, bifid scrotum, small phallus, and proximal hypospadias. It is diagnosed by low levels of cholesterol and high levels of 7-dehydrocholesterol and genetic analysis.

Treatment for DSD associated with Smith-Lemli-Opitz Syndrome (SLOS) is masculinizing genitoplasty (orchiopexy, hypospadias repair, Penoscrotal transposition correction) and testosterone supplementation at puberty. Other metabolic defects due to mineralocorticoid and glucocorticoid deficiency may be needed to be treated appropriately.

25.4 StAR Deficiency and P450scc Deficiency

Steroidogenesis is initiated with conversion of cholesterol to pregnenolone mediated by cholesterol side-chain cleavage enzyme cytochrome P450scc, also known as CYP11A1 (previously known as 20–22 desmolase). Steroidogenic acute regulatory (StAR) protein facilitates the transport of cholesterol within the mitochondria [6]. Though the chronic regulation of steroidogenesis by ACTH occurs at the level of gene transcription, the acute regulation occurs by the StAR-mediated transport of substrate from the outer mitochondrial membrane into the inner mitochondrial membrane. This is the mechanism that induces the formation of testosterone in response to LH pulse [7].

Absence of StAR is an autosomal recessive condition characterized by severe deficiency of all adrenal and gonadal steroids and leads to a condition known as Lipoid Congenital Adrenal Hyperplasia (LCAH). It is seen more frequently in Arabs, East Asians, and Swiss [8]. P450scc enzyme suboptimal quantity or activity leads to the same deficiency in steroidogenesis and presentation similar LCAH, except that adrenal cells are not oversaturated by lipid droplets as seen in StAR deficiency. Absence of testosterone in embryonal life leads to a complete female external phenotype and absence of internal genitalia. A severe deficiency of glucocorticoids and mineralocorticoids can cause death in neonatal period due to salt wasting hypovolemia, hyponatremia, hyperkalemia, acidosis. Typical newborn has a female phenotype, features of salt wasting and hyperpigmentation due to ACTH stimulation. Despite a female phenotype, there is absence of uterus and Mullerian structures as anti mullerian hormone (AMH) is normally secreted by the Sertoli cells. However, some amount of StAR-independent steroidogenesis can still occur and may sustain life till child is older. However soon, the chronic stimulation by ACTH, angiotensin and then LH at puberty causes the adrenal cells to be saturated with unused cholesterol esters and cholesterol oxidation products and causes the "second hit" by disrupting the cells. This destroys the ability of cell to secrete any StAR-independent steroidogenesis. Thus, milder cases may present as an undervirilized male with late onset adrenal insufficiency [9]. These patients are prone to develop testicular adrenal rest tumors leading to late onset testicular failure.

Diagnosis is suspected in a case of child with ambiguous genitalia, hyperpigmentation, and salt wasting and confirmed by low levels of hydrocortisone, aldosterone, and testosterone with elevated LH, ACTH, and renin and normal male range AMH. Testosterone levels fail to rise in response to HCG stimulation too.

Treatment has to focus first on the glucocorticoid and mineralocorticoid replacement as lifesaving measures. Usually these children are brought up as females and may undergo gonadectomy due to the risks of malignancy. Some type of feminizing genitoplasty in form of vaginal reconstruction of a rudimentary vagina or clitoral reduction may be done if indicated. Milder variants may decide to rear as females but would need extensive male genitoplasty in form of orchidopexy, hypospadias repair and also testosterone supplementation at puberty.

25.5 Combined 17-Alpha Hydroxylase, 17,20-Lyase Deficiency (P450c17D)

This is a rare autosomal recessive disease due to mutations in CYP17A1 gene, which causes defect in conversion of pregnenolone to 17 hydroxy pregnenolone, dihydroepiandrosterone, androstenedione, thus decreasing further formation of cortisol and testosterone. Aldosterone is excessively produced as all substrates are directed towards the mineralocorticoid pathway, leading to hypertension and hypokalemia. Thus, this condition is characterized by undervirilized 46 XY neonate with female phenotype (undescended gonads, very small clitoris-like phallus, and perineal hypospadias), absent Mullerian structures due to normal AMH, hypertension, and low potassium.

The phenotype results due to high aldosterone and low testosterone levels not responding to HCG stimulation. Deoxycorticosterone and corticosterone levels are high, whereas deoxycortisol and cortisol levels are low, thus demonstrating high deoxycorticosterone/deoxycorticosterol and corticosterone/cortisol levels The female may look absolutely normal and can either present in neonatal period due to hypertension and hypokalemia or in adolescence due to primary amenorrhea and absence of breast development.

These children are mostly brought up as girls and would require gonadectomy and estrogen supplementation. Control of hypertension is achieved by supplementing glucocorticoids.

25.6 Isolated 17-20 Lyase Deficiency

Isolated 17-20 lyase deficiency is a rare autosomal recessive disease that is characterized by absence of only the last step of conversion of progesterone to androgens. Thus, there are normal to high levels of progesterone, pregnenolone, and their 17 hydroxy derivatives (since 17-alpha hydroxylase is normal), normal aldosterone and cortisol but low to absent dihydroepiandrosterone, androstenedione, testosterone, and estrogen.

The neonate has a female phenotype with 46XY karyotype or are severely undervirilized male phenotype. Though there is absence of any secondary sexual characteristics, breast development may occur at puberty, so those brought up as males may have gynecomastia.

As in other conditions with enzyme deficiencies, children are better brought up as females and undergo gonadectomy with estrogen supplementation at puberty.

25.7 17 Beta Hydroxy Steroid Dehydrogenase Deficiency [10, 11]

This condition closely resembles other causes of partial male undervirilization, notably PAIS and 5-alpha reductase deficiency. The final enzyme responsible for conversion of androstenedione to testosterone, i.e. 17 beta hydroxysteroid dehydrogenase, is missing, leading to deficiency of testosterone and high levels of androstenedione. Since androstenedione itself has reasonable testosterone-like properties, these children are partially virilized unlike other enzyme deficiencies where they are totally unvirilized. In this deficiency, there are no metabolic abnormalities seen as other steroid synthesis is not affected. Also, at puberty under gonadotropin stimulus, large amounts of androstenedione may collect which may cause virilization by itself or by non-adrenal conversion to testosterone. These children may have undescended but palpable gonads, presence of penis but of small size, varying degree of hypospadias, and bifid scrotum. Some degree of male pattern brain imprinting also occurs. If they are brought up as girls, then they may have sudden virilization at puberty including increase in phallus size, descent of gonads, and male pattern hair development. For this reason, diagnosis is better confirmed at birth and decision about sex of rearing carefully taken, considering both present level of virilization and virilization spurt expected at puberty.

Diagnosis is confirmed by low level of testosterone but reasonably good response to HCG stimulation. Ratio of testosterone/androstenedione is <0.8 [11]. Genetic analysis confirms the diagnosis.

Those brought up as females need bilateral gonadectomy well before the pubertal spurt so that any possibility of virilization is avoided. In addition, they need feminizing genitoplasty, clitoral reduction and vaginal reconstruction, plus estrogen supplementation at puberty.

Those brought up as males may need orchiopexy and hypospadias repair. Testosterone supplementation may be considered at puberty depending on the level of virilization achieved and the pubertal spurt expected [11].

25.8 3 Beta Hydroxysteroid Dehydrogenase 2 Deficiency

This disease is similar to 17 beta hydroxy steroid dehydrogenase deficiency as it also causes partial undervirilization unlike the other enzymes involved in testosterone biosynthesis which lead to complete absence of virilization. It is a rare disorder with incidence <1 in 1,000,000 [8]. 3 beta hydroxysteroid dehydrogenase (3 BHSD) type 2 converts pregnenolone to progesterone, 17-alpha hydroxy pregnenolone to 17-alpha hydroxyprogesterone, and dehydroepiandrosterone (DHEA) to testosterone. 3BHSD type 1 is responsible for conversion to testosterone in peripheral tissues like liver, brain, placenta, and mammary glands. 3BHSD2 deficiency leads to excess of 17 hydroxy progesterone and DHEA. Since DHEA has some androgenic action, and is also converted to testosterone peripherally by 3 BHSD1, it leads to partial virilization in the child, resulting in a male phenotype but who is undervirilized. Thus, the child may have undescended testis, hypospadias, bifid scrotum, penoscrotal transposition, and vaginal pouch. Pubertal spurt with the attendant LH stimulus results in large amounts of DHEA and appearance of male secondary sexual characteristics. These children are also prone to testicular adrenal rest tumors.

Diagnosis is based on high ratios of 17 hydroxy pregnenolone to 17 hydroxy progesterone and DHEA to testosterone in urine specially after ACTH stimulation [5microgram/kg ACTH injected IM or IV, followed by sample collection at 1–2 h]. 17-hydroxypregnenolone levels above 100 nmol/L (3300 ng/dl) either baseline or after ACTH stimulation are the single best criteria for diagnosis of 3BHSD 2 deficiency. In addition, a 1000 fold elevation of 17-hydroxypregnenolone to cortisol ratio is almost confirmatory of 3BHSD [12].

Neonatal diagnosis may be difficult because of effect of 3BHSD 1 activity. Genetic testing provides confirmation.

25.9 Leydig Cell Hypoplasia

Leydig cell hypoplasia is a rare autosomal recessive disorder caused by mutation of LHCGR gene which codes for the LH/HCG receptor on Leydig cells. Mutation of this gene causes a defective receptor which makes the Leydig cell nonresponsive to

LH/HCG stimulation. It does not secrete testosterone and gradually undergo atrophy. The cells are also unresponsive to placental gonadotrophins so that the fetal secretion of testosterone is absent. Similarly response to exogenously administered gonadotrophins like in HCG stimulation test is also absent.

Two types are identified—Type 1 where Leydig cells are totally unresponsive and Type 2 where partial response may be present. Type 1 children are completely feminized at birth but with absence of Mullerian structures as AMH secretion is normal. Type 2 may present as undervirilized males with undescended testis and hypospadias [13, 14].

Diagnosis is usually suspected at puberty when a phenotypic female fails to develop secondary sexual characteristics. Rarely it may be suspected in a neonate with Type 2 Leydig cell hypoplasia who presents with ambiguous genitalia. Diagnosis is suggested by testicular biopsy which shows solid tissue with arrested spermatogenesis, Leydig cell hypoplasia, and normal Sertoli cells. It is confirmed by testing for LHCGR mutation on chromosome 2p. Unless genetic testing is available, testicular biopsy is essential to clinch the diagnosis.

Treatment depends on sex of rearing and in case of children brought up as females, it involves gonadectomy and estrogen supplementation. Those brought up as males need orchiopexy, hypospadias repair, and testosterone supplementation.

25.10 Diagnosis

Most of the disorders of testosterone synthesis share common clinical features in terms of phenotype and presence of associated adrenal dysfunction.

Since testosterone is absent since fetal life, both internal and external genitalia do not masculinize. Thus, formation of vas, epididymis, and seminal vesicles is affected, testis do not descend, phallus does not increase in size and appears as clitoris; there is unfused urogenital folds appearing as labia and penile urethra does not develop leading to perineal urethral meatus. All this gives rise to a normal female appearing neonate. Ambiguity is not seen and DSD is not suspected unless associated with other symptoms of adrenal insufficiency like electrolyte imbalance. DSD may be suspected later in a female child with enlarged clitoris, inguinal hernia showing prolapsing gonad, or a virilization spurt during puberty.

Some partial enzyme deficiencies may result in formation of testosterone albeit in very decreased amount or testosterone-like action of its precursors like DHEA and androstenedione. These children show partial virilization with undescended gonads, microphallus, and hypospadias. In these children, DSD may be suspected at birth due to ambiguous genitalia. Hormonal assays in blood or urine are the first step for diagnosis but are plagued by many difficulties. Diagnosis at birth by hormonal assays is slightly difficult due to confounding effect of peripheral activity of alternate enzyme isoforms [7]. However, there are some periods of normal gonadotropin surge during neonatal period and infancy which can aid in diagnosis. 6–36 h after birth is a normal testosterone surge under the influence of placental gonadotrophins which can help in assessing the presence of functioning testicular tissue, though normal levels for this age group need to be standardized. Again during mini puberty (15–90 days after birth) there is a normal activation of hypothalamus–pituitary–gonadal axis which causes a LH led surge in testosterone. It can help in differentiating between DSD due to defects in testosterone synthesis in which case, testosterone levels would remain low despite high LH and disorders of testosterone action in which case testosterone would show a concomitant increase with LH despite the presence of ambiguous genitalia [15, 16].

Most hormonal assays are afflicted by low specificity due to cross reactivity between the various steroids and their metabolites [7]. Also, only one steroid can be measured at a time and many assays may be required to arrive at a diagnosis. Mass spectrometry, liquid chromatography and gas chromatography are useful for testing multiple steroids and metabolites simultaneously but these are not routinely available [17].

Genetic testing is the next step in arriving at diagnosis in cases of testosterone synthesis disorders. Despite the limitations, if confirmed on genetic testing, one can get a lot of clarity regarding prognosis and possible clinical course of disorder and this makes the counseling that much more effective. However, many of these disorders are multigenomic and not all genetic mutations for all have been characterized. Additionally, the multitude of complex tests required for accurate mutation testing like chromosomal microarray, gene sequencing, gene slicing panels, etc., are expensive and not easily available. Owing to all this, only about 20–50% of 46XY DSD receive a genetic diagnosis. [18–20].

Table 25.1 presents a synopsis of the biochemical and genetic defects seen in the various disorders of testosterone synthesis [15].

Table 25.1 Hormonal and genetic profile of the various disorders of testosterone synthesis [15]

	Defects of Androgen Production									
		LHCGR Defects								
DSD Etiology	Smith-Lemli-O PTIZ Syndrome	Complete Form	Partial Form	20,22-Desmoloase Deficiency	Star Deficiency	3B-HSD2 Deficiency	17a-Hydroxylase Deficiency	POR Deficiency	17,20-Lyase Deficiency	17B-HSD3 Deficiency
Inheritance	Autosomal recessive	Autosomal recessive	Autosomal recessive	Autosomal recessive	Autosomal recessive	Autosomal recessive	Autosomal recessive	Autosomal recessive	Autosomal recessive	Autosomal recessive
Most common sex assignment	Male	Female, male-mid form	Male/female	Female, male-mid form	Female, male-mid form	Male	Female in most patients	Male	Female	Female
Gender change	Rare	Rare	Rare	Rare	Rare	Rare	Rare	Rare	Rare	Not so rare
External genitalia	Micropenis and/or hypospadias, hypoplasic or bifid scrotum; female	Female micropenis (mid form)	Atypical to male	Female micropenis (mid form)	Female micropenis (mid form)	Atypical	Female-like to atypical	Atypical	Atypical	Atypical, frequently female-like at birth
Mullerian duct derivatives	Rarely present	Absent	Absent	Absent	Absent	Absent	Absent	Absent	Absent	Absent
Wolffian duct derivatives	Absent to male	Absent to hypoplastic	Male	Absent to hypoplastic	Absent to hypoplastic	Male	Hypoplastic to normal	Hypoplastic to normal	Normal	Normal
Gonads/usual position	Testis/inguinal or scrotal region	Testis/inguinal Or intra-abdominal	Testis/scrotal or inguinal region	Testis/inguinal, intra-abdominal or scrotal region	Testis/inguinal, intra-abdominal or scrotal region	Testis/ inguinal, intra-abdominal or scrotal region	Testis/ intra-abdominal or inguinal	Testis/ inguinal, intra-abdominal or scrotal region	Testis/ inguinal, intra-abdominal or scrotal region	Testis/ inguinal, or intra-abdominal region
Associated clinical features	Mental retardation, syndromic features	None	None	Early adrenal failure	Enlarged adrenal glands Early adrenal failure	Adrenal failure or not	Hypertension (low renin)	Antley-Bixler sydreome	None	None

(continued)

Table 25.1 (continued)

	Defects of Androgen Production									
	Smith-Lemli-O	LHCGR Defects								
DSD Etiology	PTIZ Syndrome	Complete Form	Partial Form	20,22-Desmoloase Deficiency	Star Deficiency	3B-HSD2 Deficiency	17a-Hydroxylase Deficiency	POR Deficiency	17,20-Lyase Deficiency	17B-HSD3 Deficiency
Puberty	Normal	Absence of spontaneous development	Partial virilization absence gynecomastia	Absence of spontaneous development	Absence of spontaneous development	Partial virilization absence gynecomastia	Absent or slight virilization variable gynecomastia	Partial virilization	Poor virilization, Variable gynecomastia	Virilization, Variable Gynecomastia
Gene location	11q12-q13	2p21	2p21	15q24.1	8p11.2	1p13.1	10q24.3	7q11.2	10q24.3	9q22
Molecular defect	Inactivating mutations in *DHCR7*	Inactivating mutations in *LHCGR* (complete inactivation	Mutations in *CYP11A1*	Mutations in *HSD332*	Mutations in *CYP17a1*	Mutations in *CYP11A1*	Mutations in *CYP17A1*	Mutations in *POR*	Mutations in the redox partner binding site of CYP17A1	Mutations in *HSD17B3*

25.11 Management

The most important decision in management of 46XY DSD children is regarding sex of rearing, especially when the child is diagnosed in neonatal period.

Enzyme deficiencies like StAR, SF1, 17-20 lyase and combined 17-20 lyase and 17-alpha hydroxylase deficiency, Leydig cell hypoplasia lead to a totally external female phenotype with undescended gonads, perineal urethra, vaginal pit, and clitoris-like phallus. These children present at puberty due to primary amenorrhea with or without breast development. Since these conditions have close to zero virilization potential even at puberty with the testosterone surge, these children are better brought up as females. Despite absence of uterus, these females are potentially sexually active though infertile. Gender dysphoria is minimal as there is no testosterone even during fetal life to cause male pattern brain imprinting. These children require gonadectomy to prevent testicular malignancies, incidence of which rises to the tune of 20–30%, especially after puberty. Some form of feminizing genitoplasty like clitoral reduction or vaginal augmentation may be required. Estrogen supplementation is required at puberty for normal female secondary sexual characteristics.

Other conditions where there is partial testosterone synthesis or testosterone-like action due to other androgens (DHEA) are trickier to manage. These children present with partially virilized external genitalia and depending on the extent of masculinization may appear as undervirilized males or overvirilized females. Diagnosis needs to be established as early as possible so that sex of rearing can be decided. The issue is compounded because many conditions like 17 beta hydroxy steroid dehydrogenase deficiency and 3 beta hydroxysteroid dehydrogenase 2 deficiency have partial androgen action. In addition to fetal brain male imprinting, there is reasonable masculinization at puberty. Literature supports rearing these children as males. Orchiopexy, hypospadias repair, and penoscrotal transposition correction constitute the male genitoplasty. Testosterone supplementation is required at puberty to assist development of male secondary sexual features. Options for testosterone supplementation include intramuscular short-acting testosterone esters and local transdermal testosterone preparations for direct action on external genitalia [21, 22].

Both estrogen and testosterone supplementations require careful monitoring regarding timing and dose such that it mimics natural puberty as far as possible. Bone growth, physical growth, psychological and sexual maturity all need to be considered before initiating and continuing testosterone or estrogen replacement.

Disorders of testosterone synthesis due to adrenal enzyme defects have variable insufficiency of glucocorticoids and mineralocorticoids. These children also need to be supplemented with glucocorticoids and mineralocorticoids to prevent life-threatening symptoms due to hypertension and salt wasting.

References

1. Chakraborty S, Pramanik J, Mahata B. Revisiting steroidogenesis and its role in immune regulation with the advanced tools and technologies. Genes Immun. 2021;22:125–40. https://doi.org/10.1038/s41435-021-00139-3.
2. King SR, Manna PR, Ishii T, Syapin PJ, Ginsberg SD, Wilson K, et al. An essential component in steroid synthesis, the steroidogenic acute regulatory protein, is expressed in discrete regions of the brain. J Neurosci. 2002;22:10613–2.
3. Hannen RF, Michael AE, Jaulim A, Bhogal R, Burrin JM, Philpott MP. Steroid synthesis by primary human keratinocytes; implications for skin disease. Biochem Biophys Res Commun. 2011;404:62–7.
4. Li J, Papadopoulos V, Vihma V. Steroid biosynthesis in adipose tissue. Steroids. 2015;103:89–104.
5. Nowaczyk MJM, Wassif CA. Smith-Lemli-Opitz Syndrome. 1998 Nov 13. In: Adam MP, Everman DB, Mirzaa GM, et al., editors. GeneReviews® [Internet]. Seattle, WA: University of Washington; 1993-2023.
6. Arakane F, Sugawara T, Nishino H, Liu Z, Holt JA, Pain D, et al. Steroidogenic acute regulatory protein (StAR) retains activity in the absence of its mitochondrial import sequence: implications for the mechanism of StAR action. Proc Natl Acad Sci USA. 1996;93:13731–6.
7. Miller WL. Disorders of Androgen biosynthesis. Semin Reprod Med. 2002;20(3)
8. Finkielstain GP, Vieites A, Bergadá I, Rey RA. Disorders of sex development of adrenal origin. Front Endocrinol (Lausanne). 2021;12:770782. https://doi.org/10.3389/fendo.2021.770782.
9. Kolli V, Kim H, Torky A, Lao Q, Tatsi C, Mallappa A, et al. Characterization of the CYP11A1 nonsynonymous variant P.E314K in children presenting with adrenal insufficiency. J Clin Endocrinol Metab. 2019;104:269–76. https://doi.org/10.1210/jc.2018-01661.
10. Galli-Tsinopoulou A, Serbis A, Kotanidou EP, Litou E, Dokousli V, Mouzaki K, Fanis P, Neocleous V, Skordis N. 46,XY Disorder of sex development due to 17-beta hydroxysteroid dehydrogenase type 3 deficiency in an infant of Greek origin. J Clin Res Pediatr Endocrinol. 2018;10(1):74–8. https://doi.org/10.4274/jcrpe.4829.
11. Grimbly C, Caluseriu O, Metcalfe P, Jetha MM, Rosolowsky ET. 46,XY disorder of sex development due to 17-beta hydroxysteroid dehydrogenase type 3 deficiency: a plea for timely genetic testing. Int J Pediatr Endocrinol. 2016;2016:12. https://doi.org/10.1186/s13633-016-0030-x.
12. Mermejo LM, Elias LL, Marui S, Moreira AC, Mendonca BB, de Castro M. Refining hormonal diagnosis of type II 3beta-hydroxysteroid dehydrogenase deficiency in patients with premature pubarche and hirsutism based on HSD3B2 genotyping. J Clin Endocrinol Metab. 2005 Mar;90(3):1287–93. https://doi.org/10.1210/jc.2004-1552.
13. Çömlek FÖ, Yıldız R, Seyrek F, Tütüncüler F. Leydig cell hypoplasia type 1 diagnosed in early childhood with inactivating mutation in LHCGR gene, Oxford Medical Case Reports, Volume 2021, Issue 4, April 2021, omab015, doi:https://doi.org/10.1093/omcr/omab015
14. Wu SM, Leschek EW, Rennert OM, Chan WY. Luteinizing hormone receptor mutations in disorders of sexual development and cancer. Front Biosci. 2000;5:D343–52.
15. Wisniewski AB, Batista RL, Costa EMF, Finlayson C, Sircili MHP, Dénes FT, Domenice S, Mendonca BB. Management of 46,XY differences/disorders of sex development (DSD) throughout life. Endocr Rev. 2019;40(6):1547–72.
16. Paris F, Gaspari L, Philibert P, Maïmoun L, Kalfa N, Sultan C. Disorders of sex development: neonatal diagnosis and management. Endocr Dev. 2012;22:56–71. https://doi.org/10.1210/er.2019-00049.
17. Kulle A, Krone N, Holterhus PM, Schuler G, Greaves RF, Juul A, de Rijke YB, Hartmann MF, Saba A, Hiort O, Wudy SA, EU COST Action. Steroid hormone analysis in diagnosis and treatment of DSD: position paper of EU COST Action BM 1303 "DSDnet". Eur J Endocrinol. 2017;176(5):P1–9.
18. Achermann JC, Domenice S, Bachega TA, Nishi MY, Mendonca BB. Disorders of sex development: effect of molecular diagnostics. Nat Rev Endocrinol. 2015;11(8):478–88.

19. Barseghyan H, Delot EC, Vilain E. New technologies' to uncover the molecular basis of disorders of sex development. Mol Cell Endocrinol. 2018;468:60–9.
20. Hiort O, Birnbaum W, Marshall L, Wunsch L, Wünsch L, Werner R, Schröder T, Döhnert U, Holterhus PM. Management of disorders of sex development. Nat Rev Endocrinol. 2014;10(9):520–9.
21. McGriff NJ, Csako G, Kabbani M, Diep L, Chrousos GP, Pucino F. Treatment options for a patient experiencing pruritic rash associated with transdermal testosterone: a review of the literature. Pharmacotherapy. 2001;21(11):1425–35.
22. Dunkel L, Quinton R. Transition in endocrinology: induction of puberty. Eur J Endocrinol. 2014;170(6):R229–39.

Disorders of Anti-Mullerian Hormone and Its Receptors: Persistent Mullerian Duct Syndrome

26

Mrinal Arora and Aparajita Mitra

Abstract

Persistent Mullerian duct syndrome (PMDS), first described by Nilson in 1939, is a rare disorder of internal male sex development. It is considered a very rare disorder and around 300 cases have been published in literature till date. PMDS can be idiopathic or the fallout of an autosomal recessive (85%), X linked or autosomal dominant sex-limited trait. It is estimated that mutation in the anti-Müllerian hormone (AMH) gene on the nineteenth chromosome (type I PMDS, 45%) and an alteration in the AMH receptor (AMHR) gene on the twelfth chromosome (type II PMDS, 40%) are the most common causes. In the remaining cases, the cause is unknown (idiopathic PMDS).

Children with PMDS have normal virilized male external genitalia but may have unilateral or bilateral inguinal hernia or cryptorchidism. The Müllerian derivatives are incidentally discovered during the surgery for an inguinal hernia or cryptorchidism. On the basis of pathoanatomy, three types have been described in literature. In the **female type PMDS** (60–70%), both testes are adherent to the uterus and lie in the ovarian position inside the abdominal cavity. Blind-ending vas and vessels abut the fallopian tube-like structures. The next most common variant, **hernia uteri inguinalis** (20–30%) is also known as the "male-type" PMDS. It presents as an inguinal hernia with the ipsilateral testis and Mullerian structures as contents of the sac. The contralateral testis is usually found to be undescended on physical examination. The **transverse testicular ectopia (TTE) variant** is the least common (10%), with both testes and the Müllerian structures descended into one hemiscrotum.

M. Arora · A. Mitra (✉)
Department of Pediatric Surgery, ESI-PGIMSR and ESIC Medical College & Hospital, Faridabad, Haryana, India

S. K. Ratan (ed.), *Children with Differences in Sex Development*, https://doi.org/10.1007/978-981-97-1639-5_26

The basic principles to be followed in the surgical management of PMDS are (1) Early detection and confirmation of PMDS, (2) Preservation of gonadal function and fertility potential, and (3) Facilitation of testicular self-examination by early orchidopexy. Persistent Mullerian ducts should be excised, or the mucosa should be destroyed completely at the least, to prevent malignant degeneration and recurrent urinary tract infection (UTI). In case it is necessary to leave them in situ, an annual ultrasound is recommended for surveillance to assess any changes in size or a new mass lesion.

Keywords

Persistent Mullerian duct syndromes (PMDS) · Transverse testicular ectopia (TTE) · Anti-Mullerian hormone (AMH)

26.1 Introduction

Persistent Mullerian duct syndrome (PMDS), first described by Nilson in 1939, is a rare disorder of internal male sex development [1]. It is described as the presence of Müllerian duct structures, the uterus, vagina, and the fallopian tubes in an otherwise normal male phenotype and genotype.

26.2 Embryology and Genetics

While the incidence of PMDS cannot be accurately determined, there are around 300 published cases available in literature. The various hormones involved in the growth of male sexual organs are testosterone, dihydrotestosterone, and the anti-Müllerian hormone (AMH)/Müllerian inhibiting substance (MIS) (Tables 26.1, 26.2, 26.3, and 26.4). The differentiation of Wolffian duct structures into the epididymis, vas deferens, seminal vesicles, and ejaculatory ducts is orchestrated by testosterone. AMH is produced by the fetal Sertoli cells at the end of the seventh week of gestation which drives the retrogression of the Müllerian duct structures. In the absence of this hormone, Müllerian structures transform into the uterus, fallopian tubes, and upper vagina by default. The pathoanatomy includes a cervix which is attached to posterior urethra at the verumontanum and a short free segment of vas deferens before it embeds into the mesosalpinx and lateral wall of uterus and cervix and eventually opens into the vaginal fornix. Chromosomal analysis of all cases of PMDS reveals a 46 XY disposition.

AMH is an offshoot of the TGFβ (transforming growth factor β) family of growth factors and drives cellular degeneration. Any defect in the AMH receptor and its synthesis causes a persistence of the Müllerian duct and hence, this particular condition. PMDS may be idiopathic or as a consequence of an autosomal recessive (85%), X linked or autosomal dominant sex-limited trait. It is estimated that a mutation in the AMH gene on the nineteenth chromosome (type I PMDS, 45%) and an

Table 26.1 Testosterone hormone levels in PMDS vs normal individuals

Hormone	Age	PMDS	Normal reference range
Testosterone	Newborn	75–400 ng/dl	75–400 ng/dl
	1–5 months	14–363 ng/dl	14–363 ng/dl
	6–24 months	<37 ng/dl	<37 ng/dl
	2–3 years	<15 ng/dl	<15 ng/dl
	4–5 years	<19 ng/dl	<19 ng/dl
	6–7 years	<13 ng/dl	<13 ng/dl
	8–9 years	2–8 ng/dl	2-8 ng/dl
	10–11 years	2–165 ng/dl	2–165 ng/dl
	12–13 years	3–619 ng/dl	3–619 ng/dl

Table 26.2 Serum FSH levels in PMDS and normal individuals

Hormone	Age	PMDS	Normal range
FSH	Before puberty	0–5 IU/L	0–5 IU/L
	During puberty	0.3–10 IU/L	0.3–10 IU/L

Table 26.3- Serum LH levels in PMDS and normal individuals

Hormone	Age	PMDS	Normal range
LH	<1 year	<0.02–5 IU/L	<0.02–5 IU/L
	1–8 years	<0.02–0.5 IU/L	<0.02–0.5 IU/L
	9–10 years	<0.02–3.6 IU/L	<0.02–3.6 IU/L
	11–13 years	0.1–5.7 IU/L	0.1–5.7 IU/L

Table 26.4 Serum AMH levels in PMDS and normal individuals

Hormone	Age	PMDS	Normal range
AMH	<14 days		35–140 ng/ml
	15 days–6 months		55–210 ng/ml
	6 months–2 years		85–320 ng/ml
	2–9 years		55–250 ng/ml

alteration in the AMH receptor (AMHR) gene on the twelfth chromosome (type II PMDS, 40%) are the most common causes. In the remaining cases, the cause is unknown (idiopathic PMDS) [2]. Hence, it is considered necessary to screen the immediate kin and second-degree relatives of children with PMDS.

One might query as to whether there is any role of AMH measurement in the management of PMDS. Serum AMH levels reflect the amount of functional Sertoli cells in prepubertal boys, except in the first 2 weeks of life. This declines later owing to the inhibitory effect exerted by androgens and meiotic germ cells. It has been seen that the predictive value of serum AMH for the existence of normal testicular tissue is higher than that of testosterone elevation response to human chorionic gonadotropin (hCG) [3]. In a recent study, extensive genetic and biochemical investigations in a series of 12 PMDS patients showed that all but two patients with AMH defects showed low serum AMH concentrations, although all those with AMHR2 defects showed elevated AMH levels [4]. There are not any guidelines as yet regarding the hormonal or genetic profiling of children with PMDS.

26.3 Clinical and Anatomical Features

Children with PMDS have normal virilized male external genitalia but may have unilateral or bilateral inguinal hernia and/or cryptorchidism. The Müllerian derivatives are likely to be incidentally discovered during the surgery for an inguinal hernia or cryptorchidism. The exact incidence of PMDS is as yet unknown and the US National Institute of Health estimates stand at around less than two million cases in their country. There is significant under-reporting of this condition in India as a great number of cryptorchidism cases remain untreated and, even if they are, most general surgeons are unaware with this condition, leading to misdiagnosis and unsatisfactory surgical management [2]. While ultrasonography (US), computed tomography (CT), and magnetic resonance imaging (MRI) have all been described as tools for diagnosis, the gold standard still remains surgical exploration with a recent drive towards minimally invasive methods.

In this condition, the testes are firmly adherent to the uterus which is held in its anatomical position by the broad ligament which stretches to the lateral wall of the pelvis. This ensures that the testes remain in a high "ovarian" position, and bilateral cryptorchidism ensues leading to type 1 PMDS. In some individuals, the broad ligament is poorly formed or is nonexistent [5], which allows one or both testes to descend, dragging the uterus and tubes along with them. When one testis, together with the uterus and fallopian tube, descends into the inguinal canal while the other set remains intra-abdominal, the "hernia uteri inguinalis" type of PMDS results. Sometimes both testes descend into the same hemiscrotum together with the Müllerian derivatives, resulting in type 3 PMDS or "PMDS Transverse Testicular Ectopia" (as there may be other reasons for TTE).

The vas deferens and vessels are often narrow or blind ending and included in the wall of the fallopian tube and uterus. Vaso-epididymal dissociation from the testis is frequent and is often associated with cryptorchidism [6].

With that background in the pathoanatomy of PMDS, the three types described in literature are as follows [2]:

1. **PMDS female type (60–70%)**: This is the most common variant. In this type, both testes are adherent to the uterus and lie in the ovarian position inside the abdominal cavity. Blind-ending vas and vessels abut the fallopian tube-like structures (Fig. 26.1).
2. **PMDS Hernia Uteri Inguinalis (20–30%):** This is the next most common variant, also known as the "male-type" PMDS, presents as an inguinal hernia with the ipsilateral testis and Müllerian structures as contents of the sac. The other testis is usually undescended. (Fig. 26.2).

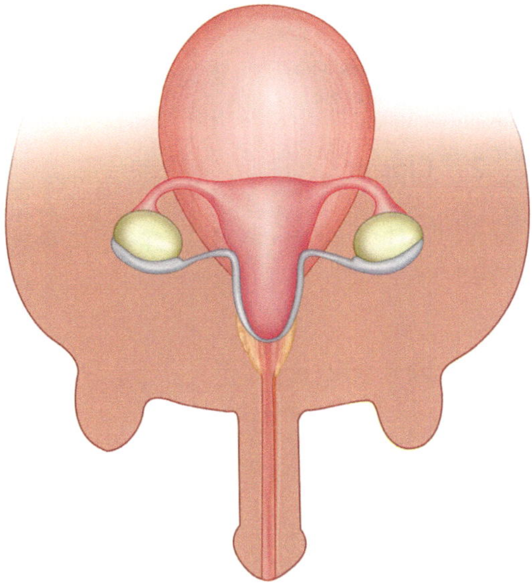

Fig. 26.1 PMDS female type—Both testes adherent to the uterus with a clinical presentation of bilateral cryptorchidism

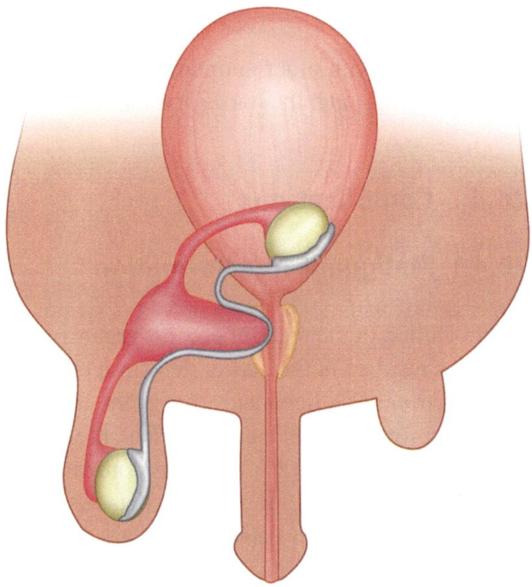

Fig. 26.2 PMDS hernia uteri inguinale—One testis with Müllerian structures is palpable within the hernia sac

3. **PMDS Transverse Testicular Ectopia (TTE) (10%):** This is the least common variant, with both testes and the Müllerian structures descended into one hemiscrotum. Although, all cases of TTE are not mandatorily PMDS (Fig. 26.3).

   Type 1 TTE accounts for roughly 50% of cases and is related only with inguinal hernias. **Type 2 TTE** comprises up to a third of the total cases and has associated persistent Müllerian structures. **Type 3 TTE** is found in 20% and is

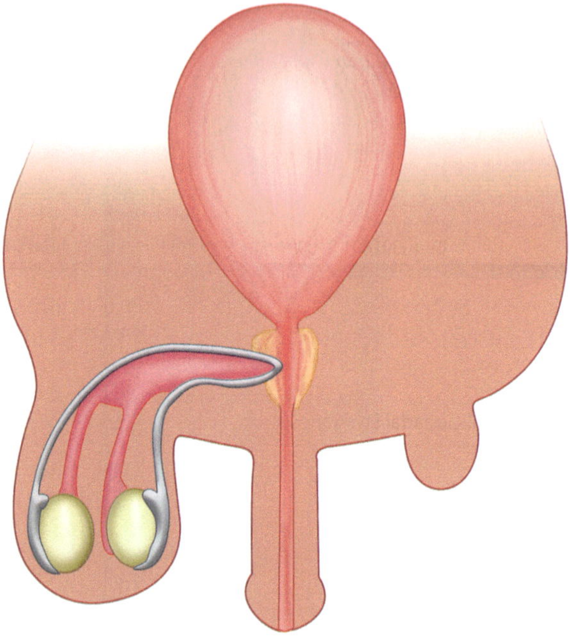

Fig. 26.3 PMDS transverse testicular ectopia—Both testes with Müllerian structures are palpable within one hernia sac

associated with other anomalies such as scrotal anomalies, hypospadias, fused vas deferens, and the horseshoe kidney.

26.4 Complications

26.4.1 Malignant Transformation

The evolution of the thought process regarding the early management of cryptorchidism has resulted in the earlier detection of PMDS. The diagnosis in adults follows the detection of malignancies due to the malignant degradation of the testes or the Müllerian remnants. In earlier studies, the incidence of testicular cancer in PMDS had been reported as 18%, which is comparable to the risk for cryptorchid testes in general [7]. However, more recent datasets have reported a much higher rate with around 33% of PMDS patients 18 years and older experiencing some form of unilateral or bilateral malignant testicular malignancy [8].

Early orchidopexy is desirable to facilitate the early detection of malignant transformation by making the testicle amenable for self-examination. The exceptionally high prevalence of testicular malignancy in PMDS compared to that seen in cryptorchid patients suggests that other than the aberrant positioning of the testis, there are other factors at play which favor malignant transformation in this disorder. This is supported by experimental studies that looked into the deletion of the AMH and

AMHRII genes in mice [9, 10]. It seems that the mechanisms of inhibin suppression or WT1/β-catenin imbalance, respectively, facilitate the neoplastic process.

26.4.2 Infertility

The most common complication of PMDS is infertility. Normal spermatogenesis has also been described in the literature provided at least one of the testes with intact male excretory duct is fixed in the scrotal sac. As per published literature, around 19% of reported adult patients fathered one or more children [8].

26.4.3 Congenital Malformations

There are multiple reports of congenital abnormalities seen in patients with PMDS including [8]

- Intestinal defects (Urioste syndrome)
- Hirschsprung's disease
- Renal anomalies including horseshoe kidney, renal Polycystosis, and hydronephrosis
- Mental deficiency, deafness
- Lipodystrophy and vitamin D-resistant rickets
- Prematurity or small for gestational age syndrome

26.5 Management

The surgical management of PMDS varies within the individual types but some basic principles to be followed are

A) **Early diagnosis and confirmation of PMDS**

PMDS is usually a surgical surprise during orchidopexy. The early recognition of female type PMDS can be attributed to the increasing use of diagnostic laparoscopy for undescended testis. This modality allows us to completely assess the Wolffian and Müllerian derivatives from the point of view of disorders of sex development. In case there is a clinical suspicion, a gonadal biopsy and karyotyping is recommended prior to proceeding with the orchidopexy. The staged approach of yore has been replaced by a single-stage procedure as the presence of the epididymis and vas is incontrovertible proof that the gonad is a testis [11]. Another advantage of laparoscopy is the ability to visualize deep into the pelvis till the prostatic utricle which helps with the thorough excision of the remnants.

In childrenwith Hernia Uteri Inguinalis (HUI) and TTE, the Müllerian structures are discovered during herniotomy. Opinion is divided as regards the Müllerian structures encountered in such cases. Some surgeons advocate that in order to preserve testicular vascularity and the ductal structures, the Müllerian structures should be left intact. A few other groups state that these structures should be completely removed while prioritizing the preservation of the vas and vascularity of the testis so as to preserve gonadal function. The watershed time period was the late nineties when a huge number of case reports appeared in literature citing a variety of malignancies from the Müllerian remnants [2]. The current understanding is that at least the mucosa should be destroyed assiduously while averting thermal injury to the vas even if the Müllerian structures cannot be stripped away in their entirety. This serves the dual purpose of obliterating the cavity of the prostatic utricle remnant and precludes recurrent urinary tract infections and urolithiasis which may otherwise result [2].

In children with HUI, it is advised that the opposite undescended testis should be fixed within the scrotal sac. In children with TTE, the testis can either be placed in the contralateral hemiscrotum by transseptal orchidopexy or can be placed within the ipsilateral side. The Müllerian remnants should be addressed as stated previously.

The perils of ignoring this condition and delaying surgery can be gauged from the reports in adult urologic literature. The most common presentation stems from the malignant degeneration of the testis and Müllerian remnant. Certain reports of hematuria (cyclical or otherwise) could also be a herald for this condition [12]. This happens secondary to a hormonal imbalance as the testes progressively produce less androgens with a proportionate increase in the estrogen levels.

B) **Preserving Gonadal Function and Fertility Potential:** In children with PMDS, the undescended gonads show normal testicular tissue on biopsy. Early orchidopexy is now well established across the globe for the maximal preservation of fertility. This though, might not always ensure good results as regards fertility in PMDS patients. Absolute care should be taken to preserve the gonadal vessels and limit any injury to the Wolffian ductal structures during hernia repair and orchidopexy (Fig. 26.4).

The precise excision of Müllerian structures is dependent upon the experience of the individual surgeon and the possible impact on the vas and testicular vascularity. There should be regular clinicoradiological surveillance in case the remnants are being left in situ. If the descent of the testes is limited due to the short length of the fallopian tubes even after the gonadal vessels have been divided, the main body of the Mullerian structure may be split sagittally (Fig. 26.4) to allow each half to add to the cord length without compromising the vasa or testicular vasculature [13]. Microvascular autotransplantation is yet another viable option.

C) **The Management of Malignancy:** Adenocarcinomas and adenosarcoma are the described malignancies in PMDS [14]. The risk of malignancy is considered to be much lower in the Mullerian derivatives than the risk of malignancy in the associated undescended gonad. PMDS has been described in a child as young

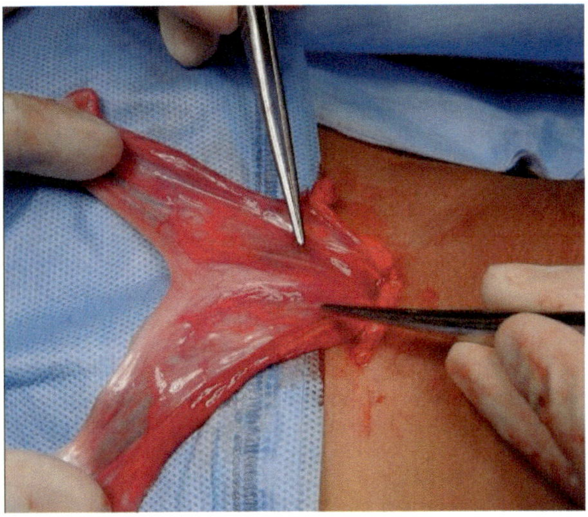

Fig. 26.4 The vas deferens (indicated by the tips of the forceps) is closely applied to the body of the uterine structure on both sides in this boy with female type PMDS

as 4 years and old to 68 years of age. If the Müllerian remnant is left in situ, an annual ultrasound surveillance is recommended to assess for change in size or new mass lesions. MRI is also a useful modality to screen for changes in the Müllerian remnants.

26.6 Hormonal Levels According to Age

In PMDS, serum AMH can be low/normal/elevated and depends on the genetics AMH mutation or AMHRII mutations.

26.7 Learning Points

- PMDS may be sporadic or inherited and there is no correlation between the phenotype and genetic basis.
- PMDS is not an innocuous condition due to the potential for malignancy in the Müllerian remnants.
- Clinical suspicion helps plan imaging and subsequent management.
- There is a definite role of AMH in the context of DSD and differentiating the type of PMDS but it has limited clinical applications as of now.
- Very limited data is available regarding the prostatic utricle and how to deal with it intra-operatively but it can be the reason for recurrent UTI and urolithiasis.
- A detailed discussion about the potential of future fertility is important in the pre-operative counseling of the family.

References

1. Nilson O. Hernia uteri inguinalis bein Manne. Acta Chir Scand. 1939;83:231–49.
2. Manjunath BG, Shenoy VG, Raj P. Persistent Müllerian duct syndrome: How to deal with the Müllerian duct remnants — a review. Indian J Surg. 2010;72(1):16–9.
3. Josso N, Rey R, Picard JY. Testicular anti-Müllerian hormone: clinical applications in DSD. Semin Reprod Med. 2012;30(5):364–73.
4. Tian HJ, Wu DH, Ru W, Wu DW, Tao C, Chen GJ, et al. Surgical management and molecular diagnosis of persistent Müllerian duct syndrome in Chinese patients. Asian J Androl. 2022;24(1):78–84.
5. Miller A, Hong MKH, Hutson JM. The broad ligament: A review of its anatomy and development in different species and hormonal environments. Clin Anat. 2004;17(3):244–51.
6. Sharma S, Sen A. Complete testicular epididymal dissociation in the abdominal cryptorchid testis. J Pediatr Urol. 2013;9(6, Part B):1023–7.
7. Bilateral testicular carcinoma in patient with the persistent müllerian duct syndrome [Internet]. [cited 2023 Jan 26]. Available from: https://arts.units.it/handle/11368/2918823
8. Picard JY, Cate RL, Racine C, Josso N. The persistent müllerian duct syndrome: an update based upon a personal experience of 157 cases. Sex Dev Genet Mol Biol Evol Endocrinol Embryol Pathol Sex Determ Differ. 2017;11(3):109–25.
9. Behringer RR, Finegold MJ, Cate RL. Müllerian-inhibiting substance function during mammalian sexual development. Cell. 1994;79(3):415–25.
10. Tanwar PS, Commandeur AE, Zhang L, Taketo MM, Teixeira JM. The Müllerian inhibiting substance type 2 receptor suppresses tumorigenesis in testes with sustained β-catenin signaling. Carcinogenesis. 2012;33(12):2351–61.
11. Loeff DS, Imbeaud S, Reyes HM, Meller JL, Rosenthal IM. Surgical and genetic aspects of persistent müllerian duct syndrome. J Pediatr Surg. 1994;29(1):61–5.
12. Smith-Harrison LI, Patel MS, Smith RP, Schenkman NS. Persistent Müllerian duct structures presenting as hematuria in an adult: Case report of robotic surgical removal and review of the literature. Urol Ann. 2015;7(4):544–6.
13. Brandli DW, Akbal C, Eugsster E, Hadad N, Havlik RJ, Kaefer M. Persistent Mullerian duct syndrome with bilateral abdominal testis: surgical approach and review of the literature. J Pediatr Urol. 2005;1(6):423–7.
14. Adhya AK, Pradhan MR. Persistent mullerian duct syndrome with mixed germ cell tumor of undescended testis: a case report. Indian J Pathol Microbiol. 2019;62(1):178.

Disorders of Testosterone Action in 46 XY DSD

27

Nidhi Sugandhi

Abstract

Disorders of androgen action constitute the largest subgroup of 46XY DSD. These children suffer from inadequate testosterone action on the developing gonadal structures in the embryo. The defects in androgen action leads to defective masculinization of internal and external genitalia and functional deficits later. The most common cause of defective testosterone action is androgen insensitivity syndrome due to anomalous testosterone receptors and deficiency of 5-alpha reductase which converts testosterone to its active form, dihydrotestosterone. These defects can have a variable range of severity leading to a wide spectrum of phenotypic manifestations. Thus, the timing and mode of presentation in these children are highly variable. The most severe form of 46 XY DSD can present as a completely female phenotype. They may be investigated and further diagnosed due to inguinal hernia with prolapsing testis or primary amenorrhea at puberty. On the other end of the phenotypic spectrum is an undervirilized male with small clitoris-like phallus, undescended gonads, unfused bifid scrotum, and possibly variable degree of hypospadias and/or a blind vaginal pouch. These children may be suspected to have DSD in neonatal period or infancy and may be investigated with karyotyping and further hormonal studies. Decision about the sex of rearing in these children need multifactorial consideration, including extent of virilization, male pattern genetic imprinting, and virilization potential at puberty. In general, children with complete androgen insensitivity (CAIS) are better reared as females, whereas those with partial androgen insensitivity (PAIS) and 5-alpha-reductase deficiency have better potential for virilization and may be assigned

N. Sugandhi (✉)
Department of Paediatric Surgery, Vardhaman Mahavir Medical College, New Delhi, India
e-mail: drnidhisugandhi@gmail.co

S. K. Ratan (ed.), *Children with Differences in Sex Development*,
https://doi.org/10.1007/978-981-97-1639-5_27

the male gender. Estrogen or testosterone supplementation at puberty and feminizing or masculinizing genitoplasty are required according to the gender assignment.

Keywords

Testicular feminization syndrome · 5-alpha reductase · Decision about the sex off deficiency · Androgen insensitivity syndrome · Pseudovaginal penoscrotal hypospadias

27.1 Introduction

Disorders of androgen action constitute the largest subgroup of 46XY DSD. These genotypically 46XY males suffer from a variable degree of inadequate testosterone action on the developing gonadal structures in the embryo leading to defective masculinization of internal and external genitalia and functional deficits later. Most of these cases are identifiable at birth due to abnormal external genitalia. Milder variants may have minimally undervirilized external genitalia such as hypospadias or undescended testis and may miss the diagnosis unless specifically investigated However, the most severe form of 46XY DSD can present as a completely female phenotype with absolute absence of masculinization. These may be evident only at puberty, due to sudden virilization or primary amenorrhea.

The exact incidence of 46XY DSD due to testosterone action is difficult to assess since it is a heterogenous group. It is known though that androgen insensitivity syndrome (AIS), which constitutes the major clinical group in disorders of testosterone synthesis and action, occurs in 1–5 per 100,000 live female neonates. This incidence may be higher if undervirilized males are also considered [1]. Many of the milder cases of undervirilization such as isolated hypospadias may remain undiagnosed unless specifically investigated. This is more so in conditions which may present with variable severity such as 5-alpha-reductase deficiency.

There is a distinct geographical variation seen in the prevalence of disorders of testosterone action, being much more common in the Caribbean and African continents and the Middle east. In fact, these disorders are so commonly prevalent here that there are distinct local names such as "Guavadoces," "Machihembras," and "Turnims" assigned to such undervirilized males or females who suddenly show masculinization at puberty [2].

Apart from the evident morbidity of ambiguous genitalia, undecided sexual role, ill developed secondary sexual characteristics and infertility, these children also suffer from hidden morbidity in form of metabolic consequences. Reduced bone mineral density, increased obesity, insulin resistance, and lipid abnormalities are some of the issues faced. They are also more prone to development of gonadoblastoma and germ cell tumors though this risk is higher in 46XY DSD due to gonadal dysgenesis. Finally, there are also reports that defective androgen action on the brains of 46XYDSD individuals makes them more prone to depression, addictions, and eating disorders [3].

Though a large number of these people are able to identify and function as males, but there is a significant subgroup who are completely unvirilized, are brought up as females, and may prefer to continue as functional though infertile females. Sex determination in these as with all DSDs is a complex issue, complicated by the effect of male pattern imprinting on developing fetal brain by testosterone.

27.2 Embryology and Classification

The presence of functioning SRY region on short arm of Y chromosome ensures the sexual determination of the primitive gonad to form into a testis at 5–6 weeks . This process is tightly orchestrated by molecular and genetic signals such as SRY gene, SF1 (steroidogenic factor1), SOX 9, and FGF9. Leydig cells and Sertoli cells soon begin to form in this destined testis and the secreted testosterone by Leydig cells and anti-Mullerian hormone (AMH) by Sertoli cells control further sexual differentiation to males during weeks 8–12. In particular, whereas AMH is responsible for regression of the default female Mullerian structures, testosterone initiates formation of male internal genitalia including the epididymis, vas deferens, and seminal vesicles. In the normal course, testosterone is converted further to dihydrotestosterone (DHT) by 5-alpha-reductase enzyme which exerts its action through androgen receptors and this is responsible for the development of phallus, scrotum, and urethra from genital tubercle, genital folds, and urethral folds, respectively thus leading to complete male external and internal sex organs by 16 weeks. Figure 27.1 summarizes a flowchart of the male sexual differentiation and the various potential points of development of DSD due to defect in testosterone synthesis and action.

Based on the embryology and the defect in the pathway, the disorders of testosterone synthesis and action can be classified as in Table 27.1. Out of this heterogenous group, the androgen insensitivity syndrome (partial or complete) forms the majority, followed by 5-alpha-reductase deficiency. The disorders of synthesis are encountered rarely, though this relatively less incidence may also be due to failure in diagnosing these cases as they require detailed enzymatic, molecular, and genetic analysis.

Absence of testosterone due to defects in synthesis or Leydig cell hypoplasia can result in completely unvirilized child who would present as a female during neonatal period. Absence or deficiency of DHT can cause a partially virilized boy with small phallus and/or undescended testis. The phenotype in androgen insensitivity can be highly variable with almost a normal looking boy with trivial deficiencies such as hypospadias to a completely female phenotype.

Thus, the timing and mode of presentation in these children are highly variable. The most common appearance is of an undervirilized male with small clitoris-like phallus, undescended gonads, unfused bifid scrotum, and possibly variable degree of hypospadias or a blind vaginal pouch. These children may be suspected to have DSD in neonatal period or infancy and may be investigated with karyotyping and further hormonal studies. The other common clinical setting in 46XY DSD is of a completely normal looking female presenting with inguinal hernia with a

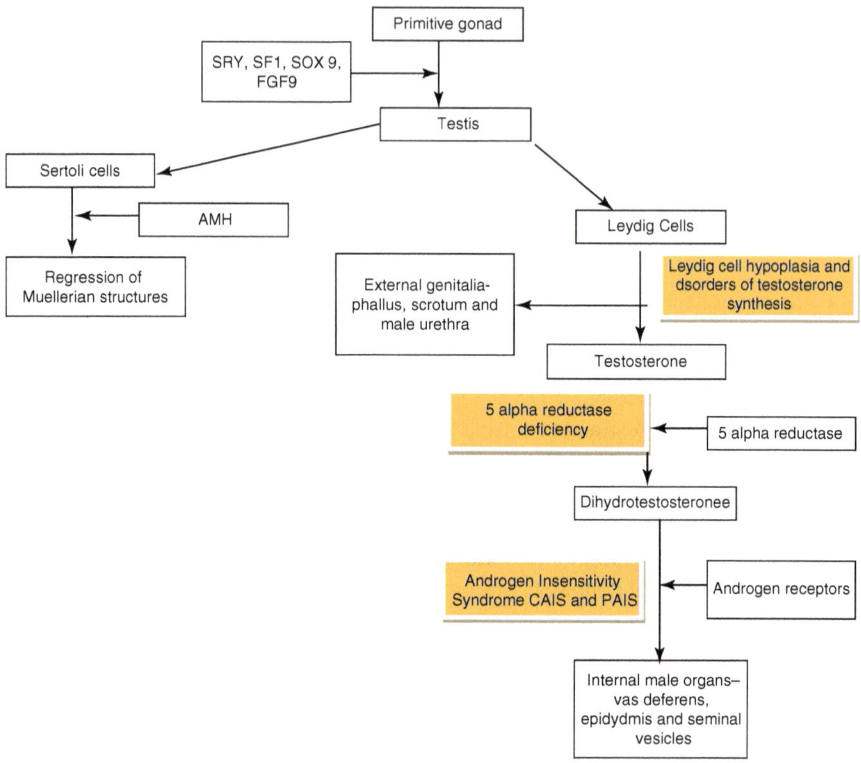

Fig. 27.1 Flowchart of male sexual differentiation with defects causing 46 XY disorders of testosterone synthesis and action

Table 27.1 Classification of disorders of androgen synthesis, metabolism, and action [4]

46,XY DSD ASSOCIATED WITH CHOLESTEROL SYNTHESIS DEFECTS Smith–Lemli–Opitz syndrome
46 XY DSD due to testosterone production defects
Impaired Leydig cell differentiation (*LHCGR* defects)
Enzymatic defects in testosterone synthesis *Defects in adrenal and testicular steroidogenesis* StAR deficiency
P450scc deficiency
3-β-hydroxysteroid dehydrogenase II deficiency
17α-hydroxylase and 17,20 lyase deficiency
P450 oxidoreductase defect (*electron transfer disruption)*
Defects in testicular steroidogenesis
Isolated 17,20-lyase deficiency *7β-hydroxysteroid dehydrogenase III deficiency*
Alternative pathway to DHT
3α-hydroxysteroid dehydrogenase deficiency due to AKR1C2 and AKR1C4 defects
46,XY DSD DUE TO DEFECTS IN TESTOSTERONE METABOLISM 5α-reductase type 2 deficiency
46,XY DSD DUE TO DEFECTS IN ANDROGEN ACTION
Androgen insensitivity syndrome Complete and partial

prolapsing gonad, primary amenorrhea, or sudden virilization at puberty. Again karyotyping and radiological and hormonal studies are required to clinch the diagnosis.

27.3 46 XY DSD:5-Alpha-Reductase Deficiency

27.3.1 Epidemiology

5-alpha-reductase deficiency, previously also known as "pseudovaginal penoscrotal hypospadias" is one of the most dramatically presenting DSD, mainly due to sudden virilization seen during the pubertal spurt [5]. It is an autosomal recessive condition caused by deficiency of 5-alpha-reductase enzyme, isoform 2, which is coded by a gene on short arm of chromosome 2. It has 254 base pairs and almost 67 mutations have been identified, probably causing the wide spectrum of severity and symptoms.

Its exact incidence is not known but it has a definite geographical predilection being very commonly found in Papua New Guinea and the Dominican Republic, probably due to consanguineous marriages. They are locally known as Guavadoces and Machihembras meaning girls turning into boys alluding to sudden virilization that happens at puberty [2].

27.3.2 Pathophysiology

Though testosterone secreted by Leydig cells is the primary masculinizing hormone and causes differentiation of male internal genitalia, it cannot cause formation of the external genitalia in the fetus in its native form and requires to be converted to dihydrotestosterone. This is achieved by an enzyme called 5-alpha reductase. It has two isoforms—5-alpha reductase 1 and 2. Isoform 2 is more active during fetal life and present in large quantities in urogenital tubercle and urogenital folds [6]. It converts the circulating testosterone into dihydrotestosterone which stimulates the differentiation and development of phallus and scrotum from the genital tubercle and folds, respectively. Absence or underactivity of this enzyme thus leads to undeveloped male external genitalia.

Few factors need to be considered, which makes the management of this condition complex. First, though the genital tubercle and folds are not very responsive to direct testosterone action during fetal life and require DHT for differentiation, this roadblock is surpassed at puberty. This could be due to very high levels of circulating testosterone, which may act directly on the external genitalia or result in a relatively greater levels of DHT due to presentation of large quantity of testosterone substrate to partially functioning 5-alpha-reductase enzyme. Isoform 1 of 5-alpha reductase also seems to be activated and induced during the pubertal surge, further neutralizing the effect of isoform 2 deficiency [7]. It is also thought that most pubertal changes including initiation of spermatogenesis, increased muscle mass, deepening of voice, etc., are directly brought about by testosterone. DHT is responsible for male pattern hair and enlargement of prostate at puberty, though this seems to be more due to Isoform 1 [8].

Thus, a normal appearing female with this condition may show sudden virilization at puberty. Also, it needs to be emphasized that these children have normal functioning testis, and testosterone levels during fetal life are normal or on higher side. Thus, the fetal brain is fully exposed to and affected by the male hormones which cause a male gender imprinting on these children even though they may have female phenotype. Consequently, these female appearing children may experience gender identity issues, especially at puberty, even if they have undergone gonadectomy earlier.

27.3.3 Clinical Features

The severity of undervirilization of external genitalia in the fetus depends on the extent of underactivity of the 5-alpha-reductase enzyme.

Severely affected kids may be born as normal looking females. Few features which may raise a suspicion of DSD are enlarged clitoris, fused labial folds, single urogenital opening, or presence of a prolapsing gonad in an inguinal hernia. These children present with primary amenorrhea with absence of breast development or sudden features of virilization including deepening of voice, increased clitoris size (phallus), male pattern hair, though scarce or descent of gonads into the labioscrotal folds. On the other end of scrotum are undervirilized males but with identifiable male external genitalia. They may have undescended testis (usually symmetrical and palpable), severe hypospadias (usually penoscrotal variety), or a micropenis. In these children, the diagnosis may be suspected at birth, whereas in those with female phenotype, it may only be thought of, during the virilization phase at puberty.

27.3.4 Diagnosis

The diagnostic feature of 5-alpha-reductase deficiency is increased testosterone/DHT ratio in a child with 46 XY karyotype. Normal ratio is <16, with ratios more than 30 indicating a severe deficiency [9]. Circulating testosterone, LH and AMH levels are usually normal. Interpretation of this ratio may be difficult since the normal ratio varies according to age and severity of 5-alpha-reductase deficiency. This test is most diagnostic post puberty, when adequate testosterone secretion is established. In prepubertal children, baseline testosterone levels are very low and this ratio can be assessed only after HCG stimulation. Thus, baseline testosterone and DHT are measured and repeated after HCG stimulation (age appropriate dosage of 500 IU <3 years, 1000 IU at 3–5 years, and 1500 IU >5 years at days 1, 3, and 5) and the ratio assessed. HCG stimulation is not required in the neonatal period when there is a natural LH and testosterone surge but it may be confounded by higher than normal activity of isoform 1 of 5-alpha reductase [10].

Similar to testosterone/DHT ratio, the ratio between other 5 alpha to 5 beta reduced steroids in urine can be profiled by gas chromatography or mass spectrometry. Thus, an increased ratio of androstenedione to etiocholanolone, 5-alpha tetrahydrocortisol to tetrahydrocortisol and 5-alpha tetrahydrocorticosterone to tetrahydrocorticosterone is indicative of deficiency of 5-alpha-reductase enzyme deficiency [11].

Molecular and genetic testing of the 5-alpha-reductase locus on chromosome 2(SRD5A2) is recommended to confirm the diagnosis in neonates especially because of difficulties and fallacies encountered in assessing the testosterone/DHT ratio [12].

5-alpha-reductase deficiency can also be assessed in the skin fibroblasts, particularly genital foreskin fibroblasts, since the activity of this enzyme may vary according to the region. However, there is considerable variability in the normal range of enzyme activity even in controls with significant overlap between normal individuals and those with 5-alpha-reductase deficiency. Hence, this test has high specificity but low sensitivity, i.e. normal activity rules out deficiency but low activity may be present in normal individuals too [6, 13].

Radiological investigations like ultrasound and MRI are useful only to confirm the absence of Mullerian remnants.

27.3.5 Differential Diagnosis

5-alpha-reductase deficiency can be confused with other conditions causing undervirilized males with symmetrical gonads or in normal looking females virilized at puberty. These include partial androgen insensitivity syndrome (PAIS), 17 beta hydroxysteroid dehydrogenase deficiency, and Leydig cell hypoplasia. It is important to differentiate these as early as possible as PAIS and 5-alpha-reductase deficiency are better raised as males due to the propensity for virilization at puberty whereas the other two conditions would most likely continue as females.

A combination of hormonal levels and molecular/genetic testing may help in differentiating between these conditions (Fig. 27.2).

27.3.6 Management

The management depends on the age at diagnosis and the degree of virilization in a child. Also, significant factors to take into account in management are the effect of testosterone on brain imprinting of gender identity in the fetal life as well as definite surge of virilization to be experienced at puberty. These factors should be clearly

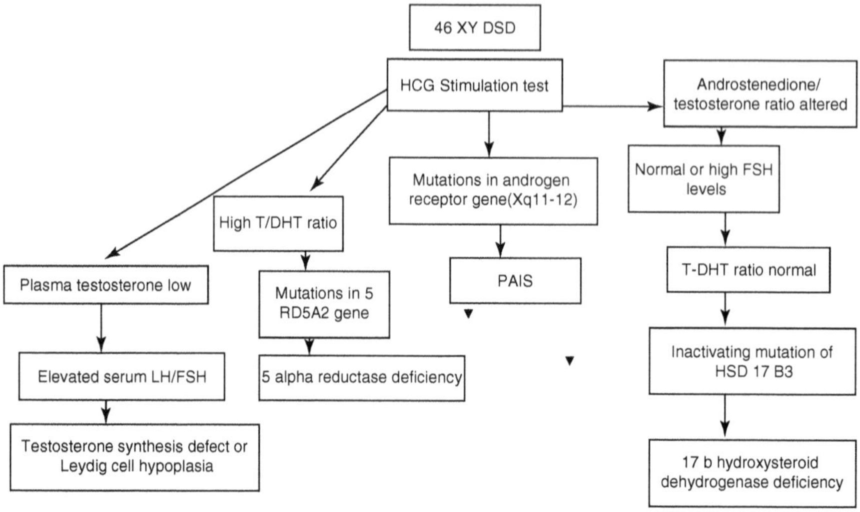

Fig. 27.2 Differential diagnosis and approach to 46XY DSD. Adapted from A handbook of Disorders of Sex Development by VR Ravikumar Printed by Coimbatore Association of Paediatric Surgeons 2022. Ed 1 [14]

informed to the child and guardians and decision about sex of rearing should be done only after adequate consideration. While early surgery has been traditionally recommended to prevent social and psychological repercussions [15, 16]. It is also suggested by some to postpone the decision about the gender assignment till puberty so that the child may participate in the decision-making and the virilization surge is also experienced [17–19].

Undervirilized males may continue to be reared as males and undergo corrective surgeries for masculinizing genitoplasty. These include orchiopexy, hypospadias repair, and penoscrotal transposition correction, to be done at the usually recommended pre-school age. During puberty, local application of 5 DHT cream or testosterone injections may assist the normal growth of phallus to adequate size. These individuals are usually able to lead a satisfactory sexual life, though fertility may be compromised mainly due to the effect of undescended testis on spermatogenesis.

Those with fully female phenotypes or those females diagnosed at puberty may choose to be reared as females, though there is a possibility of later gender dysphoria due to brain imprinting. These children need to undergo gonadectomy to prevent virilization and minimize risk of malignancy with constant estrogen replacement at puberty. Cyclical progesterone is not required as uterus is absent. Phallic reduction (clitoral reduction or recession) may be required. Vaginal augmentation may be done with molds or serial dilatation in those with blind vaginal pit. Others with complete absence of vagina may require vaginal reconstruction with colon or ileum. Again, acceptable sexual function is achievable though they will obviously be infertile females.

27.4 46XY DSD Androgen Insensitivity Syndrome

27.4.1 Epidemiology

Androgen insensitivity syndrome also known as testicular feminizing syndrome is a condition described by absence of virilization despite normal testis and normal testosterone levels due to end organ resistance to testosterone. It is the most common cause of 46XY DSD, with an incidence of 4 in 100,000 female newborns and 1 in 20,000–64,000 male newborns [20]. It results from mutations of the androgen receptor gene which is coded by the long arm of chromosome X at Xq12.

27.4.2 Pathophysiology

Androgen receptor insensitivity is an X linked recessive condition due to defects in the androgen binding receptors on the target tissues. Mutations in the androgen receptor gene leads to decreased number of receptors, absent or decreased binding affinity of receptors to testosterone or DHT, or defects in downstream signalling pathways. It is characterized by high levels of circulating testosterone and DHT but no action on the target organs. Absence of testosterone action starts during fetal life with deficient formation of internal genitalia. The testis are formed appropriately as the factors involved in testicular differentiation including SRY, SF1, and SOX9 are functioning correctly. However, these are not followed by formation of vas, epididymis, or seminal vesicles. Further, even after conversion to DHT, it cannot exert its action on the urogenital tubercle and folds and this results in absent or partial virilization of the external genitalia. A wide variety of mutations are seen in the gene which results in a wide spectrum of severity of undervirilization. The clinical severity depends on the binding affinity of the receptors and can vary according to the type of mutation. Based on clinical severity, this condition is classified as [20, 21]

1. Complete androgen insensitivity syndrome (CAIS): Exemplified by complete female looking genitalia and no signs of virilization even at puberty.
2. Partial androgen insensitivity syndrome (PAIS): Distinguished by partially or undervirilized external genitalia.

Some people also like to recognize a third variety called as MAIS or minimal androgen insensitivity syndrome in which there are minimal defects in the male external genitalia like distal hypospadias. However, the pathology and existence of this entity is controversial and not well established.

Mutations in the androgen receptor gene is also associated with male breast cancer, spinal and bulbar muscular atrophy (Kennedy syndrome), and prostate cancer [22, 23].

27.4.3 Clinical Features

Table 27.2 summarizes the clinical presentation of different types of AIS [24]

CAIS children are born as normal looking females, with unfused labial folds, clitoris-like phallus, and a perineal urethral opening. They mostly have a variable but short length vagina, though other Mullerian structures like uterus and fallopian tubes are absent due to action of anti-Mullerian hormone which is normally secreted. They usually seek medical advice at puberty for primary amenorrhea. Normal breast development may occur due to peripheral conversion of testosterone to estrogen. The testis which are normally functioning are usually intra-abdominal in location but may present as a prolapsing gonad in an inguinal hernia, in which case these children may seek medical attention earlier. These children usually have no gender dysphoria as complete absence of testosterone action ensures no brain imprinting in fetal life. Ancillary effects may be reduced muscle mass and decreased bone density. Chances of malignancy in undescended intra-abdominal gonads shoot up dramatically after puberty to the tune of almost 30%.

PAIS has a more variable presentation, and the neonate may either be an undervirilized male or an overvirilized female, depending on the extent of receptor activity. The undervirilized male may have a micropenis with or without hypospadias. The testis are bilaterally symmetrically undescended, either abdominal or inguinal. Bifid scrotum and penoscrotal transposition may also be seen. At puberty, variable but inadequate degree of masculinization occurs with enlargement of phallus and descent of testis, sparse hair may also develop. At the same time, gynecomastia may also develop due to peripheral conversion of the high levels of testosterone. These children though may identify with male gender due

Table 27.2 Clinical features of AIS

Type	External genitalia (synonyms)	Findings
CAIS	Female ("testicular feminization")	• Absent or rudimentary Wolffian duct derivatives • Absence or presence of epididymis and/or vas deferens • Inguinal or abdominal testes • Blind-ending vagina • Scant or absent pubic and/or axillary hair • Gynecomastia at puberty • Minimal testosterone imprinting on brain
PAIS	Ambiguous/ predominantly female	• Microphallus (<1 cm) with clitoris-like underdeveloped glans; labia majora-like bifid scrotum • Inguinal or labial testis • Perineoscrotal hypospadias or urogenital sinus • Gynecomastia (development of breasts) in puberty
	Predominantly male	• Severe (perineal) hypospadias with normal-sized or micropenis, bifid scrotum, either descended or undescended testes • Gynecomastia in puberty

to partial effect of testosterone on the developing brain, but they are infertile because of defective spermatogenesis.

The overvirilized female may have an enlarged looking clitoris, urogenital sinus or a separate short or pit-like vagina and labial or inguinal gonads. They may present due to the palpable gonads or inguinal hernia or even seeking correction of the enlarged clitoris. At puberty these "females" have some breast development but primary amenorrhea along with appearance of male pattern hair.

The MAIS individuals have minimal defects and present mainly for deficient secondary sexual characteristics, infertility due to impaired spermatogenesis or gynecomastia.

27.4.4 Diagnosis

In 46 XY individuals with ambiguous genitalia gene sequencing of the androgen receptor gene at Xq12 focus can provide definitive diagnosis of AIS. However, not all mutations have been characterized yet and in many individuals no identifiable mutations may be found. Biochemically, these individuals show normal or raised LH/FSH, raised testosterone and DHT, mildly altered ratios of testosterone/DHT.

Androgen receptor activity in cultured genital fibroblasts may be tested and found deficient [25]. Also, the decline in sex hormone binding globulin (SHBG) in response to exogenous administration of artificial androgen stanozolol is mitigated [24].

27.4.5 Management

The management depends on age at presentation, severity of undervirilization, sex of rearing, and gender role identified with. There are many more issues in managing patients with PAIS than CAIS, who for all practical purposes look and identify themselves as females.

CAIS children with female phenotype, continue to be reared as females. They need to undergo gonadectomy sometime before adulthood as incidence of malignancy in the undescended gonads is unacceptably high. The timing of this gonadectomy can be variable. Some recommend it on diagnosis or in neonatal period to allay anxiety of parents and to avoid the risk of malignancy, which is albeit low in prepubertal age, though higher than in general population. These children will require estrogen supplementation at puberty to stimulate breast development. Others recommend gonadectomy at puberty as the surging testosterone at that time results in breast development through peripheral aromatization to estrogen. Vagina though blind ending, is usually adequate for sexual function, or can be self-dilated by vaginal molds.

PAIS children need to be examined objectively to assess whether they are more masculinized or feminized. Quigley scale (Fig. 27.3) was a grading proposed in 1995 to evaluate the degree of masculinization which may help in decision-making

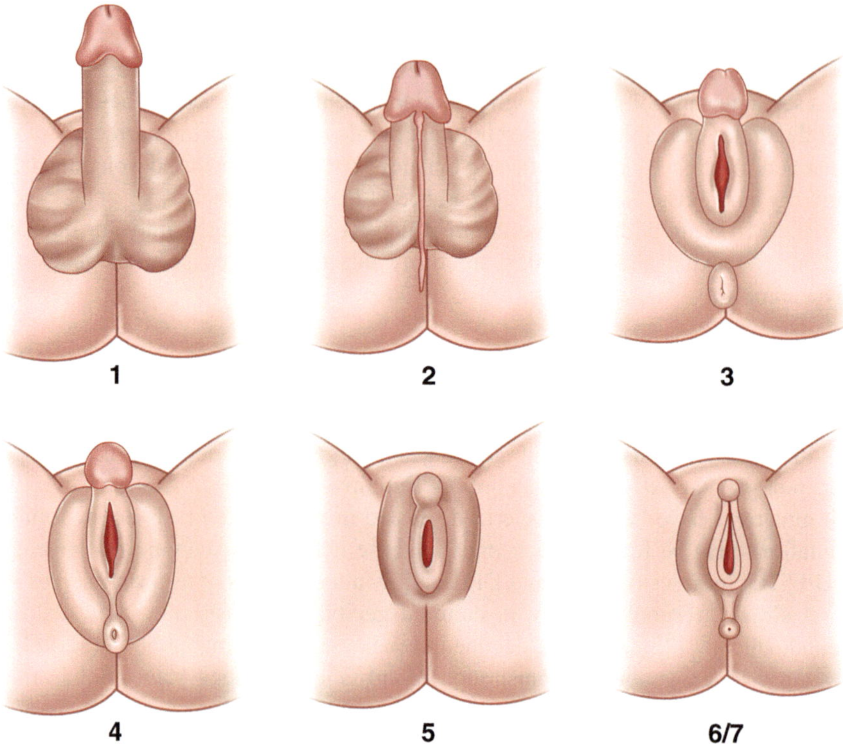

Fig. 27.3 Quigley scale for degree of masculinization of external genitalia [26]

in PAIS [26]. It has grades 1–7 with 1 being completely male appearing external genitalia and Grade 7 being completely female appearing genitalia.

Those with grades 1–3, reasonable sized phallus may be raised as males and undergo corrective male genitoplasty. This would include orchiopexy and hypospadias repair. Usually the perineal or penoscrotal hypospadias requires staged correction. Orchiopexy is done at usual age. The extent of secondary sexual characteristics and virilization at puberty is unpredictable due to variable activity of androgen receptors. In the event of inadequate virilization, the boys may suffer from depressive tendencies, dissatisfaction, and suicidal tendencies. Local 5 DHT creams may result in some increase in phallic length but it is nowhere comparable to response in 5-alpha-reductase deficiency. Though spermatogenesis is variably affected, there have been reports of PAIS males fathering children.

The children with external genitalia of grade 4–7 on Quigley scale may be offered to continue as females as virilization is poor to begin with and not expected to dramatically change during puberty. Clitoral reduction or recession is carried out. The vaginal pit may be adequate or can be self-dilated with vaginal molds. In cases where there is not even a rudimentary vagina, vaginal reconstruction can be done using colon or ileum. This is best done at puberty, as it requires regular dilatations to maintain its adequate size. These children need gonadectomy and the timing in case of PAIS is less clearly defined as compared to CAIS. In CAIS, due to complete absence of testosterone activity, the high levels at puberty entirely contribute to feminization by peripheral conversion to estrogen. In PAIS, along with peripheral conversion to estrogen, there will be variable direct virilization on the genital structures and male secondary sexual characteristics depending on the extent of functional androgen receptors—this can be very distressing to the child brought up and destined to be females. Estrogen replacement is required to induce adequate feminizing changes at puberty. These girls have a satisfactory sexual life though obviously will not be fertile.

In order to ease the decision-making of gender assignment, investigators have attempted to quantify the level of functionality of the androgen receptors in an attempt, to predict the level of virilization expected at puberty. It is believed that the decrease of sex hormone binding globulin in response to external anabolic steroid Stanozolol may give some idea of the functioning of androgen receptors [27]. The level of androgen receptor activity in genital fibroblasts, HCG stimulation test can give some idea about the virilization potential. If genetic analysis is available then somatic and inherited mutation in the androgen receptor needs to be differentiated as somatic mutation individuals may have some functioning androgen receptor activity.

27.5 Key Points

Disorders of testosterone action are the most common cause of 46XY DSD. They have a spectrum of clinical presentation varying from fully feminized to different degrees of masculinization. Sex determination usually needs to be done early except in 5 alpha reductase deficiency where dramatic changes in phenotype and sudden virilization may be seen at puberty. Masculinizing or feminizing genitoplasty with gonadectomy may be required depending on the phenotype and virilization potential. Gender dysphoria may be seen in 5-alpha-reductase deficiency and PAIS, rarely in CAIS. A dedicated team of endocrinologists, pediatric surgeons, and psychologists should collaborate for optimum management of these patients.

References

1. Berglund A, Johannsen TH, Stochholm K, Viuff MH, Fedder J, Main KM, Gravholt CH. Incidence, prevalence, diagnostic delay, and clinical presentation of female 46,XY disorders of sex development. J Clin Endocrinol Metab. 2016;101(12):4532–40.
2. Imperato-McGinley J, Miller M, Wilson JD, Peterson RE, Shackleton C, Gajdusek DC. A cluster of male pseudohermaphrodites with 5 alpha reductase deficiency in Papua New Guinea. Clin Endocrinol(Oxf). 1991;34(4):293–8.
3. Wisniewski AB, Batista RL, Costa EMF, Finlayson C, Sircili MHP, Dénes FT, Domenice S, Mendonca BB. Management of 46,XY differences/disorders of sex development (DSD) throughout life. Endocr Rev. 2019;40(6):1547–72. https://doi.org/10.1210/er.2019-00049.

4. Lee PA, Hughes I. Consensus statement on management of intersex disorders. Pediatrics. 2006;118(2):e488–500. https://doi.org/10.1542/peds.2006-0738.
5. Pc W, Madden JD, Harrod MJ, Goldstein JL, Mac Donald PC, Wilson JD. Familial incomplete male pseudohermaphoditism, Type 2. Decreased dihydrotestosterone formation in pseudovaginal perineoscrotal hypospadias. N Eng J Med. 1974;291(18):944–9.
6. Thigpen AE, Silver RI, Guileyardo JM, Linette IM, Connell M. Tissue distribution and ontogeny of steroid 5 alpha reductase isozyme expression, vol. 92. Am Soc Clin Investig; 1993. p. 903–10.
7. Wilson JD, Griffin JE, Russel DW. Steroid 5 Alpha reductase2 deficiency. Endocr Rev. 1993;14(5):577–93.
8. Randall VA. Role of 5 alpha-reductase in health and disease. Bailliere Clin Endocrinol Metab. 1994;8(2):405–31.
9. Mendonca BB, Batista RL, Domenice S, et al. Steroid 5α-reductase 2 deficiency. J Steroid Biochem Mol Biol. 2016;163:206–11. https://doi.org/10.1016/j.jsbmb.
10. Clarkson J, Herbison AE. Hypothalamic control of the male neonatal testosterone surge. Philos Trans R Soc B Biol Sci. 2016;371(1688):20150115. https://doi.org/10.1098/rstb.2015.0115.
11. Chan AO, Kei BW, Lee M, Lam CY, Ng YY, LeungTung K, Ling JY. Diagnosis of 5α-reductase 2 deficiency: is measurement of dihydrotestosterone essential? Clin Chem. 2013;59(5):798–806. https://doi.org/10.1373/clinchem.2012.196501.
12. Achermann JC, Domenice S, Bachega TASS, Nishi MY, et al. Disorders of sex development: effect of molecular diagnostics. Nat Rev Endocrinol. 2015;11(8):478–88. https://doi.org/10.1038/nrendo.2015.69.
13. Odame I, Donaldson MD, Wallace AM, Cochrane W. Early diagnosis and management of 5 alpha reductase deficiency. Arch Dis Child. 1992;67(6):720–3.
14. A handbook of disorders of sex development by VR Ravikumar printed by Coimbatore Association of Paediatric Surgeons 2022. Ed 1.
15. Hughes IA, Houk C, Ahmed SF, Lee PA; LWPES Consensus Group; ESPE Consensus Group. Consensus statement on management of intersex disorders. Arch Dis Child. 2006;91(7):554–63.
16. Braga LH, Lorenzo AJ. Cryptorchidism: a practical review for all community healthcare providers. Can Urol Assoc J. 2017;11(1–2 Suppl 1):S26–32.
17. Mouriquand P, Caldamone A, Malone P, Frank JD, Hoebeke P. The ESPU/SPU standpoint on the surgical management of disorders of sex development (DSD). J Pediatr Urol. 2014;10(1):8–10.
18. Mouriquand PD, Gorduza DB, Gay CL, Meyer-Bahlburg HF, Baker L, Baskin LS. Surgery in disorders of sex development (DSD) with a gender issue: if (why), when, and how? J Pediatr Urol. 2016;12(3):139–49.
19. Wolffenbuttel KP, Crouch NS. Timing of feminising surgery in disorders of sex development. Endocr Dev. 2014;27:210–21.
20. Galani A, Kitsiou-Tzeli S, Sofokleous C, Kanavakis E, Kalpini-Mavrou A. Androgen insensitivity syndrome: clinical features and molecular defects. Hormones (Athens). 2008;7(3):217–29.
21. Ahmed SF, Cheng A, Hughes IA. Assessment of the gonadotrophin-gonadal axis in androgen insensitivity syndrome. Arch Dis Child. 1999;80(4):324–9. https://doi.org/10.1136/adc.80.4.324.
22. Evans BA, Harper ME, Daniells CE, Watts CE, Matenhelia S, Green J, Griffiths K. Low incidence of androgen receptor gene mutations in human prostatic tumors using single strand conformation polymorphism analysis. Prostate. 1996;28(3):162–71.
23. Lobaccaro JM, Lumbroso S, Belon C, Galtier-Dereure F, Bringer J, Lesimple T, Namer M, Cutuli BF, Pujol H, Sultan C. Androgen receptor gene mutation in male breast cancer. Hum Mol Genet. 1993;2(11):1799–802.
24. Sinnecker GH, Hiort O, Nitsche EM, Holterhus PM, Kruse K. Functional assessment and clinical classification of androgen sensitivity in patients with mutations of the androgen receptor gene. German Collaborative Intersex Study Group. Eur J Pediatr. 1997;156:7–14.

25. Weidemann W, Linck B, Haupt H, Mentrup B, Romalo G, Stockklauser K, Brinkmann AO, Schweikert HU, Spindler KD. Clinical and biochemical investigations and molecular analysis of subjects with mutations in the androgen receptor gene. Clin Endocrinol. 1996;45(6):733–9. https://doi.org/10.1046/j.1365-2265.1996.8600869.
26. Quigley CA, De Bellis A, Marschke KB, el-Awady MK, Perinea Wilson EM, French FS. Androgen receptor defects: historical, clinical, and molecular perspectives. Endocr Rev. 1995;16(3):271–321.
27. Holterhus PM, Sinneckrer GH, Hiort BO. Phenotypic diversity and testosterone induced normalization of mutant L712F androgen receptor function in a kindred with androgren insensitivity. J Clin Endocrinol Metab. 2000;85:3245–50.

Part VII

46XXDSD

Special thanks to Dr. Shilpa Sharma, Professor, Pediatric Surgery, AIIMS New Delhi for assisting and aiding me in editing this section.

Classification of Congenital Adrenal Hyperplasia

28

Shilpa Sharma

Abstract

Congenital adrenal hyperplasia (CAH) is basically of two broad types: Classic and nonclassic. A subtle cryptic type also exists as the third type. The classic is further divided into salt-wasting (SW) and simple virilizing (SV). This chapter describes the basic types and their subclassifications for a better understanding of the subject.

Keywords

Classic CAH · Nonclassic CAH · Simple virilizing · Salt wasting

28.1 Introduction

Congenital adrenal hyperplasia (CAH), an inherited disease, is basically of two main types based on the presentation; Classic and nonclassic and a third less common type known as Cryptic [1, 2]. The classic is further divided into salt wasting (SW) and simple virilizing (SV). Severe classical form occurs in about 1/15,000 live births [3]. The incidence of carriers (recessive gene) is about 1 in 60. The SW accounts for two-third of classic CAH, the SV account for one third. The incidence of nonclassic CAH is about 1 in 1000 live births among Caucasians. The atypical form usually occurs in certain races, like people of European ancestry. The classification based on the presentation is depicted in Fig. 28.1. There have been recent developments that view CYP21A2 allelic variants and their phenotypic manifestations as a continuum [1].

S. Sharma (✉)
Department of Pediatric Surgery, All India Institute of Medical Sciences, New Delhi, India

S. K. Ratan (ed.), *Children with Differences in Sex Development*,
https://doi.org/10.1007/978-981-97-1639-5_28

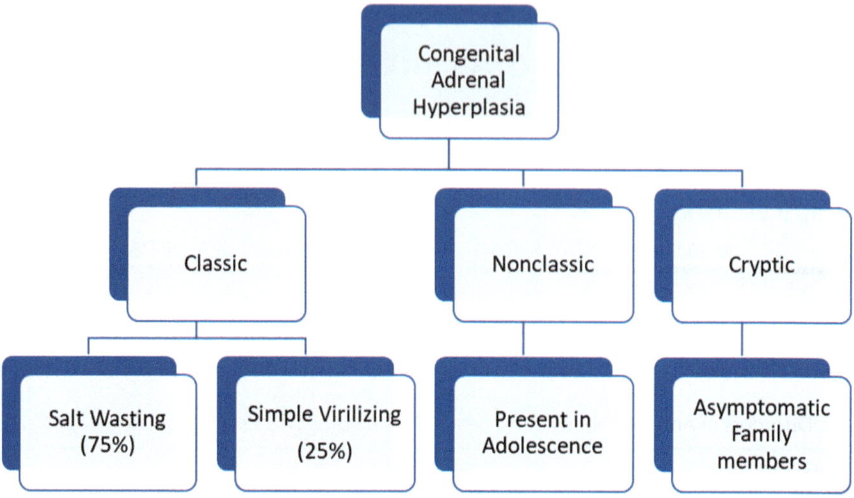

Fig. 28.1 Classification of congenital adrenal hyperplasia based on presentation

28.2 Classic CAH

Classic CAH is the more severe type. It is thus usually diagnosed during the first year of life usually in early infancy.

28.3 Salt Wasting (SW)

SW CAH is the more common type of classic CAH. Affected children lose too much water and salt in their urine. They have very low levels of cortisol and aldosterone, which can lead to a dangerous drop in blood pressure and require immediate treatment. Boys may have an enlarged penis, and girls may have ambiguous or masculine-looking genitals. The usual presentation is an infant with diarrhea, salt wasting, dehydration, virilization, and shock. Salt wasting comprises hypotension, vomiting, diarrhea, hyponatremia, hyperkalemia, hypoglycemia and circulatory collapse.

28.4 Simple Virilizing (SV)

Here, boys and girls appear more masculine, but it does not have a dramatic effect on blood pressure. Children with simple virilizing adrenal hyperplasia have decreased levels of cortisol, normal levels of aldosterone, and increased androgens. The usual presentation is a girl with ambiguous genitalia. The degree of virilization can range from simple clitoromegaly to a complete penile urethra.

28.5 Nonclassic Congenital Adrenal Hyperplasia

Nonclassic, or late-onset and mild, CAH is more common than the classic form. It is usually diagnosed in early adolescence or in younger children who have early signs of puberty. Children with nonclassic congenital adrenal hyperplasia have normal levels of aldosterone and excess amounts of androgens. Symptoms are often far milder than those associated with the classic condition, and many children do not need treatment. The nonclassic form does not progress to the classic form. The usual presentation is a post pubertal girl with hirsutism, acne, and menstrual irregularities. Young women may present with oligomenorrhea, polycystic ovaries, and hirsutism (polycystic ovarian syndrome) [4]. In rare instances, ladies may present with androgenic alopecia.

28.6 Cryptic CAH

This is a hidden variety that is present in the family members who are asymptomatic. It is diagnosed only on active diagnostic screening.

28.7 Pathogenesis

The adrenal cortex that comprises 80% of the adrenal gland produces about 50 steroids of which only 10 have hormonal activity which divided into 3 main groups:

1. Glucocorticoids—Cortisol, cortisone, and corticosterone
2. Mineralocorticoid—Aldosterone, 11-desoxycorticosterone (DOC)
3. Sex hormones—Dehydroepiandrogen (DHEA), Androstenedione, 17 Cetoandrostendions (androstentrion) and 11 β-hydroxyandrostendions

The severity and phenotypic presentation of the CAH cases vary depending on the location and extent of gene mutations or deletions [5].

28.8 Classification Based on Etiology

- Inherited mutation or deletion of CYP21A2 gene leading to deficiency of 21-hydroxylase enzyme (accounting for 90–99% of cases), disrupting 2 metabolic pathways to form Cortisol and Aldosterone. CYP21A2 gene is found on 6p21.3 as part of the human leukocyte antigen (HLA) complex. This is the most common type of CAH. CAH due to 21-hydroxylase deficiency is found in all populations.
- Mutations in CYP11B1 cause deficiency of 11-beta-hydroxylase enzyme, which disrupts the metabolic pathway to form Cortisol in the biosynthesis of adrenocortical hormone. Approximately up to 5% of CAH cases are due to this deficiency.

CAH due to 11-beta-hydroxylase deficiency is more common in those with Iranian, Jewish, or Moroccan descent [6, 7]. Mutations in CYP11B2 (aldosterone synthase) selectively impair aldosterone synthesis, causing hyponatremia and hyperkalemia with normal cortisol production.

- CYP17A1 deficiency: The enzyme CYP17A1 encodes 17α-hydroxylase and 17,20-lyase. There is a compensatory increase in ACTH that causes high level of corticosterone and DOC.
- 3β-Hydroxysteroid dehydrogenase type 2 (3β-HSD2) deficiency: This enzyme is involved in steroidogenesis.
- Mutations in steroidogenic acute regulatory (StAR) protein lead to Lipoid Congenital Adrenal Hyperplasia. It is an autosomal recessive disorder.
- P450 Oxidoreductase deficiency: P450 Oxidoreductase (POR) gene is located on chromosome 7q11-12. It encodes a protein which functions as a cofactor for several steroidogenic enzymes.

28.9 Clinical Features Based on the Various Types

SW classic CAH is an acute life-threatening form presenting with adrenal crisis in early age with poor feeding, weight loss, vomiting, diarrhea, hyperpigmentation, poor peripheral perfusion, hypotension, metabolic acidosis, and hypoglycemia. If not treated promptly, the electrolyte imbalance can lead to cardiovascular collapse, arrhythmia, cardiac arrest, and death.

SV classic CAH is usually diagnosed at birth due to virilized external genitalia, precocious pubarche (pubic hair), precocious axillary hair or precocious facial hair. Males in addition have enlargement of penis (with pre-pubertal testis). Increased linear growth and advanced bone age with compromised adult height are common if diagnosed late or with inadequate treatment. If treatment is begun at a bone age of more than 10 years, patients can go into true precocious puberty.

Nonclassic CAH is the mildest form of CYP21A2 deficiency presenting later than classic CAH. This condition is characterized by absence of virilization of external genitalia prenatally and little or no signs of hyperandrogenism during childhood. Females have acne and hirsutism, rapid growth, frontal baldness, menstrual irregularity, polycystic ovaries, and decreased fertility. Those assigned females at birth may have excess facial or body hair, permanent male body traits, irregular periods, and infertility. While there may be enlarged penis but small testicles, early puberty and short stature in those assigned males at birth.

CYP11B1 (11β-hydroxylase) deficiency: Females are born with ambiguous genitalia, whereas males have normal genitalia but can develop gynecomastia prepubertally.

CYP17A1 deficiency: Usually manifests symptoms in adolescents. Affected males have ambiguous genitalia or female-like external genitalia (few patients) and absence of pubertal changes. Females (46XX) have normal external genitalia but do not enter puberty due to estrogen deficiency and have scanty pubic and axillary hair because of adrenal androgen deficiency.

3β-HSD2 deficiency: Here the testosterone level is high in girls but low for boys. Genital ambiguity is therefore seen in 46XX and 46XY individuals with this type of CAH in addition to mineralocorticoid and glucocorticoid deficiency.

Congenital lipoid adrenal hyperplasia: All children have female external genitalia due to deficient testosterone production.

P450 Oxidoreductase (POR) deficiency: Most patients with this disorder have dysmorphism (Antley–Bixler syndrome) characterized by craniosynostosis, radio-ulnar synostosis, arachnodactyly, and bowed femur. Boys may show mild effects in testicular function whereas girls can develop hypogonadism and ovarian cysts.

References

1. Claahsen-van der Grinten HL, Speiser PW, Ahmed SF, Arlt W, Auchus RJ, Falhammar H, et al. Congenital adrenal hyperplasia-current insights in pathophysiology, diagnostics, and management. Endocr Rev. 2022;43:91–159.
2. Levine LS, Dupont B, Lorenzen F, Pang S, Pollack M, Oberfield SE, et al. Genetic and hormonal characterization of cryptic 21-hydroxylase deficiency. J Clin Endocrinol Metab. 1981;53:1193–8.
3. Milyani AA, Al-Agha AE, Al-Zanbagi M. Initial presentations and associated clinical findings in patients with classical congenital adrenal hyperplasia. J Pediatr Endocrinol Metab. 2018;31(6):671–3.
4. Rushworth RL, Torpy DJ, Stratakis CA, Falhammar H. Adrenal crises in children: perspectives and research directions. Horm Res Paediatr. 2018;89:341–51.
5. Neocleous V, Fanis P, Phylactou LA, Skordis N. Genotype is associated to the degree of virilization in patients with classic congenital adrenal hyperplasia. Front Endocrinol (Lausanne). 2018;9:733.
6. Speiser PW, Arlt W, Auchus RJ, Baskin LS, Conway GS, Merke DP, Meyer-Bahlburg HFL, Miller WL, Murad MH, Oberfield SE, White PC. Congenital adrenal hyperplasia due to steroid 21-hydroxylase deficiency: an endocrine society clinical practice guideline. J Clin Endocrinol Metab. 2018;103:4043–88.
7. Apóstolos RAC, Canguçu-Campinho AK, Lago R, Costa ACS, Oliveira LMB, Toralles MB, Barroso U. Gender identity and sexual function in 46,XX patients with congenital adrenal hyperplasia raised as males. Arch Sex Behav. 2018;47:2491–6.

Genetics of Congenital Adrenal Hyperplasia

29

Sudha Rao and Shagun Walia

Abstract

Congenital adrenal hyperplasia (CAH) is a disorder that affects cortisol biosynthesis due to autosomal recessively inherited monogenic defects in the 5 enzymes involved in adrenal steroid synthesis pathway as described below. CAH 21OHD is the commonest form and accounts for more than 90% of cases. Genotyping can help understand the phenotype and prompt life-threatening adrenal crisis. Molecular characterization of CAH 21OHD is complex with highly variable genotype.

Keywords

Congenital adrenal hyperplasia · Autosomal recessive · Genotype

29.1 Introduction

Congenital adrenal hyperplasia (CAH) is a common monogenic, autosomal recessive inherited disorder of glucocorticoid synthesis. It consists of a family of disorders caused due to the loss or severely decreased activity of one of the five enzymes required for cortisol synthesis in the adrenal cortex. These enzymes are desmolase

S. Rao (✉)
Division of Pediatric Endocrinology, Department of Pediatrics, Bai Jerbai Wadia Hospital for Children, Mumbai, Maharashtra, India

Tata Memorial Hospital, Mumbai, Maharashtra, India

Jupiter Hospital, Thane, Maharashtra, India

S. Walia
BJWHC, Mumbai, Maharashtra, India

S. K. Ratan (ed.), *Children with Differences in Sex Development*,
https://doi.org/10.1007/978-981-97-1639-5_29

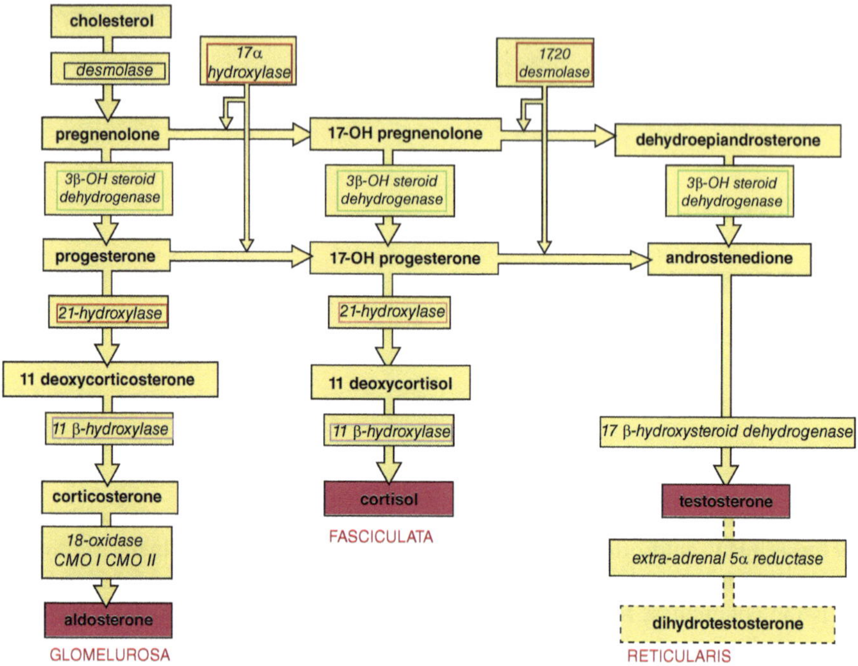

Fig. 29.1 Adrenal steroidogenesis pathway

(StAR), 3ß-hydroxysteroid dehydrogenase (*HSD3B2*), 21-hydroxylase (*CYP21A2*), 11-ß-hydroxylase (*CYP11B1&2*), and 17α-hydroxylase (*CYP17*) (Fig. 29.1).

The phenotype of all these conditions, albeit continuous, are gender specific and heterogenous. The clinical presentation is due to cortisol deficiency with either mineralocorticoid deficiency and/or androgen deficiency or excess which is based on the type of enzyme deficient in the adrenal steroidogenesis pathway. CAH due to 21-hydroxylase deficiency (*CYP21A2*) accounts for over 95% of all cases affecting 1 in 10,000 to 18,000 live births in its classic form. 11-β-hydroxylase CAH (*CYP11B1 &B2*) is the second most common type of CAH, affecting 1 in 100,000 to 200,000 live births and accounts for 5–8% of all cases of CAH.

In this chapter, we briefly describe the genetics of the various forms of CAH.

29.2 Genetics of CAH

All defects in the steroidogenesis enzyme are inherited as autosomal recessive.

- **CAH 21 – Hydroxylase deficiency (CAH 21OHD) (OMIM 201910)**

  CAH 21 OHD is the commonest form of CAH manifesting with features of cortisol deficiency and androgen excess with or without manifest mineralocorticoid deficiency. The phenotype is classified as classic salt wasting form or simple virilizing form and the nonclassic form of CAH (NCCAH). Nearly 75% of the classic form present with adrenal crisis (salt wasting and cortisol deficiency) which is life threatening and needs prompt management. Newborn screening for CAH 21OHD helps prevent life-threatening salt wasting adrenal crisis in infancy and features of hyperandrogenism in childhood and adolescence by early diagnosis and preemptive therapy.

Although the overall incidence of CAH 21OHD in the general population is 1 in 13,000 to 1:16000 live births, the prevalence in Indian population is much higher at 1:6000 individuals [1] as compared to 1:10,000 to 1:23,000 in the USA and Europe [2].

CYP21A2 gene (previously called *P450c21B, CYP21B, or CYP21*) is present on the long arm of chromosome 6, within the major human histocompatibility complex (HLA) and encodes 21-hydroxylase, a cytochrome P450 type II enzyme of 494 amino acids.

Approximately 30 kb from the *CYP21A2* gene there is a nonfunctional pseudogene *CYP21A1P*. The pseudogene *CYP21A1P* is inactive due to various insertions, deletions which prevent the formation of a functional protein. Both these genes are composed of ten exons and have a high level of nucleotide sequence homology of 98% in their exons and 96% in their introns (Fig. 29.2) [3, 4]. Next to *CYP21A2* and *CYP21A1P* genes, there are three other genes, *C4, TNXB, RP1*,encoding the fourth component of complement, an extracellular matrix protein tenascin-X23, and a serine/threonine nuclear protein kinase, respectively, and two truncated pseudogenes, RP2 and TNXA, which jointly, compose a genetic unit labelled as RCCX module (RP-C4-CYP21-TNX) and correspond to an extremely variable part of DNA of ~30Kb [5]. The standard organization is bimodular, and is formed of two RCCX modules, one with the *CYP21A1P* pseudogene and the other with the *CYP21A2* gene, where the usual sequence of genes from telomere to centromere is: RP1-C4A CYP21A1P-TNXA-RP2- C4B-CYP21A2-TNXB (Fig. 29.3) [6]. This bimodular haplotype is seen in about 69% of the Caucasian population.

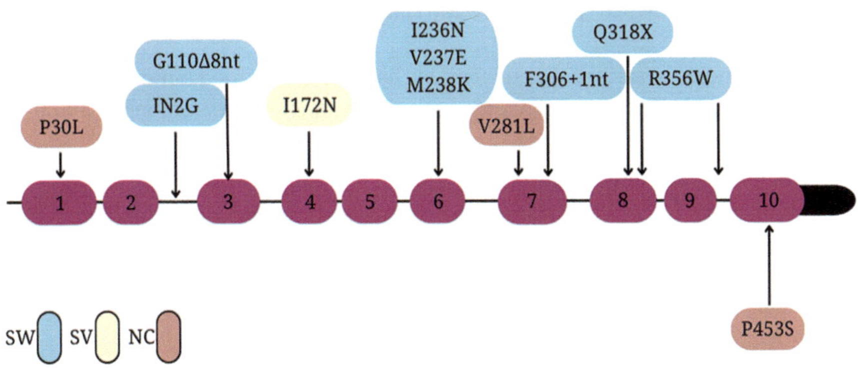

Fig. 29.2 Schematic diagram representing the common mutations in the 10 exons of *CYP21A2* gene and their clinical phenotype. (Adapted from [3]) Salt Wasting CAH; Simple Virilizing CAH; Nonclassical CAH

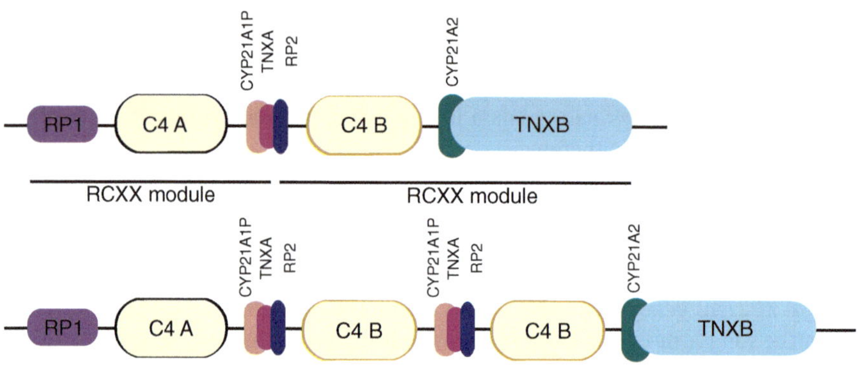

Fig. 29.3 Schematic diagram of the organization of the RCCX modules, one with the *CYP21A1P* pseudogene and the other with the *CYP21A2* gene (Adapted from [5])

29.3 Molecular Characterization in CAH 21 OHD

CYP21 gene and pseudogene have high sequence homology and being located in complex polymorphic region, misalignment and recombination events are the most common cause of CAH 21OHD. Intergenic recombination is the cause of 95% of pathogenic variants. 75% of the intergenic recombinations are confined to a single point alteration in genes, which are called "micro-conversions." In this type, a fragment of the functional *CYP21A2* gene is replaced by a segment of the pseudogene *CYP21A1P*. Nearly 20–25% of CAH result due to gross misalignment due to unequal crossing over during meiosis which can cause gene deletions, gene duplications and deletions involving *CYP21A2* and other contiguous genes. Rarely, uniparental disomy may also lead to CAH 21OHD. Approximately 1–2% of cases are caused by de novo mutations not inherited from either parent due to high variability of the *CYP21A2* locus [7].

Commonly described pathogenic variants of *CYP21A2* gene are shown in Table 29.1 [7].

Till date more than 1300 genetic variants have been recorded but only 230 affect human health. Most of the pathogenic variants are missense mutations resulting in classic form of CAH 21OHD.

Unpublished genotype data from the author's institute, of 53 children with CAH 21OHD (Table 29.2) showed intron 2g splice site mutation as the commonest as is also shown in other Indian studies [6]. Although many population do not have high Q318X mutation, a study from Tunisia reported a large number [8]. This needs to be further studied for a founder effect.

Table 29.1 Molecular characterization and clinical spectrum of common CYP21A2 mutations

Mutation	Region of CYP21A2 involved	Percentage 21OH enzyme activity	Clinical phenotype/ severity
Gene deletions and large gene conversions	Unequal cross over between CYP21A2 and CYP21P	0%	Severe form salt wasting
In2G	End of intron 2	1–2%	Severe form salt wasting
G110D8nt	Exon 3	Less than 1%	Severe form salt wasting
I236N V237E M238K	Exon 6 cluster	Less than 1%	Severe form salt wasting
F306+1nt	Exon 7	Less than 1%	Severe form salt wasting
Q318X	Exon 8	Less than 1%	Severe form salt wasting
R356W	Exon 8	Less than 1%	Severe form salt wasting
I172N	Exon 4	1–2%	May be simple virilizing or salt wasting phenotype
P30L	Exon 1	20–60%	Nonclassical or mild form
V281L	Exon 7	20–60%	Nonclassical or mild form
P453S	Exon 10	Not clearly explained	May be simple virilizing or mild form

Table 29.2 Common mutations found in a study of 53 children with CAH at the Author's institute (Unpublished data)

Mutation found	No of allele (n = 106)	%	Salt wasting form %	Simple virilizing form %
Q318X	25	24	100	
Del	26	25	100	
I 2g splice	**30**	**28.7**	**78**	**22**
I 172N	**8**	**7.6**	**12.5**	**87.5**
R356W	4	3.8	100	
Novel	5	4.8	60	40
Cluster	1	0.9	100	
FS mutation	1	0.9		100
Ins 7	2	1.8	100	
Undetermined	4	3.7	64	26

29.4 Genotype–Phenotype Correlations in CAH 21 OHD

CAH 21OHD is a clinically heterogenous disorder having mostly a compound heterozygous genotype. The genotype–phenotype correlation is seen in 90–95% of cases [9]. Being a compound heterozygous state with two different disease-causing mutations, phenotype is decided by the mutation retaining most enzyme activity. Genotype–phenotype discordance or concordance may exist across all mutation groups. In CAH 21OHD most of the phenotypic variability is seen in the simple virilizing form of CAH. In author's experience, phenotypic variability was seen in intron 2g splice site mutations and in I172N (Table 29.2). Apart from the type of mutation, several factors may be responsible for the clinical heterogeneity including length of the CAG repeats of the androgen receptor, other genes encoding proteins other than cytochrome P450 type II enzyme that have 21-hydroxylase activity. A large spectrum of phenotypes occur and genotype–phenotype discordance needs further research to find definite answers.

A contiguous gene deletion syndrome, involving both the CYP21A2 and TNXB genes, is called CAH-X syndrome in which patients of CAH 21 OHD may present with joint hypermobility, joint pain, or joint dislocations. Overall, 9% of patients with CAH due to 21-OHD are found to have CAH-X syndrome [10].

29.5 Genetic Diagnosis of CAH 21OHD

Due to the extreme variability of active gene CYP21A2 and its pseudogene CYP21AP, genetic diagnosis of 21-hydroxylase deficiency is more complicated when compared to any other monogenic disorder. Molecular analysis should be performed only in a certified laboratory with sufficient experience in analysis of CAH genotypes with standard quality control checks [11]. Southern blot techniques have been replaced by real-time quantitative polymerase chain reaction (PCR), Sanger sequencing, and multiplex ligation-dependent probe amplification (MLPA) techniques for detecting large deletions or duplications. Whole exome sequencing and other techniques will be applied by an experienced laboratory which would help identify the high variability in genomic region. Thus, gene analysis from a highly experienced laboratory dealing with CYP21A2 is advised for correct molecular diagnosis of CAH 21OHD.

- **CAH 11β hydroxylase deficiency (CAH 11β-OHD) (OMIM 202010)**

  CAH 11β-OHD is the second commonest form, accounting for 5–8% of all cases of CAH. Inherited as autosomal recessive, reduced activity of 11β hydroxylase enzyme leads to decreased cortisol with accumulation of glucocorticoid and mineralocorticoid precursors and androgen excess. Excess of 11 deoxycorticosterone (DOC) causes hypertension. The cortisol deficiency may not be much pronounced as with CAH 21OHD. The phenotype is classified as classic or severe form, where residual 11β hydroxylase enzyme activity is less than 10% [12], presenting as hypertension, atypical genitals in female newborn, and features of androgen excess manifesting as rapid postnatal growth, advanced skeletal maturation, precocious pseudo puberty in both sexes. Nonclassic form is extremely rare and may present with features of hyperandrogenism later in adolescent or adulthood.

  The overall incidence of CAH 11OHD in the general population is rare with 1 in 100,000 to 1:200,000 live births. Prevalence in Indian population could not be found in available literature.

The CYP11B1 gene, located on Chromosome 8q21, encodes the steroid 11-beta-hydroxylase which functions mainly in mitochondria in the zona fasciculata of the adrenal cortex to convert 11-deoxycortisol to cortisol and 11-deoxycorticosterone to corticosterone. CYP11B1 gene consists of 9 exons and encodes the 503 amino acid protein which belongs to the cytochrome P450 system. This gene is 40 kb away from the homologous gene CYP11B2 gene encoding the aldosterone synthase enzyme expressed in zona glomerulosa. More than 50 different mutations in CYP11B1 gene have been found which are mostly nonsense or missense mutations and tend to cluster around exons 2,6,7,and 8 [11].

- **CAH 17 α hydroxylase deficiency (CAH 17 OHD) (OMIM 202110)**

  CAH 17-OHD due to CYP17A1 enzyme deficiency accounts for 1% of all cases of CAH. CYP17A1, a microsomal P450 enzyme, catalyzes both 17 α hydroxylation of pregnenolone and progesterone to 17 hydroxy (OH) pregnenolone and 17 OH progesterone as also the conversion of 17-hydroxypregnenolone to dehydroepiandrosterone and, with lesser efficiency, of 17-OH progesterone to androstenedione through the 17,20 lyase reaction (Fig. 29.1). Thus, CAH 17-OHD results in glucocorticoid and sex steroid deficiency and accumulation of the mineralocorticoid precursors leading to hypertension with hypokalemia. The typical presentation of severe CAH 17-OHD is a phenotypic female, in both 46,XX and 46,XY individuals, with hypertension and delayed puberty. In less severe forms or partial CYP17A1 deficiency, 46,XY individuals may present with undervirilization or atypical genitalia as infants.

  CYP17A1 gene, encoding the activity of both 17 α hydroxylase and 17,20 lyase enzyme, is located on chromosome 10q24.3 locus and consists of 8 exons encoding a 508-amino-acid protein. Over 70 CYP17A1-inactivating mutations exist. No hot spots in the gene have been found yet in most large populations. Therefore, sequencing of the complete coding region is necessary for molecular genetic diagnostics.

- **CAH 3β-hydroxysteroid dehydrogenase type 2 deficiency (CAH HSD3B2) (OMIM 201810)**

  CAH HSD3B2 is a rare form of CAH with unknown incidence. CAH HSD3B2 is caused by homozygous or compound heterozygous mutation in the HSD3B2 gene, located on chromosome 1p12 and is expressed in both adrenals and gonads. The severe form of HSD3B2 deficiency is characterized by decreased excretion of cortisol, aldosterone, and androgens. In newborns affected by the severe form, there are signs of mineralocorticoid and glucocorticoid deficiency which if not treated promptly is life threatening. Paradoxically, both male and female infants can be born with ambiguous genitalia, which depends on severity of enzyme deficiency.

- **P450 oxidoreductase deficiency (POR) (OMIM 613571)**

  POR deficiency is the most complex of the various types of CAH, as it affects many enzymes composing steroidogenesis pathway. It is a rare variant with unknown incidence. POR is the electron donor for all microsomal P450 enzymes, including the three steroidogenic enzymes P450c17(17α-hydroxylase/17,20-lyase) CYP17A1; P450c21 (21-hydroxylase) CYP21A2; and P450aro (aromatase) CYP19A1. Both males and females present with severe genital ambiguity due to under- and over-virilization, respectively. Female fetus and mother may present with virilization due to reduced placental activity of aromatase CYP19A1 activity. Adrenal insufficiency is present in the majority of patients due to deficiency of CYP21A2. Overlapping clinical features with Antley–Bixler syndrome like craniofacial malformations may also be present [13].

Since the first description of POR mutations in 2004, around 50 patients have been recorded. The POR gene is present on the long arm of chromosome 7 (7q11.2) and consists of 15 translated exons extending in a region of around 32.9kb. About 26 different POR mutations have been identified till now, which includes 14 missense mutations, 8 frameshift mutations, 3 splicing errors, and 1 deletion [14].

- **Lipoid Congenital Adrenal Hyperplasia (LCAH) (OMIM 201710)**

  Lipoid CAH is the most fatal form of CAH caused by defect in, steroidogenic acute regulatory protein (StAR), a transport protein regulating cholesterol transfer in the mitochondria which is the rate limiting step for adrenal and gonadal steroidogenesis.

  Lipoid CAH was first described as an inherited disorder by Prader and colleagues in 1955 [15]. This disorder was wrongly termed "20,22 desmolase deficiency" initially and was considered to involve the cholesterol side-chain cleavage enzyme P450scc *CYP11A1.*

  The pathophysiology involves an initial loss of steroidogenesis that is StAR-dependent and a subsequent loss of steroidogenesis that is StAR-independent which is due to double hit in the form of cellular damage by the accumulated cholesterol esters in the adrenal and gonads.

  The StAR protein is 285 amino acids long and weighs 31.2 kD. The first 26 amino acids of the N-terminal end are presumably implicated in mitochondrial signaling. LCAH is caused by homozygous or compound heterozygous mutation in the gene encoding steroidogenic acute regulatory protein on chromosome 8p11.23. In classic forms of congenital lipoid CAH, there is complete or near complete absence of all steroid hormones. A less severe, nonclassic form is described where 10–25% of StAR activity is retained [16].

  LCAH is common among the Japanese, Korean, and Palestinian Arab populations, but is rare in the rest of the world. Most mutations in the *STAR* gene are located in exons 5, 6, or 7: p.Q258X in Asians, and p.R182L in Palestinian Arabs. At our Center, we have 7 genetically proven children with LCAH due to StAR mutation diagnosed over last 15 years (unpublished data). Two of these children had novel mutations—homozygous deletion of a single nucleotide G in exon 7 of the *STAR* gene, and a homozygous novel nucleotide substitution in exon 4 of the *STAR* gene c.441G>A. Two other cases had p.R182C mutation and 1 child had missense mutation c.653C>T (p.A218V) in exon 6 of the STAR gene and one had homozygous deletion exon 1 c.12_36 del (p.Thr5ProfsTer8). Exact prevalence of LCAH in Indian population is not known but might be higher than the world population and this needs proper identification through genotype studies.

- **Cholesterol side-chain cleavage (P450 SCC) enzyme deficiency (OMIM 118485)**
  Once postulated as lethal, P450 SCC enzyme deficiency is an autosomal recessive disorder of CAH caused by inactivating mutations in the CYP11A1 gene located on chromosome 15q23–24, containing 9 exons. This gene encodes the 30kD, P450 SCC protein of 521 amino acid residues. 28 patients, mostly from Eastern Turkey, have been described so far with CYP11A1 mutations causing P450 SCC deficiency. The phenotype of SCC deficiency is similar to StAR deficiency although different phenotypes have been reported, probably because of variable enzyme activity due to genotypic differences.

29.6 Antenatal Congenital Adrenal Hyperplasia (CAH)

In the presence of an index case or a family history of CAH 21OHD, fetal genetic testing for CAH 21OHD is undertaken at around 10 to 12 weeks of gestation by chorionic villus sampling or later at 15–16 weeks gestation by amniocentesis. In an ongoing pregnancy, the above 2 traditional and potentially dangerous testing techniques have been replaced by noninvasive prenatal testing involving cell-free DNA from maternal blood.

The phenotypic determination of fetal external genitalia happens by ninth gestational week and hence any treatment to prevent virilization of a female fetus has to be started much before that although the modality used for prenatal diagnosis must reasonably rule out if the fetus needs prenatal treatment because of risks associated with the treatment. In 2 cases, fetus does not need therapy—1) Nonbiallelic mutations in CYP21A2 (diagnosed by parallel sequencing of cell-free DNA to genotype CYP21A2 gene); 2) Sex typing to know presence of Y chromosomes which can be done as early as 6 weeks of gestation by collection of fetal DNA samples in maternal peripheral blood (male will not be affected by excess androgens). Such techniques, however, are still expensive and often unavailable.

In long-term studies to assess side effects of antenatal dexamethasone treatment on developing fetus, cleft palate, encephalopathic changes, cognitive, behavioral deficits, impaired verbal intelligence, social anxiety, decreased working memory, potentially increased risk of metabolic and cardiovascular disturbances, pancreatic maldevelopment have been observed in varying studies [17–19]. Hence, an algorithmic approach with initial karyotyping and then genetic analysis in a fetus to prevent unnecessary treatment of a male fetus, a female fetus with NCCAH and CAH carriers. It is important to counsel parents regarding the side effects of antenatal treatment with dexamethasone.

The recently published CAH clinical practice guidelines by Endocrine Society [20] for antenatal care/prenatal therapy as summarized below state

- It is advised that clinicians regard prenatal therapy as experimental. Thus, specific treatment protocols are not recommended [21].

- In pregnant women at risk for carrying a fetus affected with congenital adrenal hyperplasia and who are considering prenatal treatment, the treatment must be in accordance with protocols approved by Institutional review boards at centers which are competent of analyzing outcomes from large number of patients so the benefits and complications of interventions are precisely recorded.
- It is advised that research protocols for prenatal therapy include looking for Y-chromosomal DNA in maternal blood to exclude these fetuses from treatment allocation.

29.7 Conclusion

CAH is a clinically heterogenous autosomal recessive inherited disorder. CAH 21 OHD is the commonest form and accounts for more than 90% of cases. Genetic evaluation helps understand the phenotype and preempt life-threatening adrenal crisis. Molecular characterization of CAH 21OHD is complex with highly variable genotype. Molecular diagnosis should always be done form a laboratory experienced in genetic analysis of CAH 21 OHD.

References

1. Kaur G, Thakur K, Kataria S, et al. Current and future perspective of newborn screening: an Indian scenario. J Pediatr Endocrinol Metab. 2016;29:5–13.
2. van der Kamp HJ, Wit JM. Neonatal screening for congenital adrenal hyperplasia. Eur J Endocrinol. 2004;151(Suppl 3):U71–5.
3. Higashi Y, Yoshioka H, Yamane M, Gotoh O, Fujii-Kuriyama Y. Complete nucleotide sequence of two steroid 21-hydroxylase genes tandemly arranged in human chromosome: a pseudogene and a genuine gene. Proc Natl Acad Sci USA. 1986;83(9):2841–5.
4. White PC, New MI, Dupont B. Structure of human steroid 21-hydroxylase genes. Proc Natl Acad Sci USA. 1986;83(14):5111–5.
5. Yang Z, et al. Modular variations of the human major histocompatibility complex class III genes for serine/threonine kinase RP, complement component C4, steroid 21-hydroxylase CYP21, and tenascin TNX (the RCCX module). J Biol Chem. 1999;274:12147–56.
6. Gangodkar P, Khadilkar V, Raghupathy P, Kumar R, Dayal AA, Dayal D, Ayyavoo A, Godbole T, Jahagirdar R, Bhat K, Gupta N, Kamalanathan S, Jagadeesh S, Ranade S, Lohiya N, Oke RL, Ganesan K, Khatod K, Agarwal M, Phadke N, Khadilkar A. Clinical application of a novel next generation sequencing assay for CYP21A2 gene in 310 cases of 21- hydroxylase congenital adrenal hyperplasia from India. Endocrine. 2021;71(1):189–98. https://doi.org/10.1007/s12020-020-02494-z.
7. Pignatelli D, Carvalho BL, Palmeiro A, Barros A, Guerreiro SG, Macut D. The complexities in genotyping of congenital adrenal hyperplasia: 21-hydroxylase deficiency [published correction appears in Front Endocrinol (Lausanne). 2020 Mar 04;11:113]. Front Endocrinol (Lausanne). 2019;10:432. Published 2019 Jul 4. doi:https://doi.org/10.3389/fendo.2019.00432.
8. Romdhane L, Kefi R, Azaiez H, Ben Halim N, Dellagi K, Abdelhak S. Founder mutations in Tunisia: implications for diagnosis in North Africa and Middle East. Orphanet J Rare Dis. 2012;7:52.

9. Krone N, Braun A, Roscher AA, Knorr D, Schwarz HP. Predicting phenotype in steroid 21-hydroxylase deficiency? Comprehensive genotyping in 155 unrelated, well defined patients from southern Germany. J Clin Endocrinol Metab. 2000;85:1059–65.
10. Morissette R, Chen W, Perritt AF, et al. Broadening the spectrum of Ehlers Danlos syndrome in patients with congenital adrenal hyperplasia. J Clin Endocrinol Metab. 2015;100:E1143–52.
11. Krone N, Arlt W. Genetics of congenital adrenal hyperplasia. Best Pract Res Clin Endocrinol Metab. 2009;23(2):181–92. https://doi.org/10.1016/j.beem.2008.10.014.
12. Mooij CF, Parajes S, Rose IT, Taylor AE, Bayraktaroglu T, Wass JA, et al. Characterization of the molecular genetic pathology in patients with 11beta-hydroxylase deficiency. Clin Endocrinol. 2015;83:629–35. https://doi.org/10.1111/cen.12834.
13. Krone N, Reisch N, Idkowiak J, Dhir V, Ivison HE, Hughes BA, Rose IT, O'Neil DM, Vijzelaar R, Smith MJ, MacDonald F, Cole TR, Adolphs N, Barton JS, Blair EM, Braddock SR, Collins F, Cragun DL, Dattani MT, Day R, Dougan S, Feist M, Gottschalk ME, Gregory JW, Haim M, Harrison R, Olney AH, Hauffa BP, Hindmarsh PC, Hopkin RJ, Jira PE, Kempers M, Kerstens MN, Khalifa MM, Köhler B, Maiter D, Nielsen S, O'Riordan SM, Roth CL, Shane KP, Silink M, Stikkelbroeck NM, Sweeney E, Szarras-Czapnik M, Waterson JR, Williamson L, Hartmann MF, Taylor NF, Wudy SA, Malunowicz EM, Shackleton CH, Arlt W. Genotype-phenotype analysis in congenital adrenal hyperplasia due to P450 oxidoreductase deficiency. J Clin Endocrinol Metab. 2012 Feb;97(2):E257–67.
14. Scott R, Miller WL. Genetic and clinical features of P450 oxidoreductase deficiency. Horm Res. 2008;69:266–75. https://doi.org/10.1159/000114857.
15. Prader A, Gutner HP. The syndrome of male pseudohermaphrodism in congenital adrenocortical hyperplasia without overproduction of androgens (adrenal male pseudohermaphrodism). Helv Paediatr Acta. 1955;10(4):397–412.
16. Metherell LA, Naville D, Halaby G, Begeot M, Huebner A, Nurnberg G, Nurnberg P, Green J, Tomlinson JW, Krone NP, Lin L. Nonclassic lipoid congenital adrenal hyperplasia masquerading as familial glucocorticoid deficiency. J Clin Endocrinol Metabol. 2009;94(10):3865–71.
17. Van der Grinten HLC, Speiser PW, Ahmed SF, Arlt W, Auchus RJ, Falhammar H, Flück CE, Guasti L, Huebner A, Kortmann BBM, et al. Congenital adrenal hyperplasia—current insights in pathophysiology, diagnostics, and management. Endocr Rev. 2022; 43: 91–159.
18. Wapner RJ, Sorokin Y, Mele L, Johnson F, Dudley DJ, Spong CY, Peaceman AM, Leveno KJ, Malone F, Caritis SN, et al. Long-term outcomes after repeat doses of antenatal corticosteroids. N Engl J Med. 2007;357:1190–8.
19. Hirvikoski T, Nordenström A, Lindholm T, Lindblad F, Ritzeén EM, Wedell A, Lajic S. Cognitive functions in children at risk for congenital adrenal hyperplasia treated prenatally with dexamethasone. J Clin Endocrinol Metab. 2007;92:542–8.
20. Speiser PW, Arlt W, Auchus RJ, Baskin LS, Conway GS, Merke DP, Meyer-Bahlburg HFL, Miller WL, Murad MH, Oberfield SE, White PC. Congenital adrenal hyperplasia due to steroid 21-hydroxylase deficiency: an endocrine society clinical practice guideline. J Clin Endocrinol Metab. 2018;103(11):4043–88.
21. Blanchong CA, Zhou B, Rupert KL, Chung EK, Jones KN, Sotos JF, Zipf WB, Rennebohm RM, Yung YC. Deficiencies of human complement component C4A and C4B and heterozygosity in length variants of RP-C4-CYP21-TNX (RCCX) modules in Caucasians. The load of RCCX genetic diversity on major histocompatibility complex-associated disease. J Exp Med. 2000;191(12):2183–96.

30 Congenital Adrenal Hyperplasia

Rajesh Joshi

Abstract

Congenital adrenal hyperplasia is a group of heritable disorders associated with inability or reduced ability to produce cortisol. Due to cortisol insufficiency, excess of adrenocorticotropic hormone (ACTH) is released from anterior pituitary causing unremitting stimulation of adrenal cortex, causing hyperplasia. In many cases, deficiency of aldosterone also causes hyponatremia and hyperkalemia. Majority of cases are due to 21-hydroxylase deficiency. The classic form is the most common cause of atypical genitalia in 46 XX newborns. Salt losing crisis which occurs early in life in many patients needs urgent treatment with fluids, replacement with hydrocortisone and fludrocortisone, and correction of electrolyte imbalance, so as to prevent a fatal outcome. Long-term therapeutic goals, unlike other forms of adrenal insufficiency, are twofold—replacing the missing adrenal cortex hormones and suppressing the excess androgen levels. Improved strategies in dosing and timing of medications and close follow-up for adequacy of treatment, growth and complications related to the disease and treatment have improved the outcome of patients with this disorder. Clinicians should also consider patients' quality of life and consult mental health professionals as deemed necessary.

Other forms of CAH are uncommon; their incidence varies with country and ethnic background. This chapter describes the pathophysiology, clinical features, and management of 21-hydroxylase deficiency in detail and briefly describes other forms of CAH like CYP11B1 (11β-hydroxylase) deficiency, CYP17A1 deficiency, 3β-HSD2 deficiency, congenital lipoid adrenal hyperplasia, and P450 oxidoreductase deficiency.

R. Joshi (✉)
Division of PED Endocrinology, Department of Pediatrics, B.J. Wadia Hospital for Children, Mumbai, Maharashtra, India

S. K. Ratan (ed.), *Children with Differences in Sex Development*,
https://doi.org/10.1007/978-981-97-1639-5_30

Keywords

Congenital adrenal hyperplasia · 21-hydroxylase deficiency · Cortisol · Aldosterone

30.1 Introduction

The adrenal cortex secretes mainly 3 steroid hormones—aldosterone from zona glomerulosa, controlled by the renin–angiotensin system; cortisol from zona fasciculata controlled by the hypothalamic–pituitary–adrenal axis (CRH stimulating ACTH which in turn stimulates cortisol release); and androgens [mainly dehydroepiandrostenedione (DHEA) and androstenedione] controlled by an as yet not defined tropic hormone. All steroid hormones are derived from cholesterol through formation of various intermediate steroidogenic products mediated by group of p450 enzymes (Fig. 30.1). Deficiency of these steroidogenic enzymes, an autosomal recessive condition causes congenital adrenal hyperplasia (CAH). The mode of presentation, clinical features, and biochemical profile are dictated by the severity of block.

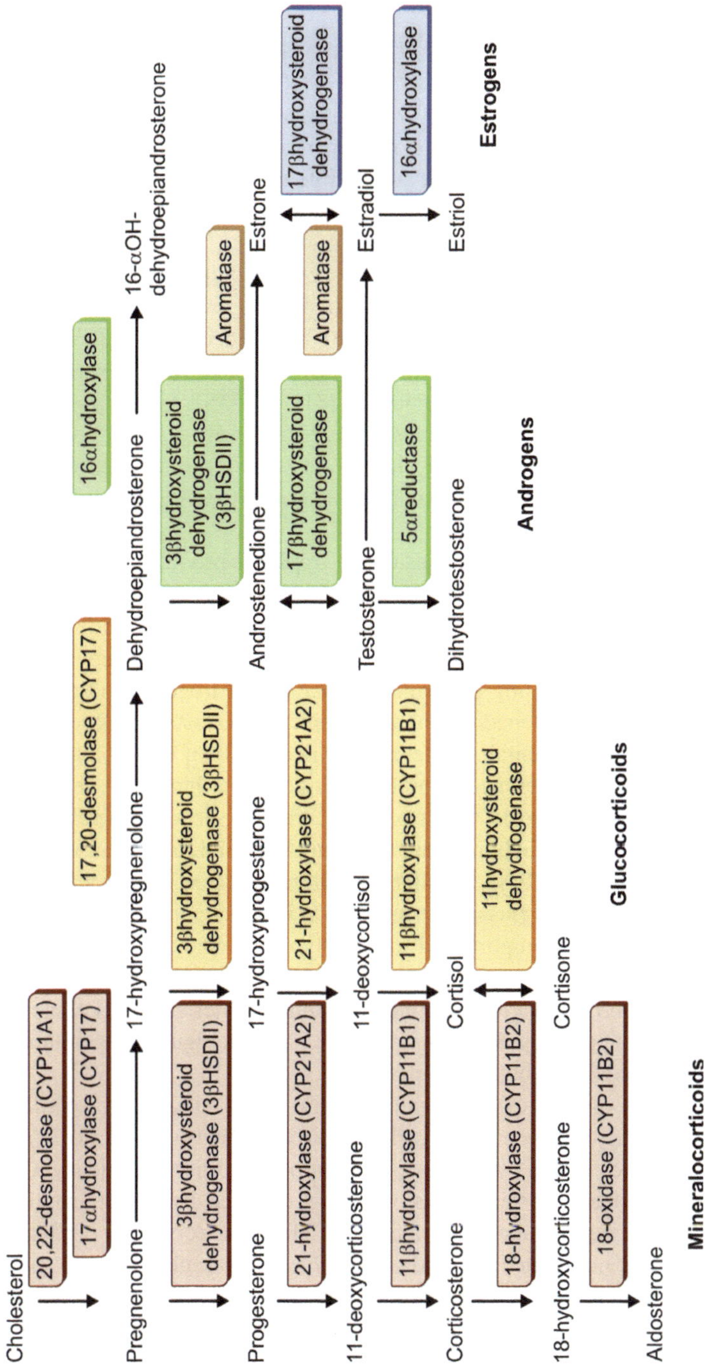

Fig. 30.1 Adrenal steroidogenesis

30.2 CYP21A2 (21-Hydroxylase) Insufficiency

CAH due to deficiency of CYP21A2 (21-hydroxylase) is the most common cause of steroidogenic defect with an incidence of 1 in 10,000 to 20,000 births. 95% cases of CAH are due to CYP21A2 deficiency with a reported screen positivity rate of 1 in 5762 babies at birth in India. The gene is located on chromosome 6, within human leukocyte antigen complex, making it susceptible to mutation.

The various clinical phenotypes of this condition are classic, i.e. salt wasting (SW) and simple virilizing (SV) CAH; and nonclassic (NC-CAH), depending on residual 21-hydroxylase activity. The genotype–phenotype correlation is seen in 80–90% cases. Severity of mutations is classified by its residual enzymatic activity: "null" (0% residual function; del/con, G110_8bp, E6 cluster, F306+t, Q318X, R356W), "A" (0–1% residual function; I2G), "B" (1–2% residual function; I172N), or "C" (20–60% residual function; P30L, V281L, P453S). Severe genotypes, i.e. null and A, correlate with SW in 97% and 91%, respectively. Usually complete loss-of-function mutations causes severe virilization and salt wasting in females. SV CAH has 1–2% residual enzyme activity as against NC-CAH which has 20–60% enzyme activity.

A decrease in p450c21 activity results in impaired production of cortisol and aldosterone, increasing the ACTH, subsequently leading to build up of steroid precursors higher in the pathway, i.e. there is insufficient conversion of 17-hydroxy progesterone (17-OHP) to 11-deoxycortisol leading to build up of 17-OHP (Fig. 30.1), ultimately leading to formation of more androgens (androstenedione, DHEA, DHEAS). Androstenedione is metabolized to testosterone leading to masculinization of external genitalia in females which makes it easy to recognize this disorder at birth due to presence of ambiguous genitalia. Increased androgen production starts between 6 and 10 weeks of intrauterine life, leading to posterior labial fusion of varying degree and hypertrophy of clitoris. Labial fusion causes formation of urogenital sinus at the base of enlarged clitoris. Degree of masculinization of female external genitalia is classified according to Prader staging (Fig. 30.2). Degree of virilization is directly proportional to severity of enzyme defect; extreme forms lead to complete fusion of labia and formation of penile urethra. Female patients with CAH have normal ovaries and internal Mullerian derivatives, i.e. uterus, fallopian tubes, and upper vagina. In males, there is no ambiguity of genitalia; however, the penile length can be at 95th percentile.

SW CAH (75% of classic CAH) is an acute life-threatening form presenting with adrenal crisis in early age (neonate or early infancy). Marked deficiency of CYP21A2 causes severe deficiency of aldosterone in addition to cortisol deficiency. Children present with poor feeding, weight loss, vomiting, diarrhea, hyperpigmentation (due to excess melanocyte-stimulating hormone (MSH)), poor peripheral perfusion, and hypotension and if not treated promptly, leads to cardiovascular collapse and death. These manifestations usually occur between 4th and 15th day of life or sometimes by 6 to 12 weeks of age. Rarely a patient may present first in childhood with salt

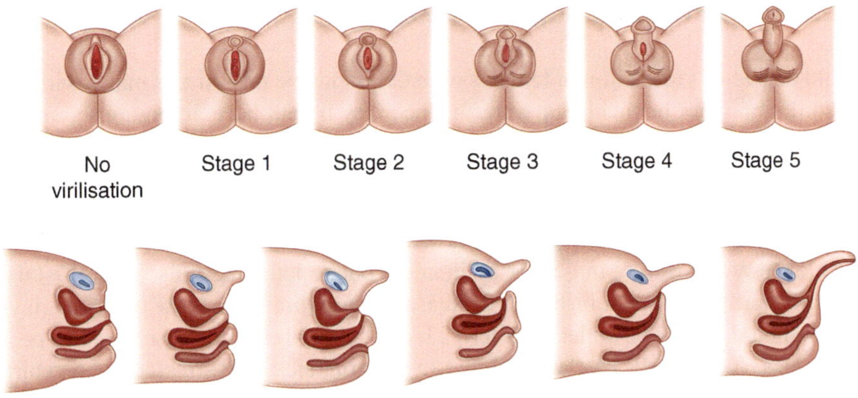

Fig. 30.2 Degrees of female genital virilization according to Prader staging

loss, during an episode of infection. Salt losing tendency may decrease after 4–5 years of age. Salt wasting presents with biochemical features of hyponatremia, hyperkalemia, and metabolic acidosis. Some children are misdiagnosed as pyloric stenosis; however, absence of characteristic biochemical findings of hypochloremia, hypokalemia, and metabolic alkalosis seen in pyloric stenosis clinches the diagnosis of CAH. History of affected sibling or their death due to salt wasting also helps in diagnosis of CAH.

SV CAH accounts for approximately 25% of classic CAH. Mineralocorticoid (MC) pathway is relatively spared unless the patient is under stress-like fluid restriction or salt deprivation. This is because the adrenal gland synthesizes more molar concentration of cortisol (100 to 1000 times) compared to aldosterone and hence glucocorticoid (GC) insufficiency is more pronounced. SV females (46XX) are usually diagnosed at birth due to virilized external genitalia. Other modes of presentation are precocious pubarche, i.e. Appearance of pubic hair between 6 months and 2 years of age, axillary hair between 2 and 4 years of age, and facial hair by 8–14 years of age. Males in addition have enlargement of penis (with pre-pubertal testis). Increased linear growth and advanced bone age with compromised adult height are common if diagnosed late or with inadequate treatment. If treatment is begun at a bone age of more than 10 years, patients can go into true precocious puberty.

NC-CAH is the mildest form of CYP21A2 deficiency presenting later than classic CAH. This condition is characterized by absence of virilization of external genitalia prenatally and little or no signs of hyperandrogenism during childhood. Females have acne and hirsutism, sometimes with other features like frontal baldness, menstrual irregularity, polycystic ovaries, and decreased fertility.

Occasionally, contiguous gene deletion can affect other genes like tenascin X in addition to 21-OHD. CAH can be associated in these cases with Ehlers–Danlos syndrome (homozygous deletion) or only joint hypermobility (if haploinsufficiency). These features should be carefully examined in all CAH patients.

30.3 Diagnostic Laboratory Findings

Mild degree of virilization in females with CAH may be missed and males with this disorder may not be picked up at birth since they do not have ambiguous genitalia and usually do not manifest salt wasting features in immediate postnatal life. Not recognizing and treating the salt wasting crisis can be fatal. Hence, newborn screening program done from dried blood sample (DBS) which measures 17-OHP in addition to other tests will help to treat these children early, thus reducing morbidity, mortality, and mistakes in gender assignment due to late diagnosis. Since 17-OHP level is normally high at birth and decreases rapidly in the first few postnatal days, DBS should ideally be collected after 72 h of life in order to decrease the false positive rates. As 17-OHP levels are higher in preterm, sick, and stressed babies, birth weight and gestational age based cut-off values for 17OHP should be used, so as to reduce the recall rate. A value of more than 90 nmol/l by fluoroimmunoassay is considered positive screen. All screen positives are confirmed by venous sampling by LC-MS/MS based assay.

In suspected patients where newborn screening (NBS) is not done, basal or if possible ACTH-stimulated 17-OHP and other steroids are used for diagnosis. In classic CAH (SW, SV), basal values are usually >20 ng/ml increasing to >50–100 ng/ml after ACTH. Patients with NC-CAH have an ACTH-stimulated values of >10–100 ng/ml. Elevated plasma renin activity is seen in SW and most SV patients (especially if sodium restricted).

Molecular genetic testing in 21-OHD is a difficult process due to occurrence of both, a gene and homologous pseudo gene. Therefore, genotyping for 21-OHD should be done in selected patients for confirming difficult cases, carrier detection and for adequate and correct genetic counseling.

30.4 Treatment of 21-OHD

Treatment of CAH is like treating any child with adrenal insufficiency. However, it is challenging and at times difficult since a delicate balance between undertreatment and overtreatment is to be maintained.

Infants presenting with SW crisis have hyponatremia, hyperkalemia, and metabolic acidosis. They require intravenous fluids (0.9% DNS) at 1.5–2 times maintenance rate with monitoring of hydration status and electrolytes. An injectable bolus of 25 mg hydrocortisone followed by 50–100 mg/m^2/day is given in divided doses till the child is out of adrenal crisis. This is followed by a gradual taper to maintenance dose of hydrocortisone orally. Hyperkalemia is managed by administering calcium gluconate, glucose-insulin drip, bicarbonate, salbutamol nebulizations depending on potassium level and acid base status. Hypoglycemia if present is treated with 4 ml/kg of 10% dextrose. Oral MC, fludrocortisone is added when the child is fit to consume orally. Newborn and young infants have a higher requirement of fludrocortisone (0.15-0.3 mg daily) due to aldosterone resistance and this dose decreases to 0.05 to 0.15 mg as the child grows. Infants also require salt supplements (1–2 gm daily) for better effect of MC at level of renal tubule.

After the acute management of adrenal crisis, children are treated with hydrocortisone at dose of 10–15 mg/m^2/day in 3 divided doses. Though the normal secretory rate of cortisol in body is 6–8 mg/m^2/day, in CAH a higher dose is required to suppress ACTH and adrenal androgens. Steroid supplement according to circadian rhythm is preferred, i.e. the first dose of the day in early morning and the last dose early in evening. Prednisone and dexamethasone should be avoided in growing children due to growth suppressant effects. Infants and small children require small doses of hydrocortisone which should be given by pulverizing tablets (available as 5, 10 mg) into small doses and then mixing them with milk or fluids just before administration. It should not be left dissolved for subsequent use. Small dose of fludrocortisone is beneficial in SV CAH. Fludrocortisone increases intravascular volume, decreasing the hypovolemic drive to ACTH secretion. This permits the lower dose of hydrocortisone in patients with SV 21 OHD.

It is important to teach stress dosing of GC during illness and emergencies. This increased dosing is provided during febrile illness (temperature > 38.5 °C), gastroenteritis, procedures with general anesthesia, and surgery. For minor illness hydrocortisone is increased to 2–3 times the usual daily dose, whereas for severe illness/surgery hydrocortisone is increased to 50–100 mg/m^2/d. Patient is also advised to increase fluid intake and consume carbohydrates frequently during illness. Care givers are also taught to administer intramuscular hydrocortisone in emergencies. The injectable dose of hydrocortisone is 25 mg (<3 years age), 50 mg (3–12 years age), and 100 mg (≥12 years age) which provides 4–6 h of coverage, allowing some time till the patient reaches health care facility. 20 mg hydrocortisone has a mineralocorticoid effect of 0.1 mg of 9α-fludrocortisone. Hence, when oral intake is not possible during severe illness, stress dose hydrocortisone and sodium chloride supplements can substitute oral MC replacement.

Choice of GC therapy in postpubertal patients is either hydrocortisone or potent steroids like prednisolone (4–6 mg/day) or dexamethasone (0.25–0.5 mg/day) which have the advantage of once or twice daily dosing.

Gender assignment is done by experts after proper counseling and discussion with parents. Most 46 XX CAH children are assigned female gender at birth. Karyotype and additional biochemical tests may be needed before this decision is made. Usually the gender identity is female and fertility is preserved. A recent review from India done in 42 SV CAH patients reported male gender assignment in one-fourth patients (median age: 2 months) as against female sex of rearing in all SW patients (median age: 0.4 months).

30.5 Monitoring

- Patients ≤ 18 months—close monitoring (monthly) in the first 3 months of life. Thereafter every 3 months
- After 18 months, evaluation every 4 months

- At all clinic visits the child should be examined for following- weight, length/height, blood pressure, skin pigmentation, cushingoid features, appearance of pubic hair, unexpected phallic or body growth, development of adult apocrine odor, and symptoms of salt craving or adrenal crisis
- MC dose titration is based on blood pressure, serum sodium, potassium, and plasma renin activity (PRA). Inadequate fludrocortisone dosing can cause salt craving, hyponatremia, and hyperkalemia. A high PRA level even with normal electrolyte level indicates need to increase fludrocortisone dose
- Serum 17-OHP, androstenedione, and testosterone are traditional indicators of the adequacy of GC replacement in CAH
- 17-OHP should be done in morning before the dose of GC. Diurnal variability is marked for 17-OHP than for androstenedione and testosterone levels. Normal value of 17-OHP indicates overtreatment. 17-OHP levels should be kept at ULN to 4–12 ng/ml. Androstenedione and testosterone levels are kept within the normal age-adjusted range and, in females, a ratio of testosterone to sex hormone binding globulin (SHBG) should be less than 0.05. However, dose adjustment should be in the context of the overall clinical picture and not solely based on a single laboratory assessment
- Laboratory data should indicate the need for dose adjustment before pathological changes in growth, bone age, and physical features occur
- Protracted undertreatment is suspected by accelerated height velocity, signs of virilization, and advanced bone maturation, whereas protracted overtreatment causes reduced height velocity, accelerated weight gain, cushingoid features, and high blood pressure
- Annual bone age is done after 2 years of age
- It is advisable to screen for testicular adrenal rest tumor (TART) in boys after 8–10 years of age by serial USG of testis
- Adolescents may require a higher and also more frequent dosing (up to 4 times daily) of hydrocortisone due to decrease in activity of 11 β-HSD type 1 enzyme and increased clearance of cortisol. Care should be taken not to exceed hydrocortisone dose above 17 $mg/m^2/day$ as it may affect growth. It may be difficult to suppress morning rise of 17-OHP in this age group which can be overcome by earlier dosing of hydrocortisone in morning

30.6 Surveillance for Long-Term Complications of CAH

Long-term complications in CAH are low bone mineral density (BMD) and fractures, cardio metabolic problems, adrenal and testicular tumors, fertility problems, altered gray and white matter microstructure, reduced working memory, and low quality of life.

One study has shown that patients with CAH had a mean adult height score of −1.4 SD (10 cm) below the population mean.

Adrenal myelolipomas may occur in patients with classic CAH. These benign and non-steroid producing tumors may grow to massive size and require surgical removal because of mass effects.

Surveillance in adults

- Annual physical examinations—BP, BMI, and assessment for cushingoid features + assess MC and GC supplement adequacy
- Counseling regarding healthy lifestyle choices at an early age to avoid metabolic syndrome and related sequelae
- Screening BMD in anyone subjected to a prolonged period of higher-than-average GC dosing, or who has suffered a non-traumatic fracture
- Reserve adrenal imaging in patients who have clinical evidence of adrenal mass, poor disease control, a lapse in treatment of several years, or lack of response to intensified therapy (no routine USGs)

30.7 Mental Health

Clinicians should be aware that individuals with congenital adrenal hyperplasia may be at risk for developing mental health problems and should have a low threshold for referral to psychological or psychiatric treatment.

Mental health practitioners should have specialized expertise in assessing and managing congenital adrenal hyperplasia-related psychosocial problems

30.8 Psychosexual Concerns

Females with CAH (especially salt-wasters) may have

- Male-typical play
- Physical aggression
- Low-interest in infants/maternal nurturing
- Increased rate of homosexuality compared to peers

Most raised as female express female gender identity, gender role, and heterosexual orientation.

30.9 Newer/Investigational Therapy

Improved glucocorticoid preparation—To mimic physiological cortisol secretion, modified-release HC, i.e. dual-release formulation (Plenadren) and delayed and sustained formulation (Chronocort).

Interference with HPA axis—ACTH is the major stimulator of adrenal cortex which increases the androgen levels. Antagonizing the CRF receptor and hence the ACTH or targeting ACTH itself may decrease adrenal androgens and allow for decreasing glucocorticoid dose. Examples are CRF1 receptor antagonists (tildacerfont), ACTH antibodies, and ACTH receptor antibodies.

Targeting adrenal steroid production—Abiraterone acetate is a potent and selective inhibitor of the enzyme CYP17A1 (17α-hydroxylase/17,20-lyase). This drug lowers the production of adrenal precursor steroids and adrenal androgens despite persistently elevated ACTH allowing for physiological glucocorticoid replacement.

30.10 Restoring Functional Anatomy by Surgery

- Minimally virilized girls-parents should be informed about surgical options, including delaying surgery and/or observation until the child is older
- Extensive discussions regarding risks and benefits, shared decision-making, review of potential complications, and fully informed consent need to occur prior to surgery. The option to forgo surgery should be considered
- Treatment of minors—all surgical decisions remain the prerogative of families (i.e., parents and assent from older children) in joint decision-making with experienced surgical consultants

30.11 Other Forms of CAH

CAH other than 21-hydroxylase deficiency is uncommon. These are discrete genetic events with founder mutations and therefore the prevalence varies with ethnic and geographic distribution. For example, the second most common CAH in Middle East is CYP11B1 (11β-hydroxylase) deficiency; in Japan, congenital lipoid adrenal hyperplasia; in Brazil, CYP17A1 deficiency. 3β-HSD2 deficiency though rare worldwide, pockets affected with this disorder are found in Amish Lancaster County, Pennsylvania.

30.11.1 CYP11B1 (11β-Hydroxylase) Deficiency

Approximately 5% cases of CAH are due to this deficiency. This deficiency impairs cortisol and cortisone production leading to increase in steroid precursor's 11-deoxycortisol, DOC, and 18-hydroxy DOC. 17-OHP also is increased though to a lesser extent. Females are born with ambiguous genitalia, whereas males have normal genitalia but can develop gynecomastia prepubertally. Elevated DOC can cause sodium retention with time, hypokalemia, and hypertension (low renin). This is the reason that salt wasting is usually not present. Severe adrenal insufficiency is uncommon because of high concentration of corticosterone which can mask cortisol deficiency. This disorder can be mislabeled as CYP21A2 deficiency; however, characteristic steroid profile, i.e. increased levels of DOC and 11-deoxycortisol, electrolyte abnormalities as mentioned above and genetic analysis can confirm the diagnosis. Treatment is with GC, with treatment of hypertension as required. Fertility can be preserved in well controlled males and females with this disorder.

30.11.2 CYP17A1 Deficiency

CYP17A1 encodes 17α-hydroxylase and 17,20-lyase. It is expressed in gonads and the adrenals leading to inability to form androgens and estrogens in addition to glucocorticoid insufficiency. Compensatory increase in ACTH causes high level of corticosterone and DOC which causes hypertension (low renin), hypernatremia, and hypokalemia. Clinical presentation usually manifests in adolescents and these individuals are labeled with other conditions till a correct diagnosis is made. Affected males have ambiguous genitalia or female-like external genitalia (few patients) and absence of pubertal changes. Females (46XX) have normal external genitalia but do not enter puberty due to estrogen deficiency and have scanty pubic and axillary hair because of adrenal androgen deficiency. Increased pigmentation of skin may be present; however, overt adrenal insufficiency may not occur due to high molar concentration of corticosterone which has a low affinity for glucocorticoid receptor. Partial forms of deficiency present with severe hypospadias in males and mild phenotype in females (46XX). Treatment is with GC and when indicated sex steroid replacement and treatment of hypertension. Most of these patients are reared as females and have a female gender identity. 46, XY patients with isolated 17,20-lyase deficiency who are mildly affected might have a male sex of rearing and gender identity. In such individuals, testosterone therapy will be required in adolescent age. 46 XX individuals have ovarian cysts. Pregnancy is reported in them after ovulation induction. 46 XY individuals however are not fertile.

30.11.3 3β-HSD2 Deficiency

Type 2 3β-HSD (3β-HSD2) is expressed in adrenals and gonads. 3β-HSD activity produces all active steroid hormones – mineralocorticoids, glucocorticoids, androgens, and estrogens and therefore its deficiency causes most profound effect on steroidogenesis. 3β-HSD2 converts Δ^5 steroids (pregnenolone, 17α-hydroxypregnenolone, DHEA) to Δ^4 steroids (progesterone, 17-OHP, androstenedione). Because of type 1 3β-HSD enzyme present in peripheral tissues, Δ^4 steroids which are one step past the block are elevated and therefore some androstenedione is converted to testosterone. This unusual metabolism causes testosterone level somewhat high in girls but low for boys. Genital ambiguity is therefore seen in 46XX and 46XY individuals with this type of CAH in addition to mineralocorticoid and glucocorticoid deficiency. Most girls (46XX) with this disorder have clitoromegaly as high DHEA are converted to androgens by peripheral 3β-HSD type 1. ACTH stimulated ratio of Δ^5 to Δ^4 steroids or urine steroid profiling helps to suspect this condition. Genetic analysis is confirmatory. Treatment is by replacement of GC and fludrocortisone (with salt supplement in infancy). Severely undervirilized 46XY cases might be reared as females. Patients require sex appropriate sex steroid replacement during puberty, and boys might develop gynecomastia if not treated.

30.11.4 Congenital Lipoid Adrenal Hyperplasia

This AR disorder is due to loss-of-function mutations in steroidogenic acute regulatory protein (StAR) gene. This defect causes impaired conversion of cholesterol to pregnenolone due to defective StAR protein function which normally moves cholesterol from outer to inner mitochondrial membrane. Complete disruption of StAR activity causes severe defect in glucocorticoid, mineralocorticoid, and sex steroid synthesis. Salt wasting usually occurs between 2 and 4 weeks of life. All children have female external genitalia due to deficient testosterone production. All gonadal and adrenal steroids are low or undetectable. This is a disease affecting infants very early in life; however, few cases with later presentation and atypical genitalia or rarely normal genitalia in 46 XY patients are described. Girls (46XX) with this defect can develop secondary sexual characters but later develop anovulatory cycles due to primary ovarian insufficiency. Treatment is by replacement of GC and fludrocortisone and salt supplement in infancy. Phenotypic girls (46XY) will require gonadectomy and estrogen replacement for life. 46 XX girls will require careful monitoring of puberty and replacement with estrogen when required. 46 XY adults are infertile whereas in 46 XX pregnancy has been achieved by ovulation induction and in vitro fertilization.

30.11.5 P450 Oxidoreductase (POR) Deficiency

POR gene located on chromosome 7q11-12, encodes a protein which functions as a cofactor for several steroidogenic enzymes. The main adrenal enzymes involved are P450c17 and P450c21 with variable involvement of aromatase, p450C51, and many hepatic enzymes. Most patients with this disorder have dysmorphism (Antley–Bixler syndrome) characterized by craniosynostosis, radio-ulnar synostosis, arachnodactyly, and bowed femur. Atypical genitalia and glucocorticoid insufficiency are seen in 46 XX and 46XY patients. During puberty, there can be variable defect in sex steroid synthesis. Boys may show mild effects in testicular function, whereas girls can develop hypogonadism and ovarian cysts. Urinary steroid profile shows a characteristic pattern of metabolites. In most of the patients, ACTH and 17-OHP are mildly elevated. Replacement with glucocorticoids is the treatment of choice. Sex steroid replacement is often needed at puberty. Fertility is more likely in females than males. Females often require ovulation induction and in vitro fertilization.

To conclude, rare forms of CAH are characterized by steroids that are deficient and in excess, uniquely elevated steroids and by knowing if the gonads are also affected by the defect.

Suggested Reading

Miller WL, Fluck CE, Breault DT, Feldman BJ. The adrenal cortex and its disorders. In: Sperling MA, Majzoub JA, Menon RK, Stratakis CA, editors. Pediatric endocrinology. 5th ed. Elsevier; 2021. p. 425–90.

Hughes CR, Man E, Achermann JC. The adrenal cortex and its disorders. In: Dattani MT, Brook CGD, editors. Brooks clinical pediatric endocrinology. 7th ed. Wiley-Blackwell; 2020. p. 335–407.

Vats P, Dabas A, Jain V, Seth A, Yadav S, Kabra M, Gupta N, Singh P, Sharma R, Kumar R, Polipalli SK, Batra P, Thelma BK, Kapoor S. Newborn screening and diagnosis of infants with congenital adrenal hyperplasia. Indian Pediatr. 2020;57(1):49–55.

Riedl S, Röhl FW, Bonfig W, Brämswig J, Richter-Unruh A, Fricke-Otto S, Bettendorf M, Riepe F, Kriegshäuser G, Schönau E, Even G, Hauffa B, Dörr HG, Holl RW, Mohnike K, AQUAPE CAH Study Group. Genotype/phenotype correlations in 538 congenital adrenal hyperplasia patients from Germany and Austria: discordances in milder genotypes and in screened versus prescreening patients. Endocr Connect. 2019;8(2):86–94.

Walia R, Singla M, Vaiphei K, Kumar S, Bhansali A. Disorders of sex development: a study of 194 cases. Endocr Connect. 2018;7:364–71.

Merke DP, Auchus RJ. Congenital adrenal hyperplasia due to 21-hydroxylase deficiency. N Engl J Med. 2020;383(13):1248–61.

Schröder MAM, Claahsen-van der Grinten HL. Novel treatments for congenital adrenal hyperplasia. Rev Endocr Metab Disord. 2022;23(3):631–45.

Nordenström A, Falhammar H, Lajic S. Current and novel treatment strategies in children with congenital adrenal hyperplasia. Horm Res Paediatr. 2022;

Speiser PW, Arlt W, Auchus RJ, Baskin LS, Conway GS, Merke DP, Meyer-Bahlburg HFL, Miller WL, Murad MH, Oberfield SE, White PC. Congenital adrenal hyperplasia due to steroid 21-hydroxylase deficiency: an endocrine society clinical practice guideline. J Clin Endocrinol Metab. 2018;103(11):4043–88.

Baumgartner-Parzer S, Witsch-Baumgartner M, Hoeppner W. EMQN best practice guidelines for molecular genetic testing and reporting of 21-hydroxylase deficiency. Eur J Hum Genet. 2020;28(10):1341–67.

Prentice P. Guideline review: congenital adrenal hyperplasia clinical practice guideline 2018. Arch Dis Child Educ Pract Ed. 2021;106(6):354–7.

Claahsen-van der Grinten HL, Speiser PW, Ahmed SF, Arlt W, Auchus RJ, Falhammar H, Flück CE, Guasti L, Huebner A, Kortmann BBM, Krone N, Merke DP, Miller WL, Nordenström A, Reisch N, Sandberg DE, Stikkelbroeck NMML, Touraine P, Utari A, Wudy SA, White PC. Congenital adrenal hyperplasia-current insights in pathophysiology, diagnostics, and management. Endocr Rev. 2022;43(1):91–159.

Mallappa A, Merke DP. Management challenges and therapeutic advances in congenital adrenal hyperplasia. Nat Rev Endocrinol. 2022;18(6):337–52.

Witchel SF. Congenital adrenal hyperplasia. J Pediatr Adolesc Gynecol. 2017;30(5):520–34.

Auchus RJ. The uncommon forms of congenital adrenal hyperplasia. Curr Opin Endocrinol Diabetes Obes. 2022;29(3):263–70.

Principles of Surgery for Congenital Adrenal Hyperplasia

31

Shilpa Sharma

Abstract

The objectives of surgical repair are to create a cosmetic appealing and functional external genitalia for girls with congenital adrenal hyperplasia (CAH) who have virilized genitalia. The surgical treatment of CAH is a controversial issue. There is no consensus regarding the timing of the repair, the indications for repair and also the surgical techniques adopted for feminizing genitoplasty. The type of repair and the need for redo procedures also vary. This chapter describes the basic principles regarding surgical repair aspects for CAH.

Keywords

Female genitoplasty · Vaginoplasty · Clitoroplasty

31.1 Introduction

The aims for surgical repair in congenital adrenal hyperplasia (CAH) are to create a feminine-appearing external genitalia that can serve the purpose of sexual, reproductive, and urinary function. The concerns remain regarding need for additional procedures, gender dysphoria in future, psychosocial effects, and sexual functional outcomes. Reconstructive genital surgery has been tried and tested since a long time. However, in 2012, the UN Report of the Special Rapporteur on degrading treatment has brought to limelight the importance of no forced gender-confirming surgery [1].

In 2017, the Council of Europe emphasized the right of physical integrity in children with unknown gender identity [2]. This point of view has been adopted in

S. Sharma (✉)
Department of Pediatric Surgery, All India Institute of Medical Sciences, New Delhi, India

S. K. Ratan (ed.), *Children with Differences in Sex Development*,
https://doi.org/10.1007/978-981-97-1639-5_31

Differences in Sexual Development (DSD) management. However, there is a tendency to exclude CAH from the term DSD [3]. On the other hand, many families of patients with CAH support early surgery [4].

31.2 Need for Surgical Intervention

The need for any surgical intervention depends on sociocultural conditions. In a country like India, there is a social emergency to offer the appropriate gender deciding surgery in an individualized and judicial manner. The patients' and parents' opinion are essential and vary from place to place.

The current recommendation in some countries is to perform surgery only in severe virilized CAH (Prader III, IV, and V) [5]. The surgery may be deferred in those with mild virilization. Recently, a study suggested deferring genital surgery into adolescence [6].

There are difficult ethical issues involved when parents' consent for surgeries lead to permanent damage to the genital sensations [7].

31.3 Timing of Surgery

The is a bimodal peak in the ages at which children undergo reconstruction, one in early age up to 2 years of age and then at puberty. Those against early surgery are concerned regarding the further surgeries and a lack of follow-up on function. On the other hand, proponents of early surgery feel there are psychosocial benefits for child and parents provided by anatomy appropriate with the gender and later minor procedures as compared to initial reconstruction [8]. In a German consensus paper, young CAH patients agreed with early surgery [9].

31.4 Aims of Surgical Repair for Female Genitoplasty

The surgical repair should focus on the following:

1. Clitoroplasty—Reduction in size of visible clitoris to make it invisible on first inspection of the genitalia and only visible on gentle labial separation.
2. Vaginoplasty—The introitus should be appropriate for age and visible as a separate opening. If the child is nearing puberty, the introital size should be adequate to allow for egress of menstrual fluid. When the child reaches adulthood, the introital size should allow intercourse without dyspareunia.
3. Vestibuloplasty—The vagina should be placed within the boundaries of the vestibule and lined by mucosa for at least the anterior 50%.
4. Labioplasty—The labia minora should preferably cover the introitus on either side for at least half the opening and mucosa lined. The labia majora should cover the introitus on either side.
5. Monsplasty—The position of the clitoris should be above the lower part of the pubic symphysis covered by the mons pubis.

31.5 Staged Versus Single Stage Surgery

The basic complete surgical procedure involves clitoroplasty, vaginoplasty, vestibuloplasty, monsplasty, and labioplasty. These procedures may be done in one stage or in different stages. The clitoroplasty is the most essential part of the procedure for female gender assignment and female genitoplasty. The vestibuloplasty with exteriorization of the vagina may just be a simple V-Y Barrows vaginoplasty.

Early surgery done in infancy and up to 2 years of age is preferred by most surgeons [10]. For clitoroplasty, partial excision of the corpora cavernosa and the skin flap or "U flap" vaginoplasty is done [10]. Routine vaginal dilatations after puberty were done by 28% [10]. More than 75% reported good outcomes [10]. In most studies, majority of the cases are done between 1 and 4 years of age [8, 11].

A large international study revealed geographical and temporal differences [12]. The outcome of early clitoral surgery at puberty has been reported as unsatisfactory (clitoral atrophy or prominent glans) [13]. Fibrosis and scarring were most evident in those who had undergone aggressive attempts at vaginal reconstruction in infancy. The long-term evaluation of feminizing genitoplasty and psychosocial aspects in these patients are important [14]. The author has described a method for assessing the results of clitoroplasty and vaginoplasty in these patients and published it earlier in 2006 [14]. These are outlined in Tables 31.1 and 31.2. The author has used size of the visible clitoris to assess cosmetic outcome for clitoroplasty and Hegar's dilators to assess the functional outcome following vaginoplasty [14].

Table 31.1 Cosmetic and functional assessment system for clitoroplasty (modified and reprinted with permission from Springer from [14])

Outcome	Excellent	Satisfactory	Poor
Clitoral visibility (cosmetic outcome)	Not visible or just visible without labial separation	May be visible without labial separation	Atrophied
Clitoral size in cm (cosmetic outcome)	0.5–1	Minor deviation from expected size 0.3–0.4 or 1–1.5	Large size >1.5
Sensations (functional outcome)	Preserved	Decreased	Absent

Table 31.2 Cosmetic and functional assessment system for vaginoplasty (reprinted with permission from Springer from [14])

Visible vaginal orifice	Adequate introitus	Stenosed introitus	Strictured (fibrosed) introitus
Cosmetic outcome	Good	Fair	Poor
Vaginal diameter assessed by Hegar dilator size (#)	>16	10–16	<10
Therapy	Nil	Dilatation	Redo vaginoplasty
Functional outcome	Satisfactory	Satisfactory	Unsatisfactory

31.6 Male Genitoplasty for 46XX CAH Reared as Boys

Cases of 46XX CAH presenting late with almost complete masculinization may be subjected to male genitoplasty. Though there are reports of gender dysphoria in some, most patients have a good long-term psychosocial outcome [15–17].

References

1. Méndez J. Report of the special rapporteur on torture and other cruel, inhuman or degrading treatment or punishment. Human Rights Council. United Nations Digital Library; 2013.
2. https://www.ohchr.org/EN/Issues/LGBTI/Pages/IntersexPeople.aspx, 2016; Parliamentary Assembly of the Council of Europe, Resolution 2191. https://assembly.coe.int/nw/xml/XRef/Xref-XML2HTML-en.asp?fileid=24232&lang=en
3. Szymanski KM, Rink RC, Whittam B, Hensel DJ. Majority of females with a life-long experience of CAH and parents do not consider females with CAH to be intersex. J Pediatr Urol. 2021;17:210.e1–9.
4. Shalaby M, Chandran H, Elford S, Kirk J, McCarthy L. Recommendations of patients and families of girls with 46,XX congenital adrenal hyperplasia in the United Kingdom regarding the timing of surgery. Pediatr Surg Int. 2021;37:137–43.
5. Hughes IA, Houk C, Ahmed SF, Lee PA. Consensus statement on management of intersex disorders. J Pediatr Urol. 2006;2:148–62.
6. Bougneres P, Bouvattier C, Cartigny M, Michala L. Deferring surgical treatment of ambiguous genitalia into adolescence in girls with 21-hydroxylase deficiency: a feasibility study. Int J Pediatr Endocrinol. 2017;2017:3.
7. Schoer MB, Nguyen PN, Merritt DF, Wesevich VG, Hollander AS. The role of patient advocacy and the declining rate of clitoroplasty in 46,XX patients with congenital adrenal hyperplasia. Clin Pediatr (Phila). 2018;57:1664–71.
8. DaJusta D, Xu L, Baker L. 535 Frequency of feminizing genitoplasty for congenital adrenal hyperplasia with the geographical distribution of surgeries in the US. J Urol. 2010;183(4S):e211.
9. Krege S, Eckoldt F, Richter-Unruh A, Köhler B, Leuschner I, Mentzel HJ, et al. Variations of sex development: The first German interdisciplinary consensus paper. J Pediatr Urol. 2019;15:114–23.
10. Yankovic F, Cherian A, Steven L, Mathur A, Cuckow P. Current practice in feminizing surgery for congenital adrenal hyperplasia. A specialist survey. J Pediatr Urol. 2013;9:1103–7.
11. Sturm RM, Durbin-Johnson B, Kurzrock EA. Congenital adrenal hyperplasia: current surgical management at academic medical centers in the United States. J Urol. 2015;193(5 Suppl):1796–801.
12. Hebenstreit D, Ahmed SF, Krone N, Krall C, Bryce J, Alvi S, et al. Surgical practice in girls with congenital adrenal hyperplasia: an international registry study. Sex Dev. 2021;15:229–35.
13. Alizai NK, Thomas DF, Lilford RJ, Batchelor AG, Johnson N. Feminizing genitoplasty for congenital adrenal hyperplasia: what happens at puberty? J Urol. 1999;161(5):1588–91.
14. Gupta DK, Sharma S, Amini AC, Gupta M, Aggarwal G, Gupta D. Congenital adrenal hyperplasia: long-term evaluation of feminizing genitoplasty and psychosocial aspects. Pediatr Surg Int. 2006;22:905–9.
15. de Jesus LE, Costa EC, Dekermacher S. Gender dysphoria and XX congenital adrenal hyperplasia: how frequent is it? Is male-sex rearing a good idea? J Pediatr Surg. 2019;8(54):2421–7.
16. Khattab A, Yau M, Qamar A, Gangishetti P, Barhen A, Al-Malki S, et al. Long term outcomes in 46,XX adult patients with congenital adrenal hyperplasia reared as males. J Steroid Biochem Mol Biol. 2017;165:12–7.
17. Sharma S. Gupta DK Male genitoplasty for 46 XX congenital adrenal hyperplasia patients presenting late and reared as males. Indian J Endocrinol Metab. 2012;16(6):935–8.

46, XX Testicular DSD

32

Medha Mittal and Vandana Jain

Abstract

46, XX testicular DSD is the presence of testes in an individual with 46, XX karyotype. In over 80%, the condition results from translocation of *SRY* to X chromosome during paternal meiosis. The molecular defect in the *SRY* negative cases can be highly variable ranging from a duplication of pro-testicular genes (such as *SOX9* or *SOX3*), deletion of pro-ovarian genes (such as *WNT4* or *RSPO1*), to defects in genes such as *NR5A1* and *WT1*. The affected individuals may have typically male genitalia, with only small testes, presenting in adolescence or adulthood, or may have atypical genitalia with hypospadias or cryptorchidism presenting in neonatal period or infancy. Azoospermia and infertility are seen in all. We describe this rare entity with case illustrations and an algorithmic approach to diagnosis.

Keywords

SRY translocation · Ovotesticular · Atypical genitalia

M. Mittal
Department of Pediatrics, Chacha Nehru Bal Chikitsalaya, Delhi, India

V. Jain (✉)
Division of Pediatric Endocrinology, Department of Pediatrics, All India Institute of Medical Sciences, New Delhi, India
e-mail: drvandanajain@aiims.edu

S. K. Ratan (ed.), *Children with Differences in Sex Development*,
https://doi.org/10.1007/978-981-97-1639-5_32

32.1 Introduction

Differences in sex development in individuals with 46, XX karyotype arise due to exposure to androgens. The majority of 46, XX DSD is constituted by congenital adrenal hyperplasia, with the adrenals being the source of androgens. However, a small but significant subset comprises individuals with ovotesticular or testicular DSD, where testicular tissue is the source of androgens.

46, XX testicular DSD is the presence of testes in an individual with 46, XX karyotype, irrespective of the functional capacity of the testes (of producing androgens and/or anti-mullerian hormone (AMH) or the structure being dysgenetic. The phenotype is quite heterogeneous and may be dependent on the underlying genetic etiology. It can range from normal male external genitalia, presenting in adolescence with concerns of small testes, with or without impaired pubertal development or gynecomastia, or in adulthood with infertility, to presentation in the neonatal period as undervirilized males with penoscrotal hypospadias and undescended testes [1]. Over 80% of individuals with 46, XX testicular DSD are *SRY* positive (have the gene that codes for testis determining factor on Y chromosome), and typically have normal male genitalia, while the remaining who are negative for *SRY* have other molecular defects (discussed later), and have varying degrees of masculinization [1, 2].

The condition was first described by de la Chapelle et al. in the year 1964 and is quite rare, with an estimated incidence of 1 in 20,000 males [3]. Smith and colleagues proposed in 1966 that the underlying mechanism could be translocation of Y chromosome material containing *SRY* to X chromosome [4]. Later studies, by de la Chapelle (in 1984), Mc Elreavey et al., and others, confirmed the presence of Y chromosome material in phenotypic males with XX karyotype [5, 6]. *SRY* negative cases were described by Berkovitz and Villain et al. in 1994 [7, 8]. Since 2010 much advance has been made into the understanding of *SRY* negative cases with role of specific genes been described. The previous nomenclatures of de la Chapelle syndrome, 46, XX male sex reversal syndrome, or true hermaphroditism were discarded in favor of 46, XX testicular DSD in the 2005 Chicago consensus [9].

32.2 Clinical Presentation

The following two cases illustrate the heterogeneity of clinical presentation.

Case 1: A 17-year-old boy presented with concerns of small testicular size and gynecomastia. He was a product of non-consanguineous marriage, born at term after an uneventful pregnancy. On examination, his weight was 67 kg, height was 162 cm, arm span 163 cm, upper segment/lower segment 0.8, penile length 9 cm, meatus at tip, bilateral testes were scrotal, 4 ml in volume and firm, pubic hair was Tanner stage 4. Bilateral nontender gynecomastia with a diameter of 5 cm was present.

Hormone analysis showed LH of 19 mIU/mL, FSH 38 mIU/mL, testosterone 99 ng/dL (suggesting hypergonadotropic hypogonadism). Ultrasonography showed absence of Müllerian structures, with normal prostate gland and seminal vesicles. Karyotype was 46, XX and FISH showed that he was *SRY* positive. A diagnosis of 46, XX testicular DSD was made, the adolescent was started on testosterone replacement and given the option of reduction mammoplasty, and family was counseled regarding infertility.

Case 2: A 14-month-old baby, first in birth order, reared as a girl, was brought with atypical genitalia noticed since birth. Growth and development were normal, and there was no family history of atypical genitalia in any member of the family. On examination, weight and length were normal, with no dysmorphism or hyperpigmentation. The genitalia consisted of symmetrical bilateral hypoplastic labioscrotal folds with rugosity, stretched genital tubercle length of 2 cm, urethral meatus and vaginal opening in perineum, and no palpable gonads (Prader 2). Systemic examination was unremarkable. 17-hydroxyprogesterone was normal (0.65 ng/ml). Hormonal profile was normal for age as the baby had crossed the age of minipuberty (LH 0.42mIU/mL, FSH 2.8mIU/mL, AMH 17.5 ng/mL, estradiol <10 pg/mL, testosterone 0.04 ng/mL). An ultrasound and subsequent MRI pelvis defined the presence of uterus, with testes anterolateral to the bladder on either side, and absence of ovaries. Karyotype was 46, XX and on PCR, *SRY* gene was absent. A diagnosis of 46, XX (*SRY* negative) testicular DSD was thus made. A whole genome array for molecular characterization of the genetic cause was planned. The family was counseled in detail regarding the need for multidisciplinary treatment. An assessment of testoserone levels after stimulation with inj. human chorionic gonadotropin was also planned. With demonstrable functionality of testicular tissue, gender assignment would need thoughtful deliberation.

Testicular and ovotesticular DSD represent a continuum of presenting features with existence of different phenotypes in the same family [10]. The testes are typically small owing to arrested meiosis and germ cell degeneration. In normal XY males, Sertoli cells proliferate and mature at puberty and trigger spermatogenesis, responsible for the increase in testicular size. In XX males, these are lacking causing the testes to remain small. They may be soft in early childhood and become firmer with age. The testicular tissue present is always dysgenetic, and azoospermia and infertility are almost universal [11]. However, this testicular tissue causes variable degree of virilization, ranging from mild to nearly complete. A combination of biochemical testing, imaging, and surgical exploration help to identify the presence of ovarian and testicular tissue.

In a review of 37 papers describing 64 patients of 46, XX testicular DSD with mean (SD) age of 33.1 (11.4) years, penis size was normal in 26/32 (81.2%) cases, erectile function was normal in 27/30 (90%) cases, and libido was preserved in 20/20 (100%) cases. However, testes volume was small in 14/14 cases, and gynecomastia was present in 17/39 (44%) cases. Hypergonadotropic hypogonadism was universal, with mean (SD) FSH 40.00 (19.7) mIU/mL, mean (SD) LH 26.5 (16.7) mIU/mL, and mean (SD) total testosterone 2.6 (1.4) ng/mL. *SRY* was present in 51/57 and absent in the rest 6 [2].

SRY positive XX testicular DSD typically has completely normal looking male genitalia. Boys may present with short stature around the age of puberty, the height being similar to that of girls of the same age. The testes fail to enlarge at puberty due to arrested spermatogenesis. Very often the condition is detected only later during work up of infertility. Since the Y chromosome present is only the short arm (Yp), there is universal azoospermia and poor semen volume. Gynecomastia is less common than in *SRY* negative cases. Testosterone levels are low and gonadotropins high. AMH is usually normal. About 10% of ovotesticular DSD also has *SRY* translocation.

XX males, who are *SRY* negative are more likely to have poorly functioning testicular tissue with incomplete virilization, with varying degrees of hypospadias and cryptorchidism, to female appearing genitalia with clitoromegaly and labial fusion, with presence of Müllerian structures. The presence of atypical genitalia brings them to medical attention at an earlier age. *SRY* negative 46, XX individuals are more likely to have ovotesticular DSD. Seminiferous tubule dysfunction leads to low inhibin B, with consequent rise in FSH due to loss of feedback inhibition, and gynecomastia at puberty due to increased conversion of testosterone to estradiol by FSH. There are typical consequences of decreased testosterone production with reduced secondary sexual characters (sparse body hair, low muscle mass), low libido and possible erectile dysfunction, and increased risk of osteopenia. The hormone levels depend on the amount of functional gonadal tissue.

32.3 Genetic Basis of 46XX Testicular DSD

In early embryonic life, the bipotential gonadal ridges differentiate from the mesonephric mesoderm and coelomic epithelium, under the regulation of *CBX2*, *SF1*, *WT1*, *DMRT1*, and *DMRT2* genes. In the 7th week of gestation, the undifferentiated gonad differentiates to form either the testis or ovary. Sex determining region Y (*SRY*) gene present on the Y chromosome is the major regulator that tips the balance of the complex interplay of various genes towards formation of the testis [10]. It regulates the expression of *SOX9* in Sertoli cell precursors which in turn activates the other testis-specific genes, *PGD2* and *FGF9* in a positive feedback loop. At the same time, the enhanced *SOX9* expression represses the pro-ovarian genes, *DAX1*, *FOXL2*, *WNT4*, and *RSPO1*, promoting testicular development. Presence of *SRY* in 46, XX fetuses is the underlying genetic basis of 46,XX testicular DSD in the majority of cases though the condition could develop due to other mechanisms as detailed below [11].

32.3.1 SRY Positive Cases

SRY maps to Yp11.2 adjacent to the pseudoautosomal region 1 of the Y chromosome. It encodes a transcription factor belonging to the high mobility group of the DNA binding proteins and is the major factor determining testicular differentiation.

Translocation of the *SRY* gene onto the short arm of X chromosome due to an unequal Yp to Xp chromosomal interchange during meiosis in the father (as a de novo mutation) is responsible for 80–90% of the XX testicular DSD cases. Rarely, *SRY* may be translocated onto an autosome. These cases usually have male phenotype with typical male genitalia. However, the genes responsible for sperm production are located on Yq; therefore, even with *SRY* translocation these individuals have small testes and azoospermia.

32.3.2 SRY Negative Cases

46, XX *SRY* negative cases have an overexpression of pro-testicular genes or an underexpression of pro-ovarian genes or certain unknown pathogenic mechanisms [10, 12]. Copy number variations in *SOX9* and *SOX3* genes account for a majority of *SRY* negative cases.

Overexpression of pro-testicular factors

1. *SOX9* (*SRY* related HMG box 9) gene is located on chromosome 17 and initiates testis differentiation and AMH production. It is key to maintaining the integrity of Sertoli cells. Duplication of this gene causes expression of male phenotype even in the absence of *SRY* [13, 14]. Duplication or triplication of *SOX9* regulatory sequences can result in similar effect.
2. *SOX3* maps to Xq27.1 and codes for a protein required for normal brain, pituitary, and craniofacial development. Although it does not initiate testicular differentiation in normal circumstances, its overexpression could cause testicular development, mainly by triggering *SOX9* overexpression [15]. Duplications of this gene have been identified in XX *SRY* negative ovotesticular DSD cases. Hypoplasia of anterior pituitary and ectopic position of posterior pituitary were also present in two patients [16].
3. *SOX10* maps to chromosome 22q13.1 and though not involved in normal gonadal development, its overexpression in cases of partial or complete duplication of chromosome 22 may result in ovotestes, ovarian dysgenesis, or testicular differentiation. The external genitalia could be male type or atypical. Affected patients have been described to have various other associated conditions such as cleft palate, heart defects, peripheral demyelinating neuropathy, central demyelinating leukodystrophy, Waardenburg syndrome, and Hirschsprung's disease [17].
4. Fibroblast growth factor (FGF) family is involved in development of gonadal ridges apart from limbs, lungs, and adenohypophysis. Located on 13q 11-q12, duplication of the locus encompassing the entire *FGF9* gene has been reported in XX *SRY* negative males with azoospermia [18].
5. *DMRT1* mapping on to 9p24.3 upregulates many pro-testicular genes. Overexpression of *DMRT1* gene in gonadal ridges has been shown to induce testicular differentiation in mice, though not yet described in humans.

Underexpression of pro-ovarian factors

1. *WNT4*—This maps to 1p36.12 and promotes müllerian duct formation. Along with *RSPO1*, it reduces the degradation of β-catenin thereby increasing its levels. This antagonizes *SOX9* (pro-testes factor) which antagonizes steroidogenic fac-

tor 1 (SF1). Heterozygous loss of function mutations have been reported in three 46, XX patients with mild virilization but normal external genitalia [10]. Inactivation of both copies of the gene in 46, XX karyotype leads to SEx Reversion, Kidneys, Adrenal and Lung dysgenesis (SERKAL) syndrome with testicular or ovotesticular DSD, ambiguous genitalia, renal agenesis, adrenal hypoplasia, and pulmonary and cardiac abnormalities that are incompatible with survival [19].
2. *RSPO1*—One of the members of the R-spondin family and located at 1p34.1, this gene synergizes with *WNT4* to stabilize β-catenin. Loss of function mutations of *RSPO1* lead to 46, XX DSD with absence of Müllerian structures, atypical genitalia and palmoplantar hyperkeratosis [20].

Other genes

1. *NR5A1*, present on 9q33.3, encodes steroidogenic factor 1(SF1). This is essential for gonadal development in both males and females. In the XY embryo, it acts synergistically with *SRY* and upregulates *SOX9*, initiating male gonadal development. On the other hand, in XX embryo, it induces *WNT4* and *RSPO1* expression, upregulates *FOXL2* and *DAX1*, thus promoting ovarian development. Mutations that result in loss of function of the gene have been identified in XX *SRY* negative testicular or ovotesticular DSD [21].
2. *NR2F2* maps to 15q26.2 and encodes a transcription factor that has an anti-testis and pro-ovarian role, besides other morphogenetic functions. Frameshift mutations in this gene have been described to cause testicular/ovotesticular DSD, with other anomalies that included congenital heart disease and congenital diaphragmatic hernia [22].
3. *WT1* maps to 11p13 and its expression is essential for normal urogenital development, lineage specification of Sertoli cells of the testis, and granulosa cells of the ovary. Its classic pathogenic variants are associated with anomalies of testis and renal development and are important causes of XY DSD, steroid resistant nephrotic syndrome, and increased propensity for Wilm's tumor and gonadoblastoma. Recently, de novo frameshift mutation *WT1*p.R485G has been identified in an XX *SRY* negative patient with clitoromegaly, single perineal opening, and a short blind-ending vagina and dysgenetic testes. Lack of *WT1* would lead to increased SF1 that upregulates *SOX9* and represses *FOXL2* [23].

32.4 Approach to Diagnosis

Often the working diagnosis in a young male with small testes, gynecomastia, and hypergonadotropic hypogonadism is Klinefelter Syndrome (KFS). However, those with KFS are relatively tall with long limbs in contrast to 46, XX males who are short. The latter may also have cryptorchidism. On the other hand, the initial diagnosis may well be congenital adrenal hyperplasia in female looking genitalia with enlarged "clitoris" and sequential investigations may finally lead to the correct diagnosis as in case 2. Hence, it is essential to carefully tread upon the trail of work up as suggested below.

1. Thorough age-appropriate history—Increase in size of penis and testis with age, history of infertility, erectile function, libido in adults.
2. Careful clinical examination—Height, arm span, any associated anomalies, presence of hyperpigmentation, hernial orifices, gynecomastia, Tanner staging.
3. Examination of genitalia—Labioscrotal folds, rugosity, stretched penile/genital tubercle length, position of urethral meatus and vaginal opening, symmetry of genitalia/gonads, presence of gonads at labioscrotal folds, testes volume and consistency, position of anal opening, anogenital distance, Prader score, external masculinization score (scoring each of glans size, labioscrotal fusion, position of gonads, and position of urinary meatus).
4. Abdominal and pelvic ultrasound and/or MRI—To see for presence and anomalies of Müllerian structures and testes.
5. Hormone profile—Serum 17-hydroxyprogesterone and other adrenal steroids (if suspicion of congenital adrenal hyperplasia), LH, FSH, testosterone, AMH, estradiol and prolactin.
6. Genetic analysis—Karyotype of 100 metaphase lymphocytes and detection of SRY by fluorescent in situ hybridization (FISH) using *SRY* gene probe or polymerase chain reaction (PCR) or on chromosomal microarray (CMA). Further, the rearrangement present (on X or autosome) is detected as that would be important for genetic counseling.

In case *SRY* is absent, molecular genetic testing can include targeted and comprehensive genetic testing (Fig. 32.1).

(a) Targeted gene testing can target single gene or multiple genes using techniques of sequence analysis, deletion/duplication analysis, and non-sequencing based tests. Sequence analysis can detect small intragenic deletions/insertions and variants but cannot typically detect exon or whole gene deletion/duplications or its chromosomal location. Also, it may be expensive to go for repeated single or multiple gene testing. Hence, comprehensive testing is the preferred modality.

(b) Chromosomal microarray (CMA) has excellent diagnostic value in detecting copy number variants and is especially useful in syndromic cases [24]. For *SRY* negative cases, it is useful to go for CMA first, as copy number variants of *SOX9* and *SOX3* are the common etiologies. Genome sequencing, typically performed by next generation sequencing, is used if CMA is not able to yield a definitive diagnosis, mainly in research settings [25]. It can detect single nucleotide variants as well as chromosome breakpoints that are in non-coding regions. However, it has a lower sensitivity to identify copy number variants as compared to CMA and it can generate a lot of data that needs careful interpretation by an experienced molecular geneticist.

7. Gonadal biopsy—Testicular biopsy may show decreased number and hyalinization of seminiferous tubules, peritubular fibrosis, absence of germ cells, and hyperplasia of Leydig cells. Immunohistochemical analysis with monoclonal

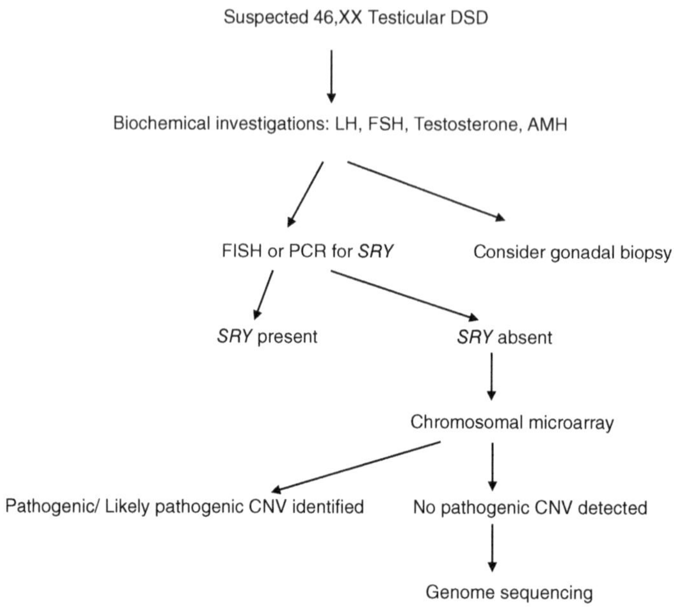

Fig 32.1 Suggested approach towards work up of 46XX testicular DSD

antibodies against inhibin and vimentin can detect presence of Leydig cells to identify testicular tissue in case of ovotestes.

8. If presentation is in adulthood for infertility, semen analysis for sperm count will be an additional test.
9. Investigations for associated urogenital anomalies and other somatic anomalies as may be required.

32.5 Management

In testicular DSD, the social sex is most often male, with genitalia being unambiguously male in a majority of cases. In ovotesticular DSD, the clinical phenotype would depend on the predominance of ovarian or testicular tissue. In testicular predominance, the genitalia is male with signs of hypovirilization (hypospadias or cryptorchidism) and the social sex is still usually male. Testosterone replacement is the mainstay of treatment for induction of development of secondary sexual characters, correct hormonal imbalance, and prevent gynecomastia. It is typically started at or after the age of puberty onset. Removal of functional ovarian tissue also helps to avoid gynecomastia and other effects of estradiol production. However, such major surgical procedure should be undertaken only after the legal age of consent. Reduction mammoplasty may be considered if gynecomastia is significant. Ultrasonography, micturating cystogram, and cystoscopy are useful to define the urogenital abnormalities. Hypospadias and cryptorchidism may need surgical intervention. Though sex assignment is usually male, counseling regarding sexual physiology, orientation, erectile function is helpful for overall well-being of the individual. In cases where gender assignment is female, the testicular tissue would need to be removed after the age of consent. Before that, medical management with pubertal suppression may be needed to avoid virilization during puberty. In view of infertility, the individuals may consider adoption or in vitro fertilization with a sperm donor. The risk of tumor development in testicular tissue is reported to be low even when the tissue is dysgenetic as the Y sequences are absent [26].

Consultation with an experienced genetic expert and psychological team are essential components of multidisciplinary management. Even though awareness about the issues faced by people with DSD is increasing, they continue to face stigma and discrimination. A sensitive and inclusive approach of the society will improve their overall well-being.

32.6 Summary

46, XX testicular DSD is quite rare but clinicians need to keep this entity in mind especially when faced with a mismatch of phenotypic, gonadal, and genetic sex. They should also keep this in mind when investigating cases of gynecomastia and infertility. FISH and PCR can accurately detect the presence of *SRY* which is present in the majority of the cases. A multidisciplinary management to address the surgical issues, hormone replacement, and psychological concerns of the patients is needed.

References

1. Vilain E. 46, XX testicular disorder of sex development. Gene Rev Last Updat. 2009;26:38–41.
2. Terribile M, Stizzo M, Manfredi C, Quattrone C, Bottone F, Giordano DR, et al. 46XX Testicular disorder of sex development (DSD): a case report and systematic review. Medicina (Kaunas). 2019;55(7):371.
3. de la Chapelle A, Hortling H, Niemi M, Wennström J. XX sex chromosomes in a human male. First case. Acta Med Scand. 1964;175(Suppl 412):25–8.
4. Ferguson-Smith MA. X-Y chromosomal interchange in the aetiology of true hermaphroditism and of XX Klinefelter's syndrome. Lancet. 1966;2(7461):475–6.
5. de la Chapelle A, Tippett PA, Wettestrand G, Page D. Genetic evidence of X-Y interchange in human XX male. Nature. 1984;307:170–1.
6. McElreavey K, Rappaport R, Vilain E, Abbas N, Richaud F, Lortat-Jacob S, et al. A minority of 46, XX true hermaphrodites are positive for the Y-DNA sequence including SRY. Hum Genet. 1992;90(1–2):121–5.
7. Berkovitz GD. Abnormalities of gonad determination and differentiation. Sem Perinatology. 1992;16:289–98.
8. Vilain E, le Fiblec B, Morichon-Delvallez N, Brauner R, Dommergues M, Dumez Y, et al. SRY-negative XX fetus with complete male phenotype. Lancet. 1994;343(8891):240–1.
9. Lee PA, Houk CP, Ahmed SF, Hughes IA. International Consensus Conference on Intersex organized by the Lawson Wilkins Pediatric Endocrine Society and the European Society for Paediatric Endocrinology. Consensus statement on management of intersex disorders. International Consensus Conference on Intersex. Pediatrics. 2006;118(2):e488–500.

10. Grinspon RP, Rey RA. Disorders of sex development with testicular differentiation in SRY-negative 46,XX individuals: clinical and genetic aspects. Sex Dev. 2016;10(1):1–11.
11. Syryn H, Van De Vijver K, Cools M. Ovotesticular difference of sex development: genetic background, histological features, and clinical management. Horm Res Paediatr. 2021;1:275–84.
12. Grinspon RP, Rey RA. Molecular characterization of XX maleness. Int J Mol Sci. 2019;20(23):6089.
13. Cox JJ, Willatt L, Homfray T, Woods CG. A SOX9 duplication and familial 46,XX developmental testicular disorder. N Engl J Med. 2011;364(1):91–3.
14. Vetro A, Dehghani MR, Kraoua L, Giorda R, Beri S, Cardarelli L, et al. Testis development in the absence of SRY: chromosomal rearrangements at SOX9 and SOX3. Eur J Hum Genet. 2015;23(8):1025–32.
15. Sutton E, Hughes J, White S, Sekido R, Tan J, Arboleda V, et al. Identification of SOX3 as an XX male sex reversal gene in mice and humans. J Clin Invest. 2011;121(1):328–41.
16. Woods KS, Cundall M, Turton J, Rizotti K, Mehta A, Palmer R, et al. Over- and underdosage of SOX3 is associated with infundibular hypoplasia and hypopituitarism. Am J Hum Genet. 2005;76(5):833–49.
17. Falah N, Posey JE, Thorson W, Benke P, Tekin M, Tarshish B, et al. 22q11.2q13 duplication including SOX10 causes sex-reversal and peripheral demyelinating neuropathy, central dysmyelinating leukodystrophy, Waardenburg syndrome, and Hirschsprung disease. Am J Med Genet A. 2017;173(4):1066–70.
18. Chiang HS, Wu YN, Wu CC, Hwang JL. Cytogenic and molecular analyses of 46,XX male syndrome with clinical comparison to other groups with testicular azoospermia of genetic origin. J Formos Med Assoc. 2013;112(2):72–8.
19. Mandel H, Shemer R, Borochowitz ZU, Okopnik M, Knopf C, Indelman M, et al. SERKAL syndrome: an autosomal-recessive disorder caused by a loss-of-function mutation in WNT4. Am J Hum Genet. 2008;82(1):39–47.
20. Parma P, Radi O, Vidal V, Chaboissier MC, Dellambra E, Valentini S, et al. R-spondin1 is essential in sex determination, skin differentiation and malignancy. Nat Genet. 2006;38(11):1304–9.
21. Bashamboo A, Donohoue PA, Vilain E, Rojo S, Calvel P, Seneviratne SN, et al. A recurrent p.Arg92Trp variant in steroidogenic factor-1 (NR5A1) can act as a molecular switch in human sex development. Hum Mol Genet. 2016;25(16):3446–53.
22. Bashamboo A, Eozenou C, Jorgensen A, Bignon-Topalovic J, Siffroi JP, Hyon C, et al. Loss of function of the nuclear receptor NR2F2, encoding COUP-TF2, causes testis development and cardiac defects in 46, XX Children. Am J Hum Genet. 2018;102(3):487–93.
23. Chen M, Zhang L, Cui X, Lin X, Li Y, Wang Y, et al. WT1 directs the lineage specification of Sertoli and granulosa cells by repressing SF1 expression. Development. 2017;144(1):44–53.
24. Délot EC, Papp JC; DSD-TRN Genetics Workgroup; Sandberg DE, Vilain E. Genetics of Disorders of Sex Development: The DSD-TRN Experience. Endocrinol Metab Clin North Am. 2017;46(2):519–37
25. Délot EC, Vilain E. Towards improved genetic diagnosis of human differences of sex development. Nat Rev Genet. 2021
26. Cools M, Looijenga LH, Wolffenbuttel KP, T'Sjoen G. Managing the risk of germ cell tumourigenesis in disorders of sex development patients. Endocr Dev. 2014;27:185–96.

33 Rarer Causes of 46 XX DSD

Mehak Sehgal, Meghna Kinjalk, and Simmi K. Ratan

Abstract

After excluding the common causes in a female child that presents with a disorder of sexual development, the less common causes, although very rare should be considered. In a newborn girl, that presents with clitoromegaly, a history of maternal virilization should be sought as this may be suggestive of a maternal tumor, exposure to androgenous drugs, or placental aromatase deficiency. In an older girl presenting with absent sexual characters and delayed onset of puberty with primary amenorrhea, the diagnosis of gonadal dysgenesis should be considered after excluding other conditions. These girls have a normal external and internal female phenotype but with premature ovarian failure, leading to bilateral streak gonads. Hormonal replacement is generally needed in these patients.

Keywords

Luteoma · Gestational hyperandrogenism · Gonadal dysgenesis · Amenorrhea

33.1 Maternal and Feto-Placental Causes

A baby girl born with ambiguous genitalia presents with a wide variety of possibilities as to what the diagnosis could be. Once the causes that require urgent management, such as the salt wasting type of congenital adrenal hyperplasia is ruled out, the baby is then evaluated for other differentials and an extensive workup is initiated to narrow in on the cause of virilization. The source of androgen causing virilization

M. Sehgal · M. Kinjalk · S. K. Ratan (✉)
Department of Pediatric Surgery, Maulana Azad Medical College, New Delhi, India

S. K. Ratan (ed.), *Children with Differences in Sex Development*,
https://doi.org/10.1007/978-981-97-1639-5_33

could be endogenous or exogenous. Exogenous source causing maternal gestational hyperandrogenism could be from:

1. Ovarian origin
2. Adrenal source
3. Exogenous exposure
4. Placental aromatization defects

The female fetus can be virilized to a variable extent if exposed to androgens between 8 and 14 weeks of gestation, as that is when the external genitalia development occurs [1]. After the 12th week, exposure to androgen may lead to clitoral hypertrophy, but not labial fusion [2]. It has been suggested that placental aromatization of androgens protects the developing fetus from excessive maternal androgens, thus serving as a "metabolic barrier" [3]. Any disturbance in this barrier or excessive androgen exposure within the sensitive time limit may serve to virilize the female fetus. After delivery, once the androgenic exposure ceases, there may be a reversal of the virilization in some cases, thus suggesting a conservative approach in some of these patients. Surgical management may be considered if the virilization persists.

Maternal androgen excess during pregnancy can lead to virilization. Some of the maternal causes include:

Ovarian origin:

1. *Polycystic ovary syndrome*: Have a high level of androgen but often not high enough to cause severe virilization [4].
2. *Hyperreactio luteinalis* or the cystic enlargement of the ovaries is caused by theca lutein cysts. These are bilateral and multiple in number. Large cysts are known to cause pressure symptoms. They may lead to virilization in the mother and fetus. These cysts are noted to regress after delivery. They are noted to be characteristically present in women with molar or multiple pregnancy [5, 6].
3. *Luteomas*: They generally present as large solid masses on ultrasound, often occurring bilaterally. These patients present with a high serum testosterone, dihydrotestosterone, and androstenedione level. They are often asymptomatic and may be detected on ultrasound. In some cases, they may present with compressive mass effect. They are the most common cause maternal virilization during pregnancy, causing virilization in the growing fetus, especially if exposed during the first trimester. These tumors can generally be observed, as they undergo spontaneous resolution in the postpartum phase [1, 7–9].
4. *Malignant ovarian tumors*: Tumors such as Sertoli–Leydig cell tumors, Krukenberg tumors, mucinous cystadenocarcinoma are rarely known to occur during pregnancy and present as a unilateral solid ovarian mass. These are often function tumors, secreting androgens and can cause fetal virilization [2, 10, 11]. The risk of malignancy is as high as 50% if a unilateral, solid mass is found in a virilized woman. The survival depends on the histology identified.

Adrenal source This includes adrenal adenoma and adrenocortical carcinoma. They are extremely rare cause of maternal and fetal virilization. These tumors are often cortisol secreting but can also secrete androgens. The effect of these hormones usually is not conducive for a pregnancy to reach term. Magnetic resonance imaging is needed to identify and confirm the diagnosis, which is often retrospective in nature [12–14].

Exogenous exposure Pregnant women receiving exogenous androgens, especially during first trimester can cause virilization of female fetus. Male fetus is generally not affected. Historically this was described in women receiving progestins such as norethindrone in high doses to prevent miscarriage [15–18].

Placental aromatase defects Resulting from an autosomal recessive disorder, mutation in P450 aromatase (CYP19A1) impairs the conversion of androgens to estrogens. This impairs the "metabolic barrier" and as the feto-placental unit cannot convert androgens to estrogens. This leads to increased androgen exposure in the developing fetus resulting in virilization. First described in Japan in 1991 [19], this rare condition should be considered especially when the mother displays signs of virilization occurring during pregnancy, which improve after delivery [20].

33.2 46XX Gonadal Dysgenesis

Introduction: Gonadal dysgenesis is characterized by incomplete or defective development of ovaries and resultant premature ovarian failure, in the presence of a female genotype and phenotype [21]. There is absence of anti-Mullerian hormone and testosterone, which allows Mullerian structure to develop and Wolffian duct to regress, respectively. These patients have female-like internal and external reproductive characteristic but lack secondary sexual features [22]. Characteristics features include:

- Phenotypically female, presenting late in adolescence with delayed or absent puberty and amenorrhea
- Chromosome 46XX
- Inheritance is autosomal recessive trait, mutations noted in BMP15 gene, FSH receptor gene, NR5A1 gene [23]
- Premature ovarian failure leading to bilateral streak gonads: Ovarian stroma without ovarian follicles
- External phenotype and internal duct structures are female
- Gonads do not have malignant potential
- Associated syndromes: Perrault syndrome (ovarian dysgenesis, deafness +/– cerebellar ataxia), lung fibrosis-immunodeficiency-gonadal dysgenesis syndrome [24]

Diagnosis: The presence of female external phenotype and normal internal structures. Karyotype of 46XX with gonadal biopsy suggestive of bilateral streak gonads.

Testosterone and estrogen levels are decreased and FSH, LH levels are raised. Clinical features may be suggestive of Turner's syndrome, differentiated on the basis of karyotype [22].

Treatment: The gonads can be left in situ. As there will be no endocrine action by streak gonads, hormone replacement (estrogen and progesterone) will be needed at puberty. These women are generally infertile [24].

References

1. Spitzer RF, Wherrett D, Chitayat D, Colgan T, Dodge JE, Salle JLP, et al. Maternal luteoma of pregnancy presenting with virilization of the female infant. J Obstet Gynaecol Can. 2007;29:835–40. https://doi.org/10.1016/S1701-2163(16)32642-1.
2. Gestational hyperandrogenism - UpToDate n.d.. https://www.uptodate.com/contents/gestational-hyperandrogenism#H3445011658 (accessed March 9, 2023).
3. Hensleigh PA, Carter RP, Grotjan HE. Fetal protection against masculinization with hyperreactio luteinalis and virilization. J Clin Endocrinol Metab. 1975;40:816–23. https://doi.org/10.1210/JCEM-40-5-816.
4. Glintborg D, Jensen RC, Bentsen K, Schmedes AV, Brandslund I, Kyhl HB, et al. Testosterone levels in third trimester in polycystic ovary syndrome: Odense child cohort. J Clin Endocrinol Metab. 2018;103:3819–27. https://doi.org/10.1210/JC.2018-00889.
5. Simsek Y, Celen S, Ustun Y, Danisman N, Bayramoglu H. Severe preeclampsia and fetal virilization in a spontaneous singleton pregnancy complicated by hyperreactio luteinalis. Eur Rev Med Pharmacol Sci. 2012;16:118–21.
6. Schnorr JA, Miller H, Davis JR, Hatch K, Seeds J. Hyperreactio luteinalis associated with pregnancy: a case report and review of the literature. Am J Perinatol. 1996;13:95–7. https://doi.org/10.1055/s-2007-994300.
7. Masarie K, Katz V, Balderston K. Pregnancy luteomas: clinical presentations and management strategies. Obstet Gynecol Surv. 2010;65:575–82. https://doi.org/10.1097/OGX.0B013E3181F8C41D.
8. Wadzinski TL, Altowaireb Y, Gupta R, Conroy R, Shoukri K. Luteoma of pregnancy associated with nearly complete virilization of genetically female twins. Endocr Pract. 2014;20:e18–23. https://doi.org/10.4158/EP13056.CR.
9. Rapisarda V, Pedalino F, Santonocito VC, Cavalli G, Zarbo G. Luteoma of pregnancy presenting with severe maternal virilisation: a case report. Case Rep Obstet Gynecol. 2016;2016:1–4. https://doi.org/10.1155/2016/3523760.
10. McClamrock HD, Adashi EY. Gestational hyperandrogenism. Fertil Steril. 1992;57:257–74. https://doi.org/10.1016/S0015-0282(16)54828-6.
11. Tinkanen H, Kuoppala T. Virilization during pregnancy caused by ovarian mucinous cystadenocarcinoma. Acta Obstet Gynecol Scand. 2001;80:476–7. https://doi.org/10.1034/J.1600-0412.2001.D01-19.X.
12. Kirk JMW, Perry LA, Shand WS, Kirby RS, Besser GM, Savage MO. Female pseudohermaphroditism due to a maternal adrenocortical tumor. J Clin Endocrinol Metab. 1990;70:1280–4. https://doi.org/10.1210/JCEM-70-5-1280.
13. Hoshino S, Obara K, Hoshii T, Kuroki H, Watanabe K, Tomita Y. Virilization of a female infant genitalia caused by a maternal androgen-producing adrenocortical tumor: a case report. Urol Case Rep. 2020:32. https://doi.org/10.1016/J.EUCR.2020.101253.
14. Morris LF, Park S, Daskivich T, Churchill BM, Rao CV, Lei Z, et al. Virilization of a female infant by a maternal adrenocortical carcinoma. Endocr Pract. 2011:17. https://doi.org/10.4158/EP10209.CR.
15. Jawetz E, Gunnison JB, King MB, Knox R, Woodroffe RC, et al. Fetal masculinization with maternal progesterone therapy. Can Med Assoc J. 1959;80:200.

16. Wilkins L. Masculinization of female fetus due to use of orally given progestins. JAMA J Am Med Assoc. 1960;172:1028–32. https://doi.org/10.1001/JAMA.1960.03020100036007.
17. Grumbach MM, Ducharme JR, Moloshok RE. On the fetal masculinizing action of certain oral progestins. J Clin Endocrinol Metab. 1959;19:1369–80. https://doi.org/10.1210/JCEM-19-11-1369.
18. Wilkins L, Jones HW, Holman GH, Stempfel RS. Masculinization of the female fetus associated with administration of oral and intramuscular progestins during gestation: non-adrenal female pseudohermaphrodism. J Clin Endocrinol Metab. 1958;18:559–85. https://doi.org/10.1210/JCEM-18-6-559.
19. Shozu M, Akasofu K, Harada T, Kubota Y. A new cause of female pseudohermaphroditism: placental aromatase deficiency. J Clin Endocrinol Metab. 1991;72:560–6. https://doi.org/10.1210/JCEM-72-3-560.
20. Alsaleem M, Miller DE, Saadeh L, Majumdar I. Aromatase deficiency: a rare cause of maternal virilisation and ambiguous genitalia in neonates. BMJ Case Rep. 2019:12. https://doi.org/10.1136/BCR-2019-231267.
21. Ambachew R, Gulilat A, Aberra T, Terefework Z, Bedilu W, Tarekegn G, et al. A rare case of 46,XX gonadal dysgenesis, Mayer–Rokitansky–Kuster–Hauser syndrome, pituitary and thyroid hypoplasia. Endocrinol Diabetes Metab Case Rep. 2022;2022 https://doi.org/10.1530/EDM-21-0103.
22. Pertusa S, Palacios A. Rare disease: 46 XX pure gonadal dysgenesis: an infrequent cause of primary amenorrhoea. BMJ Case Rep. 2009;2009 https://doi.org/10.1136/BCR.07.2008.0485.
23. Witchel SF, Lee PA. Ambiguous genitalia. In: Pediatric endocrinology. 4th ed. Elsevier; 2014. https://doi.org/10.1016/B978-1-4557-4858-7.00014-7.
24. Orphanet: 46,XX gonadal dysgenesis n.d.. https://www.orpha.net/consor/cgi-bin/OC_Exp.php?Expert=243 (accessed March 10, 2023).

Part VIII

Psycho-social Management of Individuals with DSD and their Caregivers

I whole heartedly thank Dr. Anju Gambhir, Consultant, Pediatric Surgery Fortis Hospital Shalimar Bagh, Delhi and Dr. Vishesh Jain Additional Professor, Paediatric Surgery, AIIMS New Delhi for assisting and aiding me in editing this section.

Psychological Counseling for DSD Family 34

Vijaya Raman and Suhas Chandran

Abstract

Disorders of Sexual Development (DSD) are identified at various points of time, such as at birth, early childhood, or puberty, when there is a departure from the usual course of sexual development. They are doubly challenging compared to other childhood disorders, as, in addition to medical complications they are also perceived as highly stigmatizing and alienating, with evaluations and medical assessments themselves being distressing for not only the child but the parents as well. The diagnosis itself is traumatic for parents and they have the added responsibility of assigning the gender at birth, decision-making regarding sometimes irreversible hormonal and surgical treatments for their child, and having to come to terms with child's behavior/play which may not conform to assigned or traditional gender roles. Raising a child with DSD can turn out to be a very isolating ordeal for the parents, one that can adversely impact parental mental health and their quality of life. This chapter provides an overarching view on the assessment of psychological challenges in parents of children with DSD, accompanying psychosocial and cultural factors as well as approaches for parental counseling and psychological support with a particular focus on the Indian perspective.

Keywords

Disorders of sexual development · DSD in India · Parents of children with DSD Psychosocial distress of parents · Parental counseling

V. Raman (✉) · S. Chandran
Child and Adolescent Psychiatry Unit, Department of Psychiatry, St. John's Medical College, Bangalore, Karnataka, India
e-mail: vijaya.r@stjohns.in; suhas.c@stjohns.in

S. K. Ratan (ed.), *Children with Differences in Sex Development*,
https://doi.org/10.1007/978-981-97-1639-5_34

34.1 Introduction

Disorders of Sexual Development (DSD) constitute a broad group of conditions of the urogenital tract and reproductive system, affecting human sex determination and/or differentiation [1]. These conditions can be identified and diagnosed at different points of time, such as newborns with ambiguous genitalia, or children with delayed puberty, abrupt virilization or gynecomastia or gonadal tumors. Along with physiological sexual development occurs psychosexual development, as the brain undergoes sexual differentiation [2].

In addition to medical issues, there are multiple psychosocial consequences of DSD, such as those regarding gender identity, gender assignment, early/late interventions, fertility concerns, and societal stigma [3]. Parents invariably have difficulty coping with the child's diagnosis, perceived burden of responsibility of gender assignment at birth and gender rearing as well as uncertainty about psychosocial outcomes of their child. The role of the treating team in these situations is paramount, and includes evaluating the child, parents, or other caregivers in terms of educational status, coping skills tailored to their capacity to comprehend DSD, managing social expectations, treating psychiatric disorders if any, pubertal changes, sexual orientation, and sexual functioning. Time periods of greater stress for the family include time of initial diagnosis, puberty, initiation of hormonal or surgical interventions, and when there is an onset of new relationships for the adolescent [4]. There is a paucity of research on DSD in the Indian setting, and most research is in the form of case series or retrospective studies with small sample sizes. This chapter explores DSD with a brief overview about the psychological impact of DSDs on the family that the pediatric professional needs to be cognizant of, assessments of the cultural and social challenges of the family specific to the Indian context, and psychosocial interventions for caregivers in DSD.

34.2 Psychological Assessment and Evaluation

34.2.1 Parental Challenges

Stress levels for parents of children with DSD are comparable to those described by parents of children with cancer in the phase post learning their child's diagnosis [5]. Suorsa and colleagues studied 51 parents of infants under the age of 2, who had been diagnosed with DSD in whom no decisions had been made regarding gender assignment at that stage. More than 25% of parents reported reduced quality of life, 18% had clinically significant symptoms of depression, and 25.4% had moderate to severe anxiety. A subset (15%) also reported post-traumatic stress symptoms (PTSS) [6].

Higher levels of chronic stress often predispose to more anxiety and higher risk of depression in parents of children with DSD, and a poor mental health QoL for these caregivers [7].

Table 34.1 Overview of common parental challenges

I. Treatment-based challenges
Understanding a complex medical illness which has biological, psychological, and social implications for life
Responsibility of assigning the gender at birth
Uncertainty about next course of action—Gendered rearing, medical and surgical treatments
Irreversible nature of the decisions about gender assignment and surgical procedures
Intermittent support from the medical fraternity—Support which is only forthcoming in the context of a crisis especially since most specialized centers handling children with DSD are few and far apart [9]
In some instances, careless words by professionals too can have a detrimental impact
Information available on the Internet on DSD is sparse and often unhelpful. Can have a negative impact, e.g. Parents opting for radical treatments for fear that the children "don't grow up to be gay" [10]
II. Parenting-based challenges
Ambivalence with use of gendered pronouns
Dealing with child's preference for cross-gendered activity and play contrary to the assigned gender
Adjusting to child's preferred gender—Allowing the child to explore without foisting their own views on their children
III. Psychosocial challenges
Difficult to rely on the extended family and community circles which in turn can create isolation for the parents
Contending with religious taboos
Birth certificate—Record of gender being mandatory

Parents are invariably at risk for cultivating dysfunctional parenting practices that are generally related to negative behavioral and social outcomes for their children. For example, parents of children with DSD are at higher risk for overprotective parenting and perceived child vulnerability. Perceived child vulnerability is linked with elevated anxiety [8].

A set of diverse factors that can influence parental stress and warrants evaluation by all professionals coming in contact with families of children with DSD are described in Table 34.1.

34.3 Consideration of Special Issues in the Indian Sub-continent During Evaluation

Different cultures view DSD differently, and management differs according to cultural and societal preferences for particular gender. In addition, resource poor settings necessitate the diagnosis of DSD without adequate investigations, such as hormone assays, genetic testing, and radiological imaging. Also, overcrowded hospitals and healthcare centers mean that there is little privacy for either detailed examination or discussion with the family members, and rural families are known to visit hospitals only after traditional/herbal/alternative medicine facilities are exhausted. Discussion about sexual issues is taboo, and there are many myths and misconceptions which often go unclarified and can play a major role in marginalization of individuals with DSD.

Secrecy about the diagnosis is widely prevalent, due to stigma, and fear of being considered a part of the *Hijra* community, who are often condemned as belonging to a lesser social status [11]. The other concern is the preference for a male child, as a male child is seen as a source of support, income, and care during old age. A female child, on the other hand, is viewed as a liability unless married. Many parents prefer raising the child as a male, as females with infertility may not be respected in the society compared to males with infertility. They are also resistant to suggestions of raising the child as female, or undergo hormonal or surgical treatments to reduce virilization. Consanguinity is common in many communities, and conditions such as Congential Adrenal Hyperplasia (CAH), that is autosomal recessive, is more likely to occur in several members in the family [12].

There are few Indian studies on DSD, most of which are prevalence data in the form of retrospective chart reviews or case reports and series. Kulkarni et al. evaluated the phenotype–karyotype correlation in a subset of North Indian children, noted a prevalent gender bias favoring males in that, a majority of the children had been assigned male gender, possibly consequent to the prevalent socio-cultural preference for male children. Of a total of 58 children, 43 were raised as males [13]. Kumar et al. also found similar bias towards male, with some patients undergoing male genitoplasty at the request of parents [14].

Few other common cultural challenges that professionals need to be aware of are listed in Table 34.2.

Table 34.2 Challenges rooted in culture

Cultural themes	Common challenges
Traditional belief systems	• Sexual and related issues are inescapably "forbidden" subjects in most societies and parents hesitate to discuss them even with the medical professionals. • Reaching out to faith healers, shrines, and magico-religious supports is more reflexive and comfortable for them [15].
Family survival, preference for sons, the marriage imperative	• Children are perceived to be a mode of insurance against old age. Across several cultures, the eldest son has significant unique responsibilities to upkeep the physical and spiritual well-being of the parents [16]. • Infertility, notably for a woman, can make it inconceivable for her to marry and if unmarried she could even be reduced to a social expatriate.
Rumor & discrimination	• Inappropriate use and understanding of the term "Hijra" [12].

34.4 Management

34.4.1 Psychological Management of Caregivers

34.4.1.1 General Principles

The medical and mental health requirements for a child with DSD is very much interconnected and complimentary. Medical management in DSD is invariably contingent on aiding psychological adjustment of the entire family rather than a physical health need of the child alone. For children when decisions are primarily driven by the physical considerations, the potential effect on psychological well-being on the child and family needs to be carefully considered. It is hence vital for psychological well-being and adaptation to be considered as integrated to the medical management and for this purpose psychological interventions should be provided as a quintessential part of multidisciplinary care, which must include a child mental health professional. The presence of a psychologist, or similar professional, will not pathologize DSD, but rather is appreciation of the unprecedented psychological and social challenges that a diagnosis of DSD warrants.

The educating process merits an individualized approach. Personal attitudes, past experiences, and socio-cultural background should be taken into account while planning interventions [4, 17]. The counseling process should be framed as a standard part of healthcare and not as separate "therapy." Discussions should revolve around day-to-day priorities in addition to the medical priorities. The distinction between "gender identity," "gender role," and "sexual orientation" needs to be clearly made along with an assessment of the family's understanding. There is likely to be a denial of the condition, or circumvention of medical care due to perceived negative experiences of the medical interventions, which may contribute to adverse patient outcomes, and needs to be addressed through psychoeducation and reassurance. Increased psychosocial support is required during the susceptible progression phase from adolescence to adulthood [18]. Other important time periods include at the time of birth, in the first year after birth, period of planned surgery, adolescence, and marriage. It is important to find the language or kind of discussion which the child is comfortable using, and activities such as role-play may be helpful in understanding the process of sharing information with others.

Two essential ways in which they needed to be supported:

- The physical needs—Mental health professionals need to educate themselves on medical aspects as well.
- The emotional needs—Similarly, pediatric surgeons and endocrinologists need to educate themselves on psychological aspects.

What parents need to be educated on:

- Medical facts: Sex and gender should be explained as non-binary concepts and in the background of sex determination and differentiation; Information on unavoidable vs elective medical procedures, including risks and benefits should be provided along with other alternatives. Parents should be explained to avoid sentimentally guided trigger finger decisions, delay noncritical decisions (such as those on surgery) for a few sessions and till acceptance is better.

- Gender affiliated: Differentiation between play behavior of childhood and adult gender identity should be made. Parents should also be primed that the preliminary decisions on social gender role could be subject to change later based on the self-expression of the developing child, which will help them accept the transition more openly [12].
- The difference between privacy and secrecy should be clearly delineated. Secrecy entails that the child is told not to tell anyone about their condition, which engenders shame and poor self-esteem in the child.
- Strategies to convey all the above information to other family members—Parents experience extreme stress about the revelation of the problem to family and friends and find this ambiguity very disconcerting. This challenge is often more difficult if they have already disclosed to the relatives the expected sex and/or have set up the baby's room or purchased clothes accordingly [19].

Along with the above information, parents should be assisted in applying the information successfully into everyday life and especially in how they can have an open non-judgmental conversation with their children on a regular basis. They should also be provided with tools they can use to educate others, and metaphors to help understand the condition better. It is imperative that parents should be supported with continuous psychological and psychosocial care irrespective of whether a surgical management option has been chosen or not. Professionals will need to target facilitation of encouragement as well as empowerment. For this, multidisciplinary team meetings with the family to answer questions that come up in discussions with the mental health professional regarding the DSD, and possible medical procedures, are fundamental to maintain consistent information and prevent uncertainty [20].

Parents can help the child in the following ways:

- Dealing with difficult emotions (such as shame) that keep people silent
- Developing a vernacular that they are comfortable using
- Helping them to figure out when and with whom it is useful to talk
- Role-playing and coaching the method of divulging information to peers or teachers
- Connecting them to peers who have had similar experiences.

The key is for parents to help the child acknowledge diversity, complexity, and uniqueness of the condition. It is imperative that parents discuss bodily divergences and sexual intimacy in means that will assist the child to find a comfortable way of making sense of and articulating their experience. The topics discussed with the child will also warrant a developmentally appropriate approach (Table 34.3). The professionals will need to help create an environment of appreciation and acceptance that needs to be created for the child. Informing and educating the child sometimes is met with opposition from parents who wrongly believe that holding back details will protect the children from information they regard as harmful or painful. It is important to emphasize to parents that being better informed about the medical

Table 34.3 Timing and topics based on developmental age [4, 21]

Developmental stage	Topics to be discussed
Early childhood	• Principles of body functioning including normal anatomy and physiology, role of hormones • Naturally occurring variations of the above • Play behavior • Normality of emotions such as shame, guilt, and secrecy
Early adolescence	• Puberty-related changes • Sexuality • Fertility • Gender contentedness • Peer relationships
Late adolescence and early adulthood	• Quality of sexual life • Fertility • Gender contentedness

Table 34.4 Psychological strategies and pointers for professionals (adapted from McCauley, 2017) [22]

Strategy	Description
Normalize the experience	Reassure: "We do see children with DSD and similar problems like this on a regular basis" can give a sense of respite.
Check overall understanding	Explore gently what the child and parent already know, probing for any compelling concerns they have, and listening in carefully to the way they describe the issue at hand. What is of value to the family and what they already know at baseline will determine how they hear and process the information you are about to share with them.
Educate	Educate regarding sexual development generally, utilizing images to build understanding, and enunciating in a manner that normalizes diversity in the context of development.
Discuss	Discuss forthrightly about medical information, minimum medical jargon, use pictures. Discuss small chunks of information at a time, ensure that they do not get overwhelmed.
Repeat	Begin each session by getting people to recall and summarize what they understood from the previous session. Naming and detailing a set of symptoms or medical disorder is an established technique to lower anxiety and avoidance.
Generate scripts	Develop scripts and templates for children to use when talking with peers or relevant others: a one- or two-line account they feel comfortable sharing and rehearse this with them, so that the vernacular feels natural.
Link with support	Link both parents and children with appropriate support groups which has individuals with similar experiences. Introduction to support groups can be via both offline and online forums.

condition is connected to improved psychological adaptation for the child. Informing children in a well-timed manner will cultivate circumstances to foster adaptive coping skills, nurturing a positive sense of self, and creating prospects for a rewarding adulthood in spite of the possibility of medical constraints such as infertility.

(*A few additional strategies that can be useful for professionals in their interactions with both parents are briefly discussed in Table 34.4*)

34.5 Summary and Conclusion

Our medical system does not necessarily expedite the emotional support that families invariably require. This must be granted due consideration in addition to the routine medical care which is furnished for children with DSD. Decreasing the isolation among parents of children with DSD must be a priority and in the proper environment, liaison and sharing with fellow parents appears to promote recognition and understanding. Multidisciplinary care by pediatrics, pediatric surgery, child psychology, child psychiatry, endocrinology, urology, gynecology, and social work is essential for holistic care of the child. Clinical settings which provide specialist DSD clinics should facilitate channels by which parents can be in touch with other parents of children with DSD. This may be accomplished through a "pairing" basis or the conduct of parent support groups. Research related to the effects of timing, nature of information (and a possible interaction of these two), the process of conveying information, and the impact of cultural, family, and child factors on how the information is handled is practically non-existent and needs to be evaluated in detail. It would also be sad if the courtroom were to replace a caring, holistic environment within which a positive approach by parents and doctors could effectively deal with the medical and other decisions that need to be made rather than a legal system.

References

1. Witchel SF. Disorders of sex development. Best Pract Res Clin Obstet Gynaecol. 2018;48:90–102. https://doi.org/10.1016/j.bpobgyn.2017.11.005.
2. Markosyan R, Ahmed SF. Sex assignment in conditions affecting sex development. J Clin Res Pediatr Endocrinol. 2017;9(Suppl 2):106–12. https://doi.org/10.4274/jcrpe.2017.S009.
3. Cools M, Nordenström A, Robeva R, Hall J, Westerveld P, Flück C, Köhler B, Berra M, Springer A, Schweizer K, Pasterski V. COST Action BM1303 working group 1. Caring for individuals with a difference of sex development (DSD): a Consensus Statement. Nat Rev Endocrinol. 2018;14(7):415–29. https://doi.org/10.1038/s41574-018-0010-8.
4. Wisniewski AB. Psychosocial implications of disorders of sex development treatment for parents. Curr Opin Urol. 2017;27(1):11–3. https://doi.org/10.1097/MOU.0000000000000344.
5. Wolfe-Christensen C, Fedele DA, Mullins LL, Lakshmanan Y, Wisniewski AB. Differences in anxiety and depression between male and female caregivers of children with a disorder of sex development. J Pediatr Endocrinol Metab. 2014;27(7-8):617–21. https://doi.org/10.1515/jpem-2014-0102.
6. Suorsa KI, Mullins AJ, Tackett AP, Reyes KJS, Austin P, Baskin L, et al. Characterizing early psychosocial functioning of parents of children with moderate to severe genital ambiguity due to disorders of sex development. J Urol. 2015;194:1737.
7. Ravendran K, Deans R. The psychosocial impact of disorders of sexual development. J Sexual Med Reprod Health. 2019;2:4.
8. Kirk KD, Fedele DA, Wolfe-Christensen C, Phillips TM, Mazur T, Mullins LL, et al. Parenting characteristics of female caregivers of children affected by chronic endocrine conditions: a comparison between disorders of sex development and type 1 diabetes mellitus. J Pediatr Nursing. 2011;26:e29–36.
9. Wechsung K, Marshall L, Jürgensen M, Neumann U, On Behalf of the Empower-Dsd Study Group. Diagnosis of DSD in children-development of new tools for a structured diagnos-

tic and information management program within the empower-DSD study. J Clin Med. 2022;11(13):3859. https://doi.org/10.3390/jcm11133859.

10. Ernst MM, Chen D, Kennedy K, et al. Disorders of sex development (DSD) web-based information: quality survey of DSD team websites. Int J Pediatr Endocrinol. 2019;2019:1. https://doi.org/10.1186/s13633-019-0065-x.
11. Bajpai M. Disorders of sex development: the quintessence of perennial controversies-III - DSD, transgenders and the judgment by the Hon'ble Supreme Court of India. J Indian Assoc Pediatr Surg. 2015;20(2):60–2. https://doi.org/10.4103/0971-9261.151544.
12. Krone N, Arlt W. Genetics of congenital adrenal hyperplasia. Best Pract Res Clin Endocrinol Metab. 2009;23(2):181–92. https://doi.org/10.1016/j.beem.2008.10.014.
13. Kulkarni KP, Panigrahi I, Das R, Kaur S, Marwaha RK. Pediatric disorders of sex development. Indian J Pediatr. 2009;76(9):956–8. https://doi.org/10.1007/s12098-009-0193-z.
14. Kumar J, Kumar V, Bhatia V, Dabadghao P, Chaturvedi S, Kapoor R, Ansari MS. Managing disorder of sexual development surgically: a single center experience. Indian J Urol. 2012;28(3):286–91. https://doi.org/10.4103/0970-1591.102703.
15. Chadda RK, Deb KS. Indian family systems, collectivistic society and psychotherapy. Indian J Psychiatry. 2013;55(Suppl 2):S299–309. https://doi.org/10.4103/0019-5545.105555.
16. Weidler EM, Peterson KE. The impact of culture on disclosure in differences of sex development. Semin Pediatr Surg. 2019;28(5):150840. https://doi.org/10.1016/j.sempedsurg.2019.150840.
17. Lampalzer U, Briken P, Schweizer K. Psychosocial care and support in the field of intersex/diverse sex development (dsd): counselling experiences, localisation and needed improvements. Int J Impot Res. 2021;33:228–42. https://doi.org/10.1038/s41443-021-00422-x.
18. Magritte E. Working together in placing the long term interests of the child at the heart of the DSD intervention. J Ped Urol. 2012;8:571–5.
19. Sanders C, Carter B, Goodacre L. Parents need to protect: influences, risks and tensions for parents of prepubertal children born with ambiguous genitalia. J Clin Nurs. 2012;21:3315–23.
20. Vora KA, Hanna CA, Neville KA, Bergman PB, Grover SR, O'Connell MA, Srinivasan S. Role of cross-campus multidisciplinary team meetings in decision-making for children and adolescents with differences of sex development/intersex. J Paediatr Child Health. 2021;57:1402–7. https://doi.org/10.1111/jpc.15474.
21. Wisniewski A, Sandberg D. Parenting children with disorders of sex development (DSD): a developmental perspective beyond gender. Horm Metab Res [Internet]. 2015;47(05):375–9. https://doi.org/10.1055/s-0034-1398561.
22. McCauley E. Challenges in educating patients and parents about differences in sex development. Am J Med Genet C Semin Med Genet. 2017 Jun;175(2):293–9. https://doi.org/10.1002/ajmg.c.31563.

35 Teenaged DSD Child: Psychosocial Issues and Decision-making—Parental Perspectives

Anju Gambhir and Mansi Chadha

Abstract

DSDs (Differences (Disorders) of Sex Development) are a vast set of congenital diseases of the urinary tract and reproductive system that largely affect individual sex determination and/or differentiation and can result in ambiguous genitalia. Without proper medical care and understanding of the disorder, some types of DSD—but not all—can be fatal. Moreover, some types of DSD and their management are linked to impaired reproductive potential, unsatisfactory sexual function, and gender dysphoria. In light of this, parents may find these facets of their child's DSD, its diagnosis and the possible treatment options upsetting. Parents can have a broad range of reactions and perspectives to their child's development and health due to the natural history associated with DSD which tends to vary between diagnoses. The DSD Consensus Statement (Lee PA, Houk Christopher P, Ahmed S, Pediatrics 118:e488–e500, 2006) acknowledges that DSD condition can place significant strain on the families of people with DSD and these issues demand long-term follow-up. Parents of young children with DSD report clinically significant parenting stress and reduced adaptive coping skills. Some of the primary concerns expressed by parents are concerns over their children's puberty, marriage, and fertility. Experiences of taboo, shame, or secrecy are common. Researchers

A. Gambhir (✉)
Fortis Hospital, Shalimar Bagh, Delhi, India

Maharaja Agrasen Hospital, Punjabi Bagh, Delhi, India

Saroj Hospital, Rohini, Delhi, India

M. Chadha
Department of Paediatric Surgery, All India Institute of Medical Sciences, New Delhi, Delhi, India

S. K. Ratan (ed.), *Children with Differences in Sex Development*,
https://doi.org/10.1007/978-981-97-1639-5_35

have only lately given priority to better understanding of how raising a child with DSD influences parents' psychosocial state. In fact many parents of children with DSD wish to receive mental health support because of the psychological sequelae as a result of raising a DSD affected child.

Keywords

Gender dysphoria · Sexual function · Ambiguous genitalia

Introduction

Differences (Disorders) of Sex Development (DSDs) are rare congenital conditions of the urinary tract and reproductive system that affect sex differentiation and/or differentiation and can result in ambiguous genitalia [1–6]. Discordance in genetic chromosomal sex, gonadal sex, and/or the appearance of internal/external reproductive organs typify DSD [1, 7]. Certain types of DSD are diagnosed prenatally or briefly after birth, presenting caretakers with a series of difficult, stressful, and challenging medical or surgical decisions, which may be largely aggravated by the absence of understanding about DSD's long-term consequences and potential interventions [7].

Different cultures have very different perspectives and approaches to the management of sexual development disorders (DSD). They are highly complicated conditions that even well-educated laymen struggle to understand, but when families are impoverished and the health systems are underfunded, traditional beliefs, folk remedies, and prejudice combine to make the lives of these children and adults exceedingly difficult, painful, and sorrowful. Rumor, stereotypes, and discrimination isolate them from their communities, devaluing them [7].

As rightly elicited by Warne & Raza, people with DSDs desire the same things in life as everyone else—to find someone who will love them, to be valued as human beings, to feel at home in their bodies, to be able to have satisfactory sexual relations, to be able to trust their medical advisers, and to be integrated into the general community [8].

DSDs are never easy to manage. Some kinds of DSD are linked to ambiguous external genitalia at birth. When the genitalia is ambiguous, it is difficult to determine the optimal gender. Genital surgery is frequently required in these cases of DSD, but both the type and timing of the procedure present contentious problems. Most instances have underlying endocrine abnormalities that necessitate long-term medication. Almost all DSDs impede fertility. In most of the DSD groups, gonadal development is impaired, which is usually associated with an increased risk of neoplasia. The afflicted person's childhood and adolescence are generally hampered by gender dysphoria and other psychosexual issues, which may persist into adulthood [8]. Without proper medical care, some types of DSD—but not all—can be fatal. Some types of DSD are linked to impaired reproductive potential [9]. In the light of this, parents may find many facets of their child's DSD diagnosis and the possible treatment options upsetting. Parents can have a broad range of reactions to their child's development and health due to the natural history associated with DSD which tends to vary between diagnoses. Researchers have only lately given priority to better understanding of how raising a child with DSD influences parents' psychosocial state, despite the fact that many parents of children with DSD wish to receive mental health support [10].

This chapter provides an overview of the research on parents' reactions and perspectives to their teenaged child with DSD and decision-making regarding available treatments, as well as recommendations for how to use these findings to help parents and their affected children better adapt.

35.1 Psychosexual Issues Faced by DSD Adolescents and Adults

For many years, researchers have examined DSD patients to look into the relative role of biological and psychological factors on psychosexual differentiation. Due to these factors and social factors, individuals with DSD face many issues related to gender identity, gender roles, and sexual behavior as shown in Fig. 35.1.

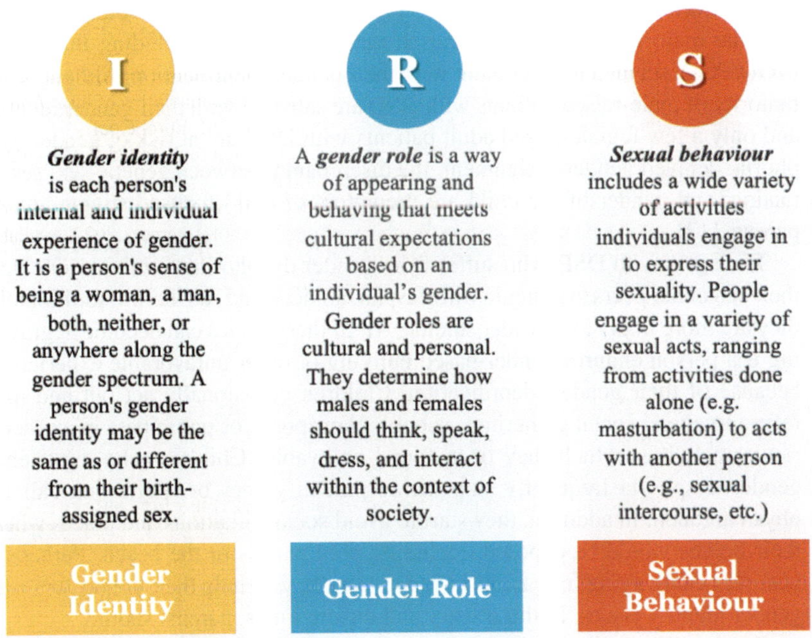

Fig. 35.1 Psychosexual issues faced by DSD adolescents and adults

- ***Gender Identity/Gender Dysphoria***

  The vast majority of men and women with mixed gonadal dysgenesis (MGD), partial gonadal dysgenesis (PGD), or congenital adrenal hyperplasia (CAH) appear to have a normal gender identity that is mostly in accordance with their sex of raising [11]. It has long been considered that choosing the female sex for raising a child with genital ambiguity due to MGD or PGD is the safest option. Even though surgical and hormonal treatments are effective, the risk of gender dysphoria is increased in DSD patients who were reared as females and were exposed prenatally to testosterone levels that were higher than those of unaffected normal females [12–14]. Reiner [15] recently reported that six (55%) of 11 patients with MGD reared as female "claimed" a male gender identification; however, it is uncertain if these patients, as in certain case reports, were indicative of MGD patients in general due to referral bias. In a study undertaken by Szarras-Czapnik, Lew-Starowicz, and Zucker, it was discovered that all ten patients raised as males had a male gender identification at follow-up. None of them had gender dysphoria or a desire to switch their gender. Whereas about two of the nine patients reared as girls (with MGD and PGD) were found to exhibit gender dysphoria, the remaining seven were found to have a female gender identity [12]. As for the two females who were dissatisfied with their gender, neither wished to change their sex nor had they considered sex reassignment. Nonetheless, they refused to accept their gender. They disliked their female attributes (e.g., breast development, body shape, etc.) [12]. According to Meyer-Bahlburg et al. [16] the majority of adult male-raised patients of DSD, including those with MGD/PGD, seemed to be content with their gender identification. Although, the majority of male-raised patients with DSD are satisfied with their gender identity and only a few female-raised adult patients with DSD are at risk of gender dysphoria, delayed gender assignment, the discrepancy between genetic sex, sex of raising, and gender of the child are predictors of child-focused stigma among parents [17].

  Teenagers with DSD who suffer from gender dysphoria are more prone than their cisgender peers to struggle with despair, anxiety, and self-harm. Additionally, they are more likely to consider suicide. All of these issues can become aggravating if a person endures gender-based bullying or other unfavorable experiences because of their gender identification. Children occasionally act out and they refuse to attend social gatherings, school, play sports, or participate in extracurricular activities which they once found enjoyable. Children who experience gender dysphoria frequently stop taking part in games or sports that call for physical touch. In addition, they start to avoid social situations and places where their bodies would be exposed, including pool parties or the beach. Perhaps it becomes challenging for parents trying to learn how to help their youngsters with gender identity issues. In these scary and unsure times, parents should

  - *Educate Themselves:* Educate themselves about gender dysphoria and mental health issues related to it and actively learn how they can help their children with gender identity issues.

- *Offer Support:* Parents should be supportive by valuing their children for who they are and loving them wholeheartedly. Parents should talk openly with their child about what they are experiencing.
- *Allow (Encourage) Gender Exploration:* Parents should allow and encourage gender exploration by allowing and encouraging the child to try new styles of clothing, hairstyles, names, and makeup that align with the gender spectrum they identify with.

- ***Gender Role***

  Research shows that there is behavioral masculinization in relation to gender role behavior in chromosomal females with DSD, such as congenital adrenal hyperplasia (CAH). Girls and women with CAH, for example, are more likely than unaffected controls to exhibit a "masculine" gender role behavior (e.g., toy interests, career preferences) [18]. Szarras-Czapnik, Lew-Starowicz, and Zucker [12] found that in childhood, five out of the ten girls demonstrated masculine gender role behavior (preference for masculine toys and games and sports, such as judo and football). Although at follow-up, these women worked in female-typical occupations, they were unhappy and uncomfortable with their situation and just appeared to adhere to social expectations. They had masculine interests and pastimes (e.g., occupations and sports). They favored male toys as children, took on masculine roles in games, and dressed themselves in men's clothing. One of these females worked as a shop clerk but longed to be a taxi driver [12]. The remaining five females appeared to be more tolerant of their gender roles as women. In childhood and at the follow-up, all ten males demonstrated masculine gender role behavior [12]. However, some females appear to be uncomfortable with their gender roles. The greatest need for psychological support is reported by parents of these individuals with DSD, and this requirement is exacerbated when the youngsters display transgender behavior [10]. It can be challenging to comprehend the particular requirements and concerns of parents of youngsters with DSD because they are more likely to refuse to take part in research studies. In addition, compared to parents of females, parents of males with DSD report greater illness uncertainty [19].

- ***Sexual Behavior***

  Aside from gender identity and gender roles, there are significant disparities in sexual behavior between DSD patients reared as male vs. female [12]. Although none of the patients in the research by Szarras-Czapnik, Lew-Starowicz, and Zucker participants who identified as a bisexual or homosexual sexual orientation had significant disparities in their experience of sexual intercourse. At the follow-up, all ten patients reared as boys experienced sexual dreams, desires, and sensitivity to sexual behavior. The sexual function of men with PGD/MDG who had hypospadias surgery was unaffected. They all reported having erections and ejaculations [12]. Furthermore, 7 (70%) of the 10 men had experienced heterosexual intercourse and were content with their sexual lives. Only two males with MGD showed an inferiority complex related to their low height and coitophobia. A small penis and an unsatisfactory appearance of genitalia were factors that contributed to dissatisfaction. One male patient felt he had a tiny penis; however, these three patients' well-being and sexual function were unaffected [12]. The majority of women, on the other hand, despite having good breast development and menstruation bleeding while on hormone replacement medication, still expressed significant sex anxiety. They indicated a lack of sexual thoughts, desire, and sensitivity to sexual action. Also, the women who had undergone successful genital gender transition surgery were less likely to have had sexual relations with male partners. They had no sexual experiences (such as dating, kissing, or genital petting) because they were disinterested. They revealed a lack of sexual desire, resulting in a lack of interest in sex. Based on clinical observations, it may be deduced that this was due, in part, to cultural issues, such as patients' unwillingness to discuss their sex lives with their respective physicians [12]. Although most males with MGD or PGD were satisfied with their sexual lives, this is not always true for individuals raised as males with androgen-resistance illnesses [20]. Though teenagers are hesitant to share their sex lives with their parents due to cultural issues and societal norms, parents should take time to inquire about whether their child has developed or is developing romantic feelings towards someone or/and also inquire about their child's physical or sexual experiences/encounters, by this parents can demonstrate to their children that they are open to talking about these things and will help in matters or issues related to them. This will help the children believe that the parents honestly care and the children will be able to share their apprehensions/issues related to their sex lives due to their DSD with the parents and seek necessary help.

In conclusion, the degree of masculinization of the external genitalia or the sex of upbringing has limited predictive value for later psychosexual qualities. Gender is not simply determined by the appearance of the external and internal genitalia [21]. Many psychosocial elements (social, familial, educational, and emotional) can impact gender development. Individuals' social status and sexual identity are influenced by their cultural environment and education. As a result, DSD patients with Y chromosomes are more prone to identify as males in a male-dominated society [12]. Thus, parents of children with DSD report stigma directed more toward the child than the parent. Although a sizable portion of both parents reports moderate to high levels of this stigma, mothers report experiencing it more frequently than fathers do. Delayed gender assignment, unusual genital appearance at birth (especially in boys), and the discrepancy between genetic sex and sex of raising are predictors of child-focused stigma among caregivers [17].

35.2 Cultural and Social Forces in the Management of DSD

Although there is insufficient data demonstrating the magnitude of social and cultural effects on management and quality of life in DSD.

- **Structural Forces (Poverty, Lack of Knowledge, and Inaccessible Medical Care):** Poverty, lack of knowledge, and inaccessible medical care all have a severe influence on CAH (DSD) patients and their families. These findings are best supported by studies carried out in underdeveloped nations, where financial issues, such as having to purchase expensive vital medications or travel long distances to see a specialist, are primary stressors for affected families [22]. According to a study conducted at a Malaysian tertiary facility, roughly one-third of CAH patients had limited knowledge regarding their diagnosis or were unaware that their medications were essential to preserve life [23]. Other difficulties affecting CAH families in these nations include a lack of newborn screening, which may result in late diagnosis, and gender allocation without proper and thorough evaluation. Aside from financial disparities, differences in healthcare delivery patterns in developed nations may influence access to proper medical care. National healthcare systems, such as those in Europe, promote the formation of multidisciplinary teams of experts [22], which can be remarked as the optimal setting for the management of DSD patients. While recognizing the specific benefits of such centralized specialized care in DSD is limited, data so far exhibit better medical management in CAH and fewer early genital procedures in DSD when care is delivered by a multidisciplinary team [24–26], though it is not clear whether these things truly improve outcome.

- **Religious and Cultural Forces:** Medical decisions and quality of life in CAH are influenced by religious and cultural influences. For instance, a study from Vietnam found that parents of CAH children were more concerned about future marriage and procreation [22]. According to reports, sex assignment in DSD is influenced in the Middle and Far Eastern cultures' by the preference for male over female infants [27, 28].
- **Social Forces (like Taboo, Shame, Social Exclusion, Secrecy Related to DSD):** Interviews with parents of adolescent children with DSD have shown that they frequently keep their child's condition a secret from others and refrain from engaging in social situations where their child's genital ambiguity or history of genital surgery might be discovered [29]. Forcible exclusion from common activities like childcare and playgroups is reportedly caused by parents fearing the threat—real or imagined—of embarrassment for their affected child. Therefore, stigmatizing children has a detrimental effect on carers' social support and increases isolation within families.

35.3 Psychological Health and Quality of Life of Persons with DSD

According to certain research, people with DSD have a higher prevalence of psychological distress, self-harm, and suicide ideation [29]. Experiences of taboo, shame, or secrecy are known causes; however, living with chronic illness and DSD-related physiological sequelae may also contribute to heightened distress [30, 31]. Parents of children with abnormal genital anatomy are frequently displeased with the way their newborns look [18], and as a consequence, these parents believe that their children are more prone to being ridiculed [32]. An educational program for caregivers of children with anomalous genitalia which explains to them how to discuss their child's anatomy in a way that both safeguards their child's privacy while properly educating other family members and friends so that it might lessen overprotective parenting and the perception that their children are vulnerable. One approach to achieve this is by listening to other parents' experiences of how they share their experiences like changing diapers, in vicinity of people outside of their immediate family [33]. Peer assistance has been demonstrated in studies to improve psychological well-being [34]. Factors connected to medical care, such as hormonal status and level of information sharing, might either limit or facilitate condition acceptance [35, 36].

Rather than Quality of Life (QoL), most studies on DSD have primarily concentrated on aspects of contentment with the sex of raising, sexual orientation, gender identity, and gender role. There have only been a few studies which studied the effect of DSD on physical and mental health. The majority of research regarding the Quality of Life (QoL) in children with DSDs has been done on 46, XX DSD (CAH) [37], which have found that individuals with DSD tend to have a low health-related quality of life (HRQoL) [38].

In a study conducted by Prematilake et al., it was found that children with 46, XY DSD have a potentially low QoL [39]. Further, study conducted by Jurgensen et al. [37] exhibited reduced QoL. It was also highlighted in the study by Prematilake et al. that in school-going children aged more than 5 years, 65.4% of school children with DSD were not drinking water in school, 57.7% of them did not use washrooms, and 65.4% of children did not participate in sports due to fear of exposure of their disease condition. The stigma and associated disgrace have a greater impact on psychological prognosis than the actual severity of the condition [39].

It can therefore be inferred that DSD largely affects an individual's overall physical, mental, emotional, and social satisfaction. Thus, these issues/concerns must be addressed as quickly as possible to improve the QoL of children/adolescents with DSDs.

35.4 Psychological Health of Caregivers of Children with DSD

The DSD Consensus Statement [40] acknowledges that DSD condition can place significant strain on the families of people with DSD and these issues demand long-term follow-up; yet, there have been very few systematic studies of how early interventions and interactions between healthcare practitioners and families of these individuals can enhance the quality of life of affected people and/or their parents [41, 42]. One study of parents of young children with DSD (mostly 46, XY) indicated clinically significant parenting stress and reduced adaptive coping skills [43]. Similarly, a Sri Lankan study identified depression symptoms in parents of children with CAH that did not improve over time. The study further elicited that parents with 46, XY DSD children have significant psychological morbidity and stigma [39]. Some of the primary concerns expressed by parents in this study were concerns over their children's puberty, marriage, and fertility [39]. This study further discovered that this condition also resulted in self-negligence and family conflict. The parents were pleased with the health care services delivered to them, but they most urgently required a support group [39]. This study also found that parental components of QoL were inversely associated with child age (p 0.01), indicating that the QoL of children rated by parents decreased as the child grew older. The study found no significant relationship between parental education level and family income and QoL in children ($p > 0.05$). This demonstrates that any intervention should be applied equally to parents regardless of their education level or income [39]. There was no statistically significant relationship between the number of procedures, pubertal changes, and Paeds QoL (Paediatric Quality of Life Scale). This is in line with the results of studies by Jurgensen et al. and Selveindran et al. [37, 44]. Similar studies on 46, XY DSD illustrate that there is significant parenting stress (19%) and decreased adaptive coping capacity (13%) in parents of children with DSD [45]. The impact of this medical condition on families was demonstrated by the fact that over 60% of these parents felt embarrassed while sharing their child's condition with relatives, friends, and acquaintances, and 68% were concerned that DSD would stigmatize their child [45].

Although some parents seem to cope well with their child's DSD and clinical management, a subset of these parents experience feelings of despair, anxiousness, and depression [46, 47] and significant stress regarding their child's future. Nearly one-quarter of parents experience clinically significant symptoms of depression and anxiety in the days following the birth of the afflicted child and the DSD diagnosis and which tend to gradually fade away [46–48]. Although these studies have begun to characterize parental experiences, relatively little is known about the nature of carers' psychosocial functioning [49], and much less is known about the determinants of such functioning. Parents of children with DSD may face anxiety, distress and anxiety concerning the lack of a formal diagnosis, unknown long-term results of surgery or timing of surgery, ambiguity about future gender identity, the necessity for life-long pharmacological therapy, unpredictable responses from friends and family, and probable challenges with sexual functioning and fertility [50].

Recent research indicates that caregiver perceptions of illness uncertainty are important for understanding early parent psychosocial adjustment to their child's DSD [18, 46, 47]. Specifically, cross-sectional or correlational studies utilizing portions of the current dataset demonstrate that caregiver-perceived illness uncertainty is associated with increased parent post-traumatic stress, anxious symptoms, and depressive symptoms following a child's DSD diagnosis [18, 46, 47]. Furthermore, caregiver perceptions of illness uncertainty within the first 2 years of their child's DSD diagnosis are associated with greater parental decisional regret 12 months later [46].

According to a study of post-traumatic stress symptoms (PTSS), the most evident markers of stress among parents of children with DSD were perplexity and uncertainty about what caused their child's DSD and how to proceed with medical therapy [51]. Other research on parents and the affected individuals has also shown that one of their key concerns is how to maintain secrecy and privacy while sharing their child's or their own diagnosis [50, 52]. As a result, the medical community is coming to appreciate the need of educating persons with DSD and their caregivers about their condition. However, there is a paucity of evidence regarding the best methods for disseminating this information [33, 53].

35.5 Attitudes, Beliefs, and Issues Related to Clinical Management of DSDs

Clinical care of patients with DSD conditions has become increasingly contested in the last decade. One of the most contentious issues is determining the gender of a newborn with ambiguous genitalia [15]. Some intersex individuals perceive themselves as notably non-female/non-male and may have a hermaphroditic identity [54]. Some authors propose the inclusion of a third or even more gender categories [55, 56]. The Intersex Society of North America supports deferring all genital surgery until the patient is mature enough to give informed consent [57]. However, genital anomalies are embarrassing and undesirable to the majority of patients and parents. If the young child has ambiguous genitalia and has not yet had surgery to "normalize" the appearance of the genitalia, mothers report higher levels of stress [18], whereas both mothers and fathers report lower levels of stress 6 months after surgery [58]. Also, delays in internal genital reconstruction surgery may result in persistent urine infection and renal failure and it is also vital to remember that patients with gonadal dysgenesis and the Y chromosome have an increased chance of developing gonad germ cell neoplasia [59]. This does not necessarily imply that genital surgery is the best course of action for reducing stress in parents of young children with atypical genitalia, but it does highlight the necessity of providing support to parents in order to help them cope with their negative feelings about their child's genital phenotype.

Quality-of-life studies can provide us with critical information regarding the optimal age for medical/surgical intervention as well as the value of various therapeutic care options [14]. Identifying the parameters that indicate a good sexual quality of life for intersex patients is the essence of knowledge that is most significant and crucial for appropriate therapeutic care [54]. Many of the patients, particularly those who were raised as females, struggled with their sexual lives as young adults. This highlights the importance of improving counseling services for these patients and their families not only during childhood and adolescence but throughout their lives [11].

35.6 Long-Term Social and Sexual Effects of Surgery

Clinicians who care for infants with anomalous external genitalia linked to intersexuality or structural issues with the lower genitourinary system have a massive challenge. In addition to being technically challenging, providing surgical care for these children also involves addressing and treating a number of additional problems that may be connected to the disease processes, such as endocrinological, social, and psychological disorders. Few studies have addressed the psychological development of these children, despite the fact that others have categorized and described reconstructive management in these children. The next task is to evaluate long-term outcomes not just with respect to physical appearance but also with respect to social, sexual, and psychological development as the ability to diagnose and treat these children is improved. There is limited data available regarding the long-term social and sexual adjustment of patients who have undergone surgical repair of ambiguous genitalia or congenital defects of the vagina and lower genitourinary tract. However, these anomalies have a considerable impact on later social and sexual development because they impair the most private parts of bodily development in childhood such as sexual, bladder, and bowel function. DSD is a condition characterized by several abnormalities which are chronic in nature that requires multiple surgical treatments as well as long-term follow-up. As a result, these children have a high risk of social and sexual underachievement due to the prolonged nature of their congenital abnormalities' treatments.

35.7 Conclusion

Parents' psychosocial adaptations and parenting skills can deteriorate due to the stressful and isolating nature of raising a child with DSD. Parents and at times immediate families of children with DSD are at a higher risk of experiencing elevated stress, anxiety, and depression, as well as a decline in quality of life. Parents typically cite perceived or actual child-focused stigma due to not seeking support and assistance from their family and friends regarding managing their child's condition. Evidence is beginning to mount to support individualized interventions to support and educate carers of children with DSD from the mental health community.

References

1. Hughes, I. A., Houk, C., Ahmed, S. F., Lee, P. A., & Society, L. W. P. E. Consensus statement on management of intersex disorders. J Pediatr Urol. 2006;2:148–62.
2. Lee PA, Nordenström A, Houk CP, Ahmed SF, Auchus R, Baratz A, et al., the Global DSD Update Consortium. Global disorders of sex development update since 2006: Perceptions, approach and care. Horm Res Paediatr. 2016;85:158–80.
3. Szymanski KM, Whittam B, Kaefer M, Frady H, Cain M, Rink R. What about my daughter's future? Parental concerns when considering female genital restoration surgery in girls with congenital adrenal hyperplasia. J Pediatr Urol. 2018a;14:417.e1–5.
4. Szymanski KM, Whittam B, Kaefer M, Frady H, Casey JT, Tran VT, et al. Parental decisional regret and views about optimal timing of female genital restoration surgery in congenital adrenal hyperplasia. J Pediatr Urol. 2018b;14:156.e1–7.
5. Thyen U, Lanz K, Holterhus PM, Hiort O. Epidemiology and initial management of ambiguous genitalia at birth in Germany. Horm Res Paediatr. 2006;66:195–203.
6. Cools M, Nordenström A, Robeva R, Hall J, Westerveld P, Flück C, Köhler B, Berra M, Springer A, Schweizer K, Pasterski V. Caring for individuals with a difference of sex development (DSD): a consensus statement. Nat Rev Endocrinol. 2018 Jul;14(7):415–29.
7. Wisniewski AB, Chernausek SD, Kropp BP. Disorders of sex development: A guide for parents and physicians. The Johns Hopkins University Press; 2012.
8. Warne GL, Raza J. Disorders of sex development (DSDs), their presentation and management in different cultures. Reviews in Endocrine and Metabolic Disorders. 2008 Sep;9(3):227–36.
9. Lee PA, Houk CP, Ahmed SF, Hughes IA. International Consensus Conference on intersex organized by the Lawson Wilkins Pediatric Endocrine Society and the European Society for Pediatric Endocrinology. Consensus statement on management of intersex disorders. Pediatrics. 2006;118(2):e488–500.

10. Bennecke E, Werner-Rosen K, Thyen U, Kleinmeier E, Lux A, Jürgensen M, et al. Subjective need for psychological support (PsySupp) in parents of children and adolescents with disorders of sex development (DSD). Eur J Pediatr. 2015;174:1287–97.
11. Dessens AB, Slijper FME, Drop SLS. Gender dysphoria and gender change in chromosomal females with congenital adrenal hyperplasia. Arch Sex Behav. 2005;34:389.
12. Szarras-Czapnik M, Lew-Starowicz Z, Zucker KJ. A psychosexual follow-up study of patients with mixed or partial gonadal dysgenesis. J Pediatr Adolesc Gynecol. 2007;20(6):333–8.
13. Meyer-Bahlburg HFL. Intersexuality and the diagnosis of gender identity disorder. Arch Sex Behav. 1994;23:21.
14. Meyer-Bahlburg HFL. Gender identity outcome in female raised 46,XY persons with penile agenesis, cloacal exstrophy of the bladder, or penile ablation. Arch Sex Behav. 2005;34:423.
15. Reiner WG. Gender identity and sex-of-rearing in children with disorders of sexual differentiation. J Pediatr Endocrinol Metab. 2005;18:549.
16. Meyer-Bahlburg HFL, Migeon CJ, Berkovitz GD, et al. Attitudes of adult 46, XY intersex persons to clinical management policies. J Urol. 2004;171:1615.
17. Rolston AM, Gardner M, Vilain E, Sandberg DE. Parental reports of stigma associated with child's Disorder of Sex Development. Intl J Ped Endocrinol. 2015;
18. Suorsa KI, Mullins AJ, Tackett AP, Scott Reyes KJ, Austin P, Baskin L, Bernabé K, Cheng E, Fried A, Frimberger D, Galan D. Characterizing early psychosocial functioning of parents of children with moderate to severe genital ambiguity due to disorders of sex development. J Urol. 2015 Dec;194(6):1737–42.
19. Cohen-Bendahan CCC, van de Beek C, Berenbaum SA. Prenatal sex hormone effects on child and adult sex-typed behavior: methods and findings. Neurosc Biobehav Rev. 2005;29:353.
20. Bouvattier C, Mignot B, Lefevre H, et al. Impaired sexual activity in male adults with partial androgen insensitivity. J Clin Endocrinol Metab. 2006;91:3310.
21. Ahmed SF, Khwaja O, Hughes IA. The role of a clinical score in the assessment of ambiguous genitalia. BJU Int. 2000;85:120.
22. Armstrong KL, Henderson C, Hoan NT, Warne GL. Living with congenital adrenal hyperplasia in Vietnam: a survey of parents. J Pediatr Endocrinol Metab. 2006;10:1207–23.
23. Zainuddin AA, Grover SR, Shamsuddin K, Mahdy ZA. Research on quality of life in female patients with congenital adrenal hyperplasia and issues in developing nations. J Pediatr Adolesc Gynecol. 2013;6:296–304. https://doi.org/10.1016/j.jpag.2012.08.004.
24. Pasterski V, Prentice P, Hughes IA. Consequences of the Chicago consensus on disorders of sex development (DSD): current practices in Europe. Arch Dis Child. 2010;8:618–23. https://doi.org/10.1136/adc.2009.163840.
25. Schaeffer TL, Tryggestad JB, Mallappa A, Hanna AE, Krishnan S, Chernausek SD, Chalmers LJ, Reiner WG, Kropp BP, Wisniewski AB. An evidence-based model of multidisciplinary care for patients and families affected by classical congenital adrenal hyperplasia due to 21-hydroxylase deficiency. Int J Pediatr Endocrinol. 2010;2010:692439.
26. Parisi MA, Ramsdell LA, Burns MW, Carr MC, Grady RE, Gunther DF, Kletter GB, McCauley E, Mitchell ME, Opheim KE, Pihoker C, Richards GE, Soules MR, Pagon RA. A gender assessment team: experience with 250 patients over a period of 25 years. Genet Med. 2007;6:348–57.
27. Bhansali A, Walia R, Sing R, Sriram M, Rao K, Dhaliwal L, Dutta P. Disorders of sex development: a study of ninety-five patients. Endocrinologist. 2009;9:98–101.
28. Uslu R, Oztop D, Ozcan O, Yilmaz S, Berberoğlu M, Adiyaman P, Cakmak M, Kerimoğlu E, Ocal G. Factors contributing to sex assignment and reassignment decisions in Turkish children with 46, XY disorders of sex development. J Pediatr Endocrinol Metab. 2007;9:1001–15.
29. Schützmann K, Brinkmann L, Schacht M, Richter-Appelt H. Psychological distress, self-harming behavior, and suicidal tendencies in adults with disorders of sex development. Arch Sex Behav. 2009;38(1):16–33.
30. Nordenström A, Thyen U. Improving the communication of healthcare professionals with affected children and adolescents. Endocr Dev. 2014;27:113–27.
31. Audí L. Past experiences of adults with disorders of sex development. Endocr Dev. 2014;27:138–48.

32. Sanders C, Carter B, Goodacre L. Parents need to protect: influences, risks and tensions for parents of prepubertal children born with ambiguous genitalia. J Clin Nurs. 2012;21:3315–23.
33. Magritte E. Working together in placing the long term interests of the child at the heart of the DSD intervention. J Ped Urol. 2012;8:571–5.
34. Baratz AB, Sharp MK, Sandberg DE. Disorders of sex development peer support. Endocr Dev. 2014;27:99–112.
35. Schweizer K, Brunner F, Gedrose B, Handford C, Richter-Appelt H. Coping with diverse sex development: treatment experiences and psychosocial support during childhood and adolescence and adult well-being. J Pediatr Psychol. 2017;42(5):504–19.
36. Lundberg T, Roen K, Hirschberg AL, Frisén L. "It's part of me, not all of me": young women's experiences of receiving a diagnosis related to diverse sex development. J Pediatr Adolesc Gynecol. 2016;29(4):338–43.
37. Jürgensen M, Lux A, Wien SB, et al. Health-related quality of life in children with disorders of sex development (DSD). Eur J Pediatr. 2014;173:893–903.
38. Gilban DLS, Alves Junior PAG, Beserra ICR. Health related quality of life of children and adolescents with congenital adrenal hyperplasia in Brazil. Health Qual Life Outcomes. 2014;12:1–9.
39. Prematilake DC, Koralage BM, Atapattu N, de Silva V, de Silva S. Assessment of psychosocial wellbeing of parents & Quality of Life (QoL) of children with 46, XY Differences of Sex Development (46, XY DSD) followed up in endocrine clinics at Lady Ridgeway Hospital for Children (LRH), Colombo, Sri Lanka.
40. Lee PA, Houk Christopher P, Ahmed S, Hughes F, Ieuan A. In collaboration with the participants in the International Consensus Conference on Intersex: Consensus statement on management of intersex disorders. Pediatrics. 2006;118(2):e488–500.
41. Frader J, Alderson P, Asch A, Aspinall C, Davis D, Dreger A, Edwards J, Feder EK, Frank A, Hedley LA, et al. Health care professionals and intersex conditions. Arch Pediatr Adolesc Med. 2004;158(5):426–8.
42. Zeiler K, Wickström A. Why do "we" perform surgery on newborn intersexed children? The phenomenology of the parental experience of having a child with intersex anatomies. Fem Theory. 2009;10(3):359–77.
43. Duguid A, Morrison S, Robertson A, Chalmers J, Youngson G, Ahmed SF. The Scottish Genital Anomaly Network: the psychological impact of genital anomalies on the parents of affected children. Acta Paediatr. 2007;96(3):348–52.
44. Selveindran NM, Syed Zakaria SZ, Jalaludin MY, et al. Quality of life in children with disorders of sex development. Horm Res Paediatr. 2017;88:324–30.
45. Duguid A, Morrison S, Robertson A, et al. The psychological impact of genital anomalies on the parents of affected children. Acta Paediatrica. 2007;96:348–52.
46. Ellens RE, Bakula DM, Mullins AJ, Scott Reyes KJ, Austin P, Baskin L, Bernabé K, Cheng EY, Fried A, Frimberger D, Galan D. Psychological adjustment of parents of children born with atypical genitalia 1 year after genitoplasty. J Urol. 2017;198(4):914–20.
47. Perez MN, Delozier AM, Aston CE, Austin P, Baskin L, Chan YM, Cheng EY, Diamond DA, Fried A, Greenfield S, Kolon T. Predictors of psychosocial distress in parents of young children with disorders of sex development. J Urol. 2019;202(5):1046–51.
48. Sharkey CM, Bakula DM, Wolfe-Christensen C, Austin P, Baskin L, Bernabé KJ, Chan YM, Cheng EY, Delozier AM, Diamond DA, Ellens RE. Parent-rated severity of illness and anxiety among caregivers of children born with a disorder of sex development including ambiguous genitalia. Horm Res Paediatr. 2018;90(5):308–13.
49. Sandberg DE, Gardner M, Callens N, Mazur T, DSD-TRN Psychosocial Workgroup, the DSD-TRN Advocacy Advisory Network, and Accord Alliance. Interdisciplinary care in disorders/differences of sex development (DSD): The psychosocial component of the DSD-Translational research network. Am J Med Genet C Semin Med Genet. 2017;175(2):279–92.
50. Crissman HP, Warner L, Gardner M, Carr M, Schast A, Quittner AL, Kogan B, Sandberg DE. Children with disorders of sex development: a qualitative study of early parental experience. Int J Pediatr Endocrinol. 2011;2011(1):1–1.

51. Pasterski V, Mastroyannopoulou K, Wright D, Zucker KJ, Hughes IA. Predictors of posttraumatic stress in parents of children diagnosed with a disorder of sex development. Arch Sex Behav. 2013;2014(43):369–75.
52. Liao LM, Tacconelli E, Wood D, Conway G, Creighton SM. Adolescent girls with disorders of sex development: a needs analysis of transitional care. J Ped Urol. 2010;6:609–13.
53. Loughlin E. The family. In: Hutson JM, Warne GL, Grover SR, editors. Disorders of sex development: an integrated approach, Chapter 19. Berlin: Springer; 2012. p. 193–200.
54. Chase C. Hermaphrodites with attitude: mapping the emergence of intersex political activism. GLQ. 1998;4:189–211.
55. Fausto SA. The five sexes. Sciences. 1993;33(2):20–4.
56. Kessler SJ. Lessons from the Intersexed. Rutgers University Press; 1998.
57. Gooren LJG. Normal and abnormal sexual differentiation: from genes to patient: psychological consequences. Semin Reprod Med. 2002;20:285.
58. Fedele DA, Kirk K, Wolfe-Christensen C, Phillips TM, Mazur T, Mullins LL, et al. Primary caregivers of children affected by disorders of sex development: mental health and caregiver characteristics in the context of genital ambiguity and genitoplasty. Int J Ped Endo. 2010;
59. Miller J, Ritzen EM, Ivarsson S-A, et al. Management of males with 45, X/46, XY gonadal dysgenesis. Horm Res. 1999;52:11.

Essentials of Counseling of a Teenaged Child with DSD at the Threshold of Decision-Making

36

S. Sulakshana and Smita N. Deshpande

Abstract

One of the first social categories learned from early childhood is gender stereotypes and the traditional perception of masculine–feminine traits encircling strong cultural and societal norms gets internalized around what real men and women should be like. Even though there is no single definition for all men and all women, the social roles, behaviors, and meanings prescribed for men and women based on sex in any society at any time, across cultures diverge, however, universally they have been institutionally organized, structured, elaborated, and experienced through interactions and expressed on the axis of identity including class, race, ethnicity, age, and sexuality. If so, then should DSD be considered an abnormality in anatomy needing correction or a social acceptance of human diversity, with the idea that differences are not equal to disease? Empirical data, mainly retrospective observations and case studies, highlighted DSDs as rare (exotic), focusing on diagnostic improvement and corrective measures (phenotypic sex), however, this congenital condition is not suggestive of a disorder or abnormality and does not imply a person's identity and therefore this chapter explores the psychological management and normalized inclusion of adolescents with DSD on the juncture of decision-making with regard to stabilizing their identity, positive acceptance of body variance, a better quality of life and sexual health.

S. Sulakshana
Ponakam, Mavelikara P.O, Alappuzha, Kerala, India

S. N. Deshpande (✉)
Department of Psychiatry, St Johns Medical College and Hospital, Bengaluru, Karnataka, India

S. K. Ratan (ed.), *Children with Differences in Sex Development*, https://doi.org/10.1007/978-981-97-1639-5_36

Keywords
Identity · Sex · Gender · Counseling · Adolescents · Re-assignment · Normalization

36.1 Historical Binarization and Sociological Evolution

Sustaining life, through the process of reproduction and genetic inheritance, is one of the most unsophisticated and rudimentary obligations among species of variance, on planet Earth and hermaphroditism is one specific way that evolution has played a role in sexual reproduction [1]. Hermaphrodite word emerged from Greek mythology; Hermaphroditus, son of Hermes and Aphrodite, fused with a woman, resulting in one (both male and female). Across the globe, this cult of duality is homogeneous in cultures like India, Syria, Cyprus, and Greece, considering such individuals as manifestations of sexual profusion, however, today such disorders are designated as genetic differentiation of the genital system [2]. Divergence in one's anatomical sex and related roles have been reported in the medical literature and until the 1950s, homosexuality, fetishism, and gender identity disorder were regarded as objectionable types of sexual perversions [3]. The optimal-gender policy practice, developed in the middle-1950s, involved assigning the female sex to congenital adrenal hyperplasia, even if severely or completely masculinized [4, 5]. In 1952, (a woman kept in the body of a man's case), the first sex-reassignment surgery (transsexualism) was performed and subsequently in 1957, John William Money advocated the concept of gender while studying their sexual consciousness [3]. Thereafter, this psychopathological condition (discrepancy between anatomical sex and gender identity) was referred and diagnosed to as Gender Identity Disorder (GID) and was introduced into DSM-III (Diagnostic and Statistical Manual of Mental Disorders) in 1980, encompassing a spectrum of conditions. Later, DSM-IV defined intersex as an exclusion criterion for GID. In DSM-V, the diagnostic name of GID changed to Gender Dysphoria (GD), referring to assigned gender rather than sex, and included disorders of sexual development. However, the new diagnosis of gender dysphoria defines DSD, as a specifier of GD since it subsumes the physical condition under the mental disorder conceived as disproportionate in their diagnosis and treatment [6, 7]. Even though, some individuals with DSD prefer to use the term intersex, irrespective of whether or not they have been diagnosed with DSD, the term Intersex has been replaced by DSDs, especially in the medical literature [8]. The concept of intersexuality subsumes, an atypical development, during the prenatal sex differentiation process which does not meet the normative definition of male and female since it encompasses conditions in which biological sexual characteristics differ and a person cannot easily be assigned to one sex [9].

In 2005, at the International Consensus Conference on Intersex in Chicago, an umbrella term, Disorders of Sex Development (DSD) was introduced, however, because of its pathologizing implication, terms such as Variations in Sex Characteristics (VSC) or Diverse Sex Development or Differences Sex Development (DSD) are used as alternatives. Disorder of Sex Development (DSD), a diagnosis mostly made at birth or infancy and in many cases requires necessary and immediate interventional actions, is referred to as a heterogeneous group of congenital conditions in which chromosomal, gonadal, and anatomical sex are atypical [10]. Anomalies in sex determination without sex ambiguity (sometimes infertility) include dysgenesis (gonadal and testicular) polysomes, male XX, and Klinefelter syndrome (dysgenesis and polysomes female phenotype; and sex reversal and Klinefelter male phenotype) and ovotesticular disorders. Among the anomalies in sex differentiation are, in Males with Androgen Deficiency (MAD) with male karyotype and gonads (XY and testes), but female phenotype (hormonal abnormalities); Females with Androgen Excess (FAE); patients having ovaries and a 46, XX karyotype, with varying degrees of external genital virilization (congenital adrenal hyperplasia); and Kallmann syndrome [11].

36.2 Cultures and Ethical Tensions

Even though clinical management of DSDs has evolved significantly (especially in the United States and other Western nations), however, complexities in decision-making concerning whether to pursue corrective genital surgery or appropriate gender of rearing, raises ethical issues and tensions, counting to roles of various medical professionals and parameters appropriate for the process as well as issues concerning patient's right to bodily integrity, informed consent, and disclosure, still exist, especially in places with social, political, and economic inequity between the sexes and may influence clinical decision-making [12]. Even though for most of Western history intersex people were integrated into the general population, adopting either a male or a female identity following binary gender norms, they maintained their ambiguous genitalia without significant psychological harm or social stigma [13].

Mounting anxiety over homosexuality, which was considered abnormal sexual behavior, by the late eighteenth century, corrective surgeries were performed on abnormal genitalia (enlarged clitorises and hypospadias), however, until the 1950s it became the medical norm with the postulation of optimum gender rearing model by John Money (biased toward female gender assignment). Prime importance was placed on genital aesthetics and to limit possible psychological distress, decisions on the child's correct sex were taken with no informed consent [14]. As a result, a large numbers of intersex infants were assigned female, since the vagina merely needed to be a hole large enough to fit a penis, and was considered less sensitive to sexual stimulation [15], thus leading to the idea that it was easier to make a hole than build a pole [16]. However, Milton Diamond's (the mid-1990s) findings discredited the claimed success of gender-reassignment protocol, debunking the theory of gender neutrality and flexibility at birth [17], and therefore corrective gender-reassignment surgery, today, is considered a violation of bodily integrity and reproductive rights, because the surgery may particularly impair reproductive ability or sexual sensitivity [18].

In many cultures across the globe, DSD management can be surprisingly influenced by the economic concerns of the families, for example, among Indian Malaysian families, daughters are considered a major financial burden (dowry) as opposed to sons who increase the family's wealth promoting male sex assignment, whereas in stark contrast, females in Muslim Malaysian families are perceived with increased economic power and independence [19]. In many countries (India, Egypt, Saudi Arabia), for economic motives, where unmarried males and females are isolated, infertile males are preferred over females [20] echoed in the family's decisions and desirability for independent, uneducated, and single working men over females who are illiterate, unmarried, or divorced, as they are considered unemployable and economically dependent [21, 22].

Fertility is valued over sexual function, among female DSDs, irrespective of religious background and can become grounds for immediate annulment of marriages, thereby pressuring parents to be the sole decision makers during sex assignment [21]. In Nigeria, the idea that a male child can become female itself could be problematic and difficult for parents to accept and perceived as a spiritual rather than a medical problem [23]. It has been observed that among DSDs, female fertility is often given higher priority than preserving male fertility since motherhood is considered more important to females than fatherhood is to males [24].

To ensure heterosexuality in cultures where homosexuality is considered a punishable crime [25], XX individuals are not reared a male, since they might get attracted to males due to female hormones [26] and in countries like Iran, to eliminate homosexuality from the country, gives out heavily subsidized sex-change operations [27]. The transition from female–male is completely different from male–female, since in some countries, like Saudi Arabia, restrictions on being female are difficult to cope with [28] prompting societal prejudices toward female reassignment [4].

36.3 India: Cultural Permanence; Social Denial

Historical evidence of intersex individuals' presence in Indian society can be traced back centuries. The concept of "tritiyaprakriti" [29] or "napumsaka" had been an integral part of the Hindu mythology, folklore, epic, and early Vedic and Puranic works of literature, denoting the absence of procreative ability, presented by differences in masculine and female roles [30]. In the Indian subcontinent, Hijra are males who adopt a female gender identity by having part or all of their genitals removed, in addition to dressing as female [31] (eunuchs, intersex people, or transgender people) live in an organized community [32] that follow a kinship system [33] and therefore, children born with DSDs are handed over to the community leaders by their parents [33]; so also to find recruits, hijras coerce [34] or take away children with DSDs (a popular belief in some regions of India), preventing the fearful parents from avoiding medical consultation during infancy [35]. In India, they are also known as Kothis, Aravanis, Jogappas, Shiv-Shakthis, Aruvanis, and Kinnar; however, in the nineteenth century, British colonial authorities criminalized hijras, to eradicate them [36]. The Supreme Court of India, in 2014, directed the Central and State Governments of India to treat Hijra (South Asia) as a third gender with full human rights, since they disconnect with the binary concept of male/female, either somatically (born hijras, individuals with a DSD) or behaviorally (made hijras, like transgenders in the West) [37, 38].

DSDs are usually diagnosed at birth, however, not all childbirths are supervised by healthcare workers (in developing countries), and few of these children remain undiagnosed until puberty or much later during adulthood, nearing marriage or presenting with infertility complaints [39]. Adolescence patients with gender-atypical pubertal development present with a lack of sexual satisfaction with their gender of rearing; hirsutism in females with CAH; gynecomastia in males with partial AIS (PAIS) and gender dysphoria. Parents of and occasionally patients with DSD prefer male gender assignment in 46, XY (excluding complete AIS [CAIS]), since they have access to greater socioeconomic independence in a patriarchal society (in India) [40]. Also, infertility in females decreases their prospects of marriage burdening the family with constant protection and economic support.

DSDs are congenital conditions and therefore psychoeducational counseling [41] is required to decrease discrimination, stigmatization, and bullying which render both, their families [42] as well as their positive psychosocial adaptation. They struggle with access to education, employment, identity documents, and marriage, especially in developing countries with worrying reports of infanticide [43]. Reactions to infertility (bereavement) are that of a major circumstantial loss (restricted experience and unnecessary loss) which facilitates understanding the vicious circle of stigmatization and learned helplessness [44]. Parents, especially in developing countries, from varied demographics of socioeconomic status, perceived children with DSD as a major medical and social problem, expecting considerable social discrimination, throughout life.

36.4 Decision-Making: Predictability over Uncertainty

Multiple disciplines of medical, political-economic, organizational, and business fields, have been involved in studying the process of decision-making [45]. Decision-making, in psychology, is a cognitive process, of making choices by identifying and defining the nature of the decision needed, from several alternatives, to achieve a desired (rational or irrational) result [46]. While deciding, the focus is either on how decisions should be made (normative approach) or by assessing the conditions of uncertainty and available alternatives (descriptive approach) [47, 48]. To reduce uncertainty [49, 50], human decisions are based both on hedonic motivations [51] linked to factors like personality [52] (people high in openness take more risks than neurotics), level of stress [53] (neurotics perform worse under pressure) anxiety [54] (decisions influenced by excessive fear and concern) emotions [55] and contexts [56] rather than rational [50] choices.

In most of the medical literature, important decisions are made by experts, who seek consensus and patients are asked to choose from a set of options excluding activities central to life choices; influenced by context, social interaction, and individual differences can have effects and repercussions on the entire life ahead [57, 58]. A one-time decision, for adolescents, is a crucial phase which will leave its deep imprints on their physical (irreversible surgeries) and emotional states, doubts concerning body image, sex life, gender role and identity, and with lifelong consequences (hormone replacement therapy) or loss of sexual sensation and body integrity [59].

The concept of shared decision-making [60] is a better alternative, where both patients and psychologist share responsibilities for their final agreement, increasing decisional competence [61] and patient satisfaction, and lowering decisional conflict, rather than putting the burden of decision-making solely on the teenagers [62]. Professional psychological health care for DSD was proposed in the mid-1990s, with the emergence of intersex patient advocacy and support groups, which stressed bringing changes to clinical management, including access to customized and individualized psychological support and psychosocial care to promote positive adaptation [63].

Health literacy is a key attribute to aid decision-making among adolescents with DSD, as it helps in building informed discussions, through psychosocial support and psychological care, promoting long-term interests, especially about non-surgical options [64]. DSD is not only a congenital condition with the incongruence of sex characteristics, but also uncertainties concerning health-related, medical, psychosocial, and legal issues that raise controversial discussions around surgical interventions and the consent of minors. An important prerequisite to ensure the improvement in well-being among DSDs is to inform the importance of fetal hormone exposure, genital development, gonadal differentiation and potential of fertility and full disclosure [65], and to promote sex education and positive acceptance of bodily diversities from a sexual health perspective. Timely psychological assessment and interventions that support informed decision-making actions are an essential aspect of holistic care toward individuals with DSD to clear their uncertainties regarding the psychosocial sphere, as most of them suffer from social isolation, secrecy, shame, self-identity questioning, and mental disorders that negatively impact their quality of life [66].

The International Consensus Conference on Intersex strongly supports gender assignment, only after a detailed examination by a multidisciplinary team with complete medical disclosure, specifically emphasizing the importance of psychologists and psychiatrists, especially when an individual is recovering from the traumatic discovery of their DSD conditions and allowing DSD patients to decide once they have sufficiently developed for a psychological assessment of gender identity [67].

36.5 Adolescence: Work Under Construction!!!

Even though DSDs, due to atypical external genitalia, are most often diagnosed during the neonatal period, however, underlying symptoms can remain hidden becoming evident only during puberty, which encompasses rapid somatic growth, brain development, sexual maturation, and attainment of reproductive capacity [68], as well as final maturation and major changes of multiple organ systems, central nervous system, and psychosocial behaviors [69]. Adolescence is the phase of life, involving biological growth and major social role transitions, stretching between childhood and adulthood, occupying a greater course for life [70], and therefore, can become overwhelmed with emotions and sensations and develop tendencies to resort to bodily centered and sensation-dominated protections (anorexia, claustrophobia, agoraphobia) [71–73]. This scenario places young DSD patients in a completely new situation, that may be difficult for them to comprehend and adjust to the complexities of either absent, delayed, incomplete, or atypical sex [74] (mostly impairment of gonadal function [75]). The impact may be presented with lack or minimal pubertal development; menarche; vaginal, uterine, or breast agenesis and inappropriate sexual development such as virilization in females or feminization (gynecomastia) in males [76].

Many of the psychological aspects of management need to be stressed during adolescence, since most DSDs require lifelong therapy, as there is never going to be a perfect outcome, with or without treatment in individuals with DSD. And therefore, management aims to give the person the best possible chance of freedom from illness, healthy psychosexual development, a positive sense of social well-being and to maximize opportunities for satisfying sexual relationships and if possible, fertility [77]. Emotional [78] (depression and anxiety), behavioral disorders (attention deficit hyperactivity disorder, conduct disorder), eating disorders, psychosis [79, 80], and risk-taking [81, 82] behaviors, suicide and self-harm are a few of the leading causes of disabilities among adolescents and failing to address adolescent mental health, may consequently extend to adulthood, impairing both physical and mental health and limiting opportunities to lead fulfilling lives as adults [83]. The purpose of adolescent counseling is to help them transition to adulthood by navigating through their complex feelings, behaviors, and thoughts while experiencing mental and physical changes affecting their moods and confidence levels [84]. Teenagers tend to withdraw as puberty sets in and become aware, confused, and sometimes overwhelmed by their bodies and sexuality therefore adolescent counseling continues until therapy goals are achieved or fully recovered [85].

36.6 Paradigm Shift: Ambiguity to Normalization

36.6.1 Essentials of Psychological Counseling:

36.6.1.1 Biological and Medical

Puberty is intricately linked to the process of becoming an adult, and by nature, discussions on genitals and pubertal development are private and intimate, more so among individuals with a DSD therefore maintaining an individual's integrity is imperative during psychological interventions [86]. During puberty, the gonadal hormones (concentration of estrogen and testosterone) organize sociosexual behaviors, which becomes the decisive course for the development of secondary sex characteristics in the body. Females typically produce more estrogen than males and thus behaviors related to energy balance and maternal care are organized in females. Similarly, males typically produce more testosterone than females leading to sexual and aggressive behaviors in males [87, 88]. Puberty is increasingly recognized as a time of distinct transitional physiology [89], encompassing stages of adrenarche (early phase puberty) with few phenotypic signs (adrenal androgen) contributing to the structural and functional development of the brain and associated behaviors in adolescence, and gonadarche phase with sexual maturation and achievement of reproductive capacity; breast buds, and onset of menarche in girls and testicular enlargement in boys [90], with the production of gonadal steroids stimulating the growth and development of secondary sexual characteristics [91].

Young girls seek medical attention due to the absence of breast development or primary amenorrhea or with increasing virilization in puberty and boys may have gynecomastia or a slowly or non-progressing pubertal development, and in rare cases, present with recurrent macroscopic hematuria as a sign of menstrual bleeding into the urethra [92]. Androgen deficiency, especially in vaginal hypoplasia, leads to sexual problems due to negative psychological adaptation and therefore treatments for vaginal hypoplasia need to be multidisciplinary and aim to optimize the patient's physical and psychological health [93].

Reversing surgically constructed genitals (during birth or infancy), is extremely perplexing, as compared to the general population, DSD children have markedly higher rates of gender transition (with or without treatment) and therefore it is one of the decisive reasons to avoid unjustified medical surgeries without patient's consent [94]. Initial gender of rearing remains a better prognosticator of gender identity than prenatal androgen exposure, even though it has large effects on gendered activities and interests, and partially on sexual orientation [95]. Being raised as female after clitoral surgery (early childhood), their sexual functioning can be compromised and may experience difficulty in engaging in sexual relations [96]. Subsequently, they might find it hard to comprehend and incorporate their diagnosis (for example, MRKH syndrome with vaginal agenesis), accompanied by the feeling of uncertainty and isolation, emotional distress, and experienced loss recurring over time in different ways [97], which is why they need to be fully informed about the potential risks of their sexual and reproductive functions [93]. An increasing number of DSDs are being identified, enhancing a better understanding of malignancy risk, which prompts stronger medical decision-making for surgery during infancy (with major genital ambiguity) and improved fertility outcomes in Turner's syndrome, Congenital adrenal hyperplasia, and Swyer's syndrome. It is also encouraged that for markedly virilized 46, XX infants with CAH to be reared as female, while an infant with a micropenis as male as well as to prevent malignancy, and removal of testes [98].

36.6.1.2 Brain Development

In adolescence, a dramatic spurt in the brain begins with marked development of both cortical and subcortical structures [99], involving two key processes: significant growth and change in regions of the prefrontal cortex [100], and improved connectivity between regions of the prefrontal cortex and regions of the limbic system [101]. During adolescence, depending on age, experience, and hormonal changes, there is an intense remodeling of the brain and mechanisms include synaptic rearrangement and the addition of new neurons and glia, establishing this as a period of significant growth and development inside the teenage brain [102]. The pruning process (use it or lose it principle) in the brain, initiates at the back and becomes more efficient as it reaches the prefrontal cortex. Teenagers rely on the amygdala to make decisions and solve problems, associated with emotions, impulses, aggression, and instinctive behavior, and therefore systematic guidance is required in the process of decision-making by weighing potential consequences [103]. These changes are thought to underpin higher-order cognitive functions, such as reasoning, interpersonal interactions, the perception of both short- and long-term risk and reward, and the regulation of behavior and emotion [104]. Dual-system and imbalance models postulate that adolescence is a particularly vulnerable period because of the imbalance (developmentally normal) between early maturation of the limbic motivational and emotional systems (intense affective and behavioral reactions) and slower, or later, development of the regulatory regions of the prefrontal cortex [101], with limited capacity to regulate them [104].

36.6.1.3 Cognitive Development

Cognitive domains, including learning, reasoning, information processing, and memory, improve as adolescents develop. Executive functioning capabilities, which facilitate self-regulation of thoughts, actions, and emotions, continue to develop in parallel with changes in the prefrontal cortex. These increases in self-regulatory control are thought to support deductive reasoning; information processing; efficiency; and the capacity for abstract, planned, hypothetical, and multidimensional thinking [105]. Highly anxious adolescents make risk-averse personal choices and avoid risks and punishments more than obtaining distant positive outcomes [106, 107].

36.6.1.4 Gender Related

Today, gender and gender identity are increasingly recognized as a continuum, rather than a binary concept, however, even though societal acceptance of non-stereotypic gender presentations and non-binary gender identities has increased in the past decade, there has also been an increase in the prevalence of psychological distress, self-harm, and suicidal tendencies in individuals with DSD and therefore gender contentedness is put forward as the ultimate goal [108]. By engaging in regular enquiries about gender experience and gender-related contentedness, sensitivity toward the secondary distress arising from one's gender discomfort could be assessed. At the same time, the assessment should be framed in a way that does not stigmatize or pathologize any particular gendered presentation, by inquiring with open-ended questions so as not to raise a typical or atypical dichotomy [95]. Gender-related developmental variance among DSDs is more compared to other populations, which could be due to complex interaction of genes, varied exposure to sex hormones in the perinatal as well as at puberty and the constant psychological, and emotional conflict throughout their lives [109]. The majority of individuals with DSDs tend to identify with the gender designated at birth, (male or female) [110], and gender identity development among DSD children, having undergone successful surgical genitoplasty (phenotype), with the same sex of rearing, have better outcomes than that of the neutral upbringing of a child with ambiguous genitalia [111].

Compared to the general population, gender dissatisfaction (unhappiness with assigned sex) occurs more frequently in DSDs, however, it is difficult to predict from karyotype, prenatal androgen exposure, degree of genital virilization, or assigned gender [112]. Compared to the general population, rates of gender dysphoria, intergender identities, and gender change are higher among individuals with DSDs, especially among self-reported gender dysphoria from an early age [95], emphasizing the need for long-term follow-up and support.

Sometimes medical and psychosocial factors (gender assignment, information management and communication, timing of medical interventions, consequences of surgery, and sexuality) may jeopardize psychological development affecting the quality of life and a desire to change gender later in life [113].

36.6.1.5 Communication: Disclosure and Consent

The importance of full disclosure (communicating diagnosis) and informed consent in DSD is a universal human right, nurturing complete participation in decision-making and informed choice and helping to undo the unnecessary burden of deception, rage, sorrow, and regret, associated with non-disclosure [114]. Adolescents with DSD have the right and responsibility to know about their conditions as confidentiality and lack of information lead to shame and trauma, leading to avoidance of future health care (women with AIS may avoid medical care including needed Hormone Replacement Therapy (HRT) and psychiatric morbidity, specifically anxiety) [115, 116]. Due to a lack of long-term information regarding the medical, surgical, and psychosexual outcome, controversy still exists regarding appropriate treatment guidelines in DSD [117]. Psychological intervention enhances informed consent among DSD adolescents to enhance medical decisions (whether to have or not to have specific interventions), and to foster self-determination and self-advocacy, placing the youth as the guide, while also encompassing psychoeducation around psychosexual concerns (fertility, sex, gender, intimacy, and sexuality) [118].

36.6.1.6 Psychological Counseling

Professional psychological health care for DSD was proposed in the mid-1990s, with the emergence of intersex patient advocacy and support groups, which stressed bringing changes to clinical management, including access to customized and individualized psychological support and psychosocial care to promote positive adaptation [119]. Psychologists, as part of specialized healthcare professionals' teams, help to explore and provide psychological support to adolescents with DSD in the areas of sexuality and body image to understand their current condition, since some may eventually need hormonal therapy or might proceed with altering their appearance of their existing genitals or to the extremes of complete genitals surgery [120]. Psychologists, to build self-efficacy among adolescents with DSD, should foster a holistic view of the specific variations (unique to each one) in DSDs, by providing support toward treatment through evaluating their mental health issues (adjustment, anxiety, depression) and psychological functioning; also addressing their quality of life by identifying factors (bio-psycho-socio) that impact their psychological health and life satisfaction [121].

Psychoeducation among DSD adolescents increases control of decisions relating to treatment, as some of them continue to remain distressed by previous non-disclosure (below the age of consent) of atypical development of the reproductive-genital system and its implications, which often poses concerns in establishing and overcoming difficulties in partnership and sexual relationships in future [122]. To challenge the concepts of normalcy and to explore different aspects of DSD, psychological therapies can offer opportunities, especially in support groups, to explore different interpretations and solutions, to help reduce isolation [85]. A cognitive-behavioral counseling approach focuses on exploration and coping with sexual problems, among DSDs, since cognition and emotions are related to sexual self-concept, and fundamental in sexual functioning [123].

During psychological management, utmost sensitivity toward their complex feelings should be analyzed, possibly by questioning their concerns regarding gender identity, sexuality, and fertility, with no hesitation or doubts about the individual's sense of gender identity because sometimes not only do they have transgender feelings, or a desire for gender change but may also want to be identified completely different from male or female [124]. Even though it is rare for individuals with a DSD to change gender, however, the atypical hormone may impact (levels of androgens is higher females and males) psychosexual development, influencing gender identity and sexual orientation, and henceforth one needs to have an awareness of the complexities of their feelings [122].

36.7 Future Scope: Way to Go

The diagnosis of DSD expressed as differentiation of urogenital tract and clinical phenotypes, requires a structured approach, involving a multidisciplinary team and different molecular techniques [125], championed to bring balance between intervention outcomes and differing opinions [126]. In cases of incompatibility, between genetic, gonadal, genital, and somatic sex, formulating individualized medical procedures for official sex determination, fertility, and hormonal activity of gonads and responsible surgical decisions (genitals and gonads) is essential. However, for appropriate sexual development with ambiguous genitalia, limited empirical evidence is available regarding the future development of their gender identity, sexual orientation, and sexual functioning [96]. Systematically introducing brief structured evidence-based psychological interventions [127] facilitating coping and recovery from surgery is crucial, to reducing future complications [128]. There is an increasing need for multidisciplinary medical care by specialized teams for improving outcome-focused research with appropriate assessment of partnership, sexuality, and fertility as well as the physical quality of life among DSDs [129]. Accordingly, the goal of clinical management is not only gender assignment and normalized genital appearance, but also ensuring the patient's quality of life encompassing the ability to form intimate relationships, and sexual functioning and sensation. The gender among DSD conditions should be based on the identity and future fertility of the patient as well as psychological evaluation for children experiencing persistent (early childhood) gender dysphoria and are in the process of gender transitioning to the desired gender [67], which is why the desire for transition should be respected, especially of children's experiencing gender dysphoria from early stages of development.

DSD is associated with myriad complex diagnoses and the physical treatment demands can vary from nothing to multiple life-saving operations, invariably impacting them emotionally and socially, requiring repetitive services from psychologists for appropriate social functioning among adolescents with DSD (especially among recently diagnosed DSDs). To avoid emotionally driven decisions, psychological counseling, including gender and sexual services is recommended to promote individualized development [130]. Compared to the general population, long-term physical health conditions are associated with higher levels of psychological problems, detrimentally affecting the capacity to self-manage health maintenance and lifestyle changes, leading to poorer health outcomes and higher usage of health care services, suggestive of longitudinal studies on gender-normative and gender-variant developments.

To improve the decision-making process regarding genital surgery in DSD, shared decision-making (SDM) is encouraged for addressing emotions; identify options, concerns, and values while making important healthcare decisions. Psychosocial support involving the participation of trained peers and online resources (psychological services) in the decision-making process, strengthens positive effects on psychological well-being among DSDs [131].

Long-term studies are required, especially in Africa and East Asia, having very different and unique cultures and languages [130], since legal and policy considerations are diverse in different countries or even by states and regions within a country. Even though knowledge of local and national laws and regulations is important, however, based on ethical and cultural considerations, the primary factor in decision-making should always be the patient's autonomy and right to an open future. Like the landmark judgment of the Madras High Court in the state of Tamil Nadu (India), in 2019 [132, 133], which led to the ban on unnecessary medical normalizing (sex-selective) surgeries on children born with DSD variations. This judgment is likely to have far-reaching consequences and it is proposed that similar stances should be taken by other governments and the medical communities, around the globe, to develop standards by upholding and recognizing the individuals' (especially children's) informed rights to consent, bodily integrity, privacy, and to determine one's gender identity.

References

1. Ghiselin MT. The evolution of hermaphroditism among animals. Q Rev Biol [Internet]. 1969;44(2):189–208. Available from: http://www.jstor.org/stable/2819439
2. Androutsos G. Hermaphroditism in Greek and Roman antiquity. Hormones (Athens) [Internet]. 2006;5(3):214–7. Available from: https://doi.org/10.14310/horm.2002.11187
3. Koh J. The history of the concept of gender identity disorder. Seishin Shinkeigaku Zasshi. 2012;114(6):673–80.
4. Diamond D, Sytsma S, Dreger A, Wilson B. Case study: Culture clash involving intersex. Hastings Cent Rep [Internet]. 2003;33(4):12. Available from: https://doi.org/10.2307/3528374
5. Thyen U, Richter-Appelt H, Wiesemann C, Holterhus P-M, Hiort O. Deciding on gender in children with intersex conditions: considerations and controversies. Treat Endocrinol [Internet]. 2005;4(1):1–8. Available from: https://doi.org/10.2165/00024677-200504010-00001
6. Kraus C. Classifying intersex in DSM-5: Critical reflections on gender dysphoria. Arch Sex Behav [Internet]. 2015;44(5):1147–63. Available from: https://doi.org/10.1007/s10508-015-0550-0
7. American Psychiatric Association. Diagnostic and statistical manual of mental disorders: DSM-5. Washington, DC: American Psychiatric Association; 2013.
8. Hughes IA. Intersex. BJU Int [Internet]. 2002;90(8):769–76. Available from: https://doi.org/10.1046/j.1464-410x.2002.02920.x
9. Richter-Appelt H, Discher C, Gedrose B. Gender identity and recalled gender related childhood play-behaviour in adult individuals with different forms of intersexuality. Anthropol Anz [Internet]. 2005;63(3):241–56. Available from: https://doi.org/10.1127/anthranz/63/2005/241
10. Bessiène L, Lombès M, Bouvattier C. Controverses et challenges sur le développement sexuel différent (DSD). Annales d'Endocrinologie. 2018;79:S22–30.
11. Acién P, Acién M. Disorders of sex development: Classification, review, and impact on fertility. J Clin Med [Internet]. 2020;9(11):3555. Available from: https://doi.org/10.3390/jcm9113555
12. Dreger AD. "Ambiguous sex": or ambivalent medicine? Ethical issues in the treatment of intersexuality. Hastings Cent Rep [Internet]. 1998;28(3):24–35. Available from: https://doi.org/10.2307/3528648
13. Dreger AD. How sex changed: a history of transsexuality in the United States (review). J Hist Sex [Internet]. 2006;15(1):148–51. Available from: https://doi.org/10.1353/sex.2006.0051
14. Dreger AD, Herndon AM. Progress and politics in the intersex rights movement. GLQ [Internet]. 2009;15(2):199–224. Available from: https://doi.org/10.1215/10642684-2008-134
15. Dreger AD, Vasey PL. Sex, gender, and sexuality diversity: Introduction. Perspect Biol Med [Internet]. 2007;50(4):479–80. Available from: https://doi.org/10.1353/pbm.2007.0045
16. Gorman C, Cole W. Between the sexes. Time [Internet]. 2004 [cited 7 Feb 2023]. Available from: https://content.time.com/time/magazine/article/0,9171,593551,00.html
17. Diamond M, Sigmundson K. Sex reassignment at birth: Long-term review and clinical complications. Arch Pediatr Adolesc Med. 1997;151(3):298–304.
18. Lloyd E. From the hospital to the courtroom: A statutory proposal for recognizing and protecting the legal rights of intersex children. Cardozo J Law Gender. 2005;12(1):155–95.
19. Kuhnle U, Krahl W. The impact of culture on sex assignment and gender development in intersex patients. Perspect Biol Med [Internet]. 2002;45(1):85–103. Available from: https://doi.org/10.1353/pbm.2002.0011
20. Taha SA. Male pseudohermaphroditism: factors determining the gender of rearing in Saudi Arabia. Urology [Internet]. 1994;43(3):370–4. Available from: https://doi.org/10.1016/0090-4295(94)90082-5
21. Warne G, Bhatia V. Intersex, East and West. In: International library of ethics, law and the new medicine. Dordrecht: Kluwer Academic; 2006. p. 183–205.

22. Shawky RM, Nour El-Din SM. Profile of disorders of sexual differentiation in the Northeast region of Cairo. Egypt. Egypt J Med Hum Genet [Internet]. 2012;13(2):197–205. Available from: https://doi.org/10.1016/j.ejmhg.2012.03.003
23. Jaja T, Yarhere I, Anochie IC. Ambiguous external genitalia in childhood in Port Harcourt, Nigeria. Niger Pediatr Ther. 2011;1(3):1–5.
24. Diamond M, Beh HG. Changes in the management of children with intersex conditions. Nat Clin Pract Endocrinol Metab [Internet]. 2008;4(1):4–5. Available from: https://doi.org/10.1038/ncpendmet0694
25. Fathi N. Despite president's denials, gays insist they exist, if quietly, in Iran. New York Times; 2007.
26. Systma S. Intersexuality, cultural influences, and cultural relativism. In: Systma S, editor. Ethics and intersex. Berlin: Springer; 2006. p. 259–70.
27. Tait R. Sex change funding undermines no gays claim. The Guardian [Internet]. 2007 [cited 7 Feb 2023]. Available from: https://www.theguardian.com/world/2007/sep/26/iran.gender
28. Usher S. Gender correction for Saudi girls. BBC News. BBC [Internet] [cited 7 Feb 2023]. Available from: http://news.bbc.co.uk/2/hi/middle_east/3814041.stm
29. Das A. Tritiya-Prakriti: people of the third sex. Philadelphia: Xlibris Corporation; 2010.
30. Pattanaik D. Shikhandi and other tales they don't tell you. New Delhi: Zubaan and Penguin Books India; 2014.
31. Nanda S. The hijras of India: cultural and individual dimensions of an institutionalized third gender role. J Homosex. 1985;11(3–4):35–54.
32. Nanda S. Gender diversity: crosscultural variations. 2nd ed. Waveland Press; 2014.
33. Seow L. Cultureshock! India. Singapore: Times Editions; 2017.
34. Mukherjee JB. Castration—A means of induction into the Hijirah group of the eunuch community in India: A critical study of 20 cases. Am J Forensic Med Pathol [Internet]. 1980;1(1):61–6. Available from: https://doi.org/10.1097/00000433-198003000-00010
35. Dalal JS, Gorea RK, Rai H, Chanana A, Kaur P, Rai G, et al. Conversion to Hijrah (Eunuch) is it need based or a crime. JIAFM. 2004;26(4):971–3.
36. Jagadish P. Mainstreaming third-gender healers: the changing perceptions of South Asian Hijras. Vanderbilt Undergrad Res J. 2009;9:1–8.
37. Hinchy J. Governing gender and sexuality in colonial India: the Hijra, c.1850-1900. Cambridge, UK: Cambridge University Press; 2019.
38. Nanda S. Neither man nor woman: The Hijras of India. 2nd ed. Belmont, CA: Wadsworth; 1999.
39. De Clercq E, Starke G, Rost M. "Waking up" the sleeping metaphor of normality in connection to intersex or DSD: a scoping review of medical literature. Hist Philos Life Sci. 2022;44
40. Syme GB. Facing the unacceptable: the emotional response to infertility. Hum Reprod. 1997;12(11 Suppl):183–7.
41. Kalra G, Shah N. The cultural, psychiatric, and sexuality aspects of hijras in India. Int J Transgend [Internet]. 2013;14(4):171–81. Available from: https://doi.org/10.1080/15532739.2013.876378
42. Lal V. Not this, not that: The hijras of India and the cultural politics of sexuality. Soc Text [Internet]. 1999;61:119–40. Available from: http://www.jstor.org/stable/488683
43. Rolston AM, Gardner M, Vilain E, Sandberg DE. Parental reports of stigma associated with child's disorder of sex development. Int J Endocrinol [Internet]. 2015;2015:980121. Available from: https://doi.org/10.1155/2015/980121
44. Statement of OII Europe on intersex, disability and the UN convention on the rights of people with disabilities [Internet]. Oiieurope.org [cited 7 Feb 2023]. Available from: https://oiieurope.org/wp-content/uploads/2015/05/CRPD_2015_Statement_OII_Europe.pdf
45. Eisenfuhr F. Decision making. New York: Springer; 2011 [Internet] [cited 7 Feb 2023]. Available from: http://www.sciepub.com/reference/264313
46. Savioni L, Triberti S, Durosini I, Pravettoni G. How to make big decisions: A cross-sectional study on the decision making process in life choices. Curr Psychol [Internet]. 2022. Available from: https://doi.org/10.1007/s12144-022-02792-x

47. Beach LR, Connolly T. The psychology of decision making: People in organizations. Thousand Oaks: Sage; 2005.
48. Kahneman D, Tversky A. Prospect theory: An analysis of decision under risk. In: MacLean LC, Ziemba WT, editors. Handbook of the fundamentals of financial decision making: Part I. Hackensack: World Scientific; 2013. p. 99–127.
49. Cabanac M. Pleasure: the common currency. J Theor Biol [Internet]. 1992;155(2):173–200. Available from: https://doi.org/10.1016/s0022-5193(05)80594-6
50. Bentham J. An introduction to the principles of morals and legislation (1789). Burns JH, editor. London: Athlone Press; 1970.
51. Kahneman D, Frederick S. Representativeness revisited: attribute substitution in intuitive judgment. In: Gilovich T, Griffin D, Kahneman D, editors. Heuristics and biases. New York: Cambridge University Press; 2002. p. 49–81.
52. Tosun C, Yıldırım M, Altun F, Yazıcı H. Personality traits and attachments styles in romantic relationships deciding. J Fam Issues [Internet]. 2022;43(5):1219–34. Available from: https://doi.org/10.1177/0192513x211022383
53. Byrne KA, Silasi-Mansat CD, Worthy DA. Who chokes under pressure? The Big Five personality traits and decision-making under pressure. Pers Individ Dif [Internet]. 2015;74:22–8. Available from: https://doi.org/10.1016/j.paid.2014.10.009
54. Hartley CA, Phelps EA. Anxiety and decision-making. Biol Psychiatry [Internet]. 2012;72(2):113–8. Available from: https://doi.org/10.1016/j.biopsych.2011.12.027
55. Mazzocco K, Masiero M, Carriero MC, Pravettoni G. The role of emotions in cancer patients' decision-making. Ecancermedicalscience [Internet]. 2019;13:914. Available from: https://doi.org/10.3332/ecancer.2019.914
56. Tversky A, Kahneman D. Judgment under uncertainty: heuristics and biases: biases in judgments reveal some heuristics of thinking under uncertainty. Science. 1974;185:1124–31.
57. Galotti KM. Approaches to studying formal and everyday reasoning. Psychol Bull [Internet]. 1989;105(3):331–51. Available from: https://doi.org/10.1037/0033-2909.105.3.331
58. Tversky A, Kahneman D. The framing of decisions and the psychology of choice. Science [Internet]. 1981;211(4481):453–8. Available from: https://doi.org/10.1126/science.7455683
59. Lampalzer U, Briken P, Schweizer K. Dealing with uncertainty and lack of knowledge in diverse sex development: controversies on early surgery and questions of consent. Sex Med [Internet]. 2020;8(3):472–89. Available from: https://doi.org/10.1016/j.esxm.2020.03.002
60. Wray LD, Stone ER. The role of self-esteem and anxiety in decision making for self-versus others in relationships. J Behav Decis Making. 2005;18(2):125–44.
61. Shay LA, Lafata JE. Where is the evidence? A systematic review of shared decision making and patient outcomes. Med Decis Making [Internet]. 2015;35(1):114–31. Available from: https://doi.org/10.1177/0272989X14551638
62. Siminoff LA, Sandberg DE. Promoting shared decision making in disorders of sex development (DSD): Decision aids and support tools. Horm Metab Res [Internet]. 2015;47(5):335–9. Available from: https://doi.org/10.1055/s-0035-1545302
63. Liao L-M, Simmonds M. A values-driven and evidence-based health care psychology for diverse sex development. Psychol Sex [Internet]. 2014;5(1):83–101. Available from: https://doi.org/10.1080/19419899.2013.831217
64. Magrite E, Williams J, Amyot E, Usipuik M, Sanders C. Listening to individuals with differences in sex development or intersex and their families: "not doing surgery doesn't mean doing nothing". Horm Res Paediatr. 2022;96(2):228–37.
65. Nabhan ZM, Lee PA. Disorders of sex development. Curr Opin Obstet Gynecol [Internet]. 2007;19(5):440–5. Available from: https://doi.org/10.1097/GCO.0b013e3282ecb13d
66. Moreno-Begines M, De Las N, Arroyo-Rodríguez A, Borrallo-Riego Á, Guerra-Martín MD. Intersexuality/differences of sex development through the discourse of intersex people, their relatives, and health experts: a descriptive qualitative study. Healthcare. 2022;10(4)
67. Littman L. Individuals treated for gender dysphoria with medical and/or surgical transition who subsequently detransitioned: A survey of 100 detransitioners. Arch Sex Behav [Internet]. 2021;50(8):3353–69. Available from: https://doi.org/10.1007/s10508-021-02163-w

68. Viner RM, Allen NB, Patton GC. Puberty, developmental processes, and health interventions. In: Bundy DAP, de Silva N, Horton S, Jamison DT, Patton GC, editors. Child and adolescent health and development, 3rd ed. Disease control priorities, vol 8. Washington, DC: World Bank; 2017. p. 107–18.
69. Patton GC, Hemphill SA, Beyers JM, Bond L, Toumbourou JW, McMorris BJ, et al. Pubertal stage and deliberate self-harm in adolescents. J Am Acad Child Adolesc Psychiatry [Internet]. 2007;46(4):508–14. Available from: https://www.sciencedirect.com/science/article/pii/S0890856709617032
70. Sawyer SM, Azzopardi PS, Wickremarathne D, Patton GC. The age of adolescence. Lancet Child Adolesc Health [Internet]. 2018;2(3):223–8. Available from: https://doi.org/10.1016/S2352-4642(18)30022-1
71. Glenn R. The post-traumatic stress disorder sourcebook: a guide to healing, recovery, and growth. 2nd ed. New York: McGraw-Hill; 2009.
72. Carvalho R II. A brief introduction to the thought of Armando B. Ferrari. Br J Psychother [Internet]. 2012;28(4):413–34. Available from: https://doi.org/10.1111/j.1752-0118.2012.01300.x
73. Romano F III. Anorexia and bulimia: Two aspects of adolescence. Br J Psychother [Internet]. 2013;29(1):5–24. Available from: https://doi.org/10.1111/j.1752-0118.2012.01301.x
74. Ahmed SF, Achermann JC, Arlt W, Balen A, Conway G, Edwards Z, et al. Society for Endocrinology UK guidance on the initial evaluation of an infant or an adolescent with a suspected disorder of sex development (Revised 2015). Clin Endocrinol (Oxf) [Internet]. 2016;84(5):771–88. Available from: https://doi.org/10.1111/cen.12857
75. Gomes NL, Chetty T, Jorgensen A, Mitchell RT. Disorders of sex development-novel regulators, impacts on fertility, and options for fertility preservation. Int J Mol Sci [Internet]. 2020;21(7):2282. Available from: https://doi.org/10.3390/ijms21072282
76. Lee PA, Houk CP. Disorders of sexual differentiation in the adolescent. Ann N Y Acad Sci [Internet]. 2008;1135(1):67–75. Available from: https://doi.org/10.1196/annals.1429.028
77. Raza J, Zaidi SZ, Warne GL. Management of disorders of sex development—With a focus on development of the child and adolescent through the pubertal years. Best Pract Res Clin Endocrinol Metab. 2019;33(3)
78. Roza SJ, Hofstra MB, van der Ende J, Verhulst FC. Stable prediction of mood and anxiety disorders based on behavioral and emotional problems in childhood: a 14-year follow-up during childhood, adolescence, and young adulthood. Am J Psychiatry [Internet]. 2003;160(12):2116–21. Available from: https://doi.org/10.1176/appi.ajp.160.12.2116
79. Kapur S. Psychosis as a state of aberrant salience: a framework linking biology, phenomenology, and pharmacology in schizophrenia. Am J Psychiatry [Internet]. 2003;160(1):13–23. Available from: https://doi.org/10.1176/appi.ajp.160.1.13
80. Pencer A, Addington J, Addington D. Outcome of a first episode of psychosis in adolescence: a 2-year follow-up. Psychiatry Res [Internet]. 2005;133(1):35–43. Available from: https://doi.org/10.1016/j.psychres.2004.10.004
81. Diclemente RJ, Hansen WB, Ponton LE. Handbook of adolescent health risk behavior. Berlin: Springer; 2013.
82. Steinberg L. A social neuroscience perspective on adolescent risk-taking. Dev Rev [Internet]. 2008;28(1):78–106. Available from: https://doi.org/10.1016/j.dr.2007.08.002
83. Adolescent mental health [Internet] [cited 7 Feb 2023]. Available from: https://www.who.int/news-room/fact-sheets/detail/adolescent-mental-health
84. Ramakrishnan VK, Jalajakumari VT. Significance of imparting guidance and counselling programmes for adolescent students. Asia Pac J Res. 2013;2(9):102–12.
85. Lampalzer U, Briken P, Schweizer K. Psychosocial care and support in the field of intersex/diverse sex development (DSD): counselling experiences, localisation and needed improvements. Int J Impot Res [Internet]. 2021;33(2):228–42. Available from: https://doi.org/10.1038/s41443-021-00422-x
86. Nordenström A. Puberty in individuals with a disorder of sex development. Curr Opin Endocr Metab Res [Internet]. 2020;14:42–51. Available from: https://www.sciencedirect.com/science/article/pii/S2451965020300545

87. Kulkarni KP, Panigrahi I, Das R, Kaur S, Marwaha RK. Pediatric disorders of sex development. Indian J Pediatr [Internet]. 2009;76(9):956–8. Available from: https://doi.org/10.1007/s12098-009-0193-z
88. Sisk CL. Hormone-dependent adolescent organization of socio-sexual behaviors in mammals. Curr Opin Neurobiol [Internet]. 2016;38:63–8. Available from: https://doi.org/10.1016/j.conb.2016.02.004
89. Rosenfeld RG, Nicodemus BC. The transition from adolescence to adult life: physiology of the "transition" phase and its evolutionary basis. Horm Res Paediatr. 2003;60(1):74–7.
90. Whittle S, Simmons JG, Byrne ML, Strikwerda-Brown C, Kerestes R, Seal ML, et al. Associations between early adrenarche, affective brain function and mental health in children. Soc Cogn Affect Neurosci [Internet]. 2015;10(9):1282–90 [cited 7 Feb 2023]. Available from: https://academic.oup.com/scan/article/10/9/1282/1675969?login=true
91. Mundy LK, Romaniuk H, Canterford L, Hearps S, Viner RM, Bayer JK, et al. Adrenarche and the emotional and behavioral problems of late childhood. J Adolesc Health [Internet]. 2015;57(6):608–16. Available from: https://doi.org/10.1016/j.jadohealth.2015.09.001
92. Sani AM, Arif II, Arshad MM, Mungadi IA, Soh KG, Soh KL. Women with amenorrhea and men with menstruation: The qualitative experiences of people with disorders of sex development in Nigeria. J Nurs Res [Internet]. 2020;28(1):e67. Available from: https://doi.org/10.1097/jnr.0000000000000313
93. Minto CL, Liao L-M, Woodhouse CRJ, Ransley PG, Creighton SM. The effect of clitoral surgery on sexual outcome in individuals who have intersex conditions with ambiguous genitalia: a cross-sectional study. Lancet [Internet]. 2003;361(9365):1252–7. Available from: https://doi.org/10.1016/S0140-6736(03)12980-7
94. Hughes IA, Houk C, Ahmed SF, Lee PA, LWPES Consensus Group, ESPE Consensus Group. Consensus statement on management of intersex disorders. Arch Dis Child [Internet]. 2006;91(7):554–63. Available from: https://doi.org/10.1136/adc.2006.098319
95. Callens N, Van Kuyk M, van Kuppenveld JH, Drop SLS, Cohen-Kettenis PT, Dessens AB, et al. Recalled and current gender role behavior, gender identity and sexual orientation in adults with disorders/differences of sex development. Horm Behav [Internet]. 2016;86:8–20. Available from: https://doi.org/10.1016/j.yhbeh.2016.08.008
96. Gooren LJG. Psychological consequences. Semin Reprod Med [Internet]. 2002;20(3):285–96. Available from: https://doi.org/10.1055/s-2002-35375
97. Holt R, Slade P. Living with an incomplete vagina and womb: an interpretative phenomenological analysis of the experience of vaginal agenesis. Psychol Health Med [Internet]. 2003;8(1):19–33. Available from: https://doi.org/10.1080/1354850021000059232
98. Kolesinska Z, Ahmed SF, Niedziela M, Bryce J, Molinska-Glura M, Rodie M, et al. Changes over time in sex assignment for disorders of sex development. Pediatrics [Internet]. 2014;134(3):e710–5. Available from: https://doi.org/10.1542/peds.2014-1088
99. Goddings A-L, Burnett Heyes S, Bird G, Viner RM, Blakemore S-J. The relationship between puberty and social emotion processing: Puberty and social brain development. Dev Sci [Internet]. 2012;15(6):801–11. Available from: https://doi.org/10.1111/j.1467-7687.2012.01174.x
100. Paus T, Keshavan M, Giedd JN. Why do many psychiatric disorders emerge during adolescence? Nat Rev Neurosci [Internet]. 2008;9(12):947–57. Available from: https://doi.org/10.1038/nrn2513
101. Casey BJ. Beyond simple models of self-control to circuit-based accounts of adolescent behavior. Annu Rev Psychol [Internet]. 2015;66(1):295–319. Available from: https://doi.org/10.1146/annurev-psych-010814-015156
102. Spear LP. Adolescent brain development and animal models. Ann N Y Acad Sci [Internet]. 2004;1021(1):23–6. Available from: https://doi.org/10.1196/annals.1308.002
103. Natterson-Horowitz B, Bowers K. Wildhood: the epic journey from adolescence to adulthood in humans and other animals. New York: Scribner; 2019.

104. Gardner M, Steinberg L. Peer influence on risk taking, risk preference, and risky decision making in adolescence and adulthood: an experimental study. Dev Psychol [Internet]. 2005;41(4):625–35. Available from: https://doi.org/10.1037/0012-1649.41.4.625
105. Kesek A, Zelazo PD, Lewis MD. The development of executive cognitive function and emotion regulation in adolescence. In: Allen NB, Sheeber LB, editors. Adolescent emotional development and the emergence of depressive disorders. Cambridge: Cambridge University Press; 2008. p. 135–55.
106. Lauriola M, Levin IP. Personality traits and risky decision-making in a controlled experimental task: an exploratory study. Pers Individ Dif [Internet]. 2001;31(2):215–26. Available from: https://doi.org/10.1016/s0191-8869(00)00130-6
107. Tversky A, Kahneman D. Judgment under uncertainty: Heuristics and biases. In: Kahneman D, Slovic P, Tversky A, editors. Judgment under uncertainty. Cambridge: Cambridge University Press; 1982. p. 3–20.
108. Meyer-Bahlburg HFL, Dolezal C, Baker SW, Ehrhardt AA, New MI. Gender development in women with congenital adrenal hyperplasia as a function of disorder severity. Arch Sex Behav [Internet]. 2006;35(6):667–84. Available from: https://doi.org/10.1007/s10508-006-9068-9
109. Zucker KJ, Lawrence AA, Kreukels BPC. Gender dysphoria in adults. Annu Rev Clin Psychol [Internet]. 2016;12(1):217–47. Available from: https://doi.org/10.1146/annurev-clinpsy-021815-093034
110. Meyer-Bahlburg H. Gender identity outcome in female-raised 46, XY persons with penile agenesis, cloacal exstrophy of the bladder, or penile ablation. Arch Sex Behav. 2005;34:423–38.
111. Nihoul-Fékété C. Does surgical genitoplasty affect gender identity in the intersex infant? Horm Res [Internet]. 2005;64(Suppl. 2):23–6. Available from: https://doi.org/10.1159/000087749
112. Cohen-Kettenis PT. Gender change in 46, XY persons with 5α-reductase-2 deficiency and 17β-hydroxysteroid dehydrogenase-3 deficiency. Arch Sex Behav. 2005;34:399–410.
113. Cohen-Kettenis PT. Psychosocial and psychosexual aspects of disorders of sex development. Best Pract Res Clin Endocrinol Metab [Internet]. 2010;24(2):325–34. Available from: https://doi.org/10.1016/j.beem.2009.11.005
114. D'Alberton F. Disclosing disorders of sex development and opening the doors. Sex Dev [Internet]. 2010;4(4–5):304–9. Available from: https://doi.org/10.1159/000317480
115. Slijper FM, Frets PG, Boehmer AL, Drop SL, Niermeijer MF. Androgen insensitivity syndrome (AIS): emotional reactions of parents and adult patients to the clinical diagnosis of AIS and its confirmation by androgen receptor gene mutation analysis. Horm Res [Internet]. 2000;53(1):9–15. Available from: https://doi.org/10.1159/000023506
116. Engberg H, Strandqvist A, Nordenström A, Butwicka A, Nordenskjöld A, Hirschberg AL, et al. Increased psychiatric morbidity in women with complete androgen insensitivity syndrome or complete gonadal dysgenesis. J Psychosom Res [Internet]. 2017;101:122–7. Available from: https://doi.org/10.1016/j.jpsychores.2017.08.009
117. Wisniewski AB, Migeon CJ, Meyer-Bahlburg H, Gearhart JP, Berkovitz GD, Brown TR. Complete androgen insensitivity syndrome: long-term medical, surgical, and psychosexual outcome. J Clin Endocrinol Metab. 2000;85(8):2664–9.
118. Cizek S, Fluharty E. The continuous textbook of women's. Professor Judith Simms-Cendan. 2022;2.
119. Liao LM. Learning to assist women born with atypical genitalia: journey through ignorance, taboo and dilemma. J Reprod Infant Psychol [Internet]. 2003;21(3):229–38. Available from: https://doi.org/10.1080/0264683031000154971
120. Differences in sex development [Internet]. nhs.uk [cited 7 Feb 2023]. Available from: https://www.nhs.uk/conditions/differences-in-sex-development/
121. Cizek S, Fluharty E, Talreja KL, Weigel N, Muscarella M, Shaw RJ, et al. Differences of sex development [Internet]. The global library of women's medicine [cited 7 Feb 2023]. Available from: https://www.glowm.com/article/heading/vol-2%2D%2Dadolescent-gynecology%2D%2Ddifferences-of-sex-development-intersex-care/id/418293

122. Jürgensen M, Kleinemeier E, Lux A, Steensma TD, Cohen-Kettenis PT, Hiort O. Psychosexual development in adolescents and adults with disorders of sex development-results from the German clinical evaluation study. J Sex Med. 2013;10(11):2703–14.
123. de Neve-Enthoven NGM, Callens N, van Kuyk M, Verhaak CM, van der Ende J, Drop SLS, et al. Sexual Self-concept in women with disorders/differences of sex development. Arch Sex Behav [Internet]. 2022;51(4):2213–29. Available from: https://doi.org/10.1007/s10508-021-02188-1
124. Kreukels BPC, Köhler B, Nordenström A, Roehle R, Thyen U, Bouvattier C, et al. Gender dysphoria and gender change in disorders of sex development/intersex conditions: results from the DSD-LIFE study. J Sex Med [Internet]. 2018;15(5):777–85. Available from: https://doi.org/10.1016/j.jsxm.2018.02.021
125. García-Acero M, Moreno O, Suárez F, Rojas A. Disorders of sexual development: Current status and progress in the diagnostic approach. Curr Urol [Internet]. 2020;13(4):169–78. Available from: https://doi.org/10.1159/000499274
126. Lee PA. A perspective on the approach to the intersex child born with genital ambiguity. J Pediatr Endocrinol Metab [Internet]. 2004;17(2):133–40. Available from: https://doi.org/10.1515/jpem.2004.17.2.133
127. Shaw E, Butler CA, Langdridge D, Gibson S, Barker M, Lenihan P. Guidelines and literature review for psychologists working therapeutically with sexual and gender minority clients. London: British Psychological Society; 2012.
128. Ridgeway V, Mathews A. Psychological preparation for surgery: a comparison of methods. Br J Clin Psychol [Internet]. 1982;21(Pt 4):271–80. Available from: https://doi.org/10.1111/j.2044-8260.1982.tb00565.x
129. Amaral RC, Inacio M, Brito VN, Bachega T, Domenice S, Arnhold I. Quality of life of patients with 46, XX and 46, XY disorders of sex development. Clin Endocrinol. 2014;82(2):159–64.
130. Warne GL. Long-term outcome of disorders of sex development. Sex Dev [Internet]. 2008;2(4–5):268–77. Available from: https://doi.org/10.1159/000152043
131. Tonkin-Hill G, Hanna C, Bonelli R, Mortimer R, O'Connell MA, Grover SR. Identifying the resource needs of young people with differences of sex development. J Clin Med [Internet]. 2022;11(15):4372. Available from: https://doi.org/10.3390/jcm11154372
132. Indiankanoon.org [cited 7 Feb 2023]. Available from: https://indiankanoon.org/doc/188806075/
133. Thenewsminute.com [cited 7 Feb 2023]. Available from: https://www.thenewsminute.com/article/ban-sex-reassignment-surgeries-intersex-infants-madras-hc-tells-tn-govt-100565

Inclusion and Well-Being of Children with DSD: Challenges and Strategies

37

Meenu Anand

Abstract

Well-being is a multifaceted paradigm that incorporates physical, psychosocial, and emotional elements. The experience of well-being for children with Disorder of Sexual Development can be challenging due to the lack of understanding, the prevalence of biases and stereotypes, and the paucity of research evidence from a psychosocial standpoint. The current chapter describes the difficulties experienced by children with DSD and their families across the multilayered systems context. The chapter also attempts to delineate a few eclectic strategies for creating an inclusive ecological system for children with DSD.

Keywords

Well-being · Mental health and Children with DSD

37.1 Introduction

World Health Organization [1] expounds well-being as a "*resource for healthy living*" and a "*positive state of health*" and being beyond the notion of skiving an ailment. It is the sustainable condition that describes the experience of constructive emotions including joy, happiness, and tranquility, the development and realization of one's inner strengths, having a sense of agency and resolve in one's life while experiencing positive relationships [2]. It enables an individual to function well in a holistic manner—psychologically, physically, emotionally as well as socially. Shah and Marks [3], 2004 explicate:

M. Anand (✉)
Department of Social Work, University of Delhi, Delhi, India

S. K. Ratan (ed.), *Children with Differences in Sex Development*,
https://doi.org/10.1007/978-981-97-1639-5_37

> Well-being is more than just happiness. As well as feeling satisfied and happy, well-being means developing as a person, being fulfilled, and making a contribution to the community (p. 2)

In other words, well-being can be described as a state of holistic living that enables a person to explore and develop the potential, form positive and prosocial relationships in life and also be creative and productive thereby making significant contributions to the community and society at large. It connotes the cognitive and affective evaluations in one's life [4] and can play a significant role in predicting positive life outcomes for an individual. As an eclectic concept, it is influenced by everything around an individual, the varied and exclusive experiences he/she has at diverse stages in life. Well-being depends on various factors deeply embedded in the person's socio-cultural context including family, community, and the larger society.

The concept of well-being can also be understood in the context of Bronfenbrenner's [5] perspective on human development which is based on the ecological model as presented in Fig. 37.1.

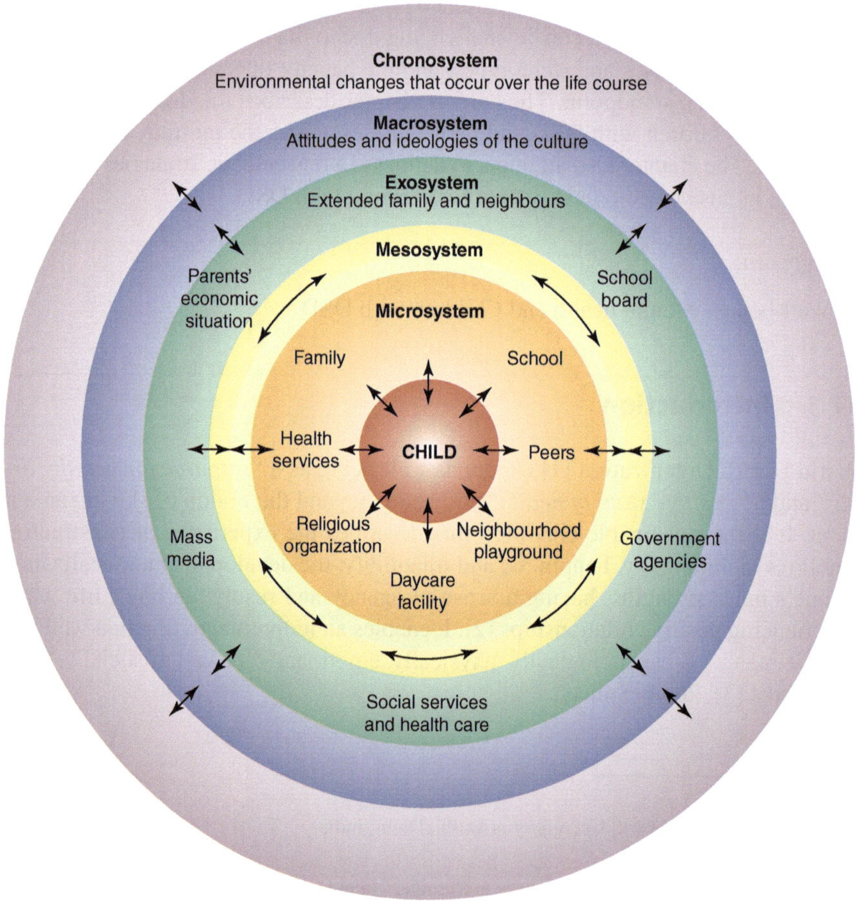

Fig. 37.1 Bronfenbrenner's ecology of human development (1977)

According to the ecological systems theory given by Bronfenbrenner, there is a focus on the interaction of individuals, integrating their social and cultural opportunities that are the determining factors contributing to their fulfillment with life. The perspective reinforces the imperative and resolute impact of the *Microsystem* (the direct milieu of the person that involves each person with whom he/ she engages on a consistent basis); the *Mesosystem* (the interface among affiliates/constituents of the microsystem); the *Exosystem* (the larger milieu that unswervingly impacts the direct setting of the person); *the Macrosystem (*the principal structure comprising of values, ethos, laws, economy, policies, etc.*); and the Chronosystem* (how certain influences impact persons over a time-period, entailing lifecycle-based incidents and alterations in the sociological as well as economic stature) on a person's behavior. The well-being of a person is therefore directly related to his/ her immediate environment; their interrelationships, the larger community as well as the society. Since no individual can exist in seclusion, the social relations, interpersonal dynamics, and the larger systems and their sub-systems impact an individual and his/ her sense of well-being.

37.2 Deciphering the Well-Being of a Child with Disorders of Sex Differentiation

Disorders of Sex Differentiation (DSD) sometimes also known as intersex variations, is a comprehensive term for 40+ inbred medical conditions that affect the development of sex characteristics and sometimes the genitals. Differences in sex development/disorders of sex development (DSD), earlier designated as intersex conditions, are inherited states in which the development of the chromosomal, gonadal, or anatomical sex is uncharacteristic [6].

There is significant research evidence concerning the medical aspects of children with DSD [7, 8] including the lengthy ethical and cultural discussions regarding the treatment of children born with anomalous genitalia. However, there is a scarcity of literature available on the significant psychological and sociological impact of DSDs on the affected children and their caregivers/ families. It is imperative to add that early childhood years lay the foundation of one's personality and are critically significant for the psychosocial well-being of children. Psychosocial well-being includes all aspects of children's psychological development and social adjustment. Since physical and mental health have a bidirectional relationship, the impact of biological development (or associated issues) is likely to impact the psychological well-being of the child diagnosed with DSD. These include cognitive and cultural capabilities, individual refuge, social amalgamation and abilities, individual's distinctiveness and assessment, sense of autonomy, and the emotive and corporeal countenances of well-being [9].

Roen [10] expounds:

> ....Firstly it is relevant to conceptualize child development in a way that questions normative imperatives. Second, it is useful to regard sex, gender and sexuality as more nuanced than is implied by binary categories such as male/female or normal/abnormal. Third, it is important to be wary of pathologizing emotion and relying on "the fiction of the rational individual who is positioned as 'in control'"...

The study by Waehre [11] to reconnoiter the mental health, psychological as well as social functioning and assess the quality of life of children and adolescents with a Difference in Sex Development (DSD) instituted that teenagers who were allocated as females with DSD probably experienced greater mental health challenges and a lower level of overall functioning in contrast to a vigorous contrasting assemblage. Since the psychosocial issues among these children and families have received little attention with trifling research [11] it is indeed imperative to explore this area from a psychosocial lens with a thrust on the well-being of the child with DSD.

37.3 Unveiling the Challenges for Children with DSD Through the Ecological Lens

The lifespan or developmental approach is a popular paradigm in mental health that was initially propounded by Erik Erikson [12]. Different stages of development are identified along with various *life crises* and expected developmental tasks. The earliest signs and symptoms of a disorder may occur at any time throughout the lifespan and the nature and timing of prevention and early intervention depend not just on the individual's age, but also on the identified pathways to mental health concerns and the identified risk factors and critical transition points that characterize those pathways. Moreover, each developmental phase with the occurrence of recurrent life events leaves behind a residual effect of stress experienced in the psyche of an individual. Based on the ecological framework, Table 37.1 enumerates two lifespan stages along with their unique developmental tasks and probable mental health issues with focus on children/ adolescents with DSD.

Table 37.1 Developmental tasks across childhood and adolescence: Areas of concern for children with DSD

Developmental stage	Major developmental tasks	Significant concerns for children with DSD	Common mental health concerns
Childhood	Physical health and well-being, secure attachment, acquisition of language and social skills, entering school, scholastic achievement, etc.	Behavioral maladjustments, confusion regarding sex-related identity, gender dysphoria, school-related concerns, bullying, acceptance by the community at large, child abuse, incest, etc.	Conduct disorders, mood disorders, Attention Deficit Hyperactive Disorder (ADHD), other behavioral maladjustments, etc.
Adolescence	Puberty, identity formation, pressure to excel in academics and vocations, career choice, etc.	Attaining puberty, concerns about "normal" biological development, physical appearance/ pressure to look *slim/beautiful/macho,* gender-specific personal and professional aspirations, access to resources, sexuality-related concerns	Anxiety, eating disorders (anorexia nervosa, bulimia nervosa, binge-eating), depression, gender dysphoria, onset of substance abuse / alcoholism, suicide, behavioral maladjustments, etc.

Adapted from Anand (2016) Gendered aspects of mental health: Issues and strategies [13]

The child with DSD is encircled by the societal world. The same comprises proximal as well as distal contexts. In the proximal world, it comprises of one's immediate family, kinship, employers, friends, and colleagues. In the context of the distal world, it entails the larger society and ethos [14]. Research has shown that socialization of children inculcates self-confidence, security, and role modelling which further impacts their overall development of personality and provides a foundation to cope with stressors in later life and meet developmental tasks successfully. The children with DSD in addition to facing the psychological and social challenges true for all children across the life span stages, also experience unique challenges owing to their specific condition throughout each developmental phase. Some of these challenges based on the ecological perspective can be enumerated as follows:

1. *Microsystem challenges*: At the microsystem level, the child with DSD is in direct contact with his/ her immediate environment including parents, siblings, teachers, and peers.
   (a) *Parental acceptance*
   Parental expectations regarding the birth of their child provide a significant context for their parenting styles as well as the transactional associations with their children. Given the emotional burden of a diagnosis of having a child with a disorder of sex development can bring about a lot of queries, guilt, apprehension as well as anxiety among the parents. Amidst the acceptance of the "different" child, significant decisions related to surgery (if needed) can become the most challenging and controversial concerns that can strain the parents and families in multiple ways.
   (b) *Concerns related to ambiguous genitalia*
   Concerns related to ambiguous genitalia may disrupt the social expectations of the parents [15] at the time of birth. There can also be debates about abnormalities/ deficiencies being caused by "nazar," witchcraft or factors associated with the mother keeping in mind the patriarchal and gendered mindset prevalent in society. Such struggles and strains regarding prescriptive multiplicities and entitlements can occur across the world [16]. These impact the attitude toward the infant thereby impacting the parent-child relationship, not mentioning the financial strains that might get added.
   (c) *Gender socialization choices*
   Apprehension regarding varied types of gender disparities highlights the challenges in the assignment of a specific gender or "sex of rearing" choices. These are sturdily impacted by the overall socio-cultural surroundings of the persons including their spiritual and traditional credence. Keeping in mind the preference for male children, there may also be a heightened burden to raise these children as males [17].
   (d) *Gender dysphoria*
   Children with DSD may face a lack of clarity regarding their own sexual identity and feel distressed. They may experience symptoms of anxiety, depression, and irritability and are at risk for mental health conditions.

2. *Mesosystem challenges*: At the Mesosystem level, the focus is on the interface between affiliates/constituents of the microsystem) concerning the child with DSD. The most significant concern is represented by the direct relationships between the family and school and with peers as well as neighbors.
   (a) *Acceptance by the school*
   Keeping in mind the probable susceptibility of young students enrolled in educational institutes, considering children's sensitive lifespan stage, there are basic postulates about the role of schools and teachers as primary socializing agents [16]. Concerns related to divulgence of the identity of the child, issues related to bullying at school, academic performance, and growing self-confidence are critical. Since there is a lack of understanding on the subject, teachers too may find it difficult to accept a "different child."
   (b) *Dynamics of adolescence*
   Puberty marks the beginning of adolescence which in itself is considered a period of stresses and strains. For a child with DSD, discussions on puberty and biological changes can become confusing as well as embarrassing. There can also be arousal of distress and psychopathology, which may impact overall well-being and psychosocial functioning. Interactions between parents and peers or parents and schoolteachers can pose difficulties owing to uncomfortable questions and lack of awareness amidst the growing adolescents with unique biological and psychological challenges.
3. *Exosystem challenges*: At the Exosystem level, the focus is on the broader environment of the child with DSD that unswervingly impacts the direct milieu of the person), e.g., the neighborhood, parents' workplaces, parents' friends, and the mass media.
   (a) *Lack of awareness*
   There is an acute lack of awareness about what constitutes DSD. While many would connect the condition with having "abnormal" genitals, others would relate such a child with belonging to the transgender community. There can also be pressure to abandon the child owing to societal attitudes.
   (b) *Stigma*
   There are various prejudices and stereotypes related to individuals with diverse sexualities. This makes them exposed to human rights desecrations when their personal details (e.g., name and sex) in certified documents are not congruent to their expression of gender identity.
4. *Macrosystem challenges*: At the Macrosystem level, the focus is on the predominant structure that comprises larger values, ethos, economy, politics, etc., and how cultural elements affect the development are significant for children with diverse sexualities.
5. *Chronosystem challenges:* At the Chronosystem level, factors such as life transitions/incidents and variations in sociological and economic status, and various environmental changes that occur over the lifetime which influence his/ her development are important. For a child with DSD, life transitions such as starting/ change of school, changes in income or family lifestyle, violence in the family, shifting of residence, climate change, and disasters can also be a source of stress.

37.4 Building an Inclusive Ecological System for Children with DSD: Few Eclectic Strategies

- *Support for parents*

  Enabling parents to engross prolifically with new-fangled evidence about their child, and helping them with the pertaining emotional demands, provides the framework for parents to express the much-needed emotional support to their child. Furthermore, keeping in mind the societal concerns related to addressing the "unspeakable shame" [18, 19], support, education as well as counselling is imperative.
- *Breaking the silence*

  The emotional challenges presented by sex development issues and medicalization can leave people unable to talk about these topics, even with people who are close to them. Being able to talk with others about DSD and healthcare experiences can be important for well-being. Having someone to talk to, and having the capacity to talk comfortably about oneself, are crucial for building relationships throughout life. Without this, families face social isolation [10].
- *Sensitization of health care workers*

  There are various paramedical professionals in the communities including the health care workers viz. Auxiliary Nurse Midwife (ANM), Trained Birth Attendants (TBA), and Accredited Social Health Activist (ASHA). Their training and capacity building can play a substantial role in awareness generation on the subject in the community, provide parental support, and make appropriate referrals for seeking medical expertise.
- *Role of schools*

  Schools can have a significant impact on a child's ability to develop into a self-assured and content adult. They play a particularly crucial role because some DSD children may not become aware of their condition until they are in their school years, especially during critical stages of sex development like puberty. Keeping into consideration the imperative role that teachers have in the overall personality development of young children, trained school teachers can skillfully guide students, deliberately divulging specific sets of information and behavioral codes of conduct through creative mediums including storytelling, reading passages, non-judgmental discussions in the class, assignment of roles/ tasks and adopting an open and non-threatening/ stigmatizing attitude [16] during their interactions with children.
- *Sensitization of teachers*

  Gender realization cannot be commanded by law, it has to be prudently fostered at the societal level to advance cognitions among people. Therefore, in this context, the prescriptive role of teachers comes into play and becomes crucial [16]. There is a need to provide teachers and school administrators with the skills and information needed to evaluate so-called "intersex inclusive" materials and guidance provided to schools by outside agencies, e.g., being sensitive while teaching about puberty that some children (whether they have a DSD or not) will experience atypical puberties.

- *Community-based approach*

  The concept of psychological care needs to be amalgamated into the medical teams concentrating on DSD. In addition, it is also imperative that persons with DSD have, it is entrée to aptly competent community-level emotional support across the various spans of life [20]. McCauley [21] expounds psychologically conversant approaches for health professionals. These include normalizing the person's understanding and perceptions, reiterating the same to the significant others of the child regarding the normalcy of such persons with altering conditions. Understanding the existing knowledge of the person, exploring their concerns and fears by listening assiduously to them followed by appropriate sex education in an open manner can be some of the significant strategies in this regard.
- *Strengthening the protective factors*

  The understanding of well-being and mental health provides a biopsychosocial perspective, with a focus on the protective factors that possibly influence the development of mental health challenges and mental disorders in persons [22]. Table 37.2 delineates some of these protective factors.
- *Support groups*

  In addition to the aforesaid, support groups can be extremely helpful in catharsis, information sharing, problem resolution, sharing of resources, etc. These can be formed for both the children with DSD as well as their parents. Professional social workers with expertise in-group work as one of their professional methods of working, can contribute significantly in this regard.
- *Rights-based approach*

  Gender identity is built around self-determination. It is the right of the children with DSD to be treated as "normal" while being acknowledged for "being different" and unique individuals as not everyone will have *periods* or will be

Table 37.2 Protective factors for the well-being of the child with DSD

INDIVIDUAL	FAMILY & COMMUNITY	STRUCTURAL
• Genetics • Adequate nutrition • Postive self regard • Social & Emotional Skills • Sense of self worth and Mastery • Physical activity	• Good perinatal nutrition • Good parenting and counselling • Attachment with family • School based awareness & support • Strong family norms and morality • Physical security and safety • Sense of belonging • Positive social networks and social supports • Green spaces	• Economic security • Good quality infrastructure • Equal access to services • Quality natural environment • Social justice & integration • Income & social protection • Social & gender equality • Sense of connectedness

Source: Adapted from the World Mental Health Report by WHO [24] (2022)

able to have children. In the contemporary context of judicial activism, the legal system can also play a proactive role with regard to legal advocacy for children with DSD from a rights-based perspective. The same can also pave the way toward resolution of various challenges currently being experienced by medical professionals, especially related to significant decisions related to surgeries.

- *Role of media*

  The media has an authoritative impact on societal perceptions and viewpoints and therefore can play a momentous role toward creating awareness on the subject through its speckled arrangements encompassing social media, television, films/documentaries, print, and contemporary platforms like OTT. This can also help to reduce the stigma associated with DSD and build support for children and persons with the condition. It can also be used effectually to cater to discourse around bringing contextual transformation in order to improve their mental health and well-being in society, in addition to contributing to the advocacy that is required to seek policy changes [23].

- *Eclectic management*

  A multidisciplinary and multipronged approach is needed toward holistic strategies aimed at rehabilitation of the children with mental illness, prevention of mental disorders, and promotion of mental well-being. Various professionals including neonatologists, pediatricians, surgeons, social workers, psychologists, endocrinologists, urologists, gynecologists, and geneticists are needed for case management.

37.5 The Way Forward

Keeping in mind the indigenized context of continency and unequal distribution of means, there is a need to raise issues and deliberate upon the inclusion and empowerment of children with DSDs to become happier, confident people with a strong sense of agency, accountable residents of the world [16]. Multipronged efforts by diverse professionals are needed with greater awareness, sensitivity, and biopsychosocial efforts toward the inclusion and well-being of children with DSDs.

References

1. World Health Organization. Basic documents. 46th ed. Geneva: World Health Organization; 2007. Available at https://apps.who.int/iris/handle/10665/43637
2. Huppert FA. Psychological well-being: evidence regarding its causes and consequences. Appl Psychol Health Well Being. 2009;1(2):13, 137–64. https://doi.org/10.1111/j.1758-0854.2009.01008.x.
3. Shah H, Marks N. A well-being manifesto for a flourishing society. J Public Mental Health. 2004;3(4):9–15. https://doi.org/10.1108/17465729200400023.
4. Diener E. The science of well-being: The collected works of Ed Diener: 2009. Dordrecht: Springer; 2009.

5. Bronfenbrenner U. Toward an experimental ecology of human development. Am Psychol. 1977;32(7):513.
6. Lee PA, Houk CP, Ahmed SF, Hughes IA. International Consensus Conference on Intersex organized by the Lawson Wilkins Pediatric Endocrine Society and the European Society for Paediatric Endocrinology. Consensus statement on management of intersex disorders. Peadiatrics. 2006;118(2):e488–500. [PubMed: 16882788]
7. Hughes IA, Nihoul-Fekete C, Thomas B, Cohen-Kettenis PT. Consequences of the ESPE/LWPES guidelines for diagnosis and treatment of disorders of sex development. Best Pract Res Clin Endocrinol Metab. 2007;21(3):351–65. https://doi.org/10.1016/j.beem.2007.06.00.
8. Thyen U, Lanz K, Holterhus PM, Hiort O. Epidemiology and initial management of ambiguous genitalia at birth in Germany. Horm Res Paediatr. 2006;66(4):195–203. https://doi.org/10.1159/000094782.
9. Francis A. Social work in mental health: Areas of practice, challenges and way forward. New Delhi: Sage; 2014.
10. Roen K. Intersex or diverse sex development: critical review of psychosocial health care research and indications for practice. J Sex Res. 2019;56(4):511–28. https://doi.org/10.1080/00224499.2019.1578331.
11. Waehre A, Heggeli C, Hald K, Myhre AG, Diseth T. A 15-20-year follow-up of mental health, psychosocial functioning and quality of life in a single center sample of individuals with differences in sex development. Health Psychol Behav Med. 2022;10(1):837–54. https://doi.org/10.1080/21642850.2022.2116329.
12. Erikson E. Childhood and society. New York: W.W. Norton; 1950.
13. Anand M. Gendered aspects of mental health: issues and strategies. In: Aleem S, Iqbal N, editors. Positive vistas on health and well-being. New Delhi: Excel India; 2016.
14. Bhugra D, et al. What is mental health? Int J Soc Psychiatry. 2013;59(1):3–4. https://doi.org/10.1177/0020764012463315. Sage
15. Jiwanmall SA, Kattula D. Profile of siblings with disorders of sexual differentiation and gender dysphoria. Indian J Psychol Med. 2021;43(4):368–9. https://doi.org/10.1177/0253717620927895.
16. Anand M. Gender understandings among teachers and students: discourses from Delhi. South Asia Res. 2018;38(3):307–26. Sage
17. Ravendran K, Deans R. The psychosocial impact of disorders of sexual development. J Sex Med Reprod Health. 2019;2:4.
18. Engberg H, Möller A, Hagenfeldt K, Nordenskjöld A, Frisén L. The experience of women living with Congenital Adrenal Hyperplasia: Impact of the Intersex/DSD and psychosocial healthcare 47 condition and the care given. Clin Endocrinol. 2016;85:21–8. https://doi.org/10.1111/cen.13054.
19. Liao L-M, Wood D, Creighton SM. Parental choice on normalising cosmetic genital surgery: Between a rock and a hard place. BMJ (Online). 2015;351:h5124. https://doi.org/10.1136/BMJ.h5124.
20. Liao LM, Green H, Creighton SM, Crouch NS, Conway GS. Service users' experiences of obtaining and giving information about disorders of sex development. Br J Obstet Gynaecol. 2010;117:193–9. https://doi.org/10.1111/j.1471-0528.2009.02385.x.
21. McCauley E. Challenges in educating patients and parents about differences in sex development. Am J Med Genet. 2017;175:293–9. https://doi.org/10.1002/ajmg.c.31563.
22. Anand M. Deconstructing the aetiology and prevention of mental illness: A developmental perspective. Delhi Psychiatry J. 2018;21(1):19–25.
23. Anand M. Mental health and human rights: An Indian inquiry. Indian J Soc Psychiatry. 2021;37:235–41. https://doi.org/10.4103/ijsp.ijsp_187_20.
24. World Health Organization. World Mental Health Report 2022. Available at https://www.who.int/publications/i/item/9789240049338

Part IX

Surgery in and for DSD

I duly acknowledge the support of Dr. Dasmit Singh Khokhar Honorary Professor, Pediatric Surgery, B.J. Govt. Medical College, Pune and Pediatric Surgery consultant Jahangir Hospital, Pune and Dr. Prabudh Goel, Additional Professor of Paediatric Surgery, AIIMS New Delhi for editing this section.

Parental and Patient Procedural Consent for Surgical Interventions in DSD

38

Velmurugan Rajendran

Abstract

Historically and in common practice, a child born with DSD is considered a social emergency rather than a medical one. This is understandable considering parental anxiety to learn and announce the sex of their child as well as individual and societal pressure to conform to the normal binary of male and female sexes. Hence, the parents and caregivers are under understandable stress of discovery and decision-making. Further, the sex of the child has to be conveyed and documented in the birth registry of local authorities in a finite time frame. It puts the treating physician under the scanner, especially a pediatric surgeon who is often asked to opine on the sex of the child. If the child requires an emergency or elective surgical procedure, issues of consent and its legality stare us in the face in the light of increasing government legislation restricting interventions. This chapter will deal with issues of procedural consent in DSD cases and how to navigate our way through this moral and ethical quagmire so that the best interests of the child and future adult are served.

Keywords

Intersex · Disorders of sexual development · Informed consent · Guidelines · Medical disclosure · Surgery

V. Rajendran (✉)
Department of Paediatric Surgery, Institute of Child Health, Madras Medical College, Park Town, Chennai, Tamil Nadu, India

S. K. Ratan (ed.), *Children with Differences in Sex Development*,
https://doi.org/10.1007/978-981-97-1639-5_38

38.1 Informed Consent

It is the legal and ethical duty of a surgeon to obtain informed consent when providing treatment [1]. The performance of procedures on a patient without consent makes the surgeon liable to battery, negligence, or malpractice [2].

Informed consent implies that the patient voluntarily agrees to surgery or treatment with a complete understanding of their condition, the nature of the proposed surgery, the risks and benefits of the proposed surgery, available alternative treatments, and the risks of the proposed surgery.

38.2 Child and Procedural Consent

Under Indian law and as per the United Nations Children's Convention, persons below 18 years are considered as children. Hence, procedural consent should be obtained from the parent or parent surrogate. The consent in DSD cases is complicated by the fact that gender identity of the future adult is an individual wish that may not conform with that of parental wish or that of existing social norms or cultural preferences. It is increasingly being accepted that the child has a right to take part in therapeutic decisions, especially after 14 years of age and irreversible interventions be postponed till the age of legal consent, i.e., 18 years.

Surgeons are reasonably apprehensive whether they can be held legally responsible if they provide care based on the consent of a minor. If a minor is authorized by law to provide consent, there is little likelihood that a clinician would be held liable based on a failure to obtain parental consent?

This authorization of minors to consent to or refuse surgery is more prevalent in the United States where there is the concept of mature minors who are empowered to decide on issues of drug abuse, abortion, etc. [3].

Parental consent to surgeries on intersex infants is contentious on two counts. First, it is debatable whether parents by themselves have the legal sanction to consent to these surgeries if the surgery affects what may be considered a constitutionally protected right. Second, even if parents have the authority to approve these procedures on their own, current informed consent practices may need to be improved to ensure that parental consent is valid and protects the fundamental human rights of the future adult.

38.3 Informed Consent in DSD

The four essential components of an informed consent are:

1. Decision-Making Capacity of the individual giving the consent.
2. Documentation of Consent, which should be clear, seamless, and retrievable.

3. Disclosure wherein all aspects of the intervention including success and failure rates as also alternative treatments should be discussed.
4. Competency of the person giving the consent.

The person obtaining consent should be capable of performing the procedure or should have specialist training in advising patients about the procedure.

Currently, people below 18 years are not considered competent to give procedural consent. This puts the onus of giving procedural consent solely on the parent.

Hence, procedures and mechanisms have to be created to ensure that the parent, parent surrogate, and physicians act in the best interests of the child while obtaining consent [4, 5].

However, in the UK, children under the age of 18 years can give consent for their own treatment if they are believed to have enough intelligence, competence, and understanding to fully appreciate what is involved in their treatment. This is known as the Gillick competence [6]. While a child under the age of 18 years may be able to legally give their consent, should they refuse their consent to a procedure, this can be overridden by someone with parental responsibility.

38.4 Activism Against Elective Surgery in DSD

Parents and caregivers were assumed to act in the best interests of the child with DSD and had a mandate to decide on elective and emergency interventions.

This mandate is being increasingly challenged by activists, individuals, and pressure groups [7]. They maintain that the child's right to bodily autonomy should be protected as also it is right to an open future till the patient can participate meaningfully in decision-making [8].

These activists and pressure groups are against early "normalising" surgery and argue that consent for irreversible and nonurgent surgical interventions given by parents violates basic fundamental rights of the future adult. Future judicial decisions on complaints of violation of human rights by an affected adult is a matter of conjecture.

Activists also seek to postpone the decision of sex to when the child becomes an adult and can be responsible for its consent and actions. The judiciary and governments have been sympathetic to such arguments by these activists causing much trepidation amongst the surgical fraternity regarding the legality of procedural consent given by the parent.

38.5 Regulatory Rulings

Legislative and judicial institutions are increasingly against elective procedures for children with DSD. There is an assumed harm due to such surgery and attendant violation of child rights and bodily integrity.

Such intervention in children has been condemned by the United Nations Special Rapporteur on Torture [9]. UN Committee on the Rights of the Child, UN Committee on the Rights of Persons with Disabilities and the Europe Commissioner of Human Rights [10].

Due to such adverse opinions, the governments of Germany, Portugal, and Malta have prohibited such surgical interventions till the affected child is of an age and maturity to provide informed consent. Other nations may be considering similar measures. Some of such legislation could be an attempt to regulate gender conversion in normal individuals and surgeons treating DSD are caught in the crossfire besides the affected children and their parents.

The State of California passed a resolution to ban genital surgery in children till the child is able to participate in the decision-making, California Senate Concurrent Resolution SCR110 [11].

Similar bills introduced in Texas and Nevada did not come to fruition.

However, the US based CARES Foundation (Congenital Adrenal Hyperplasia Research Education and Support) declared that CAH is a life-threatening adrenal disorder and not a sexual disorder and that CAH patients are not intersex and hence SCR110 should not apply to CAH. Thus, the CARES Foundation bypassed SCR by disassociating itself from DSD.

In India, the Madras High Court issued a directive to the Government of Tamil Nādu to ban all genital surgeries in children while hearing a case related to transgenders. Subsequently, the Tamil Nādu Government issued an order in August 2019 banning such surgeries. This order was later modified in February 2020 permitting emergency drainage procedures and interventions (TN GO No:63/H&FW/dt18.02.2020). This was a landmark government legislation that mandated multidisciplinary teams at local and tertiary levels, which could decide on appropriate interventions.

38.6 Guidelines for Informed Consent in DSD

It is thus imperative that surgeons caring for such children should be aware of such rulings and the possibility of parental consent being declared bad in law in the future, especially if the future adult feels that the parent has not acted in his/her best interest.

It is difficult to predict how the mind of the future adult will react and we as professionals need to create robust mechanisms to obtain legal procedural consent to serve the best interests of the child.

The treating team should first realize that there is no urgent need to normalize the genitalia and sex of the child. Corrective elective surgery is to be postponed till such time when the child can effectively participate in the decision process.

Emergency interventions should be restricted to the correction of excretory obstructions.

There should be adequate reasoning and justification for any intervention.

All interventions that have irreversible consequences on sex, sexuality, and reproductive capacity are to be avoided. Such decisions are best left to the affected individual when the person reaches the legal age for giving consent.

There should be an honest, comprehensive, and unbiased sharing of information with the parents while obtaining consent.

A shared decision-making process should be implemented to educate and guide parents [12].

The treating team should not impose their decisions on the parent but rather help them to take decisions to protect the interests of their child [13].

The treating team should not tout the benefit of only one surgical or treatment option. There should be a complete and comprehensive disclosure of information including that of alternatives to enable the parents to make a truly informed decision. This should include a full disclosure of all knowns and unknowns of the future course.

Biased or incomplete information to sway parental decision or consent may be considered coercive or inappropriate by legal authorities.

However, a surgeon can reveal his or her preference on a treatment option and give his/her reasoning for the same.

A multidisciplinary team should be involved in the counselling process which should include a surgeon, pediatrician, psychiatrist, geneticist, bioethicist, lawyer, and a social worker. This team will provide useful inputs about the family and parent motives [14].

In cases where the sex of the child cannot be assigned clearly in early life, an interim sex of rearing has to be suggested by the multidisciplinary team so as to enable documentation with birth registries and also to enable the child to have a normal childhood.

All such information and conversations have to be documented clearly and seamlessly with the parents being cosignatories. This documentation should be preserved carefully and should be easily retrievable so that the treating surgeon can protect himself from future liability [15].

Affected persons and their parents should be entitled to inspect their medical records. All such recordkeeping should take into account patient privacy and personal dignity [16].

38.7 Conclusion

It is mainly based on the legality of parental procedural consent that the surgical treatments of DSD have entered the legislative arena. It should be our endeavor to retain the decision-making within the clinical sphere by scrupulously following an exhaustive and comprehensive informed consent procedure with meticulous documentation.

An informed procedural consent given by the parent for interventions in DSD is legal in present times. But it is in our interest as surgeons to make the right calls to avoid future liability. It is also desirable to involve a multidisciplinary team in the decision-making process and ensure that the parent acts in the best interests of the child. All non-emergency and normalization procedures, which will affect the sex, sexuality, and reproductive capacity of the future adult are to be avoided if possible.

The child and future adult should have unfettered access to all his/her medical records including deliberations of consent procedure.

It is hoped that such participative decision-making will reduce surgeon–patient conflict and be in the best interest of a child with DSD.

References

1. English A. 'Consent in adolescent health care' Treating adolescents. Legal and ethical considerations. Med Clin North Am. 1990;74(5):1097.
2. de Lourdes LM, Larcher V, Kurz R. Statement of Ethics Working Group of the Confederation of European Specialists in Paediatrics. Eur J Paediatr. 2003;162:629–33.
3. Sigman GS, O'Connor C. Exploration for physicians of the mature minor doctrine. J Pediatr. 1991;119(4):520–5.
4. Diamond M, Sigmundson HK. Management of intersexuality. Guidelines for dealing with ambiguous genitalia. Arch Pediatr Adolesc Med. 1997;151:1046–50.
5. Ford K-K. First, do no harm—the fiction of legal parental consent. Yale Law Policy Rev. 2001;19:469–88.
6. United Kingdom House of Lords Decisions >> Gillick v West Norfolk and Wisbech AHA [1985] UKHL 7; 17 Oct 1985.
7. Chase C. What is the agenda of the intersex patient advocacy movement? Endocrinologist. 2003;13:240–2.
8. Feinberg J. The child's right to an open future. In: Aiken W, LaFollette H, editors. Whose child? Children's rights, parental authority, and state power. Totowa, NJ: Rowman and Littlefield; 1980. p. 124–53.
9. United Nations Human Rights Council, Méndez JE. Report of the special rapporteur on torture and other cruel, inhuman or degrading treatment or punishment. 2013.
10. European Union Agency for Fundamental Rights. The fundamental rights situation of intersex people. Vienna; 2015.
11. SCR. Senate Concurrent Resolution No. 110 (SCR-110) sex characteristics. Available online at: https://leginfo.legislature.ca.gov/faces/billTextClient.xhtml?bill_id=201720180SCR110. Accessed 17 Oct 2018.
12. Karkazis K, Tamar-Mattis A, Kon AA. Genital surgery for disorders of sex development: implementing a shared decision-making approach. J Pediatr Endocrinol Metab. 2010;23:789–805. https://doi.org/10.1515/jpem.2010.129.
13. Katz AL, Webb SA, Macauley RC, Mercurio MR, Moon MR, Okun AL, et al. Informed consent in decision-making in pediatric practice: technical report. Pediatrics. 2016;138:e20161485. https://doi.org/10.1542/peds.2016-1485.
14. Samsel C, Ribeiro M, Ibeziako P, DeMaso DR. Integrated behavioral health care in pediatric subspecialty clinics. Child Adoles Psychiatr Clin North Am. 2017;26:785–94. https://doi.org/10.1016/j.chc.2017.06.004.
15. Gardner M, Sandberg DE. Navigating surgical decision making in DSD. Front Pediatr. 2018;6:339.
16. Wiesemann C, Ude-Koeller S, Sinnecker GHG, Thyen U. Ethical principles and recommendations for management of DSD in children. Eur J Pediatr. 2010;169:671–9.

Role of Endoscopy in Diagnosis and Management of DSD

39

Dasmit Singh Khokar and Rasik Shah

Abstract

DSD is known to be an incompletely understood condition and as such a lot still needs clarification in terms of gender identity and gender roles for each child.

The multiple determinants of sexual identity may change as the child grows into adulthood. So, investigations are needed to establish identity at a stage after birth as well as after puberty. Hence, final management plan with regard to timing, type, and extent of surgery is undergoing a flux at these times. This changing pattern of surgical practice in DSD has cast an additional responsibility on the shoulders of the team to establish a precise diagnosis in DSD as the consequences would affect the child lifelong.

In addition to the extensive work-up that is advised for all children, evaluation of internal genitalia requires two most important tools Genitoscopy and Laparoscopy, which have replaced Exploratory Laparotomy for diagnosis. Many therapeutic procedures are also done Laparoscopically and this chapter discusses the various options for the same.

Conclusions: Endoscopies have become an essential part of management of DSDs both for diagnostic as well as therapeutic purposes.

D. S. Khokar (✉)
B.J. Govt. Medical College and Sassoon General Hospital, Pune, Maharashtra, India

R. Shah
Department of Paediatric Surgery, SRCC Children's Hospital Managed by Narayana Health, Mumbai, Maharashtra, India

S. K. Ratan (ed.), *Children with Differences in Sex Development*,
https://doi.org/10.1007/978-981-97-1639-5_39

Keywords

DSD · Disorders of sexual differentiation · Endoscopy · Laparoscopy · Genitoscopy · Gonad · Biopsy · Dysgenetic · Ovotestis · Hydrometrocolpos · Hematometrocolpos · Uterus didelphys · Persistent Mullerian ducts in a male · Orchiopexy · Abdominal testes · PMDS

DSD is an enigma and as such a lot still needs clarification in terms of gender identity and gender roles for each child.

There are multiple determinants of sexual identity, but these may change as the child grows into adulthood. So, it is clear that investigations are needed to evaluate the information available for each child both at a stage after birth as well as after puberty onset. Hence, final management plan with regard to timing, type, and extent of surgery is undergoing flux in these times. This changing pattern of surgical practice in DSD has cast an additional responsibility on the shoulders of the Pediatric surgeons, Endocrinologists, and the entire team of experts to correctly evaluate the child in a holistic manner and this has increased the difficulties manifold posed by every child. Establishing a precise diagnosis of DSD is extremely important as the consequences would affect the child lifelong.

In addition to the extensive work-up that is advised for all children, evaluation of internal genitalia forms an essential cog in the armamentarium of the clinician. The two most important tools available for this evaluation are Genitoscopy and Laparoscopy.

Ken Morita presented their initial experience in evaluation stating that until 1990, children with ambiguous genitalia or incompatibility between gonad and genital expression underwent explorative laparotomy in their Institute. After laparoscopic techniques were introduced in 1991, they performed diagnostic and therapeutic laparoscopic procedures on 7 intersex children over 5 years and found them to be equally effective [1].

The risks from vaginoplasty are different for high and low confluence of the urethra and vagina and need a very careful evaluation.

It is prudent while selecting the mode of surgery to increase the chances of preservation of fertility. In patients with a symptomatic prostatic utricle, excision is best done Laparoscopically to maximize the chance of preserving the vasa deferentia. Patients with bilateral ovotestes are potentially fertile due to the functional ovarian tissue [2]. Preservation of one or the other tissue in an ovotestis although technically demanding, can be undertaken Laparoscopically with better magnification.

Patients of MGD raised as male should have their streak gonads removed laparoscopically (or by laparotomy) in early childhood [2].

Children raised as females with gonadal dysgenesis and Y chromosomal material should undergo a Bilateral gonadectomy early in their childhood since the streak gonads are prone to undergo malignant change. Usually before 14 years of age as the earliest recorded malignancy is at that age [3]. The biopsies of gonadal material removed are examined and the histopathology is determined.

39.1 Indications for Genitoscopy and Laparoscopy

Almost all cases of DSD will need one or both Endoscopic evaluations, especially in cases of:

1. Ambiguous genitalia and suspected existence of testicular elements (confirmed by human chorionic gonadotropin stimulation) despite 46, XX karyotype.
2. Chromosomal mosaicism (46,XY/45,X) despite phenotypically male external genitalia.
3. Vaginal atresia/Inseparably fused labia.
4. Y element in chromosome despite phenotypically female external genitalia [1].

39.2 Diagnostic Evaluation Is Required for:

46 XX DSD
46 XY DSD
Ovotesticular DSD
Sex chromosome DSD
Gonadal dysgenesis

39.3 Indications for Therapeutic Laparoscopy

1. Gonadal biopsy.
2. Dysgenetic gonad excision.
3. Excision of inappropriate gonad in ovotestis.
4. Drainage of hydrometrocolpos or hematometrocolpos.
5. Excision of offending horn in uterus didelphys.
6. Excision of persistent Mullerian ducts in a male.
7. Orchiopexy for abdominal testes with PMDS.

39.4 Consent for Endoscopic Evaluation

Thorough counselling followed by a detailed consent for Genito-cystoscopy and/or Laparoscopy is taken. It should be worded to include diagnostic evaluation as well as excision of dysgenetic or hypoplastic organs or biopsy procedures with consent for Laparotomy, if required.

39.5 Positioning the Patient During Endoscopic Evaluation

If only Genitoscopy is planned, then the patient is placed in a lithotomy position or some modifications thereof.

If a simultaneous Laparoscopy is planned in addition to the Genitoscopy, then the lithotomy position should be such that it leaves the abdomen free for insertion of Laparoscopic instruments also. Extended lithotomy.

39.6 Technique of Genitoscopy

Genitoscopy is done under General Anaesthesia with the patient placed in a Lithotomy position or some modification of lithotomy. A cystoscope of size 4.5F–10F with 0°–30° is used depending on the size of the patient's urethra or urogenital sinus. The scope is connected to an Endocamera and the monitor is adjusted to preclude the glare and in addition, a recording device is connected. The scope should fit in snugly into the opening, but without too much resistance to sliding of the scope within the urethra. Normal Saline is used for irrigation. Most of the time, a slight pressure on the opening of the Urogenital Sinus needs to be exerted by the Surgeon while a slight compression of the NS bottle is needed to distend the urethra/urogenital sinus for better visualization. It is important to insert the scope under vision. Initial direction of insertion of the scope is posteriorly. The UGS soon turns toward a cranial direction. A careful examination is done to see whether two openings are seen at a certain distance from the perineum. This would confirm the presence of a Urogenital Sinus. The next most important point is to note the confluence of the urethra and vagina. If the confluence is below the sphincter, then it is classified as a Low UGS and if the confluence is high, then it is classified as a High UGS. This differentiation is of paramount importance as the therapeutic approach dramatically changes for the reconstruction of the UGS and Vaginoplasty. For low confluence, a cutback of the UGS can be safely done, whereas for a high UGS, being supra-sphincteric, a cut-back would directly divide the external sphincter and render the patient incontinent for voluntary control. Hence, differentiation of Low vs. High UGS is of critical value.

The external urethral sphincter lies within the deep perineal space, situated between the pelvic diaphragm and the perineal membrane. It lies within the urogenital hiatus of the pelvic diaphragm and fills the gap between the pudendal canals. The pudendal nerve and vessels pass through this canal.

The female external urethral sphincter (Urogenital sphincter) is formed from three parts. Innermost is the female external urethral sphincter, located around the proximal two-thirds of the female urethra. Thickness is more anterior than the posterior aspect and constitutes an annular sphincter around the urethra. The second part, called the compressor urethrae muscle, extends to either side bilaterally. The third part is the urethrovaginal sphincter. These broad and flat fibers pass in a posterior direction to surround both the urethra and vagina as a single unit. Extending to the bulb of the vestibule.All these smooth and skeletal muscle components of the urethral sphincter complex are continuous and need to be preserved carefully during vaginoplasty procedures. Hence precise knowledge of the location is of paramount importance to achieve proper results.

In a study published by Morei et al in the *Journal of Pediatric Urology*, they concluded that External virilization did not correlate with the length of the proximal (pre-junctional) urethra or with the depth of the vaginal–urethral confluence. The mean length of the proximal urethra was 22 mm (range 10–32 mm), and the mean vertical depth of the vaginal–urethral confluence from the perineum was 16 mm (range 8–31 mm) [4].

The accurate anatomical assessment of length of the UGS and its relationship to the sphincter may be challenging. They recommended two important points for assessment: a) the length of the urethra proximal to the urogenital sinus, as this will impact the urinary outcome; and b) the depth (level) of vaginal entry into the urogenital sinus, as this will affect the mobilization required to exteriorize the vagina.

Their conclusion was that the depth of the vaginal–urethral confluence from the perineum is an indicator of the required mobilization for the current perineal approach. In 90% of cases in the age group they studied (1–3 years old), this depth was ≤20 mm indicating a low confluence. This supports the current understanding that partial urogenital mobilization could be suitable for most cases.

39.6.1 Tricks to Differentiate Low from High UGS

Whereas no single technique can be applied uniformly, it is mandatory to familiarize oneself with various ways of identifying the pathological anatomy intraoperatively.

1. In a High UGS, the Vagina would be seen entering the UGS at a point which would be an equivalent position of the Verumontenum in males (as seen during scopy of the posterior urethra).
2. Brisk pinching of the clitoris can cause the external sphincter to contract making it easy to identify its relationship with the UGS during Genitoscopy.
3. Using a muscle stimulator with EMG probes placed on the perineum can be used to identify the sphincter during genitoscopy.
4. In low UGS, the confluence is often seen in the initial part of the scopy when the scope is still parallel to the perineum.
5. Measurement of the distance (thickness) from the perineum to the confluence. <20 mm indicates a low confluence. This can be done by placing a finger in the rectum to palpate the tip of the scope and measuring the distance on the finger with the help of a scale [4].
6. Measurement of the distance of the proximal urethra from the internal urethral meatus to the insertion of the vagina is also found to be useful by some authors. This can be assessed by Genitography or a graded urethral catheter passed beside a scope into the UGS. Marei concludes that these pre-junctional urethral length measurements were obtained from the genitography (mean = 22.8 mm, range 10–32 mm) and confirmed by cysto-urethroscopy (mean = 22.3 mm, range 10–32 mm). It was found that the concordance between both modalities was

87%. Hence fairly reliable. The length of the proximal urethra, extending from the bladder neck to the confluence, achieved a consensus of being the most important predictor of surgical complexity, regardless of the length of the distal common urogenital sinus [4].

39.7 Technique of Laparoscopy

General Anaesthesia with endotracheal tube is given. The monitor, camera, and video stack with the insufflator and light source are positioned at the foot-end of the patient. The bladder is emptied with a catheter. Open insertion of the first trochar (5 or 10 mm) is done at the umbilicus and Carbon dioxide pneumoperitoneum is created using an Insufflator preferably with an in-built Gas warmer to avoid hypothermia. Around 5–8 mmHg intraabdominal pressure suffices and ETCO2 is preferably monitored continuously.

Umbilical port becomes the primary Camera Port attached to a 0° or 30° telescope. Usually, two or three ports of 3 or 5 mm are used as working ports placed in the flanks for instruments and occasionally a fourth port may be necessary for ablative procedures to retract a dilated bowel.

Laparoscopic inspection of the pelvis is carried out to survey the presence of gonads and to inspect the pelvic organs in detail. In some cases when the gonads may not be easily identifiable, the gonadal vessels have to be identified and traced caudally. The tissue at the end of the gonadal vessels may be dysplastic and just streaks of abnormal gonadal tissue. Such dysgenetic gonads are excised. Some special energy devices like Laparoscopic Bipolar forceps, Harmonic scalpel, or a Vessel sealer device may be used in some cases, to achieve the excision or complete the hemostasis after a sharp dissection. The biopsied specimen or the excised specimen can be extracted through the 5- or 10-mm port depending on the size. It is very important to ensure that the specimen to be biopsied is not destroyed by the heat of the energy device, so sometimes it is prudent, especially for smaller specimens, to first excise the specimen with sharp scissors and then accomplish hemostasis.

Diagnostic procedures: include morphological and histological confirmation of the type of gonads and internal ductal structures. Along with the laparoscopic evaluation, ablative treatment can be simultaneously performed such as partial or total gonadectomy and partial or total hysterosalpingectomy along with the diagnosis and gender identity.

Laparoscopy has been extremely effective in the diagnosis of most variety of patients in the DSD spectrum. The presence and maturity of internal pelvic organs can be documented objectively with a well-magnified view and any gonadal structures can be biopsied or removed if indicated. In DSD patients, a diagnostic exploration would appear to be better performed laparoscopically than by an open surgical approach [5, 6].

39.8 Special Case Scenarios (Figs. 39.1, 39.2, 39.3, 39.4, 39.5, 39.6, 39.7, 39.8, 39.9, 39.10, 39.11, and 39.12). All photographs copyright @ Dr. Dasmit Singh Khokar

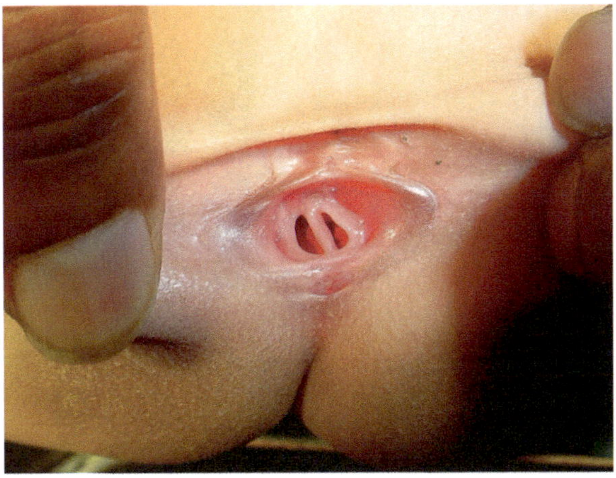

Fig. 39.1 Septate vagina—not strictly classified as a 46 XX DSD. Genitoscopy has to be done for each side of the Septate Vagina

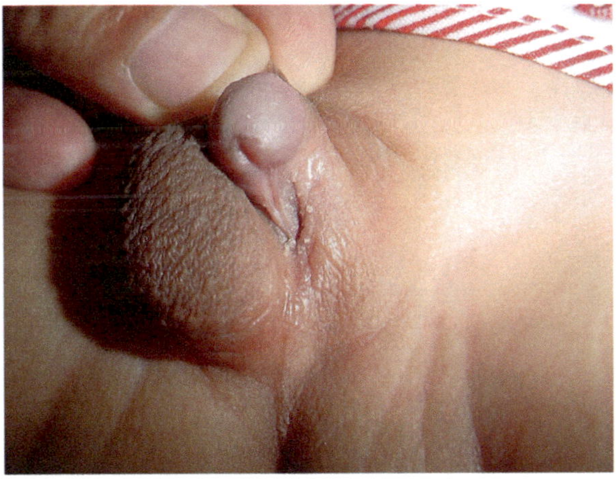

Fig. 39.2 Absent hemiscrotum—46 XX DSD. Genitoscopy revealed normal male anatomy. Laparoscopy revealed an abdominal testis which was pexed to the perineum along with single-stage hypospadias repair

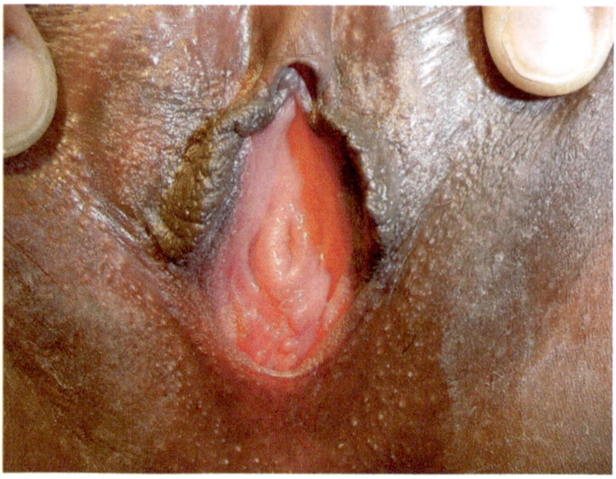

Fig. 39.3 Mullerian agenesis—a type of 46 XX DSD. Mayer Rokitansky Küster Hauser syndrome. 46 XX karyotype, with normal appearing external genitalia. Genitoscopy showed a single opening in the introitus with a blind pit of about 5 cm and no cervix. Laparoscopy showed a complete uterine aplasia with two rudimentary horns joined by a peritoneal fold and normal Fallopian tubes and normal Ovaries (MRKH type I syndrome)

Fig. 39.4 Aphallia—absent phallus—a variety of 46 XY DSD with severe penile insufficiency. Genitoscopy via an opening very close to the rectum revealed a normal proximal urinary tract of a male

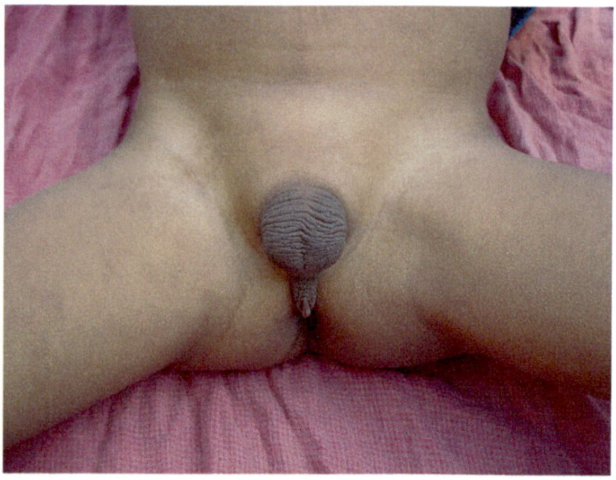

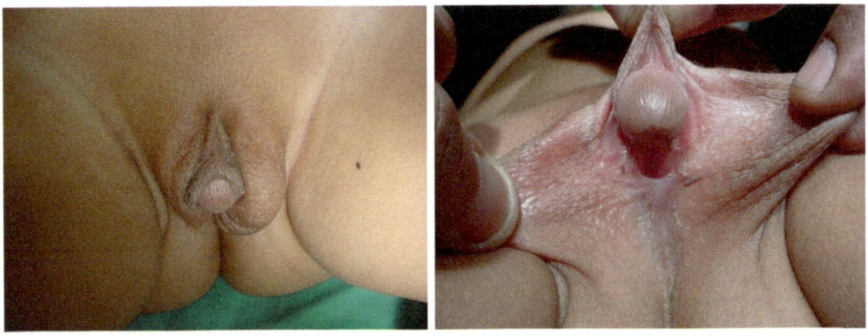

Fig. 39.5 XY DSD—partial androgen insensitivity syndrome with right non-palpable testis. Laparoscopy showed a good-sized testis intrabdominal. Laparoscopic orchiopexy done

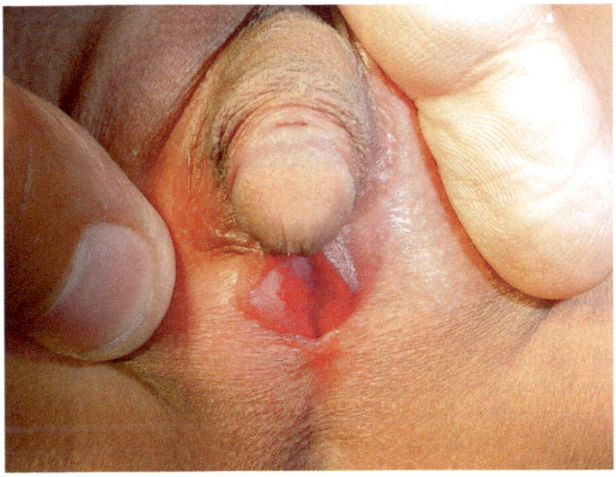

Fig. 39.6 46 XX DSD. Virilization due to congenital adrenal hyperplasia. Genitoscopy showed low confluence of the urogenital sinus

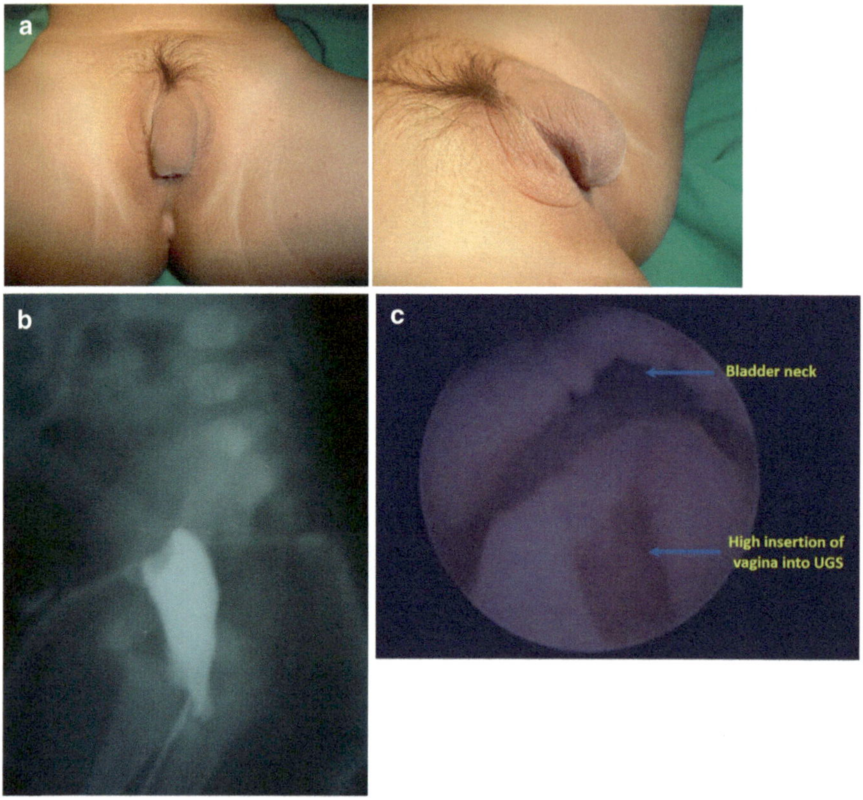

Fig. 39.7 (**a**) 46 XX DSD. Very severe virilization due to congenital adrenal hyperplasia. Genitoscopy showed a high confluence of the urogenital sinus. (**b**) 46 XX DSD. Very severe virilization due to congenital adrenal hyperplasia. Same patient as in (**a**). Genitography showed a capacious vagina with good impression of the cervix. However, the confluence could not be demonstrated on the genitogram. (**c**) 46 XX DSD. Very severe virilization due to congenital adrenal hyperplasia. Same patient as in (**a**). Genitoscopy showing a high insertion of vagina into the UGS

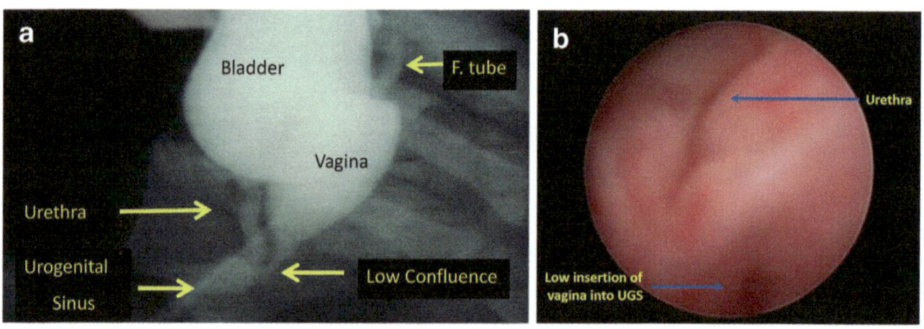

Fig. 39.8 (**a**) 46 XX DSD. Virilization due to congenital adrenal hyperplasia. Genitography showed a capacious vagina with a flow of dye into fallopian tube. The confluence seen demonstrated a low urogenital sinus. (**b**) 46 XX DSD. Same patient as in (**a**). Moderate virilization due to congenital adrenal hyperplasia. Genitoscopy showing a low insertion of vagina into the UGS

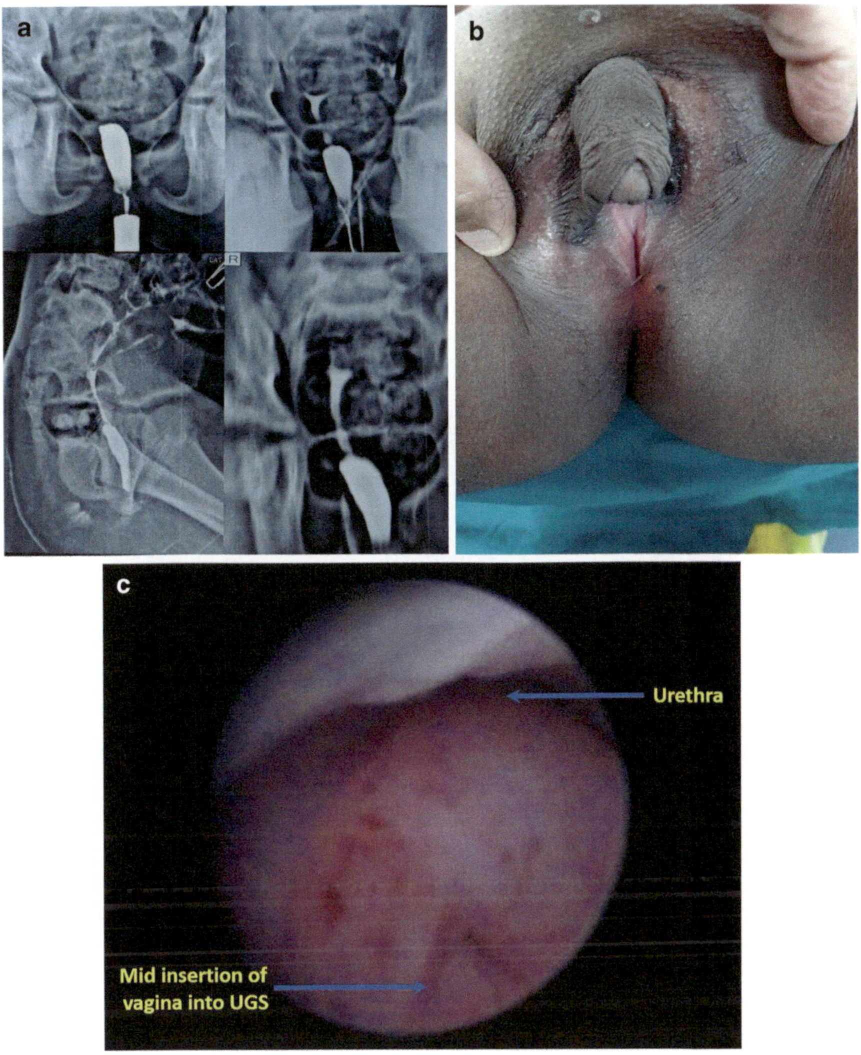

Fig. 39.9 (**a**) 46 XX DSD. Moderate virilization due to congenital adrenal hyperplasia. Genitography showing a capacious vagina with uterine cavity and fallopian tubes. (**b**) Same patent as in (**a**)—46 XX DSD. Moderate virilization due to congenital adrenal hyperplasia. Moderate clitoromegaly. (**c**) Same patent as in (**a**)—46 XX DSD. Moderate virilization due to congenital adrenal hyperplasia. Mid-confluence. Can be treated as a low confluence for surgery

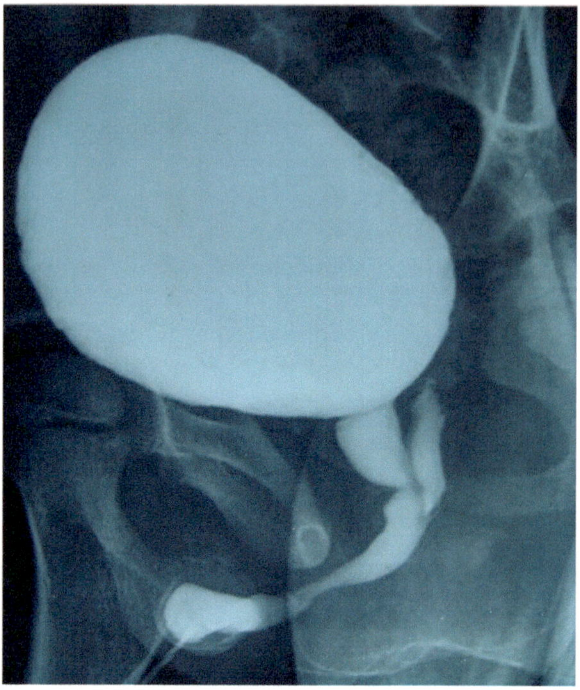

Fig. 39.10 46 XY DSD. Partial androgen insensitivity syndrome. Prostatic utricle causing difficulty in catheterization. Genitoscopy done and catheterization done over a guide wire

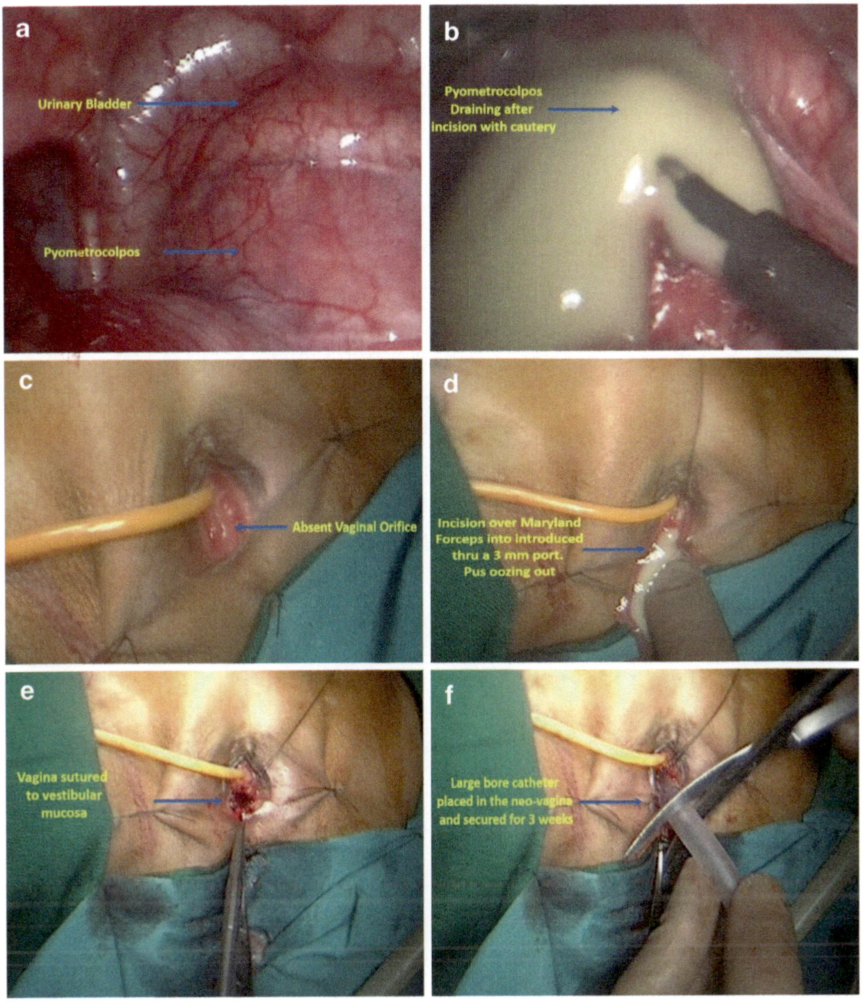

Fig. 39.11 46 XX DSD—pyometrocolpos. Laparoscopic drainage with pull-thru vaginoplasty (**a**–**f**)

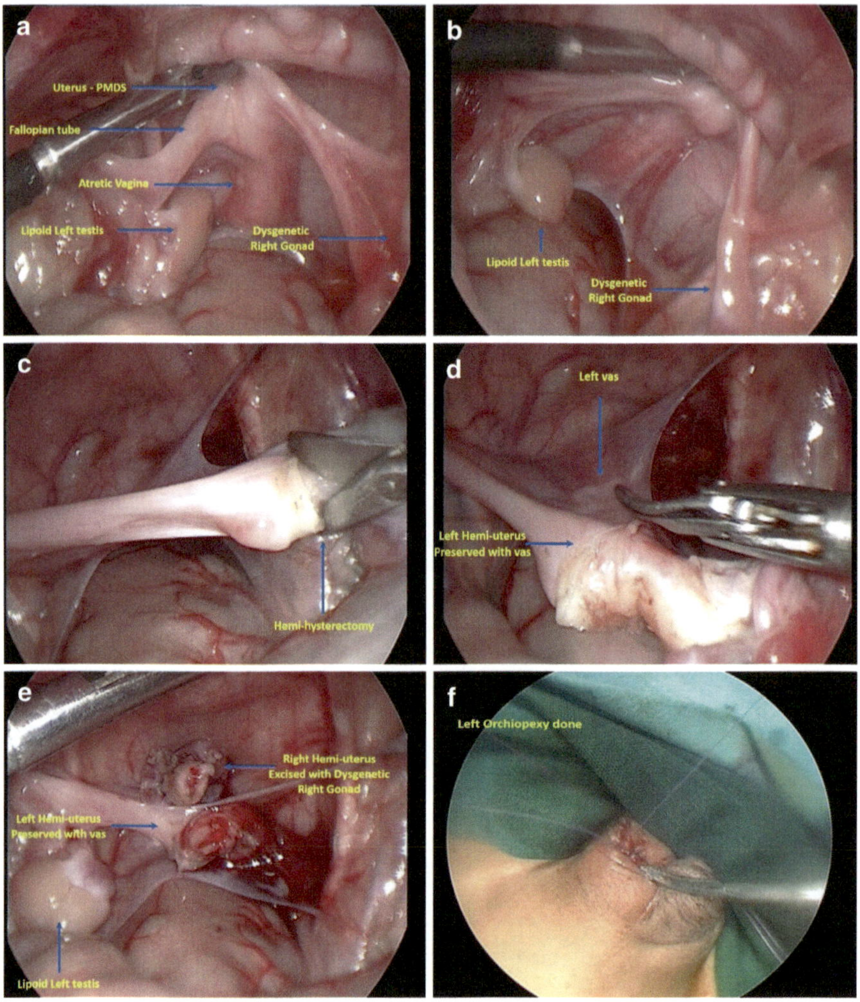

Fig. 39.12 46 XY DSD—gonadal dysgenesis, with persistent Mullerian ducts. Hemihysterectomy done. Uterus split longitudinally to preserve the vas on the left side for the orchiopexy (**a**–**f**)

References

1. Morita K, et al. Experience of laparoscopic exploration and gonadectomy in intersex children. Diagn Ther Endosc. 1998;4:127–33.
2. Grumbach MM, Hughes IA, Conte FA. Disorders of sex differentiation. In: Larsen PR, Kronenberg HM, Melmed S, et al., editors. Williams textbook of endocrinology. 10th ed. Philadelphia: WB Saunders; 2003. p. 842–1002.

3. Hurt WG, Bodurtha JN, McCall JB, et al. Seminoma in pubertal patient with androgen insensitivity syndrome. Am J Obstet Gynecol. 1989;161:530–1.
4. Marei MM, et al. Anatomical measurements of the urogenital sinus in virilized female children due to congenital adrenal hyperplasia. J Pediatr Urol. 2016;12(5):282.e1–8.
5. Clayman RV. Pediatric laparoscopy: Quo vadis? A view from the outside. J Urol. 1994;152:730–3.
6. Nihoul-Fékété C. The Isabel Forshall Lecture. Surgical management of the intersex patient: an overview in 2003. J Pediatr Surg. 2004;39:144–5.

Masculinizing Genital Surgery

40

Johannes Wirmer and Ahmed T. Hadidi

Abstract

It is important to differentiate between hypospadias and disorders of Sexual development and it is very unfair for many children with simple straight forward distal, midpenile, or even proximal hypospadias with normal descended testes to be put under the same category as patients with unclear gender differentiation. Once it is decided that the baby is a boy, it is important to proceed with surgery as early as 6 months of age to try to avoid corporeal disproportion. Ideally, the masculinizing surgery should be carried out in a well-planned staged procedure including bringing the testes into the scrotum, correction of severe chordee by excision of the hypoplastic urethral plate as well as the outer longitudinal layer of tunica albuginea (Tunica Albuginea longitudinal excision or TALE procedure) with or without corporotomy and grafting. Dorsal plication has a very high rate of complications and recurrence of chordee. The next step is to bring down healthy skin (Two-Stage Bilateral Based Flap (BILAB), Chordee Excision and Distal Urethroplasty, CEDU procedure), two-stage Byars flaps or healthy skin, prepuce or buccal mucosal graft (Cloutier or Bracka graft).

Keywords

Perineal hypospadias · Chordee · Tunica albuginea longitudinal excision · Chordee excision and distal urethroplasty

J. Wirmer
Sana Klinikum Offenbach, Goethe Frankfurt University, Frankfurt, Germany

A. T. Hadidi (✉)
Hypospadias Center & Pediatric Surgery Department, Sana Klinikum Offenbach, Goethe Frankfurt University, Frankfurt, Germany

S. K. Ratan (ed.), *Children with Differences in Sex Development*,
https://doi.org/10.1007/978-981-97-1639-5_40

40.1 Patients Selection

It is mandatory to carry out all the necessary tests that are mentioned in detail elsewhere in this book before proceeding to Masculinizing genital surgery. In most countries, the approval of the designated committee is essential beforehand.

We need to differentiate between the classic perineal hypospadias in an undisputed male with both testes in the scrotum, a reasonable-sized phallus, a moderate to severe chordee without corporeal disproportion, and a child with disorders of sexual development, unilateral or bilateral undescended testis, a very small phallus, a small glans (usually less than 10 mm in width), corporeal disproportion and almost absent ventral skin (Fig. 40.1).

Most of these children would require Pre-operative Hormone Stimulation (PHS). In most centers, PHS is given before the first surgery. In the hypospadias center in Frankfurt, there is a standard protocol where PHS is given after the first operation of chordee excision when the glans width is 12 mm or less. The philosophy is that the ventral hypoplastic tissues of the penis are less responsive to hormone therapy and it is better to excise all the hypoplastic tissues to achieve the optimum effect of PHS.

The ideal surgical technique for masculinizing genioplasty depends on two main factors: the presence/or absence of corporeal disproportion and availability of healthy ventral penile skin.

If there is corporeal disproportion and some healthy ventral penile skin, we would use the two-stage BILAB (Bilateral Based flaps from the penis and prepuce) where a dermal graft from the groin or tunica albuginea flap or graft is used to correct the corporeal disproportion and the skin from both sides are brought to the midline to reconstruct a new urethral plate (Fig. 40.2).

If there is no corporeal disproportion, we would use the TALE (Tunica Albuginea Longitudinal Excision) technique to straighten the penis and use the CEDUE

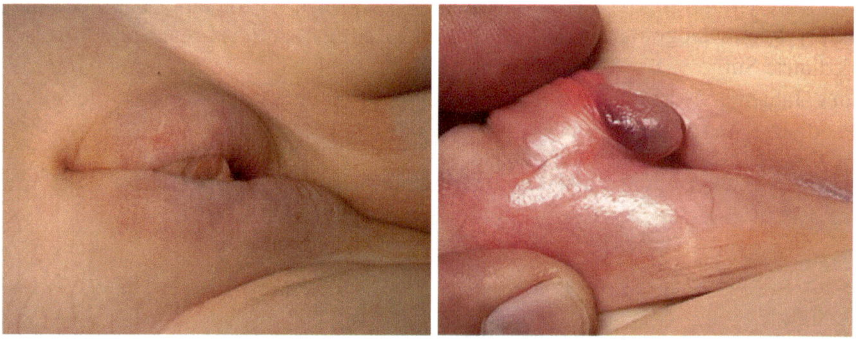

Fig. 40.1 Perineal hypospadias with very small phallus, undescended right testis, bifid scrotum, very severe chordee that the small glans is at the perineum and no perineal skin (© Ahmed T. Hadidi 2022. All Rights Reserved)

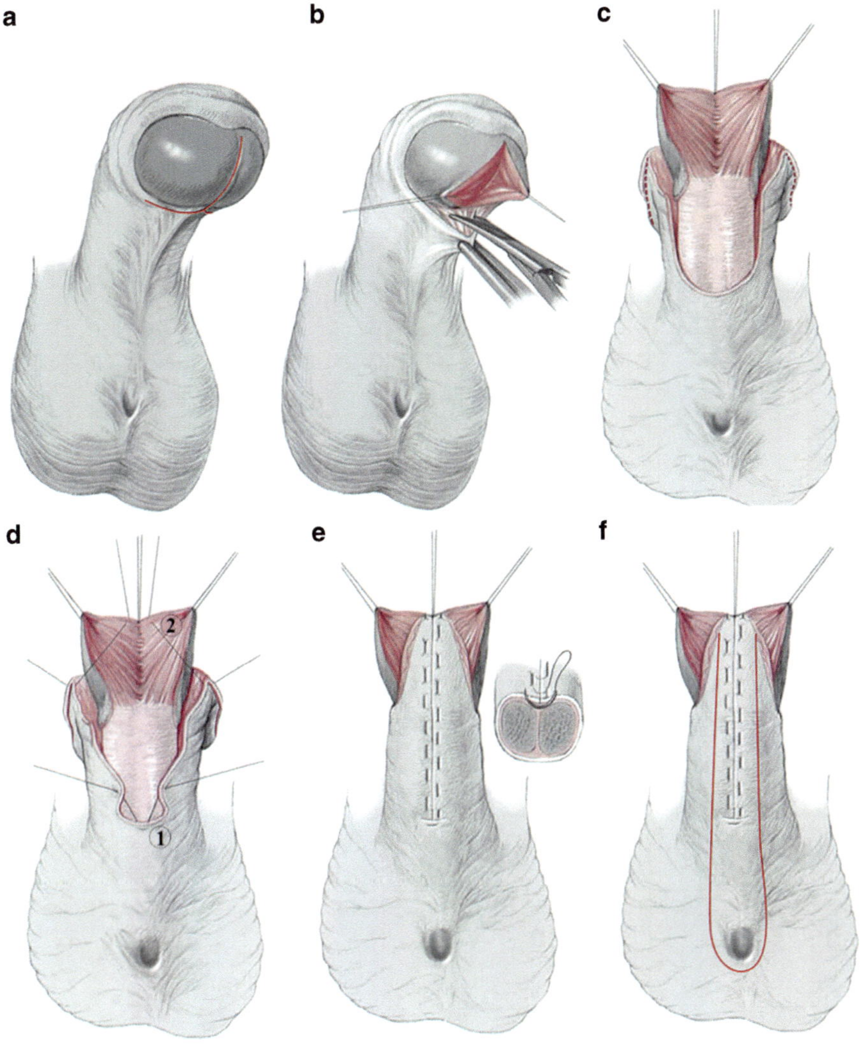

Fig. 40.2 The BILAB technique. After excision of severe chordee, the two lateral skin edges are brought to the midline and fixed to the underlying corpora cavernosa to create a "urethral plate." During the second stage, a new urethra is reconstructed using the Thiersch/ Duplay Principle (© Ahmed T. Hadidi 2022. All Rights Reserved)

technique (Chordee Excision and Distal Urethroplasty) to translocate the preputial skin ventrally and reconstruct the distal urethra (Fig. 40.3).

If there is corporeal disproportion and absent ventral penile skin, we would use a preputial skin graft after correcting the corporeal disproportion using a tunica albuginea flap (Figs. 40.4 and 40.5).

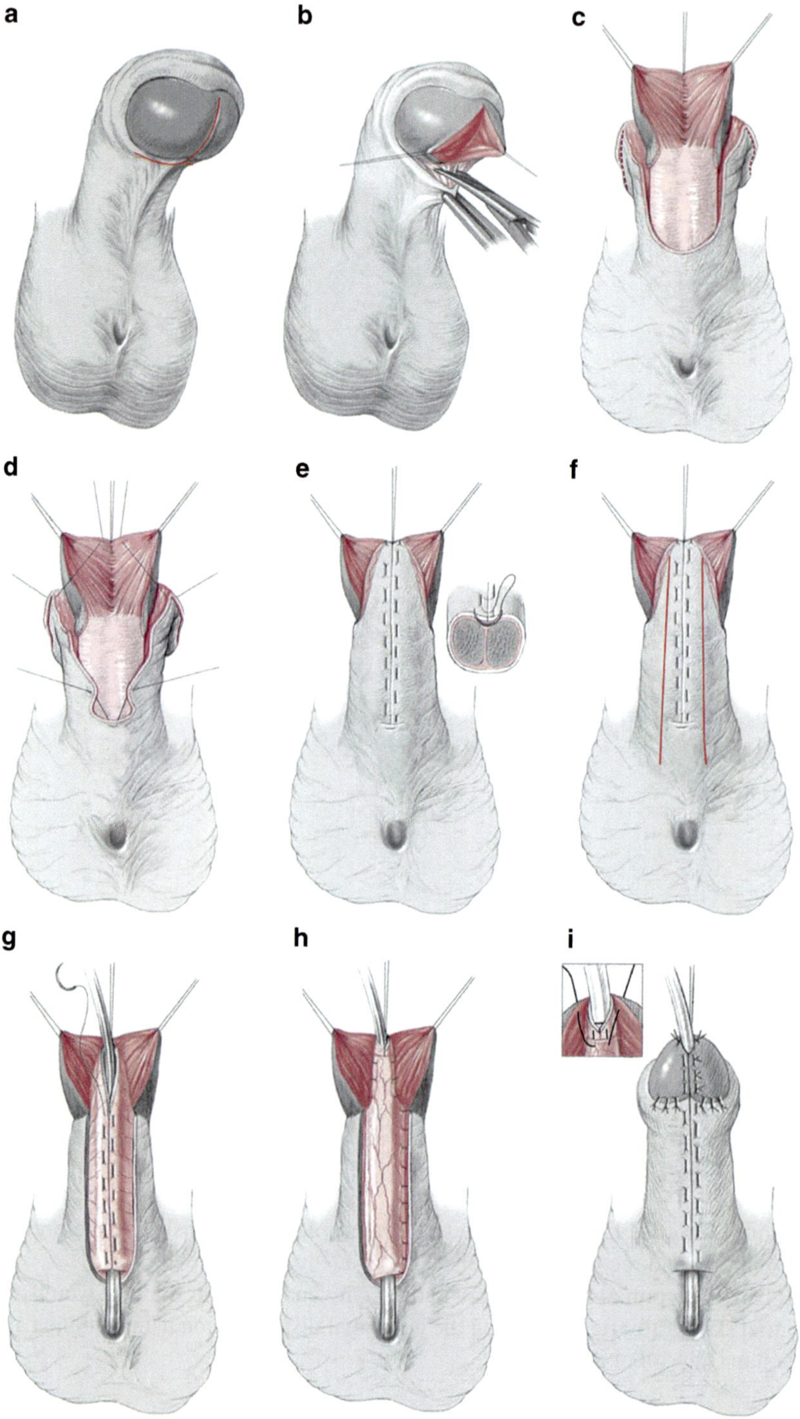
a
b
c
d
e
f
g
h
i

Fig. 40.3 The CEDU technique. After excision of severe chordee, the two lateral skin edges are brought to the midline and fixed to the underlying corpora cavernosa to create a "urethral plate." At the same first stage, distal urethroplasty is performed leaving a "perineal urethrotomy" for urinary diversion. During the second stage, the perineal urethrotomy is closed (© Ahmed T. Hadidi 2022. All Rights Reserved)

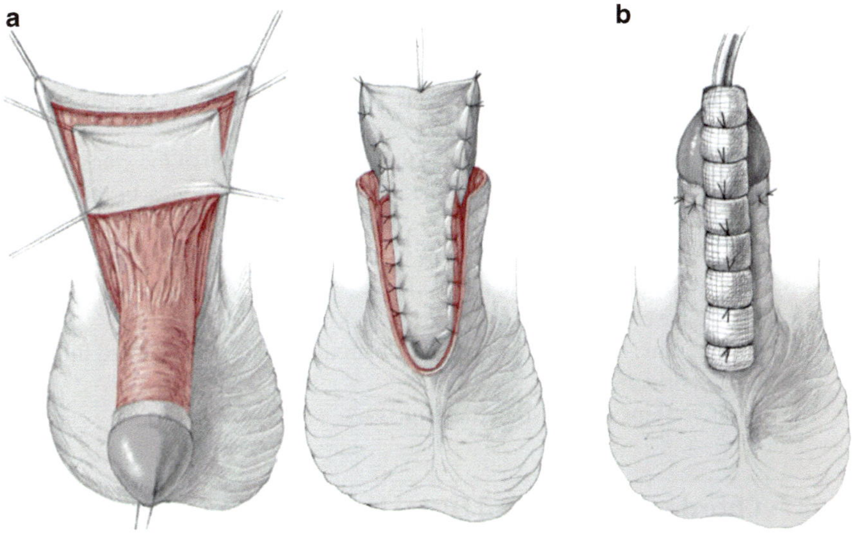

Fig. 40.4 *Preputial graft* accurately tailored to the defect with a V-shaped inset of the graft into the back wall of the urethral meatus. (**a**) The dimensions of the graft should not be overgenerous; the graft should fit snugly into the ventral defect and should not heap up into folds. (**b**) Graft may be immobilized with a "tie-over" pressure dressing prior to compressive foam dressing. The authors prefer to use "Blanket" sutures with some fenestrations to avoid hematoma formation under the graft (© Ahmed T. Hadidi 2022. All Rights Reserved)

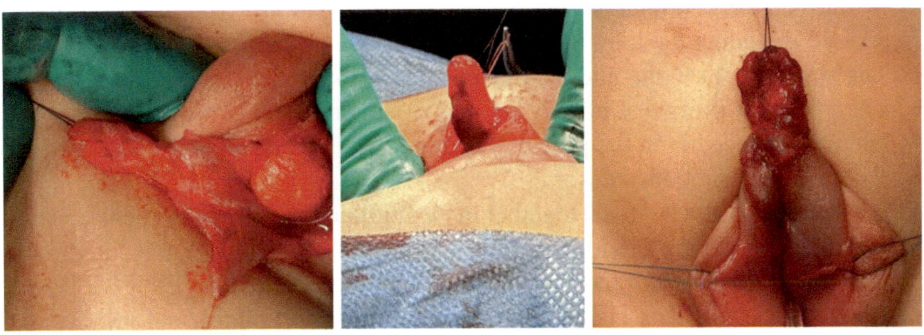

Fig. 40.5 Tunica albuginea flap was used to cover the ventral defect of the corpora cavernosa after corporotomy required to correct severe chordee with corporeal disproportion. A preputial graft was used to cover the ventral surface of the penis. The penis was fixed to the abdominal wall to maintain straightening of the penis (© Ahmed T. Hadidi 2022. All Rights Reserved)

40.2 The Two-Stage BILAB (Fig. 40.2)

- BILAB is designed to reconstruct a new urethra from the penile and preputial skin when the urethral plate has to be divided and excised in order to correct deep severe chordee (more than 30°).
- A transverse subcoronal incision is performed and all the hypoplastic tissues are removed including the longitudinal layer of tunica albuginea. The proximal edge is usually retracted back to the scrotum.
- The glans is split in the midline like an open book to provide adequate space for the new urethra.
- A transverse lateral skin incision in the prepuce is performed at the junction between the outer skin and inner mucosa on both sides.
- Both medial edges are fixed together in the midline at the tip of the glans creating a new "urethral plate" that is well fixed to the tunica albuginea in the midline (inset).
- A 4/0 Ethelon suture through the glans fixes the penis to the abdominal wall for 2–3 weeks to help to straighten the penis.
- A catheter is inserted in the bladder for a week.
- After 3–6 months, tabularization of the new urethral plate is performed following the Thiersch principle.
- The operative steps are illustrated in Fig. 40.2.

40.3 Chordee Excision and Distal Urethroplasty (CEDU) (Fig. 40.3)

The concept of CEDU operation and steps are the same. The only difference is that in CEDU, the original perineal meatus is left intact as a "***perineal fistula***" for 3 months until the wound has healed well then the perineal fistula is closed and final adjustment of the shape of the penis is performed.

- A transverse subcoronal incision is performed and all the hypoplastic tissues are excised including the longitudinal layer of tunica albuginea.
- The glans is split in the midline like an open book.
- The proximal skin edge usually retracts to the proximal penis and a transverse lateral skin incision in the preputial skin separates the outer skin from the inner mucosal layer on both sides.
- Both medial edges are fixed together in the midline to the tip of the penis creating a new "urethral plate" that is well fixed to the tunica albuginea in the midline as well as laterally.
- A catheter size F12 is inserted into the bladder. Two longitudinal incisions are made that stop 1 cm before the meatus.
- Distal urethroplasty is performed using Vicryl® 6/0 around a catheter size F12.
- Two protective layers are mobilized from both sides.

- The remaining skin is tailored to cover the glans and penis. The catheter is replaced with a silastic catheter size 10 F at the end of the procedure. The glans is fixed to the abdominal wall to stretch the penis and provide stability for 2–3 weeks after surgery.

40.4 Preputial Graft (Fig. 40.4)

Suggested Reading

1. Thiersch K. Ueber die Enstehungsweise und operative Behandlung der Epispadie. Arch der Heilkd. 1869;10:20.
2. Duplay S. Sur le traitement chirurgical de l'hypospadias et de l'epispadias. Arch Gen Med. 1880;5:–257.
3. Cloutier AM. A method for hypospadias repair. Plast Reconstr Surg. 1962; https://doi.org/10.1097/00006534-196209000-00007.
4. Bracka A. A versatile two-stage hypospadias repair. Br J Plast Surg. 1995; https://doi.org/10.1016/S0007-1226(95)90023-3.
5. Hadidi AT. Perineal hypospadias; the bilateral based (BILAB) skin flap technique. J Pediatr Surg. 2014; https://doi.org/10.1016/j.jpedsurg.2013.09.067.
6. Hadidi AT. Perineal hypospadias: back to the future chordee excision & distal urethroplasty. J Pediatr Urol. 2018; https://doi.org/10.1016/j.jpurol.2018.08.014.

Feminizing Genitoplasty for DSD

41

Shivaji Mane and Sonali Mane

Abstract

Feminizing genitoplasty is a commonly done reconstructive surgery for external genitalia of some of the cases of DSD, where there are varying degrees of virilization of external genitalia due to various causes, the commonest being congenital adrenal hyperplasia. This surgery is technically very demanding due to varying degrees of virilization and length of the common urogenital sinus.

While doing clitoroplasty, one must take all precautions to preserve normal clitoral innervation for normal sexual satisfaction.

There is still no consensus about the ideal age for surgery for feminizing genitoplasty in CAH.

Feminizing genitoplasty is commonly done for CAH patients and other conditions like Ovotesticular DSD, Gonadal dysgenesis (Mixed gonadal dysgenesis [MGD]).

Vaginoplasty may be required for Complete Androgen Insensitivity Syndrome [CAIS] at puberty if the vaginal pouch is not of adequate length for sexual intercourse.

Keywords

Cliteroplasty · Urogenital sinus · Vaginoplasty · Labioplasty · Sigmoid vaginoplasty · Congenital adrenal hyperplasia · Disorder of sexual differentiation

S. Mane (✉)
J.J. Hospital, Mumbai, Maharashtra, India

Jupiter Hospital, Thane, Maharashtra, India

S. Mane
Jupiter Hospital, Thane, Maharashtra, India

S. K. Ratan (ed.), *Children with Differences in Sex Development*,
https://doi.org/10.1007/978-981-97-1639-5_41

41.1 Historical Aspects

Schmid [1] was the first to report, preservation of dorsal neurovascular bundle with excision of corporal body.

Dissection of neurovascular bundle is often associated with diminished sensation of glans clitoris and occasional sloughing off of glans clitoris [2].

Kogan [3] described subtunical excision of erectile tissue of corporal body by incising Buck's fascia laterally.

In 1999, Baskin [4] described neurovascular anatomy of clitoris, where nerves go from dorsolateral aspect, and its branches fan out laterally to go on ventral aspect, hence lateral incision described by Kogan was changed to ventral incision on Buck's fascia just lateral to ventral urethral plate and corpus spongiosum.

Posteriorly based perineal skin flap for vaginoplasty was described by Fortunoff [5] in 1964.

Hendren [6] described pull-through vaginoplasty for high vaginal confluence in 1969.

In 1997, Alberto Pena [7] described total urogenital sinus mobilization [TUM] for repair of common cloaca, in which urogenital sinus is mobilized, by dissecting all around to get urethra and vaginal opening up to the perineum.

41.2 Preoperative Preparation

CAH patient for feminizing genitoplasty is admitted 1 day prior to surgery to start a double dose of injectable steroids with the help of a pediatric endocrinologist, which is continued for another 2 days post-surgery and then gradually tapered and started on oral normal maintenance dose. 17-Hydroxyprogesterone level should be near normal prior to surgery.

Glycerin rectal suppository is given early morning on the day of surgery to keep the rectum empty to avoid soiling of the perineum at the time of surgery.

All patients for feminizing genitoplasty should undergo cystogenitoscopy either just prior to surgery or 1 week prior to definitive surgery to assess the length of the urogenital sinus and at what distance from the bladder neck does the vagina open onto the posterior urethral wall to form the urogenital sinus distally.

The distance from the bladder neck to the confluence is important to plan surgery, especially vaginoplasty. Distance from meatus of urogenital sinus to the confluence is not as important, as the site of the meatus of the urogenital sinus will vary according to Prader's stage and severity of Prader's stage may not be associated with the level of confluence of urogenital sinus.

41.3 Surgical Technique

The child is given general anesthesia and an epidural catheter is placed for relief of post-operative pain. Extended lithotomy position is given. A stay suture is taken on the glans clitoris and then the incision is marked as follows.

Firlit's collar-like incision but taking care to leave more of inner preputial skin attached to the coronal sulcus so that it will give a better-hooded appearance for clitoris over a period. This incision is extended vertically down parallel to the urethral plate till the opening of the UG sinus. From the UG sinus incision will extend vertically down in the midline up to the apex of the Fortunoff skin flap. U-shaped incision on the inferior aspect of the Labia majora, from center of this incision extends medially to the lateral edge of the Fortunoff flap, eventually it is Y-V plasty to get labia majora down, so that the vaginal introitus remains within the labia majora (Figs. 41.1 and 41.2).

Degloving of the penile shaft is done as done in hypospadias surgery, remaining close to the Buck's fascia. Dorsally, the shaft skin is dissected till the pubic symphysis to expose the bifurcation of the corporal body going on either side toward the respective inferior pubic ramus. Each corporal body is dissected all around to put a tourniquet on either side to control bleeding while doing sub-tunical excision of erectile tissue (Figs. 41.3 and 41.4).

Vertical incision on Buck's fascia is taken on the ventral aspect just lateral to the urethral plate to expose the erectile tissue. Both edges of Buck's fascia are stretched

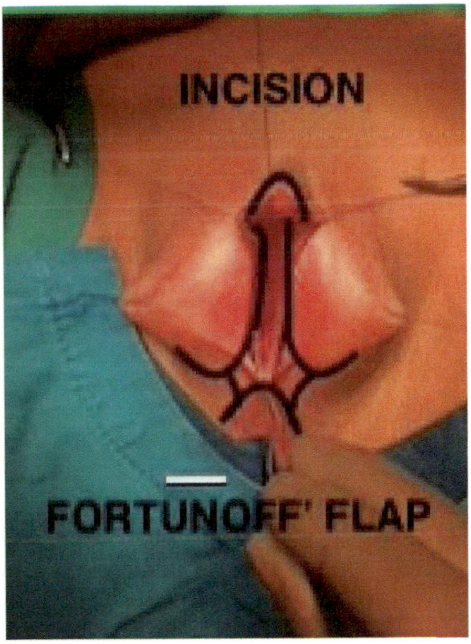

Fig. 41.1 Incision

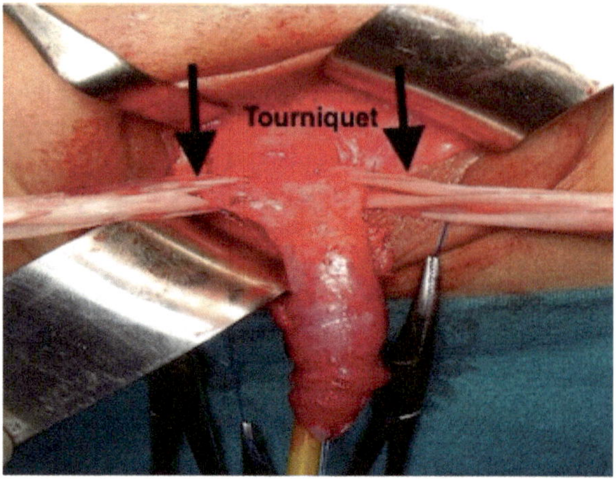

Fig. 41.2 Tourniquet around corporal body

Fig. 41.3 Ventral incision

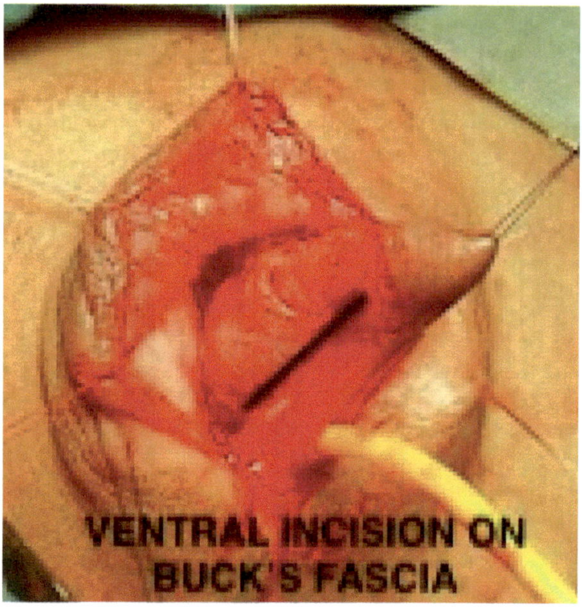

to dissect the erectile tissue, starting proximally and dissected distally to reach the glans clitoris. The edge of the glans clitoris is everted to dissect erectile tissue within the glans. This will help to reduce the bulk of the glans clitoris without taking any external incision on the glans clitoris. This will avoid unnecessary mutilation of the glans clitoris as well as it will protect the most sensitive outer surface of the glans. This is essential for sexual satisfaction. Proximally, just distal to the tourniquet, the erectile tissue is transfixed with 5/0 Polyglycolic acid suture. Edges of Buck's fascia

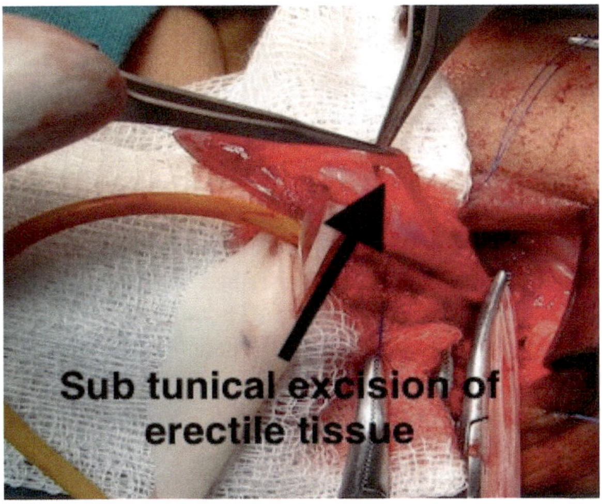

Fig. 41.4 Sub-tunical excision of erectile tissue

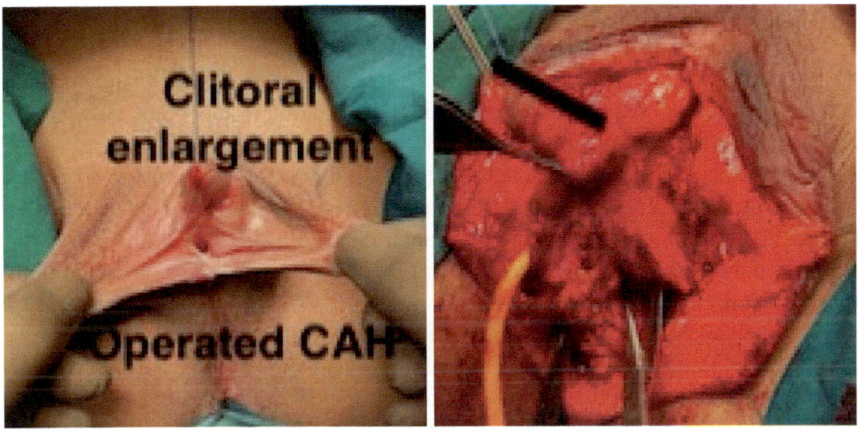

Fig. 41.5 Stump regrowth of clitoris

are sutured with 5/0 Polyglycolic acid suture and then the tourniquet is released to look for any bleeding.

Kogan in 1983 described the technique of sub-tunical excision of erectile tissue. In this technique, we do not dissect the dorsal neurovascular bundle at all, as described by Schmid, hence there is no loss of sensation of glans and there is no regrowth of stump of corporal body as described in Schmid's technique of clitoral reduction. We have seen two cases where there was regrowth of the corporal bodies following the conventional technique of excision of the corporal bodies, despite regular steroid therapy (Fig. 41.5).

41.4 Vaginoplasty

Four different types of vaginoplasties are described for feminizing genitoplasty in CAH.

1. Simple cutback.
2. Fortunoff's skin flap vaginoplasty.
3. Partial or total UG sinus mobilization as described by Alberto Pena.
4. Vaginal pull-through procedure as described by Hendren for high-vaginal confluence.
5. Sigmoid vaginoplasty, may be required for patients with very short vagina-like CAIS, and MRKH syndrome [Mayer Rokitansky Küster-Hauser] patients with absent vagina.

41.5 Operative Management of Low UGS

More than 95% of cases of CAH confluence of UG sinus is low and they are easily managed with Fortunoff's skin flap and partial UG sinus mobilization.

After clitoral reduction, diluted 2% lignocaine with adrenaline is infiltrated into the perineum where Fortunoff's skin flap is marked, as well as deep inside the perineal body to dissect the plane easily and reduce bleeding in this plane between the anterior wall of the rectum and posterior wall of the UG sinus.

Fortunoff's skin flap is dissected away from the posterior wall of the UG sinus, followed by dissection of the posterior wall of UG sinus away from anterior wall of rectum taking care to avoid injury to the anterior rectal wall. To aid this dissection, one may put one's index finger in the rectum to identify it properly and avoid injuring the rectum. After adequate dissection of the posterior wall of the UG sinus, the ventral wall of the UG sinus is opened till one can identify the vaginal opening. Preoperative placing of Fogarty's catheter in the vaginal opening genitoscopically may help to identify the vaginal opening easily. Urethra is catheterized with a Foley's catheter.

In a large number of cases where the UG sinus confluence is low, there is no need to mobilize the UG sinus all around, only posterior and lateral dissection is adequate. The vaginal opening is enlarged by cutting into the posterior wall of the vagina till one sees a good thickness of the posterior vaginal wall, and then Fortunoff's skin flap is sutured in two layers into the defect created in the posterior wall of vagina (Figs. 41.6, 41.7, and 41.8).

Some cases do need partial UG sinus mobilization to get the vaginal introitus in the perineum, but while doing so, anterior dissection is restricted till the pubic symphysis only taking care to avoid going upto the pubourethral ligament. Damaging the pubourethral ligament will lead to urinary incontinence post-surgery as observed by Rink et al. [8]. Rink described partial UG sinus mobilization to prevent damage to the pubourethral ligament (Fig. 41.9).

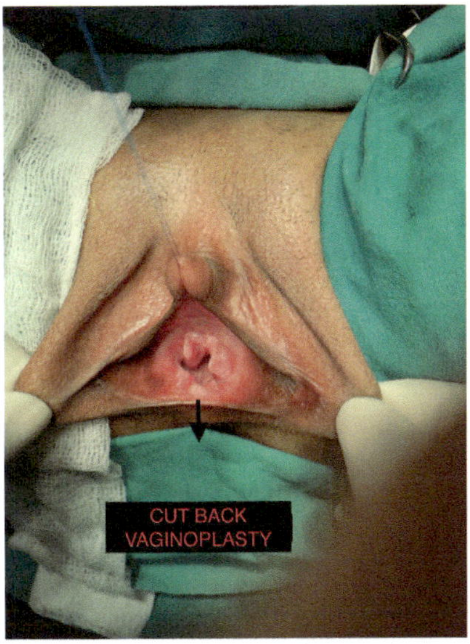

Fig. 41.6 Cutback vaginoplasty

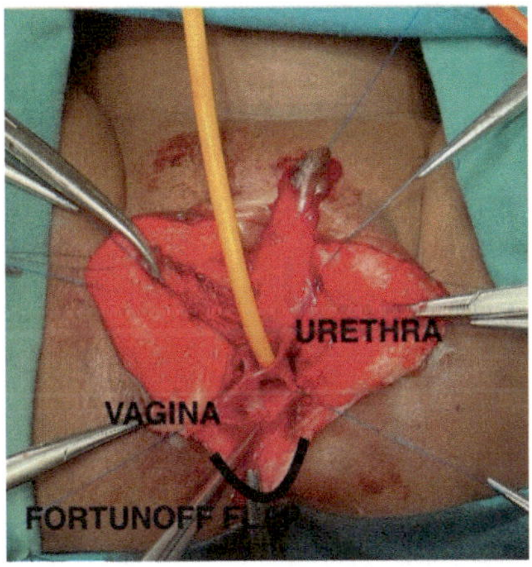

Fig. 41.7 Dissected UG sinus

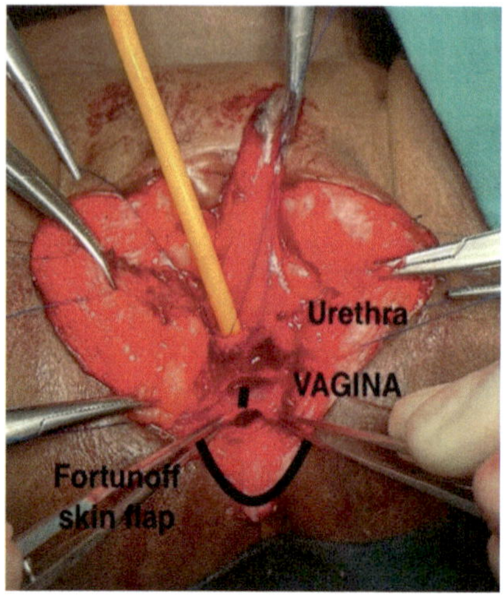

Fig. 41.8 Cut post-vaginal wall

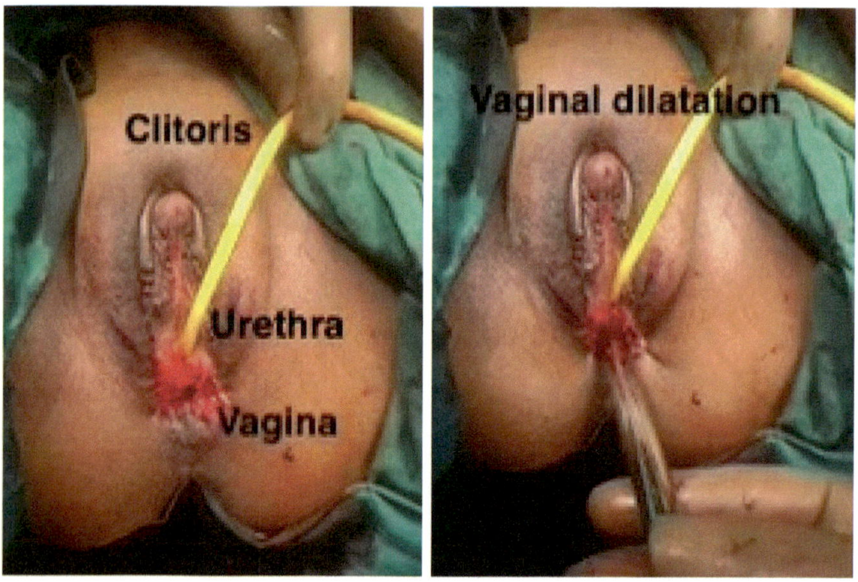

Fig. 41.9 Result of feminizing genitoplasty

41.6 Operative Management of High UGS

(A) Total Urethral Mobilization (TUM). In less than 5% of cases of CAH, urogenital confluence is high, at the level of the posterior urethra, where total urogenital sinus mobilization [TUM] may be required as described by Alberto Pena in common cloaca [1997]. For total urogenital sinus mobilization [TUM], dorsal

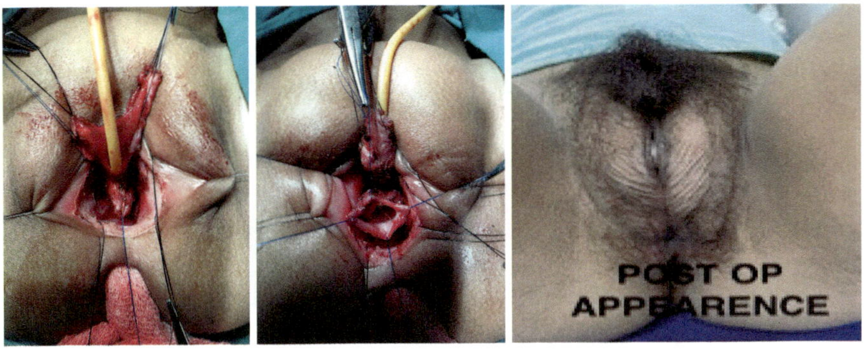

Fig. 41.10 Hendren's pull-through vaginoplasty

dissection across pubourethral ligament is required, since it has a high incidence of urinary incontinence, so we avoid doing this dissection and instead, we prefer to do a Hendren's vaginal pull-through procedure.

(B) Hendren's Vaginal Pull though: After posterior dissection of the UG sinus, the lateral wall of the UG sinus on either side is dissected. Anterior dissection between the posterior wall of the urethra and the anterior wall of the vagina is dissected after infiltrating diluted lignocaine adrenaline in this plane. This is quite a difficult dissection as the anterior wall of the vagina is very thin distally. Dissection is done taking care that the urethra is not damaged. Anterior dissection is done till the peritoneal reflection to get the vagina up the perineum. Ventral wall of the urethra is repaired over a proper-sized Foley's catheter to get the urethral opening just a little distal to the anterior vaginal wall. Posterior wall of the vagina is opened in the midline to widen the vaginal introitus and Fortunoff's skin flap is put in this defect, as described earlier.

Dorsal preputial skin is cut in the midline as for Byar's flaps in hypospadias, and used for making the labia minora, and sutured medial to the labia majora. Medially, this skin flap is sutured to the lateral edge of the urethral plate, inferiorly to the anterior wall of pulled through vagina, and laterally to medial margin of labia majora. If the labia majora are bulky due to androgenization, then it is partly excised to give a good cosmetic appearance. Y-V plasty is done at the inferior-most margin of the labia majora to locate the vagina within the labia majora. Vaginal opening is packed with paraffin gauze, compression dressing is given to the perineum and urethral catheter is left for 10 days. The vaginal pack is removed after 3 days and vaginal dilatation is started after 15 days of removal of the urethral catheter. Vaginal dilatation is continued for 2 months to prevent early vaginal stenosis (Fig. 41.10).

41.7 Substitution Vaginoplasty

Often, a short and rudimentary vagina is seen with CAIS, and a complete absence of vagina is seen with MRKH syndrome. Failure of initial gradual dilatation of the rudimentary vagina in CAIS patients and patients with MRKH will need

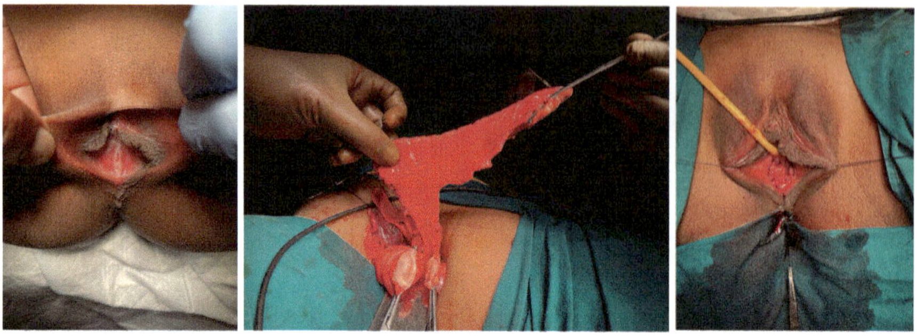

Fig. 41.11 Sigmoid vaginoplasty

vaginoplasty. Various techniques are described for this. It will be beyond the scope of this chapter to discuss technical details of all these procedures.

Most done operations for this entity are McIndoe's vaginoplasty technique and Sigmoid vaginoplasty.

McIndoe's vaginoplasty [9] technique, where a split skin graft is used, which is harvested from the inner aspect of thigh and refashioned over a vaginal mold, and placed in a space created between the urethra and the bladder anteriorly and rectum posteriorly. Common complications with this technique are, high incidence of vaginal stenosis, hence requiring regular and long-term dilatations. Dyspareunia is very common due to a dry and narrow, stenosed vagina.

Sigmoid vaginoplasty is the most done technique for patients with absent vagina in MRKH. This technique was first described by Baldwin [10] in 1904, but subsequently Pratt [11] in 1972 refocused the attention on bowel vaginoplasty, especially the use of the sigmoid colon segment (Fig. 41.11).

The ideal age for doing a sigmoid colon vaginoplasty will be post-puberty. Bowel is prepared 24 h prior to surgery by giving oral PEGLEC. A sigmoid colon segment is isolated on the inferior colic artery or the superior rectal artery. It may sometimes be difficult to get the distal end of the sigmoid colon easily to the perineum, and in such cases it is better to rotate the sigmoid colon segment through 180° in an anticlockwise direction to get the proximal end of sigmoid colon to the perineum. For this, one must preserve the marginal vessels of the sigmoid colon based on the inferior colic or superior rectal artery. The segment is brought down in a space created between the urethra and bladder anteriorly and the rectum posteriorly.

The advantages of sigmoid vaginoplasty are that it provides a very capacious vagina, with fewer chances of stenosis, not requiring vaginal dilatations and does not even need lubrication.

The drawbacks of sigmoid vaginoplasty are mucous discharge and occasional vaginal prolapse, which can be taken care of by keeping the segment of colon short up to only 10 cm, and fixing the proximal end of the sigmoid colon retroperitoneally. Syed [12] described diversion colitis in 3 patients out of 18 patients who had undergone sigmoid vaginoplasty. Mucinous adenocarcinoma of colovaginoplasty was reported by Hiroi [13] and associates from Japan 30 years after surgery.

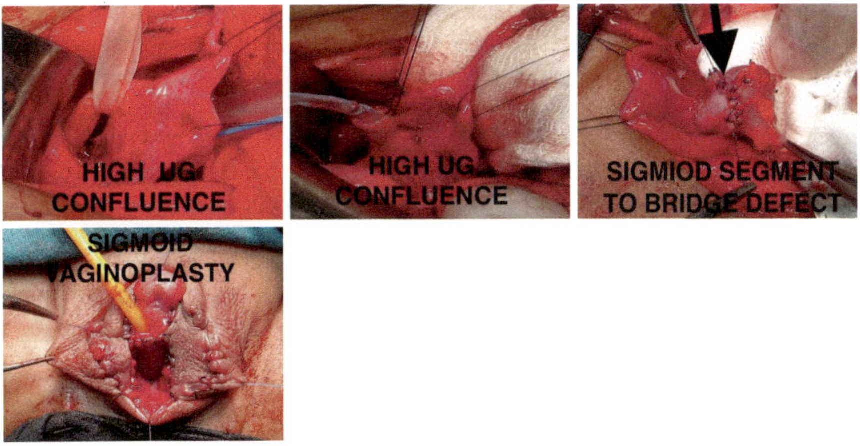

Fig. 41.12 Sigmoid vaginoplasty to bridge the gap in high UG sinus conflunce

In case of a very high confluence of UG sinus, wherein, after mobilization of the vagina as described by Hendren, if the vagina cannot be brought to perineum due to its short length, either a sigmoid colon segment or Sigmoid Monti tube [14] can be used to bridge the gap between the proximal end of vagina and the perineum (Fig. 41.12).

41.8 Ileal Vaginoplasty

Even though the preferred intestinal segment for bowel vaginoplasty is sigmoid colon, the specific indication for use of ileum for vaginoplasty may be cases of cloa cal exstrophy with vaginal agenesis where colon is absent, in these cases ileal segment is the preferred technique.

Common difficulties faced with ileal vaginoplasty are:

1. Mesentry of ileum may not reach easily in pelvis to get ileal segment easily in perineum for vaginoplasty.
2. So, to get ileal segment easily in perineum one may use a longer segment of ileum for vaginoplasty, which is usually associated with large mucus production.
3. Friability of small bowel mucosa with intercourse.

41.9 Vecchietti's [15] Technique of Vaginoplasty

This technique needs either laparotomy or laparoscopy to pass a suture through a space between anterior wall of rectum and posterior wall of bladder and urethra, across the vaginal dimple. A plastic olive is placed at the vaginal dimple, and a suture is tied to plastic olive, the other end of thread is brought out of anterior abdominal wall and attached to traction device which is strapped to the anterior abdominal wall for a period of 10 days. Gradual and steady pull-on the thread will produce progressive invagination of vaginal dimple to create a vagina. This technique is cumbersome and produces some discomfort for the patient. This newly created vagina needs regular vaginal dilatation for stenosis and can be associated with complications like bladder and rectal injury due to pressure necrosis [16].

Many such innovative techniques are available and depending upon the malformation in each child, the Pediatric surgeon has the option to choose on what is best suited for that particular child in consultation with the parents. In complex or confusing cases, the decision on the final reconstructive surgery may be deferred till the child is old enough to participate in the decision-making process. However, most of the cases, especially of CAH, can be conclusively proved and corrective surgery is done as per the existing protocols of those times.

References

1. Rink R, Kaefer M. Surgical management of intersexuality, Cloacal malformation. In: Schmid: Campbell-Walsh—Urology. 9th ed. St. Louis, MO: Saunders; 2007. p. 3854–5.
2. Yang J, Felsen D, Poppas DP. Nerve sparing ventral cliteroplasty: Analysis of clitoral sensitivity and viability. J Urol. 2007;178:1598–601.
3. Kogan SJ, Smey P, Levitt SB. Subtunical total reduction cliteroplasty: a safe modification of existing technique. J Urol. 1983;130:746–8.
4. Baskin LS, Erol A, Li YW, Liu WH, Kurzrock E, Cunha GR. Anatomical studies of human clitoris. J Urol. 1999;162:1015–20.
5. Fortunoff S, Lattimer JK, Edson M. Vaginoplasty technique for female pseudohermaphrodites. Surg Gynecol Obstet. 1964;118:545–8.
6. Hendren W, Atala A. Repair of high vagina in girl with severely masculinized anatomy from Adrenogenital syndrome. J Pedatr Surg. 1995;30:91–4.
7. Pena A. Total urogenital mobilization: An easier way to repair cloaca. J Pediatr Surg. 1997;32:267–8.
8. Rink RC, Kaefer M, Casale AJ, Meldrum KK, Cain MP. Partial urogenital mobilization: A limited proximal dissection. J Pediatr Urol. 2006;3:351.
9. McIndoe AH. Treatment of congenital absence and obliterative conditions of vagina. Br J Plast Surg. 1950;2:254.
10. Baldwin JF. The formation of artificial vagina by intestinal transplantation. Ann Surg. 1904;40:398.
11. Pratt JH. Vaginal atresia corrected by use of small and large bowel. Clin Obstet Gynecol. 1972;96:639.
12. Syed HA, Malone PS, Hitchcock RJ. Diversion colitis in children with colovaginoplasty. BJU Int. 2001;87(9):857–60.
13. Hiroi H, Yasungi T, Matsumeto K, et al. Mucinous adenocarcinoma arising in a neovagina using sigmoid colon 30 years after operation: A case report. J Surg Oncol. 2001;77:61.
14. Garcia-Roig M, Castellan M, Gonzalez J, Gorin MA, Cruz-Diaz O, Labbie A, Gosalbez R. Sigmoid vaginoplasty with a modified single Monti tube: a pediatric case series. J Urol. 2014;191(5 Suppl):1537–42.
15. Vecchitti G. Neovagina nella syndrome di Rokitansky-Kuster-Hauser. Attuat Ostet Ginecol. 1965;11:131.
16. Grosfeld JL, Coran AG. Abnormalities of female genital tract. In: Pediatric surgery, vol. 2. 6th ed. Philadelphia, PA: Elsevier; 1939.

Gynecomastia

42

Ajay Verma, Prabudh Goel, and Devendra Kumar Yadav

Abstract

Gynecomastia, the benign enlargement of male breast tissue, often emerges during puberty due to hormonal imbalances. Estrogen and testosterone fluctuations trigger mammary gland development, causing swelling beneath the nipples. While typically harmless, gynecomastia can evoke psychological distress and self-esteem issues. Treatment options vary based on severity: monitoring hormonal levels, lifestyle changes, or surgical intervention like mastectomy in persistent cases. Understanding its physiological underpinnings and psychological impacts is crucial for tailored management. Gynecomastia underscores the complex interplay of hormones and body image perceptions in male health.

Keywords

Gynecomastia · Puberty · Male breast · Mastectomy · Mammary

42.1 Introduction and Terminology

The term gynecomastia comes from *Gynae* (Female) + *Mastia* (Greek word meaning "breast"), meaning development of female-like breast in a male or enlargement of male breast. A male can have enlarged breasts physiologically at birth, due to maternal hormones, and in old age due to androgen resistance or high estrogen-to-testosterone ratio. Nearly 75% of males experience an increase in breast size at the time of puberty. It can also occur in males who are on long-term medication of drugs such as verapamil, risperidone, nifedipine, alkylating agents, PPI such as omeprazole, HIV medications (efavirenz), anabolic steroids and opioids. Common

A. Verma (✉) · P. Goel · D. K. Yadav
Department of Pediatric Surgery, All India Institute of Medical Sciences, New Delhi, India

S. K. Ratan (ed.), *Children with Differences in Sex Development*,
https://doi.org/10.1007/978-981-97-1639-5_42

drugs causing gynecomastia are digitalis, isoniazid, spironolactone, estrogens and stilbestrol. Common conditions such as cirrhosis of liver and hyperthyroidism can also cause gynecomastia. It can be unilateral but is most commonly bilateral. In young males, at the time of puberty it is accompanied by paraesthesia and pain. True gynecomastia is >4 cm in diameter and should be distinguished from excess adipose tissue, the glandular tissue is firm and has fibrous cord-like structure on palpation.

42.2 Physiology of Normal Breast Development

In normal females, at the time of puberty, the mammary ducts proliferate under the influence of estrogens, whereas progesterone helps in the development of lobules causing enlargement of breasts. This is supplemented by excess adipose tissue proliferation and deposition around mammary glands [1].

42.3 Causes of Gynecomastia

The most common cause of gynecomastia is decreased androgen-to-estrogen ratio. This can happen due to androgen deficiency or increased estrogen synthesis. The latter may happen due to increased aromatase activity by adrenal and gonadal androgens. Androgen insensitivity can also cause gynecomastia. Excess estrogens can be produced by Sertoli cells tumor. Tumors producing human Chorionic Gonadotropin (hCG) can also stimulate Leydig cell estrogen synthesis. Extra glandular androgen-to-estrogen conversion is increased when there is access to androstenedione such as in Congenital Adrenal Hyperplasia (CAH), or due to decreased catabolism of androstenedione, such as in liver disease. This shunts the estrogen precursors to aromatase in the peripheral sites. One good example is increased estrogen in obese males due to increased aromatization of androgen precursors to estrogens. Increased peripheral aromatase activity has also been described in families where it inherits as autosomal dominant or as an X-linked disorder. Some of these families were detected to have inversion of chromosome 15q21.2–3. Drugs can cause gynecomastia due to directly estrogenic substances such as oral contraceptives and digitalis, or by inhibiting androgen synthesis such as ketoconazole or decreasing androgen action such as spironolactone. Gynecomastia is a characteristic feature of children with Klinefelter syndrome and other hypo-gonadotrophic states. Ovotesticular DSD (Disorder of Sexual Differentiation) formerly known as *true hermaphrodite*, is one condition where gynecomastia can develop at puberty. Since the phenotypic presentation in these children is varied, the appearance of gynecomastia is welcomed in children who are reared as females. Similarly, patients with complete androgen insensitivity syndrome (CAIS) may develop gynecomastia and makes gender assignment easier if parents/patient wants a feminine gender identity. Other forms of 46 XY DSD such as complete gonadal dysgenesis, mixed/partial gonadal dysgenesis, and testis regression syndrome, may also present with gynecomastia. The decision for mastectomy as gender assignment is taken after considering the patient's gender identity, parental concerns, and psycho-sociological evaluation by a team comprising pediatric surgeon, pediatric endocrinologist, psychologist, and social worker.

42.4 Clinical Workup for Gynecomastia

Most of the males presenting with gynecomastia have benign disease, hence detailed investigation or intervention is not required in all cases. However, children presenting with rapid breast enlargement, tenderness, and with genital anomalies should undergo detailed evaluation. Also sometimes Disorders of Sex Development (DSD) may present unsuspectingly at puberty with gynecomastia being one sign drawing attention. Hence, all boys and young men presenting with gynecomastia should undergo a rigorous work to rule out DSD, which is detailed as under.

Clinical history should include antenatal history (maternal history of intake of ………), drug history by the children, history of infertility in family, first noticing of gynecomastia and its progress, hematuria, behavioral history, and psychological pattern, other features commensurating with male pubertal spurt (voice change included) and any urinary or other complaints. While assessment of anthropometry and general habitus for some syndromic associations is important, a careful genital examination including measurement and examination of testes and assessment of virilization is a must. Workup should also include biochemical evaluation of testosterone and other reproductive hormones, especially estradiol, assessment of their precursors, androstenedione, serum Leutinizing Hormone (LH), and hCG. A karyotype evaluation is the cornerstone investigation of such patients and so is for associated anomalies (also important for pre-anesthetic workup) In adults, the gynecomastia has been classified by BE Simon.

Simon classification [2]	
Grade 1	Small enlargement, no skin excess
Grade 2a	Moderate enlargement, no skin excess
Grade 2b	Moderate enlargement with extra skin
Grade 3	Marked enlargement with extra skin

Some patients with DSD may develop gynecomastia at puberty, and should be worked up. Children with 46 XY DSD reared as males will require gender assignment surgery and mastectomy will be a part of the process. However, children reared as females or with feminine gender identity may benefit from gynecomastia and will help in gender assignment. Children with 46 XX DSD such as CAH and Persistent Mullerian Duct Syndrome (PMDS) reared as males sometimes present with hematuria at the time of puberty. This is mainly due to hormonal effect on the endometrium producing menstrual blood mixed in urine, female gender assignment to such children will be facilitated by presence of gynecomastia. Sex chromosome DSD such as Klinefelter syndrome may present with gynecomastia and the fate of breast should be decided on the final gender assignment.

Common forms of DSD presenting with gynecomastia		
46 XY DSD	46 XX DSD	Sex chromosome DSD
• Complete gonadal dysgenesis • Partial gonadal dysgenesis • Ovotesticular DSD • Gonadal regression • Complete Androgen Insensitivity Syndrome (CAIS)	• Congenital adrenal hyperplasia • Ovotesticular DSD	• 47, XXY (Klinefelter syndrome/variants)

The workup for gynecomastia should differentiate between true glandular enlargement versus increased adipose tissue deposition, whereas the latter does not warrant any further workup the true glandular enlargement should be further evaluated for consistency (soft/hard), fixation to underlying tissue (fixed/mobile), rate of growth (rapid growing/slow growing), and duration of onset of growth (recent onset/delayed onset). Recent onset, rapidly growing, hard mass fixed to the underlying structure should be subjected to mammography with or without biopsy to rule out malignancy. Small breast mass of size less than 4 cm, peripubertal breast enlargement, neonatal breast enlargement, and associated liver disease should warrant regular follow-up with serial examination. If breast mass is more than 4 cm in diameter in a child without liver disease or there is clinical evidence of androgen deficiency such as small testis and breast tenderness the child should undergo complete biochemical analysis such as serum Testosterone (T), serum Luteinizing Hormone (LH), Follicle Stimulating Hormone (FSH), serum 17-beta Estradiol (E_2) and beta human chorionic gonadotropin (hCG). Increased beta hCG indicates hCG secreting tumors. Low T with a high E_2/T ratio suggests androgen deficiency syndrome whereas increased E_2 with normal T suggests increased aromatization of androgen to estrogen such as obesity, feminizing adrenal tumors, Sertoli cell tumors, or inherited dysregulation of aromatase.

42.5 Medical Treatment

Medical management of concerning physiological pubertal gynecomastia after detailed workup involves androgens such as danazol and aromatase inhibitors such as anastrozole; however, both these drugs have their side effects and should be used judiciously.

42.6 Surgical Treatment

Surgical treatment in physiological pubertal gynecomastia is considered when there is no scope for medical treatment, and the presence of gynecomastia is significant to cause psychological trauma. The basic principle is to remove all breast tissue while preserving the nipple-areola complex for cosmesis. When nipple–areola complex is preserved it is termed as *subcutaneous mastectomy*, the term simple mastectomy is used when complete breast tissue along with nipple and areola is excised. Patients with the Disorder of Sexual Differentiation (DSD) are referred for surgical removal of gynecomastia either as part of masculinizing genitoplasty in XY DSD or for gender reassignment in case of severely virilizing XX DSD where the patient insists on male gender identity.

The primary surgical principle is to remove all breast tissue (mastectomy) with minimal scarring. The surgical excision of breast tissue can be done by following approaches:

1. Sub-mammary incision
2. Circum-areolar incision
3. Minimally invasive axillary endoscopic excision
4. Adjunctive techniques
   (a) Liposuction (with/without US guidance)
   (b) Orthopedic arthroscopic shavers/suction-assisted cartilage shaver technique (SACST)
   (c) Microdebrider excision with liposuction technique (MELT)

In sub-mammary incision, the scar is along the normal skin crease and exposure is wide for complete breast excision. However, in patients where healing is not proper, it gives a concerning scar. In circum-areolar incision, the incision is along the curve of the areola and is cosmetically accepted. The incision is often small, and complete breast tissue excision is challenging. This approach is more frequently used along with adjunctive techniques for liposuction.

Minimally invasive endoscopic excision or Endoscopic Subcutaneous Mastectomy (ESCM) has gained popularity in recent times. In this approach, a small incision is given in axilla for insertion of endoscope. Then, two additional working ports are inserted, and a subcutaneous space is created using carbon dioxide insufflation. This space is now utilized for the complete excision of breast tissue using an endoscopic technique. Liposuction is an adjunctive procedure along with endoscopic subcutaneous mastectomy and is seldom used in isolation (Figs. 42.1 and 42.2).

Mastectomy is indicated in patients of DSD for gender assignment, cosmesis, and as well as treatment modality if carcinoma breast develops. Carcinoma breast has been reported in patients of Klinefelter's syndrome. Breast papilloma with hyperplasia of unusual type has also been reported in complete Androgen Insensitivity Syndrome (AIS) [3].

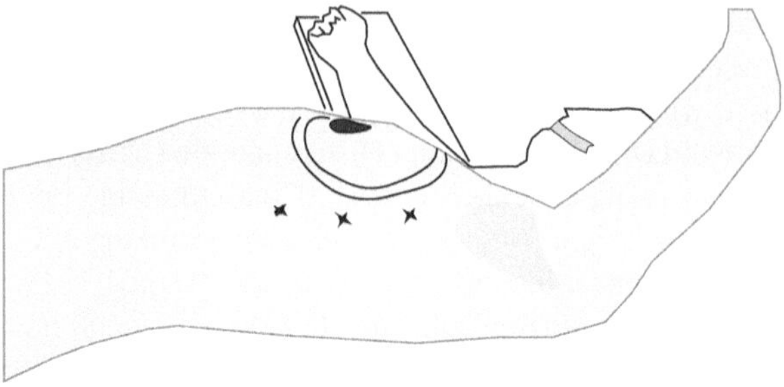

Fig. 42.1 Pencil sketch showing the port position in endoscopic subcutaneous

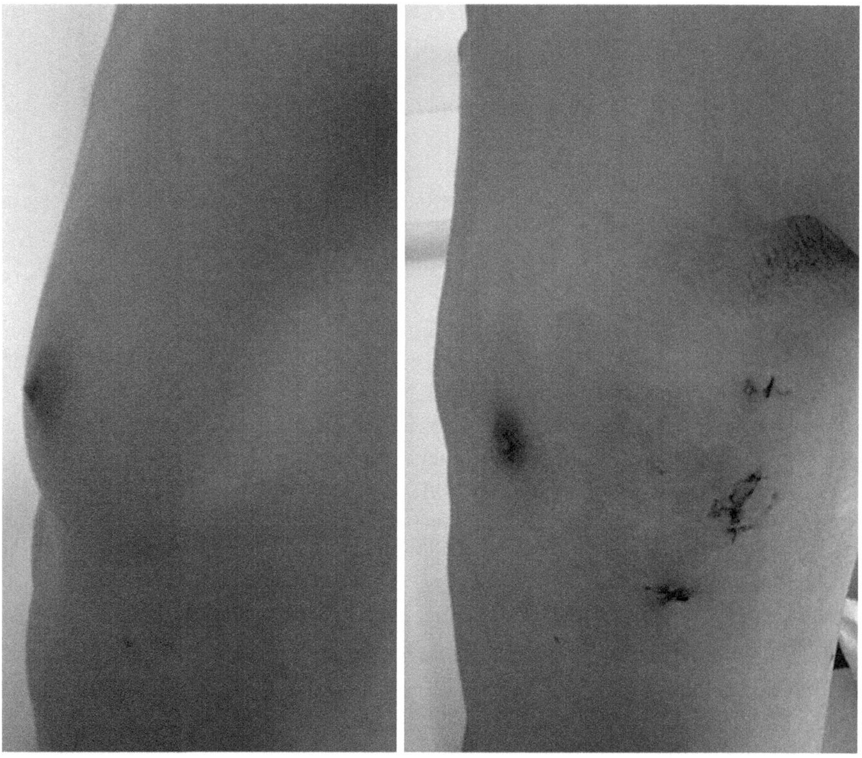

Fig. 42.2 Showing pre- and post-op photographs after endoscopic subcutaneous mastectomy [4]

References

1. Barrett KE, et al. Ganong's review of medical physiology, 26e. New York: McGraw Hill; 2019. https://accessmedicine.mhmedical.com/content.aspx?bookid=2525&sectionid=204289956
2. Simon BE, Hoffman S, Kahn S. Classification and surgical correction of gynecomastia. Plast Reconstr Surg. 1973;51:48.

3. Srinivasa Rao C, Suresh V, Sangeetha S, Rukmangada N. Breast tumour in a male with complete androgen insensitivity. J Clin Sci Res. 2016;5:144–5. https://doi.org/10.15380/2277-5706.JCSR.16.03.004.
4. Cao H, Yang ZX, Sun YH, Wu HR, Jiang GQ. Endoscopic subcutaneous mastectomy: A novel and effective treatment for gynecomastia. Exp Ther Med. 2013;5(6):1683–6. https://doi.org/10.3892/etm.2013.1032. Epub 2013 Apr 2. PMID: 23837054; PMCID: PMC3702723

Phalloplasty

43

Rajat Gupta, Richie Gupta, and P. S. Bhandari

Abstract

Disorders of sexual development are associated with a wide array of problems giving the treating clinicians a great challenge to deal. This includes Gender Dysphoria, severe penile inadequacy in length and girth, extrophy, etc. The authors in this series have extensive experience in penile reconstruction with various flaps that include radial artery forearm flap, ALT flap, Free fibula, MLD, and Groin flap in stages. Each of the procedures has its own advantages and disadvantages and is customized according to the demands of the patient explaining risks and drawbacks along with probable clinical outcomes. The goals of phalloplasty may be different for different patients and each of the procedures has to be carefully judged and offered to the patient. The gold standard procedure for phalloplasty today is Radial artery forearm flap which carries inherent drawback of exposed skin grafted donor site.

Keywords

DSD · Radial artery forearm flap · Anterolateral thigh flap · Free fibula · Groin flap · Bird wing abdominal flap · MLD · Ulnar artery flap · Penile prosthesis

R. Gupta (✉) · R. Gupta
Department of Plastic and Reconstructive Surgery, Fortis Hospital, Shalimar Bagh, Delhi, India

P. S. Bhandari
Max Superspeciality Hospital, Saket, New Delhi, India

KDH Kakkardooma New Hospital, New Delhi, India

S. K. Ratan (ed.), *Children with Differences in Sex Development*,
https://doi.org/10.1007/978-981-97-1639-5_43

Disorders of sex development are often associated with incongruence between Gender assigned at birth (i.e., what the person feels or wants the society to accept as him/her) and sex organs developed [1]. Secondly, even if there is no gender incongruence, DSD may be associated with severe penile inadequacy, undervirilization, or abnormal development of the phallus (e.g., inadequate size penis(micropenis), exstrophy, and hypospadias). Standard surgical penile lengthening techniques like metaoidioplasty do not provide patients with a phallus suitable for sexual intercourse (Erect length in normal adult less than 3 in.) leading to low self-esteem, psychosexual dysfunction, and low partner satisfaction scores. Therefore, reconstructed phalloplasty provides satisfactory genital appearance and sexual function in such cases.

43.1 The Goals of Phalloplasty [2]

In the absence of functioning corporal bodies, achieving an ideal phalloplasty is a difficult proposition. However, the goals of ideal reconstructed phallus are:

1. It should have a satisfactory, aesthetic appearance and dimensions, resembling a normal penis.
2. It should be sensate (tactile) and allow the individual to reach orgasm (erogenous sensation).
3. It should provide sufficient and lasting rigidity for vaginal penetration and sexual intercourse.
4. It should have an inbuilt neourethra with meatus at tip, which allows voiding in erect position.
5. The urethra should be of sufficient dimension and free from stenosis, so that there is a sufficient stream and no backpressure changes in urinary system.
6. When not engaged in sexual act, it should be easily concealable/ foldable in clothing appropriate for that society and era. It should also allow sports activities without getting in the way.
7. The donor area morbidity should be minimal.
8. Preferably, the procedure should be single staged (Fig. 43.1).

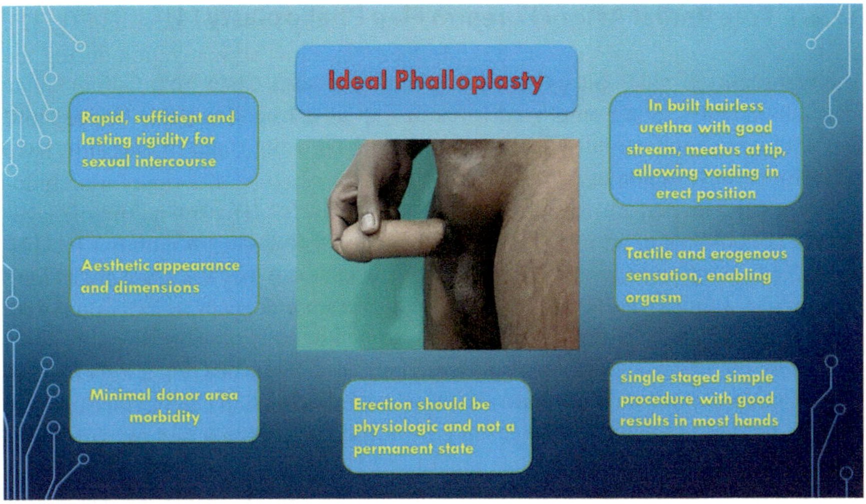

Fig. 43.1 Attributes of an ideal phalloplasty

In the present era, it is quite evident, that no single procedure today satisfies these ideals. However, modern phalloplasty techniques have evolved sufficiently through efforts of generations of surgeons, to allow for patients with penile insufficiency to function as a male.

43.2 Phalloplasty Techniques [2]

1. Free Radial Artery Forearm Flap (RAFF) (Gold Standard)
2. Anterolateral thigh flap (ALT)
3. Fibula osteocutaneous flap
4. Abdominal flaps/Bird wing flap
5. Groin flap
6. Thoracodorsal artery perforator flap (TDAP)
7. Musculocutaneous Latissimus dorsi flap (MLD) [3]
8. Ulnar artery flap

Phalloplasty comprises three components:

1. Body/Shaft with Glans
2. Urethra
3. Prosthesis/Bone

Flaps that provide adequate dimensions to form tube-in-tube technique can be used to form body and urethra of phallus. These include RAFF and ALT. Flaps that provide only body are ALT, TDAP, MLD, Fibula, Groin, and Abdominal flaps.

However, composite flaps may be used to form body and urethra like ALT, MLD, TDAP for body, and RAFF for urethra. Secondly, there may be a staged procedure where flaps can be prelaminated with skin grafts to form future penile urethra.

43.2.1 Free Radial Artery Forearm Flap Phalloplasty [4]

The first report of phalloplasty using RAFF in four cases came from Gao in China [5]. Subsequently, Song employed forearm flap as a free sensate osteocutaneous flap with wide applications. Chang in 1984 reported using forearm flap in modern form, for reconstruction of penis in single stage.

Allen's test is performed in all the cases selected for RAFF phalloplasty. It is usually not possible to perform the procedure if donor forearm circumference is less than 15 cm just above the wrist. Also, the subcutaneous tissue thickness should not be less than 4 mm or more than 8 mm. The authors prefer ulnar neourethra (Ulnar skin being non-hairy in most of the cases) and penile body along with glans being formed by the lateral part of forearm flap. The Flap dimensions depend on the size of the forearm, most of the forearm skin is recruited in the flap leaving ulnar subcutaneous border, especially in the distal part of the flap to make tube-within-a-tube construction. Minimum dimensions are 14 × 17 cm. A proximal extension of the neourethra of 2–3 cm is taken for tension-free urethral anastomosis. The radial artery with venae comitantes and cephalic vein is identified distally at the wrist and followed proximally, preserving septocutaneous perforators while ligating muscular and periosteal branches. The artery is dissected proximally until the bifurcation of the brachial artery. The medial and lateral antebrachial cutaneous nerves are identified and dissected for an additional 2–3 cm proximal to the skin paddle to allow for tensionless coaptation at the recipient site. The femoral artery is dissected in the thigh proximally for 5–6 cm in length. The great saphenous vein is dissected medially for 18–20 cm, i.e., from saphenous junction to distal third of the thigh, distal end is ligated, looped, and anastomosed end to side with femoral artery to form an arterialized segment of saphenous vein draining in femoral vein at this point. Large saphenous vein tributaries are usually available for secondary drainage. Epineural neurorrhaphy is performed between the lateral antebrachial cutaneous nerve and the dorsal nerve of the penis/clitoris or the deep pudendal nerve, allowing for erogenous sensation. If medial or posterior antebrachial cutaneous nerves are included in the flap, they can additionally be sutured with the ilioinguinal or other sensory nerves to provide tactile sensation. The urethra is anastomosed end-to-end over a Foley catheter. De-epithelialized edges around the proximal neourethra can be used to reinforce the anastomosis. Neoglans is created through the Norfolk procedure consisting of a rolled skin flap and split-thickness graft. The advantages of RAFF are:

(a) Good aesthetic outcome of the neophallus
(b) Good sensitivity
(c) Normal dimensions simulating a circumcised penis
(d) Good sexual outcomes

The main disadvantage of the RAFF is visible donor site scarring when the patient wears half-sleeve attire, as the forearm donor area, requires a large skin graft for coverage. Urethral complications are the commonest complications of phalloplasty with any technique with, urethral fistula rates of 15–77% and stricture and stenosis rates of 21–51% (Fig. 43.2).

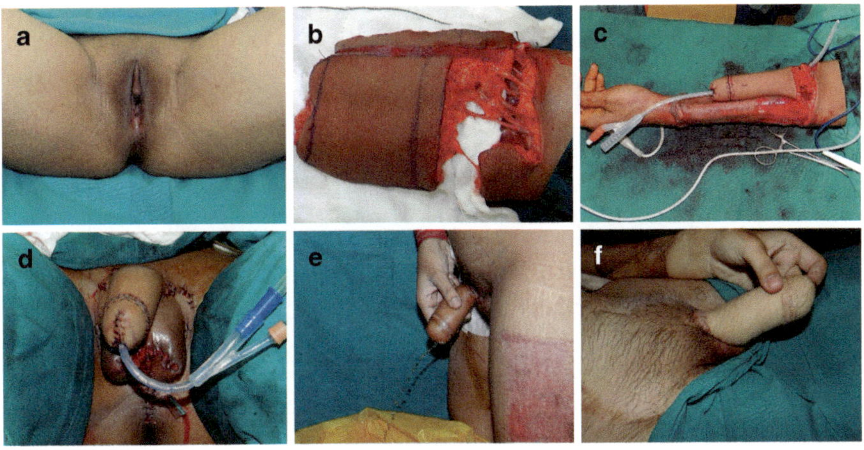

Fig. 43.2 Phalloplasty with free radial artery forearm flap (RAFF). (**a**) Preop feminine external genitalia (**b**) The RAFF including strips for penile body laterally and urethra completely dissected on vessels and nerves. (**c**) The flap rolled into a phallic configuration around a catheter. Glansplasty has been done. (**d**) Immediate postop result showing neophallus and neoscrotum. (**e**) Patient micturating via neophallus at 3 weeks postop. (**f**) Long term result

43.2.2 Anterolateral Thigh Flap Phalloplasty [6]

Preoperatively, perforator vessels from the descending branch of the Lateral Circumflex Femoral Artery (LCFA) are identified and marked on the skin with a Doppler probe. The anterosuperior iliac spine and the lateral border of the patella are marked, the line joining the two points corresponds to the intermuscular septum between rectus femoris and vastus lateralis. The perforators are usually located around the midpoint of this line. An ALT flap of dimensions 14 cm width and 16 18 cm length (in adult) is planned, inclusive of a lateral area that has been prelaminated with skin graft/vaginal mucosa and also contains the premarked perforator vessels. The medial edge of the flap is incised and lifted laterally to locate the perforator in the subfascial plane. Once the major perforator vessel is located, the dissection of the vascular pedicle is performed through the muscle or the septum to reach the origin of descending branch of the LCFA. The lateral femoral cutaneous nerve is included in the flap for providing sensitivity to neophallus. At the end of the pedicle dissection, the identified flap is incised and harvested suprafascially, leaving only a small amount of protective fascia around the perforator pedicle. Inverted "U" shaped skin incision is made on the pubis and the flap is transposed to this area by passing it under rectus femoris and sartorius muscles. The flap is tubularized to form the body of the phallus and inset over the desired area. If the flap is too thick for a tension-free primary closure, it can be thinned to avoid venous congestion during tubularization. In the case of phalloplasty, primary closure of ALT donor site is not possible due to flap dimensions. The donor area is decreased with the help of reducing sutures and then skin grafted. The flap end of the lateral femoral cutaneous nerve is coapted with ilioinguinal nerve or dorsal nerve of penis/clitoris.

Urethra may be constructed primarily using tube-in-tube technique if the flap thickness is less than 1.5 cm or can be done in two stages, in which case in first stage,

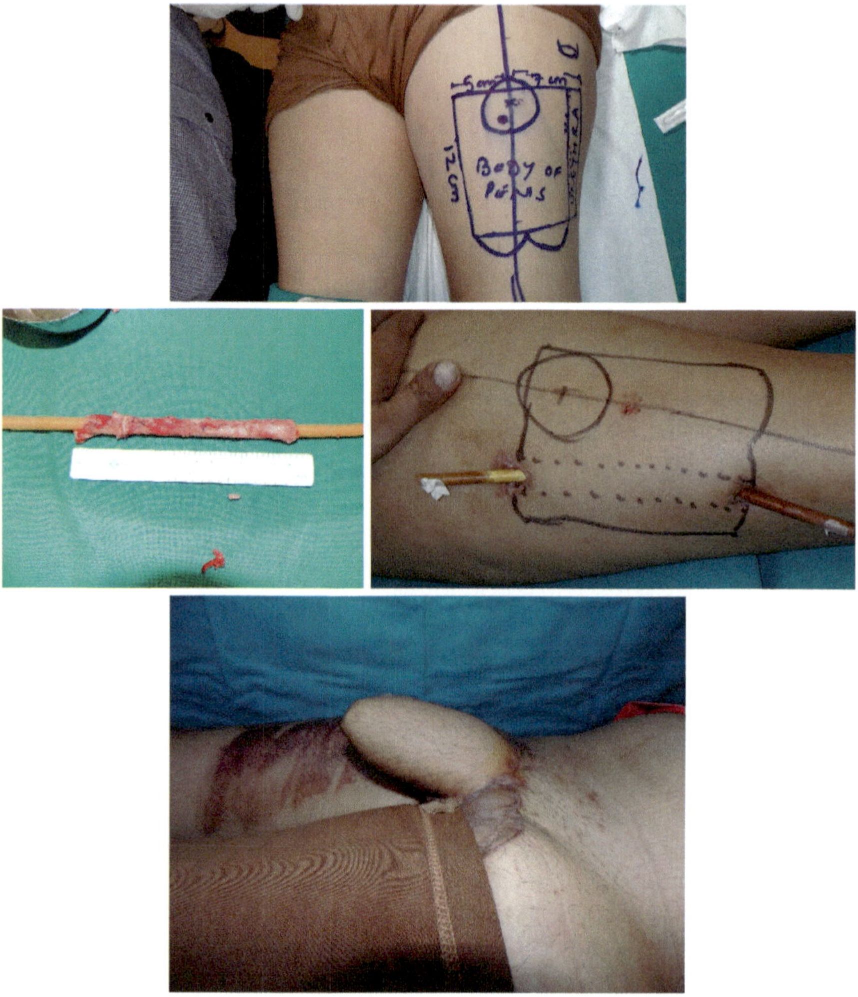

Fig. 43.3 Anterolateral thigh flap (ALT) phalloplasty. (**a**) Planning of ALT flap on left thigh. (**b**) Vaginal mucosa graft rolled around a 16G Foley's catheter. (**c**) ALT flap prelaminated with vaginal mucosa graft urethra. (**d**) ALT phallus at 3 months, prior to glans plasty

prelamination with skin graft is done in lateral part of the flap. In the second stage, the prelaminated neourethra is buried in the body of the phallus and transferred to the desired area (Fig. 43.3).

43.2.3 Osteocutaneous Free Fibula Flap Phalloplasty [7–10]

From 1992 to 1993, Sadove et al. started using Free fibula for phalloplasty relying on the intrinsic rigidity of the flap (based on an autologous bone transfer) and its superior length of the vascular pedicle. This flap is based on the peroneal artery and vein and their perforating vessels. The flap fibular bone component affords phallic rigidity without the need for penile prosthesis. However, Colebunders et al. reported that the persistent erection of the flap and a pointed distal part deformity could generate discomfort and complaints. Since the skin of the fibular flap is somewhat stiff and not free to be rolled on itself to create the neourethra, some authors have proposed to prelaminate the neourethra with STSG 6 months prior to the final flap surgery. Donor site scarring is much reduced compared to other flap options and can be easily covered by socks or long trousers. CT angiography of the leg is performed preoperatively as it was reported by Strauch that in up to 8% of cases the peroneal artery is large (arteria peronea magna) and also supplies the territory of posterior tibial artery. In such a case, the blood supply of the leg could be compromised if fibula flap based on peroneal artery is harvested. The fibula along with the skin paddle after confirming perforators on handheld Doppler are marked. The Lateral Sural Cutaneous Nerve (LSCN) is also marked out. The surgical incision is deepened up to the subfascial layer and the posterior intermuscular septum is dissected, without damaging the septocutaneous branches. Proximal and distal fibular osteotomies are performed leaving 4–5 cms of bone at either end in order to safeguard common peroneal nerve and stability of ankle joint respectively. The vascular pedicle (peroneal artery and vein) can thereafter be identified and dissected proximally up to its origin. Once harvested, the cutaneous component is rolled up to cover the fibula in a tension-free manner. The flap is transferred to the pubic area where it is anchored to the pubic periosteum and flap vessels are microsurgically anastomosed to the recipient femoral artery and saphenous vein. Similarly, LSCN is coapted to one of the dorsal penile nerves. The flap donor site is skin grafted. Donor site morbidity from fibula grafting is minimal, although published reports are limited.

Hage [11] described refinements in free fibula sensate phalloplasty to maximize the properties of neophallus as well as decrease donor area morbidity. He describes that it is particularly important to select the proper skin island innervated by LSCN, determined by LA injection. Also, a longitudinal flap design, prefabrication of urethra, and fixation of bone only to corporal tunica albuginea and not bone to allow foldability of phallus and transmit sensation through clitoris. He recommended this flap when a patient does not accept the FA donor area.

43.2.4 Thoracodorsal Axis, Musculocutaneous Latissimus Dorsi Phalloplasty

Latissimus dorsi musculocutaneous flap is raised conventionally or on perforators (TDAP) into a tube for reconstructing a neophallus. The thoracodorsal vessels are anastomosed to recipient vessels in groin/ pubic area and thoracodorsal motor nerve is anastomosed to dorsal penile/clitoral nerves. However, there is poor sensory recovery. Also, there is no provision for tube in tube urethra. The flap is grafted with buccal mucosa/ skin graft to create urethra in stages.

43.2.5 Groin Flap Phalloplasty [12]

The groin flap was more commonly used as the first choice for phalloplasty prior to the development of microsurgical techniques. However, it may still be used as a salvage flap if primary reconstruction with a free flap fails. Groin flap is pedicled and generally based on the Superficial Circumflex Iliac Vessels (SCIA). A lateral groin flap comprises thin hairless skin, which has an appearance similar to genital area skin. It may therefore have a cosmetic advantage compared to free flaps. Donor site scarring can also be minimized as the area is easily concealable. On the other hand, there is no available nerve to provide sensitivity to neophallus.

Doppler ultrasound is used to outline the course of SCIA which is marked on the skin. Ordinarily, groin flap territory does not extend beyond 5–6 cm distal to the anterosuperior iliac spine. Outline delay helps to extend the territory of groin flap in order to accommodate neophallic dimension and extend the reach of flap to pubic area. Authors prefer to perform two delays at intervals of 10 days each. 10 days after the second delay procedure, the flap along with its pedicle is gently dissected up to the SCIA origin. The distal end of the flap is transferred to the pubic area after tubularization. After around 3 weeks, flap is divided near the origin of its pedicle. The procedure could be performed in one or two steps. Likewise, other abdominal flaps, and a skin graft prefabricated urethra can be placed within the flap before its transfer. The distal part of the flap is shaped and closed while in case of prefabricated urethra, the anastomosis between the neourethra and the pars fixa is carried out with local flaps.

43.2.6 Abdominal Flap and Bird Wing Phalloplasty [13]

Phalloplasty from lower abdominal skin is described as Bird Wing phalloplasty and was published by Bajpai et al. in 2013. As the name suggests a "bird wing" incision is marked with its base in suprapubic/mons pubic location and lateral extensions up to lower abdominal skin crease extending to both flanks. The base to limb ratio of the flaps is kept at 4 or 5:1, so that adequate blood supply is ensured to the most distal end. It may be noted that a unique feature of this flap design is the common base which sustains blood supply to both the flaps. The depth of the incision reaches up to the anterior rectus sheath and the external oblique aponeurosis, from medial to lateral. Although the blood supply of this flap is provided by bilateral superficial inferior epigastric (SIEA) and bilateral superficial circumflex iliac (SCIA) vessels as described by Bajpai, a look at the configuration suggests that bilateral SIEA pedicles would need to be divided in order to reduce the pedicle width and achieve a tubular configuration.

Canadian researchers Noah Adams and François Grenier presented a poster at the 2012 CPATH conference that compiles phalloplasty research data "in an attempt to weigh the actual outcome of phalloplasty against the stereotypes and perceptions of this surgery."

The review analyzed 29 English language academic studies on female-to-male phalloplasty published since 1980. Included were charts showing the popularity of various phalloplasty techniques, the ability to stand to void, and rates of satisfaction and complications. The popularity of donor flaps was Forearm 76%, Fibula 9%, Pedicled Abdominal Flap 7%, Myocutaneous LD 5%, ALT 1%, and Groin 1%.

Surgical technique	Advantages	Disadvantages	Important tips
Radial forearm-free flap	If harvested as osteocutaneous flap, can be single staged Gives the best sensory return among all flaps Best aesthetic result Thin pliable skin with little fat, allowing easy molding and erogenous sensation Considered the gold standard No sacrifice of major blood vessel	Exposed donor area at forearm is often a giveaway Requires microsurgical techniques Stiffener is required if not osteocutaneous Over time, depigmentation occurs with poor local color match to pubic skin A major vessel in forearm is sacrificed	Forearm length should be >25 cm Forearm circumference >15 cm at distal flap margin Allen's test should indicate good perfusion of hand through the ulnar artery Forearm subcutaneous tissue thickness should be between 4 and 8 mm Veins in donor forearm should be protected from venepuncture from the time when the patient is in your care If proximal tubing is difficult, do not close. Use skin graft or local flap there
Osteocutaneous fibula flap	Gives the best sexual experience among all flaps Donor area is easily concealed	Sensation is not as good as RAFF Requires microsurgical techniques Difficult to conceal in clothing due to bulky fibula, resulting in a state of permanent erection	
Anterolateral thigh flap	May be transferred as a pedicled flap Sensate, due to incorporated lateral femoral cutaneous nerve Easily concealed donor area Major vessel is not sacrificed Microsurgical setup is not required	Often results in bulky penis, despite primary thinning Urethra is from hair-bearing area, unless prelaminated Deep, tedious perforator dissection	
Latissimus dorsi flap with nerve anastomosis to gracilis nerve	Has the capability of stiffening with achievement of sexual function in 35–40% of patients	No sensory anastomosis reported Urethra has not been reconstructed Requires microsurgical expertise	

Surgical technique	Advantages	Disadvantages	Important tips
Adjacent flaps	Easy technique, microsurgical skills are not required Good local color match Penile shaft can be longer than with RAFF No sacrifice of major blood vessel Can be used as salvage procedures	Often multistaged Urethral reconstruction is often unsatisfactory Poor sensation Stiffener is required Poor cosmesis, often bulky	

References

1. Adi K, et al. Phalloplasty and urethroplasty in adults with DSD. Urol Case Rep. 2019;26:100930.
2. Heston AL, et al. Phalloplasty: Techniques and outcome. Transl Androl Urol. 2019;8(3):254–65.
3. Jun MS, Pusica S, Kojovic V, et al. Total phalloplasty with latissimus dorsi musculocutaneous flap in female-to-male transgender surgery. Urology. 2018;120:269–70. https://doi.org/10.1016/j.urology.2018.06.032.
4. Morrison SD, Shakir A, Vyas KS, et al. Phalloplasty: a review of techniques and outcomes. Plast Reconstr Surg. 2016;138:594–615. https://doi.org/10.1097/PRS.0000000000002518.
5. Gao XS. Reconstruction of penis: report of 4 cases (author's transl.). Zhonghua Wai Ke Za Zhi 1980;18(1):67–8.
6. Wei FC, Jain V, Celik N, et al. Have we found an ideal soft-tissue flap? An experience with 672 anterolateral thigh flaps. Plast Reconstr Surg. 2002;109:2219–26.; Discussion 2227–30. https://doi.org/10.1097/00006534-200206000-00007.
7. Pototschnig H, Schaff J, Kovacs L, Biemer E, Papadopulos NA. The free osteofasciocutaneous fibula flap: clinical applications and surgical considerations. Injury. 2013;44(03):366–9.
8. Schaff J, Papadopulos NA. A new protocol for complete phalloplasty with free sensate and prelaminated osteofasciocutaneous flaps: experience in 37 patients. Microsurgery. 2009;29(05):413–9.
9. Papadopulos NA, Schaff J, Biemer E. The use of free prelaminated and sensate osteofasciocutaneous fibular flap in phalloplasty. Injury. 2008;39(Suppl 3):S62–7.
10. Chen HC, Gedebou TM, Yazar S, Tang YB. Prefabrication of the free fibula osteocutaneous flap to create a functional human penis using a controlled fistula method. J Reconstr Microsurg. 2007;23(03):151–4.
11. Kjolhede A, et al. Metoidioplasty and groin flap phalloplasty as two surgical methods for creation of neophallus in FTM confirming surgery: A retrospective study comprising 123 operated patients. JPRAS. 2019;22:1–8.
12. Hage JJ, Winters HAH, Lieshout JV. Fibula free flap phalloplasty: Modifications and recommendations. Microsurgery. 1996;17(7):358–65
13. Bajpai M. Bird wing flap abdominoplasty. J Indian Assoc Pediatr Surg. 2013;18(2):49–52.

Clitoroplasty: Points of Concern

44

Pradnya S. Bendre, Gayatri Munghate, and Arka Banerjee

Abstract

The chapter details a balanced, patient-centric approach to the management of Disorders of Sexual Differentiation (DSD) with clitoral hypertrophy through feminizing genitoplasty (clitoroplasty, labioplasty, and vaginoplasty), stressing the need for surgical precision, legal awareness, ethical sensitivity, and the adaptability to evolving practices and guidelines.

Amidst growing debates fueled by concerns over Gender Identity Disorder (GID) in children with DSD and emergence of human rights activists, the practice faces significant ethical, legal, and medical scrutiny. The situation demands a multidisciplinary approach to decision-making, involving experts across pediatric surgery, endocrinology, psychology, and other relevant fields, to individualize treatment and minimize legal and ethical risks. Emphasizing informed consent as a cornerstone, the chapter outlines the necessity of thorough communication regarding the diagnosis, potential outcomes, fertility, gender identity implications, and the details of proposed surgical interventions. This includes specific consent protocols for minors, tailored to age-appropriate comprehension levels.

Surgically, the chapter advances toward techniques that prioritize anatomical and sensory preservation, marking a shift from historical practices toward less invasive methods like clitoral recession and subtunical reduction. It underscores the critical understanding of clitoral anatomy, particularly the dorsal nerve, to ensure sensory preservation. The application of hyaluronidase for minimizing surgical trauma is discussed as a technique to enhance dissection efficiency and reduce complications.

P. S. Bendre (✉) · G. Munghate · A. Banerjee
Department of Pediatric Surgery, Bai Jerbai Wadia Hospital for Children, Mumbai, Maharashtra, India

S. K. Ratan (ed.), *Children with Differences in Sex Development*,
https://doi.org/10.1007/978-981-97-1639-5_44

Keywords

Clitoroplasty · Clitoromegaly · Feminizing genitoplasty · Informed consent

44.1 Introduction

Management of ambiguous genitalia in Congenital adrenal hyperplasia with clitoral hypertrophy by Feminizing genitoplasty has been the standard practice. Clitoroplasty, labioplasty, and vaginoplasty are the components of Feminizing genitoplasty [1]. The magnitude and timing of the surgery have always been the areas of debate [2–5]. With the identification of Gender Identity Disorder (GID) in children with Disorders of Sexual Differentiation (DSD) later in life and many human rights activists recommending a ban on such medically unnecessary normalizing surgery, it is increasingly becoming controversial. Therefore, it is important that surgeons are aware of the local laws and controversies. Full justice should be given to patients without putting the surgeons at legal risk. Some studies have reported that genital sensitivity is impaired after surgery [6]. The aim of the surgery should be restoration of normal anatomy as well as preservation of sensory function [1]. A structurally, functionally, and cosmetically acceptable genitalia appropriate to the gender should be achieved at an appropriate time by an experienced surgeon. This chapter focuses on the nuances of the clitoral and labial surgery in CAH and is based on the experience of the author along with the recent developments in the field.

44.2 Multidisciplinary Team

Western countries are increasingly imposing restrictions on genital surgery. While some have it in the form of recommendations, others have strict laws against it [7]. India too is not lagging in this direction. While there are activists protesting against the surgeons performing these surgeries in northern India, the state of Tamil Nadu has banned all medically unnecessary genital surgeries [8]. The general consensus is to individualize the management by discussing on merits of each case [2, 5]. To protect surgeons against possible legal hazards, author recommends forming a multidisciplinary team in each hospital to decide on the indications and the timing of the surgery by consensus. The Multidisciplinary team should consist of members from Pediatric urology/surgery, Endocrinology, Psychology, Psychiatry, and if pertaining to the concerned case, Neonatology, and Genetics as well.

Each case should be discussed and consensus obtained regarding the extent and timing of surgery. This will avoid the potential legal hazards to the surgeon.

44.3 Consent

Written informed consent is the most important step. The surgery should be undertaken after obtaining informed consent from the parents or the individual, who must be fully aware of all current information, both for and against. The consent should include the following details:

1. Full details of diagnosis including chromosomes, internal and external organs.
2. Long-term prognosis as mentioned in available literature including potential of malignancy.
3. Potential for fertility, together with options for assisted reproduction.
4. Gender identity outcomes and possible psychological repercussions.
5. Rationale and information for the proposed procedure(s).
6. If gonadectomy is contemplated, cryopreservation of sperm/ova.
7. Post-surgical care.
8. Alternative procedures, along with their risks and benefits.
9. The surgeon's experience.
10. Referrals to professional and support groups.

For children less than 18 years parents should sign the consent. For children 7–11 years of age a verbal assent and for those from 12 to 17 years a written assent should also be obtained. The author practices taking consent on legal paper from both parents and notarizing it by a court notary (Fig. 44.1).

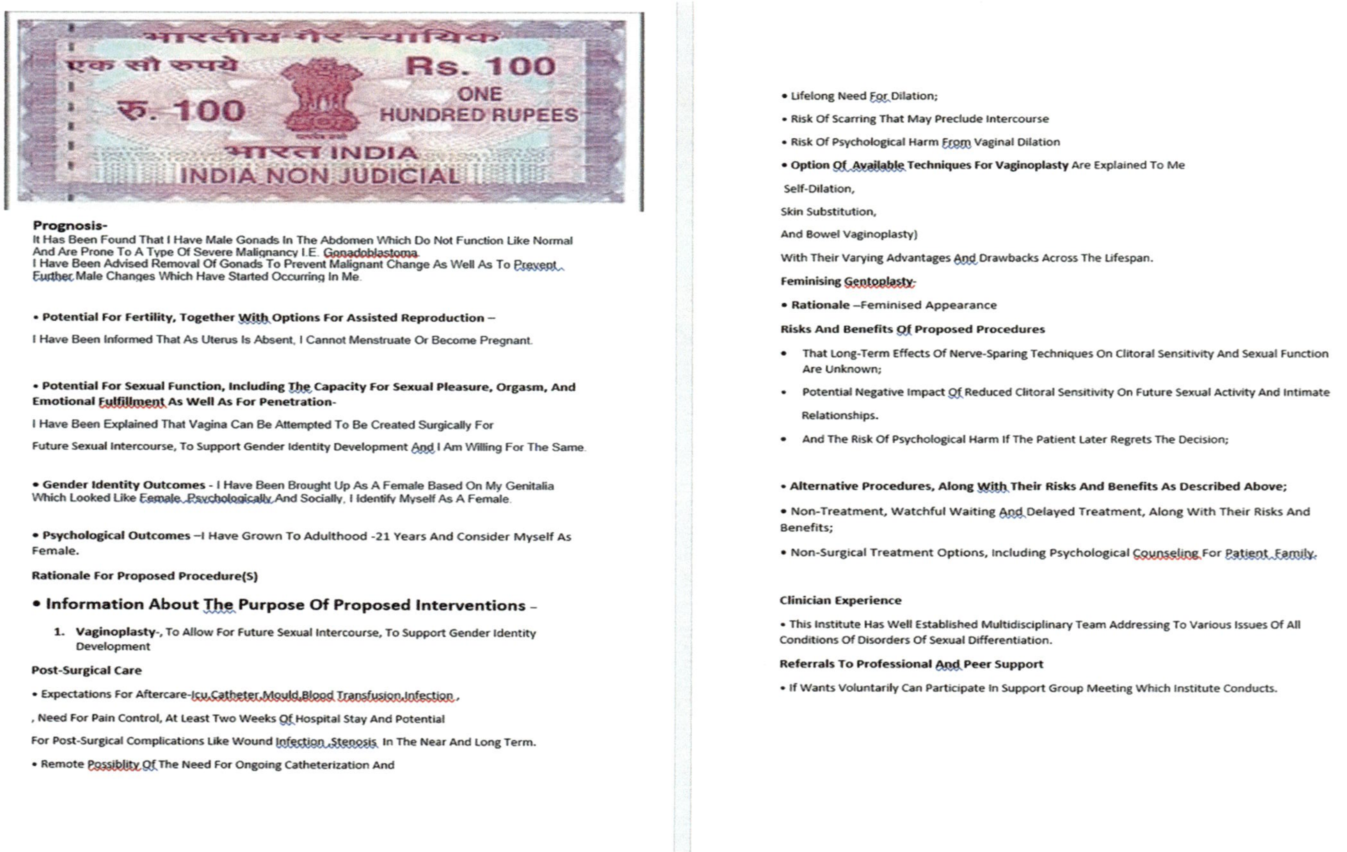
भारतीय गैर न्यायिक
एक सौ रुपये
Rs. 100
रु. 100
ONE HUNDRED RUPEES
भारत INDIA
INDIA NON JUDICIAL

Prognosis-
It Has Been Found That I Have Male Gonads In The Abdomen Which Do Not Function Like Normal And Are Prone To A Type Of Severe Malignancy I.E. Gonadoblastoma.
I Have Been Advised Removal Of Gonads To Prevent Malignant Change As Well As To Prevent Further Male Changes Which Have Started Occurring In Me.

• **Potential For Fertility, Together With Options For Assisted Reproduction –**

I Have Been Informed That As Uterus Is Absent, I Cannot Menstruate Or Become Pregnant.

• **Potential For Sexual Function, Including The Capacity For Sexual Pleasure, Orgasm, And Emotional Fulfillment As Well As For Penetration-**

I Have Been Explained That Vagina Can Be Attempted To Be Created Surgically For

Future Sexual Intercourse, To Support Gender Identity Development And I Am Willing For The Same.

• **Gender Identity Outcomes** - I Have Been Brought Up As A Female Based On My Genitalia Which Looked Like Female. Psychologically And Socially, I Identify Myself As A Female.

• **Psychological Outcomes** –I Have Grown To Adulthood -21 Years And Consider Myself As Female.

Rationale For Proposed Procedure(S)

• **Information About The Purpose Of Proposed Interventions** –

1. **Vaginoplasty**-, To Allow For Future Sexual Intercourse, To Support Gender Identity Development

Post-Surgical Care

• Expectations For Aftercare-Icu,Catheter,Mould,Blood Transfusion,Infection ,

, Need For Pain Control, At Least Two Weeks Of Hospital Stay And Potential

For Post-Surgical Complications Like Wound Infection ,Stenosis, In The Near And Long Term.

• Remote Possiblity Of The Need For Ongoing Catheterization And

• Lifelong Need For Dilation;

• Risk Of Scarring That May Preclude Intercourse

• Risk Of Psychological Harm From Vaginal Dilation

• **Option Of Available Techniques For Vaginoplasty** Are Explained To Me

Self-Dilation,

Skin Substitution,

And Bowel Vaginoplasty)

With Their Varying Advantages And Drawbacks Across The Lifespan.

Feminising Gentoplasty-

• **Rationale** –Feminised Appearance

Risks And Benefits Of Proposed Procedures

- That Long-Term Effects Of Nerve-Sparing Techniques On Clitoral Sensitivity And Sexual Function Are Unknown;
- Potential Negative Impact Of Reduced Clitoral Sensitivity On Future Sexual Activity And Intimate Relationships.
- And The Risk Of Psychological Harm If The Patient Later Regrets The Decision;

• **Alternative Procedures, Along With Their Risks And Benefits As Described Above;**

• Non-Treatment, Watchful Waiting And Delayed Treatment, Along With Their Risks And Benefits;

• Non-Surgical Treatment Options, Including Psychological Counseling For Patient, Family.

Clinician Experience

• This Institute Has Well Established Multidisciplinary Team Addressing To Various Issues Of All Conditions Of Disorders Of Sexual Differentiation.

Referrals To Professional And Peer Support

• If Wants Voluntarily Can Participate In Support Group Meeting Which Institute Conducts.

Fig. 44.1 Proforma for consent on legal paper

44.4 Preparation

Bowel preparation is done before surgery. Although rare, rectal injury remains a known and dreaded complication of surgery [4]. As children with CAH are on steroids, the stress dose must be given on the day of surgery and is repeated if surgery is prolonged beyond 3 hours or as advised by the endocrinologist.

44.5 Clitoral Reduction

This surgery is one of the components of feminizing genitoplasty and is the center of controversy. Clitoris is responsible for the sensitivity required for the sexual gratification in the intercourse. Few studies have demonstrated that genital sensitivity is impaired after surgery and this impairment is related to the difficulties in sexual function [6]. The potential damage to clitoral sensitivity and the irreversible character of this procedure are the two main criticisms made against it. There has been a paradigm shift from clitoral amputation to clitoral recession to subtunical reduction [9–11]. While there have been major advances in clitoral surgery over the last 50 years, long-term data on current techniques remains to be assessed.

Knowledge of the nerve supply of the clitoris is essential for maximizing the sensory preservation (Fig. 44.2).

The Dorsal Nerve of Penis (DNP) provides sensory innervation to the glans penis via linear projections with ramifications toward the distal end of the penis. Lateral branches of the DNP from the penile shaft linearly penetrate the spongy tissue of the distal glans with ramifications, while its dorsomedial branches give off nerve fibers to the dorsum of the mid-glans (Fig. 44.3).

The distal glans shows more compact innervations by ramifications from major branches of the DNP. Arborized nerve fibers perforate the Tunica Albuginea (TA) and form fine reticular networks in the dermis. Knowledge of anatomy of clitoris can provide insight into the finer techniques of surgery aiming for sensory preservation.

Thus, sensory preservation is achieved by sparing of nerves, sparing of the inner preputial skin, avoiding reduction of glans, avoiding tourniquets, generous diathermy and, most importantly, surgical expertise [14].

The steps of the surgery as performed by the author are:

The meatus/ Urogenital sinus (UGS) is circumscribed for hypospadias surgery, taking care to leave a redundant segment of dorsal inner foreskin to fashion a hooded prepuce, thus preserving an important source of sensation. We preserve all of the preputial skin, which has been shown to be the second most sensitive area on the female perineum. The clitoris is degloved; the ventral strip of the urethral plate is kept intact. Following degloving of the dorsal skin, the distal portion of the UGS (urethral plate) is divided below the glans clitoris. The urethral plate/UGS is mobilized off the ventral aspect of the clitoral shaft downward to below the bifurcation of the corporal bodies. Next, clitoral reduction is carried out; the author does not use a tourniquet. For dissection, hyaluronidase injection is used. It reduces bleeding and

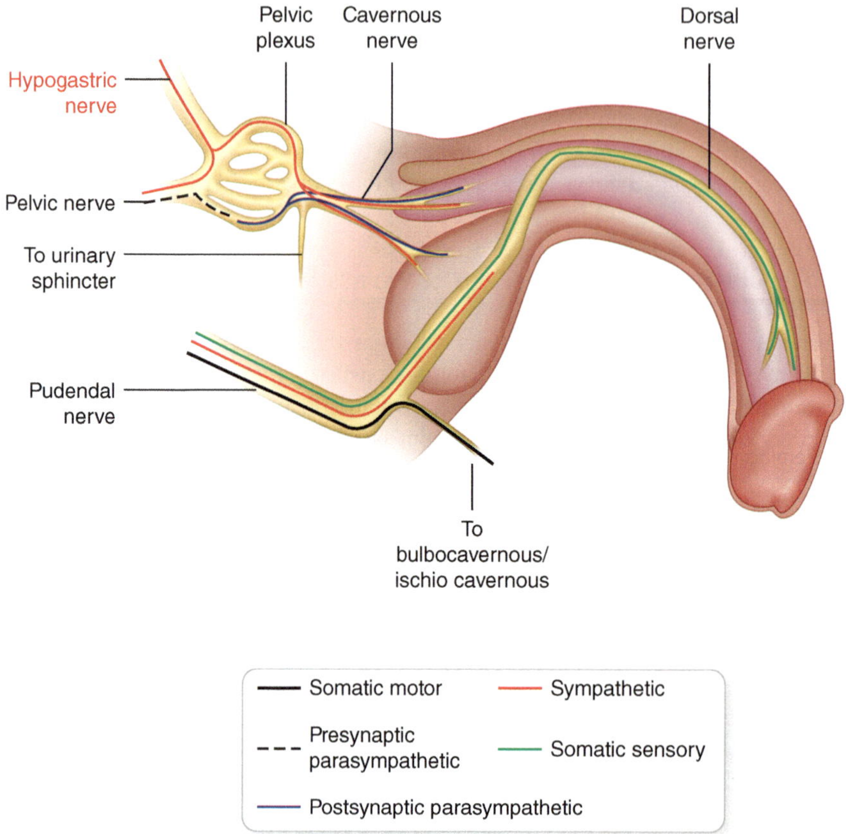

Fig. 44.2 Innervation of clitoris/penis [12]

Fig. 44.3 Innervation of the glans [13]

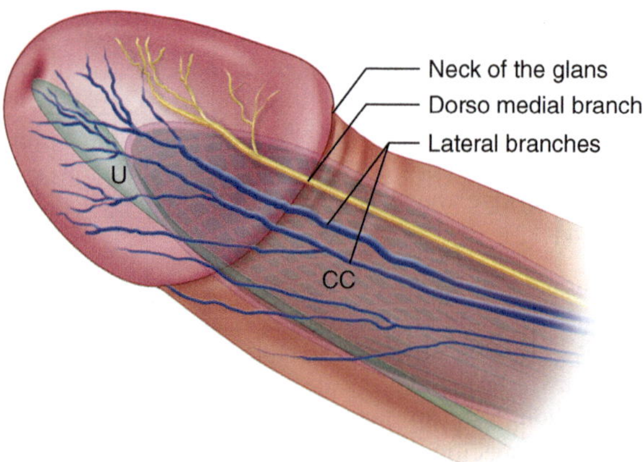

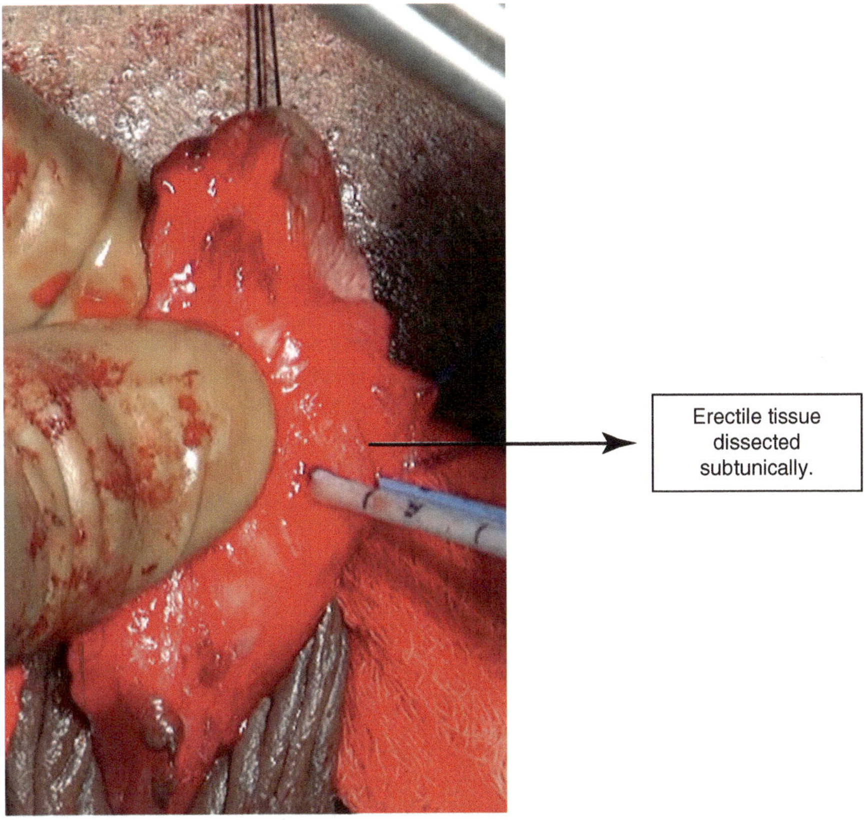

Fig. 44.4 Subtunical dissection of erectile tissue

planes are entered easily making dissection smooth and almost bloodless. Longitudinal ventral incisions are made on the corporal bodies to spare the nerves. The erectile tissue is dissected within the bodies. Hyaluronidase injection is used for this dissection. After dissection, the erectile tissues are excised carefully transfixing the base (Fig. 44.4).

The body of the glans is sutured to the pubic symphysis with absorbable sutures. The sutures are taken as vertical mattress sutures to avoid injury to neurovascular elements. The reduced clitoral shaft is covered with surrounding fat tissue, thus fashioning a normal-looking mons pubis. In most cases, the authors try to avoid reduction of the glans clitoris, but in patients with severe enlargement, a wedge of glans tissue is excised from the ventral aspect of the glans with careful re-approximation of the glans tissue with interrupted 6/0 absorbable sutures. The dorsal mucosal collar should cover the glans partially, giving it a hooded appearance. The remainder of the dorsal prepuce is divided in the midline (Byar's technique). The preputial skin wings are rotated inferiorly and incorporated lateral to the urethral plate to create labia minora.

44.6 Labioplasty

Labia majora—the labioscrotal folds have rugosities depending on the extent of virilization. The majora folds need reduction to achieve removal of rugosities for optimum cosmetic outcomes. Crescent-shaped flaps are created to achieve this (Fig. 44.5).

A stay suture is applied and pulled inferiorly to the level of inferior level of introitus to assess the extent of flaps that need excision. The edema developed intra-operatively can make rugosities less severe and can be misleading in the assessment of the extent of flaps reduction.

Labia minora—as mentioned in the clitoroplasty the remainder of the dorsal prepuce is divided by Byar's technique and wings rotated inferiorly lateral to the urethral plate to create labia minora (Fig. 44.6). The wings are pulled inferiorly to the inferior level of introitus created by opening of the common channel.

The wings are sutured in the manner of plicating sutures. These sutures ensure that the minoras remain as folds and do not become flattened in the long term (Fig. 44.7, 44.8, 44.9, and 44.10).

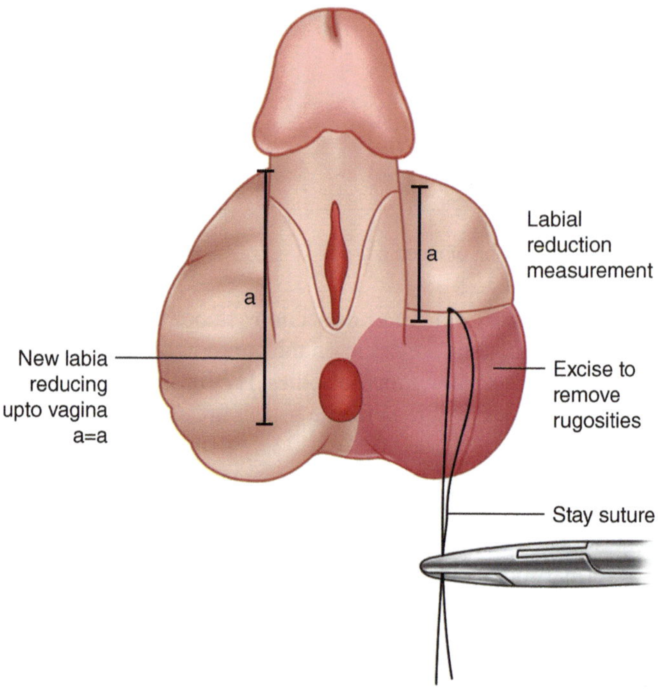

Fig. 44.5 Marking of flaps for labia majora

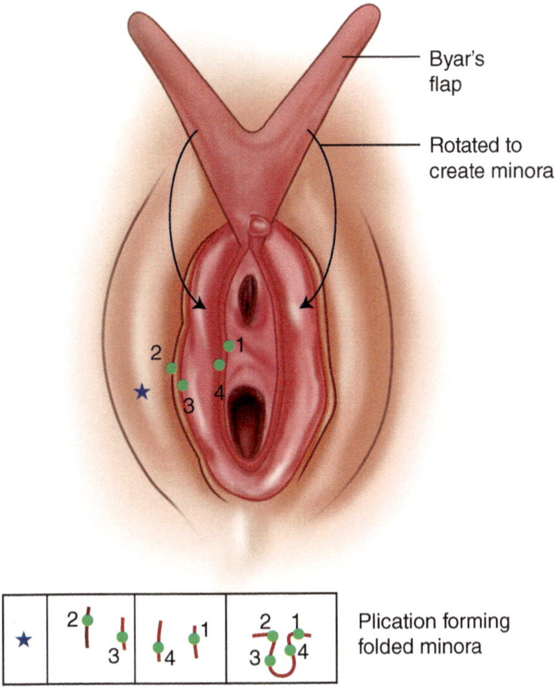

Fig. 44.6 Byar's flap for labia minora

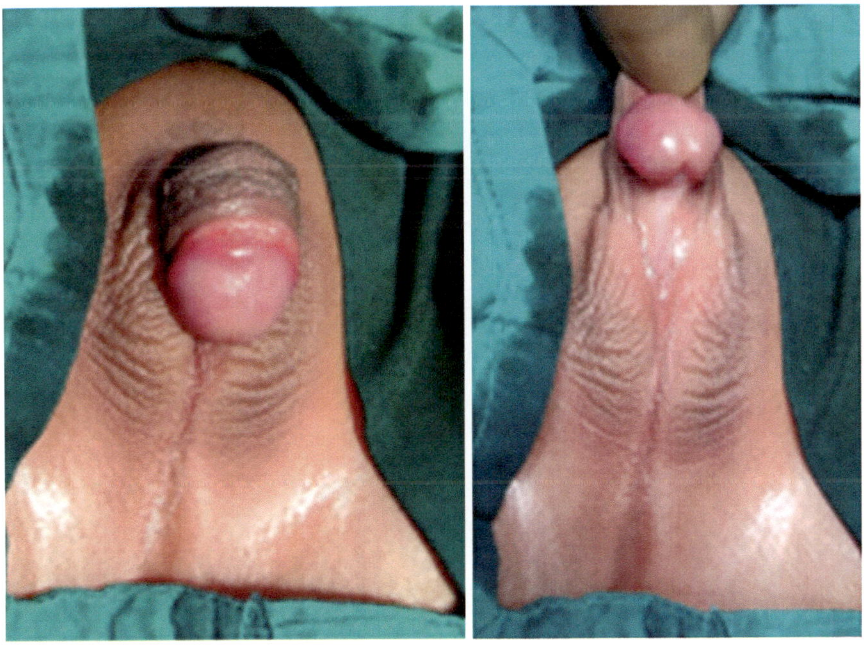

Fig. 44.7 Preoperative image

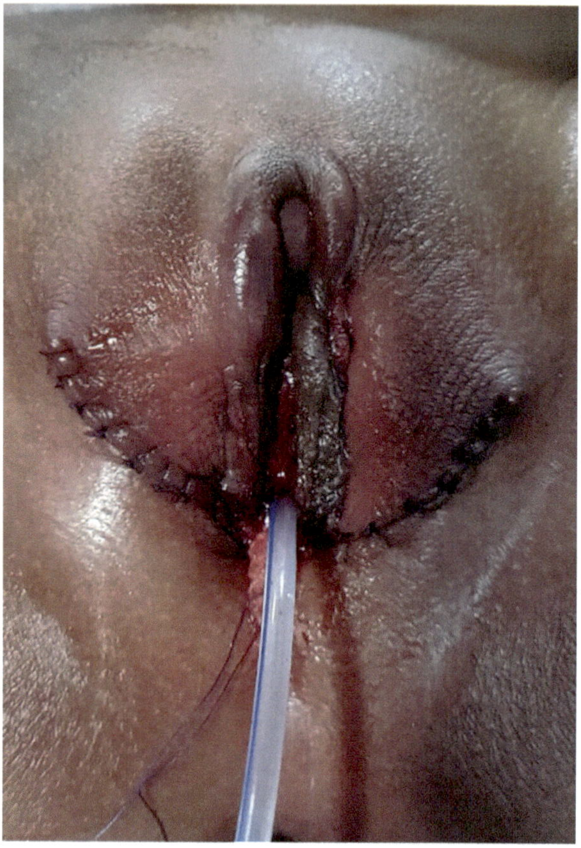

Fig. 44.8 Post-operative image

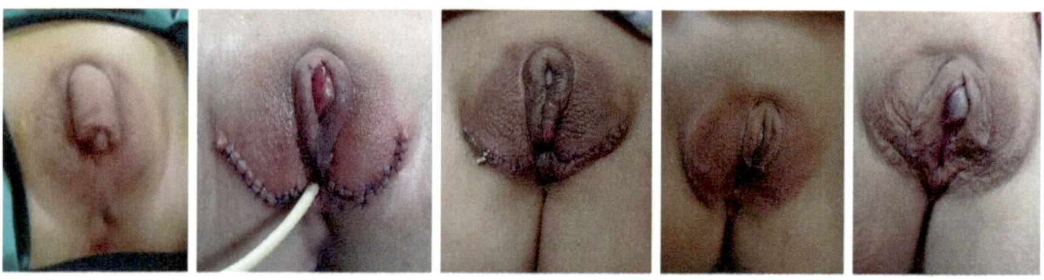

Fig. 44.9 Post-operative appearance in a 3-year-old child

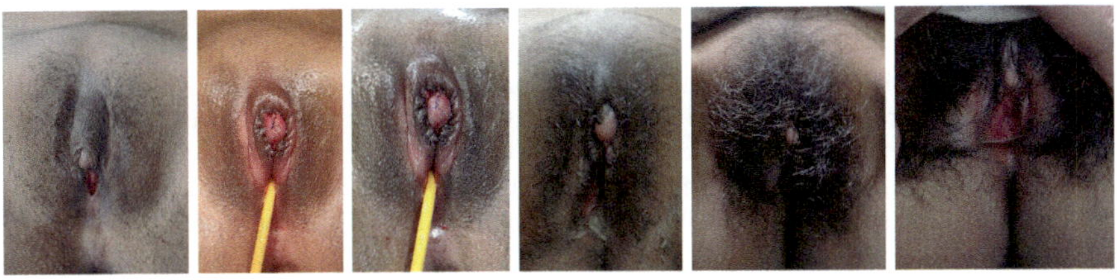

Fig. 44.10 Post-operative appearance in a 14-year-old child

44.7 Hyaluronidase Infiltration

Hyaluronidase is an enzyme that acts to make the tissue between cells more permeable. It produces liquefaction of colloid which fills tissue spaces. It lowers viscosity to the water level [15]. Hyaluronidase is used regularly in ophthalmic and plastic surgery for dissection [16, 17]. The author uses this in surgeries involving difficult dissection to facilitate tissue plane separation and minimize blood loss.

[Preparation of injection Hyaluronidase—1 amp (1500 IU) in 500 ml of Ringer's Lactate solution; Dose: 3 units/kg; Permissible amount: 50–300 units.]

The tips and tricks can be summarized as follows:

1. Multidisciplinary team
2. Informed written consent
3. Adequate preparation of patient
4. Injection hyaluronidase for dissection
5. Preservation of glans and prepuce
6. Avoid clitoral reduction
7. Avoid tourniquet/diathermy
8. Preservation of nerves
9. Plicating sutures for creating labia minora
10. Constantly updating self-regarding latest developments in surgery and legal status

Vaginoplasty and gonadectomy although components of feminizing genitoplasty are not included in this chapter. These are discussed in detail elsewhere.

References

1. Snyder HM 3rd. Feminizing genitoplasty: a proven technique. J Craniofac Surg. 2013;24(1):41–3. https://doi.org/10.1097/SCS.0b013e318270fcd8. PMID: 23321870
2. Hughes IA, Houk C, Ahmed SF, Lee PA, LWPES Consensus Group; ESPE Consensus Group. Consensus statement on management of intersex disorders. Arch Dis Child. 2006;91(7):554–63.

https://doi.org/10.1136/adc.2006.098319. Epub 2006 Apr 19. PMID: 16624884; PMCID: PMC2082839

3. Speiser PW, Arlt W, Auchus RJ, Baskin LS, Conway GS, Merke DP, Meyer-Bahlburg HFL, Miller WL, Murad MH, Oberfield SE, White PC. Congenital adrenal hyperplasia due to steroid 21-hydroxylase deficiency: an endocrine society clinical practice guideline. J Clin Endocrinol Metab. 2018;103(11):4043–88. https://doi.org/10.1210/jc.2018-01865. Erratum in: J Clin Endocrinol Metab. 2019;104(1):39–40. PMID: 30272171; PMCID: PMC6456929
4. Bennecke E, Bernstein S, Lee P, van de Grift TC, Nordenskjöld A, Rapp M, Simmonds M, Streuli JC, Thyen U, Wiesemann C, dsd-LIFE Group. Early genital surgery in disorders/differences of sex development: patients' perspectives. Arch Sex Behav. 2021;50(3):913–23. https://doi.org/10.1007/s10508-021-01953-6. Epub 2021 Mar 12. PMID: 33712989; PMCID: PMC8035116
5. Mouriquand PD, Gorduza DB, Gay CL, Meyer-Bahlburg HF, Baker L, Baskin LS, Bouvattier C, Braga LH, Caldamone AC, Duranteau L, El Ghoneimi A, Hensle TW, Hoebeke P, Kaefer M, Kalfa N, Kolon TF, Manzoni G, Mure PY, Nordenskjöld A, Pippi Salle JL, Poppas DP, Ransley PG, Rink RC, Rodrigo R, Sann L, Schober J, Sibai H, Wisniewski A, Wolffenbuttel KP, Lee P. Surgery in disorders of sex development (DSD) with a gender issue: If (why), when, and how? J Pediatr Urol. 2016;12(3):139–49. https://doi.org/10.1016/j.jpurol.2016.04.001. Epub 2016 Apr 9. PMID: 27132944
6. Minto CL, Liao LM, Woodhouse CR, Ransley PG, Creighton SM. The effect of clitoral surgery on sexual outcome in individuals who have intersex conditions with ambiguous genitalia: a cross-sectional study. Lancet. 2003;361(9365):1252–7. https://doi.org/10.1016/S0140-6736(03)12980-7. PMID: 12699952
7. https://www.reuters.com. Intersex surgery 'abuses' condemned by UN rights forum.
8. https://intersexasia.org/wp-content/uploads/2019/09/INT_CRPD_CSS_Ind_35644_E.docx
9. Jones HW Jr, Jones GE. The gynecological aspects of adrenal hyperplasia and allied disorders. Am J Obstet Gynecol. 1954;68(5):1330–65. PMID: 13207225
10. Randolph JG, Hung W. Reduction clitoroplasty in females with hypertrophied clitoris. J Pediatr Surg. 1970;5(2):224–31. https://doi.org/10.1016/0022-3468(70)90279-4. PMID: 5419072
11. Pippi Salle JL, Braga LP, Macedo N, Rosito N, Bagli D. Corporeal sparing dismembered clitoroplasty: an alternative technique for feminizing genitoplasty. J Urol. 2007;178(4 Pt 2):1796–800. https://doi.org/10.1016/j.juro.2007.03.167. Discussion 1801. Epub 2007 Aug 17. PMID: 17707426
12. Auffenberg GB, Helfand BT, McVary KT. Normal erectile physiology. Contemporary treatment of erectile dysfunction: a clinical guide. 2011;11–22.
13. Lee SH, Koh KS, Song WC. Macro/microscopic distribution of the dorsal nerve of penis in human glans penis. J Anat. 2020;237(5):849–53.
14. Kogan SJ, Smey P, Levitt SB. Subtunical total reduction clitoroplasty: a safe modification of existing techniques. J Urol. 1983;130(4):746–8. https://doi.org/10.1016/s0022-5347(17)51436-6. PMID: 6887407
15. Farr C, Menzel J, Seeberger J, Schweigle B. Klinische Pharmakologie und Anwendungsmöglichkeiten von Hyaluronidase unter Berück-sichtigung von Hylase "Dessau" [Clinical pharmacology and possible applications of hyaluronidase with reference to Hylase "Dessau"]. Wien Med Wochenschr. 1997;147(15):347–55. German. PMID: 9432805
16. Zahavi A, Grigg JR. Hyaluronidase injection for improved tissue dissection in Baerveldt tube surgery. Eur J Ophthalmol. 2018;28(3):339–40. https://doi.org/10.5301/ejo.5001065. Epub 2017 Oct 16. PMID: 29077179
17. Sharma DSC, Lahiri MA. Use of hyaluronidase in plastic surgery: A review. J Plast Reconstr Aesthet Surg. 2021;74(7):1610–4. https://doi.org/10.1016/j.bjps.2021.03.125. Epub 2021 Apr 22. PMID: 34024740

Timing of Surgery: Do We Have a Consensus?

45

Mehak Sehgal and Prabudh Goel

Abstract

Despite the multitude of literature and data, the timing of surgery for a neonate who is born with ambiguous genitalia is still in debate. Surgical interventions for these children, apart from those with outflow obstruction, are generally elective. There is an important matter of consent, and whether by providing early surgery, the parents should get to decide the genital phenotype or should the child be involved in this important process of decision-making once he can understand the consequences of it. There exist certain forms of DSD where the choice of rearing may differ from the choice of rearing, which may lead to gender dysphoria and dissatisfaction later on. A combination and interplay of several factors influence the decision-making including not only the endocrinological, psychological, and surgical aspects, but also the parental, societal, and legal points of view. All these factors must be considered, with great emphasis on parental counselling, while deciding the surgical procedure and interventions, which may often be irreversible.

Keywords

DSD · CAH · Genitoplasty · Vaginoplasty · Sex assignment

45.1 Introduction

After a child with DSD is born, an important question that the surgeon or the treating physician has to answer to the parent is—When should the child be operated on? This question, however, does not have a straightforward answer. It is one of the most

M. Sehgal · P. Goel (✉)
Department of Paediatric Surgery, All India Institute of Medical Sciences, New Delhi, India

S. K. Ratan (ed.), *Children with Differences in Sex Development*,
https://doi.org/10.1007/978-981-97-1639-5_45

discussed and controversial topics in the management of a child diagnosed with a disorders of sex development. This question is often alongside, or just after the discussion on the rearing sex of the child. Due to the social and cultural expectations to characterize into a binary system of sexual rearing, there is pressure and parental anxiety to make the child "fit" into a defined gender. This makes most parents desire early surgery, which is defined as operating on a child less than 2 years of age [1]. This also means that the parents decide and give consent for the surgery and not the child. This was, in fact, the practice prevalent earlier, when a neonate is born with a Disorders of Sex Development (DSD) was a social emergency and needed urgent "correction." Over the years, data and results were collected and evaluated, not only of the surgical outcome, but also of the psychological aspects such as gender dysphoria and patient satisfaction, ethical and social issues, and the need for reassignment, which has led to significant controversy regarding timing and type of surgery. There has also been an involvement of legal groups, with issues such as patient consent and choice being highlighted in recent years, which further adds to deliberation on this topic [2].

45.2 Type of Surgery

Surgical interventions for children born with DSD can be classified as urgent or elective. Outlet obstruction or gonadal malignancies are fairly urgent in nature. In this scenario, diversion procedures are often performed in an emergency or gonads at risk for malignancy noted on biopsy are removed urgently. This often occurs with just the parental consent, by its very nature, without waiting for the child's consent if not Gillick competent [3]. Children born with DSD conditions, although are regarded as having a very significant medical condition, the framework has shifted to obtaining a correct diagnosis at the start and thereby making a series of careful decisions about assigning the gender before surgery [4]. Thus, majority of surgical interventions, fall in the category of non-urgent issues which can theoretically wait.

45.3 Surgery but Early or Late? Both Sides of the Argument

The question to operate generally comes after the process of sex assignment. Generally, three age periods are considered relevant to the problem of sex assignment [4]:

1. Neonatal period to infancy
2. Ranging 2–10 years
3. After 11 years of age

Each period is unique on its own, as described by Bajpai et al. In the first period, the parents make a decision. In the second period, sex reassignment is not recommended generally. And after 11 years, the child can often decide on his own about what he prefers. When sex assignment has to be done in infancy, the aim is to give the child a gender that will allow a close match to his future functional gender [4]. There are a lot of factors to be considered while deciding the sex assignment. These include the anatomy of the patient, genetic constitution and its correlation with anatomy, parental considerations, societal and social implications, brain imprinting, and the psychological trauma of genital surgery once the child is older or knowing such a surgery had been done in the past without consent. Historically, there was the assumption that optimal psychosocial development could only be attained when genital anatomy was in concordance with sex of rearing [5]. Enigma does exist in the outcome of children who spend their childhood in ambiguity [1, 6]. It is to note that both, irreversible surgical feminization or masculinization in a child unwilling to be so can have adverse consequences on the psychology of the adult that the child will become, but so does inaction, especially since the parental and societal values we have at present do not support it.

Each option has its own advantages and disadvantages. The proponents of early surgery argue that there is some pressure from parents, who want to rear the child in a particular gender and prevent any form of stigmatization [6]. The potential physical and psychological morbidity of delayed surgery in an older child, along with the possibly reduced psychological impact of surgery performed in infants also justifies early surgery [7]. Some surgeons also feel that the tissues are more pliable and child is more cooperative when younger, which would help better surgical outcomes. Based on feedback from older patients, early intervention has shown positive outcomes for patients and parents [8, 9]. Data is still evolving on the outcomes of children who were operated late, as compared with early surgery, which had been performed since earlier times [2].

However, others argue that by operating early the affected individual is not informed or consulted as he or she has no understanding of his or her body, thereby withdrawing the child's right to decide for him or herself due to the irreversibility of the procedure. The decision by the parents and the surgeon is likely to be coherent with the anatomy, however, it is unrealistic to predict the acceptance of the assigned gender orientation at such a young age. A smaller child or younger adult may not understand the potential risks and benefits completely to provide consent for himself or herself. There could very well be a chance that the affected child may choose a gender other than what has been assigned in infancy. Health professionals have reported poorer outcomes of prior genital surgery and the need for additional surgeries once the child attains adolescence [2, 6]. Given the risk of subtly altering clitoral and vaginal sensitivity caused by surgical damage [2], these structures could be left for later to reconstruct. Sexual dysfunction has been reported by adults who experienced early genitoplasty. Incidence of complications and need for reoperations are also noted to be common with early surgery [10–12]. Additionally, reconstructive procedures are of long duration, and there could be deleterious effects of anesthesia on a developing brain, which is still being evaluated [13, 14].

45.4 Parental and Patient-Related Aspects

Children are represented by their parents, especially in their early years. They are not aware of stereotypes, the relevance of appearances, or the socio-cultural framework they will grow to live in. The viewpoints of the parents are diverse and are often influenced by societal pressure. Some parents want to protect their children from the challenges of their anatomy not fulfilling the societal expectation of fitting into a binary gender, while others feel that the children should be given the chance to make their own decisions once they are old enough to express their gender identity and understand the consequences of it [2]. Prevalent social practices in some societies such as the eunuch taking away children with DSD, and non-acceptance in the society plague the minds of the parents and their relatives. There are societies wherein life is very difficult for an unmarried female as compared to an unmarried male. Parental distress, which is considered to be an important cause of performing early surgery should be aimed to be corrected by counselling and support [15]. The concept of "Shared Decision Making" has also been explored which may help patients and their families make decisions that are better suited to their personal and societal values [16, 17].

45.5 Endocrinological Considerations

Endocrinological factors help us understand the physiologically correct time for surgery. Androgen receptors have been noted to decrease during adolescence and thus testosterone use earlier leads to a better response. This has been used especially in conditions such as a micropenis, but it may be difficult to implement before the correction of chordee [18, 19]. The adult height is determined by a child's biological sex, and thus it is possible that an individual could miss the opportunity for GH treatment for an unfavorable height, if decisions for male sex of rearing and surgery are deferred till adolescence [19, 20].

45.6 Legal Issues and Involvement of International Agencies

An interesting but medically important case concluded in 2017, when the Court of South Carolina, United States awarded compensation to a child who underwent feminizing genital and gonadal surgery at 16 months of age but later went on to identify himself as a boy. This was a well-publicized legal case involving a young child with ovo-testicular DSD in which plaintiffs claimed there had been inadequate informed consent for the surgery [21, 22]. This case shows us that that individuals have the right to their own to bodily autonomy and reproductive integrity as per law. This would mean that procedures which are deemed irreversible should not be performed without the individual being a part of the decision-making process, unless in an emergency [2]. State legislatures have been passed to limit genital surgery on children "until the child can participate in decision making"; e.g., California Senate Concurrent Resolution 110 (SCR-110) [23]. Recognizing the implications of these irreversible interventions and surgeries on the physical and mental health of children with DSD, as per order by the Madras High Court in India, the Tamil Nadu state government banned genital reconstructive and sex assignment surgeries in newborns [24].

Physicians need to be well-versed in these aspects. Although the age at which a child is capable of making an informed decision regarding his or her care cannot be defined uniformly, this can often be determined by a psychologist. As per Indian law, a child <12 years of age is not eligible for legal consent. An important concern is that until mid or late childhood, gender identity may be unclear in children, and so ethically leaving decisions till after this sense of gender identity has developed is in concordance with the proposition of the legal philosopher Joel Feinberg "the right to an open future" [25].

It should also be noted that the patient support groups and advocacy organizations are themselves not united in their positions toward early surgery. The CARES (Congenital Adrenal Hyperplasia Research Education and Support) Foundation, based in the United States, considers CAH to be a life-threatening adrenal disorder, not a sexual disorder, and thereby not considering CAH patients as intersex [25].

45.7 Psychological Aspects

Sometimes parents may want early surgery for their child with ambiguous genitalia because they think it might avoid the psychological trauma of living with ambiguity. But the converse is also true, as many adults with DSDs have often said that they were equally traumatized once they knew about the surgery, or maybe more as it was done without their consent [26]. This can elicit negative feelings toward their parent.

Being different is not something easily accepted by society. The lack of complete knowledge about the condition of the child, coupled with a lack of initial clear definition of the child's sex can significantly impact the new parents. There is still no clear understanding regarding the adaptation of the individual mind to the designated sex as it remains unclear the bearing one's genetic sex has, and how it is affected by the sex of rearing [19]. Once a child accepts a certain sex, it leads to less inner conflict and leads to the development of a gender identity. The parents play an important role in the child's interaction with the outside world [19]. It has been noted that a child with DSD could develop a stable identity if raised without a label [27, 28].

Atypical genitalia can lead to a variety of psychological disturbances including negative body perception, which may lead to confusion about masculinity and femininity. The child may be unsure as to which identity to adopt and thereby have mannerisms deviating from those predefined. Thus, as a "preventive" intervention, a genital surgery that is performed after a complete investigation helps in reinforcing the child's belonging to a defined sexual identity [1]. The main issue highlighted by those managing DSD cases is that the patient may challenge the originally designated sex later on. Living in a body with genitalia that does not match the sexual identity desired by the patient can be equally frustrating [29, 30].

45.8 Surgical Timing: Individual Subtypes

Main surgical interventions include dealing with—Clitoris, Gonads, Mullerian remnant, External Genitalia, and Breast.

In the 46XY DSD group with 17βHSD and 5α reductase deficits, delayed surgery for the gonads and the vagina in girls, pubertal blockade and estrogen therapy have been advocated in literature till gender confirmation. In boys, androgen stimulation should begin early [31]. In the case of PAIS reared as male, the age of hypospadias surgery in a small phallus depends on the response to androgen stimulation and a sequential staged procedure is followed. When the female sex of rearing is decided, gonadectomy along with feminization surgery has been advised [31, 32]. In CAIS, gonads are often preserved till puberty to allow aromatization during the testicular surge, but efficacy studies are lacking [33, 34].

If the male sex is assigned, the ideal surgical period would be late infancy to 2 years [35, 36]. Boys, who are operated on after 1 year of age, and adolescents, have been noted to have higher complication rates as per literature. The time for orchidopexy has decreased significantly since the 2006 Nordic consensus on the management were published (from 54.5 ± 43.0 months to 41.4 ± 45.3 months) [7, 37]. Removal of Mullerian structures is not recommended and they may be kept as long as asymptomatic, due to the risk of damaging the genital ducts in an individual with pre-existing low fertility potential [31].

If the choice of rearing is female, the ideal time for genitoplasty is a big concern. There are three options to consider: the surgical reconstruction can be done early in childhood, or as a staged procedure, reducing the clitoromegaly and external genitalia reconstruction initially and vaginoplasty later, or the surgery can be delayed till the child gives her consent and is willing for dilatations. The argument in favor of early surgery is that of better quality and pliability of genital tissues, postnatal maternal estrogens providing better vascularization, and reduced anxiety with better parent–child bonding. Vaginal dilation, however, is definitely not recommended in childhood [31, 38–40]. It is, however, noted that after vaginal reconstruction in childhood, reoperations are needed in adolescence. Those advocating late surgery feel that the procedure is better performed at a time when the child can take part in the decision process and also the dilatations without discomfort. It remains unclear how much of a psychological impact non-corrected unusual genitalia may cause. Early surgery performed in a staged manner may also discard some tissue which may be useful for later vaginal reconstruction [19, 31]. As per the patient's perspective, various studies have reported that they would prefer early surgery, had better satisfaction, good quality of life, and a lower incidence of gender dysphoria [8, 41–43].

45.9 So... Do We Have a Consensus?

The 2016 consensus guideline document provides the following points which most experts agree with [40]:

(1) the need for identifying centres of expertise with a multidisciplinary approach
(2) conservative management of the gonads in complete AIS patients
(3) avoidance of vaginal dilatation during childhood
(4) keeping asymptomatic Müllerian remnants during childhood which can be removed later if necessary
(5) removal of biopsy-confirmed streak gonads early

But unfortunately, no, we do not have clear guidelines or a consensus on this highly controversial topic. There is a need for guidelines that are specific to the different entities comprising the spectrum of possible anatomic configurations. Yet, each case is unique and deserves a personalized approach. Multidisciplinary involvement: patient, parent, physician, surgeon, geneticist, endocrinologist, psychologist, medical-social worker, lawyer, government representatives, representatives from social-support and self-help groups, etc. plays an important role in the management and is key for the successful outcome in these children. Another important aspect is to consider the prevalent local regulations. Thus, the final word is yet to be said on this topic.

References

1. Wisniewski AB, Tishelman AC. Psychological perspectives to early surgery in the management of disorders/differences of sex development. Curr Opin Pediatr. 2019;31:570–4. https://doi.org/10.1097/MOP.0000000000000784.
2. Gorduza DB, Quigley CA, Caldamone AA, Mouriquand PDE. Surgery of anomalies of gonadal and genital development in the "post-truth era". Urol Clin North Am. 2018;45:659–69. https://doi.org/10.1016/J.UCL.2018.06.012.
3. Sarin YK, Singh D, Babu R, Das K, Rao S. Indian association of pediatric surgeons guidelines on the management of differences in sex development. J Indian Assoc Pediatr Surg. 2022;27:376. https://doi.org/10.4103/0971-9261.352296.
4. Bajpai M. Disorders of sex development: the quintessence of perennial controversies-II (age, timing & type of surgery). J Prog Paediatr Urol. 2014;17:55–8.
5. Hampson JG, Money J, Hampson JL. Hermaphrodism: recommendations concerning case management. J Clin Endocrinol Metab. 1956;16:547–56. https://doi.org/10.1210/JCEM-16-4-547.
6. Crouch NS, Creighton SM. Minimal surgical intervention in the management of intersex conditions. J Pediatr Endocrinol Metab. 2004;17:1591–6. https://doi.org/10.1515/JPEM.2004.17.12.1591.
7. Lee PA, Houk CP, Ahmed F, Hughes IA, Achermann J, Baskin L, et al. Consensus statement on management of intersex disorders. Int Consens Conf Intersex Pediatr. 2006;118. https://doi.org/10.1542/PEDS.2006-0738.
8. Crawford JM, Warne G, Grover S, Southwell BR, Hutson JM. Results from a pediatric surgical centre justify early intervention in disorders of sex development. J Pediatr Surg. 2009;44:413–6. https://doi.org/10.1016/j.jpedsurg.2008.10.101.

9. Bernabé KJ, Nokoff NJ, Galan D, Felsen D, Aston CE, Austin P, et al. Preliminary report: Surgical outcomes following genitoplasty in children with moderate to severe genital atypia. J Pediatr Urol. 2018;14:157.e1–8. https://doi.org/10.1016/j.jpurol.2017.11.019.
10. Pasterski V, Zucker KJ, Hindmarsh PC, Hughes IA, Acerini C, Spencer D, et al. Increased cross-gender identification independent of gender role behavior in girls with congenital adrenal hyperplasia: results from a standardized assessment of 4- to 11-year-old children. Arch Sex Behav. 2015;44:1363–75. https://doi.org/10.1007/S10508-014-0385-0.
11. Saltzman AF, Carrasco A, Colvin A, Campbell JB, Vemulakonda VM, Wilcox D. Patients with disorders of sex development and proximal hypospadias are at high risk for reoperation. World J Urol. 2018;36:2051–8. https://doi.org/10.1007/S00345-018-2350-3.
12. Creighton SM, Minto CL, Steele SJ. Objective cosmetic and anatomical outcomes at adolescence of feminising surgery for ambiguous genitalia done in childhood. Lancet. 2001;358:124–5. https://doi.org/10.1016/S0140-6736(01)05343-0.
13. Lei SY, Hache M, Loepke AW. Clinical research into anesthetic neurotoxicity: does anesthesia cause neurological abnormalities in humans? J Neurosurg Anesthesiol. 2014;26:349–57. https://doi.org/10.1097/ANA.0000000000000117.
14. Disma N, O'Leary JD, Loepke AW, Brambrink AM, Becke K, Clausen NG, et al. Anesthesia and the developing brain: A way forward for laboratory and clinical research. Paediatr Anaesth. 2018;28:758–63. https://doi.org/10.1111/PAN.13455.
15. Streuli JC, Vayena E, Cavicchia-Balmer Y, Huber J. Shaping parents: impact of contrasting professional counseling on parents' decision making for children with disorders of sex development. J Sex Med. 2013;10:1953–60. https://doi.org/10.1111/JSM.12214.
16. Karkazis K, Tamar-Mattis A, Kon AA. Genital surgery for disorders of sex development: implementing a shared decision-making approach. J Pediatr Endocrinol Metab. 2010;23:789–805. https://doi.org/10.1515/JPEM.2010.129.
17. Siminoff LA, Sandberg DE. Promoting shared decision making in disorders of sex development (DSD): decision aids and support tools. Horm Metab Res. 2015;47:335–9. https://doi.org/10.1055/S-0035-1545302.
18. Xu D, Lu L, Xi L, Cheng R, Pei Z, Bi Y, et al. Efficacy and safety of percutaneous administration of dihydrotestosterone in children of different genetic backgrounds with micropenis. J Pediatr Endocrinol Metab. 2017;30. https://doi.org/10.1515/JPEM-2016-0400.
19. Hemesath TP, de Paula LCP, Carvalho CG, Leite JCL, Guaragna-Filho G, Costa EC. Controversies on timing of sex assignment and surgery in individuals with disorders of sex development: a perspective. Front Pediatr. 2018;6. https://doi.org/10.3389/FPED.2018.00419.
20. Sharma S, Gupta DK. Gender assignment and hormonal treatment for disorders of sexual differentiation. Pediatr Surg Int. 2008;24:1131–5. https://doi.org/10.1007/S00383-008-2232-7.
21. A landmark lawsuit about an intersex baby's genital surgery just settled for $440,000. n.d. https://www.buzzfeednews.com/article/azeenghorayshi/intersex-surgery-lawsuit-settles#.ad45lyyw1r. Accessed 15 Feb 2023.
22. M.C. v. Aaronson | Southern Poverty Law Center. n.d. https://www.splcenter.org/seeking-justice/case-docket/mc-v-aaronson. Accessed 15 Feb 2023.
23. Bill Text—SCR-110 Sex characteristics. n.d. https://leginfo.legislature.ca.gov/faces/billTextClient.xhtml?bill_id=201720180SCR110. Accessed 15 Feb 2023.
24. T.N. bans sex reassignment surgeries on intersex infants, children. The Hindu. n.d. https://www.thehindu.com/news/national/tamil-nadu/tn-bans-sex-reassignment-surgeries-on-intersex-infants-children/article29273674.ece. Accessed 15 Feb 2023.
25. Gardner M, Sandber DE. Navigating surgical decision making in disorders of sex development (DSD). Front Pediatr. 2018;6:339. https://doi.org/10.3389/FPED.2018.00339.
26. Timing of surgeries | DSD guidelines. n.d. https://dsdguidelines.org/htdocs/clinical/timing_of_surgeries.html. Accessed 14 Feb 2023.
27. Money J, Imperato-Mcginley J, Peterson RE, Gautier T. Gender identity and hermaphroditism. Science. 1979;1976(191):872. https://doi.org/10.1126/science.1251201.

28. Money J, Hampson JG, Hampson JL. Imprinting and the establishment of gender role. AMA Arch Neurol Psychiatry. 1957;77:333–6. https://doi.org/10.1001/ARCHNEURPSYC.1957.02330330119019.
29. de Vries AL, Doreleijers TAH, Cohen-Kettenis PT. Disorders of sex development and gender identity outcome in adolescence and adulthood: understanding gender identity development and its clinical implications. Pediatr Endocrinol Rev. 2007;4:343–51.
30. Yang JH, Baskin LS, DiSandro M. Gender identity in disorders of sex development: review article. Urology. 2010;75:153–9. https://doi.org/10.1016/J.UROLOGY.2009.07.1286.
31. Mouriquand PDE, Gorduza DB, Gay CL, Meyer-Bahlburg HFL, Baker L, Baskin LS, et al. Surgery in disorders of sex development (DSD) with a gender issue: If (why), when, and how? J Pediatr Urol. 2016;12:139–49. https://doi.org/10.1016/J.JPUROL.2016.04.001.
32. Mazur T. Gender dysphoria and gender change in androgen insensitivity or micropenis. Arch Sex Behav. 2005;34:411–21. https://doi.org/10.1007/S10508-005-4341-X.
33. Nakhal RS, Hall-Craggs M, Freeman A, Kirkham A, Conway GS, Arora R, et al. Evaluation of retained testes in adolescent girls and women with complete androgen insensitivity syndrome. Radiology. 2013;268:153–60. https://doi.org/10.1148/RADIOL.13121068.
34. Cools M, Drop SLS, Wolffenbuttel KP, Oosterhuis JW, Looijenga LHJ. Germ cell tumors in the intersex gonad: old paths, new directions, moving frontiers. Endocr Rev. 2006;27:468–84. https://doi.org/10.1210/ER.2006-0005.
35. Steven L, Cherian A, Yankovic F, Mathur A, Kulkarni M, Cuckow P. Current practice in paediatric hypospadias surgery; a specialist survey. J Pediatr Urol. 2013;9:1126–30. https://doi.org/10.1016/J.JPUROL.2013.04.008.
36. Creighton S, Chernausek SD, Romao R, Ransley P, Salle JP. Timing and nature of reconstructive surgery for disorders of sex development—introduction. J Pediatr Urol. 2012;8:602–10. https://doi.org/10.1016/J.JPUROL.2012.10.001.
37. Kohva E, Miettinen PJ, Taskinen S, Hero M, Tarkkanen A, Raivio T. Disorders of sex development: timing of diagnosis and management in a single large tertiary center. Endocr Connect. 2018;7:595–603. https://doi.org/10.1530/EC-18-0070.
38. Braga LH, Pippi Salle JL. Congenital adrenal hyperplasia: a critical appraisal of the evolution of feminizing genitoplasty and the controversies surrounding gender reassignment. Eur J Pediatr Surg. 2009;19:203–10. https://doi.org/10.1055/S-0029-1233490.
39. Yankovic F, Cherian A, Steven L, Mathur A, Cuckow P. Current practice in feminizing surgery for congenital adrenal hyperplasia; a specialist survey. J Pediatr Urol. 2013;9:1103–7. https://doi.org/10.1016/J.JPUROL.2013.03.013.
40. Lee PA, Nordenström A, Houk CP, Ahmed SF, Auchus R, Baratz A, et al. Global disorders of sex development update since 2006: perceptions, approach and care. Horm Res Paediatr. 2016;85:158–80. https://doi.org/10.1159/000442975.
41. Wisniewski AB, Migeon CJ, Malouf MA, Gearhart JP. Psychosexual outcome in women affected by congenital adrenal hyperplasia due to 21-hydroxylase deficiency. J Urol. 2004;171:2497–501. https://doi.org/10.1097/01.ju.0000125269.91938.f7.
42. Fagerholm R, Santtila P, Miettinen PJ, Mattila A, Rintala R, Taskinen S. Sexual function and attitudes toward surgery after feminizing genitoplasty. J Urol. 2011;185:1900–4. https://doi.org/10.1016/J.JURO.2010.12.099.
43. Binet A, Lardy H, Geslin D, Francois-Fiquet C, Poli-Merol ML. Should we question early feminizing genitoplasty for patients with congenital adrenal hyperplasia and XX karyotype? J Pediatr Surg. 2016;51:465–8. https://doi.org/10.1016/J.JPEDSURG.2015.10.004.

Part X

Long Term Outcomes and Social Rehabilitation

I whole heartedly thank Dr. Pradnya Bendre, Professor Pediatric Surgery, B J Wadia Hospital for Children, Mumbai and Dr. Pragya Sharma Professor, Community medicine Maulana Azad Medical College, New Delhi for assisting and aiding me in editing this section.

46 Endocrinological Outcomes in Subjects with DSD

Sukirti Misra, Jugal Gada, and Nikhil Bhagwat

Abstract

Endocrinological implications in subjects with disorders of sex development do not stop at the management of the basic disease but have far-reaching implications as it is linked to onset of puberty, bone health, and treatment-related adverse effects. Awareness about these outcomes is important to prepare the individuals with Disorders of Sex Development for any such outcomes, which can affect prognosis and long-term quality of life.

Keywords

Disorders of sex development · Endocrinological · Outcomes

46.1 Introduction

Disorders of Sexual Development (DSD) encompass a wide array of diagnoses with varying etiologies, and long-term endocrinological outcomes differ according to the subgroup studied. Endocrine outcomes include effects on stature, puberty, bone health, metabolic sequelae, fat distribution, treatment-related adverse effects, and secondary Polycystic Ovarian Syndrome (PCOS).

S. Misra · J. Gada · N. Bhagwat (✉)
Department of Endocrinology, T.N. Medical College and B.Y.L. Nair Ch. Hospital, Mumbai, Maharashtra, India

S. K. Ratan (ed.), *Children with Differences in Sex Development*,
https://doi.org/10.1007/978-981-97-1639-5_46

46.1.1 Congenital Adrenal Hyperplasia (CAH)

46.1.1.1 21-Hydroxylase Deficiency

The goal is to prevent female virilization, optimize growth, and protect fertility. The therapeutic spectrum of glucocorticoids is narrow, and supraphysiological doses are often needed to suppress the androgens. This creates a risk for adiposity, insulin resistance, growth suppression, and osteoporosis, while insufficient replacement, can lead to precocious puberty and adrenal crisis.

46.2 Height and Puberty

Steroid administration typically begins in early infancy at high doses, which may affect height and bone maturation. Girgis and colleagues discovered no difference in present height, height at 2 years of chronological age, or growth velocity between the tightly and poorly controlled groups, but bone maturation was significantly delayed in the tightly controlled group and advanced in the poorly controlled group, probably indicating higher androgen levels [1].

A recent systematic review and meta-analysis of 35 studies showed that the mean final height of these patients was 1.38 SD score lower than the population norm, and expected given parental height. This was not associated with age at diagnosis, gender, steroid type and dose, or age of puberty onset. Mineralocorticoid users had a better height outcome than non-users [2].

Central Precocious Puberty (CPP) is common in CAH due to androgen activation of the hypothalamic–pituitary–gonadal axis, which causes premature epiphyseal fusion. Children who are poorly controlled or diagnosed later have older bone ages than tightly controlled, early-diagnosed groups.

In a study, CAH diagnosed with CPP had faster linear development and mature bones than age-matched subjects. When they received depot leuprolide for ≥2 years, it halted puberty, slowed growth, increased predicted adult height, and decreased bone age to height age ratio [3].

Treatment with GH and GnRH agonists has been used in subjects with CPP and CAH. Patients ≥1.0 SD below target height received this combination in one study. Final height SD was much higher in the treatment group than their initial projected height and the untreated group [4].

In another study, CAH patients with height prediction 2 SD below their target height, short-term use of GH alone or in combination with GnRH agonist, improved predicted final height by 11 cm. However, the data on GH use in CAH is sparse, with few studies followed till adult height to give robust recommendations for its use. Hence, these treatments can be considered for patients with CAH and growth failure (predicted adult height –2.25 SD) in the setting of a controlled research trial. The best-established approach to optimize linear growth is judicious glucocorticoid and mineralocorticoid dosing and close monitoring [2].

46.3 Bone Mineral Density (BMD)

Supraphysiological glucocorticoid dosages can lower Bone Mineral Density (BMD). Long-acting glucocorticoids may damage BMD more than short or intermediate-acting ones.

46.3.1 Children and Adolescents

In a few studies, treatment with glucocorticoids has not been shown to affect BMD in children and adolescents [5].

Girgis et al. found that serum osteocalcin levels, procollagen peptide, collagen C-terminal telopeptide, and age-corrected vertebral BMD were comparable across groups of tightly and poorly controlled CAH patients. According to these findings, as long as these children continue to grow normally in late childhood and adolescence, bone mineral accretion should be normal, and replacement therapy should not increase the risk of osteoporosis in later years [1].

46.3.2 Adults

BMD has been reported to range from increased to decreased, but the majority of studies have found low BMD in at least one of the measured locations in both sexes and across all age groups. Increased levels of adrenal androgens, which are converted to estrogens, may offset the negative effects of glucocorticoids. This could explain why older CAH women, particularly postmenopausal, have a higher risk of osteoporosis than younger CAH patients.

A meta-analysis of nine case control studies ($n = 254$ CAH patients and 344 controls) demonstrated a reduced BMD in CAH patients, but no connection with glucocorticoid dose [6]. One study observed osteopenia in 45% of salt-wasting CAH patients, 13% of simple virilizing patients, and 11% of controls. CAH patients had lower BMDs throughout the body, including the lumbar spine. DHEAS levels in CAH patients with osteopenia were lower than in those with normal BMD [7].

BMD studies in CAH are either small or varied, lacking hormonal data, stressing the necessity for thorough prospective research. BMD should be measured at the age of peak bone mass accrual (approximately 25 years) and periodically, based on baseline values, glucocorticoid therapy, and other risk factors, to assess bone health.

Fracture risk has been rarely investigated. Glucocorticoid therapy increased fracture risk in a meta-analysis. A Swedish study found younger patients with classic CAH had more fractures, whereas those with NC-CAH had a fracture rate equivalent to matched controls. In a Korean research, only subjects over 40 years old had more fractures. Classic CAH patients born after newborn screening implementation showed a similar fracture rate to matched controls. Early diagnosis and contemporary management may reduce fractures better than chronic androgen exposure [8].

46.4 Metabolic Outcomes

Suppressive doses of corticosteroids can cause metabolic complications, which can increase cardiovascular risk. Obesity, insulin resistance, and metabolic syndrome complicate hormonal therapy in CAH because they affect glucocorticoid pharmacokinetics. Hepatic steatosis may increase cortisol clearance. Decreased insulin sensitivity increases cortisol clearance regardless of body adipose mass or hepatic steatosis. Weight loss can reverse these processes.

Studies have shown that children treated for CAH had similar heights but had higher body fat, and increased prevalence of metabolic syndrome, while 58% had systolic and 24% had diastolic hypertension [9].

Similarly, adults can have a deranged metabolic profile but many contrasting studies have been reported. Falhammar found CAH women were shorter than controls with similar waist-to-hip ratio, lean and fat mass, and insulin levels in younger patients. Older patients had higher waist-to-hip ratios, lean mass, and insulin levels than younger patients. Gestational diabetes was more common in CAH patients. There was no clear evidence of unfavorable cardiovascular risk factors [10].

In the CHASE study, 203 adult CAH females were found to have obesity (41%), hypercholesterolemia (46%), and insulin resistance (29%) [11]. A meta-analysis of CAH patients and matched controls found increased blood pressure. Hypertension risk is higher in children, and compared to controls, the risk decreases with age, notably in females. The biggest risk was found in children under 2 years, showing that high fludrocortisone dosages can cause hypertension in infants, while high BMI may cause it later [8].

A systematic review of 14 studies, including 437 patients of a wide age group (14 months to 63 years) showed they had increased systolic and diastolic blood pressures, insulin resistance, and carotid intima thickness (CIMT) compared with controls. No difference was noted in fasting blood glucose, insulin level, glucose or insulin level 2 h after a glucose load, or serum lipids. Data on cardiac events was limited, and mostly focused on surrogate outcomes [12].

Elevated heart rate is linked to cardiovascular death, especially in men. Only a few studies have found increased heart rates in CAH patients. Only adult males with CAH aged 30 or older showed higher 24-h heart rates compared to matched controls, however, obesity may confound the result. CAH children and adults had higher CIMT than controls, and adults were more affected [8].

Considering increased body fat and the potential for cardiac and metabolic consequences in treated CAH subjects, lifestyle counseling should be started early.

46.4.1 Menstrual Dysfunction, PCOS, and Hyperandrogenism

PCOS can develop in both classical and non-classical CAH. In one study, 83% of the 36 CAH patients had polycystic ovaries [13].

The pathophysiology of PCOS in women with CAH is strongly linked to poor hormonal control and inadequate suppression of ACTH secretion.

Androgen–receptor antagonists may be required for controlling hyperandrogenism in addition to GC and MC. Drospirenone-containing oral contraceptives (OCP) reduce adrenal and ovarian androgen synthesis while having no effect on cortisol, blood pressure, PRA, or serum potassium levels. Oral contraceptives do not replace glucocorticoids in classic CAH. Spironolactone is relatively contraindicated in salt-wasting CAH, as it is an MC antagonist and can result in volume depletion [14].

Management of hyperandrogenism in NC-CAH is similar to primary PCOS. Usually, they are started on OCP, and anti-androgens can be added after 6 months if necessary. Antiandrogen as initial therapy can be used after counseling about the teratogenic risks if the woman is not planning pregnancy and using contraception. For resistant cases, glucocorticoids such as prednisone 4–6 mg or dexamethasone 0.25 mg once daily can be used to treat hirsutism.

46.4.2 Testicular Adrenal Rest Tumors

In males with CAH, the prevalence of Testicular Adrenal Rest Tumors (TARTs) increases after 10 years of age. Multiple studies have shown that 30–50% of adolescent and adult males with classic 21OHD develop TART, and ultrasonography is the most sensitive method of detection. Ultrasound screening should begin in adolescence, every 1–2 years in asymptomatic, or more often in symptomatic patients. Glucocorticoid dose optimization can shrink early-stage TART and prevent a progressive increase in the size of TART, resulting in infertility. Men with large TARTs may have low morning testosterone, which indicates poor Leydig cell function.

46.4.3 Adrenals

Adrenal masses affect 1–4% of healthy people, and their frequency rises with age. One CT imaging study in adults with CAH found a high prevalence of benign adrenal masses, particularly among those on inadequate GC therapy. Adrenal carcinomas are uncommon in CAH patients, with only one pediatric case reported. Several adults with CAH have had massive adrenal myelolipomas that required surgical removal due to mass effects. There is insufficient evidence to recommend routine screening for adrenal masses [14].

46.4.4 Mortality

In countries where neonatal screening was not available, more females than males were diagnosed with SW CAH. However, both females and males died undiagnosed in the neonatal period. When the neonatal screening was implemented, more people with SW CAH were found in both sexes. Even after being diagnosed with CAH, the mortality rate increased by a hazard ratio of 1.6–5.17. Adrenal crisis was the leading cause of death, followed by cardiovascular disease. However, half of the cardiovascular deaths reported had a concurrent severe infection, implying that adrenal crisis may have contributed to the deaths [8].

A study from Sweden of 588 adults with classic 21OHD found approximately a two- to fourfold increase in mortality. The major causes of death were adrenal crisis (42%), cardiovascular (32%), cancer (16%), and suicide (10%). Mortality was highest among men and among patients with the most severe salt-wasting CAH [15].

46.5 11β-Hydroxylase Deficiency (11βOHD)

The only difference from 21-hydroxylase deficiency is hypertension which can occasionally be resistant to conventional treatment with steroids and anti-hypertensive drugs; bilateral adrenalectomy has been successfully used in several cases.

46.5.1 Height, Puberty, and Bone Health

Both girls and boys with classical 11βOHD undergo rapid postnatal growth with sexual precocity and skeletal maturation, leading to short stature. There are some case reports which suggest that GH in combination with GnRH agonists or Aromatase Inhibitors (AI) may be useful in improving height outcomes [16].

No BMD studies on 11βOHD exist. However, 21OHD data suggests that long-term glucocorticoid exposure can lower BMD.

46.5.2 Metabolic

Deoxycorticosterone (DOC) has been linked to hypertension, but it has only weak mineralocorticoid activity and does not correlate with hypertension severity or virilization. DOC is less potent than 18-hydroxylated compounds, which may be the cause of hypertension in 11OHD patients.

In a small case series, age at diagnosis was associated with systolic blood pressure. Poorly controlled hypertension and glucocorticoid overuse can impact cardio-metabolic profile. Left ventricular hypertrophy, hypertensive retinopathy, nephropathy, ischemic heart disease, cerebrovascular accidents, encephalopathy, and death are associated with 11βOHD. Obesity is a well-recognized risk factor for adverse cardiovascular outcomes. There is significant variability in BMI in patients with classic and NC 11βOHD. Spironolactone or amiloride alone or with a calcium channel blocker can be used in hypertension resistant to glucocorticoids. ACE/AR inhibitors should be avoided because the renin-angiotensin system is suppressed. Thiazide diuretics may worsen hypokalemia and should be avoided [16].

46.6 17α-Hydroxylase Deficiency

The deficient enzyme is 17-monooxygenase, which prevents the ovaries from producing estrogen. In either sex, the diagnosis may be delayed until the second decade, when puberty does not occur; and hypertension associated with hypokalemia is the critical clue.

46.6.1 Height, Puberty, Bone Health

Short stature is an unusual finding in these patients because sex steroid deficiency causes epiphyseal fusion failure, which results in average or tall stature.

To promote secondary sex characteristics, 46XY patients who are phenotypically male require testosterone replacement, whereas 46XY phenotypically female patients require estrogen therapy. To induce puberty, 46XX patients require estrogen with progesterone replacement.

Five patients in a study by Kardelene et al. had osteoporosis at the time of presentation. Most of them improved after sex steroid replacement and hydrocortisone treatment. Adolescent cortisol and estrogen deficiency have been linked to osteoporosis [17].

Wu et al., on the other hand, found that these patients were osteoporotic, and hypothesized that this was due to the negative effect of corticosteroids on bone [18].

46.6.2 Metabolic

Physiologic doses of glucocorticoids reduce mineralocorticoid excess, improve blood pressure and electrolytes. If glucocorticoids alone are not enough to control blood pressure, antihypertensives should be used. Because the underlying cause is mineralocorticoid excess, spironolactone, or other mineralocorticoid antagonists are commonly used.

46.7 Mixed Gonadal Dysgenesis

46.7.1 Height, Puberty, and Bone Health

Mixed Gonadal Dysgenesis (MGD) subjects have SHOX haploinsufficiency, which is a growth-regulating gene. Short stature is commonly seen with MGD. GH and IGF-1 levels estimations have been done in a few studies and found to be similar to patients with Turners' showing GH deficiency in some subjects and GH insensitivity in others [19].

Impaired intrauterine linear growth may be present at birth, but usually, they show average or increased birth length. Then, growth deceleration occurs in infancy, and short stature is often present during childhood when height is usually below the

50th centile. The pubertal growth spurt is severely affected, being about 4 SD below the normal mean of male growth spurt (12.3 vs. 28.0 cm). Subject with 45 X/46 XY treated with GH has shown improvement in growth at times better than Turners. However, the long-term results of this treatment in some studies seem to be disappointing [19].

If short stature is present, GH treatment should begin as soon as possible after diagnosis, or as the growth curve begins to falter. Supraphysiological GH doses are likely required, as in Turner syndrome, but trials addressing the dose/growth response relationship are lacking. Because a severely impaired pubertal growth spurt appears to contribute more to adult short stature, strategies to optimize growth during the pubertal period [e.g., increased GH doses, starting age, and doses of androgen substitutive therapy (if needed), adjunctive therapy with GnRHa or AI] should be developed.

Gonadectomy or biopsy of the underlying dysgenetic gonads is required prior to GH therapy. Subjects with a history of gonadal tumors should be closely monitored for the second neoplasm before starting GH.

Some patients may enter puberty spontaneously, while others have reduced testicular function and need hormonal treatment [20].

46.8 Androgen Insensitivity Syndrome

46.8.1 Height and Puberty

Adult women with CAIS are taller than normal women, but shorter than the male population. Infants with CAIS are on average, the same size at birth as male infants, implying that Y chromosome factors rather than prenatal androgen exposure explain the sex dimorphism.

Infancy-onset CAIS can be treated with early gonadectomy and pubertal induction or delayed until early adulthood. Estrogen should stimulate puberty if childhood gonadectomy was done. Continuous, unopposed estrogen can be used due to the absence of a uterus.

46.8.2 Bone Health

Estrogen drives chondrogenesis at the epiphyseal growth plate, and increases osteoblasts and osteoclast apoptosis from puberty to peak bone mass. Androgen may impact pubertal spurt and growth plate closure since the growth plate has AR, ERα, and ERβ. CAIS and PAIS individuals have estrogen but no androgen action. Androgen Insensitivity Syndrome (AIS) patients' vertebral BMAD decreases after gonadectomy and estrogen replacement, demonstrating androgen's direct role. PAIS patients undergoing gonadectomy had higher BMAD than CAIS, suggesting that limited androgen response may protect them. Muscle atrophy due to the absence of androgen action lowers the mechanical load and may reduce skeletal growth and modelling. Androgens activate the pituitary GH-IGF-I axis.

Compared to early postmenopausal replacement regimens, younger AIS patients may need higher estrogen doses to maintain BMD. Furthermore, the loss of skeletal integrity may be partially irreversible due to the previous failure to acquire normal bone mass [21].

Although women who had late gonadectomies are taller, there is no conclusive evidence that their BMD is less impacted. CAIS does not increase fracture risk in women. In subjects with CAIS, we can measure BMD at presentation and then 2 yearly [22].

BMI control and regular physical activity, as well as calcium and vitamin D supplementation, are recommended to improve bone health. Only in the presence of severely reduced BMD and/or fractures should bisphosphonate therapy be considered [23].

46.8.3 Metabolic

Dati et al. investigated body composition and metabolism in middle-aged CAIS patients. Body fat mass increased, resulting in high total and LDL cholesterol levels, as well as high levels of HOMA-IR. Furthermore, even though the mean BMI did not differ from the general female population of the same age, they discovered an increased rate of obesity [24].

46.9 5-α Reductase Deficiency

46.9.1 Height

5-α Reductase Deficiency (5αRD) boys grow faster before 1 year of age, growth slows between 2 and 13 years, and height varies about the median height of normal males. Localized intra-skeletal metabolism by bone tissue-specific type-1 5-α-reductase isoenzyme allows testosterone to produce biological activities in bone [25].

46.9.2 Puberty

Since most men demonstrate testicular function during puberty, testosterone replacement is usually not needed. Before surgery, dihydrotestosterone gel increases penile growth and glans faster than testosterone therapy. Hormonal therapy helps individuals raised as females establish female sexual traits. Treatment must mimic puberty with a slow escalation in estrogen dose.

46.9.3 Bone Health

In contrast to CAIS subjects, BMD was found to be preserved in subjects with 5-αRD, and the role of testosterone in maintaining bone health over DHT or the presence of 5α reductase 1 locally was proposed as the reason. No studies evaluating the metabolic and cardiovascular outcomes of 5αRD are available [26].

46.10 Gonadal Dysgenesis

46.10.1 XX

46.10.1.1 Hormonal Treatment

Gonadal Dysgenesis (GD) is treated in the same way as other causes of female estrogen deficiency, with estrogen used to induce puberty and develop secondary sexual characteristics. This is followed by cyclical estrogen and progesterone replacement therapy, which is continued until the age of 50 years when treatment can usually be discontinued. Despite adequate estrogen dosing, breast hypoplasia remains a problem in those with delayed presentation.

46.10.1.2 Bone Health

Early initiation of estrogen therapy is critical for adequate bone mass accrual during adolescence, and delaying treatment may result in decreased bone mineral density. In a study by King et al., the median T-score in the lumbar spine was −1.5, and 60% of women had osteopenia [27].

Long-term outcomes for metabolic health are currently sparse.

46.10.2 XY

Patients with XY partial GD who were raised as males and have at least one testis in the labioscrotal region have a favorable prognosis for growth and spontaneous pubertal development, but not for spontaneous fertility [28].

46.11 Conclusion

Long-term endocrinological outcomes in 21 hydroxylase deficiency CAH have been extensively studied, but are currently lacking or limited to small studies in other disorders. Adequate glucocorticoid dosing to prevent impact on height, puberty, bone, and metabolic derangements in CAH subjects is the primary concern. Among other DSDs, it is critical to determine when to initiate puberty and the need for experimental therapies to optimize height and bone health.

References

1. Girgis R, Winter JS. The effects of glucocorticoid replacement therapy on growth, bone mineral density, and bone turnover markers in children with congenital adrenal hyperplasia. J Clin Endocrinol Metab. 1997;82(12):3926–9.
2. Muthusamy K, et al. Adult height in patients with congenital adrenal hyperplasia: a systematic review and metaanalysis. J Clin Endocrinol Metab. 2010;95(9):4161–72. https://doi.org/10.1210/jc.2009-2616.
3. Soliman AT, et al. Congenital adrenal hyperplasia complicated by central precocious puberty: linear growth during infancy and treatment with gonadotropin-releasing hormone analog. Metabolism. 1997;46(5):513–7.
4. Lin-Su K, et al. Treatment with growth hormone and luteinizing hormone releasing hormone analog improves final adult height in children with congenital adrenal hyperplasia. J Clin Endocrinol Metab. 2005;90(6):3318–25.
5. Bachelot A, et al. Bone health should be an important concern in the care of patients affected by 21 hydroxylase deficiency. Int J Pediatr Endocrinol. 2010; https://doi.org/10.1155/2010/326275.
6. Rangaswamaiah S, et al. Bone mineral density in adults with congenital adrenal hyperplasia: a systematic review and meta-analysis. Front Endocrinol (Lausanne). 2020;11:493.
7. King JA, et al. Long-term corticosteroid replacement and bone mineral density in adult women with classical congenital adrenal hyperplasia. J Clin Endocrinol Metab. 2006;91(3):865–9.
8. Nordenström A, et al. Long-term outcomes of congenital adrenal hyperplasia. Endocrinol Metab (Seoul). 2022;37(4):587–98.
9. Hutson JM, Grover SR. Disorders | Differences of sex development: an integrated approach to management. 2nd ed. Singapore: Springer; 2020.
10. Falhammar H, et al. Metabolic profile and body composition in adult women with congenital adrenal hyperplasia due to 21-hydroxylase deficiency. J Clin Endocrinol Metab. 2007;92(1):110–6.
11. Arlt W, Willis DS, et al. Health status of adults with congenital adrenal hyperplasia: a cohort study of 203 patients. J Clin Endocrinol Metab. 2010;95(11):5110–21.
12. Tamhane S, et al. Cardiovascular and metabolic outcomes in congenital adrenal hyperplasia: a systematic review and meta-analysis. J Clin Endocrinol Metab. 2018;103(11):4097–103.
13. Hague WM, et al. The prevalence of polycystic ovaries in patients with congenital adrenal hyperplasia and their close relatives. Clin Endocrinol (Oxf). 1990;33(4):501–10.
14. Speiser PW, et al. Congenital adrenal hyperplasia due to steroid 21-hydroxylase deficiency: an Endocrine Society Clinical Practice Guideline. J Clin Endocrinol Metab. 2018;103(11):4043–88.
15. Falhammar H, et al. Increased mortality in patients with congenital adrenal hyperplasia due to 21-hydroxylase deficiency. J Clin Endocrinol Metab. 2014;99(12):E2715–21.
16. Bulsari K, Falhammar H. Clinical perspectives in congenital adrenal hyperplasia due to 11β-hydroxylase deficiency. Endocrine. 2017;55(1):19–36.
17. Kardelen AD, et al. A rare cause of congenital adrenal hyperplasia: clinical and genetic findings and follow-up characteristics of six patients with 17-hydroxylase deficiency including two novel mutations. J Clin Res Pediatr Endocrinol. 2018;10(3):206–15. https://doi.org/10.4274/jcrpe.0032.
18. Chaoming W, et al. 17α-hydroxylase/17, 20-lyase deficiency: clinical and molecular characterization of eight Chinese patients. Endocr Pract. 2017;23(5):576–82.
19. Bertelloni S, et al. Growth in boys with 45,X/46,XY mosaicism: effect of growth hormone treatment on statural growth. Sex Dev. 2015;9(4):183–9.
20. Martinerie L, et al. Impaired puberty, fertility, and final stature in 45,X/46,XY mixed gonadal dysgenetic patients raised as boys. Eur J Endocrinol. 2012;166(4):687–94.
21. Danilovic LS, et al. Height and bone mineral density in androgen insensitivity syndrome with mutations in the androgen receptor gene. Osteoporos Int. 2007;18(3):369–74.

22. Hughes IA, Davies JD, Bunch TI, Pasterski V, Mastroyannopoulou K, MacDougall J. Androgen insensitivity syndrome. Lancet. 2012;380(9851):1419–28. https://doi.org/10.1016/S0140-6736(12)60071-3. Epub 2012 Jun 13. PMID: 22698698
23. Lanciotti L, Cofini M, Leonardi A, Bertozzi M, Penta L, Esposito S. Different clinical presentations and management in complete androgen insensitivity syndrome (CAIS). Int J Environ Res Public Health. 2019;16(7) https://doi.org/10.3390/ijerph16071268.
24. Dati E, Baroncelli GI, Mora S, Russo G, Baldinotti F, Parrini D, Erba P, Simi P, Bertelloni S. Body composition and metabolic profile in women with complete androgen insensitivity syndrome. Sex Dev. 2009;3(4):188–93.
25. Zhao X, Song Y, Chen S, Wang X, Luo F, Yang Y, Chen L, Chen R, Chen H, Su Z, Wu D. Growth pattern in Chinese children with 5α-reductase type 2 deficiency: a retrospective multicenter study. Front Pharmacol. 2019;10:173.
26. Sobel V, Schwartz B, Zhu YS, Cordero JJ, Imperato-McGinley J. Bone mineral density in the complete androgen insensitivity and 5alpha-reductase-2 deficiency syndromes. J Clin Endocrinol Metab. 2006;91(8):3017–23. https://doi.org/10.1210/jc.2005-2809. Epub 2006 May 30. PMID: 16735493
27. King TF, Conway GS. Swyer syndrome. Curr Opin Endocrinol Diabetes Obes. 2014;21(6):504–10. https://doi.org/10.1097/MED.0000000000000113. PMID: 25314337
28. Gabriel Ribeiro de Andrade J, Marques-de-Faria AP, Fabbri HC, de Mello MP, Guerra-Júnior G, Maciel-Guerra AT. Long-term follow-up of patients with 46, XY partial gonadal dysgenesis reared as males. Int J Endocrinol. 2014; https://doi.org/10.1155/2014/480724.

Urogenital Outcome After Reconstructive Surgeries in DSD

47

Pradnya S. Bendre, Gayatri Munghate, and Arka Banerjee

Abstract

This chapter delves into the intricate realm of Disorders of Sexual Differentiation (DSD), highlighting the challenges and impacts on patients' urological, cosmetic, and sexual functions. The scarcity of standardized questionnaires for gathering long-term data hampers the evaluation of cosmetic, functional, and psychological outcomes. We have discussed surgical advancements aimed at improving physiological appearances and outcomes, such as preserving dorsal neurovascular bundles and inner preputial skin during feminizing genitoplasty, striving to achieve a more physiological appearance by creation of a mucosal lined vestibule and a natural-looking mons pubis.

It is a challenging task to assess sexual function post-DSD surgery in the Indian subcontinent where societal stigma plays a significant role. The available limited data shows widespread dissatisfaction among operated DSD patients regarding their sex lives, including issues like global sexual dysfunction, low intercourse frequency, and dyspareunia.

We have stressed the importance of addressing urinary outcomes in DSD patients by employing objective assessment tools like bladder diaries, uroflowmetry, and specific questionnaires. Furthermore, the complexities of managing proximal hypospadias are discussed, emphasizing the importance of employing self-reported measures to evaluate sexual function and satisfaction. We advocate for integrating a sexologist into the DSD care team to address potential mental health issues and enhance sexual quality of life. Highlighting the long-term psychosocial effects of DSD, the chapter acknowledges the myriad factors influencing patients' quality of life, including the development of gender identities that may not align with their initial sex assignment.

P. S. Bendre (✉) · G. Munghate · A. Banerjee
Department of Pediatric Surgery, B J Wadia Hospital for Children, Mumbai, Maharashtra, India

S. K. Ratan (ed.), *Children with Differences in Sex Development*,
https://doi.org/10.1007/978-981-97-1639-5_47

Keywords

Disorders of sexual differentiation · Feminizing genitoplasty · Lower urinary tract symptoms · Sexual dysfunction · Vaginoplasty · Gender identity

47.1 Introduction

Disorders of sexual differentiation (DSD) consist of a gamut of disorders with various etiologies and diverse long-term outcomes in different subgroups. The lack of standardized, validated questionnaires for assessment makes it extremely difficult to understand the various outcomes of DSD-affected individuals [1].

The necessity of surgery for patients with DSD has perpetually sparked controversy, even with the establishment of multidisciplinary teams (MDTs) in numerous tertiary centers. The quality of life, encompassing mental well-being, sexual function, and long-term lower urinary tract dysfunction and its associated symptoms (LUTS), remains relatively underexplored by clinical researchers in the context of DSD [2]. Notably, more than a third of caregivers for girls with Congenital Adrenal Hyperplasia (CAH) express decisional regret concerning their daughter's genital surgery [3].

The prevailing consensus guidelines advocate for multidisciplinary team (MDT) care and underscore the significance of surgical professionals adept in specific reconstructive techniques to optimize outcomes and minimize complications [4]. These guidelines stipulate the establishment of specialized centers housing MDTs capable of handling rare and intricate cases of DSD. Such a framework not only facilitates the management of complex cases but also enables the systematic collection of prospective data, from birth through adulthood and beyond, to assess long-term outcomes. However, contemporary outcome studies primarily measure the effects of past management practices rather than reflecting the current state of affairs. In an evolving socio-political landscape, marked by substantial changes in the field over the past few decades, attributed partly to enhanced understanding and knowledge among healthcare providers and largely to the efforts of patient advocacy groups and families [5], these outcome studies fall short of capturing current attitudes, knowledge, and management practices.

Insufficient long-term data on cosmetic, functional, and psychological outcomes in individuals with DSD stems from a combination of low follow-up rates and a dearth of validated questionnaires and control groups. While a desirable approach would involve annual follow-ups until adolescence and the onset of sexual activity, this proves to be exceptionally challenging. The evident need for meticulously conducted outcome studies for DSD patients persists. The limited number of patients representing various forms of DSD contributes to a scarcity of prospective, standardized, and comprehensive information crucial for informing treatment decisions. Access to high-quality outcome-based information is paramount not only for guiding patients and parents in making informed choices but also for shaping treatment strategies. Although the most robust long-term data for DSD currently emanates from females with CAH, where all aspects of outcomes have been extensively scrutinized, extrapolating this data to inform the management of other DSD types remains the primary recourse. While this approach may be deemed appropriate, particularly in the context of rarer DSD subgroups where diagnosis-specific data may be unattainable, it is crucial to recognize that DSD outcomes are predominantly influenced by socio-cultural factors, including the individual patient's personality and the capabilities and acceptance of the parents.

47.2 Urinary Outcomes

Patients with DSD may have urinary dysfunction irrespective of whether they have undergone urogenital surgery. Although the goal of genital reconstruction in any form of DSD is the restoration of phenotypically and functionally normal external genitalia congruent to the desired sex, surgery on the urogenital sinus often impacts urinary function. The main emphasis of the operations is on genital reconstruction despite a radical approach to the urinary tract. Hence, the effects on the urinary sphincters and the pelvic floor are rarely addressed in descriptions of surgical techniques [6]. As a result, the long-term outcome studies mainly focus on the genital outcome while underestimating or completely neglecting the urinary outcomes [7].

Except in cases of Complete Androgen Insensitivity Syndrome (CAIS) where the urethra is usually normal, the lower urinary tract is almost always affected in females with DSD. Prenatal androgen exposure results in the formation of a common urogenital sinus of variable length [8], most commonly seen in patients with CAH. Hence, recording urinary function in all patients at regular intervals is recommended, particularly before surgical reconstruction. Although data is lacking on lower urinary tract function in non-operated patients, there are a few inherent anatomical changes in 46 XX DSD that can lead to incomplete bladder emptying. There is pooling of urine in the common urogenital sinus and one of the indications of early surgery is to prevent UTIs and related complications [9]. The slow urethral emptying in patients with a UGS also leads to post-void dribbling. Poor urinary stream, spraying, post-void dribbling, and a feeling of incomplete emptying are the symptoms most often reported by DSD males, while an increased prevalence of Lower Urinary Tract Symptoms (LUTS) is reported in CAH patients undergoing feminizing genitoplasty [10]. There is a significant increase in urge incontinence (15% of CAH patients), nocturnal enuresis, stress incontinence (31% of CAH and 13% of AIS patients), and hesitancy [11]. Fagerholm et al. [12] reported distressing emptying symptoms (incomplete emptying, straining, post-micturition dribble or dysuria) in 19% of CAH patients and 13% of AIS patients and distressing storage symptoms (urgency, urge incontinence, stress incontinence or nocturia) in 25% of CAH and 13% of AIS patients.

Urinary incontinence after genital reconstruction surgery is rarely a neurological issue; rather, it is usually related to anatomical issues and specific surgical techniques. Continence rates after feminizing genitoplasty without total UGS mobilization range between 90 and 97% [6, 13, 14]. Our experience also reiterates that genitoplasty does not negatively impact urinary function unless the pubovesical ligaments are divided while mobilizing the bladder neck during TUM. On the other hand, separation of the urethra and vagina is not required in most pre-pubertal CAH patients with a short UGS. Extensive dissection of the bladder from the uterus below the bladder neck should be avoided to prevent any inadvertent injury to the sphincter muscles at the bladder neck.

Sometimes, vaginal voiding is confused with true urinary incontinence, as parents report daytime wetting. This usually occurs in girls who sit with their legs pressed together while voiding and that compresses the labia causing urine to backflow into the vagina. Some patients may also have a hypospadiac urethra opening into the anterior vaginal wall. We usually ask the patient to void with her legs wide apart, and at the end of voiding, we ask them to stand up for a few seconds and sit again to void. This simple maneuver of double voiding resolves daytime wetness/damping in most of these patients, saving them a genital exam and social embarrassment!

In DSD patients, hypospadias often requires staged and complex surgical procedures and specialized care according to the specific DSD subgroup. Although the immediate goals of repair include straightening the penis and ensuring socially acceptable gender-congruent functional voiding capabilities, these patients are known to have abnormal flow patterns even before surgical correction. Functional outcomes of operated hypospadias patients can be objectively assessed using uroflowmetry and post-void residual volumes. These children usually have a higher incidence of voiding dysfunction (prolonged and difficult voiding, leakage, hesitancy, spraying, and urine directed downward), a lower maximum urinary flow rate compared to normal controls [15]. Although a strong correlation exists between the severity of chordee and voiding function, post-surgical urethral function is neither determined by the primary location of the meatus nor the surgical technique employed. Obstructive urinary flow patterns following urethroplasty are probably related to the severity of hypospadias, abnormal elasticity of the neourethra, and scar tissue formation [16]. Additional factors negatively affecting the urinary flow include a lack of spongiosal support for the neourethra, the presence of a prostatic utricle and the occasional development of a megalourethra. However, we have seen that the maximum urinary flow rate shows spontaneous and progressive improvement with age and watchful waiting in these patients prevents a lot of unnecessary procedures.

There are a few standardized tools for objectively assessing lower urinary tract function after urogenital reconstruction for DSD. A well-maintained bladder diary, uroflowmetry and ultrasound assessment of post-void residual urine should be sufficient to detect any urinary dysfunction during the follow-up of these patients. Among all the validated questionnaires available to address LUTS, the ICIQ-FLUTS (International Consultation on Incontinence Questionnaire Female Lower Urinary Tract Symptoms) is arguably the most appropriate questionnaire, available with both a male and a female version, for evaluating LUTS after genital reconstructive surgery and its impact on Quality of Life (QoL) [17].

47.3 Genital (Cosmetic) Outcomes

Long-term reports of genital cosmetic outcomes in DSD are less diverse. In CAH patients, whence the majority of the data arises, early feminizing genitoplasty, in experienced hands, offers good anatomical results and very few patients require repeat surgery for cosmetic reasons [6].

To optimize surgical (cosmetic and sexual) outcomes, current consensus recommends preserving the dorsal neurovascular bundles, especially the lateral branches along the corporal tissue. These neurovascular bundles branch distally in a fan-like fashion toward the glans clitoris. Hence, reduction of the clitoral erectile tissue is better achieved with a ventral para-urethral plate incision rather than lateral incisions. The attachment of the clitoris to the pubic arch should also not be disrupted during the procedure. We also preserve the sensitive inner preputial skin during feminization genitoplasty, as it contributes to clitoral sensitivity in sexually active females.

Clitoral size is reported to be adequate in over 90% of CAH patients who have undergone surgery and separate perineal orifices are found in over 80%. However, the genital sensations depend on the surgical procedures done as those needing amputation or recession are at a higher risk of atrophy and/or neurovascular bundle injuries. Glans reduction should be avoided to preserve the glanular sensations—if deemed necessary, an effort should be made to excise as little glans tissue as possible. The dorsal nerve of the clitoris has a lateral course into the glans clitoris. Hence, we prefer to do a triangular glans de-epithelialization (reduction) at the 10–2 o'clock location on the glans clitoris. Injuring the neurovascular bundles during this step will hamper sensibility at the vaginal margin resulting in varying amounts of sexual dysfunction.

The surgical effort is meant to create a neoclitoris and at the same time, reconstruct the pubic area to make it resemble a normal female perineum. This entails the creation of a mucosa-lined vestibule which we achieve by ventrally splitting the redundant sinus tissue in the midline and folding it up. This significantly improves the cosmetic appearance and also prevents future complications from scarring and contractures of excessive skin flaps. The split preputial and phallic skin is then sutured on either side of the opened sinus to create the labia minora. The subcutaneous tissue and fat are approximated in the midline to cover the neurovascular bundles, giving rise to a natural-looking mons pubis which also plays a substantial role as a shock absorber between the male and female pubic bones during frontal intercourse movements. The added advantage of using the labial and clitoral skin to cover the new introitus is that it gives a more physiological appearance than rotational skin flaps taken from the buttocks or the perineum.

While it was previously presumed that "normalizing" the bodies of individuals with DSD should be beneficial for psychosocial development and avoid stigmatization, the advent of well-organized support groups has unearthed increasing evidence of dissatisfaction with both the timing and outcome of surgery [18]. Preves et al. [19] stated that the very approach adopted to prevent psychological maladjustment to DSD is the cause of high levels of psychological and sexual distress reported.

As a result, the accepted school of thought underwent a paradigm shift from the "optimal gender policy" to a "full consent policy" as people realized that the outcomes and perceptions of those outcomes change with time. For example, those who are prone to gender dysphoria, like PAIS, Ovotesticular DSD, or a very small proportion of CAH patients. The majority of "disgruntled former patients-turned-activists" [20] are from these subgroups who have undergone irreversible procedures such as clitoral reduction and now face severe gender dysphoria. Sometimes, the growth of local genital tissue is not proportionate to the somatic growth of the child, like in individuals with 5-ARD. Hence, a cosmetically acceptable outcome gradually becomes unacceptable over the years, especially post-puberty. On the other hand, some conditions are not affected by time, such as CAIS, Pure gonadal dysgenesis, PMDS, MRKH, etc. Hence, the timing of genital surgeries and/or gonadectomy does not affect the ultimate outcome in these patients.

The cosmetic outcome of children with proximal hypospadias is challenging to categorize, as parents' and doctors' perceptions vary. It largely depends on the surgeon's ability to achieve a slit-like meatus at the tip of a well-approximated glans and an optimal skin appearance. A few assessment tools exist, such as the Pediatric Penile Perception Score, Hypospadias Objective Penile Evaluation Score (HOPE), and the Hypospadias Objective Scoring Evaluation (HOSE) system, where pediatric surgeons, a nurse, and one of the child's parents independently assess each patient. Post-operative complications, like the formation of an urethrocutaneous fistula, glans dehiscence, neourethral breakdown, meatal stenosis, urethral stricture, urethral diverticulum, megalourethra, recurrent epididymo-orchitis secondary to a prostatic utricle or stricture, adversely affect the urinary as well as cosmetic outcomes.

47.4 Sexual Outcomes

The most challenging aspect of assessing in operated DSD patients is the genital or sexual function, i.e., the ability to achieve satisfactory painless sexual intercourse. Data suggests that the majority of DSD patients who have undergone genital surgery in childhood or later are dissatisfied with the functional results and overall sex life, including clitoral arousal as well as problems related to achieving orgasm. They also report a high degree of sexual anxiety, problems with sexual communication, desire, and dyspareunia, irrespective of gender and sexuality [21].

The studies on patients with 46XX DSD, mainly CAH, report that, although intercoital relationships start later than peers, good sex life is possible [22]. Nevertheless, these are patient-reported questionnaire-based data and the SexQoL outcomes in CAH show conflicting evidence. They have global sexual dysfunction, low intercourse frequency, non-sexuality, avoidance, anorgasmia, and penetration difficulty. A German Clinical Evaluation Study revealed diminished fertility among females with CAH, correlated with severity [23], while the offsprings were normal. However, it can be explained to be due to psychological reasons as well as their sexual behavior. It is difficult to study fertility rates in CAH patients, especially those on the more severe end of the spectrum, who are not heterosexual and hence, have not tried to get pregnant.

Vaginoplasty has undergone several modifications over the years with various tissues available for neovaginal graft. In our experience, the use of the pelvic peritoneum [24] gives a good functional outcome avoiding the inherent problems of stenosis, poor lubrication, scarring, and contracture, leading to dyspareunia, that used to be associated with the now-obsolete substitutes like skin graft, artificial dermis, skin allograft, amnion graft and ileal or sigmoid vaginoplasty. It also avoids the disastrous complications of neoplastic transformation of skin grafts to squamous cell carcinoma and adenocarcinoma in a sigmoid vaginoplasty. The amnion graft is notorious for infectious complications and is known to transmit hepatitis or human immunodeficiency virus, unless freeze-dried amnion is used [25]. Although techniques using cultured vaginal tissue, buccal mucosa, artificial material like cellulose and vaginal substitutes have been described, there is a little additional benefit of such costlier alternatives, especially after Mhatre et al demonstrated the metaplastic conversion of the peritoneum to the normal vagina in these patients [26]. Whether this translates into a better SexQoL, in spite of the peritoneal graft being less extensively innervated than the native vagina and the synchronous lack of blood supply and nervous vasoactive activity in the peritoneal neovagina, is a matter of conjecture.

Data from the "dsd-LIFE" study demonstrated a satisfactory appearance after surgery in over 90% of patients, irrespective of gender. However, the function was less satisfactory due to diminished sensation on the clitoris/glans penis and due to vaginal stenosis [27]. However, these available outcome studies pertaining to gender reassignment surgeries are unable to accurately reflect the results of the newer techniques, which are constantly evolving with our advancement of knowledge about local anatomy and surgical outcomes. Moreover, there are hardly any studies from the Indian subcontinent on the long-term sexual outcomes of DSD, primarily due to the social stigma associated with the condition and the reluctance of patients to (truthfully) divulge sensitive personal data.

The outcomes of hypospadias surgery in terms of sexuality are less diverse. More severe the hypospadias, more dissatisfactory the long-term outcome. There is also a relatively higher incidence of erectile dysfunction and premature ejaculation as the severity increases. Although glanular sensitivity is not affected by the number of procedures, patients with proximal hypospadias with chordee are more likely to be dissatisfied with the length and straightness of the penile axis during erection compared to patients with milder varieties of hypospadias [28].

A better cosmetic outcome was found to be correlating with a better sexual outcome in adults who were operated for hypospadias in their childhood. This is probably related to better penile self-perception and a lack of fear of being ridiculed for the shorter penile length and stubby appearance [29]. Although men with a small penis can achieve adequate male sexual function, they report dissatisfaction with the sexual quality of life, which is attributable to problems with erection and/or premature or no ejaculation and compromised intromission, along with a fear of infertility and rejection in marriage [30].

Men with 46XY DSD are less often in intimate relationships compared to controls but more often than women with 46XY DSD [31]. Similarly, men who underwent hypospadias repair in their childhood were sexually active but less satisfied than controls [32]. Men with Partial Androgen Insensitivity Syndrome (PAIS) report lower levels of sexual interest and sexual satisfaction compared to other subgroups of 46XY DSD [33]. This is compounded by a negative body image resulting in impaired social interactions, difficulties initiating sexual contact coupled with low sexual self-confidence, and a fear of sexual contact in undervirilized men [34].

Men with Klinefelter syndrome have erectile dysfunction, hypoactive sexual desire, and premature or delayed ejaculation resulting in fewer partnerships and delayed development of sexual interest and engagement in sexual activities [35, 36].

It is well known that sexual problems can seriously compromise the quality of life and psychological well-being. Self-reported sexuality measures (i.e., sexual activity, satisfaction with sex life, genital function, and sexual problems) are the most objective way to evaluate the outcomes of genital surgery [37]. Kreukels et al. [38] suggested the inclusion of a sexologist in the DSD MDT to anticipate mental health problems (anxiety, depression, low self-esteem) associated with sexQoL in patients with DSD.

47.5 Conclusion

Genital surgery in individuals with DSD significantly influences their lives well beyond the restoration of functional urinary and sexual capabilities. While these remain our primary surgical objectives, it is crucial not to underestimate the enduring psychosocial implications of the procedure. Complications, such as a negative body image, have been inadequately explored but wield the potential to detrimentally affect the overall outcome. A noteworthy proportion of DSD patients may eventually develop a gender identity incongruent with their initial gender assignment, with some embracing and expressing non-traditional gender identities. The timing of genital surgery lacks a consensus, and there is a dearth of long-term data to guide this decision. Endocrine factors are likely as pivotal as genetic, environmental, cultural, and educational factors in determining the quality of life for these patients. Preoperative counselling for children and their parents assumes paramount significance. Emphasizing this necessity, post-operative care should incorporate ongoing psychosocial support throughout childhood and adolescence. It is imperative for pediatric surgeons, endocrinologists, and psychologists to adopt a holistic approach to treating children with DSD throughout their developmental journey.

References

1. Bouty A, Grover SR, Hutson JM. Outcomes of individuals with DSD: a world view. In: Hutson JM, et al., editors. Disorders|Differences of sex development. 2nd ed. Singapore: Springer; 2020. p. 319–38.

2. Trachta J, Mushtaq I, Petrasova N, Hradsky O, Skaba R. Lower urinary tract symptoms in adult females after feminizing genitoplasty for congenital adrenal hyperplasia. J Urol. 2021;205(5):1483–9.
3. Fisher RS, Espeleta HC, Baskin LS, Buchanan CL, Chan YM, Cheng EY, et al. Decisional regret about surgical and non-surgical issues after genitoplasty among caregivers of female infants with CAH. J Pediatr Urol. 2022;18(1):27–33.
4. Riquelme JB, Naser NF, Puertas JB, Kalil JA, Arévalo MC. Gender affirmation surgeries in transgender women: Aesthetic, sexual, and urinary results of an initial series of vaginoplasties. Actas Urológicas Españolas (English Edition). 2021;45(3):225–31.
5. Tamar-Mattis A. Patient advocate responds to DSD surgery debate. J Pediatr Urol. 2014;10(4):788–9.
6. Wang LC, Poppas DP. Surgical outcomes and complications of reconstructive surgery in the female congenital adrenal hyperplasia patient: What every endocrinologist should know. J Steroid Biochem Mol Biol. 2017;165:137–44.
7. Berraa M, Liao LM, Creighton S, Conway GS. Long-term health issues of women with XY karyotype. Maturitas. 2010;65(2):172–8.
8. Rink RC, Adams MC, Misseri R. A new classification for genital ambiguity and urogenital sinus anomalies. Br J Urol Int. 2005;95:638–42.
9. Warne GL. Long-term outcome of disorders of sex development. Sex Dev. 2008;2(4–5):268–77.
10. Schober J, Nordenström A, Hoebeke P, Lee P, Houk C, Looijenga L, et al. Disorders of sex development: summaries of long-term outcome studies. J Pediatr Urol. 2012;8(6):616–23.
11. Kojima Y, Mizuno K, Nakane A, Kato T, Kohri K, Hayashi Y. Long-term physical, hormonal, and sexual outcome of males with disorders of sex development. J Pediatr Surg. 2009;44(8):1491–6.
12. Fagerholm R, Rintala R, Taskinen S. Lower urinary tract symptoms after feminizing genitoplasty. J Pediatr Urol. 2013;9(1):23–6.
13. Tugtepe H, Thomas DT, Turan S, et al. Does common channel length affect surgical choice in female congenital adrenal hyperplasia patients? J Pediatr Urol. 2014;10(5):948–54.
14. Palmer BW, Trojan B, Griffin K, et al. Total and partial urogenital mobilization: focus on urinary continence. J Urol. 2012;187(4):1422–6.
15. Örtqvist L, Fossum M, Andersson M, Nordenström A, Frisén L, Holmdahl G, et al. Long-term followup of men born with hypospadias: urological and cosmetic results. J Urol. 2015;193(3):975–81.
16. Andersson M, Sjöström S, Doroszkiewicz M, Ortqvist L, Abrahamsson K, Sillén U, et al. Urological results and patient satisfaction in adolescents after surgery for proximal hypospadias in childhood. J Pediatr Urol. 2020;16(5):661–8.
17. www.ICIQ.net
18. Callens N, van der Zwan YG, Drop SL, Cools M, Beerendonk CM, Wolffenbuttel KP, et al. Do surgical interventions influence psychosexual and cosmetic outcomes in women with disorders of sex development? ISRN Endocrinol. 2012; https://doi.org/10.5402/2012/276742.
19. Preves SE. For the sake of the children: destigmatizing intersexuality. J Clin Ethics. 1998;9(4):411–20.
20. Koyama E. Keynote speaker: Intersex activism at the intersection of disability and gender justice-Emi Koyama.
21. Kohler B, Kleinemeier E, Lux A, Hiort O, Gru¨ters A, Thyen U. DSD Network Working Group satisfaction with genital surgery and sexual life of adults with XY disorders of sex development: results from the German clinical evaluation study. J Clin Endocrinol Metab. 2012;97(2):577–88.
22. Crouch NS, Liao LM, Woodhouse CR, Conway GS, Creighton SM. Sexual function and genital sensitivity following feminizing genitoplasty for congenital adrenal hyperplasia. J Urol. 2008;179(2):634–8. [Epub 2007 Dec 21]
23. Hagenfeldt K, Janson PO, Holmdahl G, Falhammar H, Filipsson H, Frisen L, et al. Fertility and pregnancy outcome in women with congenital adrenal hyperplasia due to 21-hydroxylase deficiency. Hum Reprod. 2008;23(7):1607–13.

24. Origoni M, Fedele F, Parma M, Di Fatta S, Bergamini V, Candiani M, et al. The peritoneal neovagina after Davydov's laparoscopic procedure in Mayer-Rokitansky-Küster-Hauser syndrome: morphology and ultrastructure investigation of the new epithelium. J Minim Inv Gynecol. 2021;28(10):1795–9.
25. Callens N, De Cuypere G, De Sutter P, Monstrey S, Weyers S, Hoebeke P, et al. An update on surgical and non-surgical treatments for vaginal hypoplasia. Hum Reprod Update. 2014;20:775–801.
26. Mhatre PN, Narkhede HR, Pawar AP, Mhatre JP, Kumar DD. Role of progenitor cell producing normal vagina by metaplasia in laparoscopic peritoneal vaginoplasty. J Hum Reprod Sci. 2016;9:215–22.
27. Rapp M, Duranteau L, van de Grift TC, Schober J, Hirschberg AL, Krege S, et al. Self- and proxy-reported outcomes after surgery in people with disorders/differences of sex development (DSD) in Europe (dsd-LIFE). J Pediatr Urol. https://doi.org/10.1016/j.jpurol.2020.12.007
28. Winship BB, Rushton HG, Pohl HG. In pursuit of the perfect penis: hypospadias repair outcomes. J Pediatr Urol. 2017;13(3):285–8.
29. Springer A. Assessment of outcome in hypospadias surgery—a review. Front Pediatr. 2014;20(2):2.
30. Callens N, De Cuypere G, Van Hoecke E, T'Sjoen G, Monstrey S, Cools M. Sexual quality of life after hormonal and surgical treatment, including phalloplasty, in men with micropenis: A review. J Sex Med. 2013;10(12):2890–903.
31. Lee P, Schober J, Nordenstrom A, Hoebeke P, Houk C, Looijenga L, et al. Review of recent outcome data of disorders of sex development (DSD): Emphasis on surgical and sexual outcomes. J Pediatr Urol. 2012;8(6):611–5.
32. Rynja SP, de Jong TP, Bosch JL, de Kort LM. Functional, cosmetic and psychosexual results in adult men who underwent hypospadias correction in childhood. J Pediatr Urol. 2011;7(5):504–15.
33. Bouvattier C, Mignot B, Lefevre H, Morel Y, Bougneres P. Impaired sexual activity in male adults with partial androgen insensitivity. J Clin Endocrinol Metab. 2006;91(9):3310–5.
34. Schonbucher V, Schweizer K, Rustige L, Schutzmann K, Brunner F, Richter-Appelt H. Sexual quality of life of individuals with 46,XY disorders of sex development. J Sex Med. 2012;9(12):3154–70.
35. Bojesen A, Stochholm K, Juul S, Gravholt CH. Socioeconomic trajectories affect mortality in Klinefelter syndrome. J Clin Endocrinol Metab. 2011;96(7):2098–104.
36. Corona G, Petrone L, Paggi F, Lotti F, Boddi V, Fisher A, et al. Sexual dysfunction in subjects with Klinefelter's syndrome. Int J Androl. 2010;33(4):574–80.
37. Kanematsu A, Higuchi Y, Tanaka S, Hashimoto T, Nojima M, Yamamoto S. Multivariate analysis of the factors associated with sexual intercourse, marriage, and paternity of hypospadias patients. J Sex Med. 2016;13(10):1488–95.
38. Kreukels BP, Cohen-Kettenis PT, Roehle R, van de Grift TC, Slowikowska-Hilczer J, Claahsen-van der Grinten H, et al. Sexuality in adults with differences/disorders of sex development (DSD): findings from the DSD-LIFE study. J Sex Marital Ther. 2019;45(8):688–705.

Quality of Life, Psychosocial Outcomes, and Care Beyond Childhood (Transitional Care) 48

Tanu Anand and Jugal Kishore

Abstract

Disorders for sex differentiation (DSD) present as a rare congenital condition that has a significant impact on the quality of life of the affected individuals. Interestingly, the quality of life (QoL) is not only influenced by medical condition but also psychological well-being, social functionality, and cultural response. The assessment of these factors is essential to monitor the treatment outcome and its quality. Timely disclosure of DSD diagnosis and related fertility issues, with gradual reinforced education and psychological counselling, results in less likelihood of children and parents developing adverse psychosocial outcomes. Continuous communication and tripartite engagement of patients with DSD, pediatric and adult service providers are essential pillars for transitional care with successful outcomes.

Keywords

Sex differential development · Quality of Life · Psychosocial well-being

48.1 Introduction

Disorders for sex differentiation/development (DSD) present complexities in various aspects of the affected individual's life. DSD present a group of varied but rare congenital conditions with an estimated incidence of 1 in 4500–5500 births [1]. The

T. Anand
Division of Development Research, Indian Council of Medical Research, Department of Health Research, MoHFW, New Delhi, India

J. Kishore (✉)
Department of Community Medicine, VMMC & Safdarjung Hospital, New Delhi, India

S. K. Ratan (ed.), *Children with Differences in Sex Development*,
https://doi.org/10.1007/978-981-97-1639-5_48

sexual ambiguity among patients with DSD implicates serious psychosocial outcomes with adverse effect on quality of life in the long term. Further, a lack of understanding about DSD outcomes results to stigma in the lives of affected children and their families, impacting their quality of life [2].

There are major concerns associated with gender assignment, gender identity, surgical timing, consent issues, and fertility preservation measures [1]. Further, due to the rarity of these conditions, the long-term outcomes (clinical and psychosocial) have not been stratified and have often been researched among patients with congenital adrenal hypoplasia (CAH) only [3]. However, available evidence for other DSD reveals low satisfaction with medical care [4]. It, thus, implies that individuals with DSD have substantial psychosocial needs and impacted quality of life that needs consideration in management of such patients. Also, the care needs to extend beyond childhood for holistic management of affected individuals so that their quality of life can be improved.

48.2 Quality of Life

Quality of life (QoL) has been defined as "an individual's perception of their position in life in the context of the culture and value systems in which they live and in relation to their goals, expectations, standards and concerns" [5]. In simpler terms, it has been conceptualized as a measure of the successful adaptation of the individual to his/her disease or condition in life [6]. Though DSD leads to physical, social, and emotional aftermath in the patients, it is evident that the disease condition does impact the quality of life of the caregivers alike [7].

Quality of life among individuals with DSD has been assessed over the last two decades with equivocal outcomes; however, very few studies have been conducted on QoL in DSD individuals in Indian population. A large cohort of adult Brazilian patients with DSD, who were followed up by a multidisciplinary team in 2014, revealed that they had good QoL [8]. Forty-three Caucasian females with 46,XY DSD have also reported good QoL and were well-adapted [9]. On the other hand, evidence from Indonesia among 118 patients with DSD aged 6–41 years shows that while children with DSD experienced problems in social contact, adults reported negative emotions when compared with the non-affected Indonesian population [10]. A case series of 5 children aged between 6 and 15 years in Indonesia also highlighted the impact on the emotional well-being due to DSD during the COVID-19 pandemic [11]. A 15- to 20-year follow-up of children diagnosed with DSD in 2002–2004 in Norway revealed similar QoL as reported by similarly aged general population [12]. A multi-center cross-sectional clinical evaluation (dsd-LIFE) study among adults with DSD in six European countries reported patients with good QoL despite deficits in social relationships [6].

Since QOL among individuals with DSD is affected by medical, psychological, social, and cultural factors [1], it is imperative to consider the assignment of social sex in newborns with ambiguous genitalia with extreme caution [3]. Additionally, physical health, chronicity of disease condition, fertility, trauma from diagnostic and surgical procedures, gender dysphoria, sexual function in adulthood, social and psychosexual adjustment, impaired body image, social participation, and support, coping, and treatment satisfaction influence the quality of life in DSD [13, 14]. A review of international research on sexual QoL among individuals with DSD and 46,XY karyotype indicated impaired Sex QoL, particularly regarding sexual dysfunctions and sexual satisfaction [15].

In the cultural context, evidence from Asian countries revealed QoL in DSD patients to be impacted by poverty, ignorance, poor basic medical knowledge, inadequate laboratory support, lack of appropriate and affordable medications, understanding of the medical condition, and future social opportunities in relationships due to reduced fertility [10, 14]. Anthropological evidence from non-Western societies has also showcased certain cultures where more than two gender categories are recognized based on gender roles and societal norms [16]. Besides, there are many health system challenges as well. Poor awareness among practitioners regarding DSD delays diagnosis and timely interventions. Limited availability health infrastructure with sensitive and skilled multidisciplinary teams also makes the care of such individuals challenging and inaccessible [17].

Thus, to ensure good QoL of patients affected with DSD, it is imperative to adopt a multidisciplinary approach with sensitive and skilled professionals who can take care of these patients not only at the time of diagnosis but also during transition to adulthood [3]. Further, appropriate timely education to patients and their families regarding their disease condition and outcome can help in allaying anxieties related to management among them. It is further recommended to provide psychosocial support during management, inculcate family-provider communication, adoption of the patient and family-centered approach in care and promote shared decision-making for improved QoL of patients affected with DSD [1, 7, 18].

48.3 Psychosocial Outcomes

It is well established that the challenges of DSD patients range from physiological to social (discrimination and rejection) and psychological, viz. anxiety, depression, low self-esteem; thus, the care also needs to extend beyond medical and surgical management. Various studies have revealed that early surgical interventions, medical follow-ups thereafter, and frequent need for surgical revisions have resulted in psychological distress [19, 20]. Psychosocial adaptation is an important measure of psychological outcomes in DSD. It means whether the affected individual has become a functional member of society, such as part of a healthy peer group, high school, employed, and has satisfactory personal and professional relationships [1]. In a study conducted in Malaysia among 30 DSD patients and controls, psychosocial assessment using Pediatric Quality of Life Inventory Version 4.0 (PedsQL) showed patients with DSD have significantly lower PedsQL and school functioning (<0.001) [21]. The German DSD Network study reported similar findings among 86 children aged 8–12 years [22]. Adult patients with partial or complete Androgen Insensitivity Syndrome (AIS) have self-reported psychological distress, suicidal tendencies, and suicidal attempts, irrespective of the gender of rearing as male or female [4].

DSD confer a significant psychosocial impact on parents/caregivers as well. As such, there is ample literature supporting the inverted relationship between chronic illnesses among children with family functioning, family finances, and parents' psychological well-being [23]. Some types of DSD are associated with ambiguous genitals at birth, while others could be life-threatening or with decreased fertility potential [24]. Thus, DSD diagnosis itself and related aspects along with associated management options could be distressing variously for the parents. A study by Suorsa et al. in 2015 revealed that among 51 parents of DSD children aged below 2, over 25% reported reduced quality of life, 18% reported symptoms suggestive of depression, and 25.4% had moderate to severe anxiety while a substantial proportion also reported post-traumatic stress symptoms (PTSS). Uncertainty regarding the overall health of the child and related developmental issues were expressed as the cause of concern for the parents [25]. Another study in 2007 observed that a large proportion of parents face apprehensions in discussing their child's condition with their extended family and friends and also were concerned about their children getting stigmatized [26]. A study by Crissman et al. in 2011 characterized parents' experiences in the early clinical care of children with DSD. The study revealed parents' gaps in understanding regarding diagnosis and optimal management and also the expression of concern about children's future physical, social, and psychological development [27]. It is pertinent that caregivers/families understand the psychosocial implications of their decisions regarding the management of DSD [7]. Thus, in addition to an understanding of medical and surgical interventions for shared decision-making, families of individuals with DSD require access to psychological counselling for self-coping, self-empowerment, and additional resources for both medical and psychological aspects [28]. Families with poor coping skills need to be identified and offered counselling support.

Thus, psychosocial care is imperative in DSD management and accorded high priority both in clinical practice and research [29]. The 2006 consensus statement also advocates for a multidisciplinary team that includes members from psychology, psychiatry, and social work background, who would assist in decisions related to gender assignment or reassignment, the timing of surgery and sex hormone replacement, and short- and long-term assessment of outcomes for people with DSD [24]. It is also suggested that timely disclosure of DSD diagnosis and related fertility issues, with gradual reinforced education and psychological counselling, results in less likelihood of children and parents developing adverse psychosocial outcomes [29]. Children who have knowledge about their diseased condition tend to be empowered and exhibit self-confidence and better coping skills.

48.4 Care Beyond Childhood (Transitional Care)

The transition of care is defined as a meaningful passage from a family-centered and parent-reliant model of care (pediatric care) to patient-centered adult care with more emphasis on autonomy in disease management [30]. Transitional care aims at the acquisition of sufficient skills and knowledge before the transfer, patients' and families' readiness, and correct knowledge of adult healthcare professionals regarding the patient's complete clinical history [31]. It also implies to a process initiated timely to ensure continuity of care while the patient is transitioning from childhood to adulthood through adolescence.

As discussed, DSD present a complex spectrum of diseases with associated chronicity. Though the need for transitional care in DSD has been highlighted at various time points, yet the care remains suboptimal [32]. Besides, adolescents with DSD have to cope with secrecy, shame, stigma, discrimination, and diminished sexual well-being [33]. These challenges pertain to disclosure of the condition to young patients, repeated genital examinations, gonadectomy, vaginal treatments, hormonal replacement therapy, and long-term medications [33]. Young adults with DSD may not have been shared with information regarding their condition while they were in pediatric care or may have limited understanding when they were small. Thus, disclosing the information on DSD diagnosis, and treatment obtained till date and understanding the implications of decisions made in the past, might take several clinic visits in transitional care. A genital examination is pertinent for young adults with DSD prior to onset of puberty, and earlier to the time when sexual activity is contemplated. This is important as it guides further clinical management of the patient in adult life (vaginal treatment-vaginoplasty, vaginal dilation) and requires multidisciplinary team (surgeon, clinical nurse, and psychologist) to take care of. Patients with complete (46,XY karyotype) or mixed/mosaic (45XO/46 XY karyotype) gonadal dysgenesis are also at risk of developing malignancy. According to the current clinical practice, for the management of such cases gonadectomy is recommended, which is usually done during adolescence as the condition is diagnosed at that time as an investigation of delayed puberty. In female patients with no dysgenesis or virilization, gonads may have to be retained till puberty is complete. Also, some individuals may prefer to retain their gonads and want to be on long-term hormonal replacement therapies. Hence, a balance for deferring surgery against unknown risks of malignancy needs to be maintained. Patients need to be continuously monitored during transitional care and provided support for appropriate decision-making regarding their management and its impact.

It is, therefore, suggested that the transition process should begin at least a year before the intended date of transfer with the flexibility to accommodate patients' and parents' needs. Continuous communication and tripartite engagement of patients with DSD, pediatric and adult service providers are essential pillars for transitional care with successful outcomes.

48.5 Research Gaps

QoL was rarely studied as an outcome of research undertaken in DSD [6]. Even now, most of the published studies on QoL among individuals with DSD (both children and adults) report inconsistencies due to insufficient sample size [9, 11, 12], methodology [6, 12], and tool used for its assessment [3]. QoL among DSD patients has been assessed using SF-36 Health Survey test, the Psychological General Well-Being questionnaire (PGWB), the Quality of Life Assessment of GH Deficiency in Adults (QoL-AGHDA) score, the Quality of Life Scale (QOLS), the Life Situation Survey (LSS), RAND 36-Item Health Survey (RAND-36), the 15D instrument and WHOQOL-Bref, thereby making comparative analysis of studies difficult [3]. Also, the lack of DSD-specific assessment tools has resulted in challenges to inform clinical practice and evaluate the outcomes [7]. Thus, the need of the hour is to undertake research to study outcomes among DSD individuals taking large samples and using standardized QoL assessment tools with focus on psychological and social domains (relationships, partnership, sexuality, fertility, etc.) in addition to the physical dimension of health [17]. Also, the impact of clinical decisions on QoL among affected individuals is currently missing from our knowledge base and must be investigated in research mode [4]. Also, there is a requirement to elucidate diagnosis-specific differences in SexQoL and assess qualitative aspects of sexuality and non-coital sexual activities [15].

There is insufficient and inconclusive data on the long-term impact of medical and surgical interventions among individuals with DSD and their parents, more so, in the psychosocial domain [19]. Hence, it is suggested to conduct more research on optimal treatment and timing of intervention of individuals with DSD and assess long-term impacts including psychosocial interventions [4, 19]. Little research has been carried out on information management [29]. Therefore, studies are needed for the optimization of long-term care in DSD, by determining the delivery of education by whom, how, and when [34]. Also, there is a need to explore the parental outcomes in children with DSD [13]. Also, theoretical models of child development and development psychopathology need to be explored in future [13].

Consequent to suboptimal evidence on transition care in DSD [33], the data continues to be fragmented and inaccessible. Hence, research needs in transitional care encompass structure, timing of the transition, and assessment/evaluation of transitional care with specific reference to DSD.

48.6 Conclusions

DSD present as a complex spectrum of diseases with far-reaching impact on various domains of QoL. There are various issues encompassing the individuals with DSD that include gender identity, gender assignment, the timing of surgical and medical interventions, and fertility-related. All these issues incur psychological stress among affected patients and caregivers alike. Therefore, assessment of psychosocial outcomes and provision of transitional care using a multidisciplinary approach is pertinent in DSD management. There exists a paucity of data on quality of life and psychosocial outcomes among DSD patients and hence, there is a need to undertake systematic studies on these long-term outcomes. Further, a multidisciplinary team with transitional care is needed for the appropriate and holistic management of DSD to ensure unerring decision-making in terms of management of DSD for a holistic outcome.

References

1. Lee PA, Nordenstrom A, Houk CP, Ahmed SF, Auchus R, Baratz A, et al. Global disorders of sex development update since 2006: perceptions, approach and care. Horm Res Paediatr. 2016;85:158–80.
2. Khanna K, Sharma S, Gupta DK. A clinical approach to diagnosis of ambiguous genitalia. J Indian Assoc Pediatr Surg. 2019;24:162–9.
3. Amaral RC, Inacio M, Brito VN, Bachega TA, Domenice S, Arnhold IJ, et al. Quality of life of patients with 46,XX and 46,XY disorders of sex development. Clin Endocrinol. 2015;82:159–64.
4. Wisniewski AB, Mazur T. 46,XY DSD with female or ambiguous external genitalia at birth due to androgen insensitivity syndrome, 5alpha-reductase-2 deficiency, or 17beta-hydroxysteroid dehydrogenase deficiency: a review of quality of life outcomes. Int J Pediatr Endocrinol. 2009;2009:567430. https://doi.org/10.1155/2009/567430.
5. WHO. WHOQOL: measuring quality of life [Internet] 2012. Available from https://www.who.int/tools/whoqol. Accessed 24 Dec 2022.
6. Rapp M, Mueller-Godeffroy E, Lee P, Roehel R, Kreukels BPC, Kohler B, et al. Multicentre cross-sectional clinical evaluation study about quality of Life in adults with disorders/differences of sex development (DSD) compared to country specific reference populations (dsd-LIFE). Health Qual Life Outcomes. 2018;16:54. https://doi.org/10.1186/s12955-018-0881-3.
7. Alpern AN, Gardner M, Kogan B, Sandberg DE, Quittner AL. Development of health-related quality of life instruments for young children with disorders of sex development (DSD) and their parents. J Pediatr Psychol. 2017;42(5):544–58. https://doi.org/10.1093/jpepsy/jsw022.
8. Jurgensen M, Kleinemeier E, Lux A, Steensma TD, Cohen-Kettenis PT, Hiort O, et al. Psychosexual development in children with disorder of sex development (DSD)—results from the German Clinical Evaluation Study. J Pediatr Endocrinol Metab. 2010;23(3):1–16.
9. D'Alberton F, Assante MT, Foresti M, Balsamo A, Bertelloni S, Dati E, et al. Quality of life and psychological adjustment of women living with 46,XY differences of sex development. J Sex Med. 2015;12:1440–9.
10. Ediati A, Verrips GHW, Juniarto AZ, Faradz SMH, Drop SLS, Dessens AB. Quality of life in late-treated patients with disorders of sex development: insights for patient-centered care. Front Pediatr. 2019;6:434. https://doi.org/10.3389/fped.2018.00434.
11. Rochmah N, Faizi M, Setiawan D. Quality of life in congenital adrenal hyperplasia with disorders of sex development: case series. Asia Pac J Pediatr Child Health. 2021;4(2):34–8.
12. Waehre A, Heggli C, Hald K, Myhre AG, Diseth T. A 15–20-year follow-up of mental health, psychosocial functioning and quality of life in a single center sample of individuals with differences in sex development. Health Psychol Behav Med. 2022;10(1):837–54.
13. Schober J, Nordenstrom A, Hoebeke P, Lee P, Houk C, Looijenga L, et al. Disorders of sex development: summaries of long-term outcome studies. J Pediatr Urol. 2012;8:616–23.
14. Zainuddin AA, Grover SR, Shamsuddin K, Mahdy ZA. Research on quality of life in female patients with congenital adrenal hyperplasia and issues in developing nations. J Pediatr Adolesc Gynecol. 2013;26:296–304. https://doi.org/10.1016/j.jpag.2012.08.004.
15. Schönbucher V, Schweizer K, Richter-Appelt H. Sexual quality of life of individuals with disorders of sex development and a 46,XY karyotype: a review of international research. J Sex Marital Ther. 2010;36(3):193–215. https://doi.org/10.1080/00926231003719574.
16. Lang C, Kuhnle U. Intersexuality and alternative gender categories in non-Western cultures. Horm Res. 2008;69:240–50.
17. Maritska Z, Saputro ED, Pangestu R, Faulinza E, Sakinah M, Pranandita F, Satopoh FA, Ridwan I, Amalia S. Current status of disorders of sexual development in Indonesia. Adv Hum Biol. 2022;12:210–3.
18. Cools M, Nordenström A, Robeva R, Hall J, Westerveld P, Fluck C, et al. Caring for individuals with a difference of sex development (DSD): a consensus statement. Nat Rev Endocrinol. 2018;14:415–29. https://doi.org/10.1038/s41574-018-0010-8.

19. Mulkey N, Streed CG, Chubak BM. A call to update standard of care for children with differences in sex development. AMA J Ethics. 2021;23(7):E550–6. https://doi.org/10.1001/amajethics.2021.550.
20. Davis G. Normalizing intersex: the transformative power of stories. Narrat Inq Bioeth. 2015;5(2):87–9.
21. Selveindran NM, Syed Zakaria SZ, Jalaludin MY, Rasat R. Quality of life in children with disorders of sex development. Horm Res Paediatr. 2017;88:324–30.
22. Jürgensen M, Lux A, Wien SB, Kleinemeier E, Hiort O, Thyen U. Health-related quality of life in children with disorders of sex development (DSD). Eur J Pediatr. 2014;173:893–903.
23. Wisniewski AB. Psychosocial implications of disorders of sex development treatment for parents. Curr Opin Urol. 2017 Jan;27(1):11–3. https://doi.org/10.1097/MOU.0000000000000344.
24. Lee PA, Houk CP, Ahmed SF, Hughes IA, International Consensus Conference on Intersex organized by the Lawson Wilkins Pediatric Endocrine Society and the European Society for Paediatric Endocrinology. Consensus statement on management of intersex disorders. International Consensus Conference on Intersex. Pediatrics. 2006;118:e488–500.
25. Suorsa KI, Mullins AJ, Tackett AP, Reyes KJS, Austin P, Baskin L, et al. Characterizing early psychosocial functioning of parents of children with moderate to severe genital ambiguity due to disorders of sex development. J Urol. 2015;194:1737.
26. Duguid A, Morrison S, Robertson A, Chalmers J, Youngson G, Ahmed SF. The Scottish Genital Anomaly Network. The psychological impact of genital anomalies on the parents of affected children. Acta Paediatr. 2007;96(3):348–52.
27. Crissman HP, Warner L, Gardner M, Carr M, Schast A, Quittner AL, Kogan B, Sandberg DE. Children with disorders of sex development: a qualitative study of early parental experience. Int J Pediatr Endocrinol. 2011 Oct 12;2011(1):10. https://doi.org/10.1186/1687-9856-2011-10.
28. de Neve-Enthoven NGM, Callens N, van Kuyek M, van Kuppenveld JH, Drop SLS, Cohen-Kettenis PT, Dessens AB. Psychosocial well-being in Dutch adults with disorders of sex development. J Psychosom Res. 2016;83:57–64.
29. Hughes LA, Nihoul-Fekete C, Thomas B, Cohen-Kettenis PT. Consequences of the ESPE/LWPES guidelines for diagnosis and treatment of disorders of sex development. Clin Endocrinol Metabo. 2007;21(3):351–67.
30. Hait EJ, Barendse RM, Arnold JH, Valim C, Sands BE, Korzenik JR, et al. Transition of adolescents with inflammatory bowel disease from pediatric to adult care: a survey of adult gastroenterologists. J Pediatr Gastroenterol Nutr. 2009;48(1):61–5.
31. Escher JC. Transition from pediatric to adult health care in inflammatory bowel disease. Dig Dis. 2009;27(3):382–6.
32. Brain CE, et al. Holistic management of DSD. Best Pract Res Clin Endocrinol Metab. 2010;24:335–54.
33. Crouch NS, Creighton MS. Transition of care for adolescents with disorders of sex development. Nat Rev. 2014;10(7):436–42.
34. Liao LM, Simmonds M. A values-driven and evidence-based health care psychology for diverse sex development. Psychol Sexual. 2014;5:83–101.

Psychiatric Morbidity Associated in Individuals with Disorders of Sex Development

49

Alka A. Subramanyam, Mansi Somaiya, and Prerna Khar

Abstract

Disorders of sexual development (DSD) are conditions which are associated with multiple psychiatric morbidities consequent to interplay of many factors, specially related to genetic and environmental factors, especially those related to gender dysphoria. The given chapter highlights such psychiatric comorbidities in detail.

Keywords

Disorders of sex development · Genetic · Psychiatric morbidity

49.1 Introduction

Disorders of sexual development (DSD) are conditions in which an atypical chromosomal, gonadal, or phenotypic sex leads to differences in the development of the urogenital tract and hence clinically present with various phenotypic manifestations [1]. Sex development follows an interplay between genetic and environmental factors. The various points considered crucial in the emergence of a disorder are—genetic programming, chromosomal sex influences, and hormonal influences [2].

A. A. Subramanyam · M. Somaiya
Department of Psychiatry, TNMC and BYL Nair Charitable Hospital, Mumbai, India

P. Khar (✉)
Department of Psychiatry, Early Intervention and Rehabilitation Centre for Children,TNMC and BYL Nair Charitable Hospital, Mumbai, India

S. K. Ratan (ed.), *Children with Differences in Sex Development*,
https://doi.org/10.1007/978-981-97-1639-5_49

49.2 Psychosexual Development

This concept was first explained by Money et al. in 1955 [3]. The human brain undergoes sexual differentiation which is synchronous with the other sexual characteristics. It is hypothesized that androgens are influential in organizing the brain in the neurodevelopmental period. Subsequently, pubertal steroids activate the same, which leads to masculine behaviour. Sexual differentiation of the genitals occurs in the first trimester (first 2 months) and that of the brain in the latter half of pregnancy. Hence, both these processes may be influenced independently, by different factors. Thus, the extent of virilization of the external genitalia may not be in sync to the degree of masculinization of the brain. Psychosexual development is a multifaceted process which is influenced by brain structure, genetics, perinatal hormonal influences, environmental, familial, and psychosocial exposure [4].

A more recent perspective states that DSD could be defined as all the conditions in which the chromosomal, gonadal, phenotypical, or gender identity (psychological) sex are incongruent [5]. As per recent literature, few aberrant neurodevelopmental patterns have been found in individuals having Gender dysphoria, such as—a larger regional grey matter and deviances in the brain activation patterns, compared to other individuals [6, 7].

Psychosexual development comprises three components [4]:

(1) *Gender identity:* Self-representation of an individual as being male, female, or even neither.
(2) *Gender role (sex-typical behaviour)*: Behaviour, attitudes, and traits that society at large identifies as being masculine or feminine.
(3) *Sexual orientation*: The person's response to sexual stimuli, which includes the behaviour, sexual fantasies, and who one finds sexually appealing (hetero/bi/homosexual).

In cases of altered sexual development, two scenarios maybe encountered [4]:

(1) *Gender dissatisfaction:* Unhappiness with the biologically assigned sex. The exact aetiology of this has been poorly understood till date. (In individuals with DSD, it is imperative to note that having a homosexual orientation or cross-sex interest is not considered indicative of incorrect gender assignment.)
(2) *Gender dysphoria (GD):* Characterized by significant incongruence between the biologically assigned gender and the experienced/expressed gender. This is associated with clinically significant functional impairment. (It can occur either in the presence or absence of DSD)

49.3 Diagnosis

Here we may encounter two scenarios—one where there are ambiguous genitalia at birth itself (early) and the second when it becomes apparent later in life, usually due to delayed puberty or gender identity (later).

(A) **Early diagnosis:**

Approach to Diagnosis
The birth of a baby with ambiguous genitalia is often a puzzling and distressing experience for the parents and can be considered a social emergency, especially in the Indian setting. Friends, families, acquaintances, all want to know if the newborn is a boy or a girl. In case of potential DSD, it may be advisable to congratulate the parents on the birth of their offspring, show them the physical characteristics, and subsequently explain that the gonadal structures are bipotential [2].

Parental Concerns
In the hospital setting, it is important to assess and discuss the parents' existing knowledge about gender development to provide sufficient room to address their concerns. The incongruence between the child's karyotype and sex of rearing may cause significant discomfort among caregivers. It is imperative to highlight that the karyotype of the child does not define their gender. Parents need to be provided with developmentally appropriate language to explain the diagnosis and care to their child. In addition, they often need hand-holding as to how to explain their child's condition to family members and close associates. Using lucid phrases such as the baby had an alteration in genital development due to which certain assessments are required to guide sex assignment can be suggested to the new parents [8].

Family Support
It is important to note that family is the major source of strength and support to children. Hence the perspective of the family as well, in addition to the child's should be taken into account during clinical decision-making. Family-centred care improves the quality of life for the child and additionally reduces parental distress.

Psychiatric Comorbidities
Young individuals with hyperandrogenism (e.g., congenital adrenal hyperplasia—CAH) are prone to develop psychiatric disorders, due to the effect of androgen on emotional reactivity, brain structure, and function [9]. Psychosexual outcomes in DSD have been most thoroughly studied in CAH. Most of these studies have shown that girls maintain their gender identity as girl/woman throughout their lifespan. However, a weaker female identity, sometimes gender dysphoria, and masculine gender role behaviour have been observed more frequently in patients with CAH than in those without CAH. Additionally, adolescents with CAH are reported to be

exhibiting more masculine than feminine behaviour compared to those without CAH [9].

Osbarun et al. evaluated 51 children aged between 1 and 18 years, over a period of 8 years, who were referred to a DSD clinic in Turkey. Thirty-four (66.7%) children had normal intellectual functioning, 3 (5.9%) had borderline intellectual functioning, 11 children (21.6%) had mild, and 3 children (5.9%) had moderate intellectual disability. In terms of psychopathology, six children (11.8%) had depression, 3 (5.9%) had anxiety disorder, 3 (5.9%) had attention deficit hyperactivity disorder (ADHD), and one (2.0%) had adjustment disorder. One child with mild intellectual disability also had ADHD and 3 children had depression, in addition to intellectual disability [10]. ADHD is a neurodevelopmental disorder, and early exposure to androgens may thus increase the risk for ADHD, especially in susceptible males [11].

In a retrospective study done by Birsen et al., analysing data of 118 children which were followed by the DSD department for 12 years, it was found that ADHD was the most common psychiatric comorbidity (15.3%). ADHD was most commonly seen in children with CAH (22.4%) and androgen synthesis defects (ASD) (22.4%). Depression was the most common diagnosis in children having complete gonadal dysgenesis and ASD. Intellectual disability was the most common diagnosis in ASD (37.5%) [9].

In a study done by Doctur et al. on 45 children with CAH, it was found that 51.1% of the patients had at least one comorbid psychiatric disorder in their lifetime. The most common diagnoses were ADHD, anxiety disorders, tic disorders, and nocturnal enuresis. The androgen-dependent process of sexual differentiation during prenatal life may have a role in the development of tic- related disorders [11].

(B) **Later diagnosis:**

During the assessment of DSD, we must consider the phenotype of the internal and external genitals, physical health, fertility, sexual function and psychosexual adjustment, mental well-being and health and quality of life. This is from the consensus statement on management of intersex disorders from 2006 [12]. A multidisciplinary team must be involved in which a mental health professional must be present while the diagnosis is being made.

Psychiatric Comorbidities

(1) Depression and anxiety

Patients with congenital adrenal hyperplasia were found to have an increased level of psychological symptoms. There were a higher prevalence of anxiety disorders in this population. They could present as adjustment disorders, generalized anxiety disorder, social anxiety and depressive disorders (dysthymic disorders and major depressive disorder). In patients with DSD, suicidal tendencies and self-harming behaviour were also observed [13]. In adult women with DSD, previous suicidal ideations, anxiety symptoms, and psychological/psychiatric counselling for severe symptoms of problems were noted [14]. In CAH women, poor long-term quality of life and increased depressive symptoms were seen in individuals with genital malformations and corrective surgery [15].

In adult males with DSD, anxiety and depressive symptoms were higher in males with DSD conditions (KS and XY-DSD males). Anxiety and depressive symptoms were noticed in females with XY DSD with partial androgenisation [16]. Patients with DSD such as TS and KS were associated with depressive symptoms and shame [16].

(2) Substance use disorder

Alcohol misuse was observed in males with DSD [15].

(3) Psychosexual outcome

Most studies have focussed on gender dysphoria, gender change, and impact of sexual functioning on development.

The majority of research regarding DSD and gender dysphoria relates specifically to DSD.

Orientation: Data limited to a few studies. In a study, 33% patients with Complete Androgen Insensitivity Syndrome (CAIS) were bisexual and 50% Partial Androgen Insensitivity Syndrome (PAIS) were homosexual [17].

(4) Psychosis

In patients with primary adrenal deficiency, patients with P30L genotype appear to have a greater risk for psychosis [18]. A study done by Falhammar H et al. in 2014 reported an increased risk in psychosis in males with 21 hydroxylase deficiency, V281L genotype, and P30 L mutation [15].

49.4 Psychiatric Assessment of an Individual with DSD

Each clinician must be aware that from the time of evaluation and even further on, every attempt must be made for the child or adult to adjust in becoming a functional member of society. Privacy must be given foremost importance.

The psychiatric assessment forms a key and integral part of the assessment, it also facilitates and allows healthy adaptation to diagnosis and further care and intervention. Additionally, assessment of coping abilities in a child is highly essential. Comorbid psychiatric and neurodevelopmental conditions need to be assessed. In children, gender identity develops by the age of 3 years, hence a psychosexual assessment would encompass gender identity. It is essential for the mental health assessment to also include gender dissatisfaction if any. Persistence of the desire to change gender is essential.

Children and adolescents must also be offered the opportunity to speak without the presence of parents so that confidentiality is maintained. Also, this allows a comfortable space for children to discuss their concerns.

In children and adolescents, another aspect of concern is a collaborative discussion on gender, karyotype, gonadal status, and fertility issues. A regular follow-up and age-appropriate discussion must follow. Each age group will demand different aspects of discussion by the mental health care practitioner with the individual.

In adults, quality of life, intimate relationships, sexual functioning, interpersonal relationships, fertility problems need to be assessed and kept in mind while intervening in these individuals [12].

Patients must be referred for therapy and detailed mental status assessment is warranted.

49.5 Ethical Dilemmas for Sex Reassignment Surgeries

Although androgenization of the body and brain usually coincide, this may not be the case always. For instance, failed or improper androgen imprinting of the brain could explain a homosexual orientation in a healthy male not having DSD. Similarly, elevated serum testosterone has been postulated to lead to different extents of androgen imprinting in individuals having CAH. Therefore, a woman with CAH with a fully virilized brain will have a male sexual orientation (homosexual attraction towards females) and those with poor androgen imprinting will have heterosexual attraction towards males. Hence, it is imperative to ensure that psychosexual orientation and development is taken into consideration, rather than purely relying on phenotypic manifestations [19]. According to the separation individuation theory, children establish object consistency by 3 years of age. It is during adolescence when the second individuation period occurs, in which the adolescents build their identity. It is essential to prevent children with DSD from acquiring a false reference and make a false gender identity decision, which may also lead to an early decision for surgery, proving to be detrimental. Although surgery can reduce parental discomfort and enhance the child–parent/family attachment, the current consensus strongly advocates that decisions for any appearance altering interventions should not be taken hastily. It is advisable to delay surgical interventions till the child is old enough to be adequately informed and provide consent for the same [10].

49.6 Psychiatric Interventions

A key stage in the management of DSD is enhancing the individual's self-esteem and resilience to tide over emotional and behavioural difficulties. Hence a clinician must also focus on enabling the patient to develop coping mechanisms. An open and proactive communication will facilitate a sense of control and empowerment [16]. Being a part of support groups will help adults with similar issues such as shame, stigma, and social isolation address challenges.

The patients must be treated with appropriate medications depending on the diagnosed disorder. In patients with depressive disorder or anxiety, appropriate treatment with SSRIs and benzodiazepines if required. In psychosis, antipsychotics can be prescribed. ADHD can be managed with stimulants.

49.7 Lacunae in the Field

Breaking the diagnosis can be a difficult task, more so in a developing country like ours, where multidisciplinary approaches are yet to be a part of standard operating procedures. Although research has started identifying psychometric screening tools for evaluating patients and families with DSDs, standardized protocols to provide developmentally appropriate mental health care are not yet available. This has led to low satisfaction with the medical care as reported by patients/caregivers. Hence it may be necessary to provide health care providers with such protocols, as a part of a multidisciplinary approach, for enhancing patient satisfaction [8].

49.8 Legal Aspects

The Transgender person's bill, 2019 was passed to identify specific health needs in them. It also clearly prohibits non-compulsory and non-consensual sex selective surgeries/sex reassignment surgeries in a minor. The bill also recognizes the identity of transgender persons, offers provisions of welfare measures, suggestions on education, social security, and health care provisions [20].

49.9 Conclusion

The goal for any multidisciplinary team while approaching a child with potential DSD should be to ensure psychosocial well-being, sexual satisfaction, preservation of fertility options, which will all ensure a satisfactory quality of life [19]. An overview of the psychiatric approach to a child with DSD has been summarized in Flowchart 49.1.

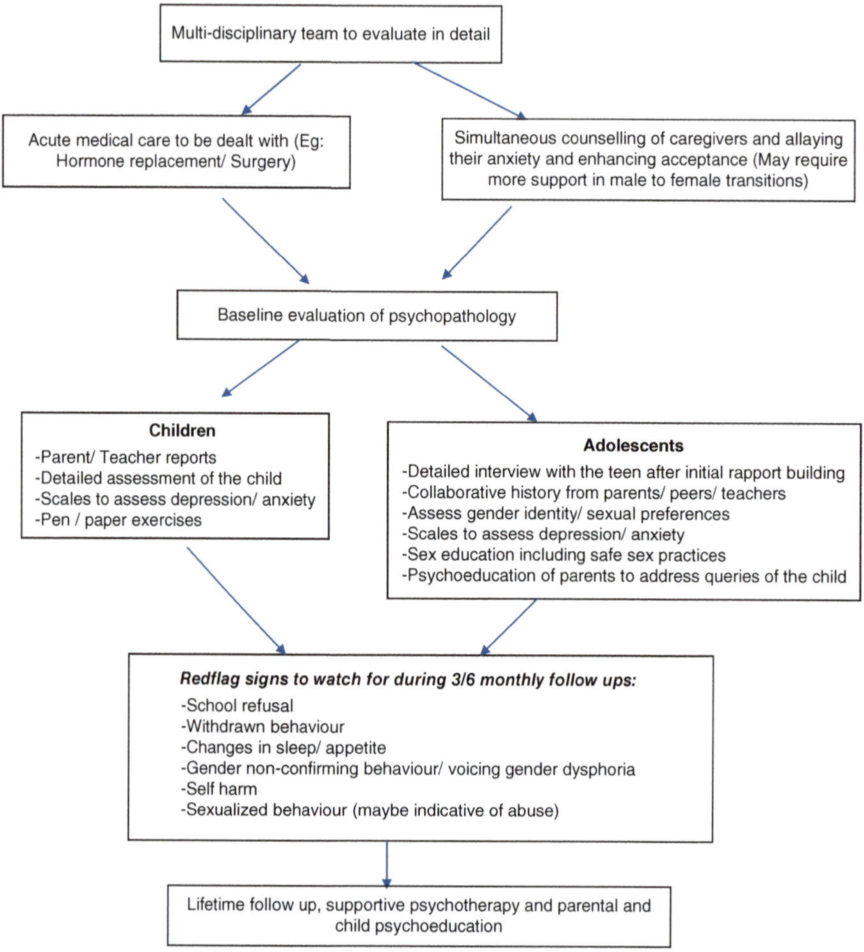

Flowchart 49.1 Bird's-Eye view: Approach to a child with DSD

References

1. García-Acero M, Moreno O, Suárez F, Rojas A. Disorders of sexual development: current status and progress in the diagnostic approach. Curr Urol. 2019;13(4):169–78.
2. Witchel SF. Disorders of sex development. Best Pract Res Clin Obstet Gynaecol. 2018 Apr 1;48:90–102.
3. Money J, Hampson JG, Hampson JL. An examination of some basics sexual concepts: the evidence of human hermaphroditism. Bull Johns Hopkins Hosp. 1955;97:301–19.
4. Sarma VP. A review of the essential concepts in diagnosis, therapy, and gender assignment in disorders of sexual development. Ann Pediatr Surg. 2022 Dec;18(1):1–2.
5. Fisher AD, Ristori J, Fanni E, Castellini G, Forti G, Maggi M. Gender identity, gender assignment and reassignment in individuals with disorders of sex development: a major of dilemma. J Endocrinol Investig. 2016 Nov;39(11):1207–24.

6. Luders E, Sánchez FJ, Gaser C, et al. Regional gray matter variation in male-to-female transsexualism. Neuroimage. 2009;46(4):904–7.
7. Schöning S, Engelien A, Bauer C, et al. Neuroimaging differences in spatial cognition between men and male-to-female transsexuals before and during hormone therapy. J Sex Med. 2010;7(5):1858–67.
8. Wisniewski AB, Batista RL, Costa EM, Finlayson C, Sircili MH, Dénes FT, Domenice S, Mendonca BB. Management of 46,XY differences/disorders of sex development (DSD) throughout life. Endocr Rev. 2019 Dec;40(6):1547–72.
9. Pilan BŞ, Özbaran B, Çelik D, Özcan T, Özen S, Gökşen D, Ulman İ, Avanoğlu A, Tiryaki S, Onay H, Çoğulu Ö. Psychiatric view for disorders of sex development: a 12-year experience of a multidisciplinary team in a university hospital. J Pediatr Endocrinol Metab. 2020 May 1;33(5):605–11.
10. Özbaran B, Özen S, Gökşen D, Korkmaz Ö, Onay H, Özkınay F, Çoğulu Ö, Erermiş S, Köse S, Avanoğlu A, Ulman İ. Psychiatric approaches for disorders of sex development: experience of a multidisciplinary team. J Clin Res Pediatr Endocrinol. 2013 Dec;5(4):229.
11. Doktur H, Tanidir C, Güneş H, Aytemiz T, Durcan G, Önal H, Kutlu E. Gender dsyphoria and psychiatric disorders in children and adolescents with congenital adrenal hyperplasia. Acta Endocrinol (Buch). 2021 Jul;17(3):365.
12. Hughes IA, Houk C, Ahmed SF, Lee PA, LWPES Consensus Group; ESPE Consensus Group. Consensus statement on management of intersex disorders. Arch Dis Child. 2006 Jul;91(7):554–63.
13. Schützmann K, Brinkmann L, Schacht M, Richter-Appelt H. Psychological distress, self-harming behavior, and suicidal tendencies in adults with disorders of sex development. Arch Sex Behav. 2009;38:16–33.
14. Johannsen TH, Ripa CPL, Mortensen EL, Main KM. Quality of life in 70 women with disorders of sex development. Eur J Endocrinol. 2006;155:877–85.
15. Falhammar H, Butwicka A, Landén M, Lichtenstein P, Nordenskjöld A, Nordenström A, Frisén L. Increased psychiatric morbidity in men with congenital adrenal hyperplasia due to 21-hydroxylase deficiency. J Clin Endocrinol Metab. 2014 Mar;99(3):E554–60. https://doi.org/10.1210/jc.2013-3707.
16. De Vries AL, Roehle R, Marshall L, Frisén L, van de Grift TC, Kreukels BP, Bouvattier C, Köhler B, Thyen U, Nordenström A, Rapp M. Mental health of a large group of adults with disorders of sex development in six European countries. Psychosom Med. 2019 Sep;81(7):629.
17. Fagerholm R, Santtila P, Miettinen PJ, Mattila A, Rintala R, Taskinen S. Sexual function and attitudes toward surgery after feminizing genitoplasty. J Urol. 2011 May;185(5):1900–4.
18. Thomsen AF, Kvist TK, Andersen PK, Kessing LV. The risk of affective disorders in patients with adrenocortical insufficiency. Psychoneuroendocrinology. 2006;31:614–22.
19. Raveenthiran V. Neonatal sex assignment in disorders of sex development: a philosophical introspection. J Neonatal Surg. 2017;6:58–64.
20. Srishti Madurai NNID Foundation. The Rights of Intersex People in India. NGO Submission for the 22nd session of the committee on the Rights of persons with disabilities (CRPD). Report accepted by office of the United Nations High Commissioner for Human Rights. Information Toolkit Human Rights of Intersex Persons in India. 1st National Intersex Human Rights Conference.

DSD: A Gynecologist's Perspective

50

Pravin Mhatre

Abstract

Disorders of sexual differentiation (DSD) is a complex entity, involving many superspecialists. Although the goal of treatment is to assign the proper sex to the unfortunate patient, the path to achieve may be very complex and difficult. Arriving at the diagnosis may be extremely easy as in Turner's syndrome or may be extremely complicated in others. The process involves gynecologists, pediatricians, pediatric surgeons, endocrinologists, genetic experts making it a cauldron of different views. Complicating it further unlike other medical conditions, patients own psychological inclination has an important role to play bringing in the psychiatrist, psychoanalyst.

Presented herewith DSD from a gynecologist's perspective and highlighted are some of the interesting and rarer cases. Ovarian transplant, a new treatment modality for Turner's syndrome, along with a novel technique of vaginoplasty developing a normal vagina is presented. Along with that, the latest research on progenitor cell and translational genomes in the vaginoplasty is also discussed.

Keywords

Disorders of sexual differentiation (DSD) · Sex assignment · Vaginoplasty

50.1 Introduction

During the evolutionary process the reproduction changed from asexual to sexual mechanism. Thus, it was a necessary biological process to have different sex assignment for specific purpose of reproduction. In the lower animals, like fish, the sex

P. Mhatre (✉)
Seth GS Medical College & KEM, Wadia Hospital, Mumbai, India

S. K. Ratan (ed.), *Children with Differences in Sex Development*,
https://doi.org/10.1007/978-981-97-1639-5_50

was interchangeable as per the need of the hour. Sharks could change the sex as per the requirement. Environmental and geographic changes also determined the sex predilection and the type of reproduction. Australian Lizards are viviparous at high altitudes, ovoviviparous at upland, and in lowland, they become oviparous. However, at evolutionary hierarchy the sex assignment is permanent. Thus, it becomes imperative that the sex determination is mandatory.

The fetal sex is determined at the time of fertilization, namely XX or XY. There are seven types of sex determinants influencing the outcome:

1: Chromosomal sex
2: Gonadal sex
3: Hormonal sex
4: Phenotypic sex
5: Sex of assignment
6: Sex of rearing
7: Psychological sex

It is important that all these determinants of sex parameters must be in unison. Any discord in these parameters will result in disorders of sexual differentiation (DSD) or Intersex. Low in the hierarchy of evolution the last three parameters do not have a role, however in the humans the addition of psychological sex along with sex of assignment and sex of rearing plays an important role. Any discrepancy in two or more parameters gets classified as "intersex" or "disorders of sexual differentiation."

In the mammalian evolutionary process, it is stated that "every fetus is destined to be a female, additional efforts are to be made to change into a male fetus." The famous experiments done by Alfred Jost [1] in 1947 finally led to the identification of SRY gene and Anti-Mullerian Hormone and their role in sex differentiation. SRY gene, present on Y chromosome, is responsible for development of the testes. The cellular fate of testis or ovary must be maintained in the adult life and is done by DMRTI (Doublesex and mab-3 related transcription factor 1) and FOXL2 (Forkhead box L2) [1]. The role of placenta is extremely important in the development of human brain and differentiation into two distinct patterns. There exist two separate patterns of male and female brain.

Allocation of sex to a newborn appears to be a simple task; however, diagnosis of DSD presents a unique challenge. This challenge is to provide immediate surgical treatment in some cases as well as long-term management. These are relatively rare conditions requiring not only multidisciplinary care but involve parental contributions and patient's psychological inclination.

Most of the DSD cases are first consulted by the pediatricians or the pediatric surgeons and approach a gynecologist much later part of childhood or in adolescent period. In fact, it should be the obstetrician who must be the first one to diagnose the DSD. The entire concentration and attention given during delivery probably results in the casual approach toward sex determination of the newborn child. Most of the pediatric surgeons do not involve the gynecologist in planning the correction surgery as these cases involve immediate management of uro-rectal problems which may be life threatening. The genital problems always seem to have a back seat approach. In the recent years the newer investigative tools like USG, antenatal screening tests, DSD may present very early in the first or second trimester of pregnancy. The advent of chorionic villous biopsy and amniocentesis can diagnose DSD and pose a problem in decision making.

50.2 McKusick-Kaufman Syndrome

McKusick-Kaufman syndrome (MKS) was first described by McKusick in the Amish population in 1978 and is characterized by polydactyly, hydrometrocolpos, and congenital heart disease. They usually present with a large cystic abdominal mass arising out of the pelvis as a result of dilated vagina and uterus with cervical secretions. Vagina is absent in lower one third. The classical finding of polydactyly where sixth digit is present on the ulnar side of the hand and the fibular side of the foot. A variety of congenital heart defects may be present in this syndrome. MKS is inherited as autosomal recessive gene.

Case capsule: *A rare case of McKusick-Kaufman syndrome presenting with hematometrocolpos with amenorrhoea.*

A 13-year-old was referred with complaints of cyclic pain in abdomen since 3 months, initially spasmodic changed to continuous type. She was born to a first-degree consanguineous couple. On the second day of life, she was operated for hydrometrocolpos with an abdominal Malecot's vaginostomy. At the age of 4 years, fistula repair was done for a discharging infraumbilical sinus and at 7 years of age, she underwent an ileal vaginoplasty.

*On examination she had frontal bossing, high arched palate and polydactyly was notable (*Fig. 50.1*). Secondary sexual characters were normal for her age with breast Tanner stage II. Her abdomen was soft but badly scarred. On local*

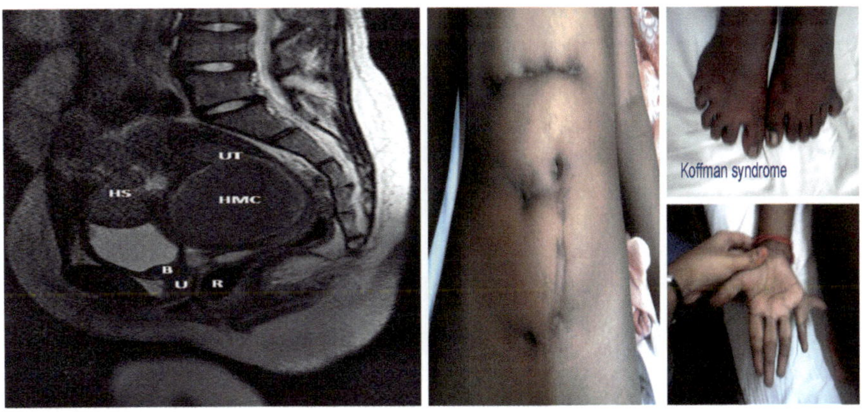

Fig. 50.1 McKusick-Kaufman syndrome

examination, her vulva, labia majora, minora, and clitoris were normal in appearance. The urethra and anus were in apposition and no vaginal orifice was seen. Single dilated opening confirmed to be hyperdilated urethra. There was no sign of vaginoplasty. On USG large hematometra, hematosalpinx, and hematocolpos were present.

*In view of multiple abdominal surgeries and a failed ileal vaginoplasty, the decision was made to operate per vaginum. We performed a unique difficult multistage operation through vaginal approach to give her full functional vagina having adequate depth. Being a perineal approach, this surgical technique was with least morbidity. She had an uneventful post-operative course and resumed her menstruation per vaginum. The patient is doing fine with 7 years of follow-up. Genetic evaluation revealed mutation in MKS gene at locus p.Arg517Cys (*Fig. 50.2*).*

Case capsule: Partial Mullerian agenesis presenting with gangrenous hematometra, hematosalpinx, hemoperitoneum, bowel obstruction and acute abdomen.

A 17-year-old unmarried girl presented with acute abdomen, pelvic mass, fever with rapid thready pulse and hypotension. She was having vomiting and colicky abdominal pain with gradual distension of the lower abdomen with a tense tender mass corresponding to 30 weeks of pregnancy. Bowel sounds were absent. The secondary sexual characters were well developed with absence of menarche.

*The resuscitation was carried out with IV fluids and blood transfusion. On laparotomy a massive hemoperitoneum of almost 1 liter with huge clots were drained. The hematometra and hematosalpinx were gangrenous and not salvageable (*Fig. 50.3*). The small bowel loops were dilated and adherent. Endometriotic cysts from both the ovaries were removed. Extirpative surgery was performed to save her life. Vaginoplasty was also done at the same sitting to drain peritoneal cavity and for future coital function. This patient got married after 4 years and had a satisfactory conjugal life.*

These cases highlight the vagaries in Mullerian agenesis, both in type and time of presentation, which poses difficult questions in terms of management. The treatment protocol cannot be generalized but is based on individual clinical scenario.

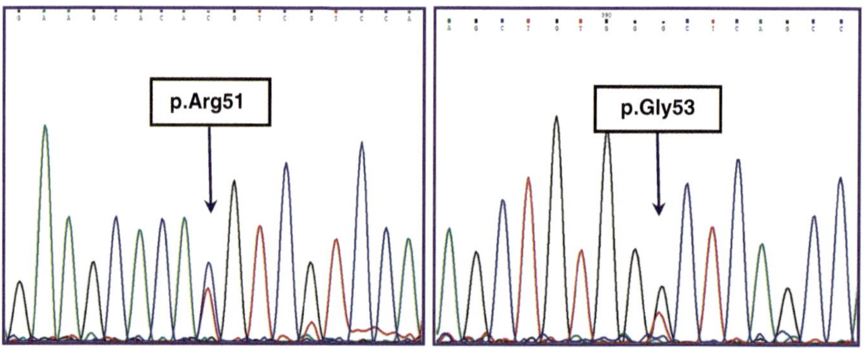

Fig. 50.2 Genetic mutation in MKS gene at locus p.Arg517Cys

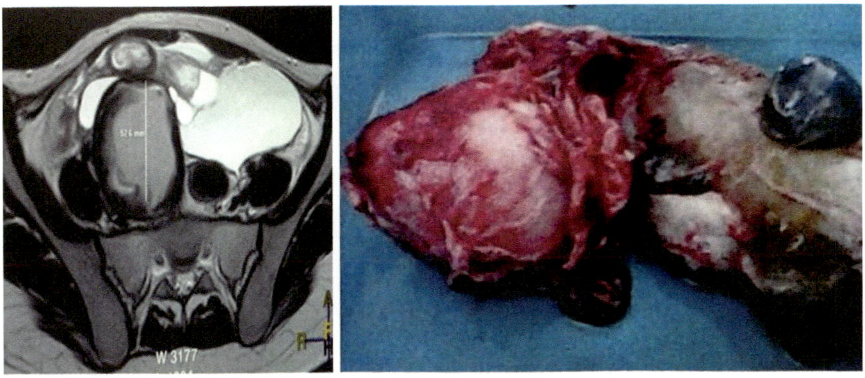

Fig. 50.3 Gangrenous hematometra and hematosalpinx

50.3 46XX DSD due to Aromatase Deficiency

The aromatase or P450 oxidoreductase deficiency is an autosomal recessive disorder having a unique pathophysiology leading to different phenotypic presentation. The world literature has 37 documented cases so far. Aromatase enzyme is mainly located in the endoplasmic reticulum of the estrogen-producing cells such as the placenta and ovary and is expressed in the other tissues such as Leydig cells of the testis, brain, adipose tissue, muscles, and hair follicles [2–7].

Aromatase deficiency can be associated with both XX and XY DSD. In XX DSD, it may present as ambiguous genitalia in newborn or in an adolescent with amenorrhea, hypergonadotropic hypogonadism with polycystic ovaries, and mild to moderate genital and skeletal malformations. They are born with clitoromegaly and the labial folds may appear scrotalized. Various degrees of posterior fusion may be present. There may be a urogenital sinus with single perineal opening.

Case capsule: *A 16-year-old girl presented with primary amenorrhea and non-development of secondary sexual characters. At birth, the infant had ambiguous genitalia where phallus was present with only one single orifice on the perineum. The baby was assigned female sex at birth. There was no evidence suggestive of adrenal insufficiency syndrome like persistent vomiting, dehydration, feeding problems, or failure to thrive.*

The phenotypic appearance was like a Turner's syndrome with short stature, cubitus valgus, and absence of secondary sexual characters. She showed failure of pubarche, adrenarche, and menarche. Gonads were not palpable in the inguinal region and a clitoromegaly of 2.5 cm was present. The urethral orifice was hidden in the pouch behind fused labia. Vaginal orifice was absent. There was complete posterior fusion of labia majora in midline with absence of labia minora. Anus was normally placed (Fig. 50.4).

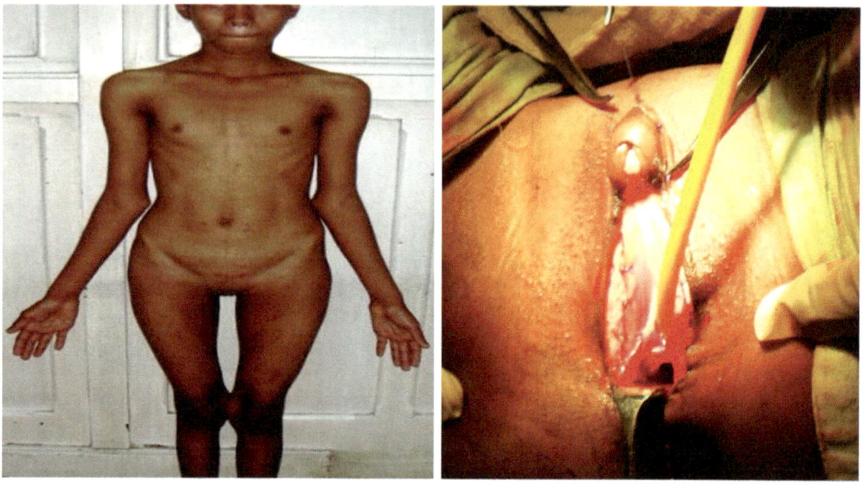

Fig. 50.4 Aromatase deficiency, clitoromegaly, absent secondary sexual characters

USG and CT pelvis showed a small-sized uterus (2.5 × 1.9 × 1.5 cm) with endometrial thickness of 1.9 mm and bilateral multiple ovarian cysts, the largest being 8.4 × 6.4 × 5.0 cm. Hormonal evaluation showed a raised serum FSH (32.93 mIU/mL) and serum LH levels (79.03 mIU/mL). However, estradiol (E2) was below 10 pg/mL (21–312 pg/mL). Hormonal analysis confirmed the presence of hypergonadotropic hypogonadism.

Karyotyping by FISH analysis was suggestive of 46XX chromosomes, negative for SRY Gene. One single gene mutation p.V80Vin CYP19A1 (Aromatase) gene was detected. CYP19A1 showed a single nucleoside polymorphism (SNP) (rs4646) while three SNPs (ID: rs3740397, rs284849, and rs10883783) were detected in the CYP17A1 (17 Hydroxylase) gene. In this case, there was no evidence of glucocorticoid or mineralocorticoid deficiency, 17-hydroxy progesterone, progesterone, testosterone, and DHEA levels were within normal limit. Once CAH (Congenital Adrenal Hyperplasia) was excluded as the cause for genital ambiguity, the diagnosis of non-adrenal female pseudo-hermaphroditism was made. All these investigations pointed toward the clinical diagnosis of 46XX DSD due to aromatase deficiency.

On laparoscopy, uterus was hypoplastic about 3.5 × 2.5 × 2.5 cm. Both ovaries showed multiple cysts filled with clear fluid. Left ovary showed the largest cyst (8 × 6 cm) with hemorrhagic fluid. Gelatinous yellowish fluid was present in pouch of Douglas. Genital examination revealed a small urethral opening 2 cm below the base of clitoris near the urogenital sinus.

The labia majora were separated in the midline and a midlevel clitoridectomy was performed. On further dissection, vagina and cervix were visualized. The edge of vagina was sutured to the free margin of labia majora. The superficial transverse Perini, bulbocavernosus, and ischiocavernosus muscles were identified. The labia majora was reconstructed over the bulbocavernosus muscle. Post-operative period was uneventful. She was started on cyclical Ethinyl-estradiol 50 mcg for 6 months with Medroxyprogesterone 10 mg twice a day to be given from day 16 to 25 of each cycle. After 6 months of follow-up, patient had normal menses and showed breast development of Tanner Stage III, the ovaries were of normal size, and uterine size had increased measuring 4.5 × 3.5 × 3.5 cm with endometrial lining of 8 mm.

50.4 Utero-Vaginal Hypoplasia or Aplasia DSD

The most common of this group is Mayer-Rokitansky-Küster-Hauser (MRKH) syndrome. Patients with ambiguous genitalia as in congenital adrenal hyperplasia or complete androgen insensitivity syndrome with vaginal hypoplasia or agenesis are relatively rare. Patients for intersex surgery also may be included in this group.

MRKH is not an uncommon syndrome and affects 1 in 4500 live born females. Although Muller described this condition almost 200 years ago, Mullerian duct has remained an enigma. Three types of MRKH have been described—Type 1 (Pure MRKH syndrome), Type 2 (MRKH with renal, cardiac, and skeletal anomalies) and Type 3 (MRKH with hyperandrogenism). Most of the parents consider these infants as transgender and try to shun away from the medical assistance. Ignorance and non-availability of specialist in the field is a major cause.

A "novel technique of vaginoplasty developing normal vagina" for MRKH syndrome and identification of the progenitor cell and role of translational genomes—Many techniques of vaginoplasty both surgical and non-surgical have been described. Unfortunately, none results in normal vagina. We present a novel technique of vaginoplasty (based on Davydov's surgery) which has shown metaplastic conversion of peritoneum to normal vagina. These changes have been recorded in a sequential form (Fig. 50.5).

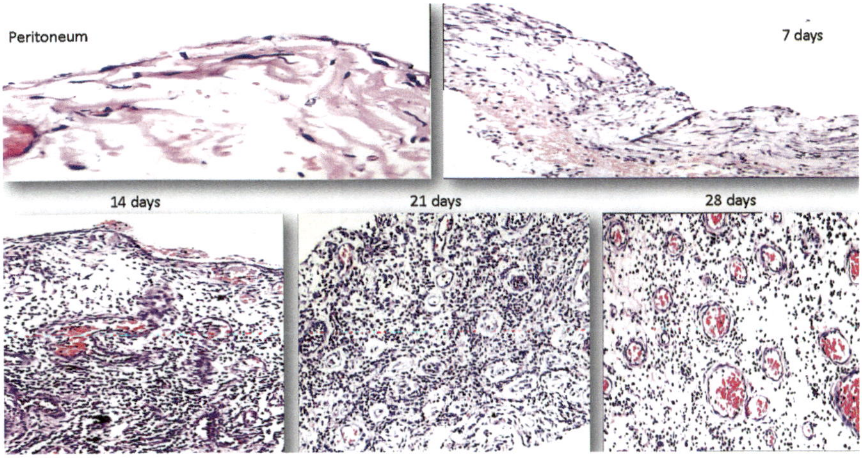

Fig. 50.5 Serial transformation of peritoneum to normal vagina

The hypothesis behind this technique is that the Mullerian ducts develop by invagination of the peritoneum and hence have the same parentage. The most important fact is the established metaplastic ability of the peritoneum to convert itself into variety of tissues like endometrium (endometriosis), cartilage, mucinous lining of intestine, and many more.

The progenitor cell producing this change has been identified. The concentration and distribution pattern of these progenitor cells at various stages has also been recorded. The expression of the specific genes (WNT4, WNT5A, and WNT7A) and their roles in the formation of neovagina have been documented. Activating these genes by genetic engineering may lead to the treatment of developmental defects of Mullerian duct and developing the rudimentary uterus to the normal size, obviating the need for a uterine transplant.

On follow-up, the neovagina was subjected to anatomical and physiological parameters confirmation, i.e., histology, pH, cytology, Doppler blood flow, and 3-D ultrasound examination. All the parameters confirmed normal vaginal development by metaplasia of peritoneum (Fig. 50.6). Normal vaginal lubrication was present. Being a laparoscopic and perineal surgery, the patient does not have any surgical scar. An excellent sexual satisfaction score is achievable and to the best of our knowledge, this study was the first to show metaplastic conversion of peritoneum to normal vagina, identifying the progenitor cell (Fig. 50.7), expression of translational stemness markers, and upregulation of specific genes [8–10].

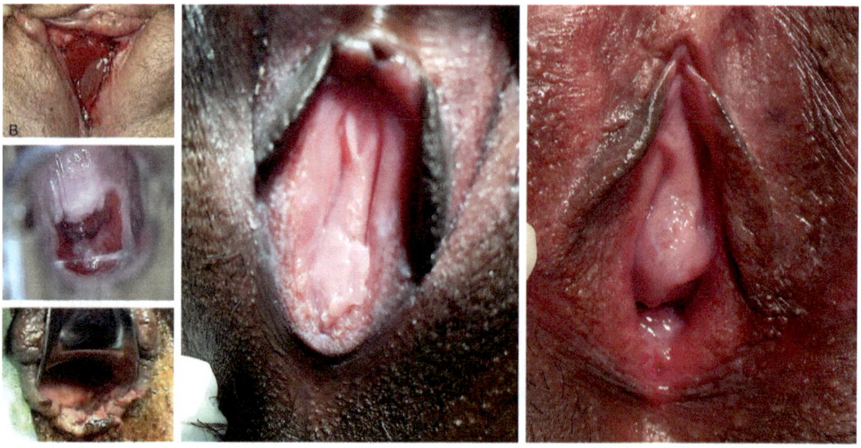

Fig. 50.6 Pre-operative and post-operative pictures showing normal vagina formation in contrast to sigmoid, ileal, and skin vaginoplasty

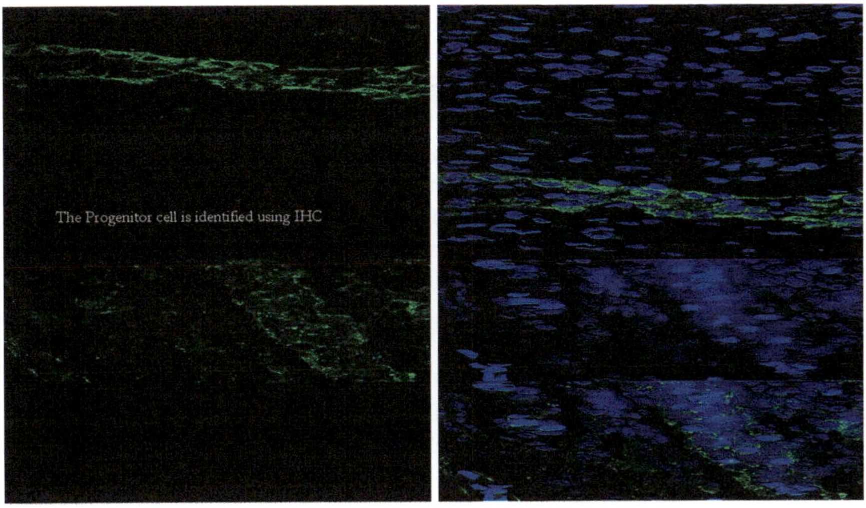

Fig. 50.7 Identification of progenitor cells in neovagina by SOX2 and OCT4

50.5 Androgen Insensitivity or Testicular Feminization 46XY DSD

Inhuman exposure to endocrine affecting chemicals is associated with androgen insensitivity syndrome. Androgen insensitivity is of three types:

Type 1—Complete having normal female external genitalia with total absence of Wolffian duct organs.
Type 2—Has normal male genitalia with gynecomastia and impaired spermatogenesis.
Type 3—Milder form.

This is a unique situation having normal chromosomes (46XY), normal gonads (testes) but due to insensitivity to the produced androgen, there is a lack of androgen response and a female phenotype. Virilization does not occur in these cases and puberty progresses normally without the need for any exogenous estrogen. There is a low malignancy risk of 5%. Most of these are reared as females. Fertility in these cases is extremely controversial due to moral and ethical dilemmas. The decision regarding gonadectomy is very difficult and controversial. Hence, it should be postponed until adolescence.

50.6 46XY Pure Gonadal Dysgenesis

In the cases of 46XY pure gonadal dysgenesis, presentation of external genitalia and gender is always female. The gonads may be testis or ovotestis. Dysgenetic gonads do not produce hormones and have no potential for fertility. Another major concern in these patients is the development of gonadoblastoma, sex cord tumors, and mixed germ cell tumors. The risk of gonadoblastoma is extremely high and increases proportionately with age. By the age of 30 years, it may be as high as 30%. Hence, prophylactic gonadectomy is advised to prevent the development of gonadoblastoma.

Gonadectomy should be performed by the age of 10 years and followed up with 6 monthly USG and tumor markers. The gonadectomy is nowadays performed laparoscopically, making it cosmetically acceptable. These patients are treated with synthetic hormone replacement to achieve secondary sexual characters, growth spurt and for maintaining bone density [11]. They have a normal uterus and vagina and can achieve fertility with in vitro techniques. However, successful full-term pregnancies are not achieved for unknown reasons, most probably due to uterine myometrial weakness.

Five such patients were managed with laparoscopic gonadectomy and were keen to have ovarian transplant. They had to be turned down due to ethical dilemmas raised by the transplant authorities. All five 46XY DSD patients were very tall, lean, athletic with significant fronto-parietal bossing and the behavioral pattern was male. All of them were reared as females [12].

50.7 45XO DSD

This is one of the commonest human chromosomal defects having an incidence of 0.8–1% of all the pregnancies. However, almost 95% of them lead to abortions. The actual incidence is about 1 in 2500 live newborn females. Many subtypes are described—45 XO, Mosaics of 45XO-46XX-46XY and 46XX with a deletion in part of one of the X chromosomes. In literature pure 45XO appears to be common form, in our series of 150 cases, mosaic form appears to be the commonest (Unpublished data).

A classical syndrome of XO or Turner's has a very typical phenotypic presentation making the diagnosis easy. However, with 3D-4D USG it is possible to diagnose these very early in the antenatal period. Lymphangiectasis resulting in to large septate cystic hygromas, nuchal fold thickening are diagnostic features. Intrauterine growth retardation, short femur, body lymphangiectasis, or cardiac defects should raise the suspicion of 45XO DSD [13, 14].

An analysis of five patients of midtrimester termination with 45XO revealed almost all of them having lymphangiectasias. The gonads were subjected for histological studies to find the primordial follicle count; however, the data was not conclusive (Unpublished data).

The commonest concern of these patients is short stature, absence of secondary sexual characters, and primary amenorrhea. If diagnosed early, growth hormone treatment is very helpful but in later part of life, it has a limited role to play. Use of sequential estrogen and progesterone therapy helps in developing secondary sexual characters and preventing osteoporosis. Theoretically, an ovarian transplant should be the ideal treatment.

We performed the world's first successful ovarian transplant on 29th March 2002 after more than 20 years of research involving animal experiments, renal transplant observership, vascular and microvascular training and finally having registered as gynecological transplant surgeon. This was done after proper government sanctions, developing a transplant unit, and following all the legal rules and regulations for such transplants. It was a vascular, orthotopic transplant from a live donor to a girl with Turner's syndrome and the results were very encouraging. The patient was given post-operative immunological suppression with methylprednisolone and cyclosporin. The transplanted ovary continued to work for nine years. Hormonal production began almost immediately and she started ovulation from second month onward. Within three years of transplant, she achieved a growth spurt and developed secondary sexual characters (Fig. 50.8).

*Surgical steps for an ovarian transplant (*Fig. 50.9*)—The donor dissection is carried out with a small Pfannenstiel incision. Extraperitoneal dissection is carried out to obtain a large ovarian pedicle. The ovary is removed after heparin injection and is washed and prepared in transplant solution. All the blood components are removed with help of catheterization. The ovarian artery and ovarian vein suitable for anastomosis are identified and the rest of venules are clipped.*

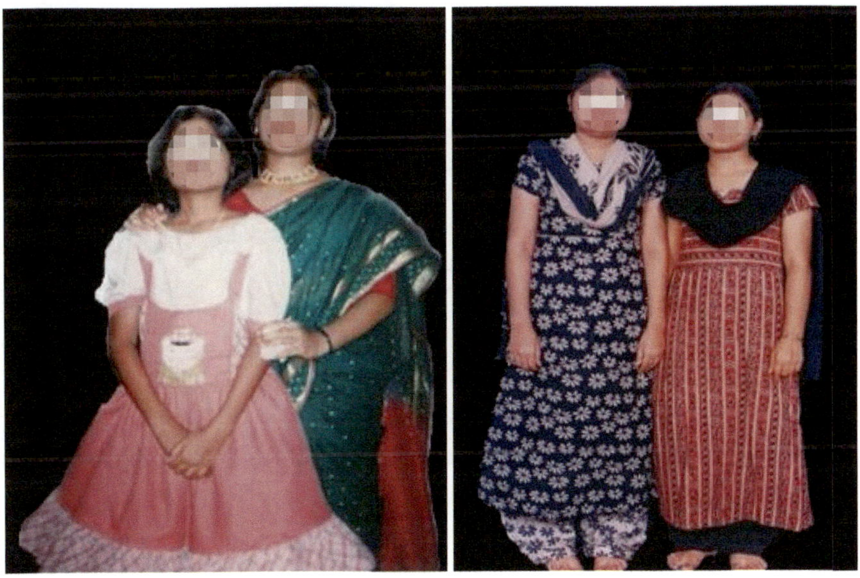

Fig. 50.8 Two years pre-transplant and three years post-transplant showing the recipient and donor

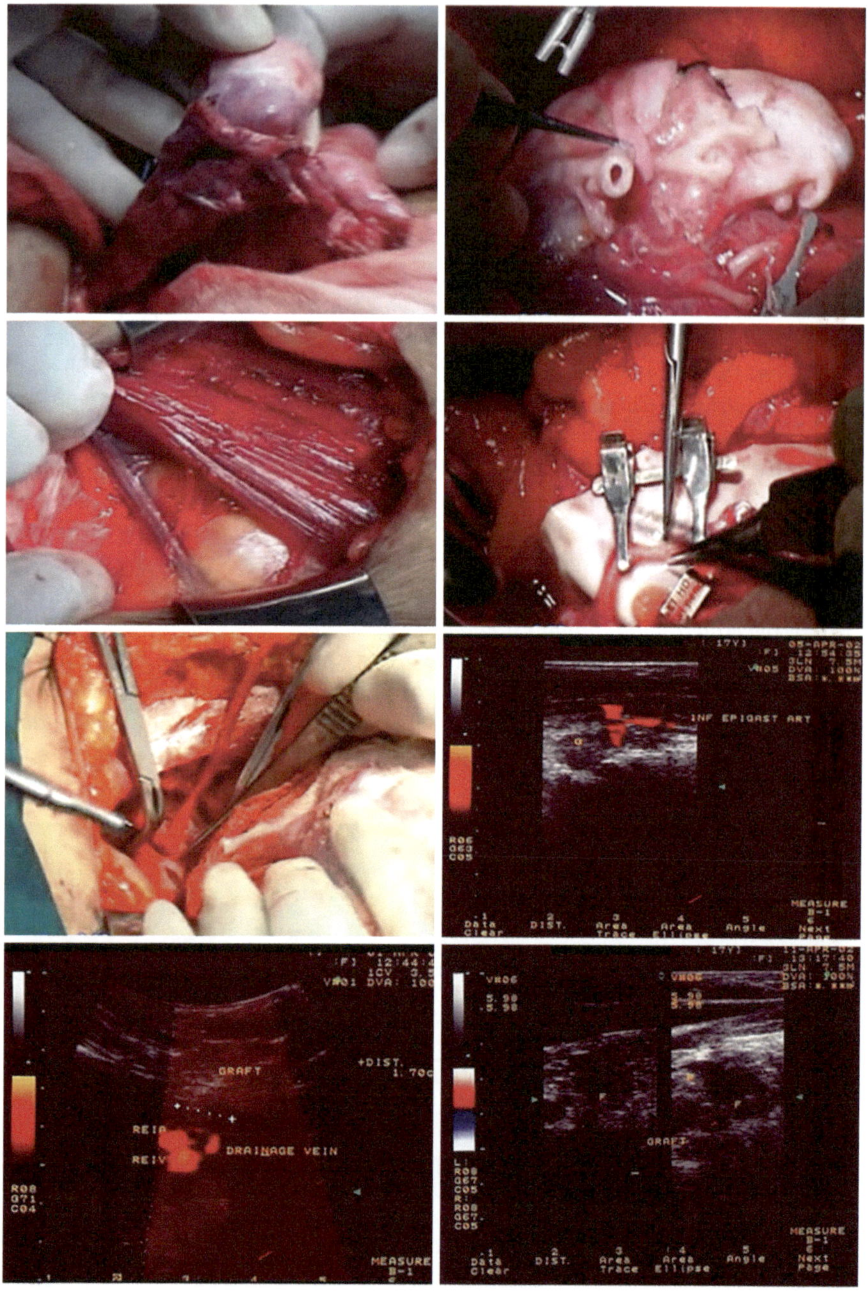

Fig. 50.9 Live donor ovarian transplant

The recipient extraperitoneal space is dissected with a reverse J-shaped incision. The inferior epigastric artery is dissected. The ovarian arteries are anastomosed side to side, interrupted 10–0 or 12–0 vascular sutures under operating microscope. The donor ovarian vein is anastomosed end to side with the recipient external iliac vein using continuous 8–0 vascular suture. After the vascular anastomosis, the peritoneum is opened and the transplanted ovary is placed orthotopically. Blood flow is assessed with Doppler and ovulation is confirmed on USG. After a successful transplant, the gonadotropic hormonal levels show downward trend and the ovarian hormones show normal values post-operatively. Another technique of avascular grafting is described for use in patients with unsuitable vasculature.

50.8 Gartner's Duct Cyst

These are remnants of Wolffian ducts and are present in normal XX females. Although they do not have any problem of DSD, they get categorized under the heading of DSD as they have a persistent Wolffian duct remnant. These are asymptomatic and vary from small anterolateral vaginal cysts to large broad ligament cysts. Simple removal suffices as a treatment modality, care to be taken to avoid damage to ureter [15–21]. In utero exposure to 2,3,7,8-tetrachlorodibenzo-p-dioxin leads to Wolffian duct remnants in rats; however, its correlation in humans is not documented.

50.9 Conclusions

This article discusses the gynecological perspective of management of DSD, a complex entity with overlapping views of multidisciplinary specialists. The multitude of problems encompassed in DSD present themselves at different age groups and require very specific management protocol. The attempt is made to highlight it by case studies. Some extremely rare cases like aromatase deficiency and McKusick-Kaufman syndrome are presented. The newer concepts of ovarian transplant and novel vaginoplasty are also discussed. The surgery of vaginoplasty must avoid the multistage surgeries using colon, small intestine, or skin and adopt the newer technique described resulting in formation of normal vagina.

The role of each specialist cannot be individualized but must be in joint consultation with others. The gender assignment and corrective surgeries require parental involvement and long-term care in childhood, adolescent, and adult life. The cases like Turner's syndrome requiring newer treatment like ovarian transplant must have consultation with endocrinologist and psychologist. The role of counsellor is very important in educating the patient and their relatives in understanding the concept of transplant, finding a suitable donor, possible complications, need for immunosuppression, and the legalities.

The counsellor plays an important role in sexual counselling of these young girls and guides them regarding possible menstrual function and fertility by using surrogacy. The subject of timing of surgery depends on the age at which the girl presents. It may be newborn with cloacal defects requiring immediate intervention or a girl in her adolescent age. Many girls present themselves at the time of marriage creating a social emergency. Surgery is advocated to any girl beyond age of 17 where the girl has attained enough maturity and before they are subjected to mental depression due to social or peer pressure. The comprehension of treatment protocol and ability for post-operative care must be judged carefully. Latest research on progenitor cells and translational genomes in the vaginoplasty and ongoing research in developing other reproductive organs are also discussed.

References

1. Huang S, Ye L, Chen H. Sex determination and maintenance: the role of DMRT1 and FOXL2. Asian J Androl. 2017 Nov–Dec;19(6):619–24.
2. Inkster SE, Brodie AMH. Immunohistochemical studies of aromatase in early and full-term human placental tissues: comparison with biochemical assay. Biol Reprod. 1989;41:889–98.
3. Kelch RP, et al. Estradiol and testosterone secretion by human, simian and canine testes, in males with hypogonadism and in male pseudohermaphrodites with feminizing testes syndrome. J Clin Invest. 1972;51:824–30.
4. Ryan KJ, Naftolin F, Reggy V, et al. Estrogen formation in brain. Am J Obstel Gynecol. 1972;114:454–60.
5. Schindler AE, et al. Conversion of androgens to estrogens by human fat tissues. J Clin Endocrinol Metab. 1972;35:627–30.
6. Longcope C, et al. Aromatization of androgens by muscle and adipose tissue in vivo. J Clin Endocrinol Metab. 1978;46:146–52.
7. Schweikert HU, et al. Aromatization of androstenendione by isolated human hairs. J Clin Endocrinol Metab. 1975;40:413–41.
8. Mhatre P, Dighe V, Das DK, Pawar A. Novel technique of vaginoplasty developing normal vagina, role of stemness markers and translational genes. J Hum Reprod Sci. 2020;13:303–9.
9. Mhatre PN, Narkhede HR, Pawar AP, Mhatre JP, Kumar DD. Role of progenitor cell producing normal vagina by metaplasia in laparoscopic peritoneal vaginoplasty. J Hum Reprod Sci. 2016;9:215–22.
10. Mhatre P, Mhatre J, Sahu R. New laparoscopic peritoneal pull-through vaginoplasty technique. J Hum Reprod Sci. 2014;7:181–6.
11. Liu AX, Shi HY, Cai ZJ, Liu A, Zhang D, Huang HF, Jin HM. Increased risk of gonadal malignancy & prophylactic gonadectomy in 102 phenotypic female patients with Y chromosome. Hum Reprod. 2014 Jul;29(7):1413–9. https://doi.org/10.1093/humrep/deu109.
12. Fedder J, Ørskov M. Pregnancies in XY women: is a weak myometrium cause of complications during pregnancy and delivery? Asian Pac J Reprod. 2013 Mar;2(1):76–9.
13. Sybert VP, McCauley E. Turner's syndrome. N Engl J Med. 2004;351:1227–38.
14. Entazami M, Albig M, Gasiorek-Wiens A. Chromosomal disorders: turner syndrome. Ultrasound diagnosis of fetal anomalies. New York, NY: Thieme Medical; 2004. p. 211–4.
15. Bedaiwy MA, Shahin AY, Falcone T. Reproductive organ transplantation: advances and controversies. Fertil Steril. 2008 Dec;90(6):2031–55.
16. Mhatre P. The first orthotopic ovarian transplant in Turner's Syndrome. Int J Gynaecol Obstet. 2003;83(3).
17. Mhatre P. The first orthotopic ovarian transplant in Turner's syndrome. Int Pediatr. 2003;18(4).

18. GUNAIKEIA –The First World Congress Ovarian Cryopreservation and Ovarian Transplantation The first orthotopic ovarian transplant in Turner's syndrome. 2003 June;8(5).
19. The Transplantation (XX International congress of Transplantation Society). The First orthotopic ovarian transplant in Turner's syndrome. 2004 July 27;78(2).
20. The Transplantation Proceedings (MEDLINE/PUBMED). Ovarian transplant - a new frontier. 2005 March;37(2):1396–8.
21. Mhatre P, Mhatre J. Orthotopic ovarian transplant--review and three surgical techniques. Pediatric Transplant. 2006 Nov;10(7):782–7.

Malignancies in Disorders of Sexual Development

51

Priti S. Mehta and Annesha Chakraborti

Abstract

The evolution of malignancies in DSD (Disorders of Sexual Differentiation) is rare and multifactorial. Aberrant germ cell development usually leads to the development of a malignancy. The gonads of children with DSD often have insufficiently differentiated supporting cells which hinders complete maturation of germ cells. Delayed gonadal development increases the risk of tumorigenesis due to the prolonged state of gonocyte immaturity. The immature germ cells abundantly express TSPY (testis-specific protein Y-encoded), which further promotes survival and proliferation. Presence of gonadal dysgenesis along with surviving germ cells places the child at a high malignant risk that increases with age.

The exact risk of malignancies in each subtype of DSD is not known. The lowest risk is seen in those with Turner Syndrome without Y chromosome, complete AIS (Androgen Insensitivity Syndrome), and ovotesticular DSD. The highest risk is seen with gonadal dysgenesis, partial AIS, and syndromic associations such as Denys Drash and Frasier syndromes.

Imaging is not very sensitive to identify masses in this population as the gonads are often aberrantly located, of heterogenous size and appearance, and maldeveloped or dysgenetic. Monthly self-examination is recommended for patients with palpable gonads, as a screening tool for testicular cancer. Tumor markers are another option for screening, but there is a lack of evidence and no set cut-offs for DSD patients. One of the latest advances in screening for Germ cell Tumor (GCT) is the development of microRNA (miRNA) testing.

P. S. Mehta (✉)
Pediatric Hematology Oncology BMT, NH SRCC Children's Hospital, Mumbai, India

A. Chakraborti
Tata Memorial Hospital, Mumbai, India

S. K. Ratan (ed.), *Children with Differences in Sex Development*,
https://doi.org/10.1007/978-981-97-1639-5_51

There are no consensus guidelines regarding the decision and timing of gonadectomy. Although ideal for the prevention or cure of an early stage invasive GCT, it has its drawbacks in the form of permanent infertility and hypogonadism. The final decision depends not only on the predicted risk of GCT but on the child's gender identity as well as desire for fertility.

Keywords

Gonads · Malignancy · Imaging

51.1 Background

Disorders of Sex Development (DSD) can predispose to germ cell tumors. However, there is not enough information available about DSD as they are not only rare but the circumstances of their discovery are not well known as well [1].

51.2 Introduction

Disorders of Sex Development (DSD) can be defined as a group of congenital conditions with abnormal development of the internal and/or external genitalia [2].

The estimated incidence of DSD is 1 in 4500–5500 births [3]. Based on Karyotyping the Chicago group has classified DSD into 3 major groups: the sex chromosome DSD, the 46 XY DSD, and the 46XX DSD [4]. The evolution of malignancies in DSD is quite rare and multifactorial, depending on the genetic makeup, morphology, histopathology, and expression of tumor markers. The most important risk factor is the expression of TSPY gene on the Y chromosome [3].

The increased risk of malignancies associated with DSD was first documented in the 1890s [5]. The manifestations range from precursor lesions, i.e., germ cell neoplasia in situ (GCNIS) and gonadoblastoma (Gb) to malignant invasive germ cell tumors (GCT), including seminomas/dysgerminomas, non-seminomas (embryonal carcinoma, yolk sac tumor, choriocarcinoma, and teratoma) [3] and rarely, stromal and epithelial tumors [6]. Data on the exact incidence of these tumors in this heterogeneous population is scarce due to the rarity of this condition. Screening and surveillance measures have to be individualized. Thus, it is difficult to devise a stratified approach to identify children at greater risk and further to recommend gonadectomy versus safe retention of their gonads [3].

51.3 Pathogenesis of Germ Cell Tumors in DSD

Primordial germ cells (PGC) are detectable at 5–6 weeks of gestation by placental alkaline phosphatase (PLAP) and OCT3/4 (octamer-binding transcription factor) staining. OCT3/4 and NANOG are transcription factors, known to enhance pluripotency, survival, and proliferation while inhibiting apoptosis. The migration of PGCs, from the yolk sac along the hindgut toward the genital ridges, is mediated primarily by SCF (stem cell factor) and its receptor c-KIT. Once within the genital ridges, the PGCs are called gonocytes and the structures are called indifferent gonads, which will differentiate into testes or ovaries depending on the karyotype [6].

Testicular differentiation is driven by SRY, SOX9 via several transcription factors including NR5A1 and ZFPM2, all of which can be mutated to cause DSD [2]. Ovarian development is enabled by absence of the SRY region as well as by signalling pathways including WNT4/β-catenin, FOXL2, and RSPO1 [2]. In the normal process of maturation, the spermatogonia and oogonia loose expression of gonocyte markers such as OCT3/4, PLAP, c-kit, NANOG, etc. [6]. However, aberrant germ cell development, at any step, can lead to the development of a malignancy. The gonads of children with DSD often have insufficiently differentiated supporting cells (Sertoli/granulosa cells), which hinders the complete maturation of germ cells [2]. As a result, these immature germ cells continue to express fetal markers like Oct3/4 and NANOG which enable their survival and proliferation [5]. Additionally, they also abundantly express TSPY (testis-specific protein Y-encoded), which further promotes survival and proliferation [6]. DSDs can also be associated with delayed gonadal development, which in turn can increase tumorigenesis due to the prolonged state of gonocyte immaturity [2].

51.3.1 Premalignant Lesions

Little is known about the natural history of gonadoblastomas, the premalignant lesions which develop from germ cells that are committed to the female pathway but get arrested in the primordial follicular stage [3]. They comprise a mixture of large germ cells resembling seminoma and small sex cord stromal cells, such as immature Sertoli and granulosa cells. The expression of TSPY gene (GBY) is the primary factor for its development [3]; an estimated 50–60% progress to dysgerminoma/seminoma and 10% progress to non-seminomatous tumors [7]. It can present in early childhood, especially in 46XY gonadal dysgenesis and mosaic Turner syndrome with Y chromosome [3].

GCNIS, on the other hand, occurs in DSDs with testicular tissue. Similar to the development of gonadoblastomas, GCNIS arises from arrested development of fetal gonocytes in seminiferous tubules [3]. It is estimated that GCNIS is present from birth and will always eventually evolve into malignancy, at a mean age of 20-35 years [2].

51.4 Risk Factors for Tumorigenesis

Presence of gonadal dysgenesis places the child at high risk as opposed to morphologically normal testis or ovaries, as does the presence of surviving germ cells in comparison with their absence (detected on biopsy or gonadectomy specimens) [2].

The risk also increases with age [2]. The expression of embryologic germ cell markers beyond infancy such as OCT3/4(POU5F1) and KITL (stem cell factor) is also associated with increased risk of malignancies [3].

51.5 Risk Stratification and Indications of Gonadectomy

The exact risk of malignancies in each subtype of DSD is not known. Data on malignant transformation of gonadoblastomas and GCNIS is also scarce. However, the risk of developing a GCT is extrapolated from data on gonadoblastomas and GCNIS identified on gonadectomy and biopsy samples [2].

The lowest risk is seen in those with Turner Syndrome without Y chromosome, complete androgen insensitivity syndrome, 46XX ovotesticular DSD, and 46XY ovotesticular DSD, wherein the overall risk is <5% [3]. The highest risk is seen in gonadal dysgenesis, partial androgen insensitivity syndrome (PAIS), and syndromic associations such as Denys Drash and Frasier syndromes [3].

Gonadectomy, while ideal for the prevention or cure of an early stage invasive GCT, has its cons in the form of permanent infertility and hypogonadism. Ideally, low risk patients can be kept on surveillance, so that hormonal and reproductive function are preserved. However, guidelines for surveillance have thus far not been unanimously developed [3]. The final decision for gonadectomy is multifactorial, depending not only on the predicted risk of GCT but also on the child's gender identity as well as their desire for fertility [2].

51.5.1 Androgen Insensitivity Syndrome (AIS)

The risk of GCT in AIS is inversely proportional to the severity of the mutation [2]. This apparent anomaly occurs because complete loss of androgen receptor function renders the testes completely devoid of germ cells. Thus, women with Complete AIS (CAIS) have pre-pubertal malignancy risk of 0.8–2% [8], which rises to 15% in adults [9]. As the gonadal tumor risk is low prior to puberty, gonadectomy can be delayed until after completion of breast development [10].

Children with PAIS, on the other hand, have some surviving germ cells, thus rendering a high risk of GCT (50%) for non-scrotal gonads [10]. Hence, patients of PAIS should undergo gonadectomy, but it can be delayed till the child has developed gender identity as malignancies generally develop after the age of 14 years [10]. However, these children should undergo orchidopexy in the first year of life. There is no role of testicular biopsy at this time as gonadal tumors tend to develop after puberty when extensive bilateral biopsies may indicate the presence of carcinoma in situ (CIS)/intra-tubular germ cell neoplasia (ITGCN). Gonadectomy is recommended only in cases with a female gender identity or if CIS is diagnosed [10].

51.5.2 46,XX DSD

The most common DSD affecting this karyotype is Congenital Adrenal Hyperplasia (CAH), most commonly due to CYP21A mutations. These children have no increased risk of DSD. Rare variants of CAH, namely mutations involving CYP17A1 and HSD17B3, which present with elevated androgens, have rarely been associated with GCTs [2].

51.5.3 DSDs Without Chromosomal Abnormalities

There are rare case reports of uterine leiomyomas in children diagnosed with Mayer–Rokitansky–Kuster–Hauser (MRKH) or Persistent Mullerian Duct Syndrome (PMDS). These children do not qualify for prophylactic removal of these remnant Mullerian structures [11].

51.5.4 46,XY Gonadal Dysgenesis

Gonadoblastomas are the most common tumors seen in children with gonadal dysgenesis. Children with complete gonadal dysgenesis have no gonadal development and present with a female phenotype, with an associated cancer risk of 15–50%. On the other hand, partial gonadal dysgenesis is characterized by incomplete testicular development, wherein the external phenotype is determined by the extent of residual testicular function. These children have a malignancy risk amounting to 16–30% [10].

Children with gonadal dysgenesis can develop tumors at any time ranging from early infancy to post pubertal period. Due to the high risk of malignant transformation, especially in intra-abdominal gonads, as well as the lack of gonadal function, it is recommended that children with 46,XY complete gonadal dysgenesis should undergo bilateral gonadectomy on diagnosis. Boys with streak gonads should also be offered early gonadectomy due to the risk of malignancy. Evidence-based recommendations on the indications and timing of gonadectomy in partial gonadal dysgenesis are yet to be developed [10]. Dysgenetic gonads can be preserved in the scrotum and kept under surveillance as intra-scrotal malignancies are extremely rare in patients with XY partial gonadal dysgeneses.

51.5.5 Turner Syndrome

Turner Syndrome patients with Y-containing material, specifically the TSPY region, have a 12–40% risk of GCNIS/GB [2]. Although non-TSPY containing Turner Syndrome is technically considered a DSD, these patients are at negligible risk of GCT. A Y-positive Turner Syndrome patient with externalized scrotal gonads (versus abdominal) also has a relatively reduced GCT risk. The risk of developing a malignant GCT increases with age in such cases, but the role of prophylactic gonadectomy (at puberty) is still controversial. A clinical and radiological follow-up without gonadectomy may be a more prudent approach rather than a bilateral gonadectomy, which was once advocated in early childhood, because of insufficient functional ovarian tissue (<10%) and high risk of malignancy [10]. The current consensus recommends prophylactic gonadectomy only if marker chromosome is demonstrated together with the cytogenetic Y chromosome or if virilization is present [2]. The option of oocyte and follicle preservation should be offered to Turner Syndrome patients who undergo gonadectomy but wish pregnancy in the future [10].

51.5.6 Ovotesticular DSD

In patients with ovotesticular DSD, the gonad sometimes comprises both ovarian and testicular tissues, and external genitalia is usually ambiguous. Most patients (87%) have a 46,XX karyotype. Histological examination of an abdominal ovotestis shows normal ovarian tissue usually with a dysgenetic testis. If the testis has normally descended into the scrotum, it may not be dysgenetic. The lack of fertility issues can be explained by the histologically normal ovarian tissue [10].

There is very little evidence to predict later gender identity in children with ovotesticular DSD. The risk of developing a germ cell tumor is determined by the degree of gonadal differentiation and the presence of Y chromosomal material. The risk for malignancy is 2–10% in 46,XY ovotesticular DSD but very low in (46,XX) ovotesticular DSD [10]. Hence, the decision for gonadectomy in children affected with ovotesticular DSD depends on patient (and parent) preferences, hormonal and histological features, and to a lesser extent, fertility issues. Ideally, this decision is based on gender identity rather than on the sex of rearing and, thus, has to be postponed until late childhood, allowing close follow-up of the developing gender identity [10].

51.5.7 Extragonadal Cancer Predisposition in Dimorphic DSDs

Extragonadal GCTs arise from germ cells that fail to migrate appropriately to the genital ridge and then fail to apoptose. Along with shared underlying disease origin, mediastinal and gonadal GCTs also share genetic signatures like hypomethylation and 12p copy number increases [2]. Unlike gonadal GCTs, mediastinal GCTs have an increased rate of resistance to platinum-based therapy secondary to high expression of TP53, resulting in survival of only ~50% [12].

Klinefelter Syndrome (47,XXY) is associated with a specific risk of mediastinal GCT. Approximately 8% of men with mediastinal GCT have Klinefelter syndrome and it is important to screen cases of mediastinal GCT for subclinical Klinefelter [2]. In addition, individuals with Klinefelter syndrome have cancer risks thought to be associated with the additional X-chromosome, including a 19-58-fold risk of breast cancer versus 46,XY men [13]. In contrast, Turner Syndrome without TSPY is associated with an increased risk of a number of non-GCT cancers, especially CNS, bladder/urethral [14]. These syndromes are unique however, in that the associated malignancies are not gonadal GCTs, thus prophylactic gonadectomy is not recommended [3].

51.6 Screening

It is extremely difficult to screen for malignancy in patients with DSD using the available screening modalities due to their poor sensitivity and specificity [3]. Although not very sensitive, monthly self-examination is recommended for patients with palpable gonads, as a screening tool for testicular cancer [10]. Hence, boys with descended testes and DSD should be encouraged and taught to palpate their testes once a month from the initiation of puberty. But many DSD patients do not have palpable gonads, limiting their ability to detect any gonadal masses themselves. Alternative options to facilitate self-examination and early detection include relocating the gonads adjacent to laparoscopy port sites or the inguinal canal [2].

Unfortunately, imaging is also not sensitive enough to identify masses in this population [2]. The gonads are often aberrantly located, of heterogenous size and appearance, and maldeveloped (e.g., streak gonads) or dysgenetic. According to the available literature, only 40–50% of gonads in DSD patients are identified with ultrasound (US) or magnetic resonance imaging (MRI). In fact, US is more sensitive than MRI at identifying dysgenetic gonads. However, gadolinium-enhanced MRI has a better sensitivity and specificity than US at localizing non-palpable testes [2]. But neither study performs well at identifying premalignant lesions, making both poor screening tools [2].

Tumor markers are another option for screening, although interpretation of these values is based on extrapolated data on testicular and ovarian cancer in the general population [3]. There are no set cut-offs for DSD patients. Another disadvantage is missing those tumors that do not secrete any of the known tumor markers (i.e., teratoma, seminoma). Although AFP, β-hCG, and LDH are associated with GCTs, there are poor evidence and unknown implications for screening in patients with DSD [3].

One of the latest advances in GCTs is the development of microRNA (miRNA) testing. First discovered in 2011, miRNA clusters appear to have elevated expression in the presence of GCTs and then decline to normal values after surgical removal of the tumor. As a screening/surveillance method, it is more sensitive than the gold standard tumor markers and/or imaging [3].

OCT3/4 is a well-established pathologic marker for GCNIS, GB, and GCT in adults, but is complicated by expression in developing fetal testes. Given the testicular dysgenesis syndrome hypothesis of developmental delay, this is of particular importance and is thought to contribute to over-diagnosis [15]. However, the expression of OCT3/4 throughout the gonad differs between premalignant/malignant cells and developmentally delayed cells, which may improve future diagnosis accuracy. Stem cell factor is also used to differentiate malignant from developing germ cells. Gonadoblastoma in particular requires the presence of a specific portion of the Y chromosome—the GBY region. The TSPY gene (testis-specific Y-encoded) within the GBY region is expressed in GB. Both these lesions—OCT3/4 and TSPY—are also expressed in GCNIS lesions [15].

51.7 Outcomes

As DSD with malignant GCT is a rare entity, data on treatment outcomes are equally rare, but can be extrapolated from data on testicular and ovarian GCTs [2]. The lifetime risk of testicular GCT is 0.4% with a 10-year survival exceeding 95%. Ovarian GCTs, with a lifetime risk of 0.07%, also have 95% 5-year survival rates. The excellent outcomes are primarily due to the exquisite sensitivity of both tumors to platinum-based chemotherapy [2]. Similar results may be expected in children with DSD who present with GCTs [3].

51.8 Conclusion

Malignancies in DSD is a well-known entity due to the inherent aberrations in gonadal development, predisposing to the same. However, the rarity of the condition has rendered it difficult to gather data on specific malignancy risks with different kinds of DSDs. There are no consensus guidelines on screening and surveillance methods for the same reason. The decision and timing of gonadectomy is also a gray area, which needs to be individualized, due to the lack of consensus recommendations. However, novel screening methods are upcoming, such as miRNA testing, which may result in betterment of screening guidelines. Further, multicentric, prospective studies need to be designed in order to study the genotype and phenotype correlations of DSDs as molecular diagnoses can better characterize the ultimate tumor risk for each kind of DSD. This would help in achieving the ultimate goal of removing the gonads for those at risk while preserving them for those who are not.

References

1. Faure-Conter C, Orbach D, Fresneau B, et al. Disorder of sex development with germ cell tumors: which is uncovered first? Pediatr Blood Cancer. 2020;67:e28169. https://doi.org/10.1002/pbc.28169.
2. Pyle LC, Nathanson KL. A practical guide for evaluating gonadal germ cell tumor predisposition in differences of sex development. Am J Med Genet C Semin Med Genet. 2017 Jun;175(2):304–14.
3. Morin J, Peard L, Saltzman AF. Gonadal malignancy in patients with differences of sex development. Transl Androl Urol. 2020 Oct;9(5):2408–15.
4. Hughes IA, Houk C, Ahmed SF, Lee PA. Consensus statement on management of intersex disorders. J Pediatr Urol. 2006;2(3):148–62.
5. Abel R. Ein Fall von Pseudohermaphroditismus masculinus mit sarcomatöser Cryptorchis sinistra. Archiv für pathologische Anatomie und Physiologie und für klinische Medicin. 1891 Dec;126(3):420–37.

6. Pleskacova J, Hersmus R, Oosterhuis JW, Setyawati BA, Faradz SM, Cools M. Tumor risk in disorders of sex development. Sex Dev. 2010;4(4–5):259–69.
7. Scully RE. Gonadoblastoma. A review of 74 cases. Cancer. 1970;25:1340–56.
8. Cools M, Drop SL, Wolffenbuttel KP, Oosterhuis JW, Looijenga LH. Germ cell tumors in the intersex gonad: old paths, new directions, moving frontiers. Endocr Rev. 2006 Aug 1;27(5):468–84.
9. Deans R, Creighton SM, Liao LM, Conway GS. Timing of gonadectomy in adult women with complete androgen insensitivity syndrome (CAIS): patient preferences and clinical evidence. Clin Endocrinol. 2012 Jun;76(6):894–8.
10. Abacı A, Çatlı G, Berberoğlu M. Gonadal malignancy risk and prophylactic gonadectomy in disorders of sexual development. J Pediatr Endocrinol Metab. 2015 Sep 1;28(9–10):1019–27.
11. Kovachev SM, Nikolov SD, Mihova A. Uterine leiomyoma in a man with persistent Müllerian duct syndrome and seminoma. Isr Med Assoc J. 2014 Nov 1;16(11):735–7.
12. Bagrodia A, Lee BH, Lee W, Cha EK, Sfakianos JP, Iyer G, et al. Genetic determinants of cisplatin resistance in patients with advanced germ cell tumors. J Clin Oncol. 2016 Nov 11;34(33):4000.
13. Brinton LA. Breast cancer risk among patients with Klinefelter syndrome. Acta Paediatr. 2011 Jun;100(6):814–8.
14. Schoemaker MJ, Swerdlow AJ, Higgins CD, Wright AF, Jacobs PA. Cancer incidence in women with turner syndrome in Great Britain: a national cohort study. Lancet Oncol. 2008 Mar 1;9(3):239–46.
15. Kathrins M, Kolon TF. Malignancy in disorders of sex development. Transl Androl Urol. 2016 Oct;5(5):794.

Role of Peer Support Groups for Individuals with Disorders of Sex Development

52

Parag J. Karkera and Pradnya S. Bendre

Abstract

In spite of the society being more open about sex today, taboo still surrounds Disorders of Sex Development (DSD). Psychoeducation and support not only provide information about the disorder, but also provide mechanisms to manage day-to-day family activities. The 2006 Consensus Statement on Management of Intersex Disorders describes peer support as integral to a comprehensive model of care for disorders of sex development (DSD). Peer support groups (PSG) bring together people with similar issues and contribute to emotional peer support and to dispel taboos, strengthen coping and assist with the process of shared and informed decision-making. Support groups also work with clinicians to improve patient care, help in research activities and provide training to peer group facilitators. The support groups may have face-to-face meetings, teleconferences or online communities. They also manage better public understanding to channel the concerns of a scattered population. Healthcare providers' awareness of and attitudes towards PSG can influence the degree to which families value such support. Support groups promote a multidisciplinary care approach for patients with DSD with the goal of improving patient/family outcomes and quality of life.

Keywords

Peer support group · Intersex · Disorders of sex development

P. J. Karkera · P. S. Bendre (✉)
Department of Pediatric Surgery, Bai Jerbai Wadia Hospital for Children, Mumbai, India

S. K. Ratan (ed.), *Children with Differences in Sex Development*,
https://doi.org/10.1007/978-981-97-1639-5_52

52.1 Introduction

A Babylonian tablet, dated approximately 700 BC, tells us that 'If a woman gives birth to an infant that has no well-marked sex, calamity and affliction will seize upon the land; the master of the house shall have no happiness'. Even in this day and age, intersex/disorders of sex development (DSD) are considered taboo, in spite of more openness in society regarding sex. Many of the ones with DSD, along with their kith and kin, find DSD conversations uncomfortable, since they are veiled in shame and secrecy [1]. Till recent times, disclosing complete information about the DSD diagnosis to the person and the family concerned was not thought appropriate by the healthcare workers, as it may lead to more psychological damage than good [2].

As a result, the affected persons and families could not clearly understand the condition and the various options of management being offered. This led to compromise of the general condition of the affected ones and their families [2] and unhappiness with the medical treatment offered, affecting the long-term psychosocial development of the child [3].

The 2006 Chicago consensus mentioned that psychosocial support services, with focus on long-term well-being of DSD individuals, is considered a primary modality of medical care for persons with DSD and their families have a major role in medical management of DSD. The Consensus stressed that the clinical teams should reveal the aetiology and discuss the various interventional modalities and the outcomes expected later at puberty with the affected ones and their families along with the ongoing psychosocial supports [4]. The transition to full disclosure has occurred at different times in different parts of the world [5, 6].

The target for people with DSD is to provide a good quality of life, consisting of good physical and mental health, body image and self-esteem, social life and relationships, ability to work, education opportunities, having a family and being an active member of the community [5, 6].

The consistency and care given to the child will depend on the parent's ability to accept and understand their child's medical and psychological needs [7]. Psychoeducation has been shown to be an important treatment tool for both physical and mental illnesses [8]. As a treatment modality, psychoeducation not only provides information about the disorder, accepts the spectrum of physical presentations that result from these variations, but also provides mechanisms to manage day-to-day family activities.

In DSD, parents are the primary caretakers and sources of psychosocial support to their children, so it becomes important that the parents get family-centred support to promote overall quality of life [9]. Peer or Parental Support groups (PSG) bring together people with similar conditions, contributing to emotional peer support and dispelling taboos thereby making these conditions easier to handle [1]. These groups take help of clinicians to improve the standard of care, for conducting research and to generate awareness and expertise. They create better public awareness via media involvement and training videos and play an advocacy role, providing one voice to channel the concerns of a scattered population with these rare conditions [10].

52.2 Psychosocial Support

The essential components of psychosocial support to the affected families imply the multidisciplinary team approach to clinical care with the involvement of a DSD clinical coordinator. The multidisciplinary team includes specialists from different disciplines (endocrinology, urology, gynaecology, psychology, nursing staff, social work, clinical ethics) to give a shared-care approach for an individual/family.

The information given to the parents should include.

- Medical: explain sex and gender as non-binary concepts and in the context of sex determination and differentiation; give proper information on the specific condition; give information on essential, functional and elective medical and surgical interventions, including risks and benefits; and offer options;
- Biological: explain the condition as a naturally occurring variation;
- Gender-related: discriminate between childhood (play) behaviour and adult gender identity; explain that the initial decisions on social gender role may be subject to later change according to the self-expression of the developing child; and put forward gender contentedness as the ultimate goal of informed decision-making.

It encourages

- Psychosocial counselling—which could be face to face, telephonic or online;
- Contact with support groups and participation of trained peers in the decision-making process or in the multidisciplinary team and referral to such groups;
- Giving a summary of the condition for the individual to share with other health professionals if needed [11, 12].

It is important to understand that learning about DSD for the first time is stepping into uncharted territory for the affected individuals and parents. Today, clinicians generally agree that children with DSD should be told about their condition early and should be allowed to continue at their own pace. If the children are explained in a simple, age-appropriate language, it improves acceptance and decreases fear and stigma. If there is clarity in communication, it instils trust in the medical team which improves compliance to treatment and combined decision-making. Delay about information provided to individuals can occur when parents don't share important clinical details with their children. It is essential that the multidisciplinary team sensitively guide the parents to accept this condition [12].

52.2.1 Peer Support Groups

Continued psychosocial support is essential for families to understand the diagnosis and make informed decisions subsequently. An old saying goes, 'a problem shared is a problem halved'. A support group acts like another home; where a person can go to speak and listen to others, to speak out about their worries and share good or bad experiences. These groups organize peer and family support meetings and educational meets that allow members to speak and share their emotions in a secure place and gain more information about their condition. Parents can speak to other parents of affected individuals, while children can meet others of a similar age group or older individuals who are in the same situation.

Such health-related support groups fill the gap between medical treatment and the need for emotional support. The individual's relationship with his doctors may not be close enough for adequate emotional support and close relatives and friends may not understand the details of the disease or treatment. Thus, it acts as a bridge between their medical and emotional needs.

Certain social aspects of DSD like infertility and relationship challenges, along with certain life-threatening aspects of conditions, such as congenital adrenal hyperplasia, in which even a simple cold may lead to serious health issues, may need some extra emotional support which is offered by these support groups [13].

Patient-centred support concentrates on the person's ability to cope, such as personal attitudes, life experiences, cultural and religious background and eventual socio-economic barriers; however, it should also address fears, concerns and questions that are relevant for individuals and families across all DSDs (for example, how parents and/or guardians can inform their child about their DSD and how they can deal with negative reactions) [12].

The long-term targets of the support groups include: to detect the specific requirements of the affected population; to give emotional, informational and appraisal support by sharing experiences; to decrease episodes of stress and isolation, improving their social life; to define distinct roles and confidentiality expectations; and to establish a screening, selection and training program for peer support facilitators [10, 14].

52.2.2 Structure of Support Groups

Support groups may be run by a nonprofit advocacy organization, clinic, hospital or community organization. They also may be run entirely by group members without any organizational support.

The format of support groups may be in the form of face-to-face meetings, teleconferences or online communities. The leader of the support group is usually a lay person—who shares or has shared the group's common experience, but a nurse, social worker or psychologist may become a professional facilitator and lead the group. Some groups may organize talks according to the needs of the group with a guest doctor, psychologist, nurse or social worker being the speaker.

52.2.3 Possible Risks

There are certain drawbacks of these groups. These problems may include: over-dependence on the facilitator to aid in solving problems, disruptive group members, personal experiences dominated by griping, no confidentiality, hyper-emotional bonds or conflicts amongst group members, improper medical advice, comparison of severity of condition among peers.

Although the benefits of the peer and family support are known, findings are diverse and not consistent and studies may not indicate the effectiveness of peer support interventions, partly because they have exploratory designs [15, 16]. Inconsistent definitions of peer support may result in mixed findings. Support groups may vary in their targets, level of professional involvement, degrees of formality, and in the training offered to facilitators, etc. Even though the results are promising with positive response towards peer support, the amount of patient involvement and the referral rates by the health care providers stay low [17]. This could be due to inadequate attention to interest determinants, to involve and retain participants, or the traits of successful peer support programs having good outcomes [18]. The involvement of the affected individuals and their families and the referral rates of the clinicians will depend on their awareness of and attitudes towards peer support. As a result, lack of understanding and prejudice will in turn lead to poor rates of clinician referral and user participation [17].

Whenever a PSG is started, the demographics of the user base should be kept in mind, as a restricted range may limit the appeal of groups and lead to lower user participation. The demographics and requirements of the participants should be well characterized, knowing that the cultural backgrounds, socio-economic status, literacy levels, age and sex of the target user base may vary. For instance, online social support is most commonly sought by mothers who are younger, well-educated, of a higher socio-economic background [19]. When new members are paired with others of a similar background in similar situations, there is an increased chance of the interactions being successful. The outcomes get better when new members are screened to optimize linking with a peer facilitator, along with training aimed to form limits and expectations of involvement. In addition to pairing new members and facilitators, the factors to consider include: detailed demographics of the user base; platform(s) to be used to provide support; confidentiality maintenance mechanisms; expectations of involvement; methods to screen making sure the support group would fulfil the desires of the user, recruitment and training of peer support facilitators; collaboration potential with clinicians, and participation charges [10].

Online PSG groups attract a lot of attention from individuals with rare medical conditions such as DSD. They offer flexible timings and personal control by allowing anytime access to personal and relevant information [18, 19]. Web-based groups provide the choice of secrecy allowing individuals to express their emotions which become difficult to share in a physical meet. Confidentiality is crucial for some

DSD patients and families who feel left out and stigmatized [20, 21]. One important goal of support groups is to reduce stress and isolation by increasing the social networks of the individuals. Although studies reveal peer support positively affects decision-making about treatment [18], the unique features of DSD still make parents hesitant to share it with their friends and family [21].

The downside of online groups is that it can't be a substitute for face-to-face interactions, as a segment of group users also want in-person contact [22]. Only written communication can lead to misunderstanding or confusion amongst participants. Some members may resort to disrespectful or ugly comments or behaviour, in the veil of anonymity. Online presence may lead to isolation from other friends or family. Online groups may also be gullible for untrue information or information overload and PSG participants may not be able to judge the quality of information received. With the expansion of social support networks, there is a risk of divulging intimate personal information and experience, which can be misused to hurt people, commit fraud or promote products for commercial benefit. Even though support groups offer a buffer against isolation [10, 23], individuals or families still may not participate as being a part of PSG may be considered as a threat to self-concept [10].

Taking part in PSG can have both positive and negative influence. Reciprocated social support is very effective, but extra support may be considered as harassment [10, 23]. Hence, the roles of PSG participants and facilitators must be clearly etched out to form realistic targets for support, e.g., fast and frequent communication, confidentiality, etc. Limits for interaction should be made to decrease the risk of the facilitators sharing conflicting ideas, being too involved emotionally, being overbearing to ones seeking help, or taking up responsibilities more than their abilities Volunteer support givers report better quality of life and validation of their personal experiences [22].

Facilitators should keep in mind the general levels of health literacy, variable opinions and quality of online health information [10]. Facilitators who have learnt techniques to provide ready to access, accurate information are the ideal personnel to make others understand the health information. Peer training and supervision of facilitators give psychological benefits [22]; guidance and feedback make sure the targets are fulfilled and required support is provided, especially early in the process of volunteering [10]. Healthcare workers can help to initiate the training process, but one should keep in mind that 'too much' formal training or structure will overshadow the feeling of 'peerness' [14]. Parents of children with DSD look up to their doctors as strong support sources and hence, their participation is encouraged [22]. Collaboration of the PSG and clinicians help the health care providers know of the PSG services and make them aware of the needs of the DSD community.

The value of PSG depends on the involvement of patients, but the clinicians' limited awareness of or attitudes towards PSG may lead to poor participation of patients and/or their families [24]. Another reason behind poor participation may be the lack of knowledge of the specific PSG services. Hesitancy to refer is sometimes intended to protect patients and families from decisional conflict or regret [24]. Participation of clinicians as PSG advisors may take care of that. Because of poor health literacy of the community, clinician-patient stakeholder partnerships improve productive dialogue and emphasize better patient-centric care. If clinicians understand and participate in PSG activities regularly, it will decrease misinformation, increase referrals and decrease the load on the strained health resources.

Peer support resources, which are well planned and implemented, promote positive changes in the people and families having chronic medical conditions. These services are cheaper than older models as they train non-healthcare workers as volunteers. Successful PSG needs to understand the requirements of the target population and train facilitators accordingly. Fortunately, the path to create effective and sustainable PSG has been systematically described, and resources for continuous quality improvement are readily available [10].

52.3 Current Status in India

For the first time in India, 7 years ago, Shyam Nair and Deepa Kannan, parents of a CAH child, started a support group called 'CAH Support India' (www.cahindia.org) involving a community of parents, grandparents and caregivers of CAH children. They also established a Facebook page describing their child's journey with Congenital Adrenal Hyperplasia to chronicle all possible events in the life of a child with CAH. The International Coalition for Endocrine Patient Support Organizations worldwide had listed this support group as the first such group for endocrine disorders in India. Soon CAH India changed officially to The Magic Foundation India with a vision to create a closed group for caregivers of all children with growth and metabolic disorders in a setting where discussions could be kept India relevant. The Magic Foundation India closed group on Facebook was created to increase support and awareness for children's Growth and Metabolic disorders, including DSD in an India-relevant and India-specific setting.

In addition, there are scattered support groups in some institutions like B J Wadia Hospital for Children, Mumbai which encourages parents of CAH patients to understand the condition better and have active involvement in decision-making.

52.4 Conclusion

Support groups put forth the hardships and difficulties in the life of an individual with DSD and his family and try to help them to come out of their shell of shame, isolation and stigma. They aid in normalizing the emotions and desires of the DSD individuals and try to change the monstrous image projected by the society. Most importantly, these support groups give them emotional and psychological support at a personal level and provide opportunities to meet and communicate with specialists in the medical profession. However, more research in PSG is required to improve the communication with doctors and to create patient-centred designs which will increase community participation. Increase in researcher and clinician involvement will decrease misguidance, improve training of facilitators and enhance funding for such groups. Ultimately, cohesive physical, medical care and emotional care offered by the hospitals/health clinics with the support groups is needed for overall physical and mental development and well-being of the patient and his family.

References

1. Cull ML. Dispelling the taboo of intersex conditions: we're only human like you! In: Balen R, Crawshaw M, editors. Sexuality and fertility issues in ill health and disability: from early adolescence to adulthood, chapter 12. London: Jessica Kingsley Publisher; 2006.
2. Thomas B. A patient's view of ambiguous genitalia. Minerva Pediatr. 2003;55(5 Suppl 1):54–7.
3. Karkazis KA. Fixing sex: intersex, medical authority, and lived experience. Durham, NC: Duke University Press; 2008. p. 2008.
4. Lee PA, Houk CP, Ahmed SF, et al. Consensus statement on management of intersex disorders. International consensus conference on intersex. Pediatrics. 2006;118(2):e488–500.
5. Nordenström A. Psychosocial factors in disorders of sex development in a long-term perspective: what clinical opportunities are there to intervene? Horm Metab Res. 2015;47(5):351–6.
6. Jones T, Hart B, Carpenter M, et al. Intersex: stories and statistics from Australia. Cambridge, UK: Open Book Publishers; 2016.
7. Schwarzer R, Leppin A. Social support and health: a theoretical and empirical overview. J Soc Pers Relat. 1991;8(1):99–127.
8. Karamanidou C, Weinman J, Horne R. Improving haemodialysis patients' understanding of phosphate-binding medications: a pilot study of a psycho-educational intervention designed to change patients' perceptions of the problem and treatment. Br J Health Psychol. 2008;13:205–14.
9. Rolston AM, Gardner M, Vilain E, et al. Parental reports of stigma associated with child's disorder of sex development. Int J Endocrinol. 2015;2015:980121.
10. Baratz AB, Sharp MK, Sandberg DE. Disorders of sex development peer support. Endocr Dev. 2014;27:99–112. https://doi.org/10.1159/000363634.
11. Hanna CA. Psychosocial support. In: Disorders|Differences of sex development. Springer; June 2020. p. 247–59. https://doi.org/10.1007/978-981-13-7864-5_19.
12. Cools M, Nordenström A, Robeva R, Hall J, Westerveld P, Flück C, Köhler B, Berra M, Springer A, Schweizer K, Pasterski V. COST action BM1303 working group 1. Caring for individuals with a difference of sex development (DSD): a consensus statement. Nat Rev Endocrinol. 2018 Jul;14(7):415–29. https://doi.org/10.1038/s41574-018-0010-8.
13. Cull ML, Simmonds M. Importance of support groups for intersex (disorders of sex development) patients, families and the medical profession. Sex Dev. 2010 Sep;4(4–5):310–2. https://doi.org/10.1159/000313889.
14. Dennis C-L. Peer support within a health care context: a concept analysis. Int J Nurs Stud. 2003;40:321–32.
15. Beatty L, Lambert S. A systematic review of internet-based self-help therapeutic interventions to improve distress and disease-control among adults with chronic health conditions. Clin Psychol Rev. 2013;33:609–22.
16. Eysenbach G, Powell J, Englesakis M, Rizo C, Stern A. Health related virtual communities and electronic support groups: systematic review of the effects of online peer to peer interactions. BMJ. 2004;328:116.
17. Van Uden-Kraan CF, Drossaert CH, Taal E, Smit WM, Bernelot Moens HJ, Van de Laar MA. Determinants of engagement in face-to-face and online patient support groups. J Med Internet Res. 2011;13:e106.
18. Lau AYS, Siek KA, Fernandez-Luque L, Tange H, Chhanabhai P, Li SYW, Elkin PL, Arjabi A, Walczowski L, Ang CS, Eysenbach G. The role of social media for patients and consumer

health: contribution of the IMIA Consumer Health Informatics Working Group. Yearb Med Inform. 2011;6:131–8.
19. Paterson BL, Brewer J, Stamler LL. Engagement of parents in on-line social support interventions. J Pediatr Nurs. 2013;28:114–24.
20. Crissman H, Warner L, Gardner M, Carr M, Schast A, Quittner A, Kogan B, Sandberg D. Children with disorders of sex development: a qualitative study of early parental experience. Int J Pediatric Endocrinol. 2011;2011:10. http://www.ijpeonline.com/content/2011/2011/2010
21. Duguid A, Morrison S, Robertson A, Chalmers J, Youngson G, Ahmed SF, Scottish Genital Anomaly Network. The psychological impact of genital anomalies on the parents of affected children. Acta Paediatr. 2007;96:348–52.
22. Schwartz CE, Sendor RM. Helping others helps oneself: response shift effects in peer support. Soc Sci Med. 1999;48:1563–75.
23. Sartore GM, Pourliakas A, Lagioia V. Peer support interventions for parents and carers of children with complex needs. Cochrane Database Syst Rev. 2021 Dec 20;12(12):CD010618. https://doi.org/10.1002/14651858.CD010618.pub2.
24. Steginga SK, Smith DP, Pinnock C, Metcalfe R, Gardiner RA, Dunn J. Clinicians' attitudes to prostate cancer peer-support groups. BJU Int. 2007;99:68–71.

Part XI

Synopsis of Special Considerations

Special thanks Dr. Vandana Jain, Professor, Pediatrics, AIIMS, New Delhi and Dr. Shashank Shekhar Panda Additional Professor Pediatric Surgery, AIIMS, Biwinagar who along with me edited this section.

Medical Management of Children with Disorders of Sex Development: An Overview

53

Jaivinder Yadav, Pamali Mahasweta Nanda,
and Vandana Jain

Abstract

Differences in Sex Development (DSD) refer to differences or discrepancies in the genetic, anatomical and gonadal sex. DSD management is a long journey of the family with a multidisciplinary team of neonatologist/paediatrician, paediatric endocrinologist, paediatric surgeon, geneticist and psychiatrist/psychologist. The management encompasses gender assignment, medical treatment with hormones, surgical corrections, assessment and counselling for gender dysphoria, and long-term follow-up for issues like marriage, sexual function, fertility and psychological problems. In children with 46 XX DSD due to congenital adrenal hyperplasia, medical management is the mainstay. Treatment and prevention of adrenal crisis and ensuring normal growth and pubertal development requires close monitoring and adjustment of steroid doses. For children with 46 XY DSD due to hypogonadotropic hypogonadism, topical or intramuscular testosterone is the mainstay for management of micropenis in infancy and for development of secondary sexual characters in adolescence. Hormonal replacement is necessary for pubertal induction in children with several types of DSD including gonadal dysgenesis and androgen biosynthetic defects. On the other hand, pubertal suppression may be needed in a few cases where pubertal development is opposite to the child's gender identity. It is important to remember that treatment with sex steroids may also lead to irreversible changes in the genital anatomy, and hence,

J. Yadav · P. M. Nanda
Paediatric Endocrinology Unit, Advanced Paediatric Centre, PGIMER Chandigarh, Chandigarh, India

V. Jain (✉)
Division of Paediatric Endocrinology, Department of Paediatrics, AIIMS New Delhi, New Delhi, India
e-mail: drvandanajain@aiims.edu

S. K. Ratan (ed.), *Children with Differences in Sex Development*,
https://doi.org/10.1007/978-981-97-1639-5_53

as for surgical management, the principles of informed consent, shared decision-making with the child's long-term interests in mind, participation of the child to the extent possible, guided by their age and maturity level, and deferring irreversible hormonal treatments to a time when child's gender identity can be assessed should be followed.

Keywords

Multidisciplinary management · Gender assignment · Topical dihydrotestosterone · Pubertal induction · Pubertal suppression

53.1 Introduction

The term DSD is an acronym for differences in sex development. It denotes the differences or discrepancies in the genetic, anatomical and gonadal sex. The various terms used in the past for these disorders, including intersex, ambiguous genitalia, hermaphrodite and pseudohermaphrodite, have been discarded in the 2006 consensus guidelines [1].

These disorders are broadly classified into chromosomal disorders, 46, XX and 46, XY disorders. The broad categories and the major disorders under the subheading of chromosomal disorders include Turner syndrome (45,X), Klinefelter syndrome (47,XXY), mixed gonadal dysgenesis (45,X/46,XY) and chimeric ovotesticular DSD (46,XX/46,XY). 46, XY disorders include disorders of testicular development (Swyer syndrome, partial gonadal dysgenesis, gonadal regression), disorders of androgen biosynthesis (17 hydroxysteroid dehydrogenase deficiency, 5 alpha reductase deficiency), or androgen action (partial or complete androgen insensitivity syndrome). 46, XX DSD disorders comprise disorders of gonadal development (ovotesticular DSD, testicular DSD, gonadal dysgenesis) and of androgen excess (congenital adrenal hyperplasia, aromatase deficiency, POR, luteoma) [1].

The diagnosis and management of DSD require the multidisciplinary involvement of a neonatologist/paediatrician, paediatric endocrinologist, paediatric surgeon, geneticist and psychiatrist/psychologist. The management includes gender assignment, medical treatment with hormones, surgical corrections, assessment and counselling for gender dysphoria and long-term follow-up for issues like marriage, sexual function, fertility and psychological problems. In this chapter, we briefly discuss the medical management of DSD.

53.2 Gender Assignment

DSD in a newborn is often a social emergency, and parents often press the doctor for early gender assignment. This is a moment of distress for the family and confusion for the doctor. The assigned gender should be in congruence with the gender identity of the baby, which is very difficult to judge at this stage. The decision should not be in haste. Some municipal corporations now offer the option for the third gender in birth certificates until the final gender assignment is done.

There is a general consensus on raising children with 46, XX CAH as girls irrespective of the degree of initial virilization. Similarly, 46, XY DSD with Complete Androgen Insensitivity Syndrome (CAIS) or Complete Gonadal Dysgenesis (CGD) are usually raised as females; although individualized decision-making is still needed. For example, when 46, XX CAH children present at peripubertal age group with extreme virilization and male gender identity (due to delayed start or poor compliance with treatment), careful evaluation with the involvement of a child psychologist may be needed to decide on the gender. Similarly, few families with children with 46,XY DSD due to CAIS or CGD insist on raising their children as boys; although based on available literature, this is not preferred. For most other DSD conditions, there is no definitive consensus, and the child's phenotype, gonadal function, the etiological diagnosis, family's expectations, surgical options, anticipated response to hormonal therapy, fertility potential and expected long-term outcome of the specific diagnosis, should be considered when counselling the parents regarding gender for rearing. Parents should be told that gender reassignment may be required as the child becomes older and greater clarity emerges in terms of medical diagnosis as well as child's preference [1, 2]. In a recent systematic review by Babu et al., the overall prevalence of gender identity disorder (GID) in patients with various DSDs was noted to be 15%, being the lowest patients with CAIS raised as females (1.7%), and the highest in those with 5-alpha reductase deficiency reared as females (53%) [3].

53.3 Consent and Ethics in DSD Care

Management of DSD is complex, and sometimes the decisions related to care, especially genital surgery and gonadectomy, are complicated and confusing. Complete information should be provided about the cause of DSD, its physical and psychosexual implications, available treatment options, short-term (like adrenal insufficiency) and long-term complications (sexual function, infertility, risk of neoplasia, gender dysphoria), and outcome of such patients from the literature. The caregiver should be emotionally stable at the decision time. Many authors recommend the child's involvement in decision-making using age-appropriate explanations about the interventions and likely outcomes. Some cases have been referred to court for inappropriate decisions related to genital surgery in DSD, and recently Madras High Court also banned early surgeries in DSD. In Australia, a special authorization was obtained from the court for carrying out special medical procedures in minors with DSD in addition to parental consent. So emotionally and cognitively informed consent should be taken in all decisions along with written documentation or video consent [4, 5]. Similar caution should be exercised while prescribing hormonal therapies that can lead to irreversible secondary sexual characters, especially when the hormonal therapy is for the opposite gender to the chromosomal and/or gonadal sex [1]. Removal of a functioning gonad in children with a gender of rearing opposite to the chromosomal sex (as in children with 46, XY DSD due to androgen insensitivity syndrome or 5-alpha reductase 2 deficiency, being reared as girls, and 46, XX DSD due to CAH being reared as boys) should be avoided below the legal age of consent, i.e., 18 years [1, 3].

53.4 General Principles of Management of DSD

The management of DSD is complex, involving hormonal replacement, surgical reconstruction and dealing with psychological concerns of family and child. Table 53.1 presents a summary of the main indications for hormonal replacement in DSD. The details are discussed in later sections.

Table 53.1 Summary of medical treatments for various DSDs

Condition	Suggested hormone replacement
Congenital adrenal hyperplasia	Hydrocortisone with or without mineralocorticoid replacement
Micropenis with or without hypospadias (46 XY DSD, mixed gonadal dysgenesis, 5α-reductase deficiency, 17β-hydroxysteroid dehydrogenase deficiency, partial androgen resistance syndrome	Intramuscular testosterone/topical dihydrotestosterone Children with 5-alpha reductase deficiency and partial androgen resistance require higher doses (up to 5 times). Small phallus interfering with urination, social anxiety are indications to treatment. Penile size comparable to age is desirable. Pubertal induction should be started at normal age of puberty if the height is normal. If the child is short, then this can be delayed.
46 XY DSD due to 5α-reductase deficiency, 17β-hydroxysteroid dehydrogenase deficiency or partial androgen insensitivity syndrome raised as girls	GnRH analogues to suppress the puberty until the final decision on sex of rearing is made by patient, parent and multidisciplinary team
46 XX DSD- turner syndrome, 46 XY with female phenotype and gender of rearing	Oestrogen replacement is started for pubertal induction

53.5 Medical Management of Chromosomal DSDs

Turner Syndrome (TS) TS is the most common chromosomal DSD in females occurring due to partial or complete loss of one X chromosome. The typical clinical phenotype is short stature, webbed neck, shield chest, high carrying angle, low posterior hairline and delayed puberty but the clinical manifestations are highly variable which often leads to a delayed diagnosis. Growth hormone (40–60 μg/kg/day) and oxandrolone are recommended for treating growth failure and result in a gain of 7–10 cm in final height [6, 7]. Patients are monitored for gonadal failure (LH and FSH estimation) from 11 years onwards, and puberty is induced with transdermal/oral oestrogens as described later in patients with hypogonadotropic hypogonadism and summarized in Table 53.6. Patients with TS should be routinely monitored for cardiovascular health (Echo and blood pressure monitoring), metabolic health (weight/BMI, HbA1c, liver function), thyroid dysfunction, hearing and ophthalmic assessments. Patients with marker chromosome and those with virilization should be screened for Y chromosome mosaicism. In those with presence of Y chromosome, prophylactic gonadectomy should be done to prevent gonadoblastoma [6, 7].

Klinefelter Syndrome (46, XXY) This is the commonest (1:600) chromosomal aneuploidy in males. The usual clinical manifestations are micropenis, cryptorchidism, clinodactyly, hypertelorism, language delay, tall stature with long legs, poor testicular growth and scant pubic hair in puberty. Patients with KS should be monitored for intellectual disability, learning difficulties, metabolic syndrome, poor bone mass and gynecomastia. Testosterone therapy is required for pubertal induction (as discussed later and summarized in Table 53.5). Mastectomy may be needed for gynecomastia [8].

Mixed Gonadal Dysgenesis (MGD) and Chimeric Ovotesticular DSD MGD refers to individuals with chromosomal mosaicism with variable internal and external reproductive anatomy. The external phenotype can be highly variable ranging from mild undervirilization to female phenotype. The evaluation includes karyotype with at least 50 cells, physical examination, gonadal and internal genitalia imaging with USG and MRI, laparoscopic and cystoscopic examination for internal anatomy, gonadal biopsy and histochemistry for markers of malignancy OCT3/4, TSPY, C-KIT ligand). The management decisions should be made through shared decision-making with parents and children after explaining the issues such as the presence of both male and female structures, infertility, malignancy risk, gender identity, dysphoria and social adaptation. Dysgenetic gonads and gonads in individuals with ambiguous genitalia carry a high risk of malignancy (15–35%) and a low threshold for gonadectomy should be kept. In patients with mild undervirilization, orchidopexy and follow-up with 3 monthly self-examinations and annual imaging with ultrasound should be done. In female phenotype, elective gonadectomy should be offered and patients not opting for it should be explained the risk of virilization during puberty. Hormonal replacement is required in gonadectomized patients and patients with delayed or arrested puberty [9].

53.6 Management of 46, XX DSD

53.6.1 Congenital Adrenal Hyperplasia (CAH)

CAH is the commonest cause of 46, XX DSD, with an estimated incidence of ~1:14,000 to 1:18,000 live births worldwide. In approximately 95% of cases, CAH is caused by 21OHD (21-hydroxylase deficiency) due to mutations in *CYP21A2*, the gene encoding adrenal P450c21. The phenotypic spectrum varies with the severity of enzyme deficiency, ranging from severe symptomatic disease presenting with acute adrenal crises and genital virilization in female infants in salt-wasting forms (SW-CAH) to only virilization in girls and precocious puberty in boys without features of adrenal insufficiency in simple-virilizing forms (SV-CAH). These two forms are collectively referred to as classical CAH. Milder degrees of enzyme deficiency can cause Non-Classical CAH (NCCAH), which may remain asymptomatic and present during late childhood or adolescence with signs of androgen excess.

53.6.1.1 Drug Therapy for CAH

Classical CAH requires replacement therapy with glucocorticoids (GC) and mineralocorticoids (MC). Besides correcting cortisol and aldosterone deficiency, treatment suppresses excessive adrenal androgen production to prevent virilization and skeletal maturation. The drug of choice for glucocorticoid replacement in children is hydrocortisone. It should be administered in physiological doses of 10–15 mg/m^2/day orally; affected infants usually require dosing at the high end of this range[9]. This dosing is expected to mimic the normal endogenous cortisol production rate (6–8 mg/m^2/day) after accounting for first-pass metabolism and bioavailability. Hydrocortisone is administered in three divided doses due to its short half-life (80–120 min). The morning dose should be administered as early in the morning as possible to simulate the body's circadian rhythm [10]. Longer-acting, more potent steroids, e.g., prednisolone and dexamethasone, are avoided in younger children because of their growth-suppressive potential, being approximately 5- and 80-times higher than hydrocortisone, respectively. However, after the closure of epiphyses, these may be used in place of hydrocortisone to permit once- or twice-daily dosing.

Mineralocorticoids should also be supplemented in all children with classical CAH. Fludrocortisone (9α-fluorocortisol) is the only oral mineralocorticoid preparation that is widely available, with a salt-retaining effect 200-fold higher than that of hydrocortisone. Mineralocorticoid replacement in 21OHD restores plasma volume, eliminating the hypovolemic drive to ACTH release. As a result, it permits lower doses of glucocorticoids in SV-CAH, thereby optimizing growth outcomes. Typical mineralocorticoid requirement in infancy is 0.1–0.2 mg/day. All infants with SW-CAH should receive salt supplements (1–2 g/day, divided into several feedings). The enhanced mineralocorticoid sensitivity in older children and adults, free access to salty foods and the stimulation of hepatic extra-adrenal 21-hydroxylating enzymes generally allow a reduction in dose and sometimes even discontinuation of fludrocortisone.

53.6.1.2 Management During Stress

During stress, all patients with CAH should be given higher doses of hydrocortisone, 50–75 mg/m^2/day (about 3–5 times the daily oral dose). Stress dosing is recommended for febrile illnesses (>38.5 °C), gastroenteritis with dehydration, major surgery requiring general anaesthesia and major trauma [11]. All patients with CAH should wear or carry medical identification indicating they have adrenal insufficiency. Their parents and guardians should be educated regarding recognizing symptoms of adrenal crisis, increasing the dose of glucocorticoids during intercurrent illnesses, and administering parenteral hydrocortisone.

53.6.1.3 Titration of Drug Therapy

Treatment of CAH is complicated and aims at preventing both hyperandrogenism and hypercortisolism. Adequate GC therapy achieves physiological cortisol replacement, prevents adrenal crisis and virilization and allows the child to reach his maximum height potential. The baby requires monthly visits in the first 3 months for dosage adjustments, followed by every 3 months till 18 months of age and every four monthly thereafter [11]. At each follow-up visit, the clinician assesses the child for weight, height/length, growth velocity, blood pressure, genitalia, signs of virilization and cushingoid features. Normal growth velocity, absence of hyperpigmentation and no progression of virilization are clinical markers of appropriate treatment. In contrast, weight gain, cushingoid habitus and poor growth velocity may indicate overtreatment with steroids and warrant a dose reduction. On the contrary, virilization or early onset of puberty, accelerated height gain and advanced bone age indicate under-treatment.

Blood pressure and serum electrolytes guide the adequacy of MC therapy. Plasma renin activity (PRA) and aldosterone-to-PRA ratio are valuable adjuncts to clinical monitoring, where feasible. Biochemical tracking should also be done in parallel, and results carefully interpreted (Table 53.2).

53.6.1.4 Genital Reconstructive Surgery in 46,XX CAH

In girls with minimal virilization, counselling the parents about delaying surgery until the child is older is advisable, as mild clitoromegaly may regress following GC therapy. For severely virilized girls (Prader stage 3 or more), early genital surgery in infancy may be required to repair the urogenital sinus [12]. The decision for surgery should not be taken in haste and must involve joint decision-making between experienced surgeons and patients' families.

Table 53.2 Monitoring of patients with CAH on follow-up and goal of therapy

Investigation	Physiology	Goal of therapy
Serum electrolytes	GC and MC replacement	Normal
17 OHP	Adequacy of GC replacement	4–12 ng/mL
Plasma renin[a]	Volume status	Low to normal unless hypertensive
Testosterone	Total androgens	At or near-normal
Androstenedione[a]	Adrenal or ovarian origin	At or near-normal
Bone age[b]	Skeletal maturation	Normal

[a]When feasible
[b]Annual assessment done after two years of age

53.6.2 Aromatase Deficiency

Microsomal cytochrome P450 aromatase enzyme is required for the conversion of androgens (androstenedione, testosterone) to oestrogens in both males and females. In the 46,XX newborns, aromatase deficiency typically presents with genital ambiguity and ovarian cysts. At puberty, there are primary amenorrhoea and delayed bone age. In males, it usually manifests as tall stature, poor bone mass, skeletal abnormalities and infertility. Maternal virilization during pregnancy, with a persistent change in the voice, is an important clue to the diagnosis. The hormonal evaluation shows high LH, FSH (often more than LH), testosterone and low oestrogen level [13]. Obesity and metabolic derangements such as insulin resistance are also common. Oestrogen supplementation is recommended in both males and females. Surgical intervention may be required for virilized external genitalia in females [14].

53.7 Management of 46, XY DSD

46, XY DSD presents with genital ambiguity in the form of micropenis, hypospadias, unfused scrotal folds and/or undescended testes. The baby requires a comprehensive evaluation of the hormonal profile and genetic testing for the etiological diagnosis of DSD. Hypospadias and undescended testes require surgical management. Micropenis due to several underlying aetiologies is amenable to medical management with local and systematic supplementation with androgens [15].

53.7.1 Micropenis

Micropenis is defined as stretched penile length (SPL) of less than 2.5 standard deviations below the mean for age in a normally formed penis, i.e., without hypospadias. It constitutes a clinical sign, not a syndrome. The normative data for SPL in healthy, full-term male neonates is primarily derived from Caucasian studies.

Table 53.3 Normative data for SPL from Indian studies

Authors, Year	Place of study	Participants	Results
Saneja et al., 2017–2018	Faridkot	800 newborns between 32–42 completed weeks in the first 72 h of life	Mean SPL in preterm group: 2.64 ± 0.40 cm, term group: 3.02 ± 0.33 cm, Micropenis (<2.5 SD) Terms: <22 mm, preterms: <16 mm
Mondal et al., 2013–2014	Kolkata	480 neonates (appropriate for gestational age) within 24–72 h of birth	3.4 ± 0.47 cm Micropenis (<2.5 SD): <24.5 mm
Prabhu SR et al., 2011	Tamil Nadu	346 term (37–41 weeks) neonates, appropriate for gestational age	Mean SPL 2.8 ± 0.5 cm
Vasudevan et al., 1995	Puducherry	135 newborns within 48 h of birth	Mean SPL 3.6 ± 0.5 cm, width 1.04 ± 0.15 cm

Worldwide studies show that there are racial and ethnic differences in penile length [16]. Therefore, it is imperative to use local normative data to interpret penile length (Table 53.3) [17–20].

53.7.2 Technique of Measurement

Penile length is traditionally measured using a transparent plastic ruler with readings from zero to 15 cm (Schonfeld's technique) [21]. In its flaccid state, the glans penis is held between the thumb and forefinger, the penis is fully stretched, and the measurement is taken from the pubic ramus to the distal tip of the glans penis over the dorsal side. The foreskin should be retracted, and the suprapubic fat pad gently compressed using one end of the ruler down to the pubic ramus.

Micropenis needs differentiation from benign variants such as buried, trapped and webbed penis. Often obese children present with the parental concern of small penile size. This is often due to excessive suprapubic fat pad (buried penis); careful examination by pressing the fatty tissue inwards as much as possible would reveal the actual penile size and avoid unwarranted evaluation and anxiety. The body of the penis may be trapped within the scarred prepubic skin following trauma or circumcision, leading to a 'trapped penis.' 'Webbed penis' refers to an abnormal attachment between the scrotum and ventral surface of the penis.

53.7.3 Etiological Factors

These are summarized in Box 53.1.

Box 53.1 Causes of Micropenis

The causes of micropenis can be broadly classified as follows:

1. Hypogonadotropic hypogonadism due to hypothalamic/pituitary insufficiency
   (a) Isolated (with anosmia: Kallmann syndrome, without anosmia: IHH)
   (b) As part of multiple pituitary hormone deficits (e.g., in Septo-optic dysplasia)
   (c) Syndromic disorders: E.g., Prader-Willi syndrome, Laurence-Moon-Biedl-Bardet syndrome
2. Hypergonadotropic hypogonadism
   (a) Anorchia
   (b) Klinefelter syndrome and other X-polysomies
   (c) Noonan syndrome, Robinow syndrome
   (d) LH receptor defect (Incomplete form)
3. 46 XY DSD conditions
   (a) Partial Androgen Insensitivity syndrome
   (b) 5-reductase deficiency
   (c) Testosterone biosynthetic defects
   (d) Partial gonadal dysgenesis
4. Miscellaneous
   (a) Idiopathic
   (b) GH/IGF-1 deficiency

53.7.4 Evaluation

1. **Hormonal evaluation:**
   (a) **LH, FSH and Testosterone:**

   First-line investigations should include measurement of gonadotropins (LH, FSH) and testosterone, preferably during the mini-puberty. These hormones are best measured in the age group between 1 and 3 months, when the gonadotropins as well as testosterone levels are physiologically at their peak, but can be measured between 2 weeks and about 9 months of age. At this time, the gonadotropins and testosterone are normally elevated and therefore it is easier to interpret low values as likely pathological. Micropenis, along with cryptorchidism, mid-line defects and the absence of mini-puberty are pointers towards hypogonadotropic hypogonadism. The evaluation for anterior pituitary hormones (cortisol, ACTH, T_4 and TSH) should be done in this setting.

   Beyond 9–10 months and before the onset of puberty, as the hypothalamo-pituitary-gonadal axis is quiescent, stimulation testing with beta hCG or

GnRH analogues is required to identify the aetiology. Hypogonadotropic hypogonadism is characterized by low gonadotropins (basal and stimulated) and low testosterone, whereas elevated gonadotropins with low testosterone levels suggest primary hypogonadism, which is most often due to testicular dysgenesis or androgen biosynthesis defects.

(b) **hCG stimulation test:** It represents a key step for assessment of functional testicular tissue (differentiated Leydig cells). This test is based on the ability of hCG to stimulate androgen production by its action on the LHCG receptor in testicular Leydig cells. Various hCG test protocols have been proposed (for the number of doses, total dose and regimes for administration of hCG), with the primary objective being assessment of testosterone, Δ4-androstenedione and dihydrotestosterone (DHT) at baseline and following stimulation [22]. A testosterone rise of more than two-three folds above the baseline is considered an adequate response [15]. The ratio of testosterone to DHT and to Δ4-androstenedione aids in the diagnosis of 5α-reductase deficiency and 17β-HSD deficiency, respectively (see Table 53.4).

(c) **Serum AMH levels:** Determination of serum levels of AMH enables assessment of functional Sertoli cells, which can further guide diagnostic testing. Serum AMH levels are normally high in infancy (127–552 ng/ml), and low levels indicate gonadal dysgenesis. On the other hand, normal AMH levels but poor rise of serum testosterone after hCG stimulation would suggest testosterone biosynthetic defect or LH receptor defect with preserved Sertoli cell function.

Table 53.4 Beta hCG Protocol and Its Interpretation

Beta hCG stimulation protocol	Normal response
Standard	
<1 year 500 IU, 1–10 years 1000 IU and > 10 years 1500 IU intramuscularly for 3 days. Samples for T, DHT and androstenedione are measured on D1 and D4. The preferred method of measurement is tandem mass spectrometry but not widely available. Routinely done with immunoassay.	Testosterone >3.5 nmol/L (100.9 ng/dl) on day 4 Below 100.9 ng/ml suggests androgen biosynthetic defect or gonadal dysgenesis.
Prolonged test	
2 additional doses in weeks two and three after the standard test. Samples are taken on day 22.	>9.5 nmol/L (274 ng/dl) on day 22 [23]
Ratios	
T: DHT Testosterone: Androstenedione	>20 suggestive of 5-alpha reductase deficiency < 0.8 suggestive of 17beta HSD3 deficiency

T testosterone, *DHT* dihydrotestosterone

2. **Genetic testing:** Karyotyping with banding studies and Y-fluorescence may be required for genetic sex determination. Specific genetic testing may be needed if syndromic causes are suspected (e.g., Prader-Willi syndrome, Noonan syndrome, etc.) or if the child has other features of genital ambiguity that mandate further molecular work-up for aetiology of DSD. The phenotypic and hormonal profile are often overlapping and molecular testing should be done whenever feasible. Targeted gene panel and whole exome/genome sequencing are being increasingly used in the molecular diagnosis of DSD.

53.7.5 Management

The chief goals of management of micropenis are to enable sufficient penile growth for the affected patient to have a normal sexual function as an adult, have a positive body image and enable him to micturate standing up. Although the data is limited, testosterone is the most commonly used drug in medical management. The data on long-term outcomes such as sexual satisfaction and fertility outcomes is also scarce [24, 25].

Testosterone Treatment

Response to androgens during infancy is an important factor for gender assignment. A short course of testosterone may be tried in newborns, and infants with micropenis. In most cases, apart from complete androgen insensitivity, some benefit is expected, *although this may not translate into a gain in the final adult penile length*. The commonly used preparations are testosterone cypionate or enanthate in oil, which are administered in doses of 25 mg intramuscularly, monthly for 3 months [24]. Side effects are usually minimal, except for a transient acceleration of growth velocity and bone age. Typically, a satisfactory response is a 100% increase in penile length after the initial treatment course and is seen in infants with low testosterone levels due to hypogonadotropic hypogonadism, gonadal dysgenesis and androgen biosynthetic defects [24]. A lack of response to testosterone trial is suggestive of androgen resistance and is likely to indicate failure to virilize at puberty. Topical application (5% testosterone cream) may also achieve penile enlargement, particularly in infancy. Various studies have demonstrated the positive effects of testosterone treatment in infancy and early childhood on penile length [25–28]. However, whether this growth continues during adolescence and adulthood is uncertain. There is a natural decline in penile androgen receptor concentration during early adulthood. Patients with hypogonadotropic hypogonadism have lowered penile androgen receptor expression. Early testosterone therapy helps to increase the expression of penile androgen receptors. Therefore, early initiation of testosterone therapy is generally recommended in children with micropenis [24].

Topical DHT therapy (0.1–0.3 mg/kg/day of 2.5% preparation) is reported to be effective in prepubertal children and presents a good alternative for children who do not respond to testosterone, e.g., those with androgen insensitivity [15, 24] and 5α-reductase deficiency [29].

In children with idiopathic GH deficiency with micropenis, GH therapy alone results in increased penile length [30]. Children with panhypopituitarism require testosterone replacement with cortisol, thyroxine and GH as early as possible.

53.7.5.1 Recombinant FSH-LH Treatment

In patients with hypogonadotropic hypogonadism, treatment with recombinant human FSH-LH has also been shown to increase penile length and testicular volumes [31]. Bougneres et al. reported continuous subcutaneous LH-FSH infusion for six months on two neonates with hypogonadotropic hypogonadism, who had micropenis and micro-orchidism, to achieve peak postnatal gonadotropin levels and replicate mini-puberty [31]. Both babies had more than a three-fold increase in penile length and testicular size.

53.7.5.2 Surgical Correction

Surgical treatment may be considered if medical therapy fails to achieve an adequate improvement in penile length.

53.8 Pubertal Induction and Suppression in DSD

Pubertal onset requires an intact hypothalamic-pituitary-gonadal axis. Hypogonadism is caused by failure of this axis, which may be hypogonadotropic (secondary to hypothalamic or pituitary defects) or hypergonadotropic (primary gonadal defects). In both conditions, pubertal induction is required to develop and maintain secondary sexual characteristics. Pubertal induction aims to mimic normal puberty in an individual, primarily targeting genital maturation, age-appropriate virilization/feminization and fertility induction. Additional goals include attaining target height per genetic potential, adequate bone mass, favourable body composition and psychosocial and mental well-being. Though the optimal timing for initiation of therapy is not well-defined, recommendations are to begin treatment no later than 14 years in boys and 13 years in girls [32].

53.9 Pubertal Induction in Boys

53.9.1 Testosterone Replacement Therapy (TRT)

Testosterone is the most common agent used for pubertal induction in boys. Various preparations are available (injectable, transdermal, oral); however, no evidence-based guidelines exist regarding the optimal formulation and regimen to be used [33]. Due to their high efficacy at inducing virilization and relatively low cost, IM testosterone esters (enanthate, cypionate, propionate) are the most commonly used.

The starting dose is 50 mg per month, gradually increasing in 50 mg increments every 6–12 months. Once a monthly dose of 100–150 mg is reached, the dosing interval can be shortened to every 2 weeks until adult dosing is achieved (up to 200 mg every 2 weeks), after the epiphyses have fused (see Table 53.5) [34]. Dose increases should be gradual to avoid premature epiphyseal fusion and a compromised adult height while allowing the gradual development of psychosexual and secondary sexual characteristics [34]. Testosterone gels (Androgel, DHT gel), oral testosterone preparations (testosterone undecanoate) and subcutaneous testosterone implants/pellets are other routes for administration; however, comparative studies are scarce.

Sex steroids are preferred agents because of their low cost, well-recognized efficacy and convenient dosing; however, the main disadvantage is their suboptimal pharmacokinetics with wide fluctuations in serum testosterone levels between injections [32]. Additionally, TRT does not induce testicular maturation or spermatogenesis. It is crucial to monitor and minimize side effects, e.g., acne, gynecomastia,

Table 53.5 Various testosterone preparations available and their doses for pubertal induction

Topical		
1. Testosterone (1%)	Cernos (Sun Pharma), Androgel	Applied over shoulders and arms (generally not used in paediatric age group)
2. DHT gel (2.5%)	Andractim gel (online via trade India, Pharmeasy)	Applied over genital skin (0.1–0.3 mg/kg/day)
Intramuscular		
1. Testosterone enanthate	Testoviron 250 mg	Start 50 mg monthly and increase dose by 50 mg 6 monthly
2. Combined [testosterone propionate (30 mg) + testosterone phenylpropionate (60 mg) + testosterone Isocaproate (60 mg) + testosterone decanoate (100 mg)]	Sustanon	Adult dose: 200 mg every 2 weeks
3. Testosterone	Aquaviron 25 mg	The exact dose equivalence of Aquaviron is not known. However, 25 mg is the adult dose. The typical dose in newborns/infants for micropenis is 1–2 mg/kg/month × 3 months
Oral		
Testosterone undecanoate	Cernos 40 mg Testoksi 40 mg	Starting dose is 40 mg once daily and increasing to adult dose (up to 120–180 mg/day) (generally not used for induction, may be used for maintenance)

aggressiveness and erythrocytosis, during TRT [32]. Periodic assessment of bone age is essential during therapy.

53.9.1.1 Gonadotropin-Based Therapy

Congenital HH is characterized by a lack of gonadal development and infertility. Pulsatile GnRH or gonadotropin treatment can also be used for pubertal and fertility induction in adolescent boys or young adults. These agents induce testosterone production by testicular Leydig cells and spermatogenesis in the seminiferous tubules.

53.9.1.2 Pulsatile GnRH

Pulsatile GnRH infusion simulates physiological gonadotropin release and induces virilization and spermatogenesis. The dose is adjusted per the testosterone response and ranges from 25 to 600 ng/kg every 2 h [35]. However, it is cumbersome and ineffective in patients with GnRH receptor mutations and pituitary damage.

53.9.1.3 Gonadotropin Therapy (GnT)

hCG has an LH-like action and is given as injections two-three times per week. hCG monotherapy can be used successively in patients with partial HH phenotype (TV ≥ 4 ml). However, those with a severe phenotype (TV < 4ml), combined gonadotropin therapy (hCG 1000-1500 IU + FSH 75-150 IU; 2–3 times per week) results in better testicular growth and improved fertility outcomes [35]. The most severely affected patients can be offered a sequential treatment approach (FSH pretreatment, followed by a combination of hCG and FSH). While hCG promotes virilization, growth spurt and psychosexual development, the early addition of FSH aids in a greater increase in TV and testicular descent and more effective spermatogenesis and therefore helps to maximize fertility potential [32]. GnT can be stopped when the patient has attained sexual maturity, with no further increase in testicular volume; this usually requires 2–3 years of therapy.

53.9.2 Pubertal Induction in Girls

Pubertal induction in females involves the administration of both oestrogens (to promote feminization) and progesterone (to prevent endometrial hyperplasia) [36].

53.9.2.1 Hormone Replacement

Oral or transdermal oestradiol preparations are commonly used. It is essential to start with low doses and gradually increase the dose (see Table 53.6) [32, 37]. The starting dose is 5–10% of the adult dose (0.25 mg oestradiol valerate, 5 μg ethinylestradiol, or 0.125 mg conjugated oestrogen), and dose increments are made by 10–20% of the adult dose every 6 months to reach the adult dose (2 mg oestradiol valerate, 30 μg ethinylestradiol and 0.625 mg conjugated oestrogen). Recent studies suggest that transdermal 17β-estradiol represents the most physiological oestrogen preparation and has a better safety and efficacy profile than ethinylestradiol or conjugated equine oestrogen [32]. The transdermal route is preferred over the oral

Table 53.6 Various oestrogen preparations available in India and their doses for pubertal induction

Preparations	Brand name	Dosage
Oral		
Conjugated Oestrogens	Premarin 0.625 mg, Conjugase, Espauz	Start with 0.3 mg, increase 6 monthly to an adult dose 1.25 mg
Ethinyl estradiol	Ethinorm 0.01 mg, Lynoral 0.01 mg	Start with 2 mcg daily and gradually advance to the adult dose of 10 mcg daily.
Estradiol valerate	Progynova 1 mg, 2 mg, Enrifol, Evadiol, Estova	Start with 0.25 mg, gradually increase to adult dose of 2 mg/day
Beta estradiol	Estrabet	Start with 0.25 mg, gradually increase to adult dose of 2 mg/day
Transdermal		
Estradiol	Estraderm Mx 50, Mx 25 μg	Start with 7 μg/day and gradually increase to 50–100 μg biweekly
Progesterone preparations		
Micronized progesterone	Miprogen 200 mg, Endogest 200 mg	Medroxyprogesterone 5–10 mg or micronized progesterone 200 mg on day 1 to day 10 of the calendar month.
Medroxyprogesterone acetate (10 mg)	Meprate 10 mg (Serum Institute)	Progesterone treatment is started after the patient experiences breakthrough bleeding or after reaching the full dose of oestrogen.

route, as the latter carries a higher risk of thrombosis and causes greater suppression of hepatic IGF-1 production [37]. Progesterone (200 μg/d micronized progesterone or 10 mg medroxyprogesterone acetate for 10 days) is added after 2 years of oestrogen replacement or when withdrawal bleeding occurs [35]. The addition of progesterone is not needed in phenotypic females with CAIS and in conditions where the uterus is not present, e.g. Mayer–Rokitansky–Küster–Hauser syndrome and MURCS syndrome (Mullerian aplasia, renal anomalies, cervicothoracic somite dysplasia syndrome).

Induction of Female Fertility

Females with HH have insufficient maturation of ovarian follicles, resulting in chronic anovulation. Ovulation induction is achieved either with pulsatile GnRH therapy or with FSH treatment followed by hCG or LH [35]. The goal is to prevent ovarian hyperstimulation and multiple pregnancies.

53.9.2.2 Adverse Effects of Hormonal Therapy and When to Stop

The testosterone preparations are well tolerated and safe. The common adverse effects reported are injection site pain, swelling, aggression and varying libido. Appearance of pubic hair (especially with topical preparation) is an adverse effect of topical testosterone and DHT gel application. Rare effects are gynecomastia and priapism, polycythaemia, low high-density lipoprotein. Higher doses of testosterone given at a younger age will result in advancement of bone age and premature

Table 53.7 Monitoring guidelines for patients on sex steroid replacement

Parameter	Male	Female
Clinical evaluation and SMR staging	3 monthly	
Bone age	Frequency to be decided based on whether statural growth is a concern. It should be assessed at baseline and 6–12 months after starting treatment, and then annually till fusion of epiphyses is complete.	
Blood pressure, liver enzymes, clotting parameters, uterine size	–	3–6 monthly
Prostate size	Baseline, 24 months then 1–2 yearly	–
LH, FSH level	Baseline, 3 months, 12, 24 months and then yearly	
DEXA	Baseline, at 24 months and then 1–2 yearly	

epiphyseal closure. The transdermal preparations may cause skin irritations and transfers to other individuals who come in contact [38, 39].

Oestrogen replacement is also safe and rare adverse events related to impaired glucose tolerance, thromboembolism and lipid profile derangements are reported. The sequential progesterone supplementation in patients with intact uterus is essential for uterine protection from cancer. Table 53.7 shows the suggested monitoring recommendations for hormone replacement therapy in male and females.

The levels of sex steroids and LH should be done six monthly and maintained in SMR stage specific range [38, 39].

Most patients with Turner Syndrome and gonadal dysgenesis require lifelong replacement. Patients with testicular regression syndrome, Klinefelter Syndrome, gonadal dysgenesis also require continuous replacement. However children with 5alpha RD and PAIS require clinical and laboratory monitoring for their needs. Most patients are satisfied and tolerate hormone replacement very well [40].

53.9.3 Pubertal Suppression in DSD

Pubertal suppression is often required in central precocious puberty developing in some DSDs like CAH. Adolescents perplexed about gender identity or developing gender dysphoria or gender identity issues at puberty may also require pubertal suppression. It is advisable to defer gonadectomy till the legal age of consent in adolescents, who otherwise don't qualify for gonadectomy due to risk of malignancy. Therefore, in adolescents with AIS raised as females who start experiencing gynecomastia, or in children with CAH with excessive virilization who identify as boys, but are concerned about the possibility of breast development, it is better to offer GnRH analogue therapy for a few years.

Conditions where children with DSD who first present to a tertiary care centre in the peripubertal or pubertal period may also benefit from pubertal suppression for some time so that the medical team gets more time to reach a diagnosis, and the patient and the family can also come to terms with the situation and decide on the course to be followed. GnRH analogues are preferred drugs for pubertal suppression.

Triptorelin pamoate and Leuprolide acetate are two GnRH analogues available in various strengths in the Indian market and are given as deep intramuscular injections at 1–3 month intervals [3].

53.9.4 Fertility in DSD

Most DSD patients have intimate partners but poor quality of sexual life. The data on fertility in DSD is very scarce [41]. In males with various DSD (5alpha reductase, AIS, hypogonadotropic hypogonadism, CAH), cases of spontaneous conception are reported along with higher success rates with newer assisted reproduction techniques. In SV, CAH live births are reported in 33–50% but the number drops to 0–10% in classical salt-wasting forms. The majority of females with TS are infertile and pregnancy rates varying from 2 to 14% are reported in classical and mosaic karyotype forms. In Klinefelter syndrome, fertility rates up to 40% are reported with newer techniques.

References

1. Lee PA, Nordenström A, Houk CP, Ahmed SF, Auchus R, Baratz A, Global DSD Update Consortium, et al. Global disorders of sex development update since 2006: perceptions, approach and care. Horm Res Paediatr. 2016;85(3):158–80.
2. Fisher AD, Ristori J, Fanni E, Castellini G, Forti G, Maggi M. Gender identity, gender assignment and reassignment in individuals with disorders of sex development: a major of dilemma. J Endocrinol Investig. 2016 Nov;39(11):1207–24.
3. Babu R, Shah U. Gender identity disorder (GID) in adolescents and adults with differences of sex development (DSD): a systematic review and meta-analysis. J Pediatr Urol. 2021 Feb;17(1):39–47.
4. Tamar-Mattis A, Baratz A, Dalke KB, Karkazis K. Emotionally and cognitively informed consent for clinical care for differences of sex development. Psychol Sex. 2014;5(1):44–55.
5. Greenberg JA. Legal, ethical, and human rights considerations for physicians treating children with atypical or ambiguous genitalia. Semin Perinatol. 2017 Jun;41(4):252–5.
6. Gravholt CH, Viuff M, Just J, Sandahl K, Brun S, van der Velden J, et al. The changing face of Turner syndrome. Endocr Rev. 2023 Jan;44(1):33–69.
7. Gravholt CH, Andersen NH, Conway GS, Dekkers OM, Geffner ME, Klein KO, International Turner Syndrome Consensus Group, et al. Clinical practice guidelines for the care of girls and women with Turner syndrome: proceedings from the 2016 Cincinnati International Turner Syndrome Meeting. Eur J Endocrinol. 2017;177(3):G1–G70.
8. Bonomi M, Rochira V, Pasquali D, Balercia G, Jannini EA, Ferlin A, Klinefelter Italia N Group (KING). Klinefelter syndrome (KS): genetics, clinical phenotype and hypogonadism. J Endocrinol Investig. 2017 Feb;40(2):123–34.
9. Weidler EM, Pearson M, van Leeuwen K, Garvey E. Clinical management in mixed gonadal dysgenesis with chromosomal mosaicism: considerations in newborns and adolescents. Semin Pediatr Surg. 2019 Oct;28(5):150841.
10. Dabas A, Vats P, Sharma R, Singh P, Seth A, Jain V, et al. Management of Infants with congenital adrenal hyperplasia. Indian Pediatr. 2020 Feb 15;57(2):159–64.

11. Speiser PW, Arlt W, Auchus RJ, Baskin LS, Conway GS, Merke DP, et al. Congenital adrenal hyperplasia due to steroid 21-hydroxylase deficiency: an Endocrine Society clinical practice guideline. J Clin Endocrinol Metab. 2018 Nov 1;103(11):4043–88.
12. Houben CH, Tsui SY, Mou JW, Chan KW, Tam YH, Lee KH. Reconstructive surgery for females with congenital adrenal hyperplasia due to 21-hydroxylase deficiency: a review from the Prince of Wales hospital. Hong Kong Med J. 2014;20(6):481–5.
13. Fukami M, Ogata T. Congenital disorders of estrogen biosynthesis and action. Best Pract Res Clin Endocrinol Metab. 2022 Jan;36(1):101580.
14. Belgorosky A, Guercio G, Pepe C, Saraco N, Rivarola MA. Genetic and clinical spectrum of aromatase deficiency in infancy, childhood and adolescence. Horm Res. 2009;72(6):321–30.
15. Menon PS, Khatwa UA. The child with micropenis. Indian J Pediatr. 2000 Jun;67(6):455–60.
16. Cheng PK, Chanoine JP. Should the definition of micropenis vary according to ethnicity? Horm Res. 2001;55:278–81.
17. Saneja V, Sethi A, Gupta H. To study of stretched penile length and testicular volume in Indian newborns. Curr Pediatr Res. 2021;25(1):313–8.
18. Mondal R, Ray S, Chatterjee K, Sabui TK, Hazra A, Das S, et al. Penile length and testicular volume in newborns. Indian J Pediatr. 2016 Nov;83(12–13):1398–404.
19. Prabhu SR, Mahadevan S, Bharath R, Jagadeesh S, Kumutha J, Suresh S. Normative data for stretched penile length in term neonates born in Tamil Nadu. Indian J Endocrinol Metab. 2014;18:585–6.
20. Manivarmane VG, Bhat BV, Bhatia BD, Kumar S. Genital standards for south Indian male newborns. Indian J Pediatr. 1995;62:593–6.
21. Schonfeld WA, Beebe GW. Normal growth and variation in male genitalia from birth to maturity. J Urol. 1942;48:759–77.
22. Bertelloni S, Russo G, Baroncelli GI. Human chorionic gonadotropin test: old uncertainties, new perspectives, and value in 46,XY disorders of sex development. Sex Dev. 2018;12(1–3):41–9.
23. Segal TY, Mehta A, Anazodo A, Hindmarsh PC, Dattani MT. Role of gonadotropin-releasing hormone and human chorionic gonadotropin stimulation tests in differentiating patients with hypogonadotropic hypogonadism from those with constitutional delay of growth and puberty. J Clin Endocrinol Metab. 2009 Mar;94(3):780–5.
24. Hatipoğlu N, Kurtoğlu S. Micropenis: etiology, diagnosis and treatment approaches. J Clin Res Pediatr Endocrinol. 2013;5(4):217–23.
25. Guthrie RD, Smith DW, Graham CB. Testosterone treatment for micropenis during early childhood. J Pediatr. 1973 Aug;83(2):247–52.
26. Main KM, Schmidt IM, Skakkebaek NE. A possible role for reproductive hormones in newborn boys: progressive hypogonadism without the postnatal testosterone peak. J Clin Endocrinol Metab. 2000;85:4905–7.
27. Arisaka O, Hoshi M, Kanazawa S, Nakajima D, Numata M, Nishikura K, et al. Systemic effects of transdermal testosterone for the treatment of microphallus in children. Pediatr Int. 2001;43:134–6.
28. Ben-Galim E, Hillman RE, Weldon VV. Topically applied testosterone and phallic growth. Its effects in male children with hypopituitarism and microphallus. Am J Dis Child. 1980;134:296–8.
29. Bertelloni S, Scaramuzzo RT, Parrini D, Baldinotti F, Tumini S, Ghirri P. Early diagnosis of 5 alpha-reductase deficiency in newborns. Sex Dev. 2007;1:147–51.
30. Levy JB, Husmann DA. Micropenis secondary to growth hormone deficiency: does treatment with growth hormone alone result in adequate penile growth? J Urol. 1996;156:214–6.
31. Bougnères P, François M, Pantalone L, Rodrigue D, Bouvattier C, Demesteere E, et al. Effects of an early postnatal treatment of hypogonadotropic hypogonadism with a continuous subcutaneous infusion of recombinant follicle-stimulating hormone and luteinising hormone. J Clin Endocrinol Metab. 2008;93:2202–5.
32. Federici S, Goggi G, Quinton R, Giovanelli L, Persani L, Cangiano B, et al. New and consolidated therapeutic options for pubertal induction in hypogonadism: in-depth review of the literature. Endocr Rev. 2022 Sep 26;43(5):824–51.

33. Han TS, Bouloux PM. What is the optimal therapy for young males with hypogonadotropic hypogonadism? Clin Endocrinol. 2010;72(6):731–7.
34. Dunkel L, Quinton R. Transition in endocrinology: induction of puberty. Eur J Endocrinol. 2014;170(6):R229–39.
35. Boehm U, Bouloux PM, Dattani MT, de Roux N, Dodé C, Dunkel L, et al. Expert consensus document: European Consensus Statement on congenital hypogonadotropic hypogonadism--pathogenesis, diagnosis and treatment. Nat Rev Endocrinol. 2015 Sep;11(9):547–64.
36. Alonso LC, Rosenfield RL. Oestrogens and puberty. Best Pract Res Clin Endocrinol Metab. 2002;16(1):13–30.
37. Kenigsberg L, Balachandar S, Prasad K, Shah B. Exogenous pubertal induction by oral versus transdermal estrogen therapy. J Pediatr Adolesc Gynecol. 2013 Apr;26(2):71–9.
38. Hewitt J, Zacharin M. Hormone replacement in disorders of sex development: current thinking. Best Pract Res Clin Endocrinol Metab. 2015 Jun;29(3):437–47.
39. Bertelloni S, Dati E, Baroncelli GI. Disorders of sex development: hormonal management in adolescence. Gynecol Endocrinol. 2008 Jun;24(6):339–46.
40. Nordenström A, Röhle R, Thyen U, Bouvattier C, Slowikowska-Hilczer J, Reisch N, DSD-LIFE group, et al. Hormone therapy and patient satisfaction with treatment, in a large cohort of diverse disorders of sex development. Clin Endocrinol. 2018 Mar;88(3):397–408.
41. Acién P, Acién M. Disorders of sex development: classification, review, and impact on fertility. J Clin Med. 2020 Nov 4;9(11):3555.

Criminal Issues Pertaining to Children and Adolescents with DSD

54

Varun Bansal and Parvathi Menon

"I am what I am, so take me as I am"

—Van Goethe

Abstracts

Disorders/Differences of Sexual Development or as commonly referred to as Intersex is a medical condition where people born with it have a different sexual anatomy than the archetypal male or female. Due to their unique identity, it becomes imperative that laws protect their vulnerability, particularly in the case of children and adolescents who have DSD. This Article attempts to delineate how far the laws in its current form accommodate DSD children and adolescents. By showcasing the lacunas in the present laws in catering the DSD Children and Adolescents, this Article advocates that the State must take steps in enacting a new law that specifically addresses the grievances of such children and adolescents. The Article further espouses that the State must amend some of the existing provisions under Criminal law that incorporate heteronormative language which tend to alienate such children and adolescents. In doing so, the State would be honouring its commitment to the Constitution laid down in the Directive Principles of State Policy and taking a positive step in promoting diverse identities.

Keywords

DSD in Children and Adolescents · Disorders of Sexual Development · Intersex · Criminal law

V. Bansal (✉) · P. Menon
Delhi High Court, New Delhi, India

S. K. Ratan (ed.), *Children with Differences in Sex Development*,
https://doi.org/10.1007/978-981-97-1639-5_54

One of the famous scientific thought experiments of quantum mechanics, developed by Erwin Schrodinger, advocates a situation where a cat is put in a sealed container along with poison and a radioactive source. The interesting thing about this experiment was that Schrodinger was only trying to establish the varied misinterpretation that results in a variety of results which may or may not relate to reality. Like in the experiment, the cat is either dead or alive depending on some presumption. Similarly, interpretation or misinterpretations pertaining to gender and sex in India have led to absurd understanding which increasingly has gone beyond reality.

However, when popular American medical shows like "Grey's Anatomy" explore the theme of Disorders/Differences of Sexual Development (DSD)*, colloquially referred to as "Intersex",[1] the issue ventures out of the scope from the activist community and medical academia to mainstream society. And, even with growing attention to the subject in the media, its substance or understanding in the jurisprudence remains largely sparse.

The situation is much worse in developing countries like India that suffer from acute misunderstanding of the needs of people with DSD. Further, the laws have grossly failed to identify this vulnerable group, which has slowed their formal integration into the society.

This article is an attempt to analyse the various issues pertaining to children and adolescents with DSD through the lens of the Criminal Laws in India. The first part of the article shall strive to articulate the concept of DSD and how it differs from LGBT. Thereafter, the article shall analyse the various provisions under the Criminal Laws to understand how far it caters to DSD and highlights the limitations in law which has furthered the adverse understanding.

54.1 What Is DSD?

As per Selma Feldman Witchel, MD, Disorder/Differences of Sexual Development is the medical terminology that is used to "*denote a group of congenital conditions associated with atypical development of internal and external genital structures.... Affected individuals may be recognized at birth due to ambiguity of the external genitalia. Others may present later with postnatal virilization, delayed/absent puberty, or infertility*" [1]. In simple terms, it refers to a condition where people born with it have a different sexual anatomy than the archetypal male or female. It is believed that around 1.7–4% of the population is born with DSDs [2].

Oftentimes it is noticed that people having DSDs are confused with the transgender or the transsexual. This confusion becomes quite glaring when one analyses the seminal judgement of the Honourable Supreme Court in the case of *NALSA v. Union of India*,[2] which amalgamated multiple identities such as intersex, Hijras, etc. under one umbrella term of "transgender". Similarly definition of transgender under Section 2(k) of the Transgender Persons (Protection of Rights) Bill, 2019 which

[1] In this Article, Intersex has been used interchangeably with DSD.

[2] WRIT PETITION (CIVIL) NO.400 OF 2012.

states—*"transgender person" means a person whose gender does not match with the gender assigned to that person at birth and includes trans-man or trans-woman (whether or not such person has undergone Sex Reassignment Surgery or hormone therapy or laser therapy or such other therapy),* **person with intersex variations,** *genderqueer and person having such socio-cultural identities as kinner, hijra, aravani and jogta.* By clubbing DSD/intersex with the transgender, law and judiciary has further perpetuated the prevailing confusion which in turn elevates the gap between the needs of people with DSD and the legal interventions necessary to regress their grievances.

In order to understand the difference between the transgender and DSD, it is pertinent to understand the difference between the term "sex" and "gender". Sex is usually determined on the basis of biological characteristics whereas gender is a more complex subject and generally refers to socially constructed roles of men, women and gender diverse people [3]. DSD/Intersex refers to a condition where the sexual anatomy of a person is different from male or female whereas the transgender may have been born with typical sexual anatomy but may identify themselves differently from the gender assigned to them at birth [4]. There may be intersection between both the terminologies wherein some people with DSD may identify themselves as transgender. However, most people with DSD identify themselves as male or female, and not as transgender [4]. Therefore, it becomes imperative for law and judiciary to have a nuanced understanding about the difference between different identities so as to cater to their needs rather than clubbing them all together.

54.2 Criminal Law Pertaining to Children & Adolescents with DSD

The Directive Principle of State Policy casts a duty upon the State under Article 39(f) of the Constitution of India. The Article states that "*the State shall direct its policy towards securing that* **children** *are given opportunities and facilities to develop in a* **healthy manner** *and conditions of freedom and dignity and that childhood and youth are* **protected against exploitation** *and against moral and material abandonment.*" Therefore the State has to ensure that all the children and adolescents, including those with DSD, shall have the opportunity to develop in a healthy manner without any discrimination. But is our law doing enough to protect the children and adolescents with DSD?

Children and adolescents with DSD face discrimination both in ways that are similar to those faced by the LGBT community, as well as in ways that are different to them. Therefore this article shall discuss the criminal issues faced by the children and adolescents with DSD under two categories—(1) Those that are specifically affecting the children and adolescents with DSD and (2) Those that are affecting the entire spectrum of LGBTQI Community as a whole.

54.3 Criminal Issues Pertaining Specifically to Children & Adolescents with DSD

The most concerning thing that children and adolescents with DSD face is being subjected to forced medical/surgical procedures that aim to "normalize" them and integrate them into the society. These include procedures such as clitoral reduction, removing sensitive erectile tissue to reduce clitoral size for cosmetic reasons, gonadectomies and forced sterilizations.[3] The World Health Organization in its report Sexual Health, Human Rights and Law has stated

> major concern for intersex people is that so-called sex normalizing procedures are often undertaken during their infancy and childhood, to alter their bodies, particularly the sexual organs, to make them conform to gendered physical norms, including through repeated surgeries, hormonal interventions and other measures. As a result, such children may be subjected to medically unnecessary, often irreversible, interventions that may have lifelong consequences for their physical and mental health, including irreversible termination of all or some of their reproductive and sexual capacity. Medical procedures may sometimes be justified in cases of conditions that pose a health risk or are considered life-threatening. Such procedures, however, are sometimes proposed on the basis of weak evidence, without discussing and considering alternative solutions...[4]

Similarly, the WHO, the Office of the High Commissioner for Human Rights ("OHCHR"), UN Women, The Joint United Nations Programme on HIV and AIDS ("UNAIDS"), the United Nations Development Programme ("UNDP"), the United Nations Population Fund ("UNFPA") and the United Nations Children's Fund ("UNICEF") jointly opposed early genital surgeries on intersex children in 2013, stating, *"intersex persons, in particular, have been subjected to cosmetic and other non-medically necessary surgery in infancy, leading to sterility, without informed consent of either the person in question or their parents or guardians."*[5]

One of the biggest criticisms that is often levelled against the Transgender Persons (Protection of Rights) Bill, 2019 is that it fails to redress the grievances of the Intersex Persons as it fails to prohibit and criminalize forced medical procedures.

The Indian Penal Code, 1860 (hereinafter IPC), which is the primary criminal law statute of India, lists out General Defences to the Crime in its Chapter IV. This Chapter on General Defences lists out certain circumstances under which if any person commits any offence, then that person shall be granted immunity for the same. One of the General Defences has been listed under Section 89 which states

> Act done in good faith for benefit of child or insane person, by or by consent of guardian.— Nothing which is done in good faith for the benefit of a person under twelve years of age, or of unsound mind, by or by consent, either express or implied, of the guardian or other

[3] Centre for Law and Policy Research & Solidarity Foundation, "Beyond The Binary: Policy Brief, Advancing Legal Recognition for Intersex Persons in India", p. 17 (2020).

[4] World Health Organization, "Sexual health, human rights and the law", p. 26 (2015).

[5] World Health Organization, "Eliminating forced, coercive and otherwise involuntary sterilization: an interagency statement, OHCHR, UN Women, UNAIDS, UNDP, UNFPA, UNICEF and WHO", p. 2 (2014).

person having lawful charge of that person, is an offence by reason of any harm which it may cause, or be intended by the doer to cause or be known by the doer to be likely to cause to that person

Under the protection of Section 89 IPC, Doctors undertake complex medical procedures on children having DSD to alter their bodies and ensure that they conform to the gender norms after taking consent from the parents. Such medical procedures are risky, immensely painful, damage nerves permanently, scarring and infections being some possible side effects.[6] Unfortunately, the parents in absence of information on the aftereffects of the surgery give consent to such procedures. Therefore, it becomes imperative that such medical procedures are prohibited and criminal Penalty is imposed.

However, not everything is dark and grim and positive steps are being undertaken in this direction as well. The Honourable Supreme Court took the first step in this direction in its judgement in ***NALSA* case**[7] by reaffirming the Yogyakarta Principle and observing—*that no one shall be forced to undergo medical procedures, including sex reassignment surgery, sterilization or hormonal therapy, as a requirement for legal recognition of their gender identity.*

Thereafter, the Honourable Madras High Court gave a landmark decision in the case of ***Arun Kumar v The Inspector General of Registration***[8] wherein it agreed with the WHO Report which stated—"*that medical and psychological professionals should be educated and trained about physical, biological and sexual diversity and integrity, and that they should properly inform patients and their parents of the consequences of surgical and other medical interventions and provide additional support... It has also been recommended that investigation should be undertaken into incidents of surgical and other medical treatment of intersex people without informed consent and that legal provisions should be adopted in order to provide remedies and redress to the victims of such treatment*". In light of this report, the Madras High Court directed the Government of Tamil Nadu to issue a Government Order enshrining the aforesaid mandate so as to effectively ban sex reassignment surgeries on intersex infants and children. Pursuant to this the Government of Tamil Nadu, vide an Order Vide Letter G.O (Ms.) No 355, banned sex reassignment surgeries on intersex infants and children except in life-threatening situations and orders accordingly.[9]

Recently the Delhi Commission for Protection of Child Rights (DCPCR) has recommended that the Delhi Government should declare a ban on medically unnecessary, sex selective surgeries on intersex infants and children except in cases of life-threatening situations.[10]

---

[6] *Supra* note 3.

[7] *Supra* note 2.

[8] W. P. (MD) No. 4125 of 2019.

[9] DCPCR, Order No. F/DCPCR/20-21/Health & Nutrition/Project File VIII/13295-95 dates 13.01.2021 at p. 2.

[10] Ibid at p. 4.

54.4 Criminal Issues Pertaining to Children and Adolescents with DSD and those Who Identify Themselves as LGBT

The Constitution of India guarantees to all the citizens of India equality before law and equal protection of law under Article 14 of the Constitution. However, despite the Constitutional mandate, the LGBTQI community has led a harsh and difficult life and faces discrimination at every step of the way. The same has been affirmed by the Honourable Supreme Court in the groundbreaking judgement of ***Navtej Singh Johar v Union of India*** in which they had categorically expressed their views about the LGBTQI Community in the following words:

> History owes an apology to the members of this community and their families, for the delay in providing redressal for the ignominy and ostracism that they have suffered through the centuries. The members of this community were compelled to live a life full of fear of reprisal and persecution

After such important judgements, one would expect that IPC, being the primary criminal law statute, would provide protection to this community without any discrimination. However, by incorporating heteronormative language, it has further perpetuated the grievances of this section of the society. Some of the provisions of the IPC that do not cater to the needs of the LGBTI community by incorporating the heteronormative language are as follows:

(a) **Definition Clause—Section 10** provides the definition of Man & Woman. It states *"The word "man" denotes a male human being of any age; the word "woman" denotes a female human being of any age"*. By incorporating such a definition, IPC has furthered the marginalization of the DSD people as it has restricted even a criminal offence to man and woman only.
(b) **Miscarriage Laws—Section 313 and 314** provide punishment for causing miscarriage to a woman. However, it is possible for Person with DSD [5] to become pregnant and by incorporating the use of the word "woman" instead of a gender-neutral terminology, such a person may be denied protection under this provision.
(c) **Sexual offences—Section 354** penalizes the act of outraging the modesty of a woman by a man. **Section 354 A, B, C and D** provides punishment for the offence of sexual harassment, disrobing a woman, voyeurism and stalking, respectively. Despite the fact that such offences are regularly and prevalently committed against the LGBTI community, IPC fails to understand that such an offence can be committed against this community as well. **Section 375** is the definition clause of rape and assumes that only a man can commit such an offence and furthermore, it assumes that it can be committed only against a woman. **Section 509** again deals with insult to the modesty of women and thereby completely excludes the LGBTI Community from its ambit.
(d) **Aggravated Sexual Offences against Children—**Section 376(3) states "*Whoever, commits rape on a woman under sixteen years of age shall be pun-*

ished with rigorous imprisonment for a term which shall not be less than twenty years, but which may extend to imprisonment for life, which shall mean imprisonment for the remainder of that person's natural life, and shall also be liable to fine".

Section 376 AB states—"*Whoever, commits rape on a woman under twelve years of age shall be punished with rigorous imprisonment for a term which shall not be less than twenty years, but which may extend to imprisonment for life, which shall mean imprisonment for the remainder of that person's natural life, and with fine or with death*".

Section 376 DA states—*Where a woman under sixteen years of age is raped by one or more persons constituting a group or acting in furtherance of a common intention, each of those persons shall be deemed to have committed the offence of rape and shall be punished with imprisonment for life, which shall mean imprisonment for the remainder of that person's natural life, and with fine.*

Section 376 DB states—*Where a woman under twelve years of age is raped by one or more persons constituting a group or acting in furtherance of a common intention, each of those persons shall be deemed to have committed the offence of rape and shall be punished with imprisonment for life, which shall mean imprisonment for the remainder of that person's natural life, and with fine, or with death.*

By not incorporating a gender-neutral terminology, the State has failed in fulfilling the duty cast upon it by the Constitution under Article 39(f) in protecting all the children without discrimination.

The Transgender Persons (Protection of Rights) Bill, 2019 has also listed out certain specific offences against the transgender persons such as compelling them to do forced labour, denying them access to public place, forcing them out of their residence and physical, economic, sexual, verbal and emotional abuse of such persons. The problem with this Act is that the punishment provided for such offences is mere 6 months-2 years which is in stark contrast with the punishment provided in IPC. This lip service paid by the new Bill has met with severe criticism, however, the same has passed by both the Houses of the Parliament. Therefore, it becomes imperative to make IPC gender neutral to cater to the needs of the persons with DSD. A writ petition was presented before the Honourable Delhi High Court in the case of ***Sanjjiiv Kumar v Union of India,***[11] for making these provisions of IPC gender neutral. However, the Court rejected the contention and observed—"*In the light of the view taken by the Union of India after receiving the report of the Law Commission and also after due deliberations, a policy decision has been taken by the Union of India as stated hereinabove, we see no reason to entertain this writ, hence, this writ petition is hereby dismissed at this stage.*"

A positive step to make the laws gender neutral was made by the State by enacting the Protection of Children from Sexual Offences Act, 2012 (hereinafter POCSO). The objective of this Act is to protect children from offences of sexual assault,

[11] W.P.(C) 8745/2017.

sexual harassment and pornography and provide for establishment of Special Courts for trial of such offences and for matters connected therewith or incidental thereto. The best part about this Act is that it incorporates the use of the word "child" without any connotation of sex or gender. This ensures protection to all the diverse forms of identities and thereby fulfils the Constitutional mandate.

54.5 Conclusion

Amartya Sen in his award winning work "Development as Freedom" forcefully argued that promoting freedom of an individual to pursue his/her choice is both the end and the means to achieve development, and concentrating on the freedom of choice of the individual agency is the key to promote development in a country.[12] Therefore, it would be apt to say that empowering people with diverse identities to pursue their own choices would be both ends and means to development of the country. By making the provisions of IPC more gender neutral and by enacting new provisions to cater to the needs of the children and adolescents with DSD would be a step forward in the direction of empowering them. By doing so, the State shall fulfil its duty as the parent of the nation (parens patriae doctrine) and reaffirm the basic structure of the Constitution.

References

1. Witchel SF. Disorders of sex development. Best Pract Res Clin Obstet Gynaecol. 2018 Apr;48:90–102. https://doi.org/10.1016/j.bpobgyn.2017.11.005.
2. Kate H. Who will make room for the intersexed? Am J Law Med. 2004 Summer;30(1):41–68.
3. Canadian Institutes of Health Research. What is gender? What is sex? https://cihr-irsc.gc.ca/e/48642.html. Retrieved 15 Feb 2023
4. Srishti Madurai. Information Toolkit: Intersex Human Rights in India. https://www.apcom.org/wp-content/uploads/2020/05/Information-Toolkit-Intersex-Human-Rights-in-India. pdf. Retrieved 15 Feb 2023.
5. Schultz BAH, Roberts S, Rodgers A, Ataya K. Pregnancy in true hermaphrodites and all male offspring to date. Obstet Gynecol. 2009 Feb;113(2 Pt 2):534–6. https://doi.org/10.1097/AOG.0b013e3181866456.

[12] Amartya Sen, "*Development as Freedom*" (New York, NY: Knopf, 1999) as cited in Mariana Prado & Michael Trebilock, "*Advanced Introduction to Law and Development* (Massachussetts, Edward Elgar Publishing Inc., 2014) at 8.

55 Neurophysiological Basis of Gender Incongruence in DSD vs Transgender Population

Asmita Patil, Simmi K. Ratan, and Devendra Yadav

Abstract

Gender is a social construct of male and female with varied responsibilities. Any deviation from optimum gender may create some problem in form of various behavioral expression which may be affected by neurophysiological changes. The gender incongruence may be due to the fronto-occipito-parietal disconnection. Brain structures like right insula, pre-cuneus, temporo-parietal junction, orbital and medial frontal, and anterior cingulate cortex have been implicated in cognitive processes. The brain dimorphism may also be due to combine genetic and epigenetic changes.

Keywords

Gender incongruence · Neurophysiology · Brain · Transgender

55.1 Introduction

Primal goals of any organism are survival of self and perpetuation of its species. Nature also favors those species that are able to produce diversification among members, an attribute critical to species survival as the nature of environmental (and other) stresses changes through time. One solution to this problem is sexual differentiation, that is, the evolution of two sexually dissimilar individuals belonging to

A. Patil
Department of Physiology, AIIMS, New Delhi, India

S. K. Ratan (✉)
Department of Pediatric Surgery, Maulana Azad Medical College, New Delhi, India

D. Yadav
Department of Pediatric Surgery, AIIMS, New Delhi, India

S. K. Ratan (ed.), *Children with Differences in Sex Development*,
https://doi.org/10.1007/978-981-97-1639-5_55

the same species, one male and one female, and each equipped with its own specific attributes necessary for its particular contribution to the process of procreation. Throughout evolution, conservation and expression of genes involved in the perpetuation of a species have clearly followed a process of "adaptation": an advantageous change in structure or function of an organ or tissue to meet the challenges of new conditions.

Sex, in simple terms, is a set of biological attributes and physiological characteristics that determine reproductive functions of a given individual. Gender, on the other hand, may be fluid and not limited to binary perceptions as is seen in Differences in Sex Development (DSD). World Health Organization defines gender as per the social constructs of male and female that include responsibilities and conventions: such as norms, roles, and relationships [1]. The mismatch between the self-identified gender and the sex assigned at birth is identified as gender dysphoria (GD) [2].

Assignment of gender has been traditionally done by health-care providers via physical inspection of the external genitalia at birth. In 99.95 percentage of newborns, it is possible to discern and assign the gender which aligns with the child's anatomical sex and phenotype. The remaining around 0.2 to 0.5 percent of the children may have ambiguous genitalia which would render the task of gender allotment a challenge, as they do not fit into the typical binary notion of gender. The further process requires multiple tests for karyotyping, hormonal assays and imaging studies to render assignment of gender. This process must be accompanied with psychosocial counselling sessions for the child and parents [3]. The ambiguity of gender assignment may lead to stigmatization and discrimination from not just the society, but also from the child's own parents. Many a time the taboo associated with Differences in Sex Development (DSD) results in infanticide, abandonment, and traumatization of the child. GD is essentially when the incongruence between the experiential or identified gender and the assigned one results in clinically significant psychological distress and major functional impairment. This discordant development of gonads and brain has been attributed to interplay of various genetic factors and molecular mechanisms requiring further evaluation [4].

"Optimal gender model" states that gender identity is more a consequence of upbringing rather than genotypic or phenotypic sex. This further adds on to the confusion in neurobiological determinants of GD wherein advent of surgical techniques and hormonal therapies that could modify the external genitalia and gender phenotype. Arguments have been put forth regarding the legality of performing "normalizing" procedures to assign decided sex vs the upbringing and social acceptance of a child left untreated until the age of consent.

55.2 Background

The sex-chromosome complement of the sperm fertilizing the ova determines the gender of the offspring. After that point, a number of molecular processes take place that cause germ cells to grow and migrate to the urogenital ridge where they

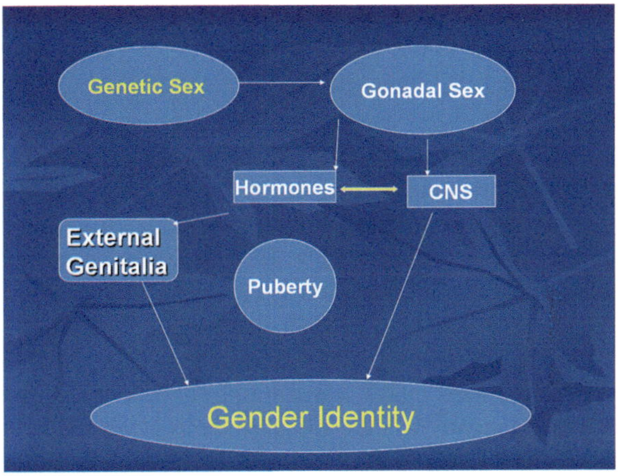

Fig. 55.1 Overview of factor influencing development of gender identity

eventually give rise to either male or female gonads. These male or female gonads produce specific hormones as they grow and develop, which pave the way for determination and differentiation of internal and external genitalia and determine the gender identity [5] (Fig. 55.1).

Differences in Sex Development (DSD) is a spectrum of developmental aberrations involving genetic, endocrine and/or receptor mediated rare conditions of the reproductive organs and genitalia. This variation in sexual development may also be classified as Variations in Sex Characteristics (VSC) or Diverse Sex Development. Gender incongruence is when a given individual identifies as a gender different from the one assigned at birth, which leads to gender dysphoria. According to the Diagnostic and Statistical Manual of Mental Disorders (DSM-5) and the American Psychiatric Association (APA), gender dysphoria is a nosological phenomenon that is defined as an individual's discontent with having been assigned to a certain gender or with being limited to a single gender [6].

The common problems associated with the gender incongruence are multiple and variable. The individuals are overtly emotionally charged and have issues related to gender identity and psychological issues around puberty [7]. These individuals have poor reproductive potential and if they procreate, there is risk of recurrence in future offspring. The genetic and hormonal disorders predispose to life-threatening metabolic complications and malignancy. The gender assignment controversy issues not only pertain to a person and family, but also resonate into the society. There is socio-cultural bias exhibited by differential disposition in social registration, legal regulations and religious edicts toward individuals with DSD [8].

55.3 History

Reisert and colleagues first investigated the role of genetics in the sexual differentiation of the diencephalon and mesencephalon in a rat embryonic cell culture model in 1991. Subsequent scientists confirmed that sex variances in vasopressin in the lateral septum, TH-ir cell density in the mesencephalon, spatial perceptive capacity, and steroid sensitivity may all be regulated genetically rather than by gonadal hormones [9].

Research using cadavers revealed an association between brain anatomy and gender identity. The researchers observed differences in the bed nucleus of stria terminalis, which is a region of the brain associated with sex and anxiety responses [10]. Additionally, no connection was discovered between sexual orientation and gender altering surgery [11]. Review studies during 2004 and 2006 verified that sexual differentiation disorder in sexually dimorphic brains could be correlated with gender incongruence [12, 13].

When it came to sexual dimorphism and gender incongruence, Garcia et al. (2008) reported that the interstitial nucleus of the anterior hypothalamus was far more significant than the stria terminalis [14].

In 2009, Luders et al. studied regional gray matter concentrations and found significantly greater volume of gray matter in the right putamen in transgender men compared to cisgender men. They concluded that transgender identity was associated with a distinct cerebral pattern [15]. The limitation of poor visibility of cerebral nuclei on MRI was remedied in 2010 by dissecting brains post-mortem and further performing diffusion tensor imaging (DTI) [16].

A 2011 review further correlated the structural differences in stria terminalis and interstitial nucleus of the anterior hypothalamus with circulating hormone levels in adults. This analysis supported the neurobiological theory about the origin of gender dysphoria, thereby attributing GD to the size of brain nuclei, the neuronal numbers, and their functional connectivity [17].

In 2019, Nguyen postulated that the GD persons' cortical thickness, gray matter volume, white matter microstructure, structural connectivity, and corpus callosum shape were more indicative of their chosen gender than their natal sex [18].

A review of brain studies published in 2021 noted that to draw more definitive conclusions or identify features associated with GD, better neuroimaging tools and larger sample sizes of representative populations were needed. This was because previous studies had yielded inconsistent results [19].

55.4 The Brain of Persons with Gender Incongruence

Sexual differentiation of the brain is an essential component of gender determination and differentiation. Hormonal regulation and deregulation of sexual differentiation of the reproductive system and the brain have been already discussed in previous chapters. Gonadal hormones act on the different neuronal systems via alteration of neurotransmitters, thereby resulting in physiological, psychological, and behavioral manifestations in gender expression.

Individuals with GD feel estrangement to their assigned physical sex, which was previously associated with prenatal or early post-natal sex hormone exposure affecting brain morphogenesis.

Recent studies propose the observed fronto-occipito-parietal disconnection as a neurobiological hallmark for GD. With advent in neuroimaging technologies attempts are being made to measure cortical thickness, gray matter volume, and structural connectivity in GD. These researches associate right-hemispheric differences particularly in the right insula, pre-cuneus, temporo-parietal junction, orbital and medial frontal, and anterior cingulate cortex with GD. The above structures have been implicated in cognitive processes involving body perception and ego-centric representation [20].

An f-MRI and DTI correlation study was conducted on 82 children and 82 teenagers to assess the effects of puberty on changes in the white-matter organization and regional fractional anisotropy (FA) as determined by diffusion MR imaging. A prominent neurological correlate of gender incongruence, according to the authors, is the bilateral inferior fronto-occipital fasciculus (IFOF), forceps major, and corpus callosum, wherein transgender adolescents showed lower FA in comparison with their cisgender peers [21].

Inhalation of androstadienone, an odorous androgen resembling artificial steroid induces hypothalamic response in heterosexual females as shown by increase in regional cerebral blood flow under positron emission tomography [22]. Another study in teenagers with GD demonstrated a hypothalamic excitatory response synchronous with their perceived gender and not the natal or assigned sex. After gender-affirming surgery, similar studies established that the cognitive, emotional, and sensorimotor regions of GD brains are more similar to their perceived gender [23].

By using the MRI data, voxel-based morphometry spatially normalizes the pictures to the same stereotactical space. This allows for the extraction of data regarding the volume of fluid, white matter, and gray matter. By applying the above-mentioned method, Simon et al. reported that, in comparison with cisgender male controls, girls and males with gender dysphoria exhibited a reduced gray matter volume in their left somatosensory and primary motor cortices [24]. Additionally, GD patients were shown to have less gray matter in the left posterior superior hemisphere of the cerebellum than in male controls, as well as less gray matter in the right inferior orbitofrontal cortex than in female controls [25, 26]. It has been demonstrated that these regions exhibit high levels of steroid receptors and sex hormones. Uncertainty surrounds the underlying causes of these neurological findings. Genetic, environmental, or endocrine variables, or a mix of these factors, may be involved of the specified etiopathogenesis.

Delving further, Clemens et al. used a specifically targeted "region-of-interest" approach to examine the most reported gray matter regions between trans- and cisgender individuals. The areas focused on were thalamus, putamen, cerebellum, angular gyrus, and precentral gyrus. They found significant pattern of neuroanatomical differences corroborated with role of putamen in cognitive processes related to body imagery and suggested it as a potential structural biomarker for GI [27].

Analyzing transgender women to the cisgender population, Hahn et al. observed a reduced intra-hemispheric connection ratio in the subcortical and limbic areas [28]. Comparing the transgender male population to the cisgender population and transgender women, they discovered a reduction in intra-hemispheric connectivity in the right frontal and temporal lobes, as well as in the right subcortical/limbic areas. The direction of brain connections as well as distinctions between transgender groups indicates that GI is characterized by unique yet heterogeneous neural signatures for both transgender groups [29].

Gender-affirming hormone treatment (GAHT) affects the morphology and functional organization of adolescent/adult brain in a discerning manner. The amygdala, corpus callosum, and nucleus putamen volumes in transgender women undergoing GAHT have been found to be comparable to that of cisgender men, corresponding to their natal sex. Nevertheless, the right insular cortex and right temporal-parietal junction are larger in these women than in the two cisgender groups. However, the nucleus accumbens and the third ventricle of transgender men differ from those of cisgender women, corresponding with the chosen gender rather than the natal sex.

Cisgender men and transgender women undergoing GAHT had higher overall gray matter volume than cisgender women in the posterior superior frontal cortex. On the other hand, compared to cisgender women, transgender men and women both showed diminished gray matter volume in the insula [29]. Another study revealed that after 4 weeks of GAHT, changes in the gray matter volume in Wernicke's and Broca's regions of transgender men were inversely correlated with changes in their testosterone levels.

Notwithstanding variations in the outcomes, these investigations generally suggest that transgender women have demasculinized patterns concerning the cortical thickness of the white-matter microstructure. Additionally, transgender men have defeminized patterns, and transgender women and men both have a distinctive, intricate sexual differentiation that takes the shape of a mosaic [30].

55.5 Genetics of Gender Incongruent Brain "Sexual" Differentiation

The pathophysiology of above has been widely studied and attributed to structure-function brain changes. Comparative studies in twin population hint on genetic basis of this incongruence and the epigenetic modifications influenced by the individual's environment.

SRY and ZFY genes are expressed in the human male hypothalamus and cerebral cortex where they are implicated in the control of sexual differentiation of the brain [31]. The P450 aromatase gene, which regulates the expression of the aromatase enzyme and converts testosterone to estradiol in the brain, is activated by this SRY gene. Male: female expression variations were discovered in 50 potential candidate genes. The expression of 36 of these genes was more in females while that of one (the inactive X specific transcript) was higher by18.5 fold in males. The female: male expression ratio of the other 35 genes ranged from 1.2 to 2.4. Males had higher levels of expression for 18 genes compared to females, including the two on the Y chromosome. Even though these sexually differentiated genes' roles are currently unclear, their existence provides a starting point for more research into the genetic regulation of brain sexual differentiation and eventually the neurological and behavioral function [32].

55.6 Hormonal Control of Sexual Differentiation of the Brain

Experiments conducted in rodents showed that females exposed to testosterone during pregnancy or early infancy had a masculinized pattern of gonadotrophin production and sexual behavior when they reach adulthood. On the other hand, males undergo feminization of the same parameters following neonatal castration. As transcription factors, the estrogen receptors α and β dimerize and form complexes in response to their ligand, bind to coactivator proteins, alter the transcription of several target genes, and eventually impact neuronal function. Coactivators in the process of brain dimorphism, such as SRC-1, SRC-2, and SRC-3, have undergone extensive research. SRC-1 and SRC-2, which differ in their binding preference to ER α or β subtypes, are located in distinct regions of the hypothalamus and hippocampus and are crucial for effective ER transcriptional activity in the brain.

The functional relevance of the nuclear receptor coactivator SRC-1 in the developing brain is indicated by the distinct patterns of coregulator expression. In order for proper male reproductive behavior and brain morphology to develop in a hormone-dependent manner, SRC-1 expression is essential. In view of the above, it has been suggested that SRC-1 and SRC-2 should be included to the list of putative "susceptibility" candidate genes for gender incongruence [33].

Taking the molecular mechanism further, epigenetic changes in DNA methylation were found to be associated with several genes involved in the genetic translational processes during sexual development. The genetic and molecular mechanism data support the hypothesis that gender incongruence is complex and multifaceted and involves intricate communications between the sex steroids, their receptors, and the further downstream pathways. A more sensible strategy to understand the causes behind brain dimorphism is to combine genetic and epigenetic techniques simultaneously. Furthermore, the current intricate "mosaic" model of brain dimorphism is compatible with this theory. Thereby demonstrating that gender incongruence is a multifactorial complicated feature with complex interactions between coactivators, sex hormones, receptors, and several other genes and polymorphisms [34].

55.7 Conclusion

Growing understanding and acceptance of GD are occurring on both an individual and societal level, making it a topic that is gaining importance. It is possible to place equal emphasis on improvements in diagnostic procedures and service availability as on the change in societal perceptions of gender nonconformity. The cognitive, emotional, and sensorimotor regions of brain of individuals with GD are more comparable to their perceived gender which is predominantly influenced by the dominant sex hormones. There is lot of heterogeneity in recording and reporting of the clinical correlates and associated psychology in GD. The standardization of a genetic, epigenetic translate, or morphogenic marker for gender determination would help in appropriate diagnosis and thereby timely management of the cases with DSD.

References

1. World Health Organization. Gender and human rights. Published May 2021. Available from https://www.who.int/genderandhealth. Accessed Feb 2022.
2. American Psychiatric association. Diagnostic and statistical manual of mental disorders (DSM 5-TR). 5th edition. March 2022.
3. Ediati A, Juniarto AZ, Birnie E, Drop SLS, Faradz SMH, Dessens AB. Gender development in Indonesian children, adolescents, and adults with disorders of sex development. Arch Sex Behav. 2015;44:1339–61.
4. Tiger RD. Sexual differentiation. Encyclopædia Britannica, Inc. Published 2018. Available from https://www.britannica.com/science/sexual-differentiation
5. Bertelloni S, Tyutyusheva N, Valiani M, D'Alberton F, Baldinotti F, Caligo MA, Baroncelli GI, Peroni DG. Disorders/differences of sex development presenting in the newborn with 46,XY karyotype. Front Pediatr. 2021 Apr 22;9:627281. https://doi.org/10.3389/fped.2021.627281.
6. Ferdinand JOB, Chinnah TI. Gender dysphoria: a review investigating the relationship between genetic influences and brain development. Adolesc Health Med Ther. 2020;11:89–99.
7. Thompson L, Sarovic D, Wilson P, Samfjord A, Gillberg C. A PRISMA systematic review of adolescent gender dysphoria literature: 1) epidemiology. PLOS Glob Public Health. 2022;2(3):e0000245. https://doi.org/10.1371/journal.pgph.0000245.
8. Luyckx K, Schwartz SJ, Goossens L, Beyers W, Missotten L. Processes of personal identity formation and evaluation. In: Handbook of identity theory and research. Taylor & Francis; 2011. p. 77–98. https://doi.org/10.1007/978-1-4419-7988-9_4.
9. Wilson CA, Davies DC. The control of sexual differentiation of the reproductive system and brain. Reproduction. 2007;133:331–59.
10. Zhou JN, Hofman MA, Gooren LJ, Swaab DF. A sex difference in the human brain and its relation to transsexuality. Nature. 1995 Nov;378(6552):68–70. https://doi.org/10.1038/378068a0.
11. Kruijver FP, Zhou JN, Pool CW, Hofman MA, Gooren LJ, Swaab DF. Male-to-female transsexuals have female neuron numbers in a limbic nucleus. J Clin Endocrinol Metab. 2000 May;85(5):2034–41. https://doi.org/10.1210/jcem.85.5.6564.
12. Gooren L. The biology of human psychosexual differentiation. Horm Behav. 2006 Nov;50(4):589–601. https://doi.org/10.1016/j.yhbeh.2006.06.011.
13. Swaab DF. Sexual differentiation of the human brain: relevance for gender identity, transsexualism and sexual orientation. Gynecol Endocrinol. 2004 Dec;19(6):301–12. https://doi.org/10.1080/09513590400018231.
14. Garcia-Falgueras A, Swaab DF. A sex difference in the hypothalamic uncinate nucleus: relationship to gender identity. Brain. 2008 Dec;131(Pt 12):3132–46. https://doi.org/10.1093/brain/awn276.
15. Luders E, Sánchez FJ, Gaser C, Toga AW, Narr KL, Hamilton LS, Vilain E. Regional gray matter variation in male-to-female transsexualism. Neuroimage. 2009 Jul;46(4):904–7. https://doi.org/10.1016/j.neuroimage.2009.03.048.
16. Rametti G, Carrillo B, Gómez-Gil E, Junque C, Segovia S, Gomez Á, Guillamon A. White matter microstructure in female to male transsexuals before cross-sex hormonal treatment. A diffusion tensor imaging study. J Psychiatr Res. 2011 Feb;45(2):199–204. https://doi.org/10.1016/j.jpsychires.2010.05.006.

17. Bao A-M, Swaab D. Sexual differentiation of the human brain: relation to gender identity, sexual orientation and neuropsychiatric disorders. Front Neuroendocrinol. 2011 Feb 18;32(2):214–26. https://doi.org/10.1016/j.yfrne.2011.02.007.
18. Nguyen HB, Loughead J, Lipner E, Hantsoo L, Kornfield SL, Epperson CN. What has sex got to do with it? The role of hormones in the transgender brain. Neuropsychopharmacology. 2019 Jan;44(1):22–37. https://doi.org/10.1038/s41386-018-0140-7.
19. Frigerio A, Ballerini L, Valdes-Hernandez M. Structural, functional, and metabolic brain differences as a function of gender identity or sexual orientation: a systematic review of the human neuroimaging literature. Arch Sex Behav. 2021;50(8):3329–52. https://doi.org/10.1007/s10508-021-02005-9.
20. Burke SM, Manzouri AH, Savic I. Structural connections in the brain in relation to gender identity and sexual orientation. Sci Rep. 2017;7:17954. https://doi.org/10.1038/s41598-017-17352-8.
21. van Heesewijk J, Steenwijk MD, Kreukels BPC, Veltman DJ, Bakker J, Burke SM. Alterations in the inferior frontooccipital fasciculus—a specific neural correlate of gender incongruence? Psychol Med. 2022;53(8):3461–70. https://doi.org/10.1017/S0033291721005547.
22. Burke SM, Veltman DJ, Gerber J, Hummel T, Bakker J. Heterosexual men and women both show a hypothalamic response to the chemo-signal androstadienone. PLoS One. 2012;7(7):e40993. https://doi.org/10.1371/journal.pone.0040993.
23. Schneider MA, Spritzer PM, Minuzzi L, et al. Effects of estradiol therapy on resting-state functional connectivity of transgender women after gender-affirming related gonadectomy. Front Neurosci. 2019;13:817. https://doi.org/10.3389/fnins.2019.0081728.
24. Burke SM, Cohen-Kettenis PT, Veltman DJ, Klink DT, Bakker J. Hypothalamic response to the chemo-signal androstadienone in gender dysphoric children and adolescents. Front Endocrinol. 2014;5:60. https://doi.org/10.3389/fendo.2014.00060.
25. Simon L, Kozák LR, Simon V, et al. Regional grey matter structure differences between transsexuals and healthy controls - a Voxel based morphometry study. PLoS One. 2013;8(12):e83947. https://doi.org/10.1371/journal.pone.008394729.
26. Hoekzema E, Schagen SEE, Kreukels BPC, et al. Regional volumes and spatial volumetric distribution of gray matter in the gender dysphoric brain. Psychoneuroendocrinology. 2015;55:59–71. https://doi.org/10.1016/j.psyneuen.2015.01.01630.
27. Clemens B, Votinov M, Puiu AA, Schüppen A, Hüpen P, Neulen J, Derntl B, Habel U. Replication of previous findings? Comparing Gray matter volumes in transgender individuals with gender incongruence and cisgender individuals. J Clin Med. 2021;10:1454. https://doi.org/10.3390/jcm10071454.
28. Hahn A, Kranz GS, Küblböck M, Kaufmann U, Ganger S, Hummer A, et al. Structural connectivity networks of transgender people. Cereb Cortex. 2015;25:3527–34. https://doi.org/10.1093/cercor/bhu194.
29. Kranz GS, Seiger R, Kaufmann U, Hummer A, Hahn A, Ganger S, et al. Effects of sex hormone treatment on white matter microstructure in individuals with gender dysphoria. Neuroimage. 2017;150:60–7. https://doi.org/10.1016/j.neuroimage.2017.02.027.
30. Guillamón A, Junque C, Gómez-Gil E. A review of the status of brain structure research in transsexualism. Arch Sex Behav. 2016;45:1615–48. https://doi.org/10.1007/s10508-016-0768-5.
31. Mayer A, Lahr G, Swaab D, et al. The Y-chromosomal genes SRY and ZFY are transcribed in adult human brain. Neurogenetics. 1998;1:281–8.
32. Ramirez K, Fernández R, Collet S, Kiyar M, Delgado-Zayas E, Gómez-Gil E, Van Den Eynde T, T'Sjoen G, Guillamon A, Mueller SC, Pásaro E. Epigenetics is implicated in the basis of gender incongruence: an epigenome-wide association analysis. Front Neurosci. 2021;15:701017. https://doi.org/10.3389/fnins.2021.701017.
33. del Valle Ramírez K, Fernandez R, Delgado-Zayas E, Gomez-Gil E. Implications of the estrogen receptor coactivators SRC1 and SRC2 in the biological basis of gender incongruence. Sex Med. 2021;9:100368.
34. Esteva I, Guillamon A, Pasaro E, Joel D. Beyond the binary: rethinking sex and the brain. Neurosci Biobehav Rev. 2021;22:165–75. https://doi.org/10.1016/j.neubiorev.2020.11.018.

Urological Issues and Morbidities in Children with DSD

56

Vanessa Coles, Azad Bhushan Mathur, and Ravindar Anbarasan

Abstract

Background

Children with a disorder of sexual development (DSD) may present with a variety of different urological pathologies. In this chapter we summarise the commonly associated urological issues seen in this patient cohort.

Content

The chapter is divided into three sections:

1. Genital anomalies
2. Gonadal anomalies
3. Renal tract anomalies

Genital anomalies are part of a wider spectrum of urological defects seen in children presenting with DSD. They usually come to light at the initial referral with varying degrees of virilisation, as graded by the Prader classification. In this section we highlight the most common features, investigations, and management strategies that may be required.

Gonadal anomalies in children with DSD can be varied and may present in different ways, including undescended testes and inguinal hernias.

Renal tract anomalies associated with DSD are numerous. We have focused on the most common associations, with highest urological morbidity.

V. Coles · A. B. Mathur · R. Anbarasan (✉)
Norfolk and Norwich University Hospital NHS Foundation Trust, Norwich, UK
e-mail: vanessa.coles@nnuh.nhs.uk; azad.mathur@nnuh.nhs.uk; ravindar.anbarasan@nnuh.nhs.uk

S. K. Ratan (ed.), *Children with Differences in Sex Development*,
https://doi.org/10.1007/978-981-97-1639-5_56

Conclusion

Children with DSD have a strong association with a variety of urological issues and morbidities. It is vital that the multi-disciplinary team that is involved in the care of these children have a good understanding of the different pathologies that may be present.

Keywords

Disorder of sexual development (DSD) · Genital anomalies · Urogenital sinus · Gonadal anomalies · Renal tract anomalies

56.1 Introduction

Disorders of sexual development (DSD) is a term that encompasses a variety of different conditions, the details and classifications of which are discussed in earlier chapters. Paediatric urologists play a key role in the multi-disciplinary team (MDT) management of these patients and therefore should have a good knowledge of urological issues that may arise.

Patients with DSD may present with a range of urological pathologies and symptoms depending on their underlying karyotype and DSD condition. These may be present at the initial referral or may be picked up incidentally during the DSD workup.

In this chapter we focus on the urological issues that patients with DSD may present with, dividing them into genital, gonadal, and renal tract anomalies.

56.1.1 Genital Anomalies

Genital anomalies in DSD present with varying degrees of virilisation from mild clitoromegaly in an overall female appearing phenotype to a more male phenotype with severe hypospadias and impalpable testes. This is classified into five stages of virilisation by the Prader classification first described in 1954 [1].

Prader 0: Normal female genitalia

Prader I: Isolated clitoromegaly

Prader II: Clitoromegaly with close, but separate urethral and vaginal openings

Prader III: Clitoromegaly with single urogenital orifice and posterior fusion of the labia majora

Prader IV: Penoscrotal hypospadias with complete fusion of the labia majora

Prader V: Complete masculinisation (normal male appearing genitalia), with no palpable testes

56.1.1.1 Clitoromegaly

Clitoromegaly is an abnormal enlargement of the clitoris secondary to excess androgens.

With increasing numbers of reported cases, we now have several case series providing us with a reference guide to what classifies as clitoromegaly in the paediatric population. Brodie et al. is the most used guide to clitoral size and uses the crosswise width of the glans to determine size range in different age groups [2].

Clitoromegaly may be seen in isolation or in conjunction with other abnormalities seen in DSD. When pronounced in the neonatal period a full DSD workup is warranted. In a subset of patients with DSD, such as non-classical congenital adrenal hyperplasia, the clitoromegaly may be more subtle and the only external suggestion of an underlying pathology. A high index of suspicion must be had when asked to review such patients and if there is any doubt a karyotype and full DSD workup should be conducted [3].

Clitoral prominence, often seen in extreme premature babies secondary to elevated 17-hydroxyprogesterone and adrenal androgen concentrations, should be distinguished from clitoromegaly as it can mimic the appearances of classical congenital adrenal hyperplasia (CAH). It is usually transient and becomes less evident as the neonate grows [4, 5]. A karyotype and DSD workup may be required before reassurance to the parents is given.

In isolated clitoromegaly, surgery is usually not indicated, and reassurance is all that is required. When in conjunction with other anomalies seen in DSD, a full multi-disciplinary workup is paramount to deciding how these patients are managed. Details of clitoral surgery, including timing, approach, and patient choice are discussed in other chapters.

56.1.1.2 Urogenital Sinus

In patients with Prader III-IV virilisation there is persistence of the urogenital sinus (pUGS) resulting in a common channel between the urethra and vagina. The length of this common channel varies depending on the stage at which the embryological separation was affected and can be subdivided into patients with a high or a low confluence [6]. pUGS can be associated with a variety of other anomalies including, rectovaginal communications and renal, bladder, or vaginal agenesis. These patients can therefore present a variety of challenges to the paediatric urologist [7].

Drainage of the bladder or vagina may be affected by pUGS and patients can present with a palpable abdominal mass, indicating urinary retention and/or hydrometrocolpos. In those with urinary retention urgent bladder drainage by way of suprapubic catheterisation (SPC), or clean intermittent catheterisation (CIC) in suitable cases, should be considered [8]. Hydrometrocolpos secondary to vaginal agenesis needs surgical drainage in comparison with the one that presents later in life due to frequent refluxing of urine into the vagina [9].

Owing to the urinary reflux, these patients are at higher risk of urinary tract infections, necessitating prophylactic antibiotics [6, 7].

Radiological imaging, in the form of an abdominal ultrasound delineates the anatomy and helps diagnose associated anomalies, such as bifid uterus, hemi vagina and renal tract abnormalities [10]. However more detailed imaging, such as a micturating urethrogram (MCUG), genitogram, and magnetic resonance imaging

(MRI), may be required [10]. Endoscopic assessment of the urogenital sinus can also help confirm the length and components of the common channel.

56.1.1.3 Hypospadias

Progressive virilisation, leading to Prader IV-V, presents as severe penoscrotal hypospadias [1]. This together with unilateral or bilateral impalpable gonads warrants a complete DSD workup, as the ability to distinguish this from severe clitoromegaly with a persistent urogenital sinus and fused labioscrotal folds is challenging [11, 12].

In such situations karyotyping becomes cornerstone to decision making. In those confirmed to be chromosomally male, a standard hypospadias repair is performed [13]. Pre-operative hormonal supplementation should be considered in cases with a small phallus. The paediatric urologist also needs to be wary of ectopic insertion of the vas deferens and the presence of persistent Mullerian duct structures such as a prostatic utricle. These may present as recurrent epididymo-orchitis following hypospadias repair.

56.1.1.4 Labial Adhesions

Some patients with DSD may present with labial adhesions, also known as labial agglutination and describe midline fusion of the labia minora. They are believed to be secondary to low levels of oestrogen and can be seen in up to 5% of prepubertal girls [14].

In most patients, including those with DSD, they do not cause any symptoms and reassurance is all that is required [15]. Over 80% of labial adhesions will resolve spontaneously once the child enters puberty and the levels of oestrogen increase. However, in some patients they can increase the risk of urinary tract infections as they can lead to pooling and refluxing of urine into either the urinary or vaginal tracts. This may be an indication for treatment [15].

Conservative management is now favoured over surgical division of the adhesions, due to the high recurrence rate with surgical intervention. The mainstay of treatment is topically applied oestrogen cream, although other methods such as topical emollient applications and labial traction techniques are described in the literature [16, 17].

56.1.1.5 Cyclical Haematuria

Cyclical haematuria is uncommon, and if present must raise suspicion for an underlying disorder of sexual development. Conditions such as ovotesticular DSD (OT-DSD) and Klinefelter's syndrome may present with an overall male phenotypic appearance but with intra-abdominal ovarian or uterine tissue [18–20]. The cyclical haematuria may indicate the initiation of menstruation and may be the only presenting symptom of DSD. There is often associated gynecomastia, and this should also be looked for [19, 20].

56.1.2 Gonadal Anomalies

Gonadal assessment is a key component in the evaluation of children with DSD. Owing to the paracrine effect of gonadal tissue, genital findings may vary, as described above. This warrants investigations to confirm the presence and type of gonadal tissue.

56.1.2.1 Undescended Testes

Gonadal anomalies may present as either unilateral or bilateral undescended testes. Bilaterally impalpable undescended testes mandate full DSD evaluation and workup. However, in patients with unilateral impalpable undescended testes, a full DSD workup may still be required, if other genital anomalies are detected [12, 21].

The use of ultrasound to locate impalpable testes is unreliable, therefore a diagnostic laparoscopy becomes the investigation of choice [22]. The presence and type of intra-abdominal gonadal tissue varies from pure testicular/ovarian tissue to a combination of both depending on underlying karyotype [19]. Ovotesticular DSD (OT-DSD) is where both ovarian and testicular tissues are present [23, 24]. Presence of hemi-uterus indicates functioning testicular tissue on the contralateral side.

Management of undescended testes in DSD is complex and controversial. Options include orchidopexy, excision of gonadal tissue or monitoring via serial ultrasounds and blood tests. The overall decision rests on the sex or rearing and the likelihood of malignant transformation [23, 24]. The complexities are covered in detail in other chapters.

56.1.2.2 Inguinal Hernias

The presence of an inguinal hernia in an otherwise phenotypically female child should raise suspicion of complete androgen insensitivity syndrome (CAIS). The incidence of underlying CAIS in patients with bilateral inguinal hernias is 1 2% [25]. Gonadal (testicular) tissue may be discovered at herniotomy. Further investigation with ultrasound reveals the absence of Mullerian structures [26].

It is paramount that in any phenotypically female patient with bilateral inguinal hernias undergo pre-operative ultrasounds and karyotype or visualisation of gonads at herniotomy. This can be achieved by diagnostic laparoscopy, hernioscopy or by opening the canal to inspect the gonad by following the round ligament.

56.1.3 Renal Tract Anomalies

Renal tract anomalies are common in patients with DSD, owing to the close relationship of renal tract and sex organ embryology. Reported incidence is as high as 21%, compared to 2–4% in the general population [27]. A baseline ultrasound of the renal tract is therefore an invaluable tool in the evaluation of these patient cohorts.

56.1.3.1 Vesicoureteric Reflux

Congenital or primary vesicoureteric reflux (VUR) appears to be the most commonly reported upper renal tract anomaly seen in patients with DSD, with an incidence of over 8% in one case series of patients with CAH [27]. It also is commonly seen patients with severe penoscrotal hypospadias. This suggests an embryonic field defect due to an ectopic ureteric budding during development.

Secondary or acquired VUR can also develop due to dysfunctional voiding after major reconstructive surgery. A high index of suspicion should be had and investigations like MCUG and ultrasound are indicated.

The management of these patients depends on the grade of VUR and the presence of other associated renal anomalies such as renal tract duplication. The main aim of treatment is to prevent renal damage from the refluxing urine and monitoring with serial imaging (ultrasound, MCUG, MAG3 scans) is warranted. The initial management is with prophylactic antibiotics and observation; however, ureteric reimplantation may be indicated when there are progressive VUR and renal impairment despite medical management [27].

56.1.3.2 Renal Tract Duplication

Renal tract duplications are reported in patients with DSD and should be picked up on the baseline renal tract ultrasound [10, 27]. If uncomplicated, they do not require any intervention or treatment; however as with any patient with renal duplication, they may have associated anomalies such as ectopic ureteric implantation, VUR, and ureterocoeles which may require intervention.

56.1.3.3 Renal Agenesis

They are several report cases of patients with DSD having associated renal agenesis [10, 27], although this occurs less commonly than renal duplication, and VUR. It is usually unilateral with a contralateral normal kidney. Management is mainly focused on protecting the remaining kidney. There has been an isolated case report of a baby born stillborn who had Klinefelter's syndrome with associated bilateral renal agenesis and subsequent severe oligohydramnios antenatally [28].

Another single case report described unilateral renal agenesis in a patient with CAH who had contralateral Mullerian structure anomalies in the form of an obstructed hemivagina. This made up a rare case of OHVIRA (Obstructed HemiVagina and Ipsilateral Renal Anomaly) in a child and is the only case known to have been reported in a patient with CAH. It further supports the importance of getting a baseline renal tract anomaly when performing the initial diagnostic workup for DSD [29].

56.1.3.4 Renal Tumours

Patients with DSD are not only at an increased risk of developing tumours like gonadal tumours such as testicular carcinoma in situ (CIS) and germ cell tumours, but also renal tumours. The most well-described condition, which is inherited in an autosomal dominant pattern and caused by mutation of the WT1 gene is Denys-Drash syndrome. This condition is characterised by the following:

Gonadal dysgenesis (XY karyotype with undescended testis and DSD genitalia)
Increased risk (up to 90%) of Wilms tumour (nephroblastoma)
Progressive nephropathy
Nephrotic syndrome
Hypertension

56.2 Lower Urinary Tract Dysfunction

Lower urinary tract dysfunction is the term used for variety of symptoms including urgency, frequency, and urge incontinence. These are most widely reported in patients with DSD who have undergone major reconstructive surgery and a large part of the pre-operative planning focuses on how to minimise these post-operative complaints. However, there is also increasing literature that suggests there is an underlying weakness of the pelvic floor muscles secondary to the associated virilisation and so these symptoms may present pre-operatively or persist post-operatively despite successful reconstruction [30, 31].

56.3 Conclusion

Children with DSD have a strong association with urological anomalies, presenting with a variety of symptoms at different time points in their life. It is vital that the multi-disciplinary team have a good understanding of this varied presentation. Careful selection of investigations such as karyotyping, ultrasound, MCUG, genitogram, MRI, and diagnostic laparoscopy allows full evaluation of anatomy and helps guide management.

References

1. Prader A. Genital findings in the female pseudo-hermaphroditism of the congenital adrenogenital syndrome; morphology, frequency, development and heredity of the different genital forms. Helv Paediatr Acta. 1954;9(3):231–48.
2. Brodie KE, Grantham EC, Huguelet PS, Caldwell BT, Westfall NJ, Wilcox D. Study of clitoral hood anatomy in the pediatric population. J Pediatr Urol. 2016;12(177):e1–5.
3. Lezzi L, Lasorella S, Varriale G, Zagaroli L, Anbrosi M, Verrotti A. Clitoromegaly in childhood and adolescence: behind one clinical sign, a Clinical Sea. Sex Dev. 2018;12:163–74.
4. Couch R, Girgis R. Postnatal virilization mimicking 21-hydroxylase deficiency in 3 very premature infants. Pediatrics. 2012;129(5):e1364–7.
5. Greaves R, Kanumakala S, Read A, Zacharin M. Genital abnormalities mimicking congenital adrenal hyperplasia in premature infants. J Paediatr Child Health. 2004;40(4):233–6.
6. Marei M, Fares A, Abdelsattar A, Abdullateef K, Seif H, Hassan M, et al. Anatomical measurements of the urogenital sinus in virilized female children due to congenital adrenal hyperplasia. J Pediatr Urol. 1016;12(5):282.28.
7. Tan H, Tam S, Shunmugan R, Zakaria R, Zahari Z. A case of persistent urogenital sinus: pitfalls and challenges of diagnosis. Sultan Qaboos Univ Med J. 2017;17(4):e455–9.

8. Kitta T, Kakizaki H, Iwami D, Tanda K. Successful bladder management for a pure urogenital sinus anomaly. Int J Urol. 2004;11(5):340–2.
9. Simonetti I, Trovato P, Verde F, Tarotto L, Della Casa R, Lonardo M, et al. A rare case of hydrometrocolpos from persistent urogenital sinus in patient affected by adrenogenital syndrome. J Ultrasound. 2018;21(3):249–52.
10. Valentini A, Giuliani M, Gui B, Laino M, Zecchi V, Rodolfino E, et al. Persistent urogenital sinus: diagnostic imaging for clinical management. What does the radiologist need to know? Am J Perinatol. 2016;33(5):425–32.
11. Abokifa A, Badawy R, Anwar H, Sagheer Y. Disorders of sexual development in proximal hypospadias in children. EJHM. 2019;77(6):5771–5.
12. Nelwan D, Mount C, Morganstern B, Chan J. Infant with severe penoscrotal hypospadias: a complex case of genital ambiguity and mistaken identity. Cureus. 2021;13(5):e15191.
13. Ahmeti H, Kolgeci S, Arifi H, Jaha L. Clinical dilemmas and surgical treatment if penoscrotal, scrotal and perineal hypospadias. Bosn J Basic Med Sci. 2009;9(3):229–34.
14. Krishna T, Sagili H, Jayalakshmi D. Labial adhesions in a pre-pubertal girl: the importance of awareness. Trop Doct. 2021;51(4):661–2.
15. Bacon J, Romano M, Quint E. Clinical recommendations: labial adhesions. J Pediatr Adolesc Gynecol. 2015;28(5):405–9.
16. Dowlet-McElroy T, Higgins J, Williams K, Strickland J. Treatment of prepubertal labial adhesions: a randomised controlled trial. J Pediatr Adolesc Gynecol. 2019;32(3):259–63.
17. Goldman R. Estrogen cream for labial adhesions in girls. Can Fam Physician. 2013;59(1):37–8.
18. Tangshewinsirikul C, Dulyaphat W, Tim-Aroom T, Parinayok R, Chareonsirisuthigul T, Korliatsakul V, et al. Klinefelter Syndrome Mosaicism 46,XX/47,XXY: a new case and literature review. J Pediatr Genet. 2020;9(4):221–6.
19. Khadilkar K, Budyal S, Kasaliwal R, Sathe P, Kanalkar B, Sanghvi B, et al. Ovotesicular disorder of sex development: a single center experience. Endocr Pract. 2015;21(7):770–6.
20. Mishra S, Ray S. A case of true hermaphrodite presenting as cyclical hematuria. Int J Sci Res. 2013;6(14):2319–7064.
21. Ostergren P, Juul A, Azawi A. Disorders of sex development presenting as unilateral cryptorchidism. Scan J Urol. 2013;47(5):433–6.
22. Tasian G, Copp H, Baskin L. Diagnostic imaging in cryptochidism: utility, indications, and effectiveness. J Pediatr Surg. 2011;46(12):2406–13.
23. Ren S, Harmas V, Stara M, Cejnova L. Description of diagnosis of 45,X/46,XY ovotesticular DSD. Ceska Gynekol. 2020;85(4):259–62.
24. Syryn H, Van De Vijver K, Cools M. Ovotesticular difference of sex development: genetic background, histological features, and clinical management. Horm Res Paediatr. 2021;2021:275–84.
25. Gil A, Salgado M. Bilateral inguinal hernia in a female child. BMJ Case Rep. 2014;2014:bcr2013202452. https://doi.org/10.1136/bcr-2013-202452.
26. Terro J, El-Helou JK, Lakkis R, Shibli A, El-Chamaa B, et al. Bilateral inguinal masses or hernias in a female teenager with delayed menarche: think of Complete Androgen Insensitivity Syndrome (CAIS), a case report. Int J Surg Case Rep. 2020;76:25–9.
27. Nabhan J, Eugster E. Upper-tract genitourinary malformations in girls with congenital adrenal hyperplasia. Pediatrics. 2007;120(2):e304–7.
28. Barroeta J, Stopyra G. 47,XXY with associated bilateral renal agenesis. Arch Pathol Lab Med. 2004;128(3):e44–5.
29. Al-Najar M, Saleem M, Alassaf A, Jubouri S, Aftan M. Congenital adrenal hyperplasia and Ohvira syndrome: first report of unique combination in a child. AACE Clin Case Rep. 2015;1(2):e136–40.
30. Bogdanska M, Morris A, Hutson J, Heloury Y, Grover A. Long-term urinary symptoms in adolescent and adult women with congenital adrenal hyperplasia. J Pediatr Urol. 2018;14(3):240.el–5.
31. Taghavi A, Anbarasan R, Mushtaq I. Distal mesonephric duct anomalies in association with penile ectopia and complex hypospadias. ANZ J Surg. 2019;89(7–8):965–6.

Algorithm for Management of Children with DSD

57

Kanika Sharma and Simmi K. Ratan

Abstract

A child with ambiguous genitalia has always been considered more of a social emergency than a medical emergency. The management of a child with disorders of sexual development or sex differentiation (DSD) is a multi-specialty consensus approach focused firstly at evaluation of the cause and thereafter to carry out the medical and surgical management along with paying due attention to the counselling of the affected children and their families. The chapter outlines the noteworthy points or working algorithm for evaluating and managing the patients with DSD.

Keywords

DSD workup · Management · Workup

57.1 Introduction

Disorder of sexual development or disorder of sex differentiation (DSD) is a broad term defining the clinical conditions where chromosomal, gonadal, and anatomical development are atypical [1]. In most of the centers, neonatologists or pediatricians or neonatal or pediatric surgeons are the first clinicians contact for these patients as there could be glaring genital ambiguity at birth itself causing dilemma in binary sex assignment. However, these patients can present anytime including adolescence

K. Sharma · S. K. Ratan (✉)
Department of Pediatric Surgery, Institute of Medical Sciences-Banaras Hindu University (IMS-BHU), Varanasi, India

Department of Pediatric Surgery, Maulana Azad Medical College, New Delhi, India
e-mail: kanikasharma@bhu.ac.in

S. K. Ratan (ed.), *Children with Differences in Sex Development*,
https://doi.org/10.1007/978-981-97-1639-5_57

and adulthood with problems related to puberty attainment and later due to fertility problems. Hence the algorithm or approach of clinicians/surgeons may vary with age and complaints at presentation.

Before formulation of any working algorithm, the uniformity in application of terminologies is very important to maintain the same direction of reference of all members of the team. The earlier anatomical based classification now has chromosomal basis hinting at a greater emphasis laid on reaching the basic etiopathogenesis of the DSD condition [1,2]. It may be highlighted that variable phenotypic expressions may sometimes occur with same chromosomal makeup necessitating the need for more sophisticated investigations to reach the diagnosis. For example, 46XY chromosomal layout with male type genitalia with unilateral or bilateral undescended testes may be seen with ovotesticular DSD or with persistent Mullerian duct syndrome. Here, a combination of investigative modalities as histopathology of the gonads and target hormonal assay help clinch the diagnosis. Hence the attending clinician must be well knowledgeable about the various investigative modalities and their combinations to be employed in a given case. Needless to mention that with advancement in knowledge and greater understanding of molecular mechanism of many hitherto unexplained phenomena, these investigations are finding a greater applicability in arriving at a correct diagnosis and in the current practices, molecular workup gets initiated at early stage itself [3]. Details about diagnostic modalities have been highlighted in the earlier section of the book.

A cascade of recent developments on legal and social fronts have led to a sea change in the sequence of steps and spectrum of disciplines to be involved in the management of DSD patients [2, 4]. More stress is being laid on patient-centric management where the decisive authority lies with family (preferably with the child), thereby making the approach more acceptable and legally oriented (see Sect. 3 on legal aspects). Clinicians, however, still hold the role for emergent life-saving procedures only for DSD children, while other surgical interventions can be postponed till they reach the legal age of giving consent for such surgeries in accordance with perceived gender identity. Though the age of consent may differ between different countries (from 16 to18 years), UNICEF has declared 18 years of age as acceptable for giving medical consent to safeguard their right to life [5]. Hence the clinicians have to be well aware of the laws of the land and weave the algorithms accordingly. The legal wrangle may take ugly turn with risk of financial liability and codal punishment on the attending clinician and surgeon too. This deterrent has affected the management algorithm greatly and led to introduction of a wider discussion and inclusion of multiple disciplines in management of such patients [2, 3]. Going a step further, there is suggestion of such multidisciplinary teams (MDT) at various levels so that the best decision for child's future can be taken [6]. Hence, along with the treating clinicians, involvement of pediatric endocrinologist, clinical and molecular genetics experts, child psychologist and psychiatrist, radiologist, pediatric urologist, gynecologist, genetics experts, plastic surgeon, social workers, and others are the backbone of DSD-management team [3]. Also, in Chicago Consensus 2006, there has been reform in the terminology pertinent to DSD in order to alleviate the social stigma associated with previous nomenclature, thus making the terminologies socially acceptable [7].

Another important development is the greater recognition of entity of gender dysphoria and its detrimental role on mental health of the individual (see Sect. 2). One of the earlier aims of the medical team is to assign one of the binary sex to DSD children to make them more acceptable member of the society, or prepare for sexually active life based on their physical (genital) assets or be able to procreate, no longer hold the ground as more emphasis is being given to acceptance of gender identity. A greater sensitivity of the medical team and the society toward this concerning issue, due to the outcry of the activists contesting for transgender rights, has led to the three-step management strategy; namely, (1) Establishing the correct diagnosis, managing the emergent life-threatening conditions, counselling for the associated anomalies and any other potential medical problems, (2) Addressing gender identity of the individual and help him establish that by employing suitable medical and surgical means after getting his due consent (not only parental), and (3) Safeguarding such individuals from any discrimination (confidentiality and support group) by repeated counselling and psychosocial assessment, thus allowing him to live a wholesome life by giving him the equal human rights at all stages of life. Ushering in of advanced surgical procedures, fertility procedures, and surrogacy have been instrumental in revolutionizing this outlook with respect to DSD. Needless to say, that parent support group and peer-support group have lent multipronged support to such individuals, with good legal help from the courts who have given landmark judgements to grant equal rights to such individuals.

Based on what has been stated above, a simplified algorithm for investigating and managing individuals can be highlighted by the given flow-diagram (Fig. 57.1). This algorithm serves as a bird's eye view of the general steps toward managing DSD individuals.

Establishing the correct diagnosis is the critical part of workup and it entails through workup through a battery of investigations along with parental counselling to allay the anxiety of the family. Generally, it is advised that the newborn be referred to as "it" or "your baby" till the diagnosis about sex is made. The early referral of suspected DSD children to the centers with facilities, expertise, and experience is of paramount importance. Besides reaching the key diagnosis, the investigations should help identify associated anomalies or syndromes known to be associated with DSD. On establishment of the diagnosis, there are directives for the attending clinicians to bring such cases to the knowledge of hospital or institutional authorities so as to associate them with MDT for further management. It is also recommended that till the time an individual decides about the gender he/she wants to be in, the name of such individual be kept based on "sex of rearing" by the parents [8]. After 7 years of age psychosexual orientation of the individual must be assessed. At every stage, parental counselling is important in view of psychosocial concerns. In some institutes, these families are introduced to DSD support groups to receive not only psychological support but also to discuss matters related to further education, lifestyle, and day-to-day upbringing concerns for these children. Later, peer-support groups help and support these individuals with personal issues such as puberty, sexual attractions, dating, marriage, and job prospects. Several NGO extend support to ensure equal human rights to these individuals including legal support.

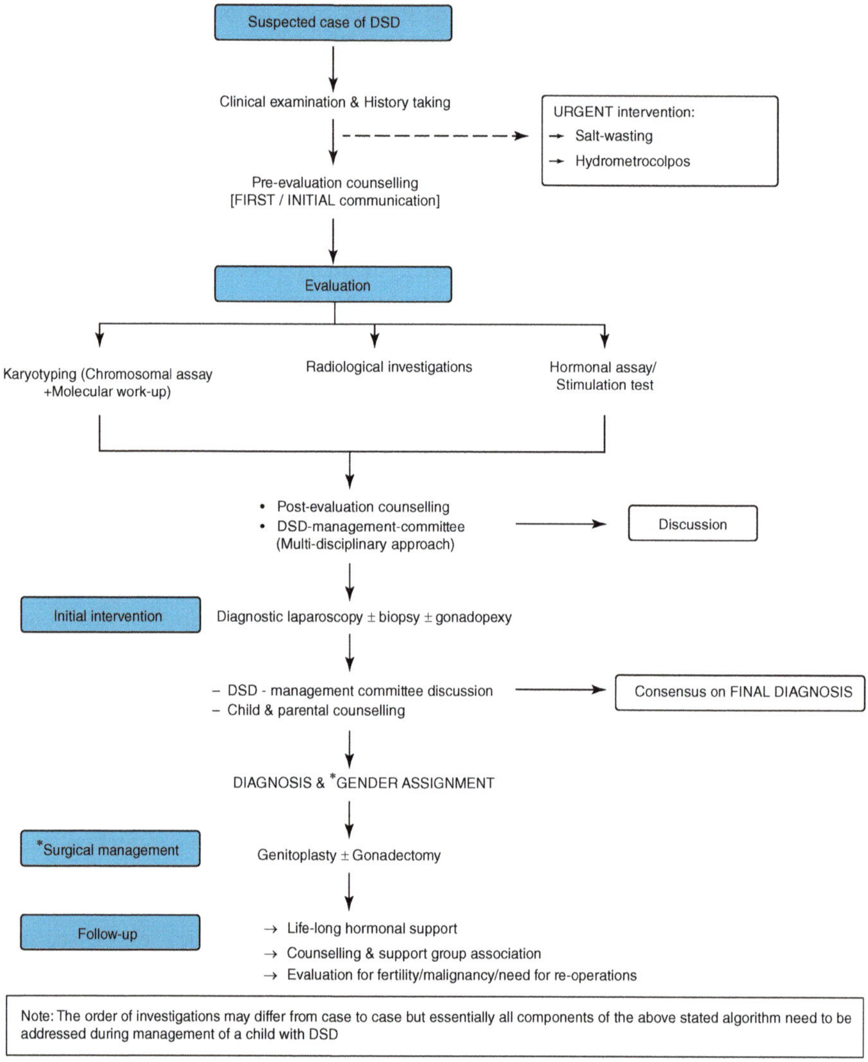

Fig. 57.1 Plan of action for working a case of DSD

57.2 When to Suspect a DSD

A child can be suspected to be lying in the spectrum of DSD from birth till adulthood. Based on the age of presentation, they can be "early presenter" or "late presenter." Table 57.1 highlights scenarios pertinent to DSD.

57.2.1 Early Presenter

Genital ambiguity (atypically development of genitalia) can be noticed immediately after birth by neonatologists or labor room staff during the general physical examination, in infancy by caretakers of the child or at a later date. Hence the normal anatomy of the normal male and female newborn should be well known and any deviation from the same, inclusive of altered parameters, hypo/hyperpigmentation or abnormal disposition of the same (virilized female external genitalia or under-virilized male) should lead to suspicion of case of DSD.

Table 57.1 Suspected case of DSD

Neonate • Overt genital ambiguity
Rarely as • Medical emergency – *Salt-wasting* • Surgical emergency – Hydrometrocolpos/hydrocolops/hematocolpos
Virilized female external genitalia • Enlarged clitoris • Posterior labial fusion • Inguinal or labial mass
Under-virilized male external genitalia • Bilateral impalpable testicles • Micropenis • Association of hypospadias and undescended testicles
Beyond infancy (10–20%) • Previously unrecognized ambiguity • Female—inguinal hernia – Labial mass or palpable gonad – Absent, delayed, or incomplete puberty – Virilization – Primary amenorrhea • Males—Gynecomastia – Cyclical hematuria/suprapubic mass – Bulge in perineum – Infertility

- Newborns can be diagnosed with genital ambiguity at birth: They need to be evaluated on an urgent basis to rule out the life-threatening variant of congenital adrenal hyperplasia (CAH) [9].
- In special cases where there has been prenatal karyotyping or genetic testing done, the presence of discordance between a prenatal karyotype and genital appearance needs to be evaluated for DSD [10].
- A subset of DSD (sex chromosomal DSD) like Klinefelter's syndrome or Turner syndrome do not present with genital ambiguity; however, they have characteristics phenotypic and behavioral features pertinent to respective syndrome [11].

57.2.2 Late Presenters

Patients with ambiguous genitalia can be missed in childhood, then they present in adolescent age or rarely in adulthood

- Other suspected scenarios for suspecting DSD beyond infancy are (1) inguinal hernia, primary amenorrhea, delayed/absent puberty, or virilization in a reared female or (2) gynecomastia, cyclic hematuria, suprapubic mass, or infertility in a reared male.

57.3 Establishment of Diagnosis

57.3.1 History

A detailed history is of importance while evaluating the case of DSD. The rearing gender, psychological sex (gender identity) of a child should be carefully noted.

Maternal history of drug intake (such as steroids, oral contraceptives, androgen, androgen antagonists, etc.) which can cause virilization of female fetus or under-virilization of male fetus should be noted. Associated maternal virilization can be attributed to aromatase deficiency or rarely, androgen secreting tumor.

Family history of similar complaints, genital ambiguity, or hypospadias should be elicited. Family history of amenorrhea, delayed/early puberty, fertility issues, or adoptions can be indirect evidence of DSD in family. Consanguinity should be documented which can be of importance in autosomal recessive conditions (CAH, 5-alpha reductase deficiency, pure gonadal dysgenesis, etc.). Unexplained neonatal death or sibling death can point toward the family history of salt-wasting CAH. A detailed family-tree should be created which should entail the details of consanguinity, similar complaints, or known syndromes in the family.

57.3.2 Clinical Examination

A detailed general physical examination should be done. Few points to be kept in mind pertinent to specific types of DSD are:

- Built and nutrition: Gigantism—Klinefelter's syndrome
  Stunted—Turner syndrome
  Overweight/obese—CAH
- *Hydration*: Dehydration—Salt-wasting CAH
- *Hypertension*: Androgen biosynthesis pathway impairment
- *Intellectual* disability: Syndromic (Turner syndrome, Klinefelter's syndrome)
- Facies: Dysmorphic features, webbed neck, skeletal deformities
  (Turner syndrome, Klinefelter's syndrome)
- Virilization: Pubic pubarche, hirsutism, hoarseness, muscle hypertrophy

Examination findings for other associated anomalies or syndromes should be documented. Sexual maturity rating (breast development, axillary and pubic hair, etc.) should be done simultaneously.

A careful inspection with palpation of external genitalia should be done in detail as it may have subtle to overt signs of genital ambiguity. The following points of genital examination should be noted and documented (Table 57.2) (See Chaps. 11 and 16 for details).

Depending upon the spectrum, for evaluation of apparent female Prader scale of virilization is utilized and for apparent male external masculinization score (EMS) is calculated [12, 13]. Ambiguity is considered when EMS < 7. A latest external genital score (EGS) has been validated in European multicentric study for

Table 57.2 Points to note in genital examination

Number of orifice in perineum
Phallus (shape, size, erectile tissue)
Meatus—shape, size, site
Labioscrotal folds—development, fusion, transposition, pigmentation
Gonads—Unilateral/bilateral (symmetry)
—palpable, impalpable
—consistency (homogenous or heterogenous)
(firm or soft)
Anogenital distance
Inguinal hernia

- Phallus is an embryological undifferentiated primodium for penis and clitoris which originates from genital tubercle. In cases of ambiguous genitalia, it is preferable to use "phallus" for penis or clitoris before genitoplasty.
- Micropenis is defined as penile length < 2 standard deviation below the mean for population and age-based penile length.
- Clitoromegaly is a hypertrophied clitoris in a virilized female which >1 cm length in term female.

ambiguous genitalia in children less than 2 years (including term and preterm newborns) [14]. EGS has gradual scale from 0 to 12 (female to male).

57.3.3 Differential Diagnosis

While the diagnosis of DSD carries a psychological burden, there are certain differential diagnoses which should be ruled out to lessen this burden. Most of these differentials can be challenged on clinical examination. A common entity which is misdiagnosed as DSD in community is labial adhesion in females, which requires day-care minor procedure of adhesiolysis and no further investigation. Another entity which is commonly mis-labelled as DSD in peripheral centers is imperforate hymen. Similarly, in males, microphallus or aphallia are labelled as DSD by the community. Therefore, it becomes important to examine the suspected case of DSD in detail before making a diagnosis of DSD. There can be isolated cases of hypospadias with undescended testis or isolated urogenital sinus, which do not fulfill the criteria for DSD. Similarly, there are other causes for clitoromegaly such as idiopathic, transient (preterm female), hormonal (masculinizing tumor, polycystic ovary syndrome), or non-hormonal (neurofibromatosis, nevus, syndrome without virilization), which do not constitute DSD. Table 57.3 outlines the common differential diagnosis of DSD.

57.3.4 Some Clinical Scenarios in DSD

Once there is a fair speculation of the existence of DSD based on clinical examination the following probable conditions (as depicted in Fig. 57.2) can be worked up.

A brief algorithm pertaining to diagnostic approach is given below though detailed workup is given in the targeted chapters within this book (Fig. 57.3).

57.3.5 Radiological Assessment

Imaging is an integral part of DSD workup as discussed in Chap. 13 [15, 16]. USG pelvis and genitogram are two common investigations ordered for cases suspected

Table 57.3 Differential diagnosis for DSD

Female patient	Male patient
Labial adhesion	Hypospadias with undescended testis
Idiopathic clitoromegaly	Bilateral undescended testis
Female epispadias	Microphallus or aphallia
Imperforate hymen	Isolated urogenital sinus
Isolated urogenital sinus	Transverse testicular ectopia
Lymphangioma perineum (e.g., clitoral/labial)	Lymphangioma perineum (e.g., scrotal/penile)

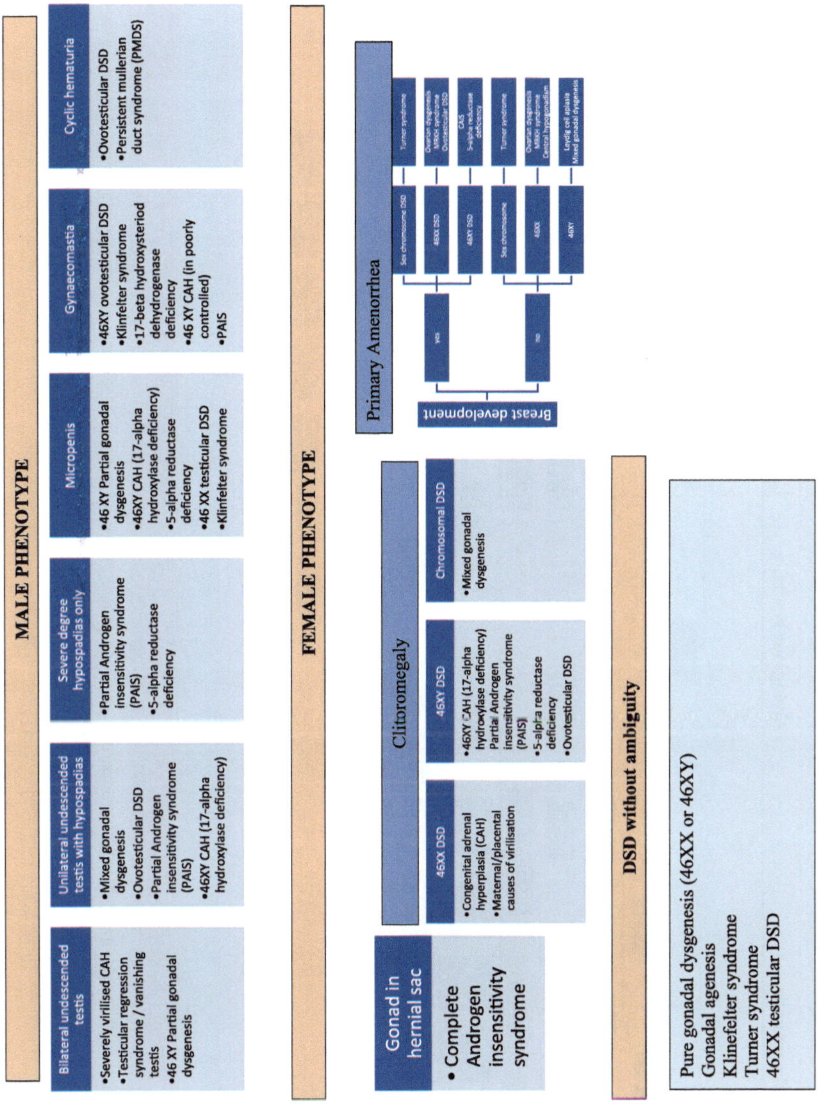

Fig. 57.2 Probable clinical scenarios in DSD based on initial examination

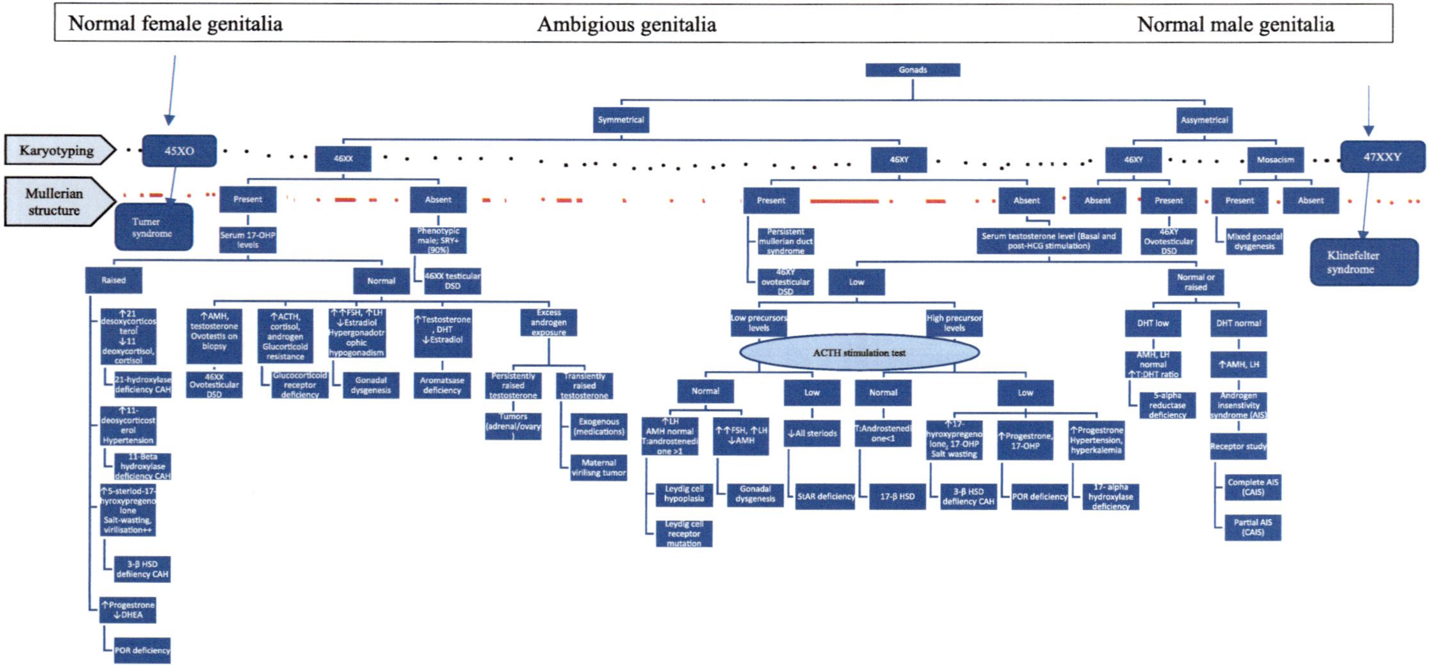

17-*OHP*, 17-hydroxyprogestrone; *β HSD*, beta hydroxysteroid dehydrogenase; *AMH,* anti-mullerian hormone; *CAH*, congenital adrenal hypertrophy; *DHT*, dihydroxy testosterone; *POR*, P450 Oxidoreductase; *StAR*, steroidogenic acute regulatory protein

Fig. 57.3 Diagnostic approach to the DSD patient

of DSD. USG pelvis, which is a non-invasive modality, aims to detect the undescended gonads and Mullerian structures.

Retrograde genitourogram (RGU), on the other hand, is an invasive investigation where radiological contrast is injected from urethral opening or suspected urogenital opening. It highlights the urethral structure along with Mullerian structures or vaginal pouch which can be interpreted based on the presence of indentation over dye filled structure behind bladder having cervical indentation or not. Limitations of these investigations and error likelihood are highlighted in Chap. 17.

Magnetic resonance imaging (MRI) can also be useful in dubious cases, where anatomy is difficult to delineate on USG or genitourogram. Both USG and MRI help reveal associated congenital problems as renal diseases, cardiac, and neurological and hence their rational use may be done in appropriate DSD scenarios. Nuclear scans are important for functional assessment of kidneys.

In cases of intra-abdominal gonads and suspected Mullerian structures, diagnostic laparoscopy with or without biopsy is still mandatory. Diagnostic laparoscopy is a more sensitive and specific modality for diagnosing Mullerian structures when compared to radiological investigations. Yet, the importance of radiological investigations cannot be under-estimated as it guides the team to plan the next step and assist in explaining the family regarding possibilities and treatment options beforehand.

57.3.6 Chromosomal/Genetics Assessment

Ever since the proposal of newer nomenclature of terminologies and groups in DSD, the newer classification of DSD is based on the chromosomal study which broadly entails three subgroups, i.e. sex chromosome DSD, 46XX DSD and 46 XY DSD (Fig. 57.4). Initial karyotype is determined using G-banding and FISH (for SRY gene). Complete karyotype should be studied to identify the mosaicism [17].

It may be also worthwhile to mention that role of geneticists and molecular work analyst has reached spiralling high for working up of cases with DSD with increasing recognition of aberrations of gene, mutations, and anomalies of downstream expressions involved in genesis of many DSD manifestations, despite a given chromosomal layout [18]. The cytogenetics of DSD has been discussed in detail in Chap. 12.

After assigning the type of chromosomal group, further characterization for specific biosynthesis or gene defect is done utilizing targeted approach of single nucleotide probes (SNP) and comprehensive genomic hybridization (CGH) arrays. Gene sequencing is also being utilized to identify the target genes. The two recent tools of molecular studies, whole genome sequencing and whole exome sequencing are being explored to establish the molecular basis of variants in DSD. DSD panels and epigenetics mechanisms have been described for workup of DSD [19].

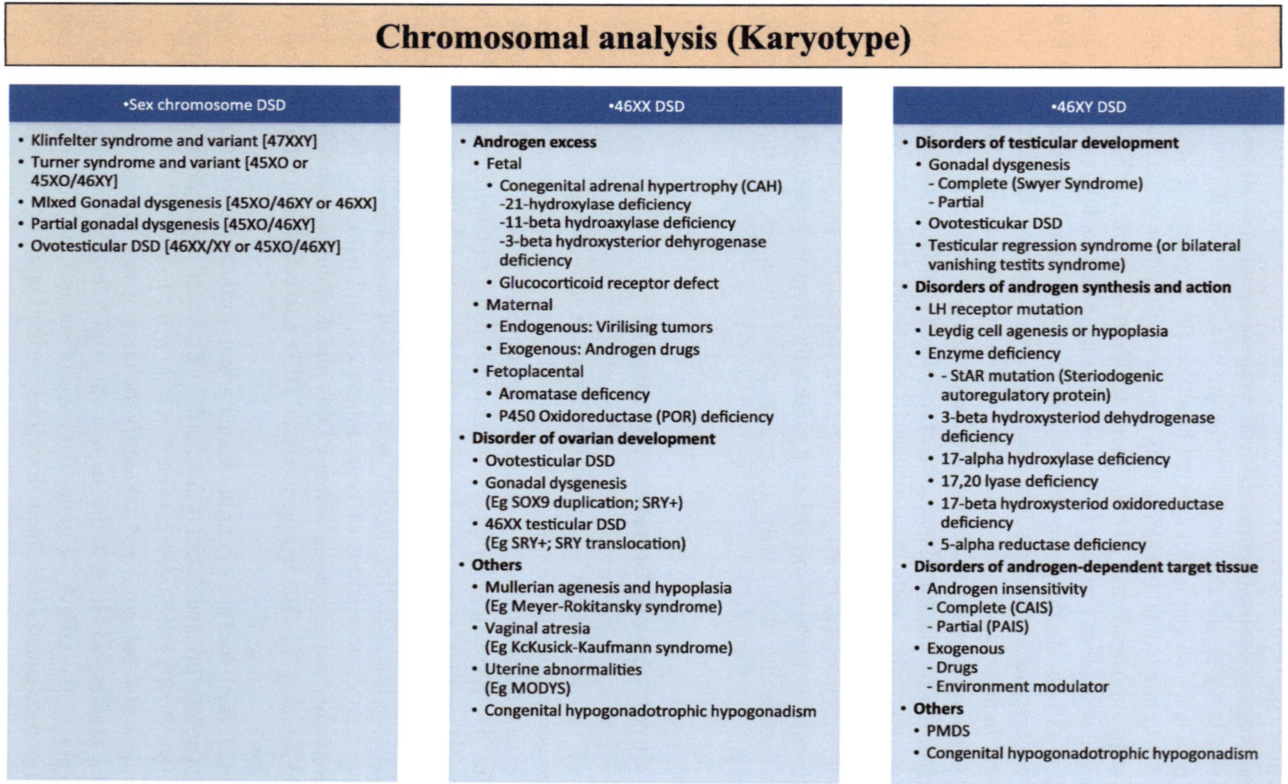

Fig. 57.4 Classification of DSD according to karyotype

57.3.7 Hormonal Evaluation

While chromosomal assay (or karyotyping) may take few weeks depending upon the testing center, hormonal evaluation with its quick result sets a working base toward diagnosing the etiology and guiding initial management [20]. The hormonal evaluation is indicative of gonadal function or integrity of various biochemical pathways involved in sexual differentiation and development. Tables 57.4 and 57.5 highlights the hormonal profile characteristics for common DSDs.

In a virilized female or suspected CAH child, the first hormonal test that is asked for is assessment of serum 17-hyroxyprogestrone levels (17-OHP) along with serum electrolytes and plasma glucose.

Initial hormonal profile includes testosterone, anti-Mullerian hormone (AMH), luteinizing hormone (LH), follicle-stimulating hormone (FSH), dihydroxytestosterone (DHT) (renin levels are also sought for specific cases) for 46XY DSD; 17-OHP for 46XX DSD [21, 22]. Further testosterone precursors level, stimulation test with HCG and ACTH are necessary to reach specific diagnosis in 46 XY DSD; and level of specific bioproducts of cholesterol pathway are sought to pin down the specific enzymatic defect in CAH. Details of endocrine profiling are mentioned in Chap. 14.

57.3.8 Diagnostic Endoscopy

As a part of initial evaluation, diagnostic panendoscopy (urethrocystoscopy + vaginoscopy) and diagnostic laparoscopy can be done [23]. Mullerian structures can be appreciated on laparoscopy (Fig. 57.5). Size (age-appropriate or infantile) and symmetry (hemi-uterus or complete uterus) of Mullerian structure should be noted. Also, laparoscopy allows simultaneous visualization of the status of bilateral gonads and associated ducts which can help identify the gonad [21]. The dual texture of ovotestis can be marked on laparoscopy. The undescended testis can also be examined, and depending upon or high or low location, staged orchiopexy can be planned. Preoperative discussion with family and team, based on chromosomal assessment and radiological investigations, should allow discussion of probable situations and possible options. In case where excision of Mullerian structure or gonadal biopsy is planned, the procedure can be conducted in a similar sitting with the help of minimally invasive methods. Some surgeons prefer laparoscopic assisted gonadal biopsy where they palpate the gonads for consistency and then decide for biopsy or orchiopexy.

Similarly, diagnostic urethrocystoscopy is utilized to define the characteristic of urethra. The presence or absence of urogenital sinus, vaginal pouch, or common channel can be outlined utilizing urethrocystoscopy. Thus, it helps in planning the next stage of genitoplasty where vaginal pouch must be excised in cases of urethroplasty and common channel maybe utilized for feminizing genitoplasty. For more details on the topic, refer to Chaps. 13 and 39.

Table 57.4 Hormonal assay in 46 XX DSD

Etiology	Gonadotrophins		Estradiol	Testosterone	17-hydroxyprogestrone (17-OHP)	Other derivatives of steroidogenesis pathway	Others
	LH	FSH					
Congenital adrenal hyperplasia	↑/N	↑/N	N/↓	↑	↑↑		
– 21-hydroxylase deficiency				↑↑		↑ 21-desoxycorticosterol ↓ corticosterone, cortisol	↑ ACTH, renin ↓ aldosterone
– 11 β hydroxylase deficiency				↑	N/↑	↑ 11-deoxycortisol, 11-deoxycorticosterone ↓ cortisol	↓ renin, aldosterone
– 3 β HSD deficiency				↑	↓	↑pregnenolone ↓cortisol, 11-deoxycortisol	↑renin ↓ aldosterone
– POR deficiency			↓↓	↑		↑progesterone ↓cortisol, 11-deoxycortisol	↓renin
Aromatase deficiency	N	N	↓	↑	N		↑ ACTH
Glucocorticoid receptor mutation	↑	↑	↓	↓	N		Glucocorticoid insensitivity
Pure gonadal dysgenesis	↑	↑	↓	N/↓	N		
Ovotesticular DSD	↑	↑	N/↓	↑	N		↑ AMH, **β HCG**
46 XX testicular DSD	↑	↑	N/↓	↓	N		Phenotypically- male

Table 57.5 Hormonal assay in 46 XY DSD

Etiology	Testosterone and precursor (basal)	Testosterone and precursor (after HCG stimulation)	Dihydroxytestosterone (DHT)	Anti-mullerian hormone (AMH)	Gonadotrophins		Others
					LH	FSH	
Defect in androgen synthesis							
Leydig cell receptor mutation or Leydig cell hypoplasia	↓	↓	↓	N/↑	↑↑	↑	Testosterone: androstenedione ratio > 1
Gonadal dysgenesis	↓	↓	↓	↓	↑/N	↑↑	Testosterone: androstenedione ratio > 1
Testosterone biosynthesis defect	↓	↓	↓	N	↑	N	
– StAR deficiency	All steroids (glucocorticoid, mineralocorticoid, androgens) decreased						Lipoidal congenital adrenal hyperplasia. ↑ ACTH, renin
– 3 β HSD deficiency	↑ 17OHPreg						↑ 17OHpreg: 17OHP ratio
– 17α hydroxylase	↑ progesterone						Low renin hypertension, hypokalemia
– 17, 20 lyase deficiency	↑ 17 OHP						
– POR deficiency	↑ progesterone, 17 OHP						
– 17β HSD deficiency.	↑ androstenedione						Testosterone: Androstenedione ratio < 1

(continued)

Table 57.5 (continued)

Etiology	Testosterone and precursor (basal)	Testosterone and precursor (after HCG stimulation)	Dihydroxytestosterone (DHT)	Anti-mullerian hormone (AMH)	Gonadotrophins		Others
					LH	FSH	
Defect in androgen metabolism							
– 5α reductase deficiency.	N/↑	↑↑	↓↓	N	N	N	↑ T:DHT ratio
Defect in androgen action							
Androgen insensitivity syndrome – Complete (CIAS) – Partial (PAIS)	N/↑	↑↑	N	↑/N	↑↑	N/↑	Receptor study

HSD, hydroxysteroid dehydrogenase; *POR*, P450 oxidoreductase; *17OHP*, 17-hydroxy progesterone; *17OHPreg*, 17 hydroxypregnenolone; *StAR*, steroidogenic acute regulatory protein

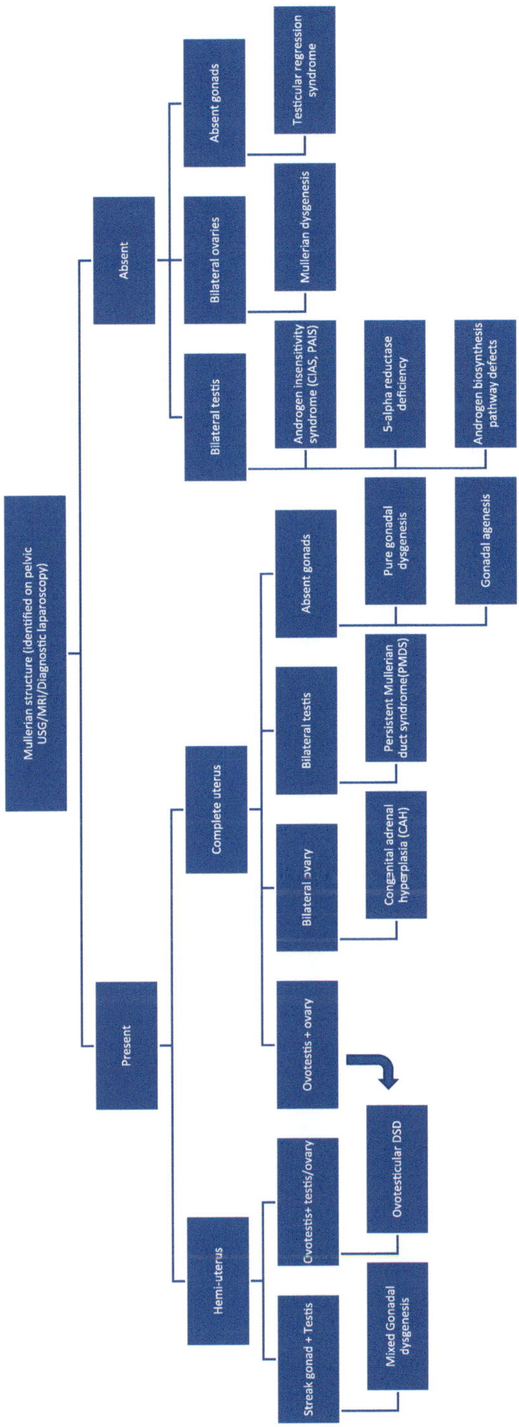

Fig. 57.5 Scheme for Mullerian structure in DSD patients

57.3.9 Histopathological Classification

In cases of 46XX CAH or 46XY with androgen biosynthesis pathway defect, the ovaries and testes are evident, respectively, therefore biopsy can be avoided in these scenarios. However, gonadal biopsy is warranted in cases where there is a dilemma pertaining to the type of gonad. Macroscopically distinct entities are (1) streak gonads, which are described as long fibrous cord-like structure, and (2) ovotestes which have dual texture and consistency with firm whitish testicular tissue and soft pinkish ovarian tissue. Histologically, the gonads can be described as [24]:

Histological description of gonads in DSD

Ovarian tissue: Normal ovarian stroma with numerous follicles
Testicular tissue: Seminiferous tubules are normal along with Leydig cells
Ovotesticular tissue: Both ovarian (at least one well-defined follicle) and testicular (ordered seminiferous tubules architecture) tissue distributed.
Streak gonad tissue: Complete absence of germ cells and supporting cells and abundance of fibrous tissue
Dysgenetic gonad tissue: Disordered germ-cell in abundant stroma

Depending upon the gonadal histology, DSDs can be described as given in Fig. 57.6. The details of histopathology are given in Chap. 15.

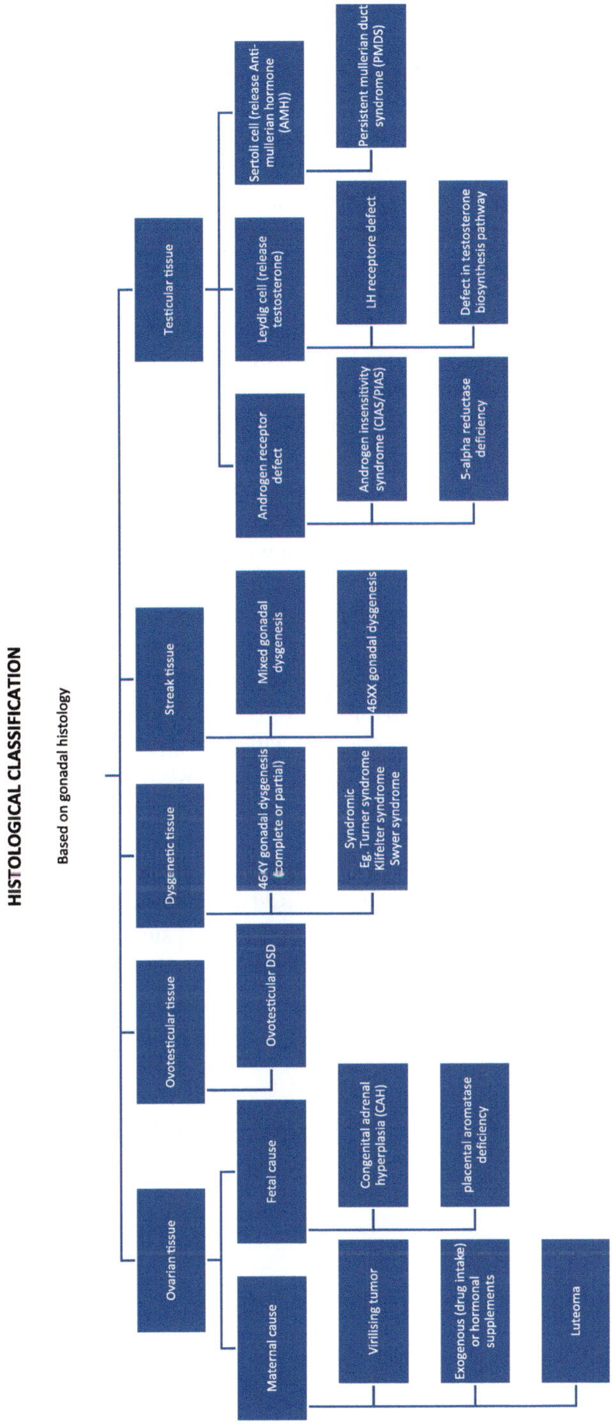

Fig. 57.6 Overview of histopathological findings and likely diagnosis

57.4 Addressing Gender Identity

57.4.1 Gender Assignment

The gender assignment is another important aspect in DSD which is extensively discussed in Sects. 2 and 8. Initial assignment of gender is done at birth based on the phenotypic appearance of genitalia (predominantly based on phallic growth). In cases of DSD, assigned gender may not be collaborative with the chromosomal sex [25].

Gender assignment is based on age at presentation, phallic growth, and rearing gender. While evaluating the case of DSD, the rearing gender should be given the preference and be the working gender of the child. One should refrain from labelling the gender purely based on karyotype.

Despite having the "third sex" as a known and legalized entity in most of the constitutions of the world, the assignment of the same to a newborn is still a stigma to families and society. A newborn with genital ambiguity should be assigned a rearing-sex at birth (as boy or girl) with the consensus of experts or DSD panel team (multidisciplinary team—MDT) in association with family to allow socially secure upbringing in society. Following this, the evaluation can be done and final gender assessment should be done after a detailed counselling and panel discussion amongst MDT and family [26].

The clinical explanation of gender assignment is guided by appearance of external genitalia, hormonal profile, and response to HCG [27]. The general rule that is followed by most of the teams, DSD children with small phallus and poor response to HCG stimulation, better be assigned female gender, while those with good phallus be given a chance to be reared as male. Figure 57.7 defines a simplified working algorithm for gender assignment in DSD.

Since the final gender assignment will be the social identity for the child for life and will affect his mental status, a well-discussed and informed decision in consultation with family, child, pediatrician, endocrinologist, surgical team along with detailed counselling with psychiatrist and psychologists is needed. Surgical corrections (i.e. genitoplasty) can be postponed till family and child are aligned with the given treatment options. The desires of child and family are to be respected [28].

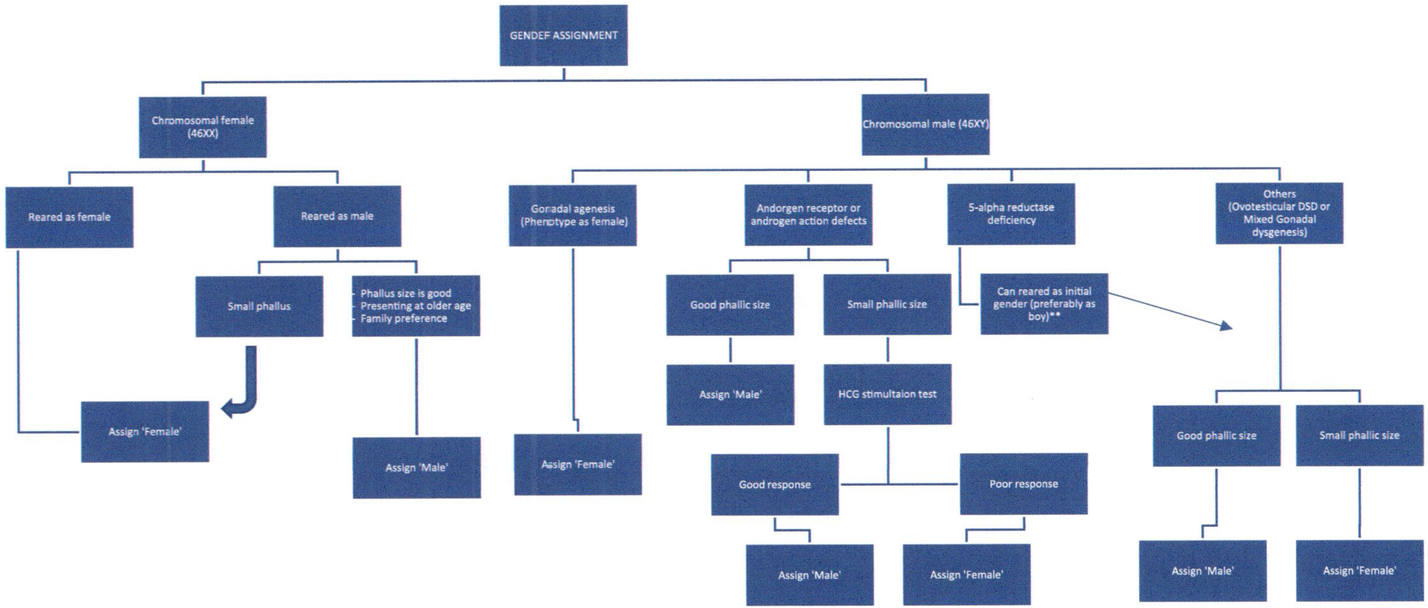

**In children with 5-alpha reductase deficiency, gender assignment can be postponed till puberty. Orchiopexy and chordee correction can be done in initial age while genitoplasty is deferred till puberty

Fig. 57.7 Flowchart for gender assignment in DSD

57.5 Preparing for Better Social Life

57.5.1 Communication and Counselling

On suspicion of DSD, the first conversation with the family is very important [29]. Clinicians should be sensitive and have a positive approach while explaining the possible medical situation and required tests for its evaluation. The families should be given privacy and humble answering to their queries regarding the needful investigation and treatment possibility. The initial counselling is very important to gain the family's confidence and build a rapport for the further management in these sensitive cases. This pre-evaluation counselling and communication can be done in conjunction with a psychological assessment and support for family to understand the need for evaluation and further plan of action better. A simultaneous psychological counselling will carefully apprehend the anxiety and psychological concerns of the child and family.

After evaluation, the reports and possible scenarios should be informed to the family [2, 3]. There should be transparency and trust on the communication. Treatment options should be discussed in a multidisciplinary roundtable discussion with the family. It should be an open discussion where the family can discuss all their concerns with the treating team. Along with the current treatment plan, the follow-up requisite (life-long hormonal support, fertility issues, risk of malignancies) should also be informed and discussed. Child and family should be allowed to make a well-informed decision pertinent to gender assignment and further treatment. The psychological counselling and social worker involvement should not be under-rated throughout the treatment.

Though timing of surgery is controversial, an informed plan can be formulated and surgery can be planned. Owing to various legal implications, some surgeons prefer to defer the genitoplasty till the child attains adulthood where child can have more active participation [28].

Despite a well-discussed and cautious gender assignment, gender dysphoria (GD) is a known issue in DSD patient, more commonly in cases of 5-alpha reductase and 17-beta hydroxysteroid dehydrogenase deficiency [30] (Chap. 6). GD is less common with CAIS. This issue needs to be tackled sympathetically, with proper counselling. For this very reason, it is ideal to wait for genital reconstruction till child ages beyond puberty.

Nowadays there is a shift of age for female genitoplasty to close to puberty. At the same time phalloplasty is also more successful in older children. There may be a need for preoperative androgen stimulation to decide the need and to interpret the success of various genitoplasties. Throughout the diagnosis, treatment, and follow-up, an open discussion should be encouraged between the DSD team, child, and family. Confidentiality of the child should be respected. An empathetic approach should be utilized to build a rapport and confidence with the child and family. The various aspects of counselling are discussed in Sect. 8.

These days various support groups and social activists are involved in the helping and supporting the families with DSD patients (for more details, please see Chap.

53). Apart from these, the institutes managing cases of DSD are encouraged to have special clinics assigned once or twice a month wherein these patients can discuss their issues and concerns not only with the multidisciplinary team but also with the fellow patients [31].

57.6 Prenatal Determination of DSD

With the advances in technology, the prenatal determination of fetal sex is possible in second trimester with 98% accuracy [32]. While owing to prenatal sex determination act, radiologists refrain from disclosing the sex of child, the utilization of sex determination can be extended to families where there has been unexplained sudden neonatal death or family history of CAH. Prenatal clinical diagnosis of ambiguous genitalia requires further molecular studies utilizing chorionic villus sampling, amniocentesis, or cell-free DNA on maternal blood to differentiate gender of child and molecular markers [33, 34]. If the salt-wasting variant of CAH is diagnosed prenatally, then with early management a neonatal death can be prevented. Use of prenatal steroids during pregnancy restricts the clitoromegaly in female fetus, which can be initiated timely in the womb and continued beyond birth. Chapter 29 may be referred to for further details.

57.7 Medical and Surgical Management

Steroids support can be extended to CAH risk pregnancies as described above.

In neonatal or infancy, salt-wasting CAH can present as a life-threatening condition which requires immediate management of shock and electrolyte imbalance. Following this, these patients require life vigilance of electrolyte balance and a mineralocorticoid supplementation (Details in Chaps. 29 and 30). The corticosteroid with or without mineralocorticoid support (depending upon the CAH variant) and maintenance is required life-long with regular monitoring of concerned enzymatic levels. The levels of corticosteroids and mineralocorticoids should be stabilized before proceeding for surgical correction. Hormone supplementation may be required to induce or sustain pubertal changes in accordance with gender. Such treatment requires long-term monitoring (physical and psychological) of side effects as well.

The surgical aspect in DSD is elaborated in Sect. 9. The only clinical scenario where surgical intervention is warranted on an urgent basis in DSD is drainage or decompression hematocolpos or hydro/hematometrocolpos.

Definitive surgical treatment entails the removal of discordant gonadal tissue and other structures to the final assigned gender. Removal of Mullerian structure can be done laparoscopically or laparotomy from Pfannenstiel incision. In cases where there is risk of gonadal tumor (e.g. 46 XY DSD with SRY-), either the gonads should be removed (gonadectomy) or a close vigilance should be kept. This is especially true for mixed gonadal dysgenesis or those with gonadal dysgenesis where chances of malignancy are high.

Timing of surgical correction of genitoplasty is under debate, which is generally resolved by consensus of family with DSD-management teams (Refer to Chap. 46). However, if child's consent is sought, then genitoplasty can be delayed till adulthood.

Male genitoplasty consists of procedures such as chordee correction, repair of hypospadias, and penoscrotal transposition correction. Gonadopexy (or orchiopexy) or gonadal prosthesis also forms a part of male genitoplasty (Details in Chap. 40). These procedures can be done in single stage or multiple stages depending upon the prevailing perineal structures and tissues.

Female genitoplasty consists of clitoroplasty, labioplasty, and vaginoplasty ± gonadectomy. Clitoroplasty can be done as clitoral reduction, clitoral recession, or clitoral resection (details in Chap. 41). In CAH, while clitoroplasty is proposed at an earlier stage, vaginoplasty can be delayed till puberty. Following vaginoplasty, a follow-up serial vaginal dilation is mandatory.

57.8 Follow-Up

DSD patients require life-long follow-up [29, 35]. Interim counselling should be promoted for them (elaborated in Sect. 10).

Patients who underwent gonadectomy will be requiring life-long hormonal support. They also need regular assessment of hormonal levels and assigned-gender-aligned-age-appropriate phenotypic response (details in Chap. 54). Not only this, childhood-to-adulthood transition care should be provided to these patients. Adult urologist should be the integral part of MDT to ease this transition and continuity of care [36].

Certain situations in DSD are at risk of developing the tumor beyond puberty, which need cautious surveillance (see details in Chaps. 52 and 64).

Some patients with CAH, 46XY, and rarely in ovotesticular DSD have fertility potential, where they need proper medical assistance and guidance for having their progeny [11] (Refer to Chap. 51).

Apart from treatment follow-up, they also need psychosocial counselling and involvement with social groups and support groups. A regular psychological evaluation should be conducted for assessment of their mental health and highlighting their concerns for proper addressal. They should be encouraged and counselled for continuation of treatment [31]. Quality of life assessment should be done at regular intervals [37] (See Chap. 49).

57.9 Multidisciplinary Team

Along with the complex and life-long management of DSD, there are various legal implications involved in DSD. Thus, involvement of a multidisciplinary team is of paramount to manage the cases of DSD [3]. It will be ideal to treat the cases of DSD at institutional level, where a cohesive DSD-management team or board including

Fig. 57.8 Constitution of DSD-management team—multidisciplinary team (MDT)

the specialist, super-specialist, and staff of inter-departmental spectrum could be organized (Fig. 57.8). This team should also involve the social workers and legal personals [31].

Every institute should organize a team of experts as Institutional committee which should harmoniously decide upon each case of DSD and suggest an individualized approach. Utilizing the expertise, the role of each member of the committee should be well-defined. The Institutional committee should also involve the concerned child and his family to discuss with them the sex assignment and possible management options.

Multiple institutional committees should have group discussions for management of complex cases. These groups or associations should form at state level and national level where a registry of all DSD cases should be maintained [6]. These groups can help in resolution of disagreement in proposed treatment by Institutional committees.

57.10 Future

With the advances in molecular and genetic studies, the future holds a bright prospective for the management of DSD. Understanding the exact genetic basis of the specific condition, we can have a gene-targeted approach for management.

With the sensitization of clinicians about DSDs, various clinical societies and social help-groups have come into organization. It is important that clinical registries and collaborative multi-institutional projects should be initiated for betterment of DSD patients.

References

1. Hughes IA, Houk C, Ahmed SF, Lee PA, LWPES Consensus Group; ESPE Consensus Group. Consensus statement on management of intersex disorders. Arch Dis Child. 2006;91:554–63.
2. Lee PA, Nordenström A, Houk CP, Ahmed SF, Auchus R, Global DSD Update Consortium, et al. Global disorders of sex development update since 2006: perceptions, approach and care. Horm Res Paediatr. 2016;85:158–80.
3. Ahmed SF, Achermann J, Alderson J, Crouch NS, Elford S, et al. Society for endocrinology UK guidance on the initial evaluation of a suspected difference or disorder of sex development (revised 2021). Clin Endocrinol (Oxf). 2021;95:818–40.
4. Babu R. Changing paradigms in intersex management: legal, ethical, and medical implications. J Indian Assoc Pediatr Surg. 2022;27:372–5.
5. https://www.unicef-irc.org/portfolios/general_comments/GC4_en.doc.html
6. Sarin YK, Singh D, Babu R, Das K, Rao S, DSD Guidelines Committee of Indian Association of Pediatric Surgeons. Indian Association of Pediatric Surgeons guidelines on the management of differences in sex development. J Indian Assoc Pediatr Surg. 2022;27:376–80.
7. Hughes IA. Disorders of sex development: a new definition and classification. Best Pract Res Clin Endocrinol Metab. 2008;22:119–34.
8. Arboleda VA, Sandberg DE, Vilain E. DSDs: genetics, underlying pathologies and psychosexual differentiation. Nat Rev Endocrinol. 2014;10:603–15.
9. Balsamo A, Baronio F, Ortolano R, Menabo S, Baldazzi L, et al. Congenital adrenal hyperplasias presenting in the newborn and young infant. Front Pediatr. 2020;8:593315.
10. Bertelloni S, Tyutyusheva N, Valiani M, D'Alberton F, Baldinotti F, et al. Disorders/differences of sex development presenting in the newborn with 46,XY Karyotype. Front Pediatr. 2021;9:627281.
11. Acién P, Acién M. Disorders of sex development: classification, review, and impact on fertility. J Clin Med. 2020;9:3555.
12. Hughes IA. Ambiguous genitalia. In: Brook CG, Clayton PE, Brown RS, editors. Brook's clinical pediatric endocrinology. Oxford, UK: Blackwell Publishing; 2005. p. 171–82.
13. Ahmed SF, Rodie M. Investigation and initial management of ambiguous genitalia. Best Pract Res Clin Endocrinol Metab. 2010;24:197–218.
14. van der Straaten S, Springer A, Zecic A, Hebenstreit D, Tonnhofer U, et al. The External Genitalia Score (EGS): a European Multicenter validation study. J Clin Endocrinol Metab. 2020;105:dgz142.
15. Moshiri M, Chapman T, Fechner PY, Dubinsky TJ, Shnorhavorian M, et al. Evaluation and management of disorders of sex development: multidisciplinary approach to a complex diagnosis. Radiographics. 2012;32:1599–618.
16. Hryhorczuk AL, Phelps AS, Yu RN, Chow JS. The radiologist's role in assessing differences of sex development. Pediatr Radiol. 2022;52:752–64.

17. O'Connell MA, Atlas G, Ayers K, Sinclair A. Establishing a molecular genetic diagnosis in children with differences of sex development - a clinical approach. Horm Res Paediatr. 2023;96(2):128–43.
18. Witchel SF. Disorders of sex development. Best Pract Res Clin Obstet Gynaecol. 2018;48:90–102.
19. García-Acero M, Moreno O, Suárez F, Rojas A. Disorders of sexual development: current status and Progress in the diagnostic approach. Curr Urol. 2020;13:169–78.
20. Grinspon RP, Castro S, Rey RA. Up-to-date clinical and biochemical workup of the child and the adolescent with a suspected disorder of sex development. Horm Res Paediatr. 2021;2021:279–90.
21. Wisniewski AB, Batista RL, Costa EMF, Finlayson C, Sircili MHP, et al. Management of 46,XY differences/disorders of sex development (DSD) throughout life. Endocr Rev. 2019;40:1547–72.
22. Alkhzouz C, Bucerzan S, Miclaus M, Mirea AM, Miclea D. 46,XX DSD: developmental, clinical and genetic aspects. Diagnostics (Basel). 2021;11:1379.
23. Tafazzoli K, Wünsch L, Bouteleux M, Lindert J, Schulz T, et al. Endoscopy and laparoscopy in disorders of sex development. Sex Dev. 2018;12:100–5.
24. de la Calle CM, Kim S, Baskin LS. Diagnosis and treatment of the intra-abdominal gonad in the pediatric population: testes, ovaries, dysgenetic gonads, streaks, and ovotestes. J Pediatr Surg. 2020;55:2480–91.
25. Gürbüz F, Alkan M, Çelik G, Bişgin A, Çekin N, et al. Gender identity and assignment recommendations in disorders of sex development patients: 20 years' experience and challenges. J Clin Res Pediatr Endocrinol. 2020;12:347–57.
26. Sarma VP. A review of the essential concepts in diagnosis, therapy and gender assignment in disorders of sexual development. Ann Pediatr Surg. 2022;18:13.
27. Markosyan R, Ahmed SF. Sex assignment in conditions affecting sex development. J Clin Res Pediatr Endocrinol. 2017;9(Suppl 2):106–12.
28. Hemesath TP, de Paula LCP, Carvalho CG, Leite JCL, Guaragna-Filho G, Costa EC. Controversies on timing of sex assignment and surgery in individuals with disorders of sex development: a perspective. Front Pediatr. 2019;6:419.
29. Cools M, Nordenström A, Robeva R, Hall J, Westerveld P, et al. COST action BM1303 working group 1. Caring for individuals with a difference of sex development (DSD): a consensus statement. Nat Rev Endocrinol. 2018;14:415–29.
30. Babu R, Shah U. Gender identity disorder (GID) in adolescents and adults with differences of sex development (DSD): a systematic review and meta-analysis. J Pediatr Urol. 2021;17:39–47.
31. Indyk JA. Disorders/differences of sex development (DSDs) for primary care: the approach to the infant with ambiguous genitalia. Transl Pediatr. 2017;6:323–34.
32. Adam MP, Fechner PY, Ramsdell LA, Badaru A, Grady RE, Pagon RA, McCauley E, Cheng EY, Parisi MA, Shnorhavorian M. Ambiguous genitalia: what prenatal genetic testing is practical? Am J Med Genet A. 2012 Jun;158A(6):1337–43. https://doi.org/10.1002/ajmg.a.35338.
33. Cera G, Locantore P, Novizio R, Maggio E, Ramunno V, Corsello A, Policola C, Concolino P, Paragliola RM, Pontecorvi A. Pregnancy and prenatal management of congenital adrenal hyperplasia. J Clin Med. 2022;11:6156.
34. Mallmann MR, Gembruch U. Clinical spectrum of female genital malformations in prenatal diagnosis. Arch Gynecol Obstet. 2022;306:1847–62.
35. Brain CE, Creighton SM, Mushtaq I, Carmichael PA, Barnicoat A, et al. Holistic management of DSD. Best Pract Res Clin Endocrinol Metab. 2010;24:335–54.
36. Auchus RJ, Quint EH. Adolescents with disorders of sex development (DSD)—lost in transition? Horm Metab Res. 2015;47:367–74.
37. Thyen U, Ittermann T, Flessa S, Muehlan H, Birnbaum W, DSD-LIFE group, et al. Quality of health care in adolescents and adults with disorders/differences of sex development (DSD) in six European countries (dsd-LIFE). BMC Health Serv Res. 2018;18:527.

Nephrological Considerations in Children with DSD

58

Sudarsan Krishnasamy and Mukta Mantan

Abstract

Disorders of Sex Development (DSD) refer to a group of disorders of sexual differentiation which lead to internal and/or external genital anomalies manifesting predominantly as ambiguous genitalia. While these disorders are associated with significant morbidity, some of these patients may have congenital anomalies of the kidneys and the urinary tract and may develop recurrent urinary tract infections, steroid-resistant nephrotic syndrome, malignancies, and/or kidney failure. Presence of any of these renal anomalies is associated with significant morbidity in the long term, hence careful evaluation for these nephrological issues is warranted in all children with DSD. Multi-disciplinary care with early referral to the nephrologist upon identification of renal anomalies, hypertension, proteinuria, or kidney dysfunction will improve the overall care offered to these patients.

Keywords

DSD · Ambiguous genitalia · CAKUT · Recurrent UTI · Wilms tumor · Gonadoblastoma

S. Krishnasamy
Department of Pediatrics, Jawaharlal Institute of Postgraduate Medical Education and Research (JIPMER), Puducherry, India

M. Mantan (✉)
Department of Pediatrics, Maulana Azad Medical College and Associated Lok Nayak Hospital, University of Delhi, New Delhi, India

S. K. Ratan (ed.), *Children with Differences in Sex Development*,
https://doi.org/10.1007/978-981-97-1639-5_58

58.1 Introduction

Disorders of sex development (DSD) are rare conditions arising due to defects in internal and/or external genitalia and may manifest anytime in childhood from birth to adolescence. The kidneys and/or the urinary tract may be affected in a subgroup of patients and may be identified while evaluating such children for other organ involvement. In this chapter, we will focus on the nephrological issues in children with DSD.

58.2 Congenital Anomalies of the Kidney and Urinary Tract (CAKUT)

CAKUT has been observed in up to 30% cases of 46,XX DSD in a large international DSD registry [1]. This frequent occurrence of anomalies of the kidneys and the urinary tracts in patients with DSD underscores the common embryological origin of the urinary and the reproductive organs from the intermediate mesoderm. Hence, defect in one system is likely to be associated with anomaly in the other system necessitating screening for both in all affected patients. The various congenital renal anomalies identified in children with DSD include:

- Renal agenesis
- Dysplastic/Hypo-dysplastic kidneys
- Ectopic kidneys
- Horse-shoe kidneys

58.2.1 Renal Agenesis

Renal agenesis refers to a complete absence of renal tissue since birth. It can be either unilateral or bilateral. While bilateral renal agenesis is incompatible with life and often associated with oligohydramnios and Potter sequence, unilateral renal agenesis has been shown to occur in 1 in 2000 individuals, based on a systematic review of 43 studies [2]. Males are slightly more commonly affected than females and the left side is more commonly involved than the right side [3]. Renal agenesis has been identified in syndromes like Turner syndrome, Kallman syndrome, Proud syndrome, Antley-Bixler syndrome, Townes-Brock syndrome, Microcephaly, Facial dysmorphism, Renal agenesis, and ambiguous Genitalia (MFRG) syndrome, etc. It is, therefore, important to identify renal agenesis as up to one-third of affected individuals may have other associated renal anomalies like vesicoureteric reflux. Also, proteinuria, hypertension, and CKD (eGFR <60 ml/min/1.73 m^2) have been identified in 21, 16, and 10% patients, respectively [2]. Therefore, unilateral renal agenesis is not a benign entity as previously thought and careful follow-up of these patients is warranted to identify complications early and institute appropriate management [3].

58.2.2 Dysplastic/Hypo-Dysplastic Kidneys

While hypoplastic kidneys are small in size on ultrasound imaging (kidney length < 2 SD from mean for age), dysplastic kidneys are small and hyperechoic. Histologically, dysplasia is characterized by distortion in glomerular and tubular architecture, metaplasia, and formation of cysts [4]. Dysplastic or Hypo-dysplastic kidneys have been reported in patients with Fraser syndrome, Smith-Lemli-Opitz syndrome, Simpson-Golabi-Behmel syndrome, Pallister-Hall syndrome, etc. This low nephron number, seen in hypo-dysplastic kidneys, predisposes to glomerular hyperfiltration and the consequent injuries leading to proteinuria, hypertension, and progression to kidney failure in the long term [5].

58.2.3 Ectopic Kidneys

Failure of the kidneys to ascend during the early stages of development leads to them being retained in the pelvis instead of their regular position in the lumbar region, leading to ectopic kidneys. Although most of them are asymptomatic and incidentally identified upon imaging for other reasons, some of them may cause nephrolithiasis or pelviureteric junction obstruction (PUJO) [6]. Also, imperforate hymen, vaginal atresia, and hypoplastic uteri have been reported in few patients [7].

58.2.4 Horse-Shoe Kidneys

Horse-shoe kidneys are seen in 1 in 500 among the general population and it has a slightly higher incidence in males [8]. Horse-shoe kidneys are among the most common renal anomalies reported in Turner syndrome (up to 30%); they are also seen in other DSDs like Klinefelter's syndrome, Denys-Drash syndrome, and lathosterolosis. Although the majority are asymptomatic, some may present with abdomen pain due to nephrolithiasis or urinary tract infection [9–11]. Earlier studies have shown a slightly increased risk of Wilms tumor and transitional cell tumors in patients with horse-shoe kidneys, hence monitoring for these tumors is recommended [12].

58.3 Recurrent Urinary Tract Infections

Patients with DSD have been reported to have recurrent urinary tract infections (UTI). Reasons identified for recurrent UTI in these children are:

(a) Defects like persistent urogenital sinus, urethral duplication, and prostatic utricle
(b) Associated constipation

(c) Improper toilet training owing to ambiguous genitalia
(d) Voiding disorders
(e) Vesicoureteric reflux (VUR)
(f) Additionally, surgical reconstruction of genitalia may predispose these patients to UTI

Hence, a proper evaluation of the kidneys, ureter, and bladder by ultrasonography and micturating cystourethrogram (MCU) is warranted in those patients who present with recurrent UTI. DMSA scan is done at least 3 months after an episode of UTI to look for scars. If there are no scars, a repeat DMSA scan may not be necessary in most cases. Although there are no specific guidelines to suggest appropriate management of recurrent UTI in this setting, antibiotic prophylaxis may be considered in those with recurrent UTIs and bowel-bladder dysfunction while carefully identifying those who may benefit from surgical intervention.

58.4 Renal Malignancy

Malignancy is an important consideration in managing patients with DSD. Such patients are at higher risk of developing gonadal malignancies like gonadoblastoma, dysgerminoma, seminoma, choriocarcinoma, Sertoli cell tumor, and yolk sac tumor [13], as covered elsewhere in this book. Apart from gonadal malignancy, DSD patients are at increased risk of Wilms tumor. The syndromes associated with Wilms tumor are WAGR (Wilms-Aniridia-Genitourinary-mental Retardation) syndrome, Denys-Drash syndrome (DDS), Frasier syndrome, Perlman syndrome, and Beckwith-Wiedemann syndrome (BWS). These syndromic forms of Wilms tumor are frequently bilateral with a much earlier age of onset (~1–2 years of age) and are often at advanced stages (stage 3 or higher) at the time of diagnosis. Past or family history of malignancy, bilateral disease, physical examination findings like aniridia, ear pits, hemihypertrophy, macrosomia, and genital abnormalities, and intellectual disability favor a diagnosis of syndromic Wilms tumor and merit a thorough search for these disorders. Although patients with BWS have 4.1% chance of developing Wilms tumor by a median age of 24 months, those with WAGR have 50% chance by 22 months; while those with DDS have 95% chance by median age of 12 months [14]. In view of the heightened risk of developing Wilms tumor, regular surveillance by ultrasonography is recommended in all these patients once every 3–4 months till the age of 5–7 years [15].

In addition, patients with BWS and hemihypertrophy are at increased risk of hepatoblastoma. Therefore, surveillance by abdominal ultrasound and serum alpha-fetoprotein (AFP) levels once every 3 months is recommended in these patients till 4–5 years of age [15].

Table 58.1 Differences between Denys-Drash syndrome (DDS) and Frasier syndrome

Clinical feature	Denys-Drash syndrome	Frasier syndrome
Type and location of mutation	Missense mutations in exons 8 and 9 of *WT1* gene	Splice site mutations in intron 9 of *WT1* gene
Onset of steroid-resistant nephrotic syndrome	Before 1 year of age	By 2–6 years of age
Kidney biopsy finding	Diffuse mesangial sclerosis (DMS)	Focal segmental glomerulosclerosis (FSGS)
Time to end stage kidney disease (ESKD)	By 1–3 years of age	By second-third decade of life
Wilms tumor risk	High (95% by 12 months of age), hence total nephrectomy recommended once ESKD attained	Low risk (5%)
Gonadoblastoma risk	Low risk (4%)	High risk (37–47%), hence prophylactic gonadectomy recommended soon after diagnosis

58.5 Kidney Failure

Patients with DDS and Frasier syndrome have a higher likelihood of developing kidney failure in childhood. The clinical manifestations include proteinuria with edema in the first few months of life, severe hypertension, hematuria, or incidental detection of Wilms tumor on radiological evaluation. Many of these patients develop steroid-resistant nephrotic syndrome and progress to kidney failure by the first decade of life. The important differences between DDS and Frasier syndrome are enumerated in Table 58.1.

58.6 Evaluation and Management

As patients with DSD are prone to multiple nephrological issues as discussed, careful evaluation is necessary in all patients. In all patients with DSD with suspected gonadal dysgenesis, assess renal function and do urinalysis for proteinuria. It is important to measure blood pressure in all patients with an appropriately sized cuff at each hospital visit to identify hypertension at the earliest. When doubtful, ambulatory blood pressure monitoring (ABPM) may be considered. Simultaneously, assessment for end-organ dysfunction like left ventricular hypertrophy by echocardiography and hypertensive retinopathy by fundus examination should carried out. Treatment with angiotensin-converting enzyme inhibitors (ACE inhibitors) like enalapril (0.2–0.6 mg/kg/day) is commenced upon identification of hypertension to keep the blood pressure around the 50th centile for age, height, and gender as per the 2017 AAP clinical practice guidelines [16]. Also, evaluation for proteinuria by a urinary dipstick at the clinic and when feasible, quantification by spot urine protein-creatinine ratio (Up/Uc) is recommended. A dipstick value of 3+ or more, or a spot Up/Uc > 2 suggests nephrotic range proteinuria and needs further evaluation, including a kidney biopsy, by a pediatric nephrologist. Nephrotic syndrome is generally resistant to steroids and other immunosuppressants like cyclosporine. Patients with proteinuria are managed with ACE inhibitors and diuretics (for edema control) while advising a high-protein and salt-restricted diet. Annual or bi-annual renal function test is done, and a glomerular function rate estimated (eGFR) by the modified Schwartz formula ([0.413 × height in cm]/serum creatinine in mg/dL) to monitor for renal dysfunction and pick up CKD (defined as eGFR <60 ml/1.73 m^2/min for 3 months or more) timely. Once CKD sets in, patients also develop metabolic acidosis, anemia, mineral bone disease, and growth failure which need to be individually addressed. It is preferable that such patients are managed in close liaison with a pediatric nephrologist to address these multiple co-morbidities. In those with recurrent UTI, evaluation by kidney ultrasound to look for hydroureteronephrosis, bladder status and post-void residual urine, micturating cystourethrogram, and DMSA scan are done.

As part of surveillance for malignancies, especially Wilms tumor and gonadal malignancies, regular ultrasounds (once every 3 months) are recommended. Empirical gonadectomy may be beneficial in some patients. In view of high risk of Wilms tumor in patients with DDS, prophylactic total nephrectomy is advised once end stage kidney failure sets in [17, 18]. These patients are subsequently put on peritoneal or hemodialysis while they complete chemotherapy, and a tumor-free interval of 1–2 years is recommended before they undergo kidney transplantation [19]. Graft outcomes following kidney transplantation are similar to other etiologies; and no recurrence of nephrotic syndrome has been reported [20].

In conclusion, patients with DSD are more prone for kidney injury as illustrated, hence a meticulous and comprehensive evaluation must be carried out in all such patients to identify problems at an early stage and offer appropriate multi-disciplinary care to improve the overall prognosis.

References

1. Cox K, Bryce J, Jiang J, Rodie M, Sinnott R, Alkhawari M, et al. Novel associations in disorders of sex development: findings from the I-DSD registry. J Clin Endocrinol Metab. 2014;99(2):E348–55.
2. Westland R, Schreuder MF, Ket JC, van Wijk JA. Unilateral renal agenesis: a systematic review on associated anomalies and renal injury. Nephrol Dial Transplant. 2013;28(7):1844–55.
3. La Scola C, Ammenti A, Bertulli C, Bodria M, Brugnara M, Camilla R, et al. Management of the congenital solitary kidney: consensus recommendations of the Italian Society of Pediatric Nephrology. Pediatr Nephrol. 2022;37:2185–207.
4. Parikh CR, McCall D, Engelman C, Schrier RW. Congenital renal agenesis: case control analysis of birth characteristics. Am J Kidney Dis. 2002;39(4):689–94.

5. Brenner BM, Mackenzie HS. Nephron mass as a risk factor for progression of renal disease. Kidney Int Suppl. 1997;63:S124–7.
6. Cinman NM, Okeke Z, Smith AD. Pelvic kidney: associated diseases and treatment. J Endourol. 2007;21:836–42.
7. Dretler SP, Olsson C, Pfister RC. The anatomic, radiologic and clinical characteristics of the pelvic kidney: an analysis of 86 cases. J Urol. 1971;105:623–7.
8. Natsis K, Piagkou M, Skotsimara A, Protogerou V, Tsitouridis I, Skandalakis P. Horseshoe kidney: a review of anatomy and pathology. Surg Radiol Anat. 2013;36:517–26.
9. Glodny B, Petersen J, Hofmann KJ, Schenk C, Herwig R, Trieb T, et al. Kidney fusion anomalies revisited: clinical and radiological analysis of 209 cases of crossed fused ectopia and horseshoe kidney. BJU Int. 2009;103(2):224–35.
10. Cascio S, Sweeney B, Granata C, Piaggio G, Jasonni V, Puri P. Vesicoureteral reflux and ureteropelvic junction obstruction in children with horseshoe kidney: treatment and outcome. J Urol. 2002;167(6):2566–8.
11. Pawar AS, Thongprayoon C, Cheungpasitporn W, Sakhuja A, Mao MA, Erickson SB. Incidence and characteristics of kidney stones in patients with horseshoe kidney: a systematic review and meta-analysis. Urol Ann. 2018;10(1):87–93.
12. Neville H, Ritchey ML, Shamberger RC, Haase G, Perlman S, Yoshioka T. The occurrence of Wilms tumor in horseshoe kidneys: a report from the National Wilms Tumor Study Group (NWTSG). J Pediatr Surg. 2002;37(8):1134–7.
13. Huang H, Wang C, Tian Q. Gonadal tumour risk in 292 phenotypic female patients with disorders of sex development containing Y chromosome or Y-derived sequence. Clin Endocrinol. 2017;86(4):621–7.
14. Liu EK, Suson KD. Syndromic Wilms tumor: a review of predisposing conditions, surveillance and treatment. Transl Androl Urol. 2020;9(5):2370–81.
15. Kalish JM, Doros L, Helman LJ, Hennekam RC, Kuiper RP, Maas SM, et al. Surveillance recommendations for children with overgrowth syndromes and predisposition to Wilms tumors and hepatoblastoma. Clin Cancer Res. 2017;23:e115–22.
16. Flynn JT, Kaelber DC, Baker-Smith CM, Blowey D, Carroll AE, Daniels SR, et al. Clinical practice guideline for screening and management of high blood pressure in children and adolescents. Pediatrics. 2017;140:e20171904.
17. Auber F, Jeanpierre C, Denamur E, Jaubert F, Schleiermacher G, Patte C, et al. Management of Wilms tumors in Drash and Frasier syndromes. Pediatr Blood Cancer. 2009;52:55–9.
18. Gariépy-Assal L, Gilbert RD, Žiaugra A, Foster BJ. Management of Denys-Drash syndrome: a case series based on an international survey. Clin Nephrol Case Stud. 2018;6:36–44.
19. Knoll G, Cockfield S, Blydt-Hansen T, Baran D, Kiberd B, Landsberg D, et al. Canadian Society of Transplantation: consensus guidelines on eligibility for kidney transplantation. CMAJ. 2005;173(10):S1–25.
20. Kist-van Holthe JE, Ho PL, Stablein D, Harmon WE, Baum MA. Outcome of renal transplantation for Wilms' tumor and Denys-Drash syndrome: a report of the North American Pediatric Renal Transplant Cooperative Study. Pediatr Transplant. 2006;9:305–10.

Anaesthesiology Considerations in Children with DSD

59

Anju R. Bhalotra and Rahil Singh

Abstract

The birth of a baby with ambiguous external genitalia is often a cause of surprise and dismay for the doctors and the family and poses a problem of assignment of sex. Various controversial terms have been used to describe these children and in response to this, the Chicago Consensus recommended a new terminology called Disorders of Sex Differentiation (DSD) which is a more inclusive definition and includes all problems noticed after the child is born in which the genitalia are atypical in relation to the gonads or chromosomes. Management of children with DSD should be individualized and multidisciplinary, and all aspects need to be considered including the psychological implications. These children may present to the anaesthesiologist for a wide plethora of diagnostic and corrective surgical procedures and may require single or multiple anaesthesia exposures. Anaesthetic implications when providing anaesthesia to children with DSD include concerns related to the paediatric age group and the associated congenital syndromes/ anomalies which may be present in a large number of these children. Older children often have emotional, psychological or behavioural problems and adequate counselling and allaying of fear and anxiety must be done preoperatively. Children with congenital adrenal hyperplasia constitute almost half the children with DSD and it is important to be able to recognise and manage these children in whom clinical manifestations are mainly due to adrenocortical insufficiency, hyperandrogenism and the adverse effects of glucocorticoid supplementation. Patients with adrenal insufficiency or those receiving long-term corticosteroid therapy are unable to respond normally to surgery or other forms of stress and require "stress dosing" or supplementation with corticosteroids during surgery.

A. R. Bhalotra (✉) · R. Singh
Department of Anaesthesiology, Maulana Azad Medical College, Delhi, India

S. K. Ratan (ed.), *Children with Differences in Sex Development*,
https://doi.org/10.1007/978-981-97-1639-5_59

The aim of anaesthesia is to maintain hemodynamic stability along with adequate perioperative analgesia which may be achieved with the use of multimodal analgesia along with epidural analgesia or USG-guided nerve blocks.

Keywords

Ambiguous genitalia · Congenital adrenal hyperplasia · Anaesthesia

The terms disorders of sexual differentiation and differences in sex development or DSD encompass a heterogenous group of conditions in which there is an atypical development of the genitalia. They may be grouped into different categories based on anatomic, gonadal and chromosomal abnormalities. Affected children may be identified very early in life due to their ambiguous external genitalia or may present at a later age with virilization, absent or delayed puberty, or infertility.

There are three main categories of DSD [1–3]:

- *46 XX DSD*
  1. Excessive androgen levels as in congenital adrenal hyperplasia (CAH) or maternal exposure.
  2. Gonadal dysgenesis (disordered ovarian development).
- *46XY DSD*
  1. Disordered androgen synthesis due to defects of biosynthesis and conversion (5 alpha reductase deficiency, 17-beta hydroxysteroid dehydrogenase 3 deficiency).
  2. Complete and partial androgen insensitivity syndromes (CAIS and PAIS).
  3. Gonadal dysgenesis (disordered testicular development).
- *Sex chromosome DSD*
  1. Mixed gonadal dysgenesis; 45XO/46XY.
  2. Ovo-testicular DSD 46XX/46XY.

Ambiguous genitalia and syndromes of multiple congenital malformations
Children with ambiguous genitalia can have other congenital malformations as many transcription factors for development and differentiation of sex are involved in development of other functions [4]. The WAGR, Denys-Drash and Frasier syndromes are due to mutations of the WT-1 gene (11p3) [4]. Whilst children with Denys-Drash syndrome have ambiguous genitalia with nephropathy and bilateral Wilms tumour [5], those with Frasier Syndrome have 46 XY gonadal dysgenesis, non-specific focal and segmental glomerulosclerosis and gonadoblastoma [6] and WAGR Syndrome consists of genitourinary malformations, Wilms tumour, aniridia and mental retardation [7]. Other mutations result in urogenital anomalies along with campomelic dysplasia and developmental delay and abnormalities of adrenal glands, anterior pituitary and hypothalamus [4]. In the Smith Lemli-Opitz syndrome there is hypocholesterolaemia and elevated 7-dehydrocholesterol levels due to mutations affecting 7-dehydrocholesterol reductase. These children can have micrognathia, cleft palate and an

abnormally hard tongue [8]. In the VATER and CHARGE syndromes, there may be abnormal development of the lower abdominal wall with pubic diastasis resulting in urogenital anomalies and bladder and cloacal exstrophy [4]. A recent retrospective review of 128 DSD patients revealed cardiac anomalies in 28 patients (22%), skeletal in 19 patients (15%) and growth problems in19 patients (15%). There were other anomalies in 53 patients (41%). Thus, it seems reasonable to screen DSD patients for cardiac and other congenital anomalies [9].

59.1 Congenital Adrenal Hyperplasia

Congenital Adrenal Hyperplasia (CAH) and mixed gonadal dysgenesis are found in almost 50% of all patients with DSD who present with ambiguous genitalia [10]. CAH is the commonest cause of 46XX DSD. Classic CAH accounts for most (>90%) of all cases and may be salt-wasting or simple virilizing type. The salt wasting type is more severe and these children present early with hyponatraemia, hyperkalaemia, and hypotension (hypovolaemia). The virilising form is a more moderate type. However, during periods of physical stress (illness, surgery, trauma) classic CAH patients of both types may develop an "adrenal crisis". Infant girls with either type of classic CAH are usually detected in the early neonatal period due to their ambiguous genitalia. Cardiovascular collapse may occur 4–14 days after birth and should be anticipated in all children with ambiguous genitalia until CAH has been excluded [11]. On the other hand, infant boys with classic CAH will have normal male sexual development apart from some scrotal hyperpigmentation and the diagnosis may be missed unless neonatal screening is done. But these neonates may also present with poor feeding and failure to thrive, hypotension, dehydration, lethargy, hyponatremia and hyperkalaemia at 2–3 weeks of age.

The non-classical form of CAH (NC-CAH) is the mildest and is often called "late onset" CAH, as symptoms occur at a later age. Newborns with NC-CAH have no genital changes and the disorder is only diagnosed when the effects of excessive androgens manifest in childhood as precocious puberty or rapid growth or during teenage or early adulthood with irregular periods, excessive body and facial hair and severe acne [12].

During their growing years, children with classic CAH should be on maintenance hydrocortisone therapy [13]. Newborns and infants should receive fludrocortisone and sodium chloride in addition to hydrocortisone. Adults with classic CAH usually receive daily hydrocortisone and/or long-acting glucocorticoids and mineralocorticoids, as indicated clinically [13].

Usual maintenance therapy in growing children with CAH [13]

- Glucocorticoid therapy as hydrocortisone tablets in a dose of 10–15 mg/m^2/day given in three divided doses.
- Mineralocorticoid therapy as fludrocortisone tablets in a dose of 0.05–0.2 mg/day given in 1 or 2 divided doses.
- Sodium chloride in a dose of 1–2 g/day in infancy given over several feedings.

Stress dosing

Patients with CAH on maintenance glucocorticoid treatment should increase glucocorticoid dosage during major trauma, major surgery under general anaesthesia, febrile illness (>38.5 °C) and gastroenteritis with dehydration, [13]. This is called stress dosing.

Stress doses of glucocorticoids to prevent an adrenal crisis [13]

- Intravenous hydrocortisone first dose: 25 mg for infants and preschool children, 50 mg for school going children and 100 mg for adults.
- Intravenous hydrocortisone successive doses: One-fourth of the initial dose (above) given every 6 h.
- Mineralocorticoids are not required with stress doses of hydrocortisone as hydrocortisone itself has a relatively high mineralocorticoid activity.
- Once stabilised, patients should resume maintenance doses of hydrocortisone.
- Patients should be instructed to avoid fasting during acute illnesses.

59.2 Surgery in Children with DSD

There is still no unanimity regarding the indications, appropriate time and extent of surgery for individuals with DSD [14]. Every child requires individual assessment and consideration by a team of multidisciplinary experts at centres which are regularly dealing with such children.

When do Anaesthesiologists encounter these children?

1. *For investigation* into cause and type of DSD. May require anaesthesia/sedation in radiology suite for CT/MRI scan or Ultrasonography.
2. *For diagnostic procedures* in the operation theatre like examination under anaesthesia (EUA), cystoscopy, retrograde genitogram or open or laparoscopic exploration with gonadal biopsy.
3. For *major genital reconstruction.*
4. For *follow up*—dilatation, EUA, cystoscopy, genitoscopy etc. after genital reconstruction.
5. For *incidental surgeries* like tonsillectomy and appendicectomy unrelated to the DSD.

Aims of surgery in patients with DSD [15]

1. To restore genital anatomy and allow penetrative sexual intercourse as a male or female when the child grows up.
2. To allow the patient to grow up and be able to reproduce if possible.
3. To minimise urinary tract infections and urinary incontinence due to abnormal genitourinary anatomy.
4. To prevent collection of blood or fluid in the uterine or vaginal cavities.
5. To avoid virilization at puberty of patients raised as girls and prevent breast development in patients who have been raised as boys.

6. To minimise risks of developing gonadal cancers.
7. To encourage these children to develop "individual" and "social identities".
8. To prevent these children from suffering shame and humiliation due to their atypical genital anatomy.
9. To allow the parents to bring up their child in society and be accepted by society.

Types of surgery in patients with DSD

Various surgical interventions can be done in these patients depending on their age and gender. These have been specified in the DSD Guideline [16].

1. *The genital tubercle*: The clitoris or glans may be left intact or reduced in size in a patient assigned as a female or may be modified in a male assigned patient. In the post-pubertal patient who chooses a male gender, phalloplasty may be required [15].
2. *The vaginal cavity:* If there is a reasonable-sized genital dimple, it can be progressively dilatated to create a full-size vagina. If there is a complete absence of the genital tract, a new vagina can be created using intestine, peritoneum, buccal mucosa etc. Surgical techniques used include partial or total urogenital sinus mobilization or creation of a mucosal funnel (introitoplasty) [15].
3. *Gonads:* These become a problem if the assigned gender is different from the gonadal sex or when there is a risk of tumour development. Gonads (testis) can be partially or completely removed, brought into the scrotum (orchidopexy), or just monitored by regular clinical examination, imaging or biopsy [15].
4. *Mullerian remnants*: These are usually asymptomatic but may become an issue when a male assigned patient develops dysuria, urinary tract infections, cyclic pain, stone formation etc. The Mullerian pouch can then be surgically removed [15].
5. *External genitalia*: Procedures include perineoplasty and genital tubercle refashioning.
6. *Breasts:* At or after puberty, bilateral mastectomy in male assigned patients and breast enhancement using hormones, prosthesis etc. may be required in female assigned patients.

59.3 Anaesthetic Considerations

59.3.1 Pre-operative Evaluation

The anaesthesiologist first comes in contact with the child with DSD when the child is posted for a procedure under general anaesthesia or monitored anaesthesia care. By this time, the child would have already been investigated and a diagnosis of the DSD made. A detailed pre-anaesthetic assessment must be done with a focus on the overall clinical condition and the presence of other congenital anomalies.

History and Examination of the neonate/infant

- *Maternal History.*
  - h/o maternal virilisation (suggests androgen secreting tumour or aromatase deficiency in the mother).
  - h/o drug intake by the mother during pregnancy which may lead to virilisation of a female foetus.
  - h/o any previous neonatal deaths (may be due to undiagnosed adrenal crisis).
  - h/o consanguineous marriage in parents (increased possibility of an autosomal recessive disorder).
  - h/o ambiguous genitalia in any family member (suggests possibility of X-linked recessive conditions like androgen insensitivity).
- *Examination of child.*
  - General appearance: sick or well looking (adrenal crisis?).
  - Developmental milestones.
  - Hydration status and blood pressure (salt loss with hypotension/hypertension in CAH).
  - Presence of dysmorphic facial features (associated genetic syndromes).
  - Jaundice (due to concomitant thyroid hormone or cortisol deficiency).
  - Airway examination (micrognathia, macroglossia, cleft palate, short neck).
  - Skeletal malformations (campomelic dysplasia?).
  - Anatomic malformations (VATER/VACTERL, cloacal exstrophy?).
  - Evidence of other anomalies—cardiac, renal.
- *Genital examination* [17].
  - Appearance of phallic structure (micropenis, clitoromegaly).
  - Presence of bilateral, unilateral or non-palpable gonads.
  - Any hypospadias.
  - Labioscrotal folds for symmetry, presence of gonads.
  - Anogenital distance.
  - Examination of overlying skin (hyperpigmentation suggests adrenal failure and possible ACTH elevation).

History and Examination of the older child/adolescent

Older children may present with delayed puberty, primary amenorrhoea or virilisation.

- Detailed family history.
- Evaluation of growth pattern and appearance of secondary sexual characteristics.
- Examination—external genital appearance, effects of oestrogens like development of breasts and effects of androgens such as pubic hair and virilisation.
- Absent or incomplete puberty in girls, with primary amenorrhoea or oligomenorrhoea and hirsutism, precocial puberty and marked acne suggest a diagnosis of CAH.

Investigations

I. *First-tier investigations* in infants with genital ambiguity include [2]:
   - Ultrasonography to identify gonads, uterus and any features of obstruction in the genitourinary tract like hydrometrocolpos and hydronephrosis.
   - Serum 17OH-progesterone (17OH-P).
   - Serum electrolytes.
   - Blood sugar. Hypothalamic, pituitary or adrenocortical insufficiency may lead to deficiency in cortisol levels and subsequent hypoglycaemia.
   - Karyotyping.
   - (Elevated levels of serum 17OH-P, hyponatraemia, hyperkalaemia, 46XX Karyotype and presence of a uterus suggests a diagnosis of CAH)

II. *Further evaluation.*
   - If karyotype is 46XY, Human Chorionic Gonadotrophin (HCG) stimulation test is done to differentiate subtypes of XY DSD.

III. *Identification of gonads and Mullerian structural derivatives.*
   - USG/MRI scan.
   - Genitogram.
   - EUA and, if indicated, genitoscopy, laparoscopy and biopsy of gonadal structures.

IV. *Other investigations if other congenital anomalies are suspected.*
   - Echocardiography, extended renal profile or thyroid function test in children with associated syndromes.
   - In conditions with a high incidence of associated spinal abnormalities, for example cloacal exstrophy, imaging of spine before considering central neuraxial blocks.
   - Electrocardiography (peaked T waves from hyperkalaemia).

59.3.2 Preoperative Preparation

Counselling

- The psychological aspects of these conditions are very important and by the time these children come to the operation theatre, they would have already received extensive counselling. Any residual doubts must be cleared in the preoperative period [18].
- These children may have emotional difficulties and are likely to be very anxious if they require repeated surgeries. Older children may also develop behavioural issues due to the nature of their underlying disorder. Adequate counselling and allaying of fear and anxiety must be done.
- Pre-operative endocrinology consultation should be obtained to optimize perioperative steroids and mineralocorticoid therapy.
- Options for managing post-operative pain should be discussed like the use of post-operative epidural analgesia and patient controlled intravenous or epidural analgesia. The associated feelings of numbness and motor weakness associated

with the use of some neuraxial blocks must also be explained so children and parents are not alarmed in the post-operative period.
- For major procedures, children and parents should be counselled and explained about post-operative concerns including the need to nurse the child in an intensive care or high-dependency unit after surgery.

59.3.3 Orders

- Adequate blood and blood products should be arranged depending on the magnitude of the surgery and anticipated blood loss.
- Standard instructions should be given regarding preoperative fasting, i.e. 2 h for clear fluids, 4 h for breast milk, 6 h for infant formula or a light meal and 8 h for a proper meal [19].
- Detailed written informed consent should be obtained from parents and written or verbal assent from older children who are able to comprehend the need and plan of the anticipated surgery. Children below 12 years of age are not eligible to give legal consent in Indian Law. Children of ages 7–12 years can, however, give their verbal assent. Older children aged 12–18 years can give written assent. This must be obtained for these procedures along with written consent from the parents [2].
- Children with adrenal insufficiency/suppression should be kept first/early on OT list in order to minimise fasting and dehydration. If there is a delay of more than 2 h in taking the child to theatre, IV maintenance fluids should be started. Morning values of S. electrolytes should be ordered.

Premedication
- Most patients (>6 months of age) will benefit from sedative and/or anxiolytic premedication. Oral midazolam is most commonly used.
- Children with CAH on maintenance doses of hydrocortisone and fludrocortisone must receive their usual morning doses.

59.4 Intraoperative Management

- Adequate intravenous access should be secured.
- Standard American Society of Anaesthesiology monitoring should be established consisting of electrocardiography, non-invasive blood pressure, pulse oximetry, end tidal carbon dioxide and temperature monitoring. More intensive monitoring like invasive blood pressure, depth of anaesthesia, central venous pressure, neuromuscular junction monitoring and monitoring of blood sugar and S. electrolytes may be considered depending on the magnitude of the surgical procedure and associated comorbidities.

- Operation theatre should be prepared. A difficult airway cart is made ready, especially if the child has an anticipated difficult airway, i.e. dysmorphic facial features, cleft palate and short neck.
- The patient's position for surgery may be lithotomy, supine or prone, depending upon the nature of surgery and the surgeon's preference.
- General anaesthesia (GA) with a regional anaesthetic technique is usually preferred for most urogenital procedures.
- In children with CAH, a "stress dose" of hydrocortisone is given at induction of anaesthesia. An additional dose administered as one-fourth of the initial dose should be given if the surgery extends beyond 4–6 h.
- Anaesthesia may be induced by the intravenous or inhalation route depending on the age of the child, presence of an intravenous cannula and preference of the anaesthesiologist.
- For intravenous induction of anaesthesia, propofol is the most widely used agent. Though there is no absolute contraindication to the use of other induction agents, etomidate should be avoided in cases of adrenal insufficiency as it inhibits steroid synthesis and this may be undesirable in haemodynamically unstable patients or patients on pre-operative steroid therapy [13, 20].
- Sevoflurane is the most recommended inhalational agent for induction and maintenance of anaesthesia in children.
- For neuromuscular blockade, any intermediate acting muscle relaxant can be used.
- In children with CAH, rocuronium may be preferred as its cardiovascular side effects are minimal [20]. Succinylcholine should be avoided in patients with preexisting hyperkalaemia. There may be a prolonged duration of action of nondepolarizing muscle relaxants in patients with uncorrected hypokalaemia.
- If there are associated renal abnormalities or renal dysfunction, atracurium or cis-atracurium should be used for muscle relaxation, short-acting opioids for analgesia and NSAIDs should be avoided. The metabolism of cisatracurium and atracurium is not affected by renal or hepatic dysfunction as they are metabolised by Hofmann elimination and ester hydrolysis. Cisatracurium has the added advantage over atracurium of not causing histamine release.
- The airway may be secured with a supraglottic airway device or endotracheal tube depending on the comfort and usual practice of the anaesthesiologist. Use of an endotracheal tube may be preferred in infants undergoing prolonged/prone position procedures or when a buccal mucosa graft is planned in hypospadias repair in which case an oropharyngeal pack should also be inserted.
- GA can be maintained using total intravenous anaesthesia (TIVA) with propofol or a balanced anaesthesia technique (oxygen in air/nitrous oxide and sevoflurane and muscle relaxant).
- Multimodal analgesia should be provided consisting of NSAIDs, paracetamol, opioids and a regional anaesthetic technique wherever possible. Short-acting opioids like fentanyl or remifentanil are preferred. USG-guided nerve blocks or central neuraxial blockade given after induction of anaesthesia provides excellent intraoperative and postoperative analgesia.

- Whilst significant bleeding does not occur from erectile tissues in infants, it may be considerable in older children and transfusion of blood and blood products may be required.
- Prophylactic antiemetics should be given, especially in patients undergoing testicular or laparoscopic surgery [21].

59.5 Post-operative Care

- At the end of surgery, most children can be extubated and shifted to PACU/ICU depending upon the duration and magnitude of the procedure.
- Patients who have undergone major procedures, children with CAH and children who develop an adrenal crisis require post-operative ICU stay.
- Adequate post-operative analgesia should be provided. After reconstructive surgery, a "mermaid dressing" may be required to provide surgical immobilisation for 4–5 days after surgery. In this, soft plastic foam or cotton is placed between the knees and ankles of the child and the legs are loosely wrapped with an elastic bandage. These children usually require continuous epidural analgesia and ICU monitoring.
- Children with CAH should receive intravenous hydrocortisone administered as one-fourth of the initial dose until clinically stable after which they can revert back to maintenance doses. A paediatric endocrinology consultation can be taken for deciding dosing of steroids [22].
- Vital signs should be monitored frequently to detect perioperative adrenal crisis.

Adrenal crisis

Adrenal crisis is defined as a major impairment of general health with at least two of the following: Hypotension (SBP < 100 mmHg), nausea/vomiting, severe fatigue, fever, somnolence, hyponatraemia (<132 mmol/l) or hyperkalaemia, and hypoglycaemia [23].

Treatment [23]

1. Stress dose of hydrocortisone (50–100 mg/m^2); repeated if necessary and followed by intravenous hydrocortisone 50–75-100 mg/m^2/24 h divided into 6 h doses OR given as a continuous intravenous infusion of 2–3 mg/m^2/h.
2. Intravenous 0.9% saline; 10–20 ml/kg bolus which is repeated as required and followed by maintenance fluids plus deficit fluids for dehydration (0.9% saline with 5% dextrose).
3. Hourly blood sugar monitoring; 2–5 ml/kg of 10% dextrose if blood sugar <3.5 mmol/l.
4. Correction of dyselectrolytaemia.
5. Monitoring of blood pressure and cardiac function.
6. Switch to oral regimen once clinically stable starting with hydrocortisone 30–50 mg/m^2/24 h and then gradually reduce to maintenance doses of hydrocortisone + fludrocortisone.

Emergency surgery

If children with CAH require emergency surgery, they should be fluid resuscitated and started on stress doses of IV hydrocortisone at the time of admission.

Problems of latex allergy in children undergoing urological surgery [21].

Early and prolonged exposure to latex containing products in children undergoing repeated surgery, especially during infancy, may lead to latex sensitization. Perioperatively a type 1 hypersensitivity reaction with bronchoconstriction and cardiovascular collapse may occur which may be delayed or even occur postoperatively. Alternatively, a type-IV hypersensitivity reaction may occur resulting in watering and itching of the eyes, sneezing, and coughing. This is delayed and usually occurs 6–48 h after exposure. Latex products should be avoided in these patients.

Options for perioperative analgesia

- Minor/intermediate procedures: simple analgesics, multimodal analgesia, wound infiltration with local anaesthetics, ilioinguinal nerve block, caudal analgesia.
- Laparoscopic surgery: multimodal analgesia, port site infiltration.
- Major procedures: multimodal analgesia, acetaminophen, intravenous opioids as continuous infusion or nurse- or patient-controlled analgesia, regional/neuraxial analgesia with or without additives like an opioid or clonidine (Additives improve analgesia, but may have side-effects like sedation, hypotension, nausea, vomiting, itching and respiratory depression etc.).

Regional blocks for intra- and post-operative analgesia

Caudal epidural: Used for infra-umbilical surgery. A major limitation is the short duration of analgesia (4–6 h) provided by a single-shot caudal injection. The use of additives or a catheter threaded into the caudal epidural space can help prolong the duration of post-operative analgesia. The catheter should be tunnelled to reduce the risk of bacterial contamination [24].

Lumbar epidural analgesia: Is recommended for all major pelvic and abdominal procedures. Relative contraindications include malformations of spine, hydrocephalus, increased intracranial pressure, unstable epilepsy or previous surgery on spine [25]. It must be performed by experienced anaesthesiologists in small children.

Trunk blocks. For most surgery involving truncal blocks the visceral pain element will be present for which systemic analgesic adjuncts must be given. Truncal blocks include:

- *Ilioinguinal and iliohypogastric nerve block*: Suitable in unilateral procedures. The iliohypogastric nerve innervates skin above inguinal ligament and the suprapubic region. The ilioinguinal nerve provides sensory supply to skin of upper medial thigh, upper part of scrotum and root of the penis/skin overlying labium majus and the mons pubis in males and females, respectively. Complete analgesia for orchidopexy repair requires additional injection of the genitofemoral nerve or local infiltration of the testicular sack.

- *Rectus sheath and paraumbilical block.* For analgesia for umbilical or midline surgical incisions and laparoscopic surgery.
- *Transverse abdominis plane block*: Has been utilized as an alternative to ilioinguinal and iliohypogastric nerve block for inguinal surgery in children. Is very popular for analgesia for abdominal wall surgery. Local anaesthetic is placed in the plane between the transverse abdominal and internal oblique muscles.

Penile block. Penis is mainly innervated by the dorsal nerve of the penis, a terminal branch of the pudendal nerve. The usual indications are for circumcision or analgesia in distal hypospadias repair. It results in lesser urinary retention compared to caudal block [26]. Epinephrine containing solutions should not be used as severe arterial vasoconstriction can cause ischaemia or necrosis.

Pudendal nerve block. The pudendal nerve provides sensory and motor innervation to the pelvic cavity and its contents, including the external genitalia. This block has been used to provide analgesia in hypospadias repair and circumcision.

References

1. Hughes IA, Houk C, Ahmed SF, Lee PA. Lawson Wilkins pediatric Endocrine Society/European Society for Paediatric Endocrinology Consensus Group. Consensus statement on management of intersex disorders. J Pediatr Urol. 2006 Jun;2(3):148–62.
2. Sarin YK, Singh D, Babu R, Das K, Rao S. DSD guidelines Committee of Indian Association of pediatric surgeons. Indian Association of Pediatric Surgeons Guidelines on the Management of Differences in sex development. J Indian Assoc Pediatr Surg. 2022 Jul-Aug;27(4):376–80.
3. Cools M, Nordenström A, Robeva R, Hall J, Westerveld P, Flück C, et al. Caring for individuals with a difference of sex development (DSD): a consensus statement. Nat Rev Endocrinol. 2018;14:415–29.
4. Krishnan S, Meyer J, Khattab A. Ambiguous Genitalia in the newborn. [Updated 2019 Dec 2]. In: Feingold KR, Anawalt B, Boyce A, et al., editors. Endotext [Internet]. South Dartmouth (MA): MDText.com, Inc.; 2000. Available from: https://www.ncbi.nlm.nih.gov/books/NBK279168/.
5. Yang Y, Zhao F, Tu X, Yu Z. Mutations in WT1 in boys with sporadic isolated steroid-resistant nephrotic syndrome. Genet Mol Res. 2016 Mar 11;15(1):15017559. https://doi.org/10.4238/gmr.15017559.
6. Mohan VK, Sharma A. Anesthetic management in a case of Fraser syndrome. Saudi J Anaesth. 2013 Jan;7(1):102–3. https://doi.org/10.4103/1658-354X.109854.
7. Fischbach BV, Trout KL, Lewis J, Luis CA, Sika M. WAGR syndrome: a clinical review of 54 cases. Pediatrics. 2005 Oct;116(4):984–8. https://doi.org/10.1542/peds.2004-0467.
8. Quezado ZMN, Veihmeyer J, Schwartz L, Nwokoro NA, Porter FD. Anesthesia and airway management of pediatric patients with Smith–Lemli–Opitz syndrome. Anesthesiology. 2002;97:1015–9.
9. Heeley JM, Hollander AS, Austin PF, Merritt DF, Wesevich VG, Washington University School of Medicine in St Louis DSD Team, Amarillo IE. Risk association of congenital anomalies in patients with ambiguous genitalia: A 22-year single-center experience. J Pediatr Urol. 2018 Apr;14(2):153.e1–7.
10. Thyen U, Lanz K, Holterhus PM, Hiort O. Epidemiology and initial management of ambiguous genitalia at birth in Germany. Horm Res. 2006;66:195–203.
11. Ogilvy-Stuart AL, Brain CE. Early assessment of ambiguous genitalia. Arch Dis Child. 2004 May;89(5):401–7. https://doi.org/10.1136/adc.2002.011312.

12. Adrenogenital syndrome. Bissonnette B, & Luginbuehl I, & Engelhardt T (Eds.), (2019). Syndromes: rapid recognition and perioperative implications, 2e. McGraw Hill https://accessanesthesiology.mhmedical.com/content.aspx?bookid=2674&sectionid=220520550.
13. Speiser PW, Arlt W, Auchus RJ, Baskin LS, Conway GS, Merke DP, Meyer-Bahlburg HFL, Miller WL, Murad MH, Oberfield SE, White PC. Congenital adrenal hyperplasia due to steroid 21-hydroxylase deficiency: an Endocrine Society clinical practice guideline. J Clin Endocrinol Metab 2018 Nov 1;103(11):4043–4088.
14. Witchel SF. Disorders of sex development. Best Pract Res Clin Obstet Gynaecol. 2018 April;48:90–102.
15. Mouriquand PD, Gorduza DB, Gay CL, et al. Surgery in disorders of sex development (DSD) with a gender issue: if (why), when, and how? J Pediatr Urol. 2016;12:139–49.
16. Guerrero-Fernández J, Azcona San Julián C, Barreiro Conde J, de la Vega JA B, Carcavilla Urquí A, Castaño González LA, Martos Tello JM, Rodríguez Estévez A, Yeste Fernández D, Martínez Martínez L, Martínez-Urrutia MJ, Mora Palma C, Audí Parera L. Guía de actuación en las anomalías de la diferenciación sexual (ADS) /desarrollo sexual diferente (DSD) [Management guidelines for disorders/different sex development (DSD)]. An Pediatr (Engl Ed). 2018;89(5):315.e1–315.e19. https://doi.org/10.1016/j.anpedi.2018.06.009. Spanish. Epub 2018 Jul 20
17. Indyk JA. Disorders/differences of sex development (DSDs) for primary care: the approach to the infant with ambiguous genitalia. Transl Pediatr. 2017 Oct;6(4):323–34.
18. Hughes IA, Nihoul-Fékété C, Thomas B, et al. Consequences of the ESPE/LWPES guidelines for diagnosis and treatment of disorders of sex development. Best Pract Res Clin Endocrinol Metab. 2007;21:351–6.
19. Joshi GP, Abdelmalak BB, Weigel WA, Harbell MW, Kuo CI, Soriano SG, Stricker PA, Tipton T, Grant MD, Marbella AM, Agarkar M, Blanck JF, Domino KB. American Society of Anesthesiologists Practice Guidelines for preoperative fasting: carbohydrate-containing clear liquids with or without protein, chewing gum, and pediatric fasting duration—a modular update of the 2017 American Society of Anesthesiologists Practice Guidelines for preoperative fasting. Anesthesiology. 2023;2023(138):132–51. https://doi.org/10.1097/ALN.0000000000004381.
20. Rapp HJ, Altenmueller CA, Waschke C. Neuromuscular recovery following rocuronium bromide single dose in infants. Paediatr Anaesth. 2004;14:329–35.
21. Gandhi M, Vashisht R. Anaesthesia for paediatric urology. Cont Edu Anaesth Crit Care Pain. 2010;10(5):152–7. https://doi.org/10.1093/bjaceaccp/mkq025.
22. Pieretti, Rafael V. and Patricia K. Donahoe. "Chapter 123—disorders of sexual development." (2012).
23. Nowotny H, Ahmed SF, Bensing S, Beun JG, Brösamle M, Chifu I, van der Claahsen Grinten H, Clemente M, Falhammar H, Hahner S, Husebye E, Kristensen J, Loli P, Lajic S, Reisch N. Endo ERN (MTG1). Therapy options for adrenal insufficiency and recommendations for the management of adrenal crisis. Endocrine. 2021 Mar;71(3):586–94.
24. Roberts SA, Guruswamy V, Galvez I. Caudal injectate can be reliably imaged using portable ultrasound—a preliminary study. Pediatr Anesth. 2005;15:948–52. https://doi.org/10.1111/j.1460-9592.2005.01606.x.
25. Andreas H. Taenzer; inadvertent spinal anesthesia during continuous epidural anesthesia in an infant. Anesthesiology. 2003;98:1014–5. https://doi.org/10.1097/00000542-200304000-00032.
26. Metzelder ML, Kuebler FJ, Glueer S, et al. Penile block is associated with less urinary retention than caudal anesthesia in distal hypospadias repair in children. World J Urol. 2009;28:87. https://doi.org/10.1007/s00345-009-0420-2.

Psychosocial Challenges and Strategies for Rehabilitation in Children with Differences in Sex Development

60

Angela Ann Joseph

Abstract

Individuals with Differences in Sex Development (DSD) are either identified at birth by the presence of atypical genitalia or at later stages due to amenorrhea or subsequent health-related concerns (such as salt-wasting type in Congenital Adrenal Hyperplasia, CAH). The birth of a child with DSD is often considered a psychosocial emergency. Parental reports suggest that the early days of getting acquainted with their child's DSD condition are traumatic to a degree comparable to that experienced by parents of children diagnosed with Cancer. They find themselves struggling to cope with the ambiguity regarding their child's future physical and gender development. Parents are faced with the challenge of making surgical decisions for their children based on medical information given to them while simultaneously coming to terms with their child's diagnosis. Surgical intervention during infancy and childhood is controversial with varying outcomes. Although documented literature exists for gender identity outcomes for certain DSD conditions, it is not true for all, as DSD is a rare and heterogenous group of congenital conditions. Even for those conditions where data exists on gender identity development, there are exceptions to the rule. Since gender identity cannot be predicted with utmost certainty at birth, any surgical intervention during infancy comes with an element of risk. In cases of unfavorable outcomes, wherein the child's gender identity is discordant with the surgical intervention during infancy, it can lead to profound psychological distress for the child. Therefore, clinical management not only involves the medical management of DSD but also careful consideration of several psychological and psychosocial

A. A. Joseph (✉)
Jindal School of Liberal Arts and Humanities, O.P. Jindal (Institution of Eminence, Deemed to be) Global University, Sonipat, Haryana, India
e-mail: aajoseph@jgu.edu.in

S. K. Ratan (ed.), *Children with Differences in Sex Development*,
https://doi.org/10.1007/978-981-97-1639-5_60

factors. These factors include psychoeducation, information management, gender assignment, genital surgery, gender reassignment, and sexuality. The following chapter takes a developmental perspective to discuss these psychosocial factors and the relevant psychosocial rehabilitation strategies that can be used to mitigate psychological distress among children with DSD.

Keywords

Psychosocial counseling · Gender assignment · Differences/Disorders of sex development

Differences in sex development (DSD) refer to a heterogeneous group of congenital conditions marked by a discordance between chromosomal, gonadal, hormonal, and morphological characteristics. Over the last two decades, medical management of DSD has shifted its focus from Dr. John Money's "optimal gender policy" and "sex corrective surgery during infancy" toward an "agency-based model" of healthcare [1] that emphasizes informed consent and prioritizes psychosocial care and well-being of the individual. The impetus for this paradigm shift toward a holistic and patient-centered model was due to the lack of follow-up studies and the risk of psychological harm caused by the former. Moreover, increasing recognition of the challenges that persons with DSD and their families endure, such as diagnostic ambiguity, gender assignment, disclosure, decisions regarding surgical options, and its future repercussions for fertility, has led to an emphasis on the need for interdisciplinary teams that include mental health experts such as pediatric psychologists and psychiatric social workers [2–4].

Individuals with DSD and their families are at an elevated risk for psychosocial problems due to several factors. Firstly, since DSD is not a visible condition, parents may choose secrecy over disclosure and thereby limit their opportunities for social support. In addition to that, due to the complex nature of the condition, it may be difficult for parents to convey age-appropriate DSD information to their affected child. Furthermore, decision-making about surgical options or medical interventions, especially when the child is very young can be overwhelming and complicated for parents. Lastly, research indicates heightened suicidality among adults with DSD. Therefore, the incorporation of psychosocial care and rehabilitation strategies is imperative for good health-related quality of life [5].

Psychosocial challenges particular to DSD may differ across the lifespan, though some may resurface repeatedly. Parents require psychosocial counseling during the antenatal stage (if the condition is diagnosed) to come to terms with the diagnosis, process their initial reactions, and decide on their future course of action. The neonatal/infancy stage is marked with questions related to gender assignment and considering options related to genital surgery and hormonal treatment (e.g., testosterone injections for micropenis, steroid replacement for CAH) whereas, during childhood, information management, uncertainties regarding gender reassignment may be discussed. In adolescence, issues pertaining to secondary sex characters, sexuality, and fertility may be a matter of concern [3, 6]. Therefore, psychosocial counseling is a continuous process across these developmental stages. The following subsections will discuss the psychosocial rehabilitation strategies for children with DSD that can be used to manage these challenges.

60.1 Gender Assignment in Infancy

The shock experienced at the birth of a child with DSD may be related to the parent's apprehensions in accepting the ambiguities regarding the child's development, due to the difficulty in identifying the sex of the child at birth. Gender assignment is based on the DSD diagnosis and longitudinal data of gender stability related to the diagnosis. Gender roles refer to the behavior, attitudes, and personality traits associated with a particular gender, whereas gender identity refers to the sense of identification one feels with a particular gender. Although there is some evidence to suggest that prenatal androgen influence may lead to masculinization (such as in the case of 46, XY 5-alpha reductase type 2 deficiency, 5α-RD-2, and 17β-hydroxysteroid-dehydrogenase 3, 17-HSD-3) [4] as opposed to those conditions that are characterized by an absence of androgen exposure (such as Complete Androgen Insensitivity Syndrome, CAIS) there exist conditions where female gender identity is the norm rather than the exception despite androgen exposure (such as Congenital Adrenal Hyperplasia, CAH) [7]. Since the relative contribution of biological versus environmental factors toward gender development in those with DSD is not clearly understood, the gender identity of any child with DSD cannot be predicted with complete accuracy. Data on gender development outcomes in certain DSD conditions suggest certain trends. CAH has been mostly associated with female gender identity though male gender roles may be displayed especially among those with the salt-wasting type [7]. Data indicates that those with 5α-RD-2 who are allowed to experience spontaneous puberty develop male gender identities [4], however, research has also documented exceptions [8]. Those with Complete Androgen Insensitivity Syndrome (CAIS) mostly develop female gender identity although gender roles may be more masculine [9]. There is greater variation in gender development outcomes among those with Partial Androgen Insensitivity Syndrome (PAIS) [4]. Gender assignment decisions, therefore, are more complex for certain DSD conditions such as PAIS, 5α – RD-2, and 17β-HSD-3 because of the lack of sufficient data on post-natal social influences that interact with biological predispositions to regulate gender identity outcomes [4].

Decision-making is a collaborative process between parents and the treating team, and a variety of factors must be considered during gender assignments. This includes healthy psychological development, the possibility of living a good quality of life given the gender assigned, the parent's capacity to cope with the complexities, the possibility of surgical interventions in the future if required, fertility, sexual functioning, and hormone replacement therapy. The 6-step model by Karkazis [10] outlines the process for shared decision-making that can be used in this context. To be able to fully participate in truly informed decision-making, parents must be in a psychological state to accurately comprehend unfamiliar medical concepts. When information regarding the DSD diagnosis involves a discordance between the external genitalia of the child and, with implications for fertility, it may lead to strong evocative responses from parents. There exist resources such as the modified SPIKES approach [11] that can guide practitioners on how to communicate potentially distress-provoking news in a compassionate manner [12].

Working through parental anxieties and fears, while providing a supportive space are initial strategies to facilitate acceptance of the diagnosis and its implications. This may take time, and information may have to be repeated over spaced intervals. It is useful to assess parents' capacity to deal with the future uncertainties and questions that lie unanswered, before urging them to be part of the decision-making process. This is recommended so that future harm to the child caused by the psychological distress of the parent can be avoided. Psychosocial assessment in this context must pay special attention to parental attitudes, religious beliefs, and cultural norms. Apart from their psychological capacity to cope, caregivers' ability to handle logistics, matters of day-to-day functioning relating to medical care, financial capacity, and an understanding of their child's condition must be estimated. Psychoeducational sessions that characterize DSD as normal variations in development (rather than terming it as a disorder) and its impact in brief, manageable modules may help in the process [13]. However, for acutely distressed families a psychoeducational approach alone will not suffice, they may require training in problem-solving skills. The essential goal here is to empower parents with knowledge and a realistic appraisal of the situation without causing unnecessary distress over things they cannot control. Once a sense of psychological acceptance and readiness to move forward has set in, then appropriate decision-making via the exploration of the medical options available may be discussed [4].

60.2 Information Management/Disclosure

Information management involves multiple routes: between clinicians and parents; between parents and child; between parents/child and the wider social environment. As the child grows older, questions related to multiple hospital visits, medical examinations, genital scars, and medications may arise. Developmentally appropriate DSD-related information must be shared with the child at spaced-out intervals and repeated as necessary. Sometimes parents may want to withhold sharing diagnostic information with the intention of protecting their child from emotional harm or fear of depriving them of a carefree childhood [14, 15]. In other situations, parents may not consider it appropriate to discuss genital/sexual-related information with their child. In such cases engaging in a dialogue with the parents about the religious background, cultural beliefs, family values, and social norms may provide perspective to their resistance. It is helpful to have a culturally competent healthcare professional on board accustomed to the ways and socio-cultural background of the parents if possible. Sometimes gentle encouragement to speak with other parents of children with DSD belonging to similar socio-cultural milieu may also help ease pressures and fear when contemplating sharing diagnostic information with their child. Children who are informed about their medical conditions have better psychological adjustment and develop better coping skills and positive self-image. Moreover, they have enough time to slowly process DSD diagnosis-specific implications such as infertility in their perspective of future adult life [4]. Clinicians need to be sensitive to socio-cultural and religious differences and must be willing to re-engage in the same conversation from time to time, especially if the child must give informed assent for medical interventions [3].

The decision to disclose information to the wider social context is debatable since the effects of doing so could be positive or negative depending on contextual factors. Psychosocial counselors assist families in comparing the risks and benefits of doing so given their individual contextual factors. Most families refrain from sharing information for fear of stigma, ostracization, the child being teased or bullied, and discrimination. Shrouding the diagnosis with a cloak of secrecy also imparts a sense of shame and self-stigma. The sense of "otherness" affects the child's ability to form peer relationships and sustain healthy socio-emotional functioning [16]. Moreover, this also limits their parents' opportunities for psychosocial support within their community. However, in sex and gender-binary societies, fears of discrimination are not unfounded. Therefore, the right to privacy and confidentiality of the child needs to be weighed carefully with respect to the risks associated with secrecy, isolation, and shame [17].

60.3 Genital Surgery

Discussions on possible surgical interventions aimed at normalizing the appearance of the external genitalia may take place after the birth of the child. Information regarding the long-term consequences of various surgical options on sexual functioning, fertility, and psychosexual outcomes needs to be thoroughly discussed with the family. While there is some support for certain kinds of genital surgery (e.g., neonatal surgery for severe virilization in female-reared children or gonadectomy in female-reared children with PAIS or CAIS) it does not extend to all conditions [4]. In older females with CAH, surgical intervention to create a vagina may be needed, however, the timing of the surgery depends on individual psychological preparedness [3]. The goals of surgery must be taken into consideration in the following questions. Is the surgery being done to reduce parental distress over the appearance of ambiguous genitalia and the fear of social ostracization or is it in the best interest of the child given the current information regarding long-term medical and psychosexual outcomes? Would opting for surgery close possibilities for further surgical interventions if the child's psychosexual development were to occur contrary to what is expected? Would it impact sexual functioning and future fertility?

There are differing perspectives on early genital surgery. While some see it as a means of harming the bodily integrity of an individual, others argue that early surgery is associated with minimal scarring and better functionality. Each individual case must use careful decision balancing to assess the potential risks versus harm. Shared decision-making is the process through which a healthcare provider and patient reach a mutual consensus through collaborative deliberation [18]. Shared decision-making protocols have been developed for CAIS [18] and CAH [19]. The use of shared decision-making with decision aids and support tools enables a collaborative process that aligns with the goals and values of patients and their families. Furthermore, it may improve decision satisfaction and mitigate decisional regret [20].

Another area of concern is the potential psychological trauma caused by repeated genital exams and medical photography. It is recommended that both be kept to a minimum and every attempt be made to preserve patient confidentiality. Since many DSD care centers are within teaching hospitals, genital exams may take place in front of numerous residents as these conditions are rare and thereby presenting limited learning opportunities. However, it may cause significant discomfort to the child and the parent may feel as though their child is treated as an exhibit on display. Therefore, the number of people in the room must be kept to a minimum to reduce distress for the child and parent [21]. Children and their families need to be prepared before genital exams and surgical interventions. A pediatric psychologist who is part of the interdisciplinary team may prepare the child on what to expect, who will be present, and how long the process will take. Several cognitive and behavioral interventions like visual imagery and breathing can be used to allay apprehensions. Requesting the child to rehearse the positions they may have to take during the exam in the presence of a caregiver and encouraging them to write to the clinician about how they feel about the exam may also be helpful. The child should be given a choice of choosing between a male and a female examiner and may be encouraged to make an evaluation of the examiner via a visual analog scale. Coercing a child for an exam when the child is not ready should be avoided at all costs. Adequate pre-genital exam preparation of the child, giving the child greater control, and post-exam debriefing are some strategies that can minimize discomfort and prevent potential psychological harm [21]. Other strategies include training a parent to act as a coach, self-reassurance [22], watching a participant modeling film, use of brief psychoeducational videotapes [23], bringing in a comfort object (toy), and giving a reward to the child after the exam [24].

60.4 Gender Reassignment

Gender assignment during infancy is based on the estimation of gender identity by drawing upon evidence from long-term studies of gender stability in those with the same diagnosis. It is only as the child grows older that gender development becomes clear. Cognitive components of gender identity which include thinking about oneself with respect to a particular gender begin at about 30 months of age, but the development of the affective component which refers to likes, dislikes, and preferences is thought to occur much later. It has been reported that age 10–13 years was considered significant in terms of gender identity formation by children who experienced both temporary and persisting forms of gender dysphoria [25]. Therefore, the possibility that the assigned gender during infancy does not align with the child's sense of gender identity may remain and uncertainties may emerge as the child grows older. Healthcare professionals and parents may be concerned if they see the child exhibiting behaviors that are contrary to the gender assignment and may question the initial gender assignment, wondering if the child meets the criteria for Gender Dysphoria. However, as per the Diagnostic and Statistical Manual of Mental Disorder, fifth edition (DSM 5) the presence of gender roles contrary to the gender of rearing does not qualify for Gender Dysphoria. Many children exhibit gender roles contrary to their gender identity which is not a reason for concern. At times, the child's gender atypical behavior may be more distressing to the parents than to the child and in such cases, it may be useful to examine the family environment and look for factors that contribute to such behavior. In scenarios where the child has consistently voiced discomfort with their current gender and wishes to be a different gender than the one in which they have been reared, the psychologist may want to look at the consistency in these expressions of dissent and discomfort as part of their evaluation. Standardized tools for the assessment of gender dysphoria are available and must be administered by experienced professionals [26]. The Standards of Care of the World Professional Association of Transgender Health (WPATH) guidelines may serve as a guide if the child nears adolescence with persisting dysphoria [27]. According to WPATH, there are three stages for gender reassignment. The first is the diagnosis of gender dysphoria and the assessment of psychological and physical readiness to transition, followed by hormone replacement therapy (HRT) and social transitioning. Social transitioning refers to living life in the gender of choice socially to get an idea of what it feels like and how others respond to the changed status. It also gives the individual a realistic perspective on what challenges lie ahead and equips them with coping strategies and adequate preparedness for the future. Surgical intervention is considered the last stage. Although DSD may have different mechanisms influencing gender development as compared to transgender youth, this model may serve as a useful guide. There are certain limitations associated with this approach when applied to patients with DSD which need to be discussed. Among patients who wish to transition from female to male, HRT may not accomplish full masculinization and surgical options may be limited if there have been multiple surgeries done earlier during infancy. Sometimes patients with DSD may be confused about their gender identity or may wish to transition for reasons other than their gender identity being different from the gender they have been raised in. In such cases, gender explorative therapy and careful assessment of psychosocial factors may help understand the complexities behind their psychological state. If an adolescent is still exploring their gender identity and requires time to decide on the further course of action, puberty suppressors (Gonadotropin Releasing Hormone Analogues) may be given to delay the development of secondary sex characteristics, while the adolescent is able to live life in the gender of their choice. Among agonadal patients with DSD, this will not be required, however, they may benefit from delayed initiation of HRT if they are unsure of their gender identity. Adolescents must be informed about the effects of HRT and that these effects are only partially reversible, so that they may take an informed decision [3].

60.5 Promoting Inclusivity in Educational Environments

Educators can act as agents of change by inculcating an atmosphere of inclusivity within their classrooms. Making students with DSD feel heard, understood, and validated can contribute positively to their educational experience and foster a sense of empowerment. Psychosocial strategies aimed at inclusivity must be child-centered and child-led, family inclusive, and culturally sensitive to meet the goals of safety, well-being, and healthy psychosocial development [28]. School counselors play a pivotal role as facilitators for communication between the parents, student, and teachers, should a student with DSD opt to disclose their sex variation and seek support. They can help students with academic stressors and peer-related issues to adapt to changes in their surroundings. At times they may also conduct psychoeducational programs for gender diversity and sex variations if they are adequately trained to do so. Sensitization programs may include content on respecting the privacy and confidentiality of the student, using preferred names and pronouns for the student, making provisions within the school system that ease out the process for change in enrolment records (i.e., when a transition is likely to happen), provisions of gender neutral/unisex washrooms and change rooms, building flexibility around school uniforms/dress-code, ensuring non-gendered seating arrangements and providing staff development opportunities whereby they receive knowledge and information on gender affirmation and social transition within the school context [28]. Another approach to building greater awareness about DSD within schools is by consulting with DSD support groups and inviting them to give a talk and engage with student welfare committees.

The health and physical education syllabus may need to be reviewed and revised to incorporate language that is inclusive and non-discriminatory based on binary notions of sex/gender. Students must be allowed to participate in sports according to their preferred category. Intra-school sports competitions can be flexible in terms of the freedom given to the student regarding participation via the category of their choice since it is within the school's purview. However, this may not hold true for inter-school sports events where the control lies with external organizing committees. In such a scenario, the student must be informed about the restrictions early in the process of signing up for such events such as to avoid disappointment and disqualification later [28].

Strategies to encourage and enable full participation in outdoor recreational activities (like excursions, camps, and overnight trips) must include providing options for gender-neutral toilets, avoiding assigning students into groups based on binary notions of sex, and briefing others at the site about the preferred name and pronouns to be used for the student. Most often issues regarding sex and reproduction are discussed in Biology and Sex Education classes, though not limited to them, hence it is important that these instructors use language that is not generalizing or universalizing, for example, instead of referring to "*All boys as having XY chromosomes*" reframing it as "*Most boys have XY chromoso*mes" may be a small change, but it is one step closer to making students with sex variations feel included. Similarly, when talking about puberty, secondary sex characteristics, and menstruation it would be more sensitive to avoid using universal statements but instead acknowledge variations that exist in sex development as natural [28, 29]. Building awareness through art and literature by including content from authors and artists with DSD is yet another subtle but strong way of influencing young minds to think critically about assumptions of sex and gender. Lastly, schools must formulate well-defined policies on tackling teasing, bullying, and harassment of students with sexual variations and implement them sincerely. Adequate facilities for obtaining help and reporting incidents must be made available in safe, accessible, and confidential spaces.

60.6 Sexuality

Psychosocial counseling may also need to address adolescent concerns related to the appearance of their external genitalia, stature, romantic relationships, fear of intimacy [30] body image [31], sexual adequacy, and sexual orientation which may start emerging during this stage of development. Fear of rejection may lead to avoidance of engaging with the opposite gender [27]. Among females with CAIS, amenorrhea and its future implications for fertility may come as a shock, especially for those with a late diagnosis. Anxieties about the repercussions it may have on finding a partner, especially in societies where arranged marriages are the norm, may overwhelm the young adolescent and the family. A safe and supportive space to process these emotions provided by the counselor and referral to peer support networks may help in fostering acceptance and overcoming any self-stigma associated with the information. Concerns regarding same-sex attraction and questions regarding sexuality may also present especially among CAH women [32, 33]. Assuming heterosexuality while inquiring about sexual activity and functioning would be a bias and any inquiry on this front would benefit from open-ended questions on the part of the counselor. Discussion on sexuality is best conducted by a professional who is comfortable and competent in discussing sexual behavior and preferences. The counselor also needs to be culturally sensitive with prior knowledge about the sexual norms, values, and beliefs of the patient's background. It is reported that sex education and timely preparation for adult relationships may contribute to positive health-related quality of life [2].

60.7 Legal Recognition and Protection of Rights

The legal recognition of DSD in India as "Intersex persons" is subsumed within the umbrella term Transgender (Supreme Court judgment in NALSA v Union of India, 2014). Conflating "Intersex" with "Transgender" implies that the experiences, perspectives, and kinds of discrimination are the same for both categories. This is far from reality, for example, while children with DSD may be subjected to genital surgery during infancy, and perhaps receive far more medical attention than children who identify as transgender. The Transgender Persons (Protection of Rights)

Act, 2019, provides a definition for "person with intersex variations" but there are no specific protections, prohibitions, and entitlements that acknowledge the needs and struggles unique to people with DSD. Hence for any legal provisions to be meaningful in what it intends to provide there must be a distinct legal recognition of intersex individuals and their rights that appropriately recognizes them as different from transgender. Non-governmental organizations like Solidarity Foundation and the Center for Law and Policy Research (CLPR, Bangalore) advocate for this much-needed change. There has also been progress in protecting the rights of infants and young children with DSD who are not able to give informed consent for genital surgeries. Recognizing that parental consent is not equivalent to child consent, the Madras High Court in a landmark judgment in Arunkumar and Anr. v The Inspector General of Registration and Ors. prohibited sex reassignment surgeries on infants and children with DSD. However, there are several other psychosocial issues that need to be addressed such as mandatory registration of a child's sex at birth which is required to receive a birth certificate. This creates an undue amount of stress for families and doctors to specify the sex of the child in binary terms and hence they are compelled to consider surgical options under these circumstances. If there was a provision for declaring the sex as "undecided" in the birth certificate, with the option of updating it at a later stage, it would reduce a lot of distress for the family and possibly delay surgical intervention [34]. As per the NALSA judgment changing the sex assigned at birth is allowed but our Birth Registration laws make the process an extremely tiresome task and only permit revisions for typographical errors. The Gujarat High Court judgment in the case of Mulla Faizal v State of Gujarat paved the way for an easier process to change the sex in the birth certificates if the previous entry recorded the gender of the appellant contrary to their present gender. Activists and legal researchers have been emphasizing the need for a common procedure to be established under The Registration of Births and Deaths Act, 1969 for changing the gender documented where required by an individual and not just restricted to situations where the initial entry is incorrect [34].

Psychosocial counselors must include content on legal rights and protection offered to individuals with DSD as part of their post-diagnostic counseling and psychoeducation sessions. Inviting legal researchers, activists, and support groups to address doctors and parents of children with DSD may also provide a meaningful way of exchanging information and gaining new perspectives for working toward a common goal of empowering individuals with DSD.

60.8 Conclusion

Children with DSD and their families face several psychosocial stressors across developmental stages that require adequate support and intervention provided through a multidisciplinary team. Current research and practice recommendations suggest psychosocial strategies in the form of shared decision-making models, psychoeducation, behavioral strategies, and standards of care to minimize distress and the likelihood of harm. While medical genital surgery during infancy is controversial, having no access to medical treatment and care may also be stressful in the absence of societal and familial support structures. Moreover, strategies to build resilience and psychosocial well-being in children with DSD, and practical accessible support for day-to-day challenges faced by caregivers are strongly needed.

References

1. Crocetti D, Monro S, Vecchietti V, Yeadon-Lee T. Towards an agency-based model of intersex, variations of sex characteristics (VSC) and DSD/DSD health. Cult Health Sex. 2021;23:500–15. https://doi.org/10.1080/13691058.2020.1825815.
2. Schweizer K, Brunner F, Gedrose B, Handford C, Richter-Appelt H. Coping with diverse sex development: treatment experiences and psychosocial support during childhood and adolescence and adult well-being. J Pediatr Psychol. 2017;42:504–19. https://doi.org/10.1093/jpepsy/jsw058.
3. Sandberg D, Gardner M, Cohen-Kettenis P. Psychological aspects of the treatment of patients with disorders of sex development. Semin Reprod Med. 2012;30:443–52. https://doi.org/10.1055/s-0032-1324729.
4. Cohen-Kettenis PT. Psychosocial and psychosexual aspects of disorders of sex development. Best Pract Res Clin Endocrinol Metab. 2010;24:325–34. https://doi.org/10.1016/j.beem.2009.11.005.
5. Hansen-Moore JA, Kapa HM, Litteral JL, Nahata L, Indyk JA, Jayanthi VR, Chan Y-M, Tishelman AC, Crerand CE. Psychosocial functioning among children with and without differences of sex development. J Pediatr Psychol. 2021;46:69–79. https://doi.org/10.1093/jpepsy/jsaa089.
6. Alderson J, Hamblin RP, Crowne EC. Psychological care of children and families with variations or differences in sex development. Horm Res Paediatr. 2022;96:222. https://doi.org/10.1159/000524517.
7. Dessens AB, Slijper FME, Drop SLS. Gender dysphoria and gender change in chromosomal females with congenital adrenal hyperplasia. Arch Sex Behav. 2005;34:389–97. https://doi.org/10.1007/s10508-005-4338-5.
8. de Vries ALC, Doreleijers TAH, Cohen-Kettenis PT. Disorders of sex development and gender identity outcome in adolescence and adulthood: understanding gender identity development and its clinical implications. Pediatr Endocrinol Rev. 2007;4:343–51.
9. Mazur T. Gender dysphoria and gender change in androgen insensitivity or micropenis. Arch Sex Behav. 2005;34:411–21. https://doi.org/10.1007/s10508-005-4341-x.
10. Karkazis K, Tamar-Mattis A, Kon AA. Genital surgery for disorders of sex development: implementing a shared decision-making approach. J Pediatr Endocrinol Metab. 2010;23:789–805. https://doi.org/10.1515/jpem.2010.129.
11. Wolfe AD, Denniston SF, Baker J, Catrine K, Hoover-Regan M. Bad news deserves better communication: a customizable curriculum for teaching learners to share life-altering information in pediatrics. MedEdPORTAL. 2016;12:10438. https://doi.org/10.15766/mep_2374-8265.10438.
12. Strauss RP, Sharp MC, Lorch SC, Kachalia B. Physicians and the communication of "bad news": parent experiences of being informed of their child's cleft lip and/or palate. Pediatrics. 1995;96:82–9.
13. How Sex Development Works Available online: https://pie.med.utoronto.ca/htbw/module.html?module=sex-development. Accessed 26 Jan 2023.

14. Wisniewski AB, Sandberg DE. Parenting children with disorders of sex development (DSD): a developmental perspective beyond gender. Horm Metab Res. 2015;47:375–9. https://doi.org/10.1055/s-0034-1398561.
15. Nordenström A. Psychosocial factors in disorders of sex development in a long-term perspective: what clinical opportunities are there to intervene? Horm Metab Res. 2015;47:351–6. https://doi.org/10.1055/s-0034-1398562.
16. Bukowski WM, McCauley E, Mazur T. Disorders of sex development (DSD): peer relations and psychosocial Well-being. Horm Metab Res. 2015;47:357–60. https://doi.org/10.1055/s-0035-1549884.
17. Crissman HP, Warner L, Gardner M, Carr M, Schast A, Quittner AL, Kogan B, Sandberg DE. Children with disorders of sex development: a qualitative study of early parental experience. Int J Pediatr Endocrinol. 2011;2011:10. https://doi.org/10.1186/1687-9856-2011-10.
18. Weidler EM, Baratz A, Muscarella M, Hernandez SJ, van Leeuwen K. A shared decision-making tool for individuals living with complete androgen insensitivity syndrome. Semin Pediatr Surg. 2019;28:150844. https://doi.org/10.1016/j.sempedsurg.2019.150844.
19. Chawla R, Weidler EM, Hernandez J, Grimbsy G, van Leeuwen K. Utilization of a shared decision-making tool in a female infant with congenital adrenal hyperplasia and genital ambiguity. J Pediatr Endocrinol Metab. 2019;32:643–6. https://doi.org/10.1515/jpem-2018-0567.
20. Siminoff LA, Sandberg DE. Promoting shared decision making in disorders of sex development (DSD): decision aids and support tools. Horm Metab Res. 2015;47:335–9. https://doi.org/10.1055/s-0035-1545302.
21. Tishelman AC, Shumer DE, Nahata L. Disorders of sex development: pediatric psychology and the genital exam. J Pediatr Psychol. 2017;42:530–43. https://doi.org/10.1093/jpepsy/jsw015.
22. Duncan MKW, Sanger M. Coping with the pediatric anogenital exam. J Child Adolesc Psychiatr Nurs. 2004;17:126–36. https://doi.org/10.1111/j.1744-6171.2004.tb00006.x.
23. Rheingold AA, Davidson TM, Resnick H, Self-Brown S, Danielson CK. The relationship between knowledge and child and caregiver distress during the medical examination for child sexual abuse. J Child Sex Abus. 2013;22:552–71. https://doi.org/10.1080/10538712.2013.800937.
24. Carmichael N, Tsipis J, Windmueller G, Mandel L, Estrella E. "Is it going to hurt?": the impact of the diagnostic odyssey on children and their families. J Genet Couns. 2015;24:325–35. https://doi.org/10.1007/s10897-014-9773-9.
25. Fuss J, Auer MK, Briken P. Gender dysphoria in children and adolescents: a review of recent research. Curr Opin Psychiatry. 2015;28:430–4. https://doi.org/10.1097/YCO.0000000000000203.
26. Zucker KJ. Measurement of psychosexual differentiation. Arch Sex Behav. 2005;34:375–88. https://doi.org/10.1007/s10508-005-4336-7.
27. Coleman E, Radix AE, Bouman WP, Brown GR, de Vries ALC, Deutsch MB, Ettner R, Fraser L, Goodman M, Green J, et al. Standards of Care for the Health of transgender and gender diverse people, version 8. Int J Transgender Health. 2022;23:S1–S259. https://doi.org/10.1080/26895269.2022.2100644.
28. Gender Diverse and Intersex Children and Young People Support Procedure.
29. Resources: DSD Families Available online: https://www.dsdfamilies.org/resources. Accessed 5 Feb 2023.
30. Bridi Filho CA, Cardoso SB, Soll BMB, Noal MF, Schwarz K, Rosito TE, Rosito NC, Lobato MIR. Descriptive study of gender dysphoria and sexual behavior in a disorder of sex development group. Front Psychol. 2022;13:13.
31. Liedmeier A, Jendryczko D, Rapp M, Roehle R, Thyen U, Kreukels BPC. Dsd-LIFE group the influence of psychosocial and sexual wellbeing on quality of LIFE in women with differences of sexual development. Compr Psychoneuroendocrinol. 2021;8:100087. https://doi.org/10.1016/j.cpnec.2021.100087.

32. Zucker KJ, Bradley SJ, Oliver G, Blake J, Fleming S, Hood J. Psychosexual development of women with congenital adrenal hyperplasia. Horm Behav. 1996;30:300–18. https://doi.org/10.1006/hbeh.1996.0038.
33. Hines M, Brook C, Conway GS. Androgen and psychosexual development: Core gender identity, sexual orientation and recalled childhood gender role behavior in women and men with congenital adrenal hyperplasia (CAH). J Sex Res. 2004;41:75–81. https://doi.org/10.1080/00224490409552215.
34. Beyond the Binary: Advancing Legal Recognition for Intersex Persons in India Available online: https://clpr.org.in/publications/beyond-the-binary-advancing-legal-recognition-for-intersex-persons-in-india/. Accessed 5 Feb 2023.

Formulation of Parent Support Group For DSD

61

Aanchal Anand, Meera Luthra, and Rhea Ratan

Abstract

Caregivers of children with DSD often find themselves plagued with a plethora of challenges ranging from long-term health consequences, financial burden of diagnosis and therapy as well as social implications of this rare misunderstood spectrum of disorders. Formulation of support groups for this cohort can serve as a vital platform to share knowledge and experiences, and receive guidance and emotional support empowering them to better adapt to their new role without compromising upon their own mental and emotional well-being. In this chapter, we will describe the purpose of parental support groups, discuss types of support groups, discuss their mode of operation, illustrate steps of formulation, and provide information about existing DSD groups.

Keywords

DSD · Support group · Parents · Emotional support

A. Anand
Department of Community Medicine, ANIIMS, Port Blair, India

M. Luthra (✉)
MCh Pediatric Surgery at Medanta Hospital, Gurugram, India

R. Ratan
Maulana Azad Medical College, New Delhi, India

S. K. Ratan (ed.), *Children with Differences in Sex Development*,
https://doi.org/10.1007/978-981-97-1639-5_61

61.1 Objectives

- To describe the purpose of parental support groups
- To discuss types of support groups
- To discuss their mode of operation
- To illustrate steps of formulation
- To provide information about existing DSD groups

61.2 Purpose

Caregivers of children with DSD often find themselves plagued with a plethora of challenges ranging from long-term health consequences, financial burden of diagnosis and therapy as well as social implications of this rare misunderstood spectrum of disorders. Formulation of support groups for this cohort can serve as a vital platform to share knowledge and experiences, and receive guidance and emotional support empowering them to better adapt to their new role without compromising upon their own mental and emotional well-being [1] (Fig. 61.1).

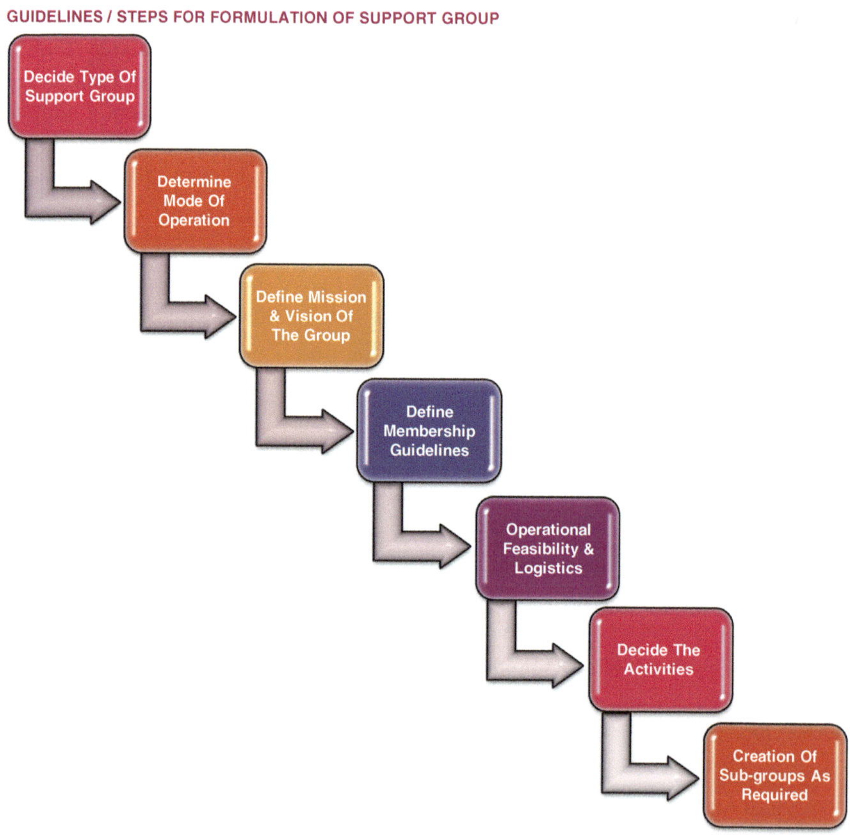

Fig. 61.1 Steps of SG formulation

A. Decide the Type of Parental Support Group

Disorders of sexual differentiation are diagnosed at birth itself when sex has to be assigned to the baby. At this stage, the complete responsibility of the child lies with the parents. Although the clinicians play their role and lend support in medical management of the child, the brutality of the fact that their child has been diagnosed with an incurable lifelong condition with physical, emotional and social consequences can leave the parents feeling devastated, confused and overwhelmed. To alleviate their anxiety and feelings of helplessness and guide them through special needs of such children, parental support groups are just not desirable but a key component of management of intersex disorders. These groups can be of several types depending upon whether they run in formal setup and are managed by professionals, by parents of children with DSD for other parents or in a hybrid mode [2, 3].

- Professional support groups—Such groups usually have licenced healthcare professionals with expertise on the subject involved as their facilitators. Also, non-governmental organizations working in this field and health facilities where DSD patients seek treatment can become initiators. Psychiatrists/behaviour therapists are roped into ease out the process of caregiving for parents. Such groups usually operate in formal settings such as hospitals and rehabilitation institutes. The meeting schedules and agendas are fixed by the facilitators.
- Peer Support Groups—Acceptance of the diagnosis and challenges that lie ahead can be very overwhelming for the parents, hence peer-support groups in the community comprising parents who have undergone similar experiences or have children who have currently been diagnosed can reduce feeling of isolation from the community and as everyone has undergone the same crisis, all the members share health and psychosocial challenges experienced with each other and how best to deal with these challenges without any hesitation or shame. Peer groups can help parents in myriad of ways to deal with physical and emotional demands of caregiving ranging from assistance in medical needs of children to maintain their social quality of life.
- Hybrid groups—The hybrid group has characteristics of both professional and peer-support groups. They bridge the gap between medical awareness and social needs of parents. Like professional groups, hybrid groups are facilitated by formal organizations and have doctors, counsellors and social workers in the loop to participate in specialized roles. Like peer groups, experiences shared by peers are highly valued alongside medical knowledge in coping with emotional issues such as denial, stress, depression anger and social isolation.

B. Determine the Mode of Operation

After deciding the type of support group, modus operandi of the group has to be decided. It can function in both offline and online modes depending upon members' preferences [4, 5].

- Offline mode—The core purpose of support groups is to develop a sense of belonging and hope amongst caregivers. When support groups operate in offline mode, it inculcates more trust and confidence amongst members. Verbal nods, empathetic touches and seeing other parents living a normal life post-diagnosis build self-confidence and reduce stress amongst members.
- Online mode—Online platforms serve as pragmatic avenues as caregivers come from diverse demographic and geographic profiles. This mode of operation offers more feasibility as demonstrated during COVID pandemic, flexibility with schedules and tends to be more affordable. Queries could be posted in groups and they can be answered by other members as per their convenient schedules. They also provide more scope for anonymity and confidentiality which some parents might be more comfortable with.

C. Define Mission and Vision of Support Group

The support group should have a constitution defining their vision and mission they are trying to accomplish for all its members. Since DSD is also associated with a lot of stigmas in society, it might be difficult for some parents to participate in such groups publicly and hence the group can choose to remain anonymous. It may conduct close meetings with strict clauses to ensure confidentiality of all its members [6].

D. Define Membership Guidelines

Membership may be offered on a voluntary basis in case of professional groups which also includes charging a nominal fee or it may be offered free of cost by peer support groups. Caregivers should be encouraged to be part of both professional and peer support groups. Once the members are enrolled, group leaders should be chosen to build consensus and coordinate effective group meetings. Leaders may be selected either on a rotational basis or by consensus obtained through balloting. They should ensure implementation of group guidelines, collaboration with multi-stakeholders and time-bound execution of designated activities.

E. Operational Feasibility and Logistics

- If the group is functioning in offline mode, the meeting venue should be accessible and agreeable to all the members of the group.
- The cost of transportation should be affordable to everyone.
- Timings—It has to be mutually decided and convenient to all members.
- Frequency of meetings can be monthly or quarterly depending upon the needs of caregivers.
- A contingency fund can be maintained for refreshments/stationary.

F. Decide the Activities

- *Raise Funding*—Treatment modalities in DSD involve hormonal therapy, genital reconstruction surgeries or a combination of both as well as cost of psychiatric counselling for both patients and parents which can be financially extremely taxing for caregivers. Support groups can collaborate with various NGOs and raise funding for all its members to meet the costs of procedures [7, 8].
- *IEC Program*—Sexual differentiation disorders are rare and being a parent to such children comes with a need for all kinds of information related to disease and coping mechanisms. A short course framework comprising of

series of lectures, interactive sessions, and educational videos providing biological explanations about intersex disorders and stress management techniques can be prepared and initiated for all new members joining the group on a monthly basis so that they are better equipped for their roles and responsibilities.

- *Handling inappropriate questions and negativity from relatives and community*—A condition like DSD comes with its own set of emotional turmoil and conflicts, especially in a country like India with a joint family predisposition and orthodox socio-cultural fabric. Caregivers are already exhausted with physical, mental and financial burden of diagnosis and often reach on the verge of breakdown when they do not receive support from community and extended family members. Psychologists/Psychiatrists and parents who have already dealt with this pressure can help and support the new members in navigating through this situation and establishing boundaries with unsupportive family members and neighbourhood.
- *Advocacy for children's legislative and social needs*—Unfortunately, these children often face prejudices in society and at workplace. They endure social isolation and hence these groups can act as their support systems facilitating networking. They can help generate awareness in society to alleviate all myths and stigmas associated with these patients and pave the way for their social inclusion. Patients with DSD also face mistreatment at their workplace hence advocacy for their legal rights is another domain that can be taken up by parental support groups. There is also growing demand for ban on irreversible genital surgeries in childhood without obtaining the individual's informed consent.
- *Screening of caregiver's health*—Caring for the caregiver is another important task that should be undertaken by support groups. Parents often neglect their own physical and mental health and prioritize their children's needs when it comes to time and resource constraints. Periodic health check-up camps can be organized by support groups to screen parents for physical and mental disorders on a regular basis.
- *Multi-sectoral collaboration for rehabilitation*—Parents who have children with DSD need guidance from various medical departments such as Paediatrics, Urology, Obstetrics and Gynaecology, Reproductive Medicine for future pregnancies, Psychiatrists, Genetics apart from which they also need the assistance of social workers, NGOs and social workers. It is not possible for an individual alone to establish contact with all these agencies on their own and hence parental support groups can act as collaborators for bringing their services under one unified platform.

G. Creation of Sub-groups as Required

Sexual differentiation disorders are further classified into different subgroups primarily based on their karyotype. All these subtypes have distinct clinical presentations, specific treatment components and varied outcomes. So once a principal parent support group has been established, it can be branched out into 46XY DSD, 46XX DSD and sex chromosomal DSD groups to cater to specific medical and behavioural needs.

61.3 Existing Support Groups for Disorders of Sexual Differentiation

Roles of support groups have been established all over the world for various other diseases. “Mom’s Belief India” and “DREAMS” are two well-known parental support groups catering to caregivers of autistic children and children with special needs, respectively. There are multiple support groups available for DSD both on global as well as national platforms to provide all kinds of support to parents [5, 6] (Table 61.1).

Table 61.1 Global and Indian support forums for parents of children with DSD

Support group	Web address	Brief description
Intersex India Forum [7]	www.india-intersex-forum/	IIF is a national forum that has threefold objective to provide guidance, create awareness and conduct research and advocacy on intersex human rights in general and rights of intersex children in particular. It also serves as a 24-h helpline for caregivers and patients.
Intersex Asia [8]	www.intersexasia.org	Intersex Asia is an international autonomous network of intersex-led organizations and individuals from Asian countries that work to support, educate and advocate for the rights and lives of intersex individuals and raise awareness of human rights violations and discrimination faced by intersex communities.
NORD National Organization for Rare Disorders [9]	www.rarediseases.org/living-with-a-rare-disease/	It is an UK-based international independent patient advocacy organization representing all individuals and families affected by rare diseases improving the health and well-being of people with rare diseases by driving advances in care, research, and policy.
InterACT [10]	www.interactadvocates.org	It is an international organization established in 2006 and composed of multi-stakeholders for improving the lives of children with DSD and majorly undertakes legal advocacy for these children.
InterConnect [11]	www.interconnect.support/	It is an international organization whose mission is to “work together to promote a better quality of life through connection, support, education and awareness”.
Gender Spectrum [12]	www.genderspectrum.org/	It is an international organization committed to the health and well-being of gender-diverse children and teens through education and support for families, and training and guidance for educators, medical and mental health providers and other professionals.

Table 61.1 (continued)

Support group	Web address	Brief description
CARES Foundation [13]	www.caresfoundation.org/	This international organization was founded in 1991. It works to improve the lives of children with CAH and seeks to advance quality health care for them through support, advocacy, education and research.
CLAN *(Caring and Living as Neighbours)* [14]	www.clanchildhealth.org/	It is an international Non-Governmental Organization committed to equity for children living with chronic health conditions in resource-poor settings.
Living with CAH [15]	www.livingwithcah.com/	This international CAH Support Group is run by people who have been living with CAH themselves. They have been providing support to people living with CAH as well as raising awareness for the condition since 1991.
Accord Alliance [16]	www.accordalliance.org/	It is an international group with the mission to "promote comprehensive and integrated approaches to care that enhance the health and well-being of people and families affected by DSD by fostering collaboration amongst all stakeholders".

References

1. Armstrong K, Benedict Yap A, Chan-Cua S, Craig ME, Cole C, Chi Dung V, Hansen J, Ibrahim M, Nadeem H, Pulungan A, Raza J. We all have a role to play: Redressing inequities for children living with CAH and other chronic health conditions of childhood in resource-poor settings. Int J Neonatal Screen. 2020 Sep 25;6(4):31.
2. Support group [Internet]. Wikipedia. 2022 [Cited 2023 Mar 2]. Available from:https://en.wikipedia.org/wiki/Support_group#:~:text=In%20a%20support%20group%2C%20members
3. Carolyn W, Fanelli AM. Guidelines for establishing and operating successful support groups for people living with HIV. Zimbabwe: CRS; 2009. p. 15.
4. Cools M, Nordenström A, Robeva R, Hall J, Westerveld P, Flück C, Köhler B, Berra M, Springer A, Schweizer K, Pasterski V. COST Action BM1303 working group 1. Caring for individuals with a difference of sex development (DSD): a Consensus Statement. Nat Rev Endocrinol. 2018 Jul;14(7):415–29.
5. Momsbelief.com. [cited 2023 Mar 9]. Available from: https://momsbelief.com
6. D.R.E.A.M.S [Internet]. D.R.E.A.M.S. [cited 2023 Mar 9]. Available from: https://www.dreamsfoundationaca.org/
7. India Intersex Forum. [cited 2023 Mar 9]. Available from: https://intersexindia.org/
8. Intersex Asia [Internet]. Intersex Asia. 2022 [cited 2023 Mar 9]. Available from: https://intersexasia.org/
9. NORD [Internet]. National Organization for Rare Disorders. 2023 [cited 2023 Mar 9]. Available from: https://rarediseases.org/
10. Advocates for intersex youth [Internet]. interACT. 2022 [cited 2023 Mar 9]. Available from: http://www.interactadvocates.org/

11. InterConnect Support. 2023 [cited 2023 Mar 9]. Available from: http://www.interconnect.support/
12. Prismic. [Internet]. Gender Spectrum. [cited 2023 Mar 9]. Available from: http://www.genderspectrum.org/
13. Learn about congenital adrenal hyperplasia [Internet]. CARES Foundation. [cited 2023 Mar 9]. Available from: https://caresfoundation.org/
14. CLAN. [cited 2023 Mar 9]. Available from: http://www.clanchildhealth.org/
15. Meet the team [Internet]. Living with CAH. [cited 2023 Mar 9]. Available from: https://www.livingwithcah.com/meet-the-team/
16. Accord Alliance [Internet]. National organization for rare disorders. 2023 [cited 2023 Mar 9]. Available from: https://rarediseases.org/non-member-patient/accord-alliance/

Rights and Agency of Children with Disorders of Sex Development

62

Rita Panicker

All human beings are born free and equal in dignity and rights [1]

Abstract

Children are our today and tomorrow. They have the rights to choose their sexual and gender identity irrespective of their biological anatomy, and physiology at birth. In a scientifically progressive world, our society is yet to acknowledge children who are born with intersex conditions. In a binary gender conformist society, the social pressure on the parents to conform their child into either male or female category has led to decisions taken by their parents, and medical practitioners to do genital modification surgery during infancy; which may not have been in the best interests of the affected child or be opposite to the gender identity that he/she develops as they grow older. This has raised many critical questions by human rights activists, on whether the rights and agency of intersex children are respected in our country. Human rights Activists advocate for children's right to take informed decisions on interventions to reassign their sex, and rights to change one's sex markers after their initial birth registration. Schools should initiate awareness of intersex conditions and non-binary gender, sensitizing children and teachers. Longitudinal studies of intersex persons would give us direction to work in the best interest of the child, collectively and systematically, with the key stakeholders from medical, mental health and legal fraternity, and with children with intersex conditions and their parents.

R. Panicker (✉)
Butterflies India, Delhi, India
e-mail: ritapanicker@butterfliesngo.org

S. K. Ratan (ed.), *Children with Differences in Sex Development*,
https://doi.org/10.1007/978-981-97-1639-5_62

Keywords

Children's agency · Children's participation · Intersex · Disorders of sex development (DSD)

62.1 Introduction

The Convention on the Rights of Children (UNCRC) was adopted by the United Nations General Assembly in 1989 and has since been ratified by all countries except the United States of America. India ratified it in 1992. The UNCRC is a historic human rights convention that recognizes the child as an individual with rights. The four fundamental principles of UNCRC are the rights of a child for survival, development, protection, and participation. Correspondingly, all the government policies, legislations, and programmes conform to the articles and principles of the Convention. Children have the right to a name, citizenship, right to practice their religion, a caring and protective family, and right to education, health care and nutrition, right to leisure and play, and right to their agency. UNCRC Articles, 3, 12, 13, 16, and 24 specifically guarantee these rights to children. The ratification of UNCRC brought in a paradigm shift in the national policies, legislations, especially when it came to children's well-being and their agency.

62.2 The Discussion Surrounding the Rights of Children with DSD

Children are not passive individuals but persons with their own agency. Children's participation, the right to be consulted and be part of the decision-making process have been advocated by social work practitioners, bilateral agencies such as the UN, international development agencies, and human rights activists [2].

This is a concept that most adults are unsure of. When we say children have an agency and need to be heard and consulted, where do we begin? Parents in this case are the most significant adults in a child's life. They are the constant person in their lives; sharing an emotional bond with their child moving through various life phases. Therefore, when one makes choices or makes decisions on behalf of the child it must be in the *"best interest of the child," that is the core principle* [3]. Wherever, appropriate, the child's views have to be considered, based on the context, seriousness of the issue, and the age of the child. The decision should be based on how it will affect the mental health and well-being of the child; will it reduce or prevent harm to the child's well-being should be the primary concern.

In this paper, I am putting forth the discussions surrounding intersex (DSD) children and their rights to be consulted and to decide for themselves about surgically modifying their genitals. This is a complex issue not only in medical terms but also with reference to the "*rights*" question of who decides in "sex-normalizing surgery"—the child, parents, medical practitioners, or collectively. The debate of an intersex individual's human rights versus that of it being viewed as a medical issue has been going on since the 1990s [Sutter, Petra De, (2020). DSD: A Discussion at the Crossroads of Medicine, Human Rights, and Politics. *Frontiers in Pediatrics, (8)*125, 1–10. https://doi.org/10.3389/fped.2020.00125] [4].

The global discourse and activism on the rights of the intersex-DSD child resulted in the 2017 Parliamentary Assembly of the Council of Europe's resolution "promoting the human rights of and eliminating discrimination against intersex people." The resolution also highlighted that intersex individuals' gender identity must be legally recognized and they must have access to health care and psychosocial support. The resolution is, however, not directly enforceable or legally binding on member states [4]. *Malta* government's 'Gender *Identity, Gender Expression and Sex Characteristics Act,' (2015)*, protects *all* children from non-consensual and non-urgent modifications to their sex characteristics. The law forbids sex assignment of all children and defers it until such time or age when the person can take an informed decision. Portugal also passed a law in 2018 to ban medical interventions on intersex infants until such time that the person can make an informed decision. It must be mentioned that in 2019, the European Parliament adopted a resolution in which it strongly condemned sex-normalizing treatments and surgery and encouraged Member States to adopt legislation on it.

In India, activists are advocating for legal recognition of intersex persons' rights to address their vulnerabilities. The argument is that without legal recognition and safeguarding of their rights, there will always be low awareness about their identity and experience, like genital modification surgery. The surgery is carried out without the person's consent, as it is carried out in infancy or early childhood. It is argued that such immediacy is mostly an attempt to fit the baby's gender into a socially accepted binary gender category [5].

In the case of LGBTQI community the Supreme Court of India's Judgement [6], recognized "third gender" and passed a landmark judgment, stating that people with third gender are equally entitled to fundamental rights under Articles 14, 15, 16, 19(1)(a), and 21 as per the Constitution of India.

The Supreme Court in its landmark judgment in 2014 also clarified that gender identity did not refer to biological characteristics. But rather referred to it as "an innate perception of one's gender." Thus, the court held that no "third gender" person should be subjected to any medical examination that would invade their right to privacy. It also stated that no one shall be forced to undergo medical procedures, as a requirement for legal recognition of their gender identity.

However, the recently passed Transgender Persons (Protection of Rights) Act (2019) has defined "person with intersex variations" without extending any specific legal protection for the intersex community. This Act has been widely criticized by activists for its failure to enact specific protections, prohibitions, and entitlements that are unique to intersex persons [7].

The Association of Pediatrics Surgeons in India [APSI] has developed guidelines for management of DSD [8]. The guidelines emphasize that children with DSD should receive safe, effective, and individualized medical care from a qualified multidisciplinary team.

As per the guidelines, the local multidisciplinary committee (LMDC) has a key role in this individualized decision-making. Guidelines effort is to ensure that the parents are explained the details of the DSD and line of treatment by a team of multidisciplinary experts. The aim is to allay parents' fears and reduce the stress. The guidelines also mandate that children above 12 years should be involved in all decisions, and a verbal/written assent be used for any procedures.

In India, if the child is above 12 years of age, the child has to be consulted. In the event of a child having undergone sex reassignment surgery, but identifying his/her gender as opposed to the assigned sex during puberty, it would become necessary for the child to get medical, mental health, and legal support to make an informed decision.

62.3 Child Rights and the Agency of Children

The traditional as well as modern institution of childhood is bound by attitudes, customs, and laws that define and locate children as *dependent beings,* thus, taking away the agency from children to decide what their lives should be like, and giving adults the right to make the choices for them. Even though, there is a paradigm shift in the relevance of child participation in development discourses, yet remains a problematic and contested concept [9]. When it comes to medical treatment and care, the lines are blurred, it is an unchartered domain.

Parents whose children are born with disorders or differences in their sex development, go through immense emotional stress in coming to terms with it. Furthermore, in a socially and culturally conservative society, the parents are under pressure and urgency to decide on surgically modifying the intersex child's genitals, without waiting for the child to come to an age where they can make an informed decision. As per [10, 11], age and maturity are socially constructed, it is based on various factors, such as culture, economic, class, and socialization.

We raise children on existing cultural and social norms, which straitjackets a girl and boy to the prevailing gender-based social norms. The Indian society has not evolved at the pace other societies have where gender is not considered binary and non-binary gender forms are accepted. There are children who believe in gender fluidity. A girl is comfortable dressing in male clothes, has a crew cut hair style, prefers bonding with boys, and a boy may prefer wearing colors and clothes that are defined as feminine, painting their nails, are not aggressive, do not engage in rough games, talks softly, comes across to other boys as not being "masculine." These are the children who are not only targeted and tormented by other children but are also chastised by parents and other adults for not conforming to socially accepted binary gender norms and traits.

A tragedy that happened in 2022, comes to one's mind, the horrific suicide of a 15-year-old boy, a student of an elite school, who could not bear the emotional, physical, and sexual torments of his schoolmates. He was ridiculed and tormented for his feminine traits. His mother, a single parent, was supportive of her son, but the school management and teachers did not take any proactive steps to safeguard the child, although they were aware of it [12]. For the child to take the ultimate decision to end his life must make us realize, we too are guilty for the death of the child. He was a brilliant, gentle soul. It is an illustration of lack of sensitivity and acceptance of non-conforming gender image in our society. In this case, the child was not intersex but nevertheless tormented because he did not conform to socially accepted gender norms. Physical attributes of a person make one as female, male, or intersex, but it is the cultural and social norms that rigidly define one as a girl (with feminine traits) or a boy (with masculine traits) with no place for an intersex child. This notion/idea of a male and female needs to be unpacked, and studied, so that there is a diversity of thought and acceptance of the same. The pressure to adhere to gender conformity needs to be questioned. We need to have more conversations on these issues.

On the question of the rights of intersex children to be consulted, we will have to look at the systems in place that would safeguard the child, and provide the psychosocial support required throughout their transition to adulthood. If the law guarantees the right of an adolescent or child to make informed decisions regarding sex-normalizing surgery and treatment, safety measures will have to be put in place in the school, neighborhood, and public places to protect the child. An intersex child needs to be given psychosocial and medical support at every stage of their life.

Schools are an important agency to begin with in educating and sensitizing the children about intersex and non-binary gender. Unfortunately, the biology classes taught in schools are sanitized versions of physiology which does not instruct children about sexual organs, puberty, sex, and sexuality beyond male and female categories. There are no studied discussions on intersex, to educate children about intersex persons and their rights. The lack of understanding of this critical issue results in insensitivity of the teachers, school management, in having provisions for child safeguarding policy and mechanisms for children with DSD. The school counsellors must be well aware of the issue to support an intersex child, counsel parents, and share information about support groups, if available. There is a need for educators and school management to be made aware of this issue and the supportive role the teachers could play. The child safeguarding policy should have mechanisms to protect an intersex child from being bullied, tortured, sexually abused, and ridiculed by other school children. Schools are part of the continuum of care for a child, from family, community to school. It would be important to create safe spaces in the school for children to express their feelings and thoughts without any inhibitions, prejudices, being open to listening, accepting them and their thoughts and lived experiences.

The State needs to put in place support systems for the care and protection of these children. As of now, there is a policy, law, and programmes for children with disabilities. However, children born with differences in sexual development have not yet been recognized as children who require support systems for their care, protection, and rehabilitation in the event they are neglected or abandoned. In a country with a significant number of poor and not literate population, there are chances of parents abandoning children, due to the existing social stigma and long-term medical treatment. The support system at a rehabilitation center should have to have multi-stakeholders, medical care professionals, mental health professionals, catering to neglected children and children living in families where the child and their parents can be referred to for ongoing support. Both children and parents require this support, as it has mental health implications, and we must ensure the emotional well-being of the child as well as the parents.

The other area of contention is birth certificates. Once a gender is put on the birth certificate, changing the sex/gender on birth certificates, is cumbersome as the Birth registration law does not allow for an easy change of any details unless it is a typographical error. As per "*Section 15 of the Registration of Births and Deaths Act (1969), for change in the birth certificate can be allowed only when it is proved to the satisfaction of the Registrar that any entry made by him under the act is erroneous in form or substance, or has been fraudulently or improperly made*." Only then can the original entry in the birth or death certificate be appropriately corrected or cancelled in accordance with the state government rules. Although this contradicts NALSA judgment which allows changing the sex assigned at birth.

In *South Africa*, intersex persons are allowed to change their sex marker, after their initial birth registration, under the Alteration of Sex Description and Sex Status Act, (2003). Intersex persons have the legal right to alter their sex description in the birth register by applying to the government authority. It is an opportune time in India to bring a law that allows an intersex person to change their sex marker after their initial birth registration.

62.4 Conclusion

It is time to question the process of our socialization. We need to acknowledge gender non-conformity and diversity of thought. The rigid cultural and social norms on categorizing persons as a girl and boy and ways one must socialize must be unpacked. There was a time when homosexuality was considered a mental health disorder, whereas today it is accepted as a variance in sexual orientation.

The importance of parents in giving the child support, and being part of the decision-making process, cannot be undermined as they bear the child's best interest. However, there are incidents when parents might take decisions under pressure and not necessarily in the child's best interest. All key players such as the medical practitioner, parents, mental health professionals, and the child (if older and mature) must have an open dialogue and be part of informed decision-making processes. We must remember that the decision taken will impact the life of the child.

Children are our today and tomorrow, they have rights, and their agency is to be respected [13]. However, unless there is a focus on awareness of the issue of intersex and the right of persons with intersex to live their lives with respect and dignity, these children will continue to face discrimination, and parents will continue to insist on early genital surgery to protect their children from discrimination and alienation. The right of adolescent children to give consent for sex-normalizing treatments and surgery [APSI Guidelines], would require a prolonged dialogue with the child to understand the procedures, line of treatment and to make an informed decision. It is complex for an adolescent child to make an informed decision. If the alternative to infancy sex-normalizing surgery is to wait until the child reaches late teens, i.e., 15–16 years, we should have in place the safeguards to protect the individual from mental stress, and anxiety, especially in a society that is culturally and socially binary gender conformist. The media could play an instrumental role in bringing about a change, sensitizing the state and the public about this medical phenomenon and acceptance of gender fluidity.

The way forward is to do a longitudinal sociological and psychological study of intersex persons to understand their views, lived experiences of the decision taken by their parents on the choice of their sex when they were infants and if that choice has been a challenge and has impacted their lives. The longitudinal study of the lived experiences of intersex persons will inform us about *when* and *who* decides the surgical interventions. The study should include the lived experiences of the parents. In retrospect, would they have waited for the child to grow up to decide on the intervention with the consent of their child? These are questions that need an answer for us to take an informed position.

It is important for all the key stakeholders, medical practitioners, parents, mental health professionals, older adolescents, legal professionals to have an open discussion on what steps should be taken that would be in the best interest of the child. It is a fact that all the key stakeholders speak a different language, but we need to come together so that we offer the best solution to this critical issue.

Lastly, re-emphasizing "Children are our today and tomorrow," they have rights, and their agency is to be respected. Children are to be cared for and protected. We as adults are there, to support their journey, be their mentor, but also give the space to the child to think freely, be confident, thrive, and script their own narratives. We owe this to children to live their lives unfettered by prejudices and negative social norms but to be comfortable with who they are.

References

1. United Nations. "Universal Declaration of Human Rights". United Nations. 1948. https://documents.un.org/doc/resolution/gen/nr0/043/88/pdf/nr004388.pdf?token=pmjxyRQ6G7EIJHUKyA&fe=true.
2. Hart R. Children's participation from tokenism to citizenship, Innocenti Essays, No. 4, UNICEF. 1999.
3. UNICEF. "Convention on the Rights of the Child". UNICEF, 1989. https://www.unicef.org/media/52626/file.
4. De Sutter P. DSD: a discussion at the crossroads of medicine, human rights, and politics. Front Pediatr. 2020;8(125):1–10. https://doi.org/10.3389/fped.2020.00125.
5. Kothari J, Balu K, Sarma R. Beyond the binary: advocating legal recognition for intersex persons in India. CLPR and Solidarity Foundation. 2020. CLPR_PolicyBrief_2.pdf.
6. Radhakrishnan, K.S. "National Legal Ser. Auth. vs Union of India & Ors on 15 April 2014". Indian kanoon, April 15, 2014. https://indiankanoon.org/doc/193543132/.
7. Human Rights Watch. "I Want to be like nature made me" Medically unnecessary surgeries on intersex children in the US. "I Want to be like nature made me": medically unnecessary surgeries on intersex children in the US | HRW. 2017.

8. Sarin YK, Singh D, Babu R, Das K, Rao S, DSD Guidelines Committee. Indian association of pediatric surgeons guidelines on the management of differences in sex development. J Indian Assoc Pediatr Surg. 2022;27(4):376–80.
9. Ennew J. How can we define citizenship in childhood? Working Paper Series. 2000;10(12). Harvard Center for Population and Development Studies.
10. Cussianovich A. Working Children and adolescents: culture, image and theory. NAT's Working Children and Adolescents International Review. 1995.
11. Qvortrup J. Childhood matters: an introduction. In: Qvortrup J, et al., editors. Childhood matters. Aldershot; 1994.
12. The Wire. Teenager dies by suicide over his sexuality, school ignored complaints, 'says mother. 2022. https://thewire.in/rights/dps-greater-faridabad-student-suicide.
13. Lansdown G. Promoting Children's participation in democratic decision-making. UNICEF: Innocenti Insights; 2001.

63 Surveillance, Prevention, and Management of Neoplasms in Children with DSD

Ankur Bhardwaj, Prabudh Goel, and Sandeep Agarwala

Abstract

Patients with Disorders of sexual differentiation (DSD) are at risk of various malignancies, predominantly, germ cell gonadal tumors, but also non-germ cell gonadal tumors and extra gonadal tumors. The risk differs according to the type of DSD, being the highest in patients with Y chromosome and gonadal dysgenesis. The risk also increases with age. In this chapter, we discuss the pathology, risk factors, surveillance, and management of malignancy in patients with DSD.

Keywords

Malignancy · Gonadoblastoma · Dysgerminoma · Gonadectomy · Surveillance

63.1 Introduction

Disorders of sexual differentiation (DSD) have been defined as a group of congenital conditions in which the development of chromosomal, anatomical, and gonadal sex is atypical [1]. DSD have been associated with various malignancies in all age groups. As DSD pathology begins in the gonads, most of the tumors in DSD are gonadal tumors. The process of tumorigenesis starts as early as the embryonic stage when genetic aberrations occur in sex chromosomes. From there on, the malignancy

A. Bhardwaj
Department of Pediatric Surgery, Chacha Nehru Bal Chikitsalya, New Delhi, India

P. Goel · S. Agarwala (✉)
Department of Pediatric Surgery, All India Institute of Medical Science, New Delhi, India

S. K. Ratan (ed.), *Children with Differences in Sex Development*,
https://doi.org/10.1007/978-981-97-1639-5_63

risk keeps increasing with age. A precise assessment of this risk is required in deciding further management. Early diagnosis of premalignant lesions is the key to prevention and management of neoplasms in DSD in children. Other factors like fertility, endocrine functions, gender identity, and psycho-social have to be balanced while dealing with tumor risk.

63.2 History

In 1953, Scully introduced the term "Gonadoblastoma" for tumors arising in the gonads of DSD patients [2]. As premalignant lesions came into light, gonadectomy and gonadal biopsy were advised for most of the DSD [3–5]. But studies published in first decade of the twenty-first century showed variable risks of malignancy with different DSD subtypes [6]. Thus, surgical intervention was advised in those DSDs which had high risk of malignancy [7–9]. Concept of shared decision-making (SDM) and gender dysphoria led to a shift in management guidelines. In many DSD subtypes, surveillance was favored till puberty when a child could participate in decision-making [10]. Studies in the last decade were targeted toward finding non-invasive molecular markers which can provide an exact risk of malignancy [11–13].

63.3 Terminology and Incidence

Tumors in DSD are mostly germ cell tumors (GCT) of gonads. Sometimes extra gonadal tumors have also been seen in some syndromic DSDs like Turner, Klinefelter, Denys Drash, and Frasier syndrome. Rarely non-germ cell gonadal tumors like Leydig cell tumor, Sertoli cell tumor, and granulosa cell tumor have also been found [6].

Seminomas are the most common GCT in the male gonad. Similar tumors if arising from the ovary or dysgenetic gonad are called dysgerminomas [6]. Rarely, other Germ cell tumors like Embryonal carcinoma, yolk sac tumor, and teratoma have also been reported [4].

Pre-malignant lesions include Gonadoblastoma (GB) and Germ Cell Neoplasia In Situ (GCNIS) [14].

Due to changes in terminologies of DSD over the last 50 years, a wide range of tumor incidence rates have been reported. Earlier studies have documented higher rates due to the higher rates of gonadectomies/biopsies and difficulty in differentiating pre-malignant lesions from pre-pubertal gonads. The incidence of tumors in DSD varies from one type to another. Almost all the studies have documented higher rates in gonadal dysgenesis subgroup. As new DSD classification was adopted in the 2006 consensus meeting, most of the studies after that have followed the terminologies of this classification [1].

In the last two decades, the proportion of malignant to pre-malignant lesions has increased as management protocols have shifted toward the preservation of gonads.

Age of incidence—Premalignant lesions are seen as early as infancy [6]. Tumor risk increases with age after puberty, especially in CAIS [15, 16].

63.4 Tumor Risk in Different DSD Subgroups

Cools et al. published a tumor risk-based classification in 2006 [6] (Table 63.1). Studies after that assessed the tumor risk based on new terminologies. The summary of this is given in Table 63.2.

Table 63.1 Tumor risk-based classification [6]

	DSD	Risk (%)
	GD (+Y) intra abd	15–35
	PAIS non-scrotal	15
	Frasier	60
High	Denys drash (+Y)	40
Intermediate	Turner (+Y)	12
	17β HSD deficiency	28
Low	CAIS	0.8
	Ovo testicular DSD	3
	Turner (-Y)	1
Unknown	5α Reductase deficiency	0
	Leydig cell hyperplasia	0
	GD (+Y) scrotal	Unknown
	PAIS scrotal gonad	Unknown

GD gonadal dysgenesis, *PAIS* Partial androgen insensitivity syndrome, *CAIS* Complete androgen insensitivity syndrome, *17β HSD* 17β hydroxysteroid dehydrogenase deficiency, *MGD* mixed gonadal dysgenesis

Table 63.2 Tumor risk in DSD based on new classification (from studies after 2006)

46 XY DSD		46XX DSD	
CGD **PGD**	15–50% [8, 17, 18] 5.5–30% [8, 17]	**Ovo testicular**	Very low risk
Ovotesticular	2–10% [8]	**CAH**	Rare [6]
CAIS **-pre-puberty** **-post-puberty**	<1% [19] Upto 22% [19, 20]		
PAIS	15–20% [18]	**Sex chromosome DSD**	
17 β-HSD	28 [6]	**MGD**	2.2–50% [8, 21]
PMDS	Rare [22]	**Turner with SRY-**	7–10% [23–25]
Male CAH(TART)	Nil [26]	**Klinefelter**	1% [27, 28]
5α Reductase (type 2)	Not known [8]		
17 α-Hydroxylase/ 17,20 lyase	Not known [8]		

CGD Complete gonadal dysgenesis; *PGD* Pure gonadal dysgenesis; *PMDS* Persistent mullerian duct syndrome, *CAH* Congenital adrenal hyperplasia; *TART* Testicular adrenal rest tumors; *PAIS* Partial androgen insensitivity syndrome, *CAIS* Complete androgen insensitivity syndrome, *17β HSD* 17β hydroxysteroid dehydrogenase deficiency, *MGD* mixed gonadal dysgenesis

63.4.1 46 XY Disorders of Sexual Differentiation

Overall, DSD with Y chromatin are associated with an increased risk of tumors [7, 8].

- Gonadal dysgenesis: Dysgenetic gonads in Complete gonadal dysgenesis (CGD)/ Mixed Gonadal Dysgenesis (MGD) have a higher risk of tumors than normal gonads in disorders of androgen synthesis/action [7, 25].
- Complete Androgen Insensitivity Syndrome (CAIS): Low risk in pre-pubertal age has been linked to massive apoptosis of germ cells after 1 year of age whereas in PAIS, the majority of germ cells are maintained [6]. Risk significantly increases after puberty (<1% to 22%).
- Partial Androgen Insensitivity Syndrome (PAIS): The risk in PAIS is influenced by the location of gonads whether abdominal or scrotal.
- Male CAH patients are prone to develop TART in rete testis. These tumors have no malignant risk. So, they have to be distinguished from testicular GCTs [26].

63.4.2 Sex Chromosome DSD

- Klinefelter: In Klinefelter and Turner syndrome, there is no increased risk of GCT due to apoptosis of germ cells. However, an increased risk of mediastinal tumors is seen in Klinefelter syndrome [27]. Mediastinal tumors found in Klinefelter include mainly extra gonadal GCT, Non-Hodgkin lymphoma, and breast cancer [28].
- Turner: 5–15% of patients of Turner syndrome have Y chromosome mosaicism [8, 23]. Most of the reports of tumors in Turner syndrome have come in those patients who have Y chromatin. A large cohort study from 1959 to 2002 showed a cumulative malignancy risk of 7.9% up to the age of 25 years in patients of Turner syndrome with Y chromatin [23].
- MGD: Risk in MGD patients is variable depending on the type of gonadal tissue-streak or dysgenetic. Streak gonads have a very lesser risk of GCT due to the absence of germ cells. Risk increases with age. Rare case reports of Juvenile granulosa cell tumors have been found in neonates. These tumors have no malignancy risk [22].

63.4.3 46 XX DSD

Very low risk is seen in XX ovo testicular DSD [8]. Androgen excess disorders do not have any increased risk.

63.5 Risk Factors for Malignancy

Tumor risk in DSD depends on the following factors:

1. Karyotype: Presence or absence of Y chromosome
2. Gonad: Dysgenetic or normal
3. Presence of Y chromatin
4. Location of gonad: Abdominal, inguinal or scrotal
5. Age of patient
6. Genetic mutations

63.6 Pathogenesis

Gonad formation: To understand the tumorigenesis in DSD gonads, it is necessary to understand normal gonadal development. Abnormality at various levels of fetal gonad development can lead to DSD. Germ cells migrate to the gonadal ridge by 5 weeks [29]. For this process of migration, the Stem Cell Factor (SCF-c KIT) pathway is crucial. At 6 weeks, with the expression of SRY gene, Primordial Germ Cells (PGCs) undergo maturation. In the absence of SRY gene and expression of other genes, no changes occur in undifferentiated gonads and germ cells undergo meiosis at 12 weeks. Under the hormonal action, spermatogonia, oocytes, and respective gonads are formed. Both oocytes and spermatogonia lose their embryonic characteristics and expression of markers like OCT3/4, AP, and c-KIT [30].

Genetic etiology: Multiple genes like SOX 9, WT-1, and DAX1 have been identified for gonadal development. Mutation in any of these genes not only leads to DSD but also increases the risk of gonadal as well as extragonadal tumors [6].

Histopathology: Multiple studies on histopathological findings in dysgenetic gonads show that a block in the maturation process is commonly found in such gonads [6]. Undifferentiated Gonadal Tissue (UGT) is more prone to neoplastic changes [6]. In DSD of androgen synthesis and action, germ cells are delayed in the maturation process due to inadequate support by Sertoli cells. These germ cells survive for a few years and undergo apoptosis [6]. OCT3/4 (Octamer binding transcription factor) is a gene product that is expressed in pluripotent cells. It is consistently overexpressed in pre-malignant and malignant lesions in DSD. Based on the distribution of OCT3/4 in gonadal tissue, delayed maturation is differentiated from pre-malignant tissue [6]. Stem Cell Factor (SCF)/KITLG is also a promising marker to differentiate pre-malignant lesions from immature gonadal tissue because it is not expressed in embryonic or adult tissue but is expressed in pre-malignant lesions [27]. However, on attaining invasiveness all these markers are downregulated [27].

TSPY gene: It was noted in almost all studies that tumors arise in DSD with Y chromatin. TSPY gene on the Y chromosome was linked to the pathogenesis of tumors. TSPY protein is expressed in spermatogonia of adult testis and is thought to be responsible for their proliferation. It becomes more abundant in tumors and pre-malignant lesions as compared to embryonic gonads and adult gonads [31]. It is downregulated upon the progression of the tumor toward invasiveness [27].

Pre-malignant lesions: There are two types of pre-malignant lesions—Gonadoblastoma (GB) and Germ Cell Neoplasia In Situ (GCNIS) in GCT in DSD. GB mostly arises in dysgenetic gonads but GCNIS arises exclusively in DSD of androgen action or synthesis. The development of GB or Carcinoma In Situ (CIS) depends on the microenvironment [30]. If Sertoli cells are present, GCNIS will form, and if absent, GB will form. Although of different morphology, both GB and CIS progress to the same type of tumors [27]. GBs arise as early as in infancy while GCNIS develops mostly after puberty. All GCNIS progress to malignancy but only around 50% of GB turn malignant [6, 32].

Microscopically, GCNIS cells mimic primordial Germ Cells (PGCs) [29]. It is difficult to differentiate pre-malignant lesions from immature gonadal tissue, especially in infants when immature cells in gonads are normally found.

63.7 Management

In the last two decades, management protocols have shifted toward the conservation of gonads, at least till puberty. Patient advocacy groups have raised concerns about gender dysphoria and shared decision-making by the child when he/she grows up [10]. But the balance has to be maintained between the risk of malignancy and preservation of endocrine functions, fertility, and gender identity. Present times ask for precise risk quantification so that high-risk cases can be offered prophylactic gonadectomy. In the rest of the DSD subtypes, with low-to-moderate risk, close surveillance is needed to pick up malignancy at an early stage.

63.7.1 Diagnostic Work Up

Multiple modalities are available to detect tumors at pre-malignant stage.

Gonadal biopsy: Gonadal biopsy though not representative of the whole gonad still helps in histopathological diagnosis. Adequate tissue is mandatory. One 3 × 3 × 2 mm tissue from the representative site has been advised. Dysgenetic gonads occasionally have yellow-brown spots. Biopsy should be taken from these sites [32]. When gonadal preservation is the final decision, biopsy may need to be repeated as malignancy risk increases with age. As management protocols are tilting toward gonad preservation, gonadal biopsy will be needed more frequently in the time to come.

Histopathology: Histological examination helps in assessing the level of testicularization and maturation. It differentiates the dysgenetic gonads from streak gonads. In dysgenetic gonads, it tells about the extent of dysgenesis whether complete or partial. It is the gold standard investigation to diagnose pre-malignant lesions (GB and GCNIS) [12]. Pre-malignant lesions are easily missed by radiological investigations. Multiple immunohistochemistry markers have been developed to detect pre-malignant lesions—OCT 3/54, TSPY, PLAP, and c-KIT/SCF [6, 12]. OCT3/5 is an established marker for GCNIS [22]. It is common to find immature cells normally till 1 year of life. Since these immature cells are morphologically similar to GCNIS, so gonadal biopsy or gonadectomy are sometimes delayed till 1 year of age [33]. Co-expression of TSPY and KITLG with OCT 3/5 also helps in differentiating GCNIS from immature gonocytes [12].

Radiological: Ultrasonography is used to evaluate extra-abdominal gonads. The presence of microlithiasis along with heterogeneous testicular parenchyma is highly sensitive (not specific) to GCNIS [34, 35].

MRI: Can assess abdominal gonads better than USG but has low specificity for monitoring intra-abdominal dysgenetic gonads [35].

Diagnostic laparoscopy: It helps in examining the gonad morphologically. It is much more sensitive and specific than USG in visualizing mullerian remnants (PMDS) and streak-like abdominal gonads [36]. But it needs anesthesia for performance.

Genetic markers: Some genes whose mutation has an established role in pathogenesis include WT-1, SOX 9, DAX1, SF-1, and SRY [6]. More and more susceptibility genes are being identified in genome studies. Some important ones include KITLG, DMRT1, andBAK1 [7]. Some microRNAs are specifically present in the serum of patients with GCNIS and malignant lesions. miR-371-3 and miR-302 microRNA clusters are overexpressed in GCT and GCNIS irrespective of subtypes [7, 13].These genetic markers have good prospects but are not available in routine clinical practice yet.

63.8 Surveillance

As discussed above, surveillance has a major role in the management of DSD because the latest guidelines are favoring gonadal preservation in many DSD subtypes. Most of the surveillance protocols include self-examination, imaging, and gonadal biopsies.

Self-examination of gonads should be done once monthly starting from puberty. Any suspected abnormality should be confirmed by USG [8].

Radiological assessment by USG/MRI should be done annually after puberty [8].

The gonadal biopsy is useful in monitoring extra-abdominal gonads in those DSD subgroups where tumor risk is moderate to severe. However, due to limited sampling and sampling errors, it may not always detect the tumor [8, 32].

63.9 Management Protocols

Management protocols in different DSD subtypes depend on the risk of malignancy, location of the gonad, external phenotype, sex of rearing, and family decision in the end. Management decisions for each patient should be individualized by assessment of tumor risk, fertility potential, and shared decision-making. Treatment is multidisciplinary, involving pediatric surgeon, pediatric endocrinologist, radiologist, geneticist, psychologist, and pathologist. Family needs to be counselled patiently and multiple times regarding chromosomal, gonadal, and phenotypic sex, future fertility potential, and malignancy risk.

63.9.1 46 XY DSD

CGD: No gonadal development occurs and the phenotype is completely female. Malignancy risk is high. So, early bilateral gonadectomy is recommended [8, 17].

PGD: Phenotype depends on the extent of testicular function. Malignancy risk depends on the location of the gonad. So, if the sex of rearing is not decided or decided as male, intra-abdominal gonads should be brought down with or without biopsy. Streak gonad if present should be removed. Dysgenetic gonad should be monitored closely as the report of gonadoblastoma has been found as early as infant age group. A low threshold for gonadectomy should be there for those who have intra-abdominal gonads, ambiguous phenotype, and low testicular function [8, 17]. Cools et al. recommended biopsy once during pre-puberty (during orchiopexy) and once post-puberty [7].

63.9.2 Disorders of Androgen Action

CAIS: Phenotype is completely female. Gonads may be abdominal or inguinal. Malignancy risk is low before puberty (<1%) but increases after puberty (up to 22%). Peripheral aromatization of androgen leads to the development of sexual characters, growth, and mineralization of bones. So, gonadectomy is delayed till puberty when breast development is complete [8, 16, 37]. Those who deny gonadectomy need to be counselled regarding malignancy risk and limitation of surveillance modalities.

PAIS: Management depends on the external phenotype and sex of rearing. Malignancy risk depends on the location of the gonad and is much more than CAIS. Tumor cases occur mostly after puberty. In phenotypic girls, early bilateral gonadectomy is done because virilization causes physical and mental disturbances in girls. In phenotypic boys, early orchiopexy is done. As tumors mostly occur after puberty, biopsy is recommended at puberty. When the sex of rearing is not decided, it is better to wait till a stable gender identity is established [8, 16, 17].

63.9.3 Androgen Synthesis Defects

External phenotype depends on the degree of testicular function. Adequate Anti-Mullerian Hormone (AMH) production ensures that mullerian structures do not develop. But Wolffian structures are not fully developed. Testis are usually at labioscrotal folds or inguinal region and rarely abdominal.

17 β-Hydroxysteroid Dehydrogenase(HSD) type 3 deficiency: The phenotype is usually female. Malignancy risk is 28% in limited studies. In girls, early gonadectomy + excision of Wolffian remnants is recommended during pre-pubertal period to prevent pubertal virilization. In boys, monitoring is done [6, 8].

5-α-Reductase deficiency: External phenotype varies depending on the degree of residual enzyme activity. Malignancy risk is not known. So, in girls management is same, i.e., early bilateral gonadectomy. In males, orchiopexy and reconstruction of male genitalia are done [8].

17 α-Hydroxylase/17,20-lyase deficiency: It is a disorder of adrenal and gonadal steroidogenesis. Cortisol and testosterone deficiency occur. Phenotypes can be male, female, or ambiguous. Malignancy risk is not clearly known but few reports of mixed GCT have been reported. Thus, no consensus is available presently regarding gonadectomy [8].

Ovotesticular DSD: This subgroup is the most complex one to decide regarding gonadectomy and the sex of rearing. External phenotype is generally ambiguous and depends on the dominant gonadal tissue. Internally, ovotestis can exist with the testis or ovary. The testicular component of ovotestis is usually dysgenetic. Malignancy risk is very low in 46XX ovotesticular DSD and 2–10% in 46 XY ovotesticular DSD. Malignancy risk depends on the degree of gonadal differentiation and Y chromosomal material. Due to complex anatomy, functional gonadal tissue, and low risk of malignancy, it is better to wait till gender identity is established. In females, intra-abdominal testicular tissue is removed but it's not possible always when the ovary circumscribes testis. In those with male chromosomal material, laparoscopic biopsy can be attempted to check the external and histological features of gonad. If streak gonad is present, it should be removed [8, 37].

63.9.4 Sex Chromosome DSD

MGD: Management protocol is similar to partial gonadal dysgenesis. Malignancy risk varies inversely with testicular function or external masculinization. In the child with a female phenotype, bilateral gonadectomy should be done because of low-functioning gonads and high malignancy risk. In phenotypic boys, fibrotic (streak) gonads should be removed and dysgenetic gonads brought down in the scrotum for close monitoring. In children with ambiguous genitalia when doubt in gender identity is there, gonadectomy is delayed and monitoring is done along with close psychological follow-up to detect gender dysphoria early [8, 21].

Turner syndrome: Patients with a Y chromosome have a tumor risk of about 10%. Ovarian functions are also low (<10%) but present. So, no clear consensus is there regarding gonadectomy. The majority favor bilateral gonadectomy if Y chromosome is detected cytogenetically or virilization happens. Oocyte or follicle preservation can be tried in those patients for gonadectomy [8].

Klinefelter syndrome: Malignancy risk is 1%. So, surveillance protocols have not been developed yet [28].

63.9.5 Syndromic DSD

Denys Drash syndrome (DDS) and Frasier syndrome are commonly associated with DSD. Both have a high risk of malignancy [6].

In patients with DDS, the risk of Wilms' tumor is higher than gonadal malignancy, whereas it is the reverse in Frasier syndrome, where the risk of gonadal malignancy is higher.

DDS—46 XY DSD. The phenotype is female. Usually, nephropathy occurs before gonadal malignancy. But with improvement in CKD management, gonadal malignancies are bound to increase in prevalence. Gonadal malignancy risk is low but Wilms' tumor risk is very high (about 70%). Bilateral gonadectomy is indicated. Screening for Wilms' tumor is necessary [38].

Frasier syndrome—Gonads have complete gonadal dysgenesis. Phenotype is female. Risk of gonadal malignany is high but Wilms' tumor risk is low. So gonadectomy is recommended [39].

63.10 Chemotherapy and Radiotherapy for Malignant Germ Cell Tumors

Chemotherapy The introduction of multi-agent chemotherapy in the form of Vincristine, Adriamycin, and Cyclophosphamide (VAC) along with radiotherapy improved survival rates significantly. Dramatic improvement in survival came about after the introduction of cisplatin-based therapy. The current regimens with proven efficacy and used by the COG include cisplatinum, etoposide, and bleomycin (PEb) [Table 63.3]. The UK trials have preferred to use carboplatin instead of cisplatin (JEB regime). The most recent outcome trials for children with malignant germ cell tumors at all extracranial sites have survival rates ranging from 85% to 95%. There is a difference between the pediatric PEb and the adult BEP. In PEb, bleomycin is administered *once per cycle (once in three weeks)* while it is given *once per week* in the adult BEP schedules. Thus, the adult regimen includes 30 IU of bleomycin weekly for 9–12 weeks and the pediatric regimen PEb is 15 IU/m^2 once per cycle (every 3 weeks). This possibly avoids the risk of fatal pulmonary fibrosis. The morbidity of both chemotherapy and surgery in survivors highlights the need for meticulous surgical staging and the refinement of the surgical approach. Renal impairment, neurotoxicity, and hearing loss are a few of the frequently observed toxicities in pediatric patients treated for germ cell tumors. Risk catego-

Table 63.3 PEb regimen

	Drug	Dose	Administration
1	**Inj. Cisplatin**	35 mg/m^2/day	Day 1–3 (IV infusion over 1 h)
2	**Inj. Etoposide**	120 mg/m^2/day	Day 1–3 (IV infusion over 1 h)
3	**Inj. Bleomycin**	15 IU/m^2/day	Day 1–3 (IV infusion over 1 h)
4	**IV Infusion-N/2 saline in 5% dextrose + 1 ml/100 ml Kcl**	125 mL/m^2/h	

Duration: Cycles at 3 weekly intervals (day 1- day1). Total of 4 courses

Table 63.4 Ovarian MGCT—Risk categorization and treatment

Stage	Risk category	Surgery	Chemotherapy
Stages I, II, and III	Intermediate	+	PEb x 4 courses
Stage IV	High risk	+	PEb x 4 courses

Note: There is no low risk for ovarian tumors

Table 63.5 Sensitivity to chemotherapy and radiotherapy treatment

	Histologic grading	Sensitivity to chemotherapy	Sensitivity to radiation
Seminoma/ germinoma	Malignant	+++	>24 Gy
Embryonal carcinoma	Malignant	+++	>45 Gy
Yolk sac tumor	Malignant	+++	>45 Gy
Choriocarcinoma	Malignant	+++	>45 Gy
Teratoma, mature/ immature	Benign/potential for malignant development	?	?

rization for germ cell tumors is shown in Table 63.4. Low-risk tumors are treated with surgery alone and these include stage I testicular EST and stage I ovarian tumors. However, later study results (AGCT0132) showed that Stage I ovarian tumors treated with surgery alone had an unacceptably low 3 year EFS and therefore. Stage I ovarian tumors are now considered as intermediate risk and given adjuvant chemotherapy.

Radiotherapy The current importance of radiotherapy is mainly in the treatment of the CNS germinoma, seminoma, and ovarian dysgerminoma. The other subtypes are not as responsive [Table 63.5]. This too has been diminished after the emergence of platinum-based chemotherapy. It remains the standard of treatment for CNS germinoma along with chemotherapy. Recurrent germ cell tumors (unresectable tumor following salvage chemotherapy) may also be treated with radiotherapy. Therefore, radiotherapy is not routinely advocated in the management of most malignant germ cell tumors.

63.11 Further Research

Studies in the last decade aimed toward finding new SNPs, miRNAs, and other biomarkers which would help in precise risk quantification of tumor risk in DSD [7, 11, 13, 17, 18].

63.12 Conclusion

DSD subgroups have variable malignancy risk. Within each group, individuals can have variable malignancy risk depending on their genetic makeup. So, each child with DSD needs an individual risk assessment. As DSD affects child physically and psychologically, so holistic approach to management is required. Its management does not just involve multiple departments of medical science but also has participants from other fields of society. So, we should prepare safe and socially approved surveillance protocols for monitoring gonadal tumors in DSD in children. The risk of tumor should be balanced with the preservation of fertility, endocrine function, and gender identity of the child.

References

1. Hughes IA, Houk C, Ahmed SF, Lee PA, Society LW. Consensus statement on management of intersex disorders. J Pediatr Urol. 2006;2:148–62.
2. Scully RE. Gonadoblastoma. A gonadal tumor related to the dysgerminoma (seminoma) and capable of sex-hormone production. Cancer. 1953;6:455–63.
3. Scully RE. Gonadoblastoma. A review of 74 cases. Cancer. 1970;25:1340–56.
4. Troche VL, Hernandez EN. Neoplasia arising in dysgenetic gonads. Obstet Gynecol Survey. 1986;41:74–9.
5. Müller J, Visfeldt J, Philip J, Skakkebæk NE. Carcinoma in situ, gonadoblastoma, and early invasive neoplasia in a nine-year-old girl with 46, XY gonadal dysgenesis. APMIS J Pathol Microbiol and Immunol. 1992;100:170–4.
6. Cools M, Drop SL, Wolffenbuttel KP, Oosterhuis JW, Looijenga LH. Germ cell tumors in the intersex gonad: old paths, new directions, moving frontiers. Endocrine Rev. 2006;27:468–84.
7. Cools M, Looijenga LH, Wolffenbuttel KP, T'Sjoen G. Managing the risk of germ cell tumourigenesis in disorders of sex development patients. In: Hiort O, Ahmed SF, editors. Understanding differences and disorders of sex development (DSD). Endocr Dev, vol. 27. Basel: Karger; 2014. p. 185–96.
8. Abacı A, Çatlı G, Berberoğlu M. Gonadal malignancy risk and prophylactic gonadectomy in disorders of sexual development. J Pediat Endocrinol and Metab. 2015;28:1019–27.
9. Mouriquand PD, Gorduza DB, Gay CL, Meyer-Bahlburg HF, Baker L, Baskin LS, et al. Surgery in disorders of sex development (DSD) with a gender issue: if (why), when, and how? J Pediat Urol. 2016;12:139–49.
10. Gardner M, Sandberg DE. Navigating surgical decision making in disorders of sex development (DSD). Frontiers Pediatr. 2018;6:339–48.
11. Elzinga-Tinke JE, Dohle GR, Looijenga LH. Etiology and early pathogenesis of malignant testicular germ cell tumors: towards possibilities for preinvasive diagnosis. Asian J Androl. 2015;17:381–93.
12. Spoor JA, Oosterhuis JW, Hersmus R, Biermann K, Wolffenbuttel KP, Cools M, et al. Histological assessment of gonads in DSD: relevance for clinical management. Sex Dev. 2018;12:106–22.
13. Dieckmann KP, Radtke A, Geczi L, Matthies C, Anheuser P, Eckardt U, et al. Serum levels of microRNA-371a-3p (M371 test) as a new biomarker of testicular germ cell tumors: results of a prospective multicentric study. J Clin Oncol. 2019;37:1412–23.
14. Humphrey PA, Ulbright TM, Reuter VE, Moch H. WHO classification of tumours of the urinary system and male genital organs. International Agency for Research on Cancer. 2016;70:106.
15. Manuel M, Katayama KP, Jones HW Jr. The age of occurrence of gonadal tumors in intersex patients with a Y chromosome. Am J Obstet Gynecol. 1976;124:293–300.
16. Mendoza N, Motos MA. Androgen insensitivity syndrome. Gynecol Endocrinol. 2013;29:1–5.
17. Huang H, Wang C, Tian Q. Gonadal tumour risk in 292 phenotypic female patients with disorders of sex development containing Y chromosome or Y-derived sequence. Clin Endocrinol. 2017;86:621–7.

18. Steinmacher S, Brucker SY, Kölle A, Krämer B, Schöller D, Rall K. Malignant germ cell tumors and their precursor gonadal lesions in patients with XY-DSD: a case series and review of the literature. Int J Environ Res Public Health. 2021;18:5648.
19. Chaudhry S, Tadokoro-Cuccaro R, Hannema SE, Acerini CL, Hughes IA. Frequency of gonadal tumours in complete androgen insensitivity syndrome (CAIS): a retrospective case-series analysis. J Pediat Urol. 2017;13:498–e1.
20. Deans R, Creighton SM, Liao LM, Conway GS. Timing of gonadectomy in adult women with complete androgen insensitivity syndrome (CAIS): patient preferences and clinical evidence. Clin Endocrinol. 2012;76:894–8.
21. Weidler EM, Pearson M, van Leeuwen K, Garvey E. Clinical management in mixed gonadal dysgenesis with chromosomal mosaicism: considerations in newborns and adolescents. Semin Pediat Surg. 2019;28:150841.
22. Kathrins M, Kolon TF. Malignancy in disorders of sex development. Translat Androl Urol. 2016;5:794–8.
23. Schoemaker MJ, Swerdlow AJ, Higgins CD, Wright AF, Jacobs PA. United Kingdom clinical cytogenetics group. Mortality in women with turner syndrome in Great Britain: a national cohort study. J Clin Endocrinol Metab. 2008;93:4735–42.
24. Gravholt CH, Fedder J, Naeraa RW, Müller J. Occurrence of gonadoblastoma in females with turner syndrome and Y chromosome material: a population study. J Clin Endocrinol Metab. 2000;85:3199–202.
25. Lu L, Luo F, Wang X. Gonadal tumor risk in pediatric and adolescent phenotypic females with disorders of sex development and Y chromosomal constitution with different genetic etiologies. Front Pediat. 2022;10:856128. https://doi.org/10.3389/fped.2022.856128.
26. Claahsen-Van der Grinten HL, Otten BJ, Stikkelbroeck MM, Sweep FC, Hermus AR. Testicular adrenal rest tumours in congenital adrenal hyperplasia. Best Pract Res Clin Endocrinol Metab. 2009;23:209–20.
27. Looijenga LH, Hersmus R, de Leeuw BH, Stoop H, Cools M, Oosterhuis JW, et al. Gonadal tumours and DSD. Best Pract Res Clin Endocrinol Metab. 2010;24:291–310.
28. Ji J, Zöller B, Sundquist J, Sundquist K. Risk of solid tumors and hematological malignancy in persons with turner and Klinefelter syndromes: a national cohort study. Int J Cancer. 2016;139:754–8.
29. Gondos BE. Development of the reproductive organs. Ann Clin Lab Sc. 1985;15:363–73.
30. Hersmus R, de Leeuw BH, Wolffenbuttel KP, Drop SL, Oosterhuis JW, Cools M. New insights into type II germ cell tumor pathogenesis based on studies of patients with various forms of disorders of sex development (DSD). Mol Cell Endocrinol. 2008;291:1.
31. Lau YF, Chou PM, Iezzoni JC, Alonzo JA, Kömüves LG. Expression of a candidate gene for the gonadoblastoma locus in gonadoblastoma and testicular seminoma. Cytogenetic and Genome Res. 2000;9:160–4.
32. Wolffenbuttel KP, Hersmus R, Stoop H, Biermann K, Hoebeke P, Cools M, et al. Gonadal dysgenesis in disorders of sex development: diagnosis and surgical management. J Pediat Urol. 2016;12:411–6.
33. Pyle LC, Nathanson KL. A practical guide for evaluating gonadal germ cell tumor predisposition in differences of sex development. Am J Med Genet Part C: Sem Med Genet. 2017;175:304–14.
34. Elzinga-Tinke JE, Sirre ME, Looijenga LH, Van Casteren N, Wildhagen MF, Dohle GR. The predictive value of testicular ultrasound abnormalities for carcinoma in situ of the testis in men at risk for testicular cancer. Int J Androl. 2010 Aug;33(4):597–603.
35. Ebert KM, Hewitt GD, Indyk JA, McCracken KA, Nahata L, Jayanthi VR. Normal pelvic ultrasound or MRI does not rule out neoplasm in patients with gonadal dysgenesis and Y chromosome material. J Pediat Urol. 2018;14:154–e1.

36. Steven M, O'Toole S, Lam JP, MacKinlay GA, Cascio S. Laparoscopy versus ultrasonography for the evaluation of Mullerian structures in children with complex disorders of sex development. Pediat Surg Int. 2012;28:1161–4.
37. Döhnert U, Wünsch L, Hiort O. Gonadectomy in complete androgen insensitivity syndrome: why and when? Sex Dev. 2017;11:171–4.
38. Roca N, Munoz M, Cruz A, Vilalta R, Lara E, Ariceta G. Long-term outcome in a case series of Denys–Drash syndrome. Clin Kid J. 2019;12:836–9.
39. Ezaki J, Hashimoto K, Asano T, Kanda S, Akioka Y, Hattori M, et al. Gonadal tumor in Frasier syndrome: a review and classification of Frasier syndrome. Cancer Prevention Res. 2015;8:271–6.
40. Kilberg M, McLoughlin M, Pyle L, Vogiatzi MG. Endocrine management of ovotesticular DSD, an index case and review of the literature. Pediatr Endocrinol Rev: PER. 2019;17:110.

Micropenis: DSD Perspective

64

John M. Park, Dinesh Sarda, and Hari Mangtani

Abstract

Micropenis—A purely clinical entity featured by the stretch penile length less than 2.5 SD from the normal age-specific penile length. Microphallus is a term often assigned to small phallus usually seen in patients with DSD (Disorders or Differences in Sex Development). A meticulous clinical examination and systematic approach are needed in managing such patients. Most of the cases can be managed with expectant treatment or hormonal supplementation. This chapter highlights different etiological conditions, investigations, diagnostic protocols, and different management options and psychosocial effects in patients with micropenis.

Keywords

Micropenis · Microphallus

J. M. Park (✉)
C S Mott Children's Hospital Department of Urology, University of Michigan Medical School, Ann Arbor, MI, USA
e-mail: jmpark@med.umich.edu

D. Sarda
C S Mott Children's Hospital Department of Urology, University of Michigan Medical School, Ann Arbor, MI, USA

Shridha Hospital and Research Institute, Nagpur, India

H. Mangtani
Pearl Endocrine Clinic, Nagpur, Maharashtra, India

S. K. Ratan (ed.), *Children with Differences in Sex Development*, https://doi.org/10.1007/978-981-97-1639-5_64

The term 'Micropenis" (Greek. mikros, small + Latin. penis) is frequently used interchangeably and perhaps correctly with the term "Microphallus" (Greek. mikros, small + Greek. phallos, penis) in a male child with smaller looking external genitalia. However, many experts prefer to use the term micropenis for an extremely small penis, which has a midline raphe, circumferential prepuce, as well as normally located urethral meatus opening. The scrotum is normal and fused, but it may be underdeveloped (hypoplastic) or sometimes empty without testicles. The term microphallus can also be used for ambiguous/atypical-looking male genitalia seen in DSD cases with varying levels of hypospadias and/or other genital abnormalities. In patients with micropenis, the gonads are exclusively testis and is having 46, XY chromosome. However, none of these are absolute criteria for defining micropenis, as it is known to occur in Klinefelter's syndrome (47, XXY) and other cases of DSD. While discussing the DSD perspective of the small penis, we will be using the term micropenis for all small male external genitalia.

Micropenis is a clinical sign, and the accurate diagnosis can be made by measuring the Stretched Penile Length (SPL). Micropenis is defined as stretched penile length of less than 2.5 SD below the mean for the age. In cases of micropenis, the circumference-to-length ratio can be typically normal and rarely there can be hypoplastic corpora cavernosa. Schonfeld and Beebe were the first to describe SPL for age as well as data on penile length and penile circumference in the US population. Since then, many ethnicity-specific normative data have been produced from various countries including India. Measurement of SPL is done by fully stretching the penis. Glans is held with the thumb and forefinger and scale is held straight from pubic ramus to tip of glans penis on the dorsal side. The suprapubic fat pad if present should be pushed maximally and intact prepuce should be retracted (if possible) while taking the measurement.

64.1 Embryology/Development

During the intrauterine period of a 46, XY male fetus, the placental chorionic gonadotropins stimulate fetal Leydig cells to produce testosterone, beginning at 8 weeks of gestation. By 12 weeks of gestation, differentiation of penis is complete under the influence of testosterone as well as a more potent Dihydrotestosterone (DHT), an intracellular conversion product of testosterone, acting through the Androgen Receptors (AR). The testosterone to DHT conversion is facilitated by the enzyme 5-alpha-reductase. The genital tubercle, genital fold, and genital selling differentiate into glans penis, shaft of penis as well as scrotum, respectively.

Second and third trimesters are characterized by significant and marked increase in size of penis under the influence of fetal androgens, the production of which is stimulated by the fetal pituitary gonadotropins. Thus, it can be deduced that an abnormality in the size of a completely formed penis (true micropenis) is likely caused by the abnormalities in the hypothalamic–pituitary–testicular axis beyond 12 weeks of gestation, while the micropenis associated with other structural DSD abnormalities of the genitalia (chordee, hypospadias, bifid scrotum, etc.) result from the abnormalities in synthesis, secretion, or action of the fetal testosterone or faulty differentiation of the testis during the first trimester. In addition, the severity of the

Table 64.1 Mean and calculated -2.5 standard deviation (SD) values for stretched penile length (cm)

Age Group	Mean (+/- 2SD)	Mean – 2.5 SD
Preterm newborns (30 weeks)	2.5±0.4	1.5
Preterm newborns (34 weeks)	3.0±0.4	2.0
Term newborns	3.5±0.4	2.5
Infants and children 0-5 months	3.9±0.8	1.9
6-12 months	4.3±0.8	2.3
1-2 years	4.7±0.8	2.6
2-3 years	5.1±0.9	2.9
3-4 years	5.5±0.9	3.3
4-5 years	5.7±0.9	3.5
5-6 years	6.0±0.9	3.8
6-7 years	6.1±0.9	3.9
7-8 years	6.2±1.0	3.7
8-9 years	6.3±1.0	3.8
9-10 years	6.3±1.0	3.8
10-11 years	6.4±1.1	3.7
Adults	13.3±1.6	9.3

Custer J, Rau R. The Harriet Lane handbook. In S. Ballel P. McIntosh (Eds), Endocrinology. Philadelphia: Elsevier Mosby 2009:269-300

Table 64.2 Age wise mean stretched penile length (SPL), 2 SD and Mean-2.5 SD

Age (years)	Mean SPL (cm)	2SD (cm)	Mean-2.5 SD (cm)
0-1	4.02	0.49	2.8
1-2	4.50	0.53	3.2
2-3	5.01	0.56	3.6
3-4	5.52	0.63	3.9
4-5	5.82	0.63	4.2
5-6	6.11	0.61	4.6
6-7	6.31	0.66	4.7
7-8	6.50	0.68	4.8
8-9	6.61	0.75	4.7
9-10	6.71	0.75	4.8

N Teckchandani, M Bajpai. Penile length nomogram for Asian Indian prepubertal boys. Journal of Pediatric Urology (2014) 10, 352-354)

structural abnormality of the genitalia depends on the severity of the underlying defect in the synthesis, secretion, or action of testosterone (Tables 64.1 and 64.2).

64.2 Etiology

As noted above, micropenis results when the full masculinization of external genitalia is hindered by inadequate androgen action in early fetal life and/or an inadequate penile growth secondary to a variable degree of lack of androgen exposure in late fetal life. The etiological factors can thus be broadly classified as follows:

1. Hypogonadotropic hypogonadism due to hypothalamic or pituitary failure.
2. Hypergonadotropic hypogonadism due to primary gonadal failure.
3. Partial androgen insensitivity syndrome (PAIS) with inadequate penile growth despite normal or even excessive androgen secretion.
4. Idiopathic with normal hypothalamic, pituitary, and gonadal function.

The failure of the hypothalamus/pituitary to secrete gonadotropins is the most common cause of micropenis. Hypogonadotropic Hypogonadism (HH) can be isolated, or it may be a part of combined pituitary hormone deficiency. Severe isolated Growth Hormone (GH) deficiency can cause micropenis even when not associated with HH. Similarly, Laron syndrome characterized by GH insensitivity due to GH receptor defects may also be associated with a small penis. These observations suggest the possible role of GH in facilitating the action of testosterone on penile growth. HH can be an isolated genetic disorder (Kallman syndrome KAL-1, GnRHR, etc.) or it may be a part of a multiorgan genetic syndrome (Prader-Willi and Laurence-Moon-Biedl-Bardet). The common syndromes associated with micropenis are shown in Table 64.3.

Table 64.3 Syndromes associated with micropenis

Category	Syndrome	Associated findings
Hypogonadotropic hypogonadism	Kallmann	Hyposmia, small testes ,X-linked recessive
	Prader-Willi	Perinatal hypotonia, short stature, small hands and feet, obesity, mental retardation
	Septo-optic dysplasia (de Morsier)	Hypopituitarism, hypoplastic optic discs, absent septum pellucidum, abnormal third ventricle
	Laurence-Moon-Biedl Bardet	Retinitis pigmentosa, spastic paraplegia, mental retardation, obesity, polydactyly
	Rud	Hyposmia, ichthyosis, developmental delay
Primary hypogonadism (Hypergonadotropic)	Klinefelter	Tall stature, hypospadias, small testes, mental retardation, gynecomastia
	Noonan	Short stature, mental retardation, cardiac defects, Turner-like dysmorphism
	Robinow	Fetal facies, brachymesomelic dwarfism
Other syndromes	Carpenter	Acrocephaly, polydactyly, syndactyly
	Down	Mongoloid facies, cardiac defects, renal defects, mental retardation
	Fanconi pancytopenia	Hypoplastic anemia, short stature, GHD
	Fetal hydantoin	Congenital anomalies, IUGK neuroblastoma, bleeding
	Triploidy	Prenatal growth failure, small eye~, cardiac defects
	Williams	Mental retardation, cardiac defects, short palpebral fissure, epicanthal folds, hypoplastic nails

P.S.N. Menon and Umakanth A. Khatwa. The Child with Micropenis. Indian Journal of Pediatrics, 2000; 67 (6): 455-460

Failure of gonads to secrete adequate amounts of testosterone required for complete masculinization of the male fetus can be due to gonadal dysgenesis, testosterone biosynthetic defect, or may be associated with syndromes like Klinefelter syndrome, Robinow syndrome, or X polysomies. Deficiency of the enzyme 5-alpha-reductase or mutation leading to the insensitivity of the Androgen Receptor (AR) to androgens also is responsible for variable degrees of incomplete masculinization. The disorders affecting the testicular differentiation leading to deficient synthesis, secretion, or action of the androgens very rarely have isolated micropenis. These disorders are mostly present with DSD/intersex states of poorly differentiated male external genitalia with variable degrees of hypospadias, chordee, and scrotal fusion. Similarly, 46, XX congenital adrenal hyperplasia typically presents as DSD, but its presentation as isolated micropenis has been reported.

In cases with idiopathic isolated micropenis without hypogonadism, there is no qualitative or quantitative defect of androgen production or binding. Micropenis in these children may be the result of transient defects in androgen synthesis or action or both, during a critical phase of embryogenesis and fetal development.

64.3 Pseudo-micropenis

Before subjecting a child to extensive hormonal, imaging, and genetic work up to find the cause of micropenis, it is important to exclude pseudo-micropenis first. The closest differential of micropenis includes the buried penis, concealed penis, trapped penis, and webbed penis.

A buried penis is often seen in prepubertal and obese boys where the small size of their penis is thought to exist by the overabundance of the prepubic fat on the penis, and the true penile structure can be revealed by pressing the surrounding fatty tissue inward (Fig. 64.1). A trapped penis is an acquired form of the inconspicuous

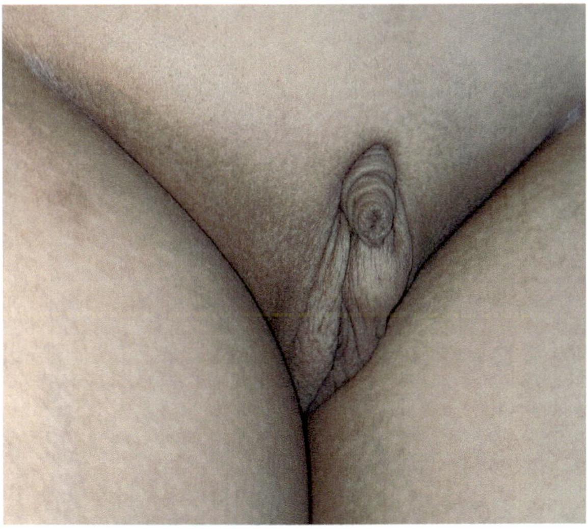

Fig. 64.1 Buried penis

penis that may become embedded in the suprapubic fat pad following circumcision. A webbed penis is characterized by skin tissue connecting the ventral aspect of the penis to the scrotum. Sometimes the loose penile skin does not stretch tightly around the body of the penis, or the penile skin is deficient, causing a concealed or inconspicuous penis (Table 64.4).

Table 64.4 Molecular basis for development disorders associated with micropenis due to hypogonadotropic hypogonadism

Gene	Phenotype
MPHD	
PROP1 (POU1F1)	Autosomal recessive GH, PRL, TSH, and LH/FSH deficiency (later onset ACTH deficiency)
HESX1 (RPX)	Autosomal recessive; and heterozygous mutations Multiple pituitary including diabetes insipidus
LHX3	Septo-optic dysplasia, Autosomal recessive, GH, PRL, TSH, FSH/LH deficiency, Rigid cervical spine
PHF6	X-linked; GH, TSH, ACTH, LH/FSH deficiency Borjeson-Lehmann syndrome: mental retardation, obesity, seizure, dysmorphic facies.
Isolated HH	
KAL1	X-linked Kallmann syndrome anosmia/hyposmia, renal agenesis
FGFR1 (fibroblast growth factor receptor)	Autosomal dominant Kallmann syndrome
KAL2	Anosmia/hyposmia, cleft lip/palate
GNRHR (GnRH receptor; G protein-coupled receptor)	Autosomal recessive
GPR54 (KiSS1-derived peptide receptor GPR54	Autosomal recessive
SNRPN (small nuclear ribonucleoprotein polypeptide SmN) Lack of function of paternal 15q11– q13 region or maternal uniparental disomy	Prader-Willi syndrome; Obesity
LEP (leptin)	Autosomal recessive Obesity
LEPR (leptin receptor	Autosomal recessive Obesity
DAX1	X-linked recessive Adrenal hypoplasia (73–75)

Grumbach MM. A window of opportunity: the diagnosis of gonadotropin deficiency in the male infant. J Clin Endocrinol Metab 2005; 90:3122-3127. Epub 2005 Feb 22).

HESX1 Homeobox gene expressed in ES cells; *PHF6* plant homeo domain-like finger gene; *FGFR1* fibroblast growth factor receptor 1; *GNRHR* GnRH receptor; *SNRPN* small nuclear ribonucleoprotein polypeptide SmN; *LEP* leptin; *LEPR* leptin receptor; *DAX1* dosage-sensitive sex reversal-adrenal hyperplasia congenita critical region on the X chromosome, gene 1; *PROP1* prophet of Pit-1; *LHX3* lim homeobox gene 3

64.4 Diagnostic Approach

Micropenis should be diagnosed as early as possible with the involvement of pediatric endocrinologist along with pediatric surgeon or urologist and geneticist form the multidisciplinary team.

A thorough physical examination and laboratory workup is required to establish the diagnosis (Table 64.5). Besides ascertaining the etiology of micropenis, it is imperative to find out any associated life-threatening abnormalities. The intersex/DSD states such as salt-losing adrenal disorder or hypogonadotropic hypogonadism may be a part of panhypopituitarism, and the associated GH and ACTH deficiency may precipitate hypoglycemia or a shock-like state due to cortisol deficiency. Serum cortisol, serum electrolytes, and plasma glucose levels should be measured to rule out acute problems. The length of the penis, the presence and size of the corporal bodies, the position of the urethral meatus, the associated chordee, and the location of the testicles should be noted. Any associated deformities and/or dysmorphological features should be noted to identify any associated genetic syndromes.

The endocrine evaluation includes Luteinizing Hormone (LH), Follicle-Stimulating Hormone (FSH), testosterone, prolactin, IGF-1, free T4, and TSH levels to identify the level of dysfunction in the hypothalamic-pituitary-testicular axis. The defect of the hypothalamus (high PRL) vs. the pituitary (low PRL) can be differentiated by serum Prolactin levels. Low LH, FSH, Testosterone with normal cortisol, IGF-1, and Free T4 and TSH suggest isolated HH whereas, a low IGF-1 along with or without low cortisol and/or a low free T4 and TSH points toward panhypopituitarism as a cause of micropenis. Isolated GH deficiency can be diagnosed on low IGF-I levels with normal cortisol, Free T4, TSH, gonadotropins, and testosterone with subnormal peak GH levels on stimulation tests. Diagnosis of Laron's syndrome is suspected with low IGF-I levels and elevated GH levels. It is important to note that fetal GnRH pulse generator activity continues in early infancy, thereby causing a postnatal surge of LH, FSH, and Testosterone. This is known as "mini-puberty" and provides a window of opportunity (at 1–3 months) for clinicians to diagnose HH. From post-infancy to the onset of puberty (prepuberty) pituitary is quiescent to secret gonadotrophs, characterized biochemically by low levels of LH, FSH, and testosterone. During this time, assessing the plasma concentration of inhibin B is helpful as its levels remain low in HH.

The testicular function may be assessed in a patient with micropenis, particularly when the micropenis is associated with cryptorchidism or when the patient presented with additional structural abnormalities of the external or internal genitalia (DSD equivalent), notably hypospadias, chordee, bifid scrotum, presence of Mullerian derivatives, or when the Wolffian derivatives are not completely formed (Figs. 64.2 and 64.3). Patients with primary hypogonadism are characterized by low levels of testosterone at all ages as well as suboptimal rise or no rise after administration of human Chorionic Gonadotropin (hCG). In contrast, normal male individuals demonstrate a significant rise in plasma testosterone after HCG stimulation.

Table 64.5 Clinical and laboratory evaluation of patients with small penile length

GENERAL EXAMINATION
Weight
Height
Stature
Nutrition (Obesity)
Muscle Tone
Neurological examination
Presence of secondary sexual characteristics in adolescent children
LOCAL EXAMINATION
Penis
- Stretched penile length
- Corporal girth
- Location of meatus
- Axis of Penis
- Associated chordee
- Adequacy ofprepuse
- Medianraphe
- Prepubicfat pad
- Penile web
- Any other abnormality

Scrotum
- Normal orhypoplastic
- Rugosities
- Bifid or fused

Testis
- Location
- Size
- Volume

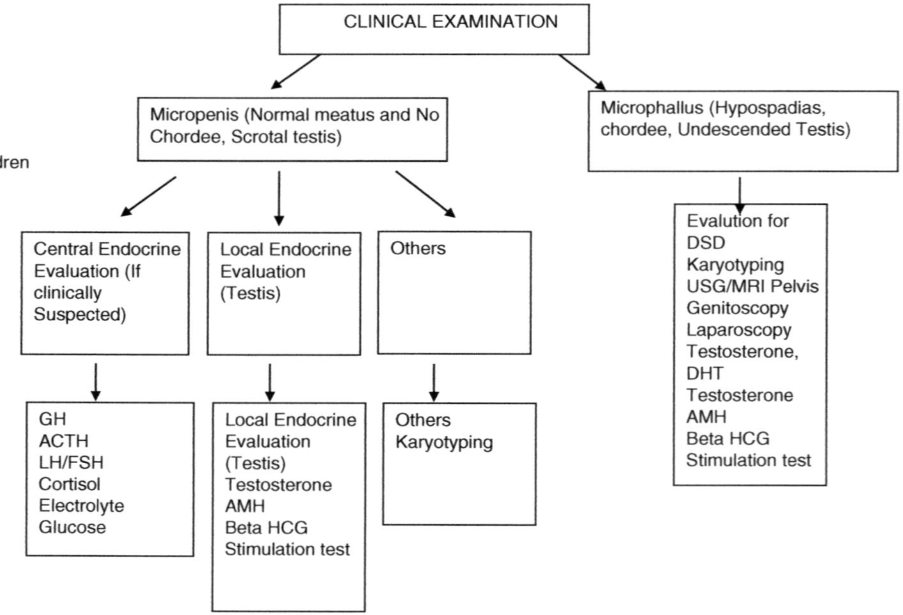

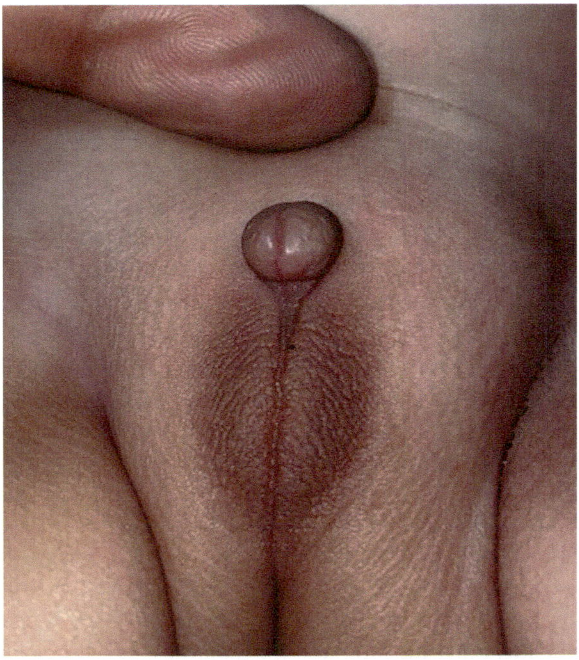

Fig. 64.2 Micropenis

Fig. 64.3 Microphallus (note associated hypospadias with unilateral undescended testis)

Along with low testosterone, primary hypogonadism is characterized by higher levels of gonadotropins (FSH/LH) due to a lack of negative feedback. The gonadotropin levels are modestly raised during early infancy and during childhood but rise significantly during the pubertal years. Functional testicular tissue can also be determined by measuring anti-Müllerian substance (AMH), also known as Müllerian-Inhibiting Substance (MIS), and inhibin B. Serum levels of testosterone and DHT can be measured before and after administration of hCG to determine the cause of micropenis/DSD in selected cases. Typically, 5-alpha-reductase deficiency can be diagnosed by normal serum testosterone levels, normal or low DHT levels with markedly increased basal and especially hCG-stimulated testosterone to DHT ratios (>17).

Patients with Androgen Insensitivity Syndrome (AIS) may demonstrate normal male levels of testosterone and DHT. However, resistance to the action of endogenous or exogenous testosterone at the cellular level may lead to absence of androgenic effects. Genetic confirmation is also possible by demonstrating a specific mutation in AR receptor genes.

Imaging studies are required to be performed in cases that are suspected to have a DSD. An ultrasound or magnetic resonance imaging of abdomen should be done to detect the Müllerian structures and also to locate the testes in patients with nonpalpable cryptorchidism. Diagnostic laparoscopy is also appropriate. Based on biochemical investigation and magnetic resonance imaging of the brain the pituitary and the olfactory tracts abnormalities can be diagnosed. A brain MRI may demonstrate a midline structural defect, such as pituitary stalk dysplasia syndrome, pituitary aplasia, or hypoplasia of the anterior pituitary with stalk abnormalities all suggestive of hypopituitarism.

64.5 Genetic Evaluations

Genetic analysis provides useful clinical insights and diagnostic applications. The genetic studies should include karyotyping with banding studies and Y-FISH to identify genetic sex and rule out Klinefelter syndrome or other forms of gonadal dysgenesis with Y chromosome cell line. Detailed molecular genetic analysis is guided by the results of the clinical and biochemical evaluation and may require the analysis of the genes that are implicated in both DSD and hypogonadotropic hypogonadism. The genetics of hypogonadotropic hypogonadism can be broadly separated into mutations associated with multiple pituitary hormone deficiencies, including gonadotropin deficiency (e.g., PROP1, HESX1, LHX3, and PHF6) and isolated hypogonadotropic hypogonadism [e.g., KAL1, FGFR1 (KAL2), GNHR, GPR54 (KiSS1 derived peptide receptor), SNRPN (Prader-Willi), LEP, LEPR, and DAX1] that can be associated with micropenis and evidence of hypogonadotropic hypogonadism in the male infant (Table 64.4). Molecular genetic analysis should be accompanied by a microarray to identify copy number variation, especially in those cases where there are associated extragenital features (Table 64.3).

64.6 Treatments

Medical therapy includes administration of androgen or gonadotropin. The surgical procedures include either correction of the associated chordee/hypospadias only or other penile lengthening procedures.

64.7 Medical Treatment

Testosterone can be administered by intramuscular route or by topical route to be applied locally on penis. Skin patches of testosterone are also available. The effect of testosterone is carried out directly by testosterone or by its metabolism to a more potent DHT. These androgens exert their effect by working through the Androgen Receptor (AR). To assess the responsiveness of the penis to respond to androgens, all patients with micropenis, irrespective of the underlying cause, should receive a short course of testosterone. Although there are no clear-cut guidelines on the dose, mode of administration, or duration of testosterone therapy for micropenis, most prefer to use testosterone cypionate or enanthate in oil with a dose of 25 mg given intramuscularly once every 3 weeks for 3 months. Temporary acceleration in growth rate and advancement of bone age are known side effects of testosterone therapy. Doubling of the SPL is considered a good response. A lack of response points toward androgen insensitivity or failure of androgen metabolism due to 5-alpha-reductase deficiency as an underlying cause of micropenis and which leads to failure to virilize at puberty. Testosterone cream (5%) applied topically also can cause penile enlargement, however, this effect is due to its systemic absorption. As the number of penile androgen receptors decrease from early infancy by 2–3 fold in early adulthood, testosterone therapy should be instituted early in infancy.

Transdermal DHT therapy is reported to have a favorable effect in achieving increase in penile length, particularly in patients with 5-alpha-reductase deficiency and less so in patients with partial AIS. DHT is nearly 10 times more potent than testosterone, so it is considered a better option for testosterone nonresponsive patients. DHT therapy is not available in some countries.

Human Chorionic Gonadotropin (hCG) stimulates the endogenous secretion of testosterone and can induce increase in penile length, particularly in hypogonadotropic hypogonadism patients. Similarly, a twice-weekly subcutaneous injection of LH and FSH was found to be effective in augmenting the penile length as well as the testicular volume.

In cases of micropenis due to idiopathic GH deficiency, growth hormone treatment alone has been shown to increase the length of penis and is the treatment of choice in such patients.

64.8 Surgical Treatment

In the early history of surgeries for micropenis, feminization of the small phallus along with female gender assignment was long considered the correct strategy prior to the 1960s. This is because, it seemed technically easier to make receptive vaginal pouch than a sensate, erect penis. However, this approach was based on the assumption that psychosexual orientation of humans was neutral during infancy. Subsequent studies have suggested that whebn 46 XY individuals with severely inadequate penis (such as penile agenesis, exstrophy and micropenis) who were assigned female sex had severly adverse physical and psychological long term outcomes. Hence many of the surgeons abandoned this approach. As per many reports, patients with micropenis are content with the male gender assignment in their adulthood and could even attain sexual function. However, they nevertheless reported significant disappointment with the quality of sexual life and genital appearance due to small size of the penis. Although hormonal treatment during infancy and childhood can improve penile length, many studies suggest that small penis size persists in adulthood, even if additional hormonal therapy has been given at puberty and beyond. It may be that early testosterone supplementation in prepubertal age stimulates growth of penis, but ultimate size of the penis does not grow more than its calculate as they reach the adulthood. Therefore, for micropenis patients whose phallus does not reach sufficient desired length despite medical treatment, surgery to increase the penile size further may be considered. In discussing surgical options, however, a humble and honest discussion, carefully balancing known pros and cons, must take place unhurriedly to establish realistic patient expectations. Available data suggest that the effect of male genitoplasty (regardless of the reported techniques) on penile length is still significantly suboptimal and typically does not significantly enhance the genital image and quality of sexual life. Phalloplasty, typically done in gender reassignment for transgender population, has resulted in a better acceptance and outcome in a small number of studies, but they are associated with a formidable risk of repeat surgeries and complications. Surgery for micropenis, therefore, should only be performed in adulthood with the patients' full autonomous consent, and never in childhood, and reserved for the most severe cases. Although the standard guidelines for indications of surgical interventions are lacking, Wessell suggested that surgical treatment should be offered only in men whose flaccid length is less than 4 cm or an SPL of less than 7.5 cm. Hinman (1970) was the first author to report the procedure and results of reconstructive surgery for micropenis. Thereafter, several techniques have been described but all of them lack noteworthy evidence to formulate guidelines for clinicians and patients.

The aim of reconstruction or penile lengthening procedures are:

1. Patient should be able to perform satisfying sexual penetrative acts during adulthood without chordee and painless erection.
2. Patient should be able to micturate in standing position with forward stream.
3. Penis should look as normal as possible to mitigate serious stigmatization and self-esteem concerns.

Penis has many unique anatomic features. The skin has a firm attachment at the glans and a loose attachment at the shaft which allows extensive stretching of skin during erection and easy mobility during sexual intercourse. Secondly, corpora cavernosa is a uniquely elastic tissue with expansile characteristics during engorgement. No other tissues in the body can act as a complete replacement or substitute for corporal reconstruction, although prosthetic devices can help. Similarly, glans and preputial skin play a key role in sexual intercourse. This area has numerous nerve terminal receipts that are distinctive and provide erotic sensations.

For a surgeon practicing penile elongation surgery, it is important to understand the facial attachments of penis. Beneath the skin are superficial dartos fascia and deep Buck's fascia which is adhered to tunica albuginea surrounding corporal bodies. On the ventral aspect of penis, penoscrotal angle is formed by blending of dartos fascia with smooth muscle fibers. At the dorsal base of penis, deep fascia has ligamentous attachment to pubic rami, which gives penopubic demarcation externally. These ligaments originate from external oblique aponeurosis and are attached to pubic symphysis. Suspensory ligaments are the extension of fibers from linea alba. Usually, they are avascular and firm but are very closely placed to penile neurovascular bundle. Scarpa's fascia layer of abdominal wall extends onto penile shaft and forms ligament of Luschka or fungiform ligament. This ligament has an attachment on pubis and extends laterally on both sides as well as ventrally in a fan-shaped manner and encircles entire penile root.

Key steps for correction of phallic length are:

1. Complete penile degloving
2. Cutting all the lateral fibrous attachment between corpora and surrounding tissue till the sphincter is exposed posteriorly and fascia covering the pubic arch laterally
3. Cutting the suspensory ligament of penis behind the pubic symphysis
4. Transection of urethral plate when required with urethral reconstruction in single or multiple stages with or without grafts
5. Corporotomies on ventral aspect if necessary
6. Partial excision of pubic fat pad

I. Techniques to increase penile length
   (i) Release of suspensory ligament of penis.

The suspensory ligament consists of the fibrous avascular attachment arising from inferior and posterior surface of pubic arch and is attached to deep fascia of penis. Complete detachment of these fibers straightens the corporal bodies and acute angulation becomes obtuse. Although the actual length of corpora is not increased there is perception of increase length of penis. The length of release of corporal bodies can be extended onto the entire pubic rami length. This can increase the length further but carries the risk of damage to neurovascular bundle.

It must be noted that this procedure increases the penile length only by 1–3 cm. Furthermore, there are few reports of recurrence in the form of post-operative penile reshortening. In order to prevent the recurrence, the placement of some filler materials such as vascularized flap or prosthesis in the space created between pubis and released suspensory ligament has been proposed. Overlying skin being stretchable, it can accommodate the released length of corpora, however, if required, dorsal V-Y advancement at the base of the penis can be combined with the above procedure.

The outcome of suspensory ligament release is not consistently encouraging. Partner and individual satisfaction rates ranges between 30 and 65%. The main disadvantage of this procedure is loss of penile support which results in dangling position of the erected penis and resulting in difficulty with sexual penetration.

(ii) Corporal elongation technique

Sliding elongation technique is a novel technique that can be used for corporal lengthening in patients with micropenis. After penile degloving, neurovascular bundle and corpora spongiosa are isolated. Two longitudinal incisions are given at 3 and 9 o'clock positions on tunica albuginea of corporal bodies along with dorsal-proximal and ventral-distal transverse incision on tunica albuginea. Subtunical dissection along with constant traction on glans elongates the corporal length. The tunical defects produced due to these incisions are covered with some prosthetic or autologous patch.

(iii) Length augmentation using the graft or pedicled flaps

Perovic and Djordjevic described placement of autologous rib cartilage graft between corpora cavernosa and glans cap and then reassembling the penis. This technique can lengthen the penis by 2–4 cm. Augmentation phalloplasty is the technique to replace the penis itself. Different techniques such as sensate osteocutaneous fibular flap, free scapular flap, suprapubic abdominal wall flap, and vertical rectus abdominis flap have been described. However, fasciocutaneous phallus reconstruction using radial artery forearm flap has remained the most widely used of all, especially in transgender patients. Prosthesis implanted after reconstruction gives better cosmetic and functional results which are more acceptable and better than other genitoplasty and penile lengthening procedures. However, complication rates remain significantly higher, even in most experienced hands.

II. Increasing the penile girth
  (i) Penile augmentation using various agents

The penile subcutaneous tissue can be injected with various agents to augment the penile girth. Various substances like liquid silicone, polyacrylamide, hyaluronic acid, and mineral oil have been described but with high rates of complications like penile distortion, granulomas, and swelling, necessitating its subsequent removal. Autologous fat injection may be a more biocompatible method of penile augmentation without foreign body reaction. However, up to 80% of the autologous fat graft will be absorbed in the first year of engraftment. Hence, multiple procedures may be required for achieving the desired effects. This procedure can be combined with one of the penile lengthening procedures.

(ii) Corporal tunica albuginea augmentation using graft

Autologous venous graft or expanded polytetrafluorethylene vessel patches can be used to augment the corporal circumference. Bilateral longitudinal incisions are placed on tunica albuginea and the grafts are placed in the defect to achieve the expansion of penile corpora cavernosa. The endothelial lining which constitutes the internal surface of veins is compatible with endothelium of corpora cavernosa and hence lesser chances of complications.

III. Increasing the perception of penile enlargement

Suprapubic lipectomy can be an additional procedure along with other techniques to increase the perceived penile length. Removal of overhanging fat pad in suprapubic region adds to the increase in exposed penile length and is more conspicuous. It also helps in the improvement of hygiene and directing the urinary stream.

IV. Penile traction devices

These devices can be used in patients who are not keen on surgery of micropenis as well as can be combined with other surgical procedures. These devices surround the base of penis and gentle tension stretches the soft tissue and increases the penile length. Devices need to be applied for many hours per day and for prolonged periods. Hence patient compliance is a critical issue in this management. Outcome is favorable as patient satisfaction has been shown to increase after use of 3–6 months. Different studies showed an increase in penile length by 1–3 cm.

Although there are significant advances in surgical procedures, the anatomy and physiology of the penis cannot be replicated, and further research may find the most suitable surgical option that can give the maximum and long-term satisfaction to these patients with the least rate of complications.

64.9 Concluding Thoughts

Micropenis represents a unique, rare congenital condition with varying etiology and syndromes. Early prompt diagnosis and appropriate hormonal treatment can significantly improve the penile size during infancy and childhood although its ultimate effect on the final adult penile size and sexual quality of life remains undetermined. It is understandable that patients with micropenis might have concerns regarding their self-esteem, sexual confidence, and overall sexual quality of life, and it is tempting to jump into surgical options to alleviate their anxiety. Although vascular pedicle phalloplasty seems to provide the best outcome in terms of penile enlargement compared to other reported techniques of genitoplasty, it is associated with multiple repeat surgeries and a high rate of complications. An honest and humble discussion with patients is paramount to establishing realistic expectations. It is also important to note that there is no clear-cut evidence that a relation exists between small length of penis and penetrative intercourse. Sircili study showed that small penile length was associated with both satisfactory and unsatisfactory sexual activities. Few of them could have sexual intercourse and some could not. Interestingly, few reports showed that men with micropenis could perform sexual acts and some men could find out other ways to reach orgasm and satisfy their partner. Micropenis, like other DSD conditions, must include psychological counselling as a routine and regular part of the overall treatment. Doing a careful job of educating and counselling is critical, and the role of parents in the case of children with micropenis cannot be overemphasized. Clinicians must work hard to provide patients with proper time for explanation, clear counselling about the diagnosis, pathophysiology, and different treatments from childhood to adulthood. Long-term psychological support is of utmost importance than the surgical results.

Further Reading

1. Schonfeld WA, Beebe GW. Normal growth and variation in the male genitalia from birth to maturity. J Urol. 1942;48:759–77.
2. Teckchandani N, Bajpai M. Penile length nomogram for Asian Indian prepubertal boys. J Pediatric Urol. 2014;10:352–4.
3. Zalel Y, Pinhas-Hamiel O, Lipitz S, Mashiach S, Achiron R. The development of the fetal penis—an in utero sonographic evaluation. Ultrasound Obstet Gynecol. 2001;17(2):129–31.
4. Evans BA, Williams DM, Hughes IA. Normal postnatal androgen production and action in isolated micropenis and isolated hypospadias. Arch Dis Child. 1991;66:1033–6.
5. Lee PA, Mazur T, Danish R, Amrhein J, Blizzard RM, Money J, Migeon CJ, Micropenis I. Criteria, etiologies and classification. Johns Hopkins Med J. 1980;146:156–63.
6. Borsellino A, Spagnoli A, Vallasciani S, Martini L, Ferro F. Surgical approach to concealed penis: technical refinementsand outcome. Urology. 2007;69:1195–8.
7. Grumbach MM. A window of opportunity: the diagnosis of gonadotropin deficiency in the male infant. J Clin Endocrinol Metab. 2005;90:3122–7. Epub 2005 Feb 22)
8. Menon PS, Khatwa UA. The child with micropenis. Ind J Pediatr. 2000 Jun;67(6):455–60.
9. Hatipoğlu N, Kurtoğlu S. Micropenis: etiology, diagnosis and treatment approaches. J Clin Res Pediatr Endocrinol. 2013;5(4):217–23.
10. Guthrie RD, Smith DW, Graham CB. Testosterone treatment for micropenis during early childhood. J Pediatr. 1973;83:247–52.
11. Levy JB, Husmann DA. Micropenis secondary to growth hormone deficiency: Does treatment with growth hormone alone result in adequate penile growth? J Urol. 1996;156:214–6.
12. Stancampiano MR, Suzuki K, O'Toole S, Russo G, Yamada G, Ahmed SF. Congenital micropenis: etiology and management. J Endocrine Soc. 2022;6(2)
13. Rajendra B, Nerli AKG, Patne PB, Hiremath MB. Penile growth in response to hormone treatment in children with micropenis. Indian J Urol. 2013 Oct-Dec;29(4):288–91.
14. Callensa N, Piet Hoebekeb Hiort O, Ahmed SF. Understanding differences and disorders of sex development (DSD). Phalloplasty: a Panacea for 46,XY Disorder of sex development conditions with Penile Deficiency? Endocr Dev, vol. 27. Basel: Karger; 2014. p. 222–33.
15. Garaffa G, Raheem AA, Ralph DJ. Asian An update on penile reconstruction. J Androl. 2011;13:391–4.

16. Custer J, Rau R. The Harriet Lane handbook. In: Ballel S, McIntosh P, editors. Endocrinology. Philadelphia: Elsevier Mosby; 2009. p. 269–300.
17. Sircili MHP, Silva F, Costa EMF, et al. Long-term surgical outcome of masculinizing genitoplasty in large cohort of patients with disorders of sex development. J Urol. 2010;184:1122–7.
18. Van Seters A, Slob A. Mutually gratifying heterosexual relationship with micropenis of husband. J Sex Marital Ther. 1988;14:98–107.
19. Stulhofer A. How (un)important is penis size for women with heterosexual experience? Arch Sex Behav. 2006;35:5–6.

65 Ovo Testicular Differences in Sexual Development

Devendra Kumar Gupta and Shilpa Sharma

Abstract

Ovo testicular type is amongst the least common of all Disorders of Sexual Differentiation. A child born with ambiguous genitalia due to Ovo testicular DSD has severe hypospadias with unilateral or bilateral undescended gonad, Mullerian structures, and 46XX chromosomes.

With recent sharing of expertise, there has been improved timely diagnosis, proper assignment of sex, and judicious approach in the surgical management with a long-term follow-up including hormonal supplements for the development of secondary sexual characters. The phallus size is adequate and thus most of these children are reared as boys. Successful management is achieved by a team approach involving the pediatric geneticist, endocrinologist, Paediatric surgeon, radiologist, and a gynecologist, especially in advanced medical centers. Long-term follow-up is necessary to monitor the outcome of surgical procedures and the need for re-do if any, titrate the hormonal requirements, and address the issues related to psychology, marriage, infertility, and rehabilitation in the society.

Keywords

True hermaphroditism · Ovotesticular DSD · Sex assignment · Genitalia reconstruction · Long-term results

D. K. Gupta (✉) · S. Sharma
Department of Pediatric Surgery, All India Institute of Medical Sciences, New Delhi, India

S. K. Ratan (ed.), *Children with Differences in Sex Development*,
https://doi.org/10.1007/978-981-97-1639-5_65

Ovo testicular Disorders of Sexual Differentiation (DSD) is amongst the least common of all types of DSD (5–10%). As the name suggests, there is the presence of both the gonads, ovary and testis, in the same individual, though the quality, size, and laterality of each gonad will vary.

The pathogenesis of Ovo testicular DSD is very poorly understood to date. The expertise is limited in the field and the clinical presentation of the disorder varies significantly; thus, it becomes quite difficult to make the initial diagnosis of the disorder in infants and children without few specialized investigations.

The development of the embryo takes place under the influence of sex chromosomes and hormones, into a male or a female foetus with normal internal and external genitalia [1]. The presence of Y chromosome induces masculinization in the female foetus, possibly under the influence of H-Y antigen, testis determining factor (TDF), Anti-Mullerian Hormone (AMH), and SRY gene. Any chromosomal abnormalities, failure of migration of the primitive germ cells from the yolk sac to the urogenital ridge, absence of the production of anti-Mullerian Substance (from the Sertoli cells) and testosterone (produced by Leydig cells), overproduction of androgens in the females and finally the lack of androgen receptors, especially in the genital tissues, may affect the chain of sequence of development and lead to an abnormal development of genitalia in the newborn.

65.1 Clinical Presentation

A child born with ambiguous genitalia due to Ovo testicular DSD has severe hypospadias with unilateral or bilateral undescended gonad. These patients may present at a later age (post-puberty) with the enlargement of breasts (due to presence of estrogens hormone secreted from the ovarian gonad), cyclic hematuria (due to menstrual flow of blood), globular mass in the lower abdomen as hematometrocolpos (due to obstruction at the vaginal introitus-urogenital sinus, resulting in collection of blood in the uterus and the vagina), or still later due to breast development in a male adult, already operated for undescended testis and hypospadias, now being investigated for infertility.

The clinical spectrum of the Ovo testicular DSD case includes the presence of ambiguous genitalia, with a severe degree of penoscrotal, scrotal, or perineal hypospadias (Fig. 65.1a) [2, 3]. The phallus is usually well-developed and triangular in shape with a narrower base. However, the phallus could be moderate in size or even tiny. It is usually very well covered with prepuce. Glans and navicular fossa are well formed. The chordee is severe, making the phallus bend significantly, restricting the growth of the phallus.

There is a single urogenital sinus, with a large, wide, and patulous orifice. The urethral meatus is lined with mucosa, which extends distally from the meatal base towards the coronal sulcus for a varying length. The vaginal orifice remains mostly hidden, covered with the skin flap (Fig. 65.1b). The Ovo testicular DSD patients do

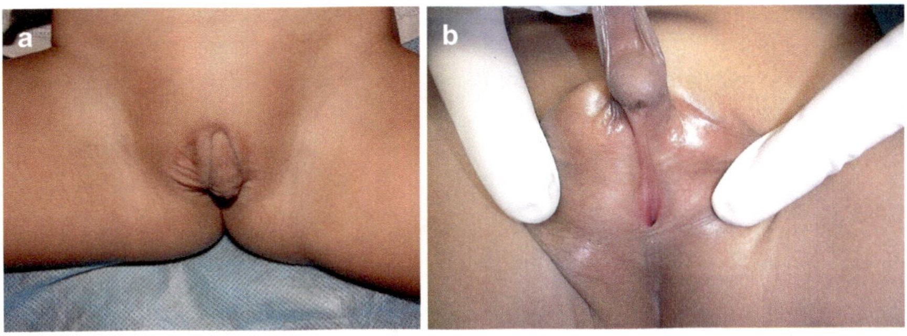

Fig. 65.1 (**a**). A 3-year boy presented with perineal hypospadias with severe chordee, poorly developed scrotal folds, and bilateral undescended gonads. Phallus is adequate though narrower towards the base. Had 46 XX on karyotype and Mullerian structures were present. Diagnosis of Ovo testicular DSD was made. (**b**). Showing a single wide meatus, lined with mucosa (arrow). A fibrous Chordee extends from the meatus to the glans

not have both the orifices (urethra and the vagina) visible in the perineum, except in a rare case - 2%), unlike in patients with severe dysgenesis (mixed gonadal dysgenesis and dysgenetic male pseudohermaphroditism).

The development of the scrotum depends on the presence or the absence of the gonad as its content. Scrotum is found well developed only if the gonad has reached it. Scrotal folds remain poorly developed if both the gonads remain in the inguinal canal or high up in the abdominal cavity. Scrotum may be well-developed unilaterally due to the descent of the gonad on that side only.

In the Ovo testicular DSD, there must be an ovarian as well as a testicular tissue in the same child. The internal gonads may be in various combinations, resulting in a pair of testis and ovary (10%), Testis/ovotestis (25%), ovary/ovotestis (30%), and bilateral ovotestis (35%). Ovotestis is the most common gonad and can be present on both sides. The ovarian gonad is always abdominal in location and is accompanied by a recognizable fallopian tube whilst the testis has a vas as the draining duct and depending on the testicular volume and maturity, descends into the scrotum. Never ever, both the ducts drain an Ovo testicular gonad. If the ovarian tissue is the major component, it will be drained by a duct resembling "fallopian tube with poorly developed fimbriae structures" and if the testicular tissue is the major volume of the Ovo testicular gonad, then it will be drained by a vas with the poorly developed epididymal structure.

The ovary remained in its original position in the abdomen, whilst the testis may descend, and the descent depends on the degree of dysgenesis. The more severe the dysgenesis, the higher the location of the testicular gonad (abdominal), and it is then smaller in volume, poor in texture, and without a firm capsule. Contrary to this, a better quality of testis may descend to the inguinal canal or even to the base of the ipsilateral scrotal fold. This can be felt clinically (Fig. 65.2a).

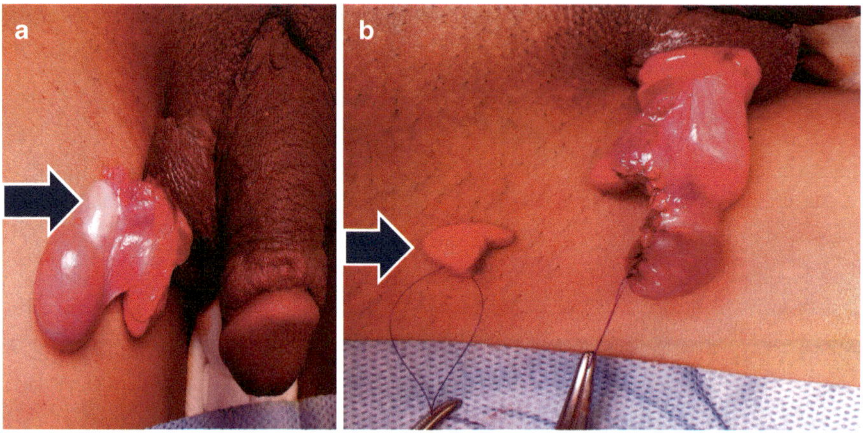

Fig. 65.2 YP—(**a**) Follow-up case of severe hypospadias and bilateral undescended gonads with 46XX, who was investigated for Ovo testicular DSD in childhood. The right-sided scrotal gonad looked testis and was biopsied. Despite all efforts, ovarian tissue could not be found even on palpation of the gonad. On the left side, it was a streak gonad which was removed (histopathology confirmed wavy stroma only, without any primordial follicles). Staged male genitoplasty was done, removing the female adnexa in childhood. This case with 46XX had remained quite intriguing. (**b**). At 16 years of age, during the follow-up, he presented with breast development. Palpation of the scrotal gonad this time confirmed a tiny firm nodule, sitting on the top of the testis (now suspected to be an ovarian tissue. Estradiol level was also found high. Thus, the labelling of the diagnosis of Ovo testicular DSD. Excision biopsy of the suspected ovarian tissue (nodule, leaving the testicular tissue behind) was done Histopathology confirmed the biopsy nodule as the ovarian tissue. Bilateral mastectomy was also completed (The ovarian component of the ovotestis gonad being so tiny was missed in early childhood)

The ovotestis may fully descend in the scrotum or become palpable anywhere along the course of the descent of the testis, with the vas deference as the draining duct. The palpable ovotestis gonad is diagnostic of Ovo testicular DSD due to the dumbbell shape of a gonad with double consistency. The ovarian part is firm whilst the testicular gonadal tissue is soft. Dual consistency confirms the diagnosis on palpation. The ovary tissue may be very tiny, like a small bead. However, the firm feel of the tissue adjoining the soft testicular gonad will clinch the diagnosis of Ovo testicular DSD, and can also be excised to retain only the testicular component (Fig. 65.2b). The ovotesticular gonads are usually positioned pole to pole to each other in the majority (90% cases) but may be hilar in location in about 10% of cases.

From 1980 to 2018, 69 cases of Ovo testicular were managed at the multidisciplinary paediatric Intersex clinic (AIIMS, New Delhi), with the following clinical details (Table 65.1).

Table 65.1 Clinical information based on the management of 69 cases of Ovo testicular DSD patients during 1980–2018 (Personal Data)

Phenotypic sex:- male severe hypospadias severe chordee well-developed phallus.
Genotypic sex:- 46,XX (94%); 46,XY (6%)
Prepuce:- Usually well developed, supple, good sized
Phallus:-Triangular phallus with a narrow base
Labioscrotal folds:- Asymmetry. Scrotal folds developed only if the testis has descended in the scrotum
Gonads:- T + ovary T + OT OT + OT, O + OT OT is the commonest gonad. There is only one draining duct, either a vas or the tube
Urethral meatus:- Severe chordee, perineal hypospadias, wide meatus, mucosal lining extending towards the base of phallus
Pubic hair:- Female pattern
Uterus (P/R):- Present; 60%—good sized, 35%—Hypoplastic; 5%—absent if XY
Breast development (Tanner's stage):-I-IIIAt puberty. Suggests the presence of good ovarian tissue
Fertility:-If reared as female, can be very rarely fertile. Yet no normal child was born.

65.2 Diagnosis

The following need to be evaluated to reach a diagnosis of an Ovo testicular DSD:

1. Chromosomal analysis. 46XX pattern is seen in 90% Ovo testicular DSD. The XY chromosomal pattern is rare though it may be present only in about 5–10% cases of Ovo testicular.
2. Retrograde Genitourethrogram (RGU). An RGU establishes not only a vaginal pouch (size and shape joining the urethra) but may also reveal a cervical indentation, suggestive of presence of uterus (Fig. 65.3). Rarely, the dye outlines not only the uterine cavity as in Fig. 65.3 like a streak, but also the fallopian tube (s) and the dye may even spill into the peritoneal cavity suggestive of existence of Mullerian structures and the patency of the fallopian tube/s. Unilateral deviation of the uterus is suggestive of hemiuterus, with ovary and fallopian tube on that side.
3. Imaging. The presence of uterus can be made per rectal examination and on ultrasonography (USG), CT, and MRI. A well-developed uterus is a feature of a normal girl or a case of Ovo testicular DSD with the presence of functional ovarian tissue at puberty. Ultrasonography is routinely performed for the evaluation of the renal tracts, internal gonads, uterus, vaginal pouch, and the adrenal glands. USG may also indicate the pattern of the gonads (ovary or testis), though it may not be very reliable. MRI scanning is rather much more useful to delineate details of various pelvic structures and to differentiate the ovary from the testis.
4. Bone age. An advanced bone age is suggestive of Congenital Adrenal Hyperplasia (CAH).

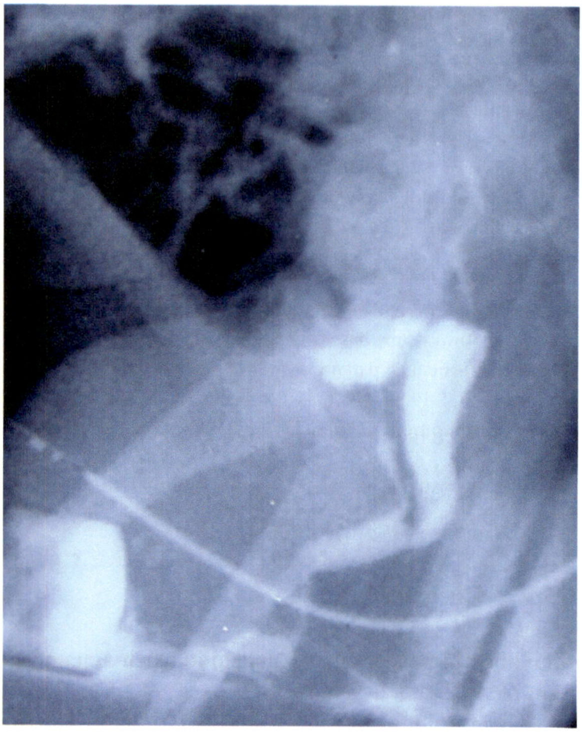

Fig. 65.3 Retrograde Genitourethrogram (RGU) in a 3-year-old child (same as above in Fig. 65.1), with perineal hypospadias, showed urogenital sinus, common channel for about 1.5 cm, well-defined vagina, cervical indentation (arrow mark—suggesting presence of uterus) and the dye actually flowing to the uterine cavity as a tiny streak (uterine cavity is very small in a 3-year-old child)

5. Hormonal assessment (17OHP, testosterone, LH and FSH, Estradiol), is done to differentiate different types of DSD. Hormonal evaluation is of particular importance in DSD cases, 4–5 times or more rise in levels of 17OHP is very suggestive of CAH. A normal value of estradiol in the presence of normal Testosterone level is suggestive of presence of an ovarian source as in Ovo testicular DSD. Similarly, T/DHT ratio is changed in MPH group of patients with 5-alfa-reductase deficiency, resulting in low levels of DHT, though T levels remain normal. In testicular feminization syndromes, part of Testosterone may be converted into estrogens and resulting in the bilateral breast development.
6. Endoscopy forms the part of the surgical procedure to evaluate the length and confluence of the vaginal pouch with the urogenital sinus. The narrower the distal opening of the pouch, higher are the chances of urinary infection and the associated hydrocolpos and hydrometrocolpos.
7. Laparotomy or laparoscopy is required for the physical examination of the internal gonads and Mullerian structures, taking a gonadal biopsy for final confirmation of the diagnosis, and also performing the surgical procedures.

65.3 Exploratory Laparotomy and Gonadal Biopsy

The Ovo testicular gonad can also be assessed at surgery by palpation. The ovarian component of the gonad is firmer and greyish whilst the testis part is softer and pinkish (Fig. 65.4). To confirm the presence of both the gonads histologically, the biopsy should include a longitudinal sleeve of the gonad through the firm and the soft parts of the gonad. The testis is usually soft and dysgenetic (prone to malignancy) whilst the ovarian gonad is firm and usually healthy and normal on histological examination. The presence of both the gonads must be confirmed histologically by an expert pathologist in the field. The ovary tissue is confirmed to show wavy stroma with the presence of primordial follicles. It is not safe to make the diagnosis based only on the presence of wavy stroma only without seeing the primordial follicles. The testis shows varying degrees of dysgenesis, depending on its location and volume. If the facility of a frozen section is available, the dysgenetic testicular tissue may be removed in the same sitting.

In Ovo testicular DSD, the uterus is present in almost all cases. It is well developed in about two-thirds of cases, and hypoplastic and chord like in the remaining ones. Very rarely, the uterus is absent. The uterus is found well developed and bulky in Ovo testicular cases presenting late, as the ovarian tissue becomes quite functional by then. Such children present with hypospadias, undescended gonad/s, cyclic hematuria, and bilateral breast development due to estrogen hormone secretion from the functional ovarian tissue present. Presence of high values of estradiol in a male child with hypospadias and undescended gonads, strongly favors the

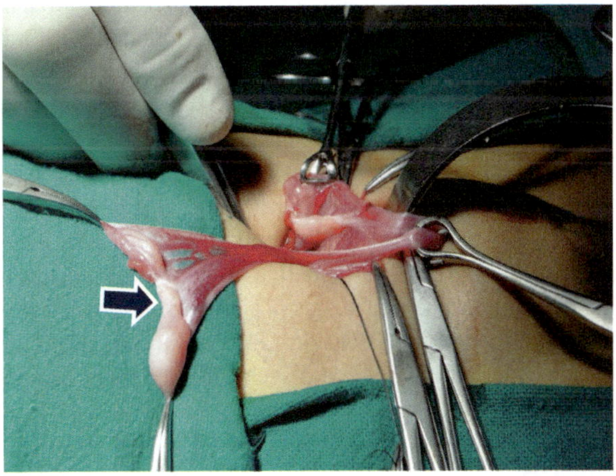

Fig. 65.4 The 3-year-old child (same child as above in Fig. 65.1,) with 46 XX, on laparotomy, showing right side ovotestis. The Ovarian tissue is smaller in component and greyish in colour (marked with arrow). The testis is a larger component (without a tunica cover, pinkish in colour), with gonadal vessels. On the left side, an ovary with adnexa. The tiny uterus is held in Babcock. Bilateral gonads were removed. Uterus and upper vagina were removed. He was assigned as a boy. Male genitoplasty was performed in stages. Bilateral testicular prosthesis was inserted later on

diagnosis of Ovo testicular DSD. Testosterone values may be normal or low, depending on the amount of normally functioning testicular tissue.

Most Ovo testicular cases are 46XX and only 5–10% of cases are 46XY. The presence of the Y cell line is associated with the presence of a testis with the vas, (rudimentary uterus), and it is dysgenesis and remains prone to development of malignancy.

65.4 Gender Assignment and Surgery

Gender assignment is the most important aspect of management.

The main problem in a child with DSD is giving the child a definite sex of rearing, male or female. After obtaining the clinical information aided by the investigations, one is in a reasonably good position to arrive at the possible diagnosis of DSD and explain the surgical approach needed to give the child a definite sex of rearing. The genetic sex does not matter and plays no role in the sex assignment of a child in most countries (except in a few Middle East and African countries).

In a developing country like ours, the plan should be discussed in confidence with the parents and grandparents. The merits and demerits of the male or the female sex assignment should be explained in detail, in terms of surgical options, fertility, phallic reconstruction, vaginoplasty, need for hormonal supplementation, insertion of testicular prosthesis, and possible risk of malignancy.

The Indian parents prefer to have an inadequate male rather than an infertile female to sustain the various types of pressures of society (marriage, livelihood, support).

It is easier to convert a DSD child into a female rather than into a male. The presence of a "Y" line (on chromosomal analysis) is associated with a high degree of malignancy. Hence, any abdominal (dysgenetic) testis with XY line must be removed at the latest by 12 years of age.

It is also easier to convince the parents and change the sex of rearing if the patient is under 1 year of age, thus the need for an early decision for sex assignment. The mental makeup of the child is extremely important and if becomes known, must be honored. Usually, the assignment of sex is dictated by the size of the phallus and the type of internal genitalia present. If the phallus is adequate, male sex is assigned and "male genitoplasty" is undertaken in stages [4]. However, if the phallus is small, and there is presence of an ovary, the fallopian tube, and a well-developed uterus, it is "desirable" to assign a female sex. "Female genitoplasty" is performed accordingly.

65.5 Follow-Up

Ovo testicular DSD patients are at an increased risk for the development of malignancy such as seminoma, gonadoblastoma, and dysgerminoma [5–7]. The risk for tumour development depends on many parameters, including gonadal morphology, germ cell location, presence of germ cells, genital morphology, patient age, and presence of TSPY [8, 9]. The malignancy risk is 3% in 46,XY Ovo testicular DSD and even lower in 46,XX Ovo testicular DSD [5].

65.6 Conclusion

With the increased awareness, the diagnosis of Ovo testicular DSD can be made with ease using clinical and investigative approaches. Child with severe hypospadias with unilateral or bilateral undescended gonad, a palpable dumbbell-shaped gonad with dual consistency, 46XX chromosomal pattern, imaging showing the presence of vaginal pouch and uterine indentation (or dye spilling to the peritoneal cavity), presence of high estradiol in boys with normal levels of testosterone and the laparoscopy and laparotomy confirming the presence of both types of gonads, clinches the diagnosis in almost 90% cases of Ovo testicular DSD. Histology finally confirms the presence of tissues from both types of gonads.

The role of laparoscopy and laparotomy is diagnostic as well as therapeutic. The laparoscopy has its limitation of tactile sensation, needed to feel the small ovarian gonadal tissue in the diagnosis of Ovo testicular DSD, which can otherwise be missed easily. Chromosomal sex is of no significance to assign the sex of rearing. Depending on the size of the phallus and the presence of the internal gonads, a male or the female sex is assigned in the Ovo testicular DSD case, after discussing the same in detail with the patient, parents, and the parents. All these patients require hormonal support to maintain their sexual secondary characters and mental makeup. A long-term follow-up is mandatory to assess the postoperative surgical outcome, marriage equations, and the psychosocial adjustment in the family and the society.

References

1. Sharma S, Kumar A, Gupta DK. Basics of human andrology: a textbook. In: Biology of male hermaphrodite and Intersex. Singapore: Springer; 2017. p. 437–50.
2. Gupta DK, Sharma S. Pediatric urology. In: Goel KM, Gupta DK, editors. Hutchison's pediatrics. 2nd ed. New Delhi: Jaypee; 2011. p. 244–57.
3. Khanna K, Sharma S, Gupta DK. A clinical approach to diagnosis of ambiguous genitalia. J Indian Assoc Pediatr Surg. 2019 Jul-Sep;24(3):162–9.
4. Sharma S, Gupta DK. Male genitoplasty for intersex disorders. Adv Urol. 2008;2008:685897.
5. Li Z, Liu J, Peng Y, Chen R, Ge P, Wang J. 46, XX Ovotesticular disorder of sex development (true hermaphroditism) with seminoma: a case report. Medicine (Baltimore). 2020;99:e22530.
6. Malik V, Gupta D, Gill M, Salvi AL. Seminoma in a male phenotype 46XX true hermaphrodite. Asian J Surg. 2007;30:85–7.
7. Kathrins M, Kolon TF. Malignancy in disorders of sex development. Transl Androl Urol. 2016;5:794–8.
8. Pleskacova J, Hersmus R, Oosterhuis JW, et al. Tumor risk in disorders of sex development. Sex Dev. 2010;4:259–69.
9. Pyle LC, Nathanson KL. A practical guide for evaluating gonadal germ cell tumor predisposition in differences of sex development. Am J Med Genet C Semin Med Genet. 2017;175:304–14.

Gender Dysphoria: Evolution of Terminologies and Psychological Aspects

66

T. S. Sathyanarayana Rao, Avinash De Sousa, Pragya Lodha, and Smita N. Deshpande

Abstract

Gender Dysphoria (GD) is a common condition seen in clinical practice. The current chapter provides the clinician with an overview of the disorders and looks at various facets of the causation, evaluation, and management of GD. Various terminology related to GD and related terms are defined. The epidemiology and clinical presentation of the disorder are discussed. The chapter provides a glance at various causative theories that have been implicated in the causation of GD. A team-based approach to the management of GD is discussed and the role of the psychiatrist and psychologist are defined. Difficulties in the management are also discussed. The chapter provides good insight for clinicians who deal with GD in their psychiatric clinics on a day-to-day basis.

Keywords

Gender · Gender identity · Gender dysphoria · Gender identity disorder

T. S. S. Rao
Department of Psychiatry, JSS Medical College, Mysore, India

A. De Sousa (✉)
Psychiatrist and Founder Trustee, Desousa Foundation, Mumbai, India

P. Lodha
Psychologist, Private Practice, Mumbai, India

S. N. Deshpande
Department of Psychiatry, St. John's Medical College Hospital and St. John's National Academy of Health Sciences, Bengaluru, India

S. K. Ratan (ed.), *Children with Differences in Sex Development*,
https://doi.org/10.1007/978-981-97-1639-5_66

66.1 Introduction

Gender Dysphoria (GD) is a term coined for "*persistent discomfort with one's biologic sex or assigned gender, replaced the diagnosis of gender identity disorder*" in the Diagnostic and Statistical Manual of Mental Disorders in 2013 [1]. The term GD has received tremendous attention in recent years because of many cases coming to the fore and the problem also emerging in adolescents in the school setting. There is some reprieve with the term GD as it replaces the earlier terms Gender Identity Disorder (GID) which gave a mental disorder-like connotation to the condition and Transgenderism which also was not liked in some quarters. The current chapter aims to provide a clinical overview of the rubric of GD and shall highlight the various facets of diagnosis and treatment of the same. It must be noted that though scientific literature commonly sets forth the idea of gender dysphoria for the trans community, the experience is not limited to the trans. Gender dysphoria can be experienced by any individual who may be exploring and/or questioning his/her gender/sexual identity, and is especially met with gender policing by the (heteronormative) society.

66.2 Overview of Clinical Terminology

Fisk first introduced the term gender dysphoria to describe "*individuals who experience discomfort with their biological sex to form the wish for sex reassignment.*" [2] DSM-IVTR referred to the condition as Gender Identity Disorder and it was included under the section of Sexual Disorders [3]. In the DSM-5, GD is defined as "*an individual's affective/cognitive discontent with the assigned gender [usually at birth and referred to as natal gender]*" (p. 451) [1].

There are many other terms abundant in literature that has been used to describe these conditions and these include "gender variant," "gender non-conforming," "gender queer," "bigender," "agender," and "non-binary," along with the terms "transmale," "transfemale," "transsexual," and "transgender." Sometimes the word "third gender" is also used [4].

66.3 Certain Other Definitions [4]

Gender: denotes "the public (and usually legally recognized) lived role as boy or girl, man or woman." Biological factors combined with social and psychological factors contribute to gender development.

Assigned gender: refers to "a person's initial assignment as male or female at birth." It is based on the child's genitalia and other visible physical sex characteristics.

Gender-atypical: refers to "physical features or behaviors that are not typical of individuals of the same assigned gender in a given society."

Gender-nonconforming: refers to "behaviors that are not typical of individuals with the same assigned gender in a given society."

Gender reassignment: denotes "an official (and usually legal) change of gender."

Sexual identity: A broad term, referring to "an individual's perception of themselves in sexual matters."

Gender Role: it is referred to as "the role or behavior learned by a person as appropriate to their gender, determined by the prevailing cultural norms."

Sexual Orientation: it is referred to "a person's identity in relation to the gender or genders to which they are sexually attracted; the fact of being heterosexual, homosexual, etc."

Gender identity: is a category of social identity and refers to "an individual's identification as male, female or, occasionally, some category other than male or female." It is one's deeply held core sense of being male, female, some of both or neither, and does not always correspond to biological sex.

Gender dysphoria: as a general descriptive term refers to "an individual's discontent with the assigned gender." It is more specifically defined when used as a diagnosis.

Transgender: refers to "the broad spectrum of individuals who transiently or persistently identify with a gender different from their gender at birth." (Note: the term transgendered is not generally used.)

Transsexual: refers to "an individual who seeks, or has undergone, a social transition from male to female or female to male." In many, but not all, cases this also involves a physical transition through cross-sex hormone treatment and genital surgery (sex reassignment surgery).

Genderqueer: blurring the lines around gender identity and sexual orientation. Genderqueer individuals "typically embrace a fluidity of gender identity and sometimes sexual orientation."

Gender fluidity: having different gender identities at different times.

Agendered: "without gender," as "individuals identifying as having no gender identity."

Cisgender: describes "individuals whose gender identity or expression aligns with the sex assigned to them at birth."

Gender expansiveness: conveys a wider, more flexible range of gender identity and/or expression than typically associated with the binary gender system.

Gender expression: "the manner in which a person communicates about gender to others through external means such as clothing, appearance, or mannerisms." This communication may be conscious or subconscious and may or may not reflect their gender identity or sexual orientation.

66.4 Epidemiology

There are no large rigorous studies on the epidemiological data of GD in the general population. There was little mention of transgender and transsexualism in the earlier studies in the 1980s, and most epidemiological studies done on it have been post the 1990s. An early study done in Taiwan in 1989 reported a lifetime prevalence of 0.3–2 per 1000 population [5]. A recent review of 25 epidemiological studies done by authors in 2009 found the prevalence to be 1–3 per 10,000 of the population across European and US studies [6]. A recent meta-analysis of 21 studies in 2015 has reported that transexual adult males were 1 in 14,700 and 1 in 38,500 in (predominantly) adult females [7]. In India, there have been no epidemiological studies done so far but we would report from personal clinical experience that there has been a marked increase in adults seeking out gender change with medical and surgical treatments. In studies, it has been noted that the prevalence of male-to-female transsexuals has been slightly more than female-to-male transsexuals in adults [8]. One may ponder based on these findings as to why is GD more common in those who were biologically male than those who were biologically female [8].

66.5 Evolution of Terminology in Psychiatric Classification [9]

"Transsexualism" as a psychiatric diagnosis (for adolescents and adults) appeared for the first time in the DSM-III in 1980 while the corresponding diagnosis for children was "gender identity disorder of childhood." [9] In DSM-IV in 1994, "*transsexualism and gender identity disorder of childhood were merged into one diagnosis, gender identity disorder (GID), with distinct criteria sets for children and adolescents/adults were established*" [9] and this was a part of the sexual disorders section. In 2013, in DSM-5, GID was renamed GD, with a chapter of its own and removed from the section on sexual disorders.

66.6 Changes in DSM-5 As Far As GD Is Concerned

Transsexualism or GID was proposed and supposed to be a normal variant of cisgender identity. Researchers felt that adding it as a disorder in DSM would stigmatize the population and having gender identity was not inherently psychopathological in nature. It was debated whether it must be included in DSM-5 or not [10]. It was retained because a legal diagnosis would be needed for health insurance and reimbursement for sex reassignment surgery (SRS) and hormonal replacement therapies. GID is a multimodal disorder that has neural and endocrinal underpinnings and not just psychiatric symptoms, but evidence to say that psychopathology does not exist is probably incorrect and hence researchers decided to keep GD as a separate disorder category in DSM-5 [11]. For inclusion, a reconceptualization was where "identity" was not considered a sign of a mental disorder was considered. The problem was an "*incongruence between one's felt gender and assigned sex/gender (usually at birth) leading to distress and/or impairment that was the core feature of the diagnosis.*" [11] Thus, a change in name from GID to GD in order to better reflect this incongruence was proposed in DSM-5 [12]. The diagnostic criteria for adolescents and adults moved to a more detailed structure which was rather skewed so far in classification systems. A 6-month duration criterion was added for clinical reasons to prevent a fast diagnosis and so that no patient with GD gets inappropriately treated and as many times the symptoms of GD may be transitory. The inclusion of disorders of sexual development (DSD) and intersex conditions under GID has always been a debatable topic. Many patients with DSD may have GD and may want to undergo SRS. The cause of GD is dependent on the core type of DSD and the disorder may be expressed by these patients in a similar manner to other patients with GD but there may be a distinct causative mechanism that underlies the process. DSD and GD may be a separate variant of GD and DSD has been included as a specifier in DSM-5 [13, 14]. The specifier of sexual orientation is now not a part of DSM-5. While sexual orientation has an interface with GD, sexual orientation has been thought to derive from a plethora of developmental and environmental factors compared to GD that has distinct causal pathways. Sexual orientation-related GD may be a subtype of GD but is not a diagnostic specifier [15]. Post-transition has been added in DSM-5 as many patients post-transition by medical or surgical means may not meet the diagnostic criteria for GD but may need chronic hormonal therapy, further surgical intervention for gender confirmation, continuous psychotherapy, and medical management for any psychopathology that may ensue [16]. There is a current debate ongoing that GD may be replaced by the term gender incongruence in further editions of DSM as well as ICD-11.

66.7 Psychiatric/Psychological Assessment of a Case of GD

Psychiatric assessment of GD includes a detailed clinical history, assessment of symptoms as per the DSM-5 criteria, and detailed history of birth, developmental, and psychosocial factors leading to GD. Psychological assessment is done using the Rorschach Inkblot test, Minnesota Multiphasic Personality Inventory-2 [17] which shall help with its Masculinity-Feminity (Mf) scales. The Feminine Gender Identity Scale for Males [18] and the Masculine Gender Identity Scale for Females [19] are aimed to provide dimension-based assessment of some indicators of GD and transsexualism. Though this may be a debated concern with the current understanding of gender and sexual orientation, from a clinical perspective, it adds clinical insight, more than a diagnostic label.

The Gender Identity/Gender Dysphoria Questionnaire for Adolescents and Adults (GIDYQ-AA) [20] has been developed recently and has 27 items that cover pertinent aspects of gender identity and GD and also cover multiple indicators of gender identity and GD, within the subjective, social, somatic, and sociolegal domains [20]. The GIDYQ-AA has parallel male and female versions. Each item is rated on a 5-point response scale from 1 (never) to 5 (always) based on the time frame of the past 12 months. The total score is derived by summing scores on the completed items and dividing by the number of marked responses [20]. Scales for body dissatisfaction the Body Image Scale [21], Body Uneasiness Test [22], and the Hamburg Body Drawing Scale [23] have been used to evaluate body image dissatisfaction with regard to sexually specific body features and general aspects of body image.

A thorough physical examination, external genital examination and karyotyping along with blood tests for routine testing and hormonal batteries may be done in patients with GD. A multidisciplinary team evaluation that consists of a psychiatrist, neurologist, endocrinologist, gynecologist, urologist, psychologist, general physician, and plastic surgeon shall all give their opinions before any take on SRS. The diagnosis of GD may be made by the opinion of the psychiatrist and endocrinologist in tandem [24].

66.8 Associated Comorbidity in GD

Mood disorders, major depression, and anxiety disorders have been reported in 20–30% cases of GD. The onset of these disorders may be before or after or with the onset of GD and need to be assessed and treated simultaneously [25]. Some studies have posited that depression may be present in 45–50% of patients with GD and panic disorder and generalized anxiety disorder may be seen in 20–25% of cases [26]. There are anecdotal case reports of GD co-occurring with schizophrenia and psychotic disorders as well as eating disorders and obesity, Hormonal factors may be at play here in these cases [27]. Suicidal thoughts have been reported in 10–20% of patients with GD with suicidal attempts being noted in 2–5% of cases. Many cases of GD may co-exist with borderline personality disorder and also demonstrate non-suicidal self-injurious behavior [28, 29]

However, evolving research literature cautions clinicians from a hasty Axis II diagnosis in patients who experience gender dysphoria and the larger LGBTQ+ community while inviting them to question their own bias. Researchers delineate that various traits of Axis II spectrum, especially, Cluster B (narcissistic, borderline, anti-social, and histrionic) tend to naturally predispose within the LGBTQ+ community as a result of the developmental trajectory, individual preferences, and a strong “bold” presentation that the individuals consciously choose for themselves to mark their identity. It is also important to use the personality disorder diagnostic framework with caution given the changing lifestyle choices observed in the millennial and Gen Z populations [30].

66.9 Some Insights into Possible Mechanisms Involved in the Developmental Trajectory of GD

The current understanding about gender and sexual orientation, away from the medical model of pathologizing the same, is that gender is a personal preference and that homosexuality is not an illness, just as heteronormativity is not the only norm. However, the existing scientific literature and understanding cannot be ignored that explain the possible factors to explain gender dysphoria. It is important to note that none of the studies are conclusive and merely present the possibility of explaining underlying factors in gender dysphoria that are *not* causative and merely correlational and predisposing. Some of them are discussed below –.

66.9.1 Genetic Factors in GD

Some genetic studies have examined whether genetics contributes to the development of gender identity and GD. There have been reports in patients with GD that 25–47% of the variance was explained by genetic factors. Heritability rates have been reported to be 0.5–0.57 in men and 0.37–0.4 in women [31]. Genes involved in sex steroid biosynthesis have also been investigated in the genetics of GD as sex steroids contribute to the sexual differentiation of the brain while in utero and early in life [32]. No single gene has been identified for GD but single nucleotide polymorphisms of the androgen and estrogen receptors have been studied [33]. The CYP19 gene concerned with the aromatization of estrogen to androgens and the CYP17 gene involved in estrogen biosynthesis has been studied in this regard. We conclude that GD is a complex condition resulting from a combination of multiple gene–environment interactions [34].

66.9.2 Sex Hormone Levels in the Pre-natal Phase

There are biological theories that explain the possibility of the role of sex hormones that play a vital role in the differentiation of the baby in utero into male and female based on the sex chromosomes. This happens at around 7 months in the uterus while the baby is developing. The early effects of sex hormones on the brain play a role in the development of male and female phenotypes. It has been proposed that gender-incongruent feelings may be due to an atypical sexual development of the brain mediated by the pre-natal hormonal peaks [35].

66.9.3 Neurobiological Underpinnings of GD

It has always been a matter of debate whether the brains of individuals with GD show atypical sexual differentiation. There have been earlier reports of female-typical hypothalamic nuclei in male-to-female transsexual individuals [36]. Many neurobiological studies have tried to elucidate whether the brains of patients with GD would show propensity to resemble the brains of their preferred gender rather than their actual sex [37]. Gray matter volumes have also been studied in patients with GD. Differences in gray matter thickness between males and females have been noted in the orbitofrontal regions, insular cortex, and medial frontal regions of the brain [38].

The putamen, a nucleus that is a part of the basal ganglia is associated with motor regulation and is the only subcortical structure with proven differences between male and female patients though the findings are inconclusive [39]. Studies of white matter imaging in GD have been done using diffusion tensor imaging (DTI). DTI studies have demonstrated a reduced inter-hemispheric connectivity ratio in subcortical/limbic regions was found in men with GD and there is evidence of increased inter-hemispheric lobar connectivity [40]. Functional connectivity studies have shown interactions between regions during a task (task-related functional connectivity) or while not engaged in a task (resting-state connectivity) [41]. Studies on functional connectivity have implicated the role of the ventral tegmental area in dimorphic genital representation and the role of the anterior cingulate cortex in conflict processing in patients with GD [42]. Most neuroimaging studies posit findings in patients with GD either "*as a consequence of atypical sexual differentiation or a mismatch between their anatomical sex characteristics and their gender identity.*" [43] Brain changes may also be triggered by psychological distress in many patients and need not be present from an early age [42].

66.9.4 Psychosocial and Cultural Factors in GD

Various factors contribute to the persistent experience of gender dysphoria. In India, it may be viewed as stigmatic to have such thoughts, particularly in sections of society where there are factions that are homophobic, heterosexist, and patriarchal. The expression of normal sexuality is itself a taboo in many quarters, let alone the expression of GD and any form of alternate sexualities. One of the most important psychological issues in patients with GD is a divide between what the self, experiences, and the expectations of a society. This disparity leads to low self-esteem, guilt, and anxiety. They experience stigma because of labelling "gender policing" from family and society they live in and face trauma of being misgendered [44].

There is a lack of psychoeducation among family members in order to understand the status of being queer which deviates from the heteronormative norms of the society. This further means that there is a lack of sensitivity to the issues that trans individuals face and feel as individuals which further leads to an unwarranted divide among them and their family members. Overall, in the larger picture, this translates into a lack of acceptance from society by large [45]. Reluctance to visit a psychiatrist and the equating of GD to major mental illness is the reason for not seeking help and these patients linger for years after being tormented by family members and relatives. Marriage is a forced treatment for GD may be advised by elders in the family as a means of curing the problem and add to the woes of patients with GD who are coerced into marriage and also closeting the problem. Thus, many people live with GD throughout their lives and rarely seek help.

66.9.5 Psychological Issues in Psychiatric Care of Persons with GD [45]

One of the nuances that often go amiss is the difference between sexual orientation and gender identity. Though the two are interlinked they are different concepts in existence and understanding. Professionals must be clear and precise about making note of the same. When talking about sex reassignment surgery, it is important to know the gender identity of the individual and not the sexual orientation. Health care needs and social experiences vary across the lifespan. Experiences of young queer people continue to remain invisible and their concerns are often neglected as they are always discriminated from the larger society. This hampers the emotional, social, and cognitive development of the youth.

The importance of informed consent, confidentiality, and anonymity is equally important for everyone irrespective of age, gender, or any other socio-demographical feature. However, the larger clinical practice sees that though these norms are maintained for the heterosexual population, there is a compromise in following these ethical principles of practice when it comes to homosexuals or trans individuals as a consequence of the inherent bias that may be held against them. Another concern that holds against patients is that there is a greater stigma attached to approaching for help in a homophobic and transphobic society. Chances are less to find a homosexual doctor or therapist as opposed to a greater chance of finding heterosexual professionals (though they may and most often, they do lack specialized care skills and knowledge to treat trans patients).

Not all adolescents who have gender dysphoria wish to undergo sex reassignment surgery, however, it is a commonly misunderstood paradigm where the dissonance between an individual's gender and sex is always looked at reconstructing and fixing them to align. In fact, another problem can arise if an adolescent may wish to undergo the sex reassignment surgery where there may be two issues- one, either there is a confusion whether the teenager is sure enough for the surgery and second is the disagreement about the surgery between the adolescent and his or her parents.

Some parents may seek psychotherapy or counseling with the misconception that it can alter or "correct" their children's sexuality. However, these beliefs are unfounded, and any practitioner adhering to such practices may be questioned on ethical grounds in health care. Unlike these misconceptions, sex reassignment therapies are intended for individuals seeking biological alignment with their psychological gender, often the opposite of their biological sex. The assessment preceding sex reassignment surgeries and hormone treatments depends on the discretion of practitioners, some of whom still diagnose Gender Identity Disorder (revised to Gender Dysphoria in DSM 5) to address distress, depression, or anxiety resulting from gender non-conformity, without attempting to dissuade personal gender identification choices.

Pre-counseling before sex reassignment surgery may overlook underlying emotional factors and, at times, discourage individuals from undergoing the surgery to conform to their biological sex. Ethically, individuals are subjected to societal norms and lifestyle adjustments before surgery to prepare for post-surgery life.

Post-surgery and hormone treatment care, along with counseling, is often overlooked, and crucial for aiding patients in adjusting to transformations. Electroconvulsive Therapy (ECT), patient restraint, and institutionalization are sometimes employed to convince individuals of the "consequences of sex change," creating an ethical dilemma. ECT is mistakenly used by some practitioners to prevent sex reassignment surgery, but it is crucial to recognize that this approach is baseless.

Due to stressors associated with gender dysphoria, an increasing number of individuals seek therapy, yet many therapists lack the necessary skills and sensitivity, remaining unaware of transgender issues. Mental health professionals and medical practitioners often lack understanding of LGBTQ spectrum health issues, leading to a deficiency in ethical considerations in treatment.

The unique needs of families with LGBTQ members, particularly those with gender dysphoria, are frequently overlooked, with limited access to family and couple's therapy. Mental health service professionals often lack training in queer and gender dysphoria issues, hindering optimal treatment. Personal biases among psychiatrists and psychotherapists may lead to the denial of necessary mental health services for queer individuals. Specialized training in counseling and psychotherapies is imperative to appropriately address the amplified mental health concerns associated with gender dysphoria.

66.9.6 Management of Patients with GD

Some very fundamental steps that can be taken up at an individual and societal level in order to ensure a just status for these patients are that GD must be understood correctly that it is the underlying depression or distress due to gender confusion or social ostracization and not a pathological identity/personality. Psychoeducation to one and all about the existing queer identities and their challenges should be addressed in order to ensure that the community acts in a sensitive manner toward them. Health care settings must offer a safer environment for queer persons to bring up mental health concerns and should make access to mental health services easy for them. Every person taken as intake for primary health care concern should include a mental health history and an assessment for active mental health concerns. Screening should include primary mental health problems, environmental and social stressors, and gender-related needs.

When there is a queer patient that a mental health professional is unable to handle, one must appropriately refer to transgender-affirming mental health services where mental health professionals specialize in the field of LGBTQ+ mental health and thus will be more equipped to treat them. Professionals should take responsible effort of understanding the community and their specific challenges in health care. There should be specialization courses run for treating individuals from the trans and LGBTQ+ community in order to better address their GD concerns and handle their specific mental health concerns with due sensitivity and responsibility. School curriculums can also include discussing the growing minority population and their challenges.

66.9.7 Role of the Psychiatrist: (Table 66.1)

The psychiatrist is part of a multidisciplinary team that manages patients of GD. He/She will serve as one of the clinicians who diagnoses GD and also helps the patient going in for hormonal therapy and SRS if needed. The monitoring of the patient at

Table 66.1 Role of the psychiatrist in the management of GD

Role of the psychiatrist in the management of GD
• Role in the diagnosis and assessment of GD.
• Role in the long-term management of GD.
• Role in the management of co-existing psychiatric disorders.
• Role in the management of sexual problems that may co-exist.
• Role in the SRS team for long-term management.
• Management of suicidality or self-injury if any.
• Psychoeducation of family members.

Table 66.2 Role of the psychologist in the management of GD

Role of the psychologist in the management of GD
• Planning out the long-term psychotherapy plan for the individual.
• Role in psychological assessment and fitness for SRS.
• Role in the psychoeducation of family members and family therapy.
• Group sessions for GD patients and the running of support groups.
• Monitoring psychiatric treatment and promoting adherence and compliance.
• Building self-esteem and overall development of the person.
• Sex education and re-education if needed.

all stages is done by the psychiatrist. The psychiatrist plays the liaison person between the endocrinologist, plastic surgeons, and other members of the treating team. The ruling out of comorbid psychiatric conditions and their treatment if present like the management of depression and anxiety is done by the psychiatrist. The psychiatrist also plays a role in ruling out psychotic disorder and ascertaining whether the patient may be fit to provide consent for all medical and surgical treatments in GD [42].

66.9.8 Role of Psychotherapy in GD: (Table 66.2)

The role of psychotherapy in GD is to help adults with GD find solace and comfort with their actual birth sex and assigned gender and is often undertaken at the behest of the patient if needed especially when doubtful about GD. Psychotherapy in literature has been shown to help cases of GD by reducing GD symptoms, and greater synchrony between gender identity and expression and assigned sex when they qualify for a diagnosis of GD [42]. Psychotherapy also aims at "long-term comfort in gender identity expression with realistic chances for success in relationships, education, and work." The patient is made to understand the "need to maintain a job, provide for the emotional needs of children, honor a spousal commitment, or not to distress a family member as currently having a higher priority than the personal wish for constant cross-gender expression." [42] Psychotherapy plays a vital role in the rehabilitation process and all the psychosocial and personal adjustments post SRS. A detailed evaluation of psychotherapy and its important facets in GD are beyond the scope of the current chapter [43].

66.9.9 Living in the Role of the Desired Gender

Earlier it was thought mandatory that one must not allow SRS without a year of full-time Real-Life Experience (RLE) in the desired gender, even when the client does not desire it. It was also posited that the patient need not undergo SRS if they are happy with changes due to hormone therapy and RLE alone. The carrying out of RLE has huge social ramifications like loss of a job, disturbed relationships, gender-based discrimination, emotional, and physical abuse along with psychological disturbances that would ensue. Many patients with GD can anyways suffice with living RLE partially, i.e., in public they may live in one gender and have a separate private life in the desired gender. The issue of mandatory RLE before SRS in patients is debatable and is believed to impinge on the autonomy and choice of the patient [44, 45].

66.9.10 Hormonal Treatments for GD

Hormonal treatments have been regarded as amongst the safest treatment options for patients with GD. Hormonal therapy for adult males with GD includes testosterone suppression using spironolactone, cyproterone acetate, or GnRH agonists. GnRH agonists are regarded worldwide as a means of testosterone suppression in adult males with GD [46].

66.9.11 Sex Reassignment Surgery in GD

There have been multiple surgical procedures that have been used in SRS for GD but studies of long-term outcomes suggest mixed results on satisfaction. The procedures carried out are different for males and females. Feminization surgeries are "surgeries that result in anatomy that is typically gendered female and include vaginoplasty, feminizing augmentation mammoplasty, orchiectomy, facial feminization surgery, reduction thyrochondroplasty (tracheal shave), and voice feminization surgery among others." [47] Masculinization surgeries are "surgeries that result in anatomy that is typically gendered male. These surgeries include chest masculinization surgery (top surgery), metoidioplasty, phalloplasty, scrotoplasty, and hysterectomy." [47]

66.10 Conclusions

In this chapter, we have provided an overview of GD in adults, including terminology, symptoms, epidemiology, diagnosis, assessment, psychopathology, causative factors, and management. As GD gets increasing recognition, there is a change with regard to acceptance and support of people with GD, including better access to health care. This chapter is aimed at providing the general adult psychiatrist with a greater understanding of the clinical issues that will help in his/her clinical work in patients with GD.

References

1. American Psychiatric Association. Diagnostic and statistical manual for the classification of psychiatric disorders—5th edition (DSM-5). New York: American Psychiatric Publishing; 2013.
2. Fisk NM. Gender dysphoria syndrome--the conceptualization that liberalizes indications for total gender reorientation and implies a broadly based multi-dimensional rehabilitative regimen. West J Med. 1974;120(5):386–91.
3. American Psychiatric Association. Diagnostic and statistical manual for the classification of psychiatric disorders—4th edition text revised (DSM-IVTR). New York: American Psychiatric Publishing; 2000.
4. Pauly IB. Terminology and classification of gender identity disorders. J Psychol Hum Sexuality. 1993;5(4):1–4.
5. Hwu HG, Yeh EK, Chang LY. Prevalence of psychiatric disorders in Taiwan defined by the Chinese diagnostic interview schedule. Acta Psychiatr Scand. 1989;79(2):136–47.
6. Zucker KJ, Lawrence AA. Epidemiology of gender identity disorder: recommendations for the standards of care of the world professional association for transgender health. Int J Transgender. 2009;11(1):8–18.
7. Arcelus J, Bouman WP, Van Den Noortgate W, Claes L, Witcomb G, Fernandez-Aranda F. Systematic review and meta-analysis of prevalence studies in transsexualism. Eur Psychiatry. 2015;30(6):807–15.
8. Judge C, O'Donovan C, Callaghan G, Gaoatswe G, O'Shea D. Gender dysphoria: prevalence and co-morbidities in an Irish adult population. Front Endocrinol. 2014;5:87.
9. Beek T, Cohen-Kettenis PT, Kreukels BP. Gender incongruence/gender dysphoria and its classification history. Int Rev Psychiatry. 2016;28(1):5–12.
10. Vance SR Jr, Cohen-Kettenis PT, Drescher J, Meyer-Bahlburg HF, Pfäfflin F, Zucker KJ. Opinions about the DSM gender identity disorder diagnosis: results from an international survey administered to organizations concerned with the welfare of transgender people. Int J Transgender. 2010;12(1):1–4.
11. Lev AI. Gender dysphoria: two steps forward, one step back. Clin Soc Work J. 2013;41(3):288–96.
12. Zucker KJ, Cohen-Kettenis PT, Drescher J, Meyer-Bahlburg HF, Pfäfflin F, Womack WM. Memo outlining evidence for change for gender identity disorder in the DSM-5. Arch Sex Behav. 2013;42(5):901–14.
13. Kraus C. Classifying intersex in DSM-5: critical reflections on gender dysphoria. Arch Sex Behav. 2015;44(5):1147–63.
14. Kreukels BP, Köhler B, Nordenström A, Roehle R, Thyen U, Bouvattier C, De Vries AL, Cohen-Kettenis PT, Cohen-Kettenis P, de Vries A, Arlt W. Gender dysphoria and gender change in disorders of sex development/intersex conditions: results from the DSD-LIFE study. J Sexual Med. 2018;15(5):777–85.
15. Nuttbrock L, Bockting W, Mason M, Hwahng S, Rosenblum A, Macri M, Becker J. A further assessment of Blanchard's typology of homosexual versus non-homosexual or autogynephilic gender dysphoria. Arch Sex Behav. 2011;40(2):247–57.
16. Davy Z, Toze M. What is gender dysphoria? A critical systematic narrative review. Transgend Health. 2018;3(1):159–69.
17. Graham JR. MMPI-2: assessing personality and psychopathology. Oxford University Press; 1990.
18. Heilbrun AB. Measurement of masculine and feminine sex role identities as independent dimensions. J Consult Clin Psychol. 1976;44(2):183–9.
19. Blanchard R, Freund K. Measuring masculine gender identity in females. J Consult Clin Psychol. 1983;51(2):205–9.
20. Deogracias JJ, Johnson LL, Meyer-Bahlburg HF, Kessler SJ, Schober JM, Zucker KJ. The gender identity/gender dysphoria questionnaire for adolescents and adults. J Sex Res. 2007;44(4):370–9.

21. Lindgren TW, Pauly IB. A body image scale for evaluating transsexuals. Arch Sex Behav. 1975;4(6):639–56.
22. Fisher AD, Castellini G, Bandini E, Casale H, Fanni E, Benni L, Ferruccio N, Meriggiola MC, Manieri C, Gualerzi A, Jannini E. Cross-sex hormonal treatment and body uneasiness in individuals with gender dysphoria. J Sexual Med. 2014;11(3):709–19.
23. Becker I, Nieder TO, Cerwenka S, Briken P, Kreukels BP, Cohen-Kettenis PT, Cuypere G, Haraldsen IR, Richter-Appelt H. Body image in young gender dysphoric adults: a European multi-center study. Arch Sex Behav. 2016;45(3):559–74.
24. Cousino MK, Davis A, Ng H, Stancin T. An emerging opportunity for pediatric psychologists: our role in a multidisciplinary clinic for youth with gender dysphoria. Clin Pract Pediatr Psychol. 2014;2(4):400–6.
25. Megeri D, Khoosal D. Anxiety and depression in males experiencing gender dysphoria. Sex Relat Ther. 2007;22(1):77–81.
26. Bergero-Miguel T, García-Encinas MA, Villena-Jimena A, Pérez-Costillas L, Sánchez-Álvarez N, de Diego-Otero Y, Guzman-Parra J. Gender dysphoria and social anxiety: an exploratory study in Spain. J Sex Med. 2016;13(8):1270–8.
27. Feder S, Isserlin L, Seale E, Hammond N, Norris ML. Exploring the association between eating disorders and gender dysphoria in youth. Eating Disord. 2017;25(4):310–7.
28. Marshall E, Claes L, Bouman WP, Witcomb GL, Arcelus J. Non-suicidal self-injury and suicidality in trans people: a systematic review of the literature. Int Rev Psychiatry. 2016;28(1):58–69.
29. García-Vega E, Camero A, Fernández M, Villaverde A. Suicidal ideation and suicide attempts in persons with gender dysphoria. Psicothema. 2018;30(3):283–8.
30. Rodriguez-Seijas C, Morgan TA, Zimmerman M. Is there a bias in the diagnosis of borderline personality disorder among lesbian, gay, and bisexual patients? Assessment. 2021;28(3):724–38.
31. Sasaki S, Ozaki K, Yamagata S, Takahashi Y, Shikishima C, Kornacki T, Nonaka K, Ando J. Genetic and environmental influences on traits of gender identity disorder: A study of Japanese twins across developmental stages. Arch Sex Behav. 2016;45(7):1681–95.
32. Henningsson S, Westberg L, Nilsson S, Lundström B, Ekselius L, Bodlund O, Lindström E, Hellstrand M, Rosmond R, Eriksson E, Landén M. Sex steroid-related genes and male-to-female transsexualism. Psychoneuroendocrinology. 2005;30(7):657–64.
33. Hare L, Bernard P, Sánchez FJ, Baird PN, Vilain E, Kennedy T, Harley VR. Androgen receptor repeat length polymorphism associated with male-to-female transsexualism. Biol Psychiatry. 2009;65(1):93–6.
34. Ujike H, Otani K, Nakatsuka M, Ishii K, Sasaki A, Oishi T, Sato T, Okahisa Y, Matsumoto Y, Namba Y, Kimata Y. Association study of gender identity disorder and sex hormone-related genes. Prog Neuro-Psychopharmacol Biol Psychiatry. 2009;33(7):1241–4.
35. Berenbaum SA, Beltz AM. Sexual differentiation of human behavior: effects of prenatal and pubertal organizational hormones. Front Neuroendocrinol. 2011;32(2):183–200.
36. Garcia-Falgueras A, Swaab DF. Sexual differentiation of the human brain in relation to gender identity and sexual orientation. Funct Neurol. 2009;24(1):17–28.
37. Savic I, Garcia-Falgueras A, Swaab DF. Sexual differentiation of the human brain in relation to gender identity and sexual orientation. Progr Brain Res. 2010;186:41–62.
38. Luders E, Sánchez FJ, Tosun D, Shattuck DW, Gaser C, Vilain E, Toga AW. Increased cortical thickness in male-to-female transsexualism. J Behav Brain Sci. 2012;2(3):357–62.
39. Manzouri A, Savic I. Possible neurobiological underpinnings of homosexuality and gender dysphoria. Cerebr Cortex. 2019;29(5):2084–101.
40. Kranz GS, Hahn A, Kaufmann U, Küblböck M, Hummer A, Ganger S, Seiger R, Winkler D, Swaab DF, Windischberger C, Kasper S. White matter microstructure in transsexuals and controls investigated by diffusion tensor imaging. J Neurosci. 2014;34(46):15466–75.
41. Kreukels BP, Guillamon A. Neuroimaging studies in people with gender incongruence. Int Rev Psychiatry. 2016;28(1):120–8.
42. Zucker KJ, Lawrence AA, Kreukels BP. Gender dysphoria in adults. Ann Rev Clin Psychol. 2016;12:217–47.

43. Coleman EJ, Bockting WO. Gender dysphoria: interdisciplinary approaches in clinical management. Routledge; 2019.
44. Venkataramu VN, Banerjee D. Gender dysphoria in psychiatric practice: understanding the clinical ambiguity and management. J Psychosexual Health. 2021;3(2):124–32.
45. Chavan K, Lodha P. Psychological issues in transgenders. Psychological issues in obstretics and gynaecology. India: CBS Publishers; 2022. p. 100–4.
46. Wylie K, Barrett J, Besser M, Bouman WP, Bridgman M, Clayton A, Green R, Hamilton M, Hines M, Ivbijaro G, Khoosal D. Good practice guidelines for the assessment and treatment of adults with gender dysphoria. Sex Relat Ther. 2014;29(2):154–214.
47. Barker H, Wylie K. Are the criteria for the 'real-life experience'(RLE) stage of assessment for GID useful to patients and clinicians? Int J Transgender. 2008;10(3–4):121–31.

Sex Chromosome Aneuploidies: DSD Perspective

67

Priyanka Gupta and Rajni Sharma

Abstract

Sex chromosome aneuploidies are a group of chromosomal disorders that are characterised by loss or gain of one or more sex chromosomes. These disorders are usually underdiagnosed despite them being more prevalent in the general population. Turner syndrome, Klinefelter syndrome, Trisomy X, Trisomy XYY and Tetrasomy XXYY are the five major sex chromosomal aneuploidies. Of these, Klinefelter syndrome is the most common with an incidence of 1 in 600–750 live births. It is typically characterised by tall stature, gynaecomastia, small firm testes, hypergonadotrophic hypogonadism and azoospermia but the presentation may vary according to the age of presentation. Multidisciplinary team approach is required for optimal care and better outcomes.

Keywords

Hypergonadotrophic hypogonadism · Trisomy XYY · Klinefelter syndrome Turner syndrome

P. Gupta · R. Sharma (✉)
Division of Pediatric Endocrinology, Department of Pediatrics, All India Institute of Medical Sciences, New Delhi, India

S. K. Ratan (ed.), *Children with Differences in Sex Development*,
https://doi.org/10.1007/978-981-97-1639-5_67

67.1 Introduction

Sex chromosome aneuploidies are characterised by abnormal number of sex chromosomes, either due to loss or gain of one or more sex chromosomes during meiosis or mitosis. There are five major sex chromosomal aneuploidies (Turner syndrome (45,XO), Klinefelter syndrome (47,XXY), Trisomy X, Trisomy XYY and Tetrasomy XXYY) as mentioned in Table 67.1 [1–5]. Their prevalence in general population is much higher than is clinically recognised. The affected individuals have a highly variable clinical appearance that differs from the typical phenotypic description given in textbooks leading to delayed diagnosis. Most of these disorders are associated with learning disabilities, physical or mental health problems and infertility adds to morbidity and possibly mortality. Clinicians should be well aware about the typical and atypical phenotypic appearance of these disorders so as to make an early diagnosis and initiate timely therapy for treatable complications [4]. Of all the sex chromosomal aneuploidies, Klinefelter syndrome (KS) is the most common sex chromosomal aneuploidy, hence will be discussed in detail in this section.

Table 67.1 Incidence, genetics, clinical features and treatment of various sex chromosomal aneuploidies

	Turner syndrome (1)	Klinefelter syndrome (2)	Trisomy X (3)	Trisomy XYY (4)	Tetrasomy (48, XXYY) (5)
Mechanism	Partial or complete loss of paternal (80%) or maternal (10–15%) single-sex chromosome (either X or Y)	Paternal/maternal non-disjunction during meiosis I/ II, or post-zygotic non-disjunction	Maternal (90%) or paternal (10%) non-disjunction during meiosis I > II or post-zygotic non-disjunction (10%)	Non-disjunction during paternal meiosis II or post-zygotic non-disjunction	Paternal double non-disjunction during meiosis
Incidence	1:2500 female live births	1:600–750 live births	1:1000 live births	1:1000 male live births	1:20,000 live births
Clinical presentation	Short stature, webbed neck, cubitus valgus, shield chest	Cryptorchidism, tall stature, hypotonia, hypertelorism, clinodactyly, pes planus, elbow dysplasia and a high-arched palate	Tall stature, epicanthal folds, pes planus, hypotonia, clinodactyly	Tall stature, macrocephaly, hypertelorism, hypotonia, clinodactyly, pes planus, dental abnormalities	Tall stature, hypertelorism, epicanthal folds, clinodactyly, pes planus, joint hyperextensibility, hypotonia, radioulnar synostosis
Other complications	Congenital heart disease (CoA, bicuspid aortic valves); genitourinary problems (horseshoe kidney); metabolic complications; autoimmune diseases including thyroiditis and celiac disease	Cardiovascular, cerebrovascular, metabolic complications and early osteoporosis	Seizures, genitourinary malformations, congenital hip dysplasia, constipation, congenital heart disease (VSD, PS, CoA)	Asthma, seizures, tremors; genitourinary abnormalities (microphallus, cryptorchidism, hypospadias)	Reactive airway disease, tremors, seizures, deep vein thrombosis, peripheral vascular disease, insulin resistance, non-Hodgkin lymphoma
Presentation with respect to DSD	Primary or secondary amenorrhoea, infertility, streak ovaries	Pubertal delay, gynaecomastia, infertility	Premature ovarian failure, rarely precocious puberty, fertility is usually normal	Decreased libido, atrophic testicles, infertility due to oligospermia or sperm chromosomal abnormalities	Microorchidism, impaired fertility, hypergonadotropic hypogonadism, gynaecomastia

(continued)

Table 67.1 (continued)

	Turner syndrome (1)	Klinefelter syndrome (2)	Trisomy X (3)	Trisomy XYY (4)	Tetrasomy (48, XXYY) (5)
Psychiatric complications	Anxiety, depression, ADHD, autistic spectrum disorder	ADHD, anxiety, depression, autistic spectrum disorder, psychotic disorders, schizophrenia	ADHD (25–35%); anxiety, depression, adjustment problems	ADHD, autistic spectrum disorder, impulse control disorders, aggressive behaviour and criminal activities	Mood disorders, anxiety, depression, autistic spectrum disorders
Cognitive phenotype	Normal verbal cognition; deficit in non-verbal IQ (visuospatial/ arithmetic abilities)	Low verbal cognition (language-based learning abilities); relatively preserved non-verbal skills	Developmental delay; IQ ranges from 55–111; verbal IQ is more affected than non-verbal IQ	Subnormal IQ, learning disabilities and speech difficulties	Decreased IQ; developmental delay (mostly language-based learning abilities)
MRI brain	Decreased grey matter volume in parieto-occipital and frontal cortex, increased volume in temporal region	High grey matter volume in parieto-occipital and sensory cortex, lower volume in temporal regions	Decreased whole brain volumes and size of amygdala; white matter high-intensity foci (HIS) may be seen in 25–30% of patients	Normal to increased white and grey matter volume in general or a pattern similar to seen in Klinefelter syndrome	Decreased brain volume, areas of white matter hyperintensity and enlarged ventricles
Treatment options	Periodic screening for the comorbidities; growth hormone therapy, hormone (combined oestrogen and progesterone) replacement therapy	Periodic screening for the comorbidities; speech therapy, physical and occupational therapy; testosterone replacement therapy in adulthood, sperm cryopreservation	Periodic screening for the comorbidities, physical and occupational therapy	Treat associated comorbidities, speech and behavioural therapy. Correct hormonal imbalance if present. For infertility- in vitro fertilisation or intracytoplasmic sperm injection	Speech and behavioural therapy, physical and occupational therapy, hormonal replacement therapy if required

CoA Coarctation of aorta; *VSD* Ventricular septal defect; *PS* Pulmonary stenosis; *ADHD* Attention deficit hyperactivity disorder; *IQ* Intelligent quotient

67.2 Klinefelter Syndrome (KS)

KS has an incidence of 1 in 600 males [6] and is characterised by the presence of an extra X chromosome in male patients. KS was first described in 1942 as a clinical phenotype which is characterised by tall stature, gynaecomastia, small firm testes, hypogonadism and azoospermia [7]. Since then, multiple researchers have elucidated the multisystem involvement in this syndrome. Moreover, implications of additional X chromosome and its impact on health, neurodevelopmental performance and human well-being are being understood. A multidisciplinary team approach is helpful in improving the longevity and quality of life of patients with KS.

67.3 Aetiology of KS

Around 80–90% of individuals with KS have XXY karyotype; the remaining 10–20% of cases may have mosaicism (46, XY/47, XXY), higher grades of aneuploidies (48,XXXY, 48,XXYY, 49,XXXXY) or structurally abnormal X chromosome (47,iXq,Y). The extra X chromosome may originate from paternal non-disjunction during meiosis I (50%), maternal non-disjunction during meiosis I (25%) or meiosis II (15%), or maybe a result of post-zygotic non-disjunction [8]. Parental age has been found to be the most important contributing factor to the non-disjunction; maternal age more than 35 years is being associated with a higher risk of errors in meiosis I. Recently, association with increasing paternal age has also been found.

Various factors including skewed X-inactivation, genetic polymorphism, gene dosage, and parental origin of extra X chromosome influence the phenotypic heterogeneity of physical and neurocognitive features in patients with KS. The presence of an extra X chromosome leads to escape of certain genes from inactivation. Of these, SHOX gene situated in the pseudoautosomal region 1 of X chromosome is the most important, the addition of which leads to tall stature, long limbs and reduced upper segment to lower segment ratio. The androgen receptor (AR) gene is another important gene influencing the genotype/phenotype correlation in patients with KS. It is located in Xq11.2-3q12 and contains highly polymorphic CAG trinucleotide repeat which influences the androgen effects. The length of CAG nucleotide repeats has been found to significantly affect phenotype of patients with KS via modulation of androgenic actions [9].

67.4 Clinical Features

The clinical presentation of patients with KS varies according to the age of presentation. The severity of presentation may vary according to the severity of sex chromosomal aneuploidy and degree of gonadal dysfunction. There are neither distinct dysmorphic features nor a classical single phenotype that can describe patients of KS. Testicular degeneration is the most consistent feature and may start as early as during the foetal life. Low testicular volume has been found to be the most sensitive feature distinguishing patients with KS from those without KS.

Majority of newborns remain asymptomatic and have a normal male phenotype. Few patients may have features of hypogonadism including cryptorchidism, reduced penile length, low testicular volume, hypotonia, hypertelorism, clinodactyly, pes planus, elbow dysplasia and a high arched palate; the actual prevalence of which is not known.

Prepubertal boys mostly come to attention during the workup of cryptorchidism, developmental delay or tall stature. Developmental delay is characterised by predominantly language-based learning difficulties, discrepancy between verbal and non-verbal cognitive abilities (visuoperceptual and non-verbal skills are better) and decreased fine motor skills. Although androgens play an important role in neurodevelopment; but androgen replacement has not been found to be beneficial in improving the cognitive abilities of children with KS.

During adolescence and after puberty, this syndrome is characterised by small firm testes, gynaecomastia, eunuchoid body habitus and features of varying degrees of androgen deficiency. Depending upon the severity of hypogonadism and the level of testosterone production, patients may have normal onset of puberty with external genitalia appearing sexual maturity stage-appropriate (Fig. 67.1); delayed or absent

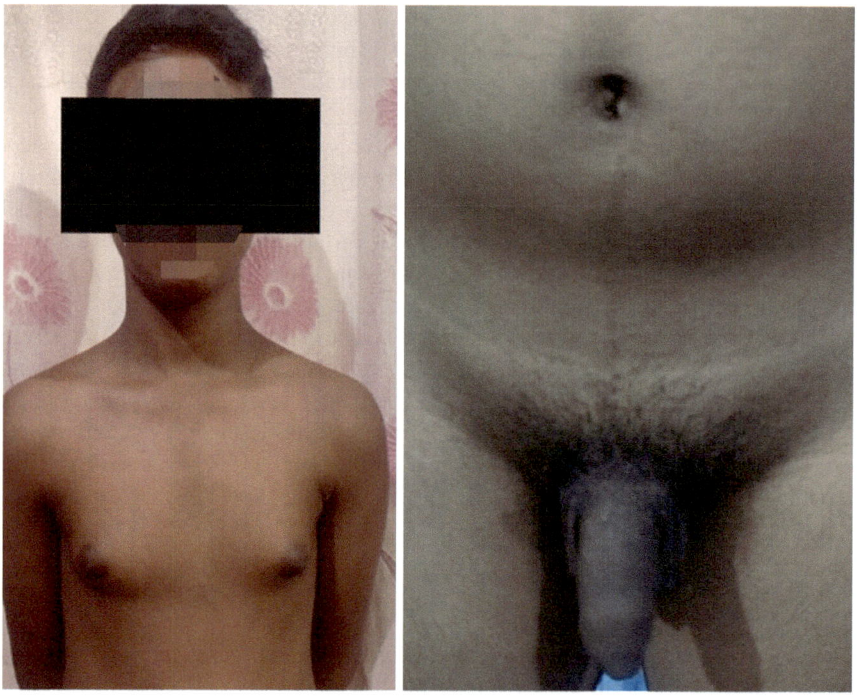

Fig. 67.1 A 17-year-old boy with gynaecomastia and delayed secondary sexual characteristics. On examination, he had absent facial hair, Sexual maturity rating 3–4 with small firm testes with volume of 4 ml. Investigations showed serum LH: 6.7 mIU/L, FSH: 44 mIU/L, testosterone: 1.99 ng/ml, oestradiol: 48 pg/ml and Karyotype 47, XXY

puberty with minimal or no signs of pubertal development; or arrested puberty with reduced level of testosterone leading to regression of secondary sexual characters. During adulthood, they mostly come to attention during evaluation of decreased libido, potency or primary infertility [10].

67.5 Comorbidities

Patients with KS are at increased risk of various comorbidities including cardiovascular, cerebrovascular, metabolic syndrome and early osteoporosis. The increased cardiovascular mortality is attributed to the presence of various structural cardiac abnormalities including left ventricular diastolic dysfunction, mitral valve prolapse, increased thickness of inner tunica of carotid arteries; aortic valve anomalies, peripheral vascular disease and deep vein thrombosis. Few studies have also found impaired cardiorhythmic integrity in patients with KS, manifesting as reduced QTc time and sudden cardiac arrest. Hence, a screening 12-lead ECG and diagnostic echocardiogram should be considered.

The increased cerebrovascular mortality is mainly due to increased risk of subarachnoid haemorrhage, rupture of intracranial saccular aneurysm or a thrombotic event. The risk of deep vein thrombosis and pulmonary embolism is increased 3–6-fold and is mainly due to higher levels of procoagulatory substance PAI-1 and low levels of testosterone. Risk of dyslipidaemia and metabolic syndrome (increased total body fat, waist circumference, insulin resistance and increased LDL and TG) is also increased and is mainly due to the intrinsic genetic background and hypogonadism.

Low bone mineral density, osteopenia and osteoporosis leading to increased risk of fracture is seen in up to 40% of patients with KS. Hypogonadism, delayed or absent puberty, hypovitaminosis D, increased fat mass to lean mass ratio and AR insensitivity are various risk factors contributing to the poor bone mass. Hence, determination of BMD by DXA scan and vitamin D levels is advised in all adult patients with KS.

Certain types of malignancies including breast cancer and extragonadal Germ Cell Tumours (GCTs) (mostly mediastinal) have been found to be increased in patients with KS. However, the overall incidence is not higher than the general population and hence routine screening is not recommended [11].

67.6 Diagnosis

Due to variability in the expression of phenotype and lack of specific signs and symptoms, KS is not only underdiagnosed, but also there is a marked delay in diagnosis. Poor knowledge and awareness of KS makes it even more difficult to diagnose at an early age. Only 25–40% of cases with KS are diagnosed in their lifetime. Of these, 50–60% of cases are diagnosed in adulthood during their fertility work-up, rest are diagnosed either prenatally (15–20%), before puberty (10%) or after puberty

(15%). Early diagnosis and timely treatment can help in decreasing morbidity and mortality associated with KS.

As there are no specific findings on antenatal ultrasonography or biochemical tests, prenatal diagnosis is usually the result of karyotyping performed for other reasons. Prenatal genetic testing using Non-Invasive Prenatal Test (NIPT) can help increasing diagnostic rate and also provide information on prenatal prevalence and natural history of KS. As prenatal screening carries the disadvantage of increasing parental anxiety, excessive medicalisation, social and ethical problems and increased rate of abortions which exceeds its potential benefits, hence, routine prenatal screening for KS is not recommended. Moreover, prenatal diagnosis of KS should be confirmed postnatally by chromosomal analysis done on peripheral blood.

Early diagnosis during the prepubertal and adolescent period provides a good opportunity for timely screening and managing the comorbidities associated with KS. Hence, a genetic diagnosis should be established in children presenting with bilateral cryptorchidism, micropenis, small testes, long legs, tall stature, speech and behavioural abnormalities. Genetic diagnosis can be established using Barr body analysis which is a quick and reliable screening test. Confirmatory diagnosis is made by chromosomal analysis using conventional karyotype in the lymphocytes. In mosaic KS, diagnosis may be achieved by chromosomal analysis performed on skin fibroblast or biopsy samples from testicular tissue.

After adolescence and pubertal age, typical biochemical abnormalities include low or low normal testosterone; increased oestradiol, sex hormone binding globulin, Luteinising Hormone (LH) and Follicle-Stimulating Hormone (FSH). Androgen sensitivity index (LH × serum testosterone) is high due to increased LH. As seminiferous tubules are damaged, hence, rise in FSH shows the best discriminatory value. Due to the damage to germ cells and Sertoli cells, inhibin B levels are significantly low specifically in the post-pubertal boys. Semen analysis typically shows azoospermia (non-obstructive).

Histological examination of testes reveals hyalinisation of seminiferous tubules, hyperplasia of Leydig cells and absence of spermatogenesis. Residual foci of normal spermatogenesis, spermatogenesis with meiotic arrest at primary spermatocyte or spermatid stage has also been reported, providing an opportunity of sperm retrieval and cryopreservation in these patients. Genetic testing and karyotype analysis are recommended in adult men presenting with non-obstructive azoospermia or oligospermia; or primary hypogonadism (low testosterone with increased LH and FSH) along with low testicular volume [12].

67.7 Management (Box 67.1)

As patients with KS have multisystem involvement, close collaboration with specialists from different domains including genetics, fertility, embryology, cardiology, neurology, psychology and endocrinology is required for optimal care and better outcomes.

During infancy and for pre-pubertal boys with KS, treatment is mainly directed towards management of cryptorchidism and micropenis, management being same as for children without KS. Growth parameters including weight, height and body proportions should be assessed periodically. Proper neurological, psychological, cognition, behavioural and speech assessment should be done and managed appropriately. Bone health status assessment using vitamin D levels and DXA scan is advised and treatment should be initiated accordingly. Testicular tissue cryopreservation, spermatogonial stem cell retrieval is not recommended in prepubertal boys with KS.

For adolescents and adults with KS, proper information regarding impact of KS on fertility and options available for fertility preservation should be imparted. Proper follow-up using testicular ultrasound and hormonal assessment should be done. If patients are willing for fertility preservation; semen analysis should be done and motile spermatozoa should be cryopreserved. In patients with azoospermia, Testicular Sperm Extraction (TESE) either using standard TESE or microdissection TESE should be done and cryopreserved. Genetic counselling is recommended for all patients with KS who opt for fertility preservation techniques and also prior to pregnancy and preimplantation.

Testosterone Replacement Therapy (TRT) is the mainstay of management in adulthood and should be started in patients presenting with delayed puberty or having features of hypogonadism after addressing fertility issues. TRT helps in improving muscle strength, libido, self-esteem, behaviour, bone mineral density and body hair. It has also been observed that TRT helps in improving anaemia, cardiovascular health and vascular endothelial function. However, TRT has not been found to affect neurocognitive functions, gynaecomastia and fertility in patients with KS; and rather found to block spermatogenesis by decreasing endogenous LH and FSH levels. Hence, TRT is usually advised once fertility issues have been addressed properly.

Lifestyle modification, yearly assessment for metabolic syndrome (BP, HbA1c, lipid profile, fasting sugars), routine assessment for bone health (Calcium, phosphate, ALP, Vitamin D, DEXA scan), cardiovascular integrity (12-lead ECG and echocardiogram) and psychosexual health is advised for all patients. Thrombosis prophylaxis is recommended for patients with KS prior to long-term flights or prior to exposure to factors known to increase risk of deep vein thrombosis and pulmonary embolism [2, 13].

67.8 Summary

KS is one of the most common chromosomal aneuploidies occurring with a frequency of 1:600. The common presenting features in childhood include cryptorchidism, micropenis and developmental delay whereas in pubertal/adult age group delayed puberty, eunachoid body habitus, tall stature, gynaecomastia and small testes are the main features. The management of KS is multidisciplinary. Testosterone replacement and fertility assistance are the mainstay of therapy.

Box 67.1 Management of Klinefelter syndrome

Investigations

1. Serum testosterone.
2. Serum oestradiol, sex hormone binding globulin, LH and FSH.
3. Androgen sensitivity index (LH × serum testosterone).
3. Serum inhibin B levels.
4. Semen analysis.

Management

1. Treat cryptorchidism and micropenis as per protocol.
2. Assess growth parameters periodically: Weight, height, body proportions.
3. Periodic assessment of speech, cognition, behaviour, psychological and neurological examination.
4. Testosterone replacement therapy in adulthood.
5. Lifestyle modification.
6. Serum calcium, phosphate, ALP, vitamin D levels and DXA scan: For bone health status.
7. Yearly screening for metabolic syndrome (BP, HbA1c, lipid profile, fasting sugars).
8. 12-lead ECG and echocardiogram: For cardiovascular health assessment.
9. Thrombosis prophylaxis: Prior to long-term flights or exposure to factors known to increase risk of deep vein thrombosis and pulmonary embolism.
10. Sperm cryopreservation: If patient is willing for fertility preservation.

Acknowledgements The authors thank Dr. Medhal Mittal, Chacha Nehru Bal Chikitsalya (New Delhi), for sharing clinical pictures.

References

1. Cui X, Cui Y, Shi L, Luan J, Zhou X, Han J. A basic understanding of turner syndrome: incidence, complications, diagnosis, and treatment. Intractable Rare Dis Res. 2018;7(4):223–8.
2. Groth KA, Skakkebæk A, Høst C, Gravholt CH, Bojesen A. Clinical review: Klinefelter syndrome-a clinical update. J Clin Endocrinol Metab. 2013;98(1):20–30.
3. Tartaglia NR, Howell S, Sutherland A, Wilson R, Wilson L. A review of trisomy X (47,XXX). Orphanet J Rare Dis. 2010;5(1):8.
4. Skuse D, Printzlau F, Wolstencroft J. Sex chromosome aneuploidies. Handb Clin Neurol. 2018;147:355–76.
5. Blumling AA, Martyn K, Talboy A, Close S. Rare sex chromosome variation 48,XXYY: An integrative review. Am J Med Genet C Semin Med Genet. 2020;184(2):386–403.
6. Bojesen A, Juul S, Gravholt CH. Prenatal and postnatal prevalence of Klinefelter syndrome: a national registry study. J Clin Endocrinol Metab. 2003;88(2):622–6.
7. Klinefelter HF Jr, Reifenstein EC Jr, Albright F Jr. Syndrome characterized by gynecomastia, Aspermatogenesis without A-Leydigism, and increased excretion of follicle-stimulating hormone. J Clin Endocrinol Metabol. 1942;2(11):615–27.
8. Lanfranco F, Kamischke A, Zitzmann M, Nieschlag E. Klinefelter's syndrome. Lancet. 2004;364(9430):273–83.

9. Bonomi M, Rochira V, Pasquali D, Balercia G, Jannini EA, Ferlin A, et al. Klinefelter syndrome (KS): genetics, clinical phenotype and hypogonadism. J Endocrinol Investig. 2017;40(2):123–34.
10. Samango-Sprouse CA, Counts DR, Tran SL, Lasutschinkow PC, Porter GF, Gropman AL. Update on the clinical perspectives and care of the child with 47,XXY (Klinefelter Syndrome). Appl Clin Genet. 2019;12:191–202.
11. Ferlin A. Strategies to improve early diagnosis of Klinefelter syndrome. Expert Rev Endocrinol Metab. 2020;15(6):375–8.
12. Bearelly P, Oates R. Recent advances in managing and understanding Klinefelter syndrome. F1000Res. 2019;8:F1000 Faculty Rev-112.
13. Espehana A, Tomlinson C. Klinefelter syndrome and the need for a multi-disciplinary approach. Front Reprod Health. 2021;3:622144.

Cloacal Exstrophy and Gender Management

68

Mehak Sehgal and Minu Bajpai

Abstract

Cloacal exstrophy is a rare condition, in the spectrum of the exstrophy–epispadias complex. There is an exposed bladder and cecal plate with prolapsed ileocecal junction and an omphalocele. The pubic bone is widely separated, which leads to the clitoris and phallus being spaced apart, creating challenges in its reconstruction. It is known to be associated with several other systemic anomalies including spinal and skeletal defects, which add to the morbidity of the condition. With improved early survival beyond the neonatal age group and success in the early surgical management separating bladder and bowel segments, gender management and genital reconstruction is a matter of dilemma and deliberation, especially for those with the male genotype. Female genital reconstruction is relatively straightforward, however, whether a child born with the XY genotype should be reared as female or should an attempt be made to reconstruct the phallus from diminished and limited phallic structures is still concerning. Despite evidence now suggesting a male gender of rearing for these patients, it remains a matter of debate and needs to be decided after discussing with the parents regarding outcomes.

Keywords

Cloacal Exstrophy · Gender assignment · OEIS complex · Genitoplasty

M. Sehgal · M. Bajpai (✉)
Department of Pediatric Surgery, All India Institute of Medical Sciences, New Delhi, India

S. K. Ratan (ed.), *Children with Differences in Sex Development*,
https://doi.org/10.1007/978-981-97-1639-5_68

68.1 Introduction

Cloacal Exstrophy falls under the most extreme of the Exstrophy-epispadias complex spectrum. Also called the vesico-intestinal fissure, it is an extremely rare abnormality, occurring in 1 in 2,00,000–4,00,000 live births varying by geographical location, with a 2:1 male predominance [1–3]. Although described as early as the 1700s, the first survivor of the condition was described by Rickham only in the 1960s [3, 4]. It is also known as the OEIS complex, first coined by Carey et al. as a synonym for this defect [5]. However, variability has been noted in the presentation. It has been noted that about 22.6% of cases diagnosed as cloacal exstrophy met the criteria for the OEIS complex, whereas 43 were classified as OEI or EIS and 54.8% were reported to have an omphalocele [6, 7]. Owing to intensive care management, nutrition options, antibiotics, and improved surgical techniques, children born with the condition now live beyond their neonatal period overcoming complications like dehydration and sepsis. This has led to a shift of focus of management from emergent surgeries and life-saving interventions to elective genital reconstructions, concepts of gender assignment, and staged reconstructions.

68.2 Etiopathogenesis and Anatomical Considerations

The exact embryology and etiology are unknown. Failure of the urorectal septum to develop and divide the urogenital sinus from the rectum, along with the failure of the mesodermal proliferation which forms the infraumbilical abdominal wall and genital tubercle, leads to the development of cloacal exstrophy. Unbalanced translocation between the long arm of chromosome 9 and the Y chromosome with resulting 9q34.1 deletion and mutation in HLXB9 and HOX family of genes are postulated causative factors [3]. It has also been seen to occur in multiple cases of monozygotic twins, implying a genetic influence [8–10].

The key components of cloacal exstrophy include (Fig. 68.1):

- Omphalocele.
- Exstrophy of the two hemi-bladders.
- Prolapsed ileocecal region which presents between the two hemi-bladders (described as the trunk of an elephant).
- Imperforate anus.
- Ambiguous genitalia.

Other anomalies are also often found to be associated with this condition, which includes:

- *Gastrointestinal*: There is associated ileocecal exstrophy with omphalocele occurring in 88–100%, along with an imperforate anus. Short bowel syndrome occurs in 25% which is either mechanical or due to a malabsorptive defect. Malrotation and duplication abnormalities may be frequently associated [11, 12].

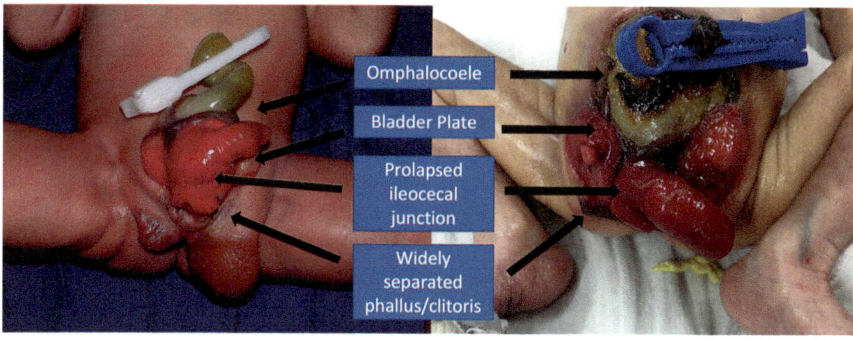

Fig. 68.1 Key components found in Cloacal Exstrophy as found in two cases seen at our institute

- *Skeletal*: Lower limb abnormalities occur in 17–26%, which include a variety of neuroorthopedic deformities. The pelvis has widened pubic diastasis with external angling of the posterior and anterior segments. The iliac wings are externally rotated and abducted. A higher incidence of bilateral asymmetry. Spinal abnormalities are seen in 22–60%, such as absent or extra vertebrae, scoliosis, and kyphosis [13, 14].
- *Central Nervous System*: Spinal dysraphism may be seen in 64%–100%. Up to one-third may have an intracranial abnormality associated [11, 15].
- *Urinary System*: Upper urinary tract abnormalities occur in 41–66%, with unilateral renal agenesis, pelvic kidney, or hydronephrosis in as high as 33% of the affected individuals [16, 17]. The exstrophy segment presents as half bladder plate and remaining hindgut plate. The bladder neck and external urethral sphincter are poorly developed [12]. Histopathological evidence of poor muscle layer with increased fibrosis, along with a scarcity of neural elements in the bladder wall explains the poor compliance of such bladders and subsequent development of incontinence and upper tract deterioration due to high pressures [18].
- *Genital*: There is complete separation, diminution, or even absence of the phallic or clitoral halves. The phallus may rarely even be located intravesical, adhered to the cecum [19]. The scrotum or labia are noted to be widely separated. Cryptorchidism with bilateral inguinal hernias can be present. There is often a failure of Mullerian duct fusion, with uterine and vaginal duplication anomalies seen in two-thirds and one-third having vaginal agenesis which develops hematometrocolpos later [20].

68.3 Perinatal Management

With recent advances in radiological imaging and fetal medicine, fetuses affected with cloacal exstrophy are often detected early on antenatal scans [21]. Parents need to be counselled about the multitude of anatomical defects which are associated with the diagnosis [22]. Termination of pregnancy can be advised based on the certainty of antenatal diagnosis. If detected beyond terminable gestational age, delivery at a tertiary care institute is to be advised for stabilization of the child, who may otherwise incur heat and evaporative losses if not adequately cared for. Post-stabilization, the child undergoes a panel of investigations for the associated anatomical anomalies [17]. This enables better prognostication of the parents after complete evaluation. Often the genitalia are small and ambiguous, for which the inclusion of karyotyping for determining the sex of the child can be done as a part of the workup [3].

68.4 Goals of Treatment and Early Surgical Considerations

The child is initially managed conservatively and adequately stabilized. Further, the challenges and goals of immediate surgical management include gastrointestinal reconstruction by either ileostomy or tubularization of the cecal plate and colostomy [23]. This enables the important aspect of the separation of bowel and bladder plates. The open bladder plate can either be repaired in a single stage (Fig. 68.2) or by a staged approach. Alternatively, it can also be left undisturbed for repair later. Another technique by which a stoma can be avoided is by performing an ileocolic anastomosis after the reduction of the prolapsed ileum and the distal colon. The end of the colon can then be pulled through the pelvis and from the neoanus and hung outside for a length of around 2–3 cm, allowing for an auto-anastomosis of the bowel to the pelvic floor. The excess hung bowel can be later trimmed after 7–10 days. This allows for a single-stage repair [19]. Meningocoele if present, is also repaired at this stage [17]. The omphalocele can be either primarily repaired or conservatively managed and epithelized.

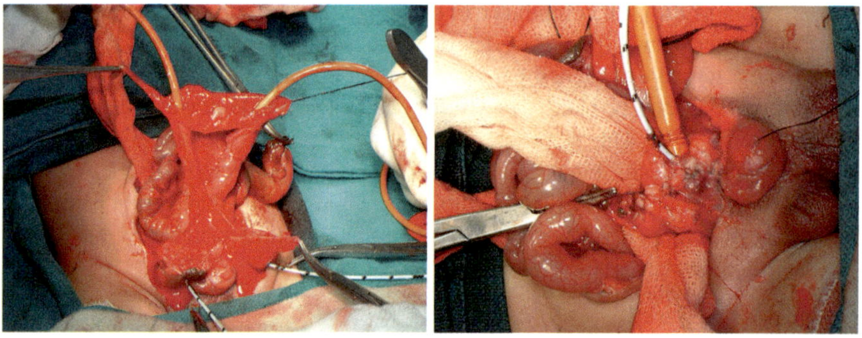

Fig. 68.2 Repair of Cloacal Exstrophy with Bladder Repaired in the same sitting

68.5 Gender Management and Genital Reconstruction

Cloacal exstrophy does not technically come under the purview of a disorder of sex differentiation. However, it is the gender assignment that is a difficult task in these patients, as the genitalia and phallic structures are small and bifid. The chromosomes and gonads do not have any ambiguity. The sex assignment to the female gender in a child with a 46XX karyotype is straightforward. However, sex assignment in a child with a 46XY karyotype is controversial. The cause of concern is the lack of adequate phallic tissue, which is small, asymmetric, and widely separated, which creates a challenge to reconstructing it into a complete function penis, although not impossible (Fig. 68.3) [24]. This led some surgeons to convert these genotypic males into females during the newborn period itself. The neonate would undergo bilateral orchidectomy to avoid the effects of testosterone, and to rear the child as female. However, it should also be considered that these children are genetically male with histologically normal testicles. They have also had an androgenic influence on the developing brain, which may potentially cause gender identity issues later, which leads some surgeons to favor the male assignment of 46XY neonates. This opinion has also been found in a recent survey of the members of the Section of Urology of the American Academy of Pediatrics [2, 25–27]. Utilization of other tissue such as grafts from the thigh and forearm may also be used later for phalloplasty [28, 29].

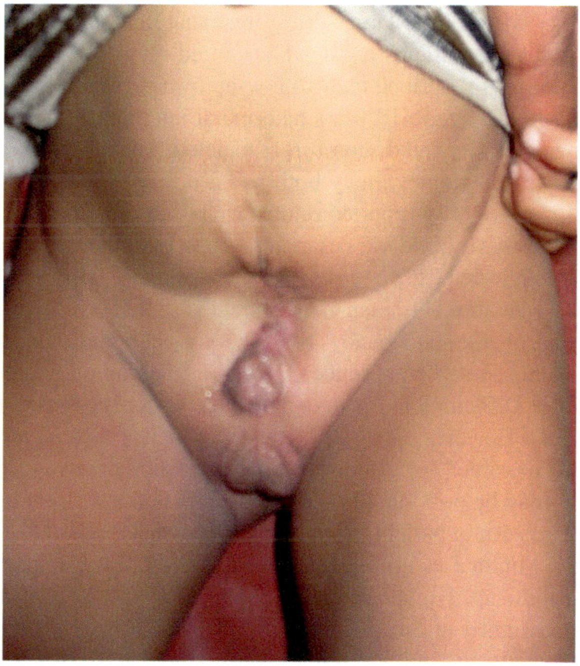

Fig. 68.3 Final appearance in a male child during follow up after complete repair of Cloacal Exstrophy and Phallic Reconstruction

There have been many studies published to question this approach and understand the best possible way forward. An important study published in the *New England Journal of Medicine* by Reiner and Gearhart followed 16 genetic male patients with cloacal exstrophy. All 16 patients, at a follow-up at 34–98 months, showed interests and attitudes typical of the male gender. Of the 14 patients who had been reassigned to the female sex during the neonatal period, eight described themselves as male. Five patients were living as females, and three were living with unclear sexual identity, of which two of these patients thought of themselves as male. During the period of their study, six of the patients reassigned themselves to the male gender [30]. Other authors have also found male patients reared as females opting later for gender reassignment [31]. Mitchell and Plaire in their study found that over half of their 46XY patients raised as female displayed male behavior patterns despite having undergone early orchiectomy [32]. Vaginal reconstruction requires a bowel segment, which is already short in length in these children. Thus, there is variability in the gender identity outcomes in 46XY males reared females.

68.6 Psycho-Sexual Outcomes

In females, vaginal reconstruction can be undertaken early. The native vagina may often be adequate to reconstruct at the perineum. Alternatively, a neovagina may be created using a large or small bowel [12, 23]. Normal menstruation occurs less frequently in these patients and there are only reports of successful pregnancies [33, 34]. Most of the large series on sexual outcomes are in combination with bladder exstrophy patients and have a limited number with cloacal exstrophy. A study by Mathews described their follow-up by a survey of cloacal exstrophy and found 3 of 5 women aged >18 years had sexual intercourse. One had stopped because of uterine prolapse and discomfort associated. The other two patients indicated sexual satisfaction. Notably, they both had a urinary stoma. The two patients who were not sexually active reported reluctance because of body image-related issues. Only one delivered a child via cesarean section, and the course of the pregnancy was rather complicated [35].

In males, Hussman et al. evaluated sexual outcomes and found feelings of sexual and phallic inadequacy in all the post-pubertal patients. Half of these individuals demonstrated an absence of erectile function on nocturnal penile tumescence testing. Difficulty in engaging in intercourse secondary to inadequate penile length [24]. However, using artificial reproductive techniques, fertility and paternity may be possible [36]. Along with the utilization of advanced techniques of phalloplasty, it may be possible to provide these patients with improved sexual outcomes, as is also seen in patients with bladder exstrophy [37].

Minimal to mild depression has been noted amongst the patient cohort during follow-up. Higher levels of depression were seen in those who had their gender reassigned at birth. In those with a stable gender identity matching their gender assignment—no differences were seen in self-esteem, social adjustment, or cognitive ability [38]. This also highlights the need to be in continued psychological counselling.

68.7 Conclusion

Cloacal exstrophy is a rare condition with a multitude of diverse associated conditions, each of which requires specialized management. Gender assignment is a difficult task, especially in cases of genetic males with cloacal exstrophy. Historically, although males have been reassigned to the female phenotype, it is possible with advancements in surgical and reproductive techniques, to avoid the reassignment and retain the male gonads. The recent 2016 guidelines thus recommend male rearing of 46XY neonates [39]. The decision, however, should be made after a detailed discussion with the parents, with both sides of the decision well explained.

References

1. Caton AR, Bloom A, Druschel CM, Kirby RS. Epidemiology of bladder and cloacal exstrophies in New York state, 1983–1999. Birth Defects Res A Clin Mol Teratol. 2007;79:781–7. https://doi.org/10.1002/BDRA.20402.
2. Tank ES, Lindenauer SM. Principles of management of exstrophy of the cloaca. Am J Surg. 1970;119:95–8. https://doi.org/10.1016/0002-9610(70)90018-8.
3. Woo LL, Thomas JC, Brock JW. Cloacal exstrophy: a comprehensive review of an uncommon problem. J Pediatr Urol. 2010;6:102–11. https://doi.org/10.1016/J.JPUROL.2009.09.011.
4. Rickham PP. Vesico-intestinal Fissure. Arch Dis Child. 1960;35:97–102. https://doi.org/10.1136/ADC.35.179.97.
5. Carey JC, Greenbaum B, Hall BD. The OEIS complex (omphalocele, exstrophy, imperforate anus, spinal defects). Birth Defects Orig Artic Ser. 1978;14:253–63.
6. Feldkamp ML, Botto LD, Amar E, Bakker MK, Bermejo-Sánchez E, Bianca S, et al. Cloacal exstrophy: an epidemiologic study from the international clearinghouse for birth defects surveillance and research. Am J Med Genet C Semin Med Genet. 2011;157C:333–43. https://doi.org/10.1002/AJMG.C.30317.
7. Vasudevan PC, Cohen MC, Whitby EH, Anumba DOC, Quarrell OWJ. The OEIS complex: two case reports that illustrate the spectrum of abnormalities and a review of the literature. Prenat Diagn. 2006;26:267–72. https://doi.org/10.1002/PD.1394.
8. Gambhir L, Höller T, Müller M, Schott G, Vogt H, Detlefsen B, et al. Epidemiological survey of 214 families with bladder exstrophy-epispadias complex. J Urol. 2008;179:1539–43. https://doi.org/10.1016/J.JURO.2007.11.092.
9. Smith NM, Chambers HM, Furness ME, Haan EA. The OEIS complex (omphalocele-exstrophy-imperforate anus-spinal defects): recurrence in sibs. J Med Genet. 1992;29:730–2. https://doi.org/10.1136/JMG.29.10.730.
10. Fullerton BS, Sparks EA, Hall AM, Velazco CS, Modi BP, Lund DP, et al. High prevalence of same-sex twins in patients with cloacal exstrophy: support for embryological association with monozygotic twinning. J Pediatr Surg. 2017;52:807–9. https://doi.org/10.1016/J.JPEDSURG.2017.01.043.
11. McHoney M, Ransley PG, Duffy P, Wilcox DT, Spitz L. Cloacal exstrophy: morbidity associated with abnormalities of the gastrointestinal tract and spine. J Pediatr Surg. 2004;39:1209–13. https://doi.org/10.1016/j.jpedsurg.2004.04.019.
12. Mathews R, Jeffs RD, Reiner WG, Docimo SG, Gearhart JP. Cloacal exstrophy–improving the quality of life: the Johns Hopkins experience. J Urol. 1998;160:2452–6. https://doi.org/10.1097/00005392-199812020-00017.
13. Loder RT, Dayioglu MM. Association of congenital vertebral malformations with bladder and cloacal exstrophy. J Pediatr Orthop. 1990;10:389–93. https://doi.org/10.1097/01241398-199005000-00018.

14. Jain M, Weaver DD. Severe lower limb defects in exstrophy of the cloaca. Am J Med Genet A. 2004;128A:320–4. https://doi.org/10.1002/AJMG.A.30103.
15. Suson KD, Novak TE, Gupta AD, Benson J, Sponseller P, Gearhart JP. Neuro-orthopedic manifestations of the omphalocele exstrophy imperforate anus spinal defects complex. J Urol. 2010;184:1651–5. https://doi.org/10.1016/J.JURO.2010.03.085.
16. Singh A, Bajpai M, Ali A. Cloacal exstrophy with vesicoureteric junction obstruction: rare association. Afr J Paediatr Surg. 2013;10:282–4. https://doi.org/10.4103/0189-6725.120891.
17. Hyun SJ. Cloacal exstrophy. Neonatal Netw. 2006;25:101–15. https://doi.org/10.1891/0730-0832.25.2.101.
18. Gupta A, Minu Bajpai S, Mallick NS, Shasanka Panda MC, Sharma. Cloacal Exstrophy: a Histomorphological analysis of the bladder plate and correlation with bladder dynamics. Journal of Progress in Paediatric. Urology. 2014;17:33–6.
19. Mitra A, Bajpai M, Dey S, Sharma N, Panda S. Cloacal Exstrophy with Intravesical phallus: an intra-operative revelation in a case of OEIS complex. J Pediatr Urol. 2014;17:41–3.
20. Hurwitz RS, Manzoni GAM, Ransley PG, Stephens FD. Cloacal exstrophy: a report of 34 cases. J Urol. 1987;138:1060–4. https://doi.org/10.1016/S0022-5347(17)43502-6.
21. Austin PF, Homsy YL, Gearhart JP, Porter K, Guidi C, Madsen K, et al. The prenatal diagnosis of cloacal exstrophy. J Urol. 1998;160:1179–81. https://doi.org/10.1097/00005392-199809020-00061.
22. Bischoff A, Calvo-Garcia MA, Baregamian N, Levitt MA, Lim FY, Hall J, et al. Prenatal counseling for cloaca and cloacal exstrophy-challenges faced by pediatric surgeons. Pediatr Surg Int. 2012;28:781–8. https://doi.org/10.1007/S00383-012-3133-3.
23. Soffer SZ, Rosen NG, Hong AR, Alexianu M, Peña A. Cloacal exstrophy: a unified management plan. J Pediatr Surg. 2000;35:932–7. https://doi.org/10.1053/JPSU.2000.6928.
24. Husmann DA, McLorie GA, Churchill BM. Phallic reconstruction in cloacal exstrophy. J Urol. 1989;142:563–4. https://doi.org/10.1016/S0022-5347(17)38816-X.
25. Diamond DA, Burns JP, Huang L, Rosoklija I, Retik AB. Gender assignment for newborns with 46XY cloacal exstrophy: a 6-year follow up survey of pediatric urologists. J Urol. 2011;186:1642–8. https://doi.org/10.1016/J.JURO.2011.03.101.
26. Raveenthiran V. Controversies of sex re-assignment in genetic males with congenital inadequacy of the penis. Indian J Pediatr. 2017;84:700–8. https://doi.org/10.1007/S12098-017-2412-3.
27. Diamond DA, Burns JP, Mitchell C, Lamb K, Kartashov AI, Retik AB. Sex assignment for newborns with ambiguous genitalia and exposure to fetal testosterone: attitudes and practices of pediatric urologists. J Pediatr. 2006;148:445–9. https://doi.org/10.1016/J.JPEDS.2005.10.043.
28. Heston AL, Esmonde NO, Dugi DD, Berli JU. Phalloplasty: techniques and outcomes. Transl Androl Urol. 2019;8:254. https://doi.org/10.21037/TAU.2019.05.05.
29. Garaffa G, Spilotros M, Christopher NA, Ralph DJ. Total phallic reconstruction using radial artery based forearm free flap phalloplasty in patients with epispadias-exstrophy complex. J Urol. 2014;192:814–20. https://doi.org/10.1016/J.JURO.2014.03.105.
30. Reiner WG, Gearhart JP. Discordant sexual identity in some genetic males with cloacal exstrophy assigned to female sex at birth. N Engl J Med. 2004;350:333–41. https://doi.org/10.1056/NEJMOA022236.
31. Diamond M, Sigmundson HK. Sex reassignment at birth. Long-term review and clinical implications. Arch Pediatr Adolesc Med. 1997;151:298–304. https://doi.org/10.1001/ARCHPEDI.1997.02170400084015.
32. Mitchell ME, Plaire C. Management of cloacal exstrophy. Adv Exp Med Biol. 2002;511:267–73. https://doi.org/10.1007/978-1-4615-0621-8_16.
33. Seat M, Boxwalla M, Hough A, Goodwin G. A case of congenital cloacal exstrophy/omphalocele-exstrophy-imperforate anus-spinal defects syndrome and a successful pregnancy. Clin Exp Reprod Med. 2022;49:215. https://doi.org/10.5653/CERM.2021.05148.
34. Dy GW, Willihnganz-Lawson KH, Shnorhavorian M, Delaney SS, Amies Oelschlager AM, Merguerian PA, et al. Successful pregnancy in patients with exstrophy-epispadias complex: a University of Washington experience. J Pediatr Urol. 2015;11:213.e1–6. https://doi.org/10.1016/J.JPUROL.2015.04.019.

35. Mathews RI, Gan M, Gearhart JP. Urogynaecological and obstetric issues in women with the exstrophy-epispadias complex. BJU Int. 2003;91:845–9. https://doi.org/10.1046/J.1464-410X.2003.04244.X.
36. D'Hauwers KWM, Feitz WFJ, Kremer JAM. Bladder exstrophy and male fertility: pregnancies after ICSI with ejaculated or epididymal sperm. Fertil Steril. 2008;89:387–9. https://doi.org/10.1016/J.FERTNSTERT.2007.03.005.
37. Harris TGW, Khandge P, Wu WJ, Leto Barone AA, Manyevitch R, Sholklapper T, et al. Sexual health outcomes after penile reconstruction in the exstrophy-epispadias complex. J Pediatr Urol. 2022;18:747–55. https://doi.org/10.1016/J.JPUROL.2022.02.016.
38. Mukherjee B, McCauley E, Hanford RB, Aalsma M, Anderson AM. Psychopathology, psychosocial, gender and cognitive outcomes in patients with cloacal exstrophy. J Urol. 2007;178:630–5. https://doi.org/10.1016/J.JURO.2007.03.144.
39. Lee PA, Nordenström A, Houk CP, Ahmed SF, Auchus R, Baratz A, et al. Global disorders of sex development update since 2006: perceptions, approach and care. Horm Res Paediatr. 2016;85:158–80. https://doi.org/10.1159/000442975.

GPSR Compliance

The European Union's (EU) General Product Safety Regulation (GPSR) is a set of rules that requires consumer products to be safe and our obligations to ensure this.

If you have any concerns about our products, you can contact us on ProductSafety@springernature.com

In case Publisher is established outside the EU, the EU authorized representative is:

Springer Nature Customer Service Center GmbH
Europaplatz 3
69115 Heidelberg, Germany

Batch number: 10091891

Printed by Printforce, the Netherlands